CALCIUM HYPOTHESIS OF AGING AND DEMENTIA

ANNALS OF THE NEW YORK ACADEMY OF SCIENCES
Volume 747

CALCIUM HYPOTHESIS OF AGING AND DEMENTIA

Edited by John F. Disterhoft, Willem H. Gispen, Jörg Traber, and Zaven S. Khachaturian

The New York Academy of Sciences
New York, New York
1994

∞The paper used in this publication meets the minimum requirements of American National Standard for Information Sciences–Permanence of Paper for Printed Library Materials, ANSI Z39.48-1984.

Cover: *The initial design of the cover logo for this volume was done by Dr. James R. Moyer, Jr. and Dr. Lucien T. Thompson.*

Library of Congress Cataloging-in-Publication Data

Calcium hypothesis of aging and dementia / editors, John F. Disterhoft
. . . [et al.].
p. cm. — (Annals of the New York Academy of Sciences, ISSN 0077-8923 ; v. 747)
Includes bibliographical references and index.
ISBN 0-89766-878-2 (cloth : alk. paper). — ISBN 0-89766-879-0 (pbk. : alk. paper)
1. Alzheimer's disease—Pathogenesis–Congresses. 2. Aging—Physiological aspects—Congresses. 3. Calcium—Physiological effect—Congresses. I. Disterhoft, John F. II. Series.
[DNLM: 1. Alzheimer's Disease—etiology—congresses. 2. Alzheimer's Disease—physiopathology—congresses. 3. Aging—physiology—congresses. 4. Calcium—physiology—congresses. W1 AN626YL v. 747 1994 / WM 220 C144 1994]
Q11.N5 vol. 747
[RC523]
500 s—dc20
[616.8'3107]
DNLM/DLC
for Library of Congress 94-41979
CIP

PCP
Printed in the United States of America
ISBN 0-89766-878-2 (cloth)
ISBN 0-89766-879-0 (paper)
ISSN 0077-8923

ANNALS OF THE NEW YORK ACADEMY OF SCIENCES

Volume 747
December 15, 1994

CALCIUM HYPOTHESIS OF AGING AND DEMENTIA [a]

Editors
JOHN F. DISTERHOFT, WILLEM H. GISPEN, JÖRG TRABER, and ZAVEN S. KHACHATURIAN

Conference Organizers
ZAVEN S. KHACHATURIAN and JOHN F. DISTERHOFT

CONTENTS

[a] This volume is the result of a conference entitled **Calcium Hypothesis of Aging and Dementia** held at the National Institutes of Health, Bethesda, Maryland, on December 15-17, 1993.

Part III. Cellular Models of Neuronal Degeneration and Alzheimer's Disease

Part IV. Calcium Channels

Financial assistance was received from:

- ALZHEIMER'S ASSOCIATION
- FARBENFABRIKEN BAYER GMBH
- FISCHER FOUNDATION
- FRENCH FOUNDATION
- NATIONAL INSTITUTE ON AGING
- NATIONAL INSTITUTES OF HEALTH

Preface

Calcium plays a pivotal role in the function of cells generally and of neurons in particular. The role of calcium in neuronal function has received considerable attention in recent years, in particular its involvement in the release of neurotransmitters, neuronal cytoarchitecture, nervous system growth and development, and intracellular communication and signal processing. The contribution of calcium has become better appreciated and theoretically more important in the area of neuronal plasticity, both in positive events such as learning and memory and in more negative manifestations such as neuronal deterioration and death. The role of calcium in these processes is important in a consideration of the cellular causes of aging and dementia, as some of the most important functional consequences of these degenerative processes in the human are on cognitive function generally and on learning in particular.

Calcium concentrations are tightly regulated in various neuronal and cellular compartments. The recognition that even very slight imbalances of calcium, when sustained over a long period of time, could lead by well described intraneuronal processes to cellular deterioration and death has led to the formulation of the "calcium hypothesis of aging." According to this hypothesis, calcium contributes significantly to the cellular alterations that occur in the processes of normal aging as well as those that underly the much more dramatic changes that occur during dementia. A corollary to this hypothesis suggests that aging neurons are especially sensitive to or predisposed to the type of cellular changes that the Alzheimer's disease process triggers. Considerable progress has been made in recent years in understanding the role of calcium in the aging process generally and in contributing to the development of demented neural states.

The material summarized in the contributions to this book was originally presented at a meeting on the **Calcium Hypothesis of Aging and Dementia** which was held on the campus of the National Institutes of Health in Bethesda, Maryland on December 15-17, 1993. Our thesis in organizing the meeting was that in order to understand the contribution of calcium to the disorders of cellular function that underly dementia, particularly Alzeheimer's dementia, we can make considerable progress by accurately characterizing the contribution of calcium to cellular functioning in normal as well as abnormal states. Because many of the most disturbing functional effects of dementia involve disruptions of cognition and learning, we concentrated considerable effort in dealing with this important aspect of brain function. Finally, as many of the most dramatic dysfunctions of the dementias are mediated through the hippocampus and associated limbic circuitry, much of the focus of the contributions to this book is on these important neural structures.

Our purpose in organizing the meeting on the Calcium Hypothesis of Aging and Dementia was to bring together a group of scientists who use a variety of approaches to study cellular, and especially neuronal, function which had relevance to the issues of aging and dementia. We dealt in some detail with the current state of the calcium hypothesis of aging and of the dementia and with the contribution of calcium to processes leading to neuronal deterioration and death. In addition, we were concerned just as importantly with the role of calcium in neuronal function in a more general fashion. So issues such as the molecular biology and molecular pharmacology of calcium channels, the role of calcium and calcium currents in behavioral learning and in models of learning such as long-term potentiation, the contribution of calcium to the development of brain systems, and the general role of calcium currents in neuronal functions were addressed. It was our hope that bringing together a group of individuals to review current thinking and neuroscientific progress on the role of calcium, in cellular and neuronal function, much of which has been gathered by scientists working in disparate fields who often do not communicate adequately, would serve to stimulate a new level of thought and progress in this pressing research area. The meeting in Bethesda was the first step in this process. We hope that this volume will summarize the scientific interactions as well as the review of the current state of the "calcium hypothesis of aging and dementia" which occurred at the meeting to the broader scientific community.

Several sponsors jointly contributed to the support of the meeting which was held in Bethesda and to this volume. These sponsors include, of course, the National Institute on Aging, the Neuroscience and Neuropsychology of Aging Program at the National Institute on Aging, and the National Institutes of Health, in general, whose facilitities were used for the conference. In addition, support for our conference was received from several nonprofit foundations including the French Foundation, the Fischer Foundation and the Alzheimer's Association. Finally, the editors would like to acknowledge the generous financial support of Farbenfabriken Bayer GmbH in Leverkusen, Germany for this project. Dr. P.-W. Ungerer, Dr. G. Fenu, and Dr. N. Schmidt-Gollas of Bayer were especially helpful in arranging this support.

The meeting in Bethesda would not have occurred without the expert and energetic organizational contributions of Nancy Rostoczy of the Neuroscience and Neuropsychology of Aging Program at the National Institute on Aging in Bethesda. Nancy worked very hard and effectively in the preparations both before and during the conference itself. Veloy Ramsey of the Department of Cell and Molecular Biology at Northwestern University Medical School carried much of the burden for the correspondence required before the conference and subsequently during the preparation of this volume. We thank both of these capable individuals for their contributions to our efforts.

JOHN F. DISTERHOFT
WILLEM H. GISPEN
JÖRG TRABER
ZAVEN S. KHACHATURIAN
Editors

Calcium Hypothesis of Alzheimer's Disease and Brain Aging [a]

Z. S. KHACHATURIAN

Neuroscience and Neuropsychology of Aging Program
National Institute on Aging
National Institutes of Health
Bethesda, Maryland 20892

The purpose of this volume, which represents the proceedings of a workshop with the same title, was to reevaluate the "Calcium Hypothesis of Brain Aging and Alzheimer's Disease." This exercise is warranted in light of new evidence that might support or refute the proposition that cellular mechanisms for maintaining the homeostasis of cytosolic calcium concentration ($[Ca^{2+}]_i$) play a key role in aging and that sustained changes in $[Ca^{2+}]_i$ homeostasis could provide the final common pathway for the neuropathological changes associated with Alzheimer's disease (AD).

The conference was held on December 15-17, 1993 in Bethesda Maryland. This was the second meeting on the same topic organized by the National Institute on Aging (NIA); the first one was held in October 1988 and published by the New York Academy of Sciences.[1] The calcium hypothesis, first formulated in 1982 and later published (1984),[2] was primarily aimed at stimulating research and redirecting the focus of studies towards cellular mechanisms of brain aging and Alzheimer's disease. In 1989, the hypothesis was revised to accommodate new data and the increasing knowledge about the aging brain and Alzheimer's disease.[3,4] The proceedings of the present conference will likely result in a further refinement of the hypothesis.

Since 1984, the calcium hypothesis has flourished by gathering stronger evidence in its support as well as several articulate proponents.[5-13] The evolution of the hypothesis has been an integral part of the recent history of the neurobiology of aging which has matured and emerged as a distinct area of study within the larger field of neuroscience. Now, there are approximately 3,000 scientists worldwide who are interested in research on the aging brain, dementia, and neurodegenerative diseases. The scientific questions being pursued currently are at the cutting edge of neuroscience; they are more concerned with the molecular mechanisms rather than descriptive studies of normal and pathological change in the aging brain. However, the neurobiology of Alzheimer's disease, now more than ever, needs an infusion of new talent and expertise; also, new concepts, and perspectives. There are a number of compelling reasons for expanding the scope and improving the vigor and the quality of research on Alzheimer's disease.

[a] Some of the discussions in this article are excerpts from previous public speeches, presentations, or publications. See references 3, 13, and 22.

RATIONALE

Several studies have indicated that "aging" is one of the most important and consistent risk factors for AD. As an increasing proportion of the population survives beyond the age of 85 years, increasing numbers of individuals will be at risk for developing a dementia. It is estimated that in the United States, by the year 2050, between 7 and 14 million individuals will be affected by some form of dementia requiring care and institutionalization. The challenge of reducing the numbers of individuals with AD is made particularly difficult by the demographic trends in the age distribution of the U.S. population. The U.S. Bureau of Census data indicate that in 1980-1990 the "65 and over age group" increased at a substantially higher rate than did any other age group.

In recent years, with the "graying of America," it has become apparent that AD as a public policy issue is very complex. The rising cost of long-term care in recent years has made the economics of health care one of the most important issues in the array of major public health concerns. Alzheimer's disease, with its long clinical course, is perhaps the most significant disability requiring various forms of long-term care services in different settings ranging from home care to special nursing homes. It is estimated that the total annual direct and indirect cost of care for AD to the country as a whole is between 90 and 100 billion dollars. The cost of disability most immediately affects the families of AD patients, but ultimately has major economic impact on society at large by reducing or impairing potential productivity.

The burden of care to the families of AD patients is not just emotional stress and physical hardship but financial considerations as well. The long clinical course of the disease, its devastating effects on the ability of the patient to function independently, the psychosocial stress on the families, and the heavy economic burden associated with care have combined to raise public awareness and concern about the lack of any means to cope with this national problem. The discovery of safe and effective treatments is an urgent public health challenge.

Strategies of developing symptomatic treatments and care programs designed to delay the disabling symptoms and postpone institutionalization will not be adequate to overcome the demographic forces increasing the total number of affected individuals. Although it might appear premature, there is a need to develop strategies for determining preventive measures. To address this problem there is an urgent need for planned systematic expansion of basic research programs on the neurobiology of AD that are already underway, such as studies designed to identify selective risk factors, the search for the cause(s), and treatments designed to halt or reverse the clinical course of the disease. The papers in this volume provide new insights into the systems regulating $[Ca^{2+}]_i$ as potential targets for developing treatments for AD.

One of the persistent scientific challenges facing AD research is the discovery of the cascade of events leading to AD and the establishment of a causal relationship between the disease and such variables as aging, genetics, clinical symptoms of dementia, neuropathological hallmarks, molecular lesions, loss of synapses, mechanisms of cell dysfunction, and mechanisms of the specificity of neuronal loss. To accommodate the rapid progress of research in AD there is a need to synthesize all the divergent pieces of data into a coherent story. The calcium hypothesis of aging and dementia as it evolves and gathers more supporting evidence has the potential of unifying and accounting for most of the available data on AD.

PARAMETERS FOR A UNIFYING HYPOTHESIS

This volume represents a compilation of recent studies buttressing the calcium hypothesis. However, the relative merits of the hypothesis and its ultimate utility need to be evaluated in light of the following essential parameters for a unifying hypothesis for AD.

First, a unifying hypothesis for AD must be parsimonious. It must provide simple explanations that are plausible on the basis of well established biological processes.

Second, the hypothesis must account for the clinical symptoms of Alzheimer's disease including cognitive and other behavioral changes on the basis of neural mechanisms of cell dysfunction and cell death. An attempt to explain the cause(s) of AD needs to address not only the biological mechanism of neural dysfunction but also the cause(s) of the behavioral manifestations of the disease. A number of the noncognitive behavioral symptoms associated with the disease, which often cause great stress to care providers, may or may not have a direct relationship to the primary cause(s) of the disease. Still, there is a need to determine the neurobiological basis of secondary behavioral and psychiatric symptoms associated with the disease. Therefore, a unifying hypothesis must seek to explain these secondary changes.

Third, the hypothesis must provide an explanation for the specificity of neuronal dysfunction, synaptic loss, and neuronal death. One of the important characteristics of AD is that specific neural systems, such as the cholinergic system, are affected. The hypothesis must provide answers to such critical questions as: why the cholinergic system is particularly vulnerable; why only specific types of cells are affected; and why particular regions of the brain and specific structures, such as synapses, and dendrites are affected?

Fourth, the hypothesis must explain the heterogeneity of AD. At present, no one can say with assurance if AD is a single disease or a complex syndrome with many subtypes and varieties of patterns in its manifestations or if it is many different diseases with similar clusters of symptoms. The heterogeneity of the disease is typified by the many aspects of its presentation including age of onset, duration, clinical course, types and patterns of neurological and psychiatric symptoms, response to treatments, and neuropathological lesions. A number of the critical scientific problems facing the field of AD research are directly associated with heterogeneity in the expression of this disease. Although during the last 14 years significant progress has been made in identifying and describing the different manifestations of AD, the underlying biological mechanisms of the heterogeneity still remain to be uncovered. The general problem of heterogeneity provides an unusually rich array of scientific opportunities for further research directions.[14]

The biological basis of heterogeneity most probably will be found in the interaction between genetic and other factors. The search for genes associated with various brain metabolic dysfunctions and abnormal processing of cytoskeletal proteins promises to be one of the most productive lines of research in uncovering the cause(s) of this disease. The recent findings of mutations in the amyloid precursor protein (APP) gene have created great excitement and given special impetus to the search for other loci and other mutations.[15] Unfortunately, identifying the locus and the nature of the mutation will not be sufficient; this field still needs to determine the functional consequences of these mutations on protein synthesis, structure, function, and behavior, that is, protein-protein or protein-membrane interactions. Ultimately, we need to determine if, and if so how, these mutations lead to cell dysfunction and/or cell death.

Presently, it is not clear if mutations in genes are a necessary and sufficient condition to cause the disease or if one or more additional biological insults are necessary to trigger the degenerative processes of AD. If a relationship exists between genetic predisposition for AD and environmental factors or systemic metabolic dysfunctions, the mechanism for the interaction between genes and such triggering factors is not well studied. We in the field need to know how changes in metabolic functions, the immune system, neuroendocrine factors, infectious agents, and exposure to toxins influence the expression of the disease or modulate its course.[14]

Fifth, the hypothesis must demonstrate a relationship between known risk factors for AD and the neurobiology of the disease. Although there is no evidence that aging per se causes AD, it is strongly associated as one of the major risk factors for this disease. A history of severe head trauma which leads to loss of consciousness has also been found to increase the risk for AD. A third risk factor is a family history of AD in a first degree relative, which increases the odds of developing AD three to four times. These three risk factors–aging, head trauma, and genetic predisposition–meet the generally accepted epidemiological criteria for causal factors, because they provide a plausible biological explanation and their effects are strong and consistent. A recently discovered gene linked to AD is the apolipoprotein E (ApoE) gene on chromosome 19, which has been associated with many late onset familial cases of AD as well as sporadic cases in the over-60 age group.[16] The ApoE gene not only appears to have a strong and consistent relationship with AD but also offers a plausible biological explanation for its role in the pathological processes of AD. It is very likely that AD is caused by complex interactions among biological variables such as genetic predisposition and environmental, cultural, or educational factors.

Cross-cultural epidemiological studies of risk factors as extensions of the search for cause(s) are essential for teasing out the complex interactions of genetic and epigenetic factors in causing AD. Recent epidemiological investigations have suggested that lack of education might be a risk factor for AD.[17] At the present it is not clear if educational attainment may be a surrogate marker for synaptic density or some other neural developmental state. If these observations are confirmed, they may provide clues to possible mechanisms of heterogeneity by linking risk factors and other life experiences to changes in synaptic density or synaptic reserve. Further systematic epidemiological investigations of comorbidity of AD with other neurodegenerative diseases, systemic metabolic disorders, vascular diseases, health history, dietary habits, occupation, exposure to toxins, and life experiences such as education may provide clues for testable hypotheses concerning the cascade of events leading to the disease.[14]

Sixth, the hypothesis must relate the mechanisms of cell dysfunction and cell death not only to the clinical symptoms of AD, but also to the key pathognomonic features of the disease and the subtle molecular lesions associated with structural and functional changes in the AD brain. The precise reasons for and the cascade of events preceding cell dysfunction and death are not known. However, it is well established that the synthesis and release of several neurotransmitters, particularly acetylcholine, are compromised in AD.[18] It is also well established that several structural and functional changes are found in AD brains including loss of synapses, pruning of dendrites, changes in membrane constituents, regulation of intracellular ions, processing of transmembrane proteins (eg, amyloid precursor protein), abnormal processing of various other cytoskeletal proteins (eg, Tau), and metabolic changes affecting cellular energy utilization. At present it is not clear which one or combination of these events initiates the process of cell dysfunction and cell death.[14]

In summary, the following observations have been linked to AD. Although some of these linkages may not be as strong as others, all should be incorporated into any hypothesis concerning AD:

- aging, head trauma, and lack of education as risk factors and ApoE as a major susceptibility gene;
- clinical, neuropathological, and genetic heterogeneity of the disease;
- specificity of cell dysfunction and loss;
- neuropathological changes including abnormal synthesis/processing/accumulation of cytoskeletal proteins; neurotransmitter deficits (particularly cholinergic); dysfunction of glucose metabolism; loss of dendrites-synapses-neurons; extra-neuronal changes involving microvessels, glia, glucose transporters, and neuro-endocrine and neuroimmune systems;
- the 5-20-year clinical course of the disease;
- symptoms of the disease including cognitive and other behavioral changes.

Any useful and testable hypothesis of AD causation needs to accommodate and account for these observations. If it cannot explain the role of any of these observations, then it must provide convincing arguments that the observation in question is an epiphenomenon.

ESSENTIAL FEATURES OF THE CALCIUM HYPOTHESIS AND CHALLENGES IT MUST MEET

At present, the fundamental biological underpinning of AD is thought to be a gradual dysfunction of neurons, eventually leading to cell death. One of the major challenges remaining in the search for etiology(ies) of AD is the proximal cause of cell dysfunction and cell death. During the last 14 years, a rich array of ideas and scientific leads has emerged, including endogenous and exogenous toxins, infectious agents, abnormal proteins, deficits in growth-promoting factors, corticosteroids, membrane changes, calcium homeostasis, and deregulation of proteolysis. Clearly there are many mechanisms each involving several steps in the process of cell dysfunction and death. Although each of these various insults or abnormal cellular processes potentially could influence cell functioning and survival, it is important to determine which ones are critical in causing the degenerative process of AD. Some of the crucial scientific questions that need to be answered are:

- What initiates the process of cell dysfunction, when does it start, and what are the key steps in the cascade of events leading to cell death?

- What factors determine or regulate the specificity of cell dysfunction and neuron death?
- What are the key interactions between various cellular components, such as changes in membrane structure affecting the behavior of membrane-bound proteins; changes in the primary or secondary structure of proteins affecting their tertiary structure and interactions with other proteins; or abnormally cleaved protein fragments that form potentially toxic aggregates?

KEY ELEMENTS OF THE CALCIUM HYPOTHESIS

The current version of the calcium hypothesis proposes six interrelated postulates. First, it proposes that cellular mechanisms that regulate the homeostasis of cytosolic free calcium ion $[Ca_2^+]_i$ play a critical role in brain aging and the neuropathology of AD, and that altered $[Ca^{2+}]_i$ might account for a number of age-related changes in neural function and AD-associated neural dysfunctions.[2–4]

Second, it suggests that the cellular mechanisms underlying AD neuropathologies are part of a continuum of molecular processes associated with the developing nervous system and aging-related changes. The neural dysfunction, dendritic pruning, loss of synapses, and neural loss observed in AD essentially involve the same molecular mechanisms as those associated with programmed cell death in the developing nervous system, regulation of neurite elongation and growth cone motility, regulation of neuroplasticity in the adult brain, and synaptic turnover/synaptogenesis in the normal aging brain.[13]

Third, it postulates that the plasticity of neuroarchitecture is regulated by a functional equilibrium between molecular mechanisms promoting growth/regeneration and those processes that control regression/degeneration. The intracellular calcium concentrations play a central role in modulating the direction of this equilibrium governing regeneration and degeneration. The $[Ca^{2+}]_i$ in particular cellular compartments, through various intermediary regulators, governs dendritic arborization, stability, regression, pruning, or complete elimination. A parsimonious explanation is that the neuronal dysfunction, dendritic regression, and loss of synapses associated with AD reflect disregulation of normal processes that are essential for the development or normal function of the adult brain.[13] The disruption of this equilibrium between growth and regression could be the consequence of the breakdown of systems that control and regulate the homeostasis of $[Ca^{2+}]_i$; thus, a small but chronic elevation in $[Ca^{2+}]_i$ could shift the equilibrium in favor of degeneration.

Fourth, it proposes a systematic interaction between the amount of the perturbation in the ($\blacktriangle[Ca^{2+}]_i$) and the duration ($\blacktriangle T$) of the deregulation in the calcium homeostasis, so that the product of the two variables is a constant:

$$K = \blacktriangle[Ca^{2+}]_i \times \blacktriangle T.$$

This relationship suggests that a small change in $\blacktriangle[Ca^{2+}]_i$ that is sustained over a prolonged period ($\blacktriangle T$) will result in similar cellular damage as will a large change in $\blacktriangle[Ca^{2+}]_i$ over a short period ($\blacktriangle T$), or;

$$\text{Large } \blacktriangle[Ca^{2+}]_i \times \text{Small } \blacktriangle T = \text{Small } \blacktriangle[Ca^{2+}]_i \times \text{Large } \blacktriangle T$$

A calcium channel that is leaky or an extrusion pump that loses its efficiency over the course of several years could potentially cause as much damage chronically as could a massive but acute insult, such as one that occurs in stroke.[3]

Fifth, it promotes the concept that the calcium-mediated signaling system and regulation of $[Ca^{2+}]_i$ homeostasis are part of the final common pathway for the cellular changes leading to neuronal dysfunction and cell death. The final common pathway concept accounts for several alternative mechanisms through which the regulation of $[Ca^{2+}]_i$ can be disrupted. These include changes in ion channel functioning or formation of new channels; changes in membrane structure altering the functioning of transmembrane proteins; and alterations in the behavior of calcium binding proteins, extrusion pumps, buffers, and sequestration.[19,20] Also it explicitly suggests that there could be many different antecedent factors, each involving a separate cascade of events, but all leading to disruptions in calcium homeostasis. The downstream consequences of destabilizing calcium homeostasis could also result in different outcomes depending on the particular cellular compartment involved and the amount and duration of the changes in the $[Ca^{2+}]_i$.[21] Some of the possible variables in the alternative cascade of events leading to neuronal dysfunction and death are listed in FIGURE 1 under the headings of: I. Antecedent Variables; II. Mechanisms for Maintaining Ca^{2+} Homeostasis; and III. Subsequent Potential Degenerative Processes.

Sixth, it proposes that age-related and AD-associated changes in the brain may not be due to a single event or insult, but are brought about by a series of different antecedent events occurring, in combination or sequence, over a long period. The concept of the "AND gate" allows for several postulated etiologic factors working in combination and sequentially to provide various plausible paths or cascades of events all leading to the final common result of disregulating calcium homeostasis.[22]

In discussing the cause(s) and effect relationship of this disorder, it might be useful to use the analogy of an "AND gate" (a digital logic element) to conceptualize the cascade of events or the process of timing and interaction of multiple potential etiological factors. As is well known, an "AND gate" as a Boolean logic element can have two or more inputs (switch) and one output. Each input as a switch can be only in one of two states, "ON" or "OFF"; the output also can be only in the "ON" or "OFF" state at any given time. To get an output it is necessary that certain well-specified rules (a truth table) be followed. That is, a discrete event has to take place at each of the inputs (all switches must be "ON") of the "AND gate" for an output to occur. The timing of the events at the various inputs can vary widely, but once an event takes place, its effect must linger on. The basic rule is that an event has to be present at all inputs at the same time for an output to occur. Now, transposing this model to AD, one can conceptualize an "AND gate" with the number of the inputs ranging between 2 and N. The simplest gate would have only two inputs, that is, a gene mutation and another event to trigger the gene. A more complex hypothetical model "AND gate" for AD could have as inputs or switches events such as:

- genetic mutation(s) predisposing one for various metabolic disorders, pathologies, or biochemical abnormalities, such as APP processing, ApoE functioning, glucose transporter, or mitochondria defect;

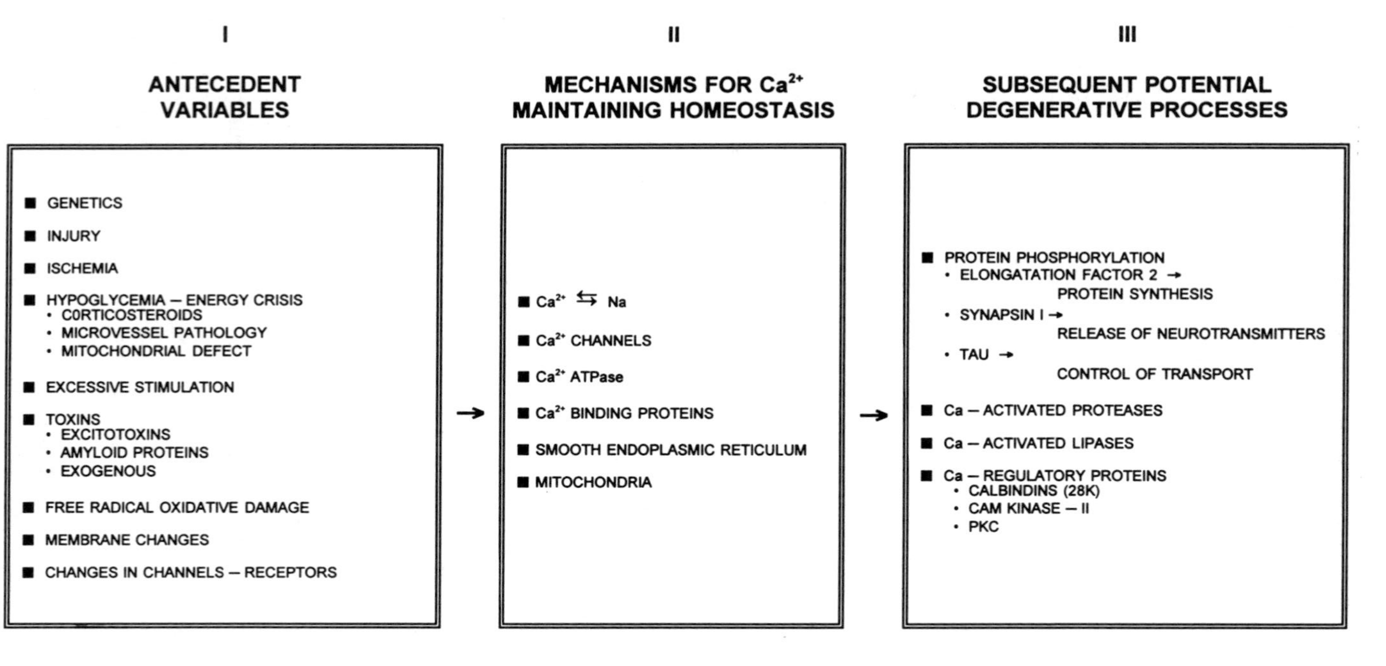

FIGURE 1. This illustrates the flow of the possible cascades of events leading to cell dysfunction and death through the final common pathway of disrupting CA^{2+} homeostasis.

- cerebral microvascular pathology which affects the transport of glucose and/or O_2, thus creating a chronic deficiency in perfusion;
- changes in oxidative metabolism due to enzymatic deficiencies; exposure to exogenous and/or endogenous toxins, such as glutamate;
- changes in the structure and function of membranes and/or proteins in systems that regulate $[Ca^{2+}]_i$ or maintain homeostasis;
- other changes in brain chemistry that affect glucose metabolism, neurotransmitter synthesis, neuroendocrine functioning, and neuroimmune system.

According to the "AND gate" model, these events do not have to occur at the same time, but each could produce a lingering effect that eventually might lead to cell loss and clinical symptoms. Therefore, to understand the etiology of AD it might be important to examine evidence showing a relationship between a particular variable, such as toxins and AD, in the context of other preceding critical physiological events during the lifetime of the patient which may have predisposed the vulnerability of the brain to the disease. For example, chronic low level hypoglycemia could modulate the toxicity of glutamate.

CONCLUSION

The Calcium Hypothesis of Aging and Dementia conference and this volume have focused on the scientific issues concerning postulated molecular mechanisms underlying the disease and potential targets for treatments. The conference and this volume have both used the calcium hypothesis as a convenient heuristic tool to help conceptualize the neurodegenerative processes in brain aging and dementia.

The calcium hypothesis has been revised twice; no doubt the proceedings of this conference and this volume will lead to a further revision.[14] Any hypothesis, to maintain its utility, must be tested constantly and revised in light of new data. The best statement about the value of scientific speculation is attributed to Henri Poincare who, in a commentary about the value of science, is to have said, "There is a value in developing an hypothesis . . . There is a hierarchy of facts. Some have no reach. They teach us nothing but themselves. The scientist who has ascertained them has learned nothing but a fact and has not become more capable of foreseeing new facts. Such facts, it seems, come once but are not destined to reappear. There are, on the other hand, facts of great yield. Each of them teaches of a new law. And since a choice must be made, it is to these that the scientist should devote himself."

It is our hope that this volume will tempt many to look for the whole story beyond the immediately available data and speculate about how the data might best fit together in solving the puzzle that is Alzheimer's disease.

ACKNOWLEDGMENTS

I wish to thank Drs. Neil Buckholtz, Creighton Phelps, and Steven Snyder and Nancy Rosztoczy for their suggestions, assistance, and patience in preparing the manuscript. My special gratitude to Dr. John Disterhoft for his gracious help in organizing the conference.

REFERENCES

1. KHACHATURIAN, Z. S., C. W. COTMAN & W. PETTEGREW. 1989. Calcium, Membranes, Aging, and Alzheimer's Disease. Ann. N.Y. Acad. Sci. **568:** 1-292.
2. KHACHATURIAN, Z. S. 1984. Towards theories of brain aging. In Handbook of Studies on Psychiatry and Old Age. D. S. Kay & G. W. Burrows, Eds.: 7-30. Elsevier. Amsterdam.
3. KHACHATURIAN, Z. S. 1989. The role of calcium regulation in brain aging: Reexamination of a hypothesis. Aging **1:** 17-34.
4. KHACHATURIAN, Z. S. 1989. Calcium, membranes, aging, and Alzheimer's disease: Introduction and overview. Ann. N.Y. Acad. Sci. **568:** 1-4.
5. LANDFIELD, P. W., L. W. CAMPBELL, S.-Y. HAO & D. S. KERR. 1989. Aging-related increases in voltage-sensitive, inactivating calcium currents in rat hippocampus: Implications for mechanisms of brain aging and Alzheimer's disease. Ann. N.Y. Acad. Sci. **568:** 95-105.
6. LANDFIELD, P. W., M. D. APPLEGATE, S. E. SCHMITZER-OSBORNE & C. E. NAYLOR. 1991. Phosphate/calcium alterations in the first stages of Alzheimer's disease: Implications for etiology and pathogenesis. J. Neurol. Sci. **106:** 221-229.
7. LANDFIELD, P. W., O. THIBAULT, M. L. MAZZANTI, N. M. PORTER & D. S. KERR. 1992. Mechanisms of neuronal death in brain aging and Alzheimer's disease: Role of endocrine-mediated calcium dyshomeostasis. J. Neurobiol. **23:** 1247-1260.
8. MATTSON, M. P. 1992. Calcium as sculptor and destroyer of neural circuitry. Exp. Gerontol. **27:** 29-49.
9. MATTSON, M. P., B. CHENG, D. DAVIS, B. BRYANT, I. LIEBERBURG & R. E. RYDEL. 1992. Beta-amyloid peptides destabilize calcium homeostasis and render human cortical neurons vulnerable to excitotoxicity. J. Neurosci. **12:** 376-389.
10. KOH, J.-Y. & C. W. COTMAN. 1992. Programmed cell death: Its possible contribution to neurotoxicity mediated by calcium channel antagonists. Brain Res. **587:** 233-240.
11. DISTERHOFT, J. F., J. R. MOYER, L. T. THOMPSON & M. KOWALSKA. 1993. Functional aspects of calcium channel modulation. Clin. Neuropharmacol. **16**(Suppl. 1): S12-S24.
12. WEISS, J. H., C. J. PIKE & C. W. COTMAN. 1994. Ca^{2+} channel blockers attenuate B-amyloid peptide toxicity to cortical neurons in culture. J. Neurochem. **62:** 372-375.
13. KATER, S. B., M. P. MATTSON & P. B. GUTHRIE. 1989. Calcium-induced neuronal denervation: A normal growth cone regulating signal gone awry(?). Ann. N.Y. Acad. Sci. **568:** 252-261.
14. KHACHATURIAN, Z. S. 1994. Scientific opportunities for developing treatments for Alzheimer's disease: Proceedings of a Planning Workshop. Neurobiol. Aging, in press.
15. GOATE, A., M. CHARTIER-HARLIN, M. MULLAN, J. BROWN, F. CRAWFORD, L. FIDANI, L. GIUFFRA, A. HAYNES, N. IRVING, L. JAMES, R. MANT, P. NEWTON, K. ROOKE, P. ROQUES, C. TALBOT, M. PERTCAK-VANCE, A. ROSES, R. WILLIAMSON, M. ROSSOR, M. OWEN & J. HARDY. 1991. Segregation of a missense mutation in the amyloid precursor protein gene with familial Alzheimer's disease. Nature **349:** 704-706.
16. CORDER, E. H., A. M. SAUNDERS, W. J. STRITTMATTER, D. E. SCHMECHEL, P. C. GASKELL, G. W. SMALL, A. D. ROSES, J. L. HAINES & M. A. PERICAK-VANCE. 1993. Gene dose of apolipoprotein E type 4 allele and the risk of Alzheimer's disease in late onset families. Science **261:** 921-923.
17. STERN, Y., B. GURLAND, T. K. TATEMICHI, M. X. TANG, D. WILDER, R. MAYEUX. 1994. Influence of education and occupation on the incidence of Alzheimer's disease. JAMA **271:** 1004-1010.
18. BARTUS, R. T., R. L. DEAN, B. BEER & A. S. LIPPA. 1982. The cholinergic hypothesis of geriatric memory dysfunction. Science **217:** 408-417.

19. MICHAELIS, M. L., C. T. FOSTER & C. JAYAWICKREME. 1992. Regulation of calcium levels in brain tissue from adult and aged rats. Mech. Ageing Dev. **62:** 291-306.
20. ARISPE, N., E. ROJAS & H. B. POLLARD. 1993. Alzheimer disease amyloid β protein forms calcium channels in bilayer membranes: Blockade by tromethamine and aluminum. Proc. Natl. Acad. Sci. USA **90:** 567-571.
21. NIXON, R. A. 1989. Calcium-activated neutral proteinase as regulators of cellular function: Implications for Alzheimer's disease pathogenesis. Ann. N.Y. Acad. Sci. **568:** 198-208.
22. KHACHATURIAN, Z. 1986. Aluminum toxicity among other views on the etiology of Alzheimer's disease. Neurobiol. Aging **7:** 537-539.

Cellular and Systems Neuroanatomical Changes in Alzheimer's Disease

GARY W. VAN HOESEN [a] AND ANA SOLODKIN

Departments of Anatomy and Neurology
University of Iowa College of Medicine
Iowa City, Iowa 52242

The neuroanatomical distribution of pathologic changes in Alzheimer's disease (AD) has received substantial scrutiny during the second half of this century, with a variety of aims motivating these efforts. One is linked clearly to advances in understanding the functional organization of the central nervous system and has clinicoanatomical correlates as a central goal.[1-9] Another aim has focused on the issue of communality of the lesions and their variance among the victims of the disease. Common to this has been a near parallel desire to characterize the range of vulnerable structures.[10-20] Finally, characterizing AD in neuroanatomical terms might highlight a pattern of changes unique to structures that share common connections, metabolic and enzymatic and/or chemical properties, thereby drawing attention to mechanisms altered or attacked by a single etiological factor destructive to all that share the phenotype.[21-32]

Although all of these are essential efforts to characterize a complicated disease, the realization of these aims is laden with difficulties. For example, the human central nervous system is highly complex and no base of experimental study exists as it does for several other species. Multiple types of pathology occur in AD, and they vary from one part of the central nervous system to another. Moreover, it is unclear how to view these changes. Are they destructive lesions, space-occupying lesions, or both? When in their evolution is function compromised? Are they, in fact, simply partial lesions which confuse the functioning of an already aging organ? Another complication centers around the fact that AD onset varies with age, and depending on this it can have a different tempo and severity.[33] Methods of staining for AD pathology have also varied, and cross correlations among them to ascertain their sensitivity have only recently been made.[34] Finally, descriptive investigations are by nature static and data points on a continuum of methodological advances and new discovery. For example, many contemporary neuroscientists have had the experience that a neuron characterized by one method today may, down the road, become a neuron with a unique property such as a neurotransmitter or a neuromodulator that greatly alters the earlier impressions of it.

[a] Address for correspondence: Dr. Gary W. Van Hoesen, Department of Anatomy, University of Iowa, Iowa City, IA 52242.

Our efforts in this chapter are aimed in part at synthesis, but equally so with providing a neuroanatomical framework with which to view AD pathology. Our focus centers largely on the cerebral cortex because it is the part of the central nervous system that bears the brunt of AD changes. Neurofibrillary tangles will receive more attention than neuritic plaques and neuropil threads. The former correlate best with cognitive changes,[9,18] and their dystrophic axons and dendrites are the likely source of the latter[35–37] as well as the abnormal density of synapses found in AD.[38] All caveats will not be obviated but should be used with the neuroanatomical framework to evaluate the conclusions that we and others have reached about the neuroanatomical and neurobehavioral correlates of AD. We begin with the most simple form of neuroanatomical/neuropathological analysis of AD, the appearance of the brain at autopsy and its lobular changes.

LOBULAR SPECIFICITY OF PATHOLOGY IN ALZHEIMER'S DISEASE

The classic characteristics of lobular changes in AD are widened sulci and flattened atrophic gyri, seen after at least a brief fixation and stripping of the meninges (FIG. 1). An irregular contour of the hemisphere may also be obvious, and brain weight may be low (usually approximating or less than 1,100 g). Discoloration of atrophic areas of the cortex may be present and the occipital pole may have an exaggerated knob-like appearance. Nearly always, the sometimes elusive central sulcus is easy to locate because the pre- and postcentral gyri that define it appear nonatrophic and appropriately colored. These characteristics typify the AD brain at endstage, usually after a long duration of illness.[39] However, in our experience they represent the less typical AD brain rather than the norm, and variants with only a subset of these appearances are far more common.

Frontal Lobe

Atrophic changes in the frontal association cortex are common and often conspicuous in the anterior parts of the superior frontal lobule, the posterior and medial parts of the orbitofrontal area, and the subgenual parts of the medial frontal region extending ventrally to the gyrus rectus. The middle frontal gyri seem less routinely affected than are the superior frontal gyri, and the inferior frontal gyri, including Broca's area, often appear spared. These changes in the frontal association areas are what contribute to the conspicuous appearance of sparing in the precentral gyrus and the premotor cortices immediately posterior to them (FIG. 1).

Parietal Lobe

Parietal lobe changes are typically characterized by atrophy in the inferior parietal lobule ventral to the intraparietal sulcus. This may involve both the supramarginal gyrus and the angular gyri, but it is not unusual for one to be more preserved than the other. The superior parietal lobule often contains atrophy and discoloration in

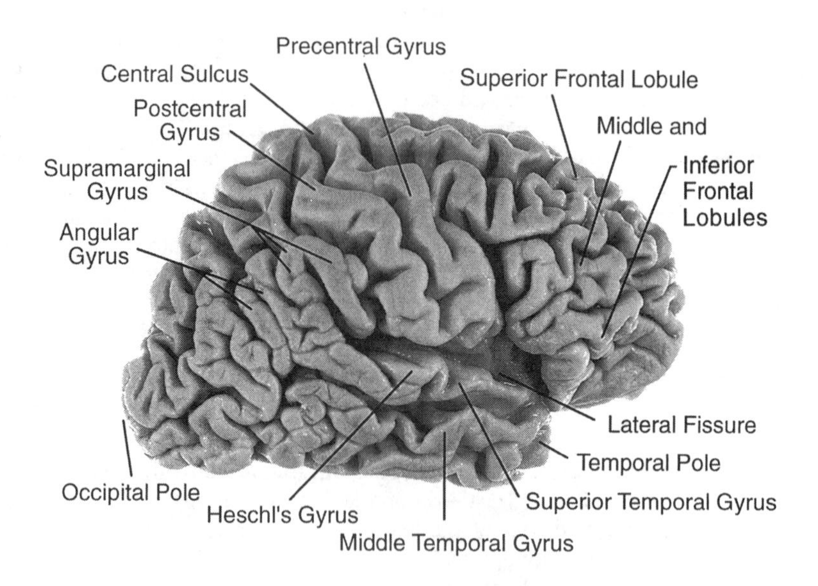

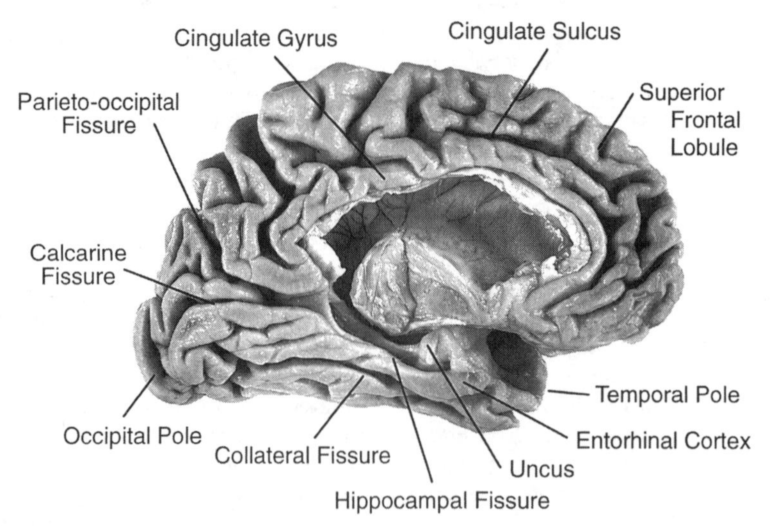

FIGURE 1. Lateral and medial photographs of the cerebral hemisphere at end-stage Alzheimer's disease after a duration of illness for 12+ years. Note the relative preservation of the primary visual cortex along the calcarine fissure, the primary auditory cortex that forms Heschl's gyrus, the primary somatosensory cortex of the postcentral gyrus, and the motor cortex of the precentral gyrus. All other association and limbic cortical areas appear affected by the disease process(es). Further discussion can be found in the text.

AD and, along with inferior parietal lobule changes, enhances the relatively normal appearance of the postcentral gyrus. The latter (Brodmann's areas 3, 1, and 2) forms the primary somatosensory cortex (FIG. 1).

Occipital Lobe

As stated earlier, the occipital pole in AD often has a bulbous or knob-like appearance. This seems due largely to the relative preservation of the primary visual cortex (Brodmann's area 17) which forms the occipital pole and the banks of the calcarine fissure, moderate atrophy of the visual association cortices, and substantial atrophy of the inferior parietal cortex formed in part by the nearby angular gyri just anteriorly (FIG. 1).

Temporal Lobe

Some of the more conspicuous and consistent lobular changes in AD occur in the temporal lobe.[8,12,14,18,40] Most notable are alterations in the temporal pole consisting of atrophy and discoloration. These give it a narrow blade-like appearance when the lateral fissure is separated. The superior, middle, and inferior temporal gyri at end stage AD are typically atrophic as well, but seldom to the degree observed at the pole. Separating the lateral fissure at the point where the central sulcus approximates it nearly always reveals a healthy appearing Heschl's gyrus traversing the posterior plane of the superior temporal gyrus. This cortex (Brodmann's area 41 and 42) is the site of the primary auditory cortex (FIG. 1).

Limbic Lobe

Broca's limbic lobe is the band of cortex that forms the medial edge of the cerebral hemisphere encircling the upper parts of the brainstem and thalamus. It is delimited by the cingulate sulcus dorsally and collateral sulcus ventrally. Bridging areas beneath the genu and splenium of the corpus callosum and across the isthmus of the insula complete the circle, providing continuity between the cingulate and parahippocampal gyri. The most invariant changes in AD occur in the limbic lobe.[8,14,30,41] Most notable are atrophic changes in the subcallosal and posterior parts of the cingulate gyrus and in the entorhinal cortex (Brodmann's area 28) which forms much of the parahippocampal gyrus along the ventromedial margin of the temporal lobe. Posterior cingulate gyrus atrophy in AD creates the appearance that the cingulate gyrus is abnormally narrow as it follows the curvature of the splenium. Entorhinal atrophic changes can range from subtle flattening and discoloration to a moth-eaten appearance reminiscent of the lunar surface.[42–44] In both cases the uncus is unusually prominent and the hippocampal fissure is uncommonly visible. Accompanying ventricular enlargement in the inferior horn of the lateral ventricle is often suggested by a thinness and softness of the anterior parahippocampal area. The anterior parts of the cingulate gyrus dorsal to the genu and body of the corpus callosum are typically spared in AD (FIG. 1).

NEUROANATOMICAL AND FUNCTIONAL SPECIFICITY IN ALZHEIMER'S DISEASE

Neuroanatomical Correlates of Cortical Areas Affected in Alzheimer's Disease

The microscopic observation of pathological changes in the cortex in Bielschowsky-stained tissue sections led Alzheimer[45] to publish his classic report on "A Singular Disease of the Cerebral Cortex." His pioneering observations have never been disputed seriously. Cortical changes quantitatively overshadow those in subcortical areas, although it is clear that the cortex is not affected uniformly. All types of cortex are affected in AD, but the changes within a given type of cortex are selective. A brief discussion of the major types of cortex and their involvement in AD follows.

Allocortical Involvement

The allocortices, a small but essential component of the cerebral cortex, are formed by two to three layers, of which one is typically acellular. Pyramidal neurons of various sizes are the major cell types. Characteristically, they have polarized apical dendrites that extend towards the pia mater and form a wide molecular layer. The dentate gyrus of the hippocampal formation, hippocampus proper, subiculum, olfactory, and periamygdaloid cortices are examples. In AD these fields are heavily but selectively compromised.[13,15,16,19,21,46–48] For example, within the hippocampal formation, the pyramids that form the subiculum and CA1 sector of the hippocampus are usually invested heavily with neurofibrillary tangles. Neuritic plaques are found among them and in their molecular layer.[49,50] In contrast, the adjoining CA3 sector of the hippocampus and the dentate gyrus may contain only an occasional neuron with neurofibrillary tangles.[21] Evidence also exists that the olfactory and periamygdaloid allocortices are damaged heavily in AD.[51–57]

Periallocortical Involvement

The periallocortical parts of the cerebral cortex are characterized by unusual laminar arrangements and cell shapes.[58] Their deep layers are related closely to the adjacent allocortex, whereas the superficial layers are reminiscent of the isocortex.[59,60] The entorhinal, parasubicular, presubicular, and retrosplenial cortices are all examples of periallocortex. As with allocortices, involvement of the periallocortices in AD is not uniform. For example, the entorhinal cortex is involved heavily with many neurofibrillary tangles.[11,16,21,61] Typically the changes are so extensive that the normal and unique cytoarchitecture of this cortex is unrecognizable in Nissl stains. The same may be said for the adjacent and closely related parasubicular cortex. In contrast, the adjacent presubicular cortex is largely spared in AD.[62] It may contain an occasional neuritic plaque and neurofibrillary tangle, but these are in random locations without

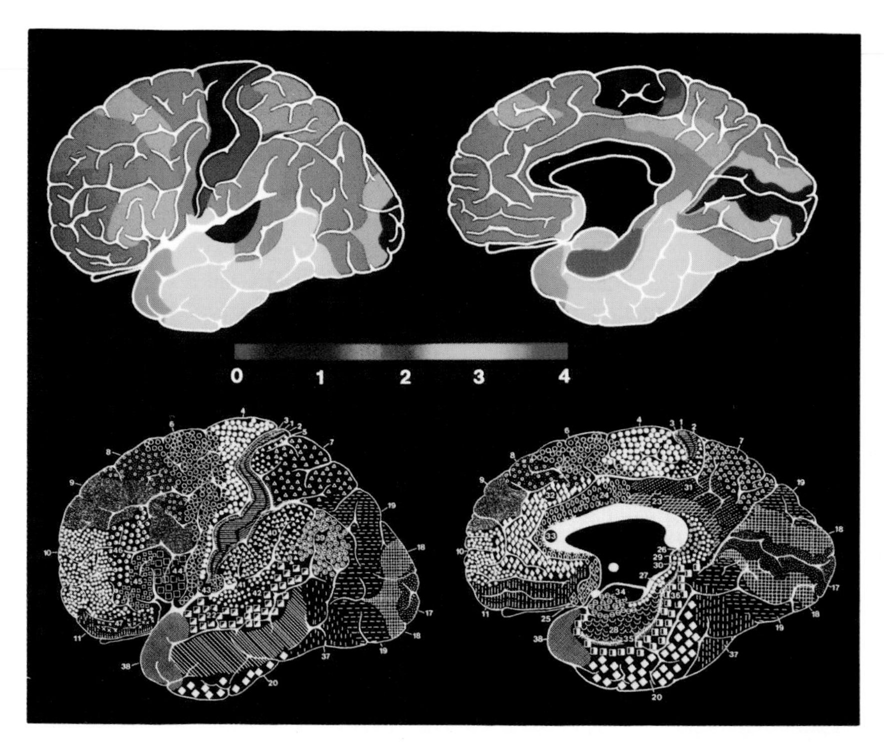

FIGURE 2. (**Top**) The lateral and medial surfaces of the human brain with cortical neurofibrillary tangle densities coded in color, as described by Arnold *et al.*,[8] for 18 hemispheres with Alzheimer's disease. (**Bottom**) The density value is plotted onto the hemisphere in accordance with Brodmann's cytoarchitectural areas. Colors towards the red end of the spectrum represent greater density than do those near the blue end of the spectrum. Note that the entorhinal cortex (Brodmann's area 28) had the greatest density of neurofibrillary tangles for all cortical areas and that the temporal lobe contains more than do the other lobes.

a predilection for a specific layer or cell type. According to our observations, the retrosplenial cortex is moderately damaged in AD (FIG. 2).

Proisocortical Involvement

The proisocortices constitute a sizable component of the cerebral cortex,[63] and together with the periallocortices and allocortices they form the limbic lobe. The posterior parts of the orbitofrontal cortex, cingulate cortex, retrocalcarine, posterior parahippocampal, perirhinal, temporal polar, and anterior insular cortices are all examples. They resemble the isocortex in that there are multiple cellular layers. However, some layers may be atypical, because they are accentuated or incipient, and generally, layers V and VI contain large and poorly differentiated pyramidal and multipolar neurons that are difficult to segregate. Judgments on the separation between these layers are typically arbitrary. The proisocortices are often uniformly affected in AD. For example, the posterior part of the parahippocampal cortex (Brodmann's area 36) is a common site of pathological change and contains an abundance of neurofibrillary tangles and neuritic plaques. These are continuous into the more anteriorly located perirhinal cortex (Brodmann's area 35) and the temporal polar cortex (Brodmann's area 38).[16,64]

The dysgranular part of the insular cortex and the posterior orbital cortex are likewise involved to a major degree, as is the subgenual part of the cingulate gyrus (Brodmann's area 25). Some investigators report that the posterior part of the cingulate gyrus (Brodmann's area 23) is one of the more heavily affected cortical areas in AD.[3] However, the anterior parts of this gyrus (Brodmann's area 24) are a major exception. This cortex does not appear affected until later in the disorder and may be spared entirely.[2,3] Interestingly, this part of the proisocortices is related closely to the motor cortices (supplementary, premotor, and primary motor cortices), which are typically spared in AD.[65,66]

Isocortical Involvement

As discussed in the first section, the nonuniform involvement of the isocortex in AD often can be detected by visual inspection of the gross brain.[39] In brief, pathological changes in the form of neurofibrillary tangles, neuritic plaques, and cell loss are abundant in the association areas of the temporal and parietal lobes in many Alzheimer's brains, and additional changes in the frontal and occipital lobes are common. In contrast, the primary sensory areas and the primary motor cortices are not affected greatly. The relative sparing of these sensory and motor areas is consistent with the paucity of motor and sensory impairment in AD. Nonetheless, the heavy and possibly early involvement of the olfactory cortex in AD[53] suggests that the olfactory modality is an exception, and behavioral data supports this idea.[67–69]

The association cortices affected in AD are of the so-called homotypical variety.[22,23] In the frontal lobe they form Brodmann's areas 9, 10, 11, 12, 13, 14, and 46. In the parietal lobe, they form Brodmann's areas 7, 39, and 40. In the occipitotemporal region, Brodmann's area 37 contains the major expanse of homotypical isocortex. Areas 20, 21, and 22 represent the major examples of this type of cortex in the temporal lobe. These regions are characterized by a similarity of architectural plan and form the largest part of the cerebral cortex in the human and nonhuman primate brain. They represent the classic six-layered pattern of isocortex. Layer II is well developed and has small neurons, mostly of the pyramidal variety. Layer III is generally wide and only slightly differentiated, with the major cell type being medium-sized pyramids. Layer IV is composed of small neurons of a pyramidal or stellate variety. Layer V is composed largely of pyramidal neurons of varying sizes with prominent apical dendrites and is distinguishable from layer VI, which is formed by a variety of neurons of varying size and shape. These areas give rise to a majority of the long corticocortical association projections that mediate interconnectivity between the various lobes and with the limbic cortical areas.[70,71] They participate prominently in both feedforward and feedback cortical connections (Figs. 2 and 3).

The degree of pathological involvement of the association cortices in AD can vary greatly, ranging from extensive involvement to only selective involvement in

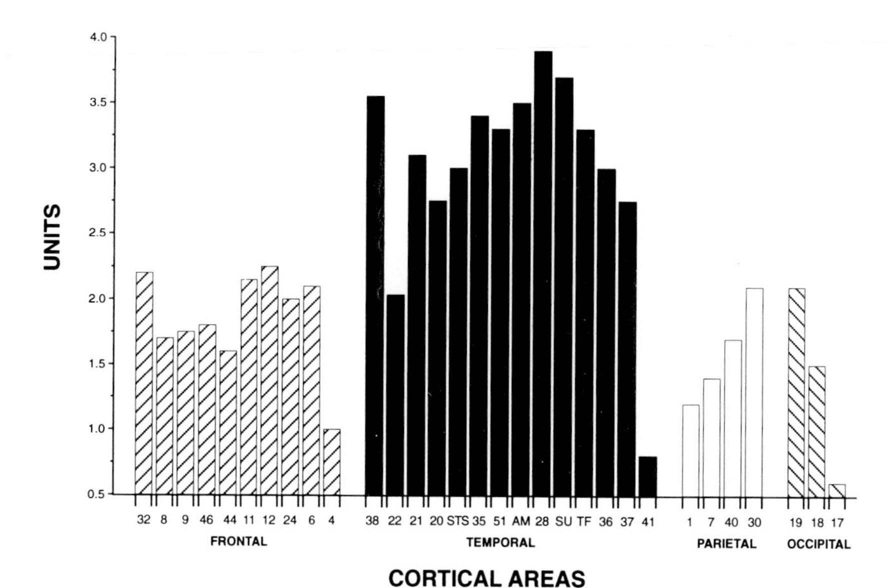

FIGURE 3. Histogram showing the mean density of neurofibrillary tangles (NFTs) per lobe for 18 hemispheres from patients with Alzheimer's disease as described by Arnold *et al.*[8] The areas sampled correspond in part to Brodmann's areas, but the amygdala (AM), subiculum (SU), and posterior parahippocampal area (TF) are included in the temporal lobe selection for comparison. Units correspond to numbers of neurofibrillary tangles per 1.6 mm^2 of tissue or per 10x microscopic field. A unit value of 1 was assigned to 1-10 tangles per field, 2 for 11-25 tangles per field, 3 for 26-50 per field, and 4 for more than 50 per field. Note the high density values for all temporal areas sampled.

one or two lobes. However, involvement of the proisocortex of the limbic lobe, the periallocortex of the anterior parahippocampal gyrus, and the allocortex of the hippocampal formation is more consistent. Whether this represents early involvement of these types of cortex or a differential degree of intensity once the disease process starts is not yet apparent, but it is likely both. Among isocortical association areas the parietotemporal and occipitotemporal regions seem affected most heavily in AD.[3]

Functional Correlates of Cortical Areas Affected in Alzheimer's Disease

Limbic Area Involvement

As shown in Figures 2, 3, and 4, our observations replicate previous observations that the allo-, periallo-, and proisocortices that form the limbic lobe represent the cortical areas affected most heavily in AD. Accordingly, one would expect functional changes in this disorder that resonate with such pathology. Clearly, memory impairment in AD, especially that of an anterograde variety, correlates well with ventromedial temporal pathology because parts of the hippocampal formation nearly always

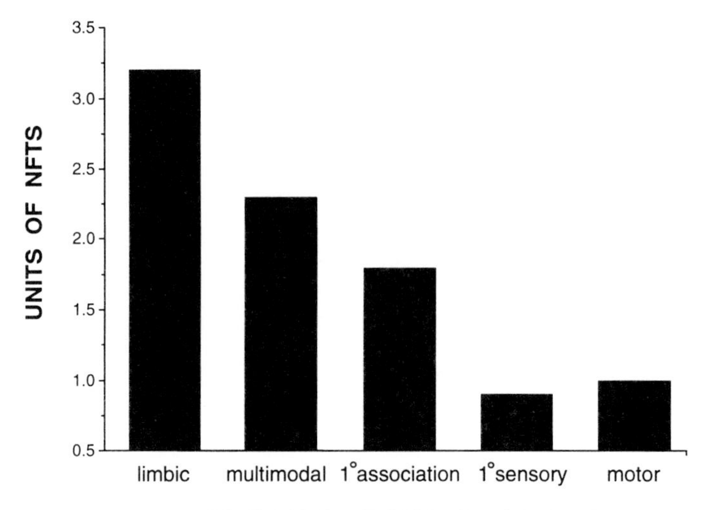

FIGURE 4. Histogram depicting the density of neurofibrillary tangles (see FIG. 3 legend for definition of units) according to functional types of cortex from 18 hemispheres of 11 brains with Alzheimer's disease. The values for limbic areas were derived from the classic cortical areas that form the limbic lobe plus the amygdala. Multimodal areas included those of the lateral frontal cortex, inferior parietal cortex, cortex in the banks of the superior temporal sulcus, and posterior parahippocampal area. Primary association areas were Brodmann's areas 18, 19, 7, and posterior area 22 adjacent to the primary auditory cortex. Primary sensory areas were Brodmann's area 17, 41, and 42 and 3, 1, and 2 for the visual, auditory, and somatosensory cortices respectively. Motor cortex sampled was Brodmann's area 4 of the precentral gyrus. Temporal limbic and multimodal areas contribute disproportionately to the high values for these functional types of cortex.

contain pathology in AD. Consistent with this is pathology in the entorhinal and perirhinal cortices inasmuch as these cortical areas are essential for cortical input to hippocampal formation and for the latter to report back via direct and indirect pathways to the association cortices. The integrity of this interplay seems essential for the assimilation of new learning and its more long-term registration elsewhere in the cortex.

Other functional correlates of limbic lobe AD pathology are less easy to characterize, but some predictions seem tenable. For example, temporal polar and orbitofrontal pathology would be expected to contribute to alterations in affect and social behavior because of their strong affiliation with the amygdala and other basal forebrain nuclei. Medial frontal pathology, particularly in Brodmann's area 25, would be expected to contribute to alterations in autonomic regulation because this cortex in all mammals projects to the hypothalamus and autonomic effector centers in the brainstem. Since area 25 also projects to sensory centers for the autonomic nervous system, general autonomic dysregulation might be predicted in AD, skewed perhaps towards sluggishness in higher order autonomic processes rather than homeostatic mechanisms per se. Finally, the pathology of the posterior cingulate cortex (Brodmann's area

23) would be consistent with attention-related disorders because these areas have widespread interactions with the frontal parietal and temporal association areas known to play roles in working memory and spatial awareness of both the personal (body) and the extrapersonal spaces.

Multimodal Area Involvement

Multimodal areas of the cortex have been defined in nonhuman primates and represent areas that receive input from several cortical areas uniquely involved with sensory processing.[72,73] The cortex that forms the banks of the superior temporal sulcus, parts of the prefrontal association cortex, the occipitotemporal cortices (Brodmann's areas 36 and 37), and Bonin and Bailey's area TF would seem to qualify as well as parts of the inferior parietal lobule. Although temporal lobe areas contribute disproportionately, these areas are damaged heavily in endstage AD (FIG. 4). The functional consequences of pathology in these areas is not understood, but it is believed by some that the integrity of the multimodal cortices is essential for nearly all forms of cognition. Certainly, these areas are expansive in man and probably essential for higher order behavior and for generating the synthetic percepts that enable the organism to capitalize maximally on the meaning of multiple ongoing environmental stimuli. These same areas are also thought to evoke combinations of percepts stored in other parts of the association cortices so that mental images of complex sensations can be retrieved.[74–76] Whatever, the cognitive handicap in AD can be profound at end stage with minimum ability to deal meaningfully with either the present or the past. Multimodal area pathology might be at the core of these behavioral changes.

Primary Association Area Involvement

These cortical areas lie adjacent to the primary sensory cortices and are dedicated functionally to whichever modality they border. The primary visual association cortices have been studied extensively in nonhuman primates, and specific areas within their boundaries serve the roles of parcellating the visual world into unique components such as form, color, and movement. It seems likely that the primary auditory and somatosensory cortices play a somewhat similar role. The primary association cortices are moderately altered in AD, and it is difficult to speculate on how this may be manifested in the behavior of victims. Certainly, one might suspect that the richness and range of sensory meaning are compromised even though the generic and rote skeleton of normal functioning persists.

Primary Sensory and Motor Area Involvement

As is apparent from gross inspection of the brain at autopsy, the cortical areas that form the primary sensory cortex and the motor, or agranular, cortex contain relatively small quantities of pathological change in AD. Functionally, this is consistent with the preservation of mobility until very late in the disease process and the

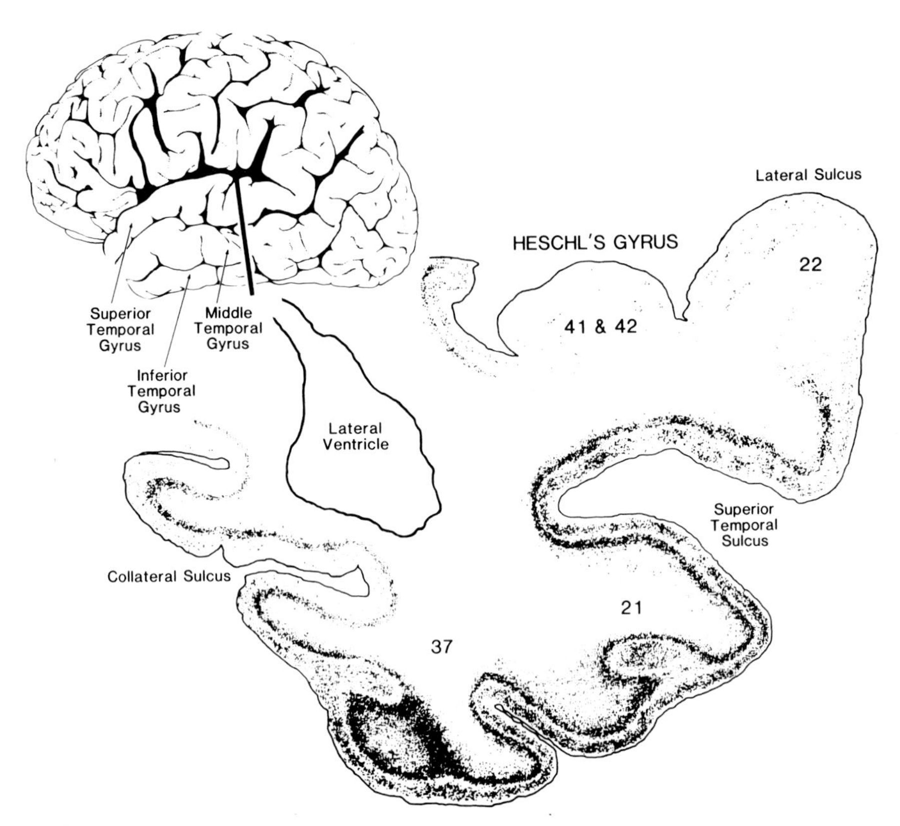

FIGURE 5. Microscope chartings of the topography and laminar distribution of neurofibrillary tangles in a temporal lobe section in Alzheimer's disease stained with thioflavin S. The *dark bar* across the temporal lobe (**top figure**) indicates the plane of section. Note that tangles are bilaminate in distribution, occurring primarily in layers III and V. Note also that they are low in density in the primary auditory cortex of Heschl's gyrus (areas 41 and 42) and the primary auditory association cortex (area 22) adjacent to it. However, their density is high in the multimodal association cortex of the superior temporal sulcus and in Brodmann's area 37 (Reproduced, with permission, from Van Hoesen and Damasio.[43])

general registry of sensory information as deduced by neurological examination (FIGS. 2 and 4).

LAMINAR SPECIFICITY OF CORTICAL PATHOLOGY IN ALZHEIMER'S DISEASE

The topographical specificity of Alzheimer's pathology in terms of cytoarchitecture is paralleled by a cellular and laminar specificity within affected areas (FIG. 5). This is clearly the case for neurofibrillary tangles. For example, neurons that form predominantly one or two cortical layers contain this cytoskeletal alteration, whereas their immediate neighbors in adjacent layers may be unaltered.[6,7,21,22,25,61,64] The same applies to some extent for neuritic plaque distribution.[23,77] Although this pathological marker may have a broader distribution in terms of laminar location,[22] there are many examples in the cerebral cortex where plaques are found largely in selective cortical layers. It is less clear whether cell loss, in addition to that attributable to neurofibrillary tangles, follows a laminar-specific pattern,[17,20,78] but it appears that larger cortical neurons are lost to a greater degree than are small neurons in both AD and normal aging. Because large neurons are located primarily in layers III, V, and VI, it would follow that at least some laminar specificity occurs in age-related cell loss whether AD is present or not.

CONNECTIONAL SPECIFICITY OF PATHOLOGY IN ALZHEIMER'S DISEASE

It is apparent that pathological changes in AD are decidedly skewed toward the multimodal association and limbic cortices (FIGS. 3, 4, and 5). As reported elsewhere,[58] cortical areas that form the association cortices are a major source of input to the limbic system and are a major recipient of limbic system output.[58,79–82] Moreover, they provide a major source of long cortical projections that link together the various parts of the cerebral cortex. Their destruction of the magnitude noted in AD would suggest that devastating cognitive change should ensue. Moreover, the laminar distribution of cortical pathology in AD correlates closely with the laminar origin of cortical projections and the laminar distribution of cortical association efferents.

The critical nature of these changes can be appreciated if one considers the connectional relation between the hippocampal formation and the cerebral cortex. The hippocampus receives few direct inputs from cortical association areas. Instead, its major cortical input arises from the entorhinal cortex (Brodmann's area 28), which forms the anterior parahippocampal gyrus. The entorhinal cortex gives rise to the perforant pathway, a major neural system of the temporal lobe, which terminates on the outer parts of the dendritic trees of dentate gyrus granule cells and hippocampal pyramidal cells. It has been stressed since the time of Ramón y Cajal's original description of the perforant pathway in Golgi-stained material that it carries the strongest source of input to the hippocampal formation and forms its exclusive link to the remainder of the cortex. The input to the entorhinal cortex thus becomes a critical issue, and many of its sources have been described in the last two decades.[58,83–88] Most afferents arise from neighboring cortical areas such as the presubiculum, parasubiculum, periamygdaloid, and prepiriform cortex and proisocortical areas of the temporal lobe. The latter includes area 35, the perirhinal cortex; area 36, the posterior parahippocampal cortex; area 37, the occipitotemporal cortex; and area 38, the temporal polar cortex. All of these areas are the recipients of powerful corticocortical association input from sensory association and multimodal cortical association areas and in turn project powerfully to the superficial layers of the entorhinal cortex. The critical relay neurons in areas 35, 36, 37, and 38 are the pyramids in layers III and V. Briefly, it is probably accurate to state that a sizable component of the feedforward system of cortical sensory connections converges onto the entorhinal cortex, where it is relayed to the hippocampal formation by the perforant pathway.

Nearly all of the key components of these neural systems are affected in AD.[21,89–91] For example, in areas 35, 36, 37, and 38, the most common distribution of neurofibrillary tangles is in cortical layers III and V. Moreover, the cells of origin for the perforant pathway, the large layer II stellate neurons and the more superficial layer III pyramids, are laden with neurofibrillary tangles (FIG. 6) and often are destroyed totally in AD.[92] There is a strong likelihood that the specific cortical neurons that form neural systems conveying cortical association input to the hippocampal formation are destroyed in AD. The reciprocal of this relationship also seems compromised greatly.[21] This feedback system arises primarily from the subicular and CA1 part of the hippocampal formation and from layer IV of the entorhinal cortex, which itself receives a large subicular projection. Both the subiculum and the CA1 zones of the hippocampal formation and layer IV of the entorhinal cortex typically contain neurofibrillary tangles in AD. Such observations led us to postulate that the pathological distribution of neurofibrillary tangles and neuritic plaques in AD dissects neural systems of the temporal lobe with a cellular precision that effectively isolates the hippocampal formation from the remainder of the cerebral cortex.[21,43,61]

MODULAR SPECIFICITY OF ALZHEIMER'S DISEASE PATHOLOGY

It is well known that certain cortical areas have a distinct vertical organization that exists in consort with the better known laminar or horizontal organization. The generic term "modular" has been used to characterize many different types of vertical organization and refers largely to periodicity, periodicity in the grouping of neurons, their axons, their transmitters, their enzyme characteristics, and importantly their functional properties.[94–110] Columns are the best characterized form of periodicity in the cortex even though exact morphological correlates to accompany and buttress the physiology can prove elusive. Nevertheless, at this time, all forms of modularity, particularly vertical organization, are of interest. Clearly, there are many forms of periodicity in the cortex, and these undoubtedly form a sandwich with many stacked elements. For these reasons, the distribution of pathological markers in AD are of great interest because they may reveal morphological and functional properties of a cortex that cannot be investigated experimentally.

Several investigators have reported evidence that neuritic plaques and neurofibrillary tangles have a modular distribution in AD cortex.[22,23,31,32,77,111] Some have argued that the modular distribution parallels the known patterns of corticocortical connections.[22,23,26,31,93,106] This would be of great interest if it could be proven, and it would indicate that AD pathology destroys the corticocortical association systems that interconnect the many areas of the cerebral cortex together.

We are favorably disposed to this hypothesis and have viewed AD pathology in a disconnection fashion since our initial report on the topic.[21] However, it is clear that AD changes in some parts of the cortex not only disconnect key structures, but also destroy certain elements of vertical or modular organization, perhaps stripping the cortex of key morphological units that support and underlie the functional operations of the cortex. This is especially apparent in the temporal cortex where AD pathology targets several types of modular organization uniquely associated with the various categories of cortex found there.[32]

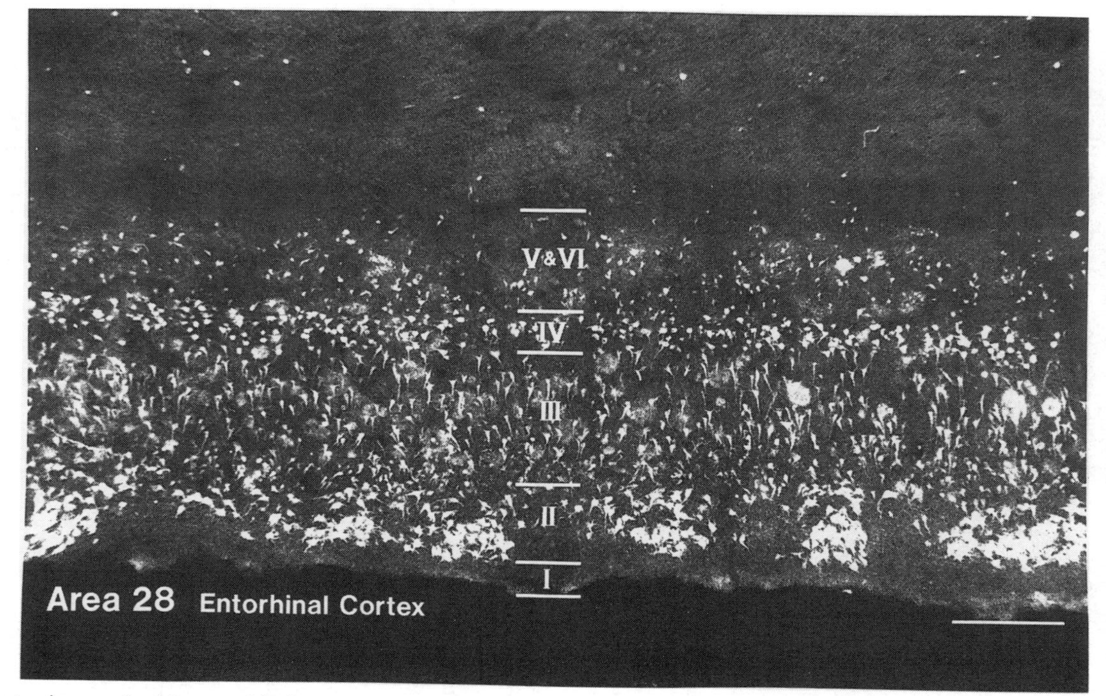

FIGURE 6. Photomicrograph of the entorhinal cortex at end stage in Alzheimer's disease stained with thioflavin S. Note the dense accumulation of neurofibrillary tangles in layers II, III, and IV. Layers II and III in other primates give rise to the perforant pathway that links the hippocampal formation to the cortex. Layer IV receives hippocampal output and projects back to these areas. Thus, entorhinal pathology of this magnitude alone can greatly compromise the interconnections between the association cortices and the hippocampal formation. Calibration bar: 200 μm.

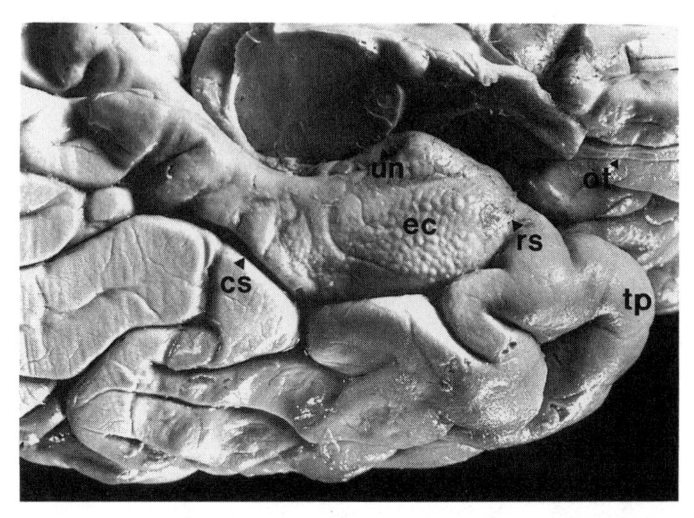

FIGURE 7. Photograph of the ventromedial temporal region in the normal human brain showing the location of the entorhinal cortex (ec). Note the mosaic of bumps or *verrucae* that can be seen on the surface of the ec. These correspond to the islands of layer II entorhinal neurons that are some of the earliest cortical neurons to develop neurofibrillary tangles in Alzheimer's disease (FIG. 6). Abbreviations: cs = collateral sulcus; rs = rhinal sulcus; un = uncus; tp = temporal pole; ot = olfactory tract.

A striking example of neuroanatomical modularity can be found in the human entorhinal cortex (FIG. 7) where the cortical surface is characterized by a mosaic of small elevations known as *verrucae*[44,88,110] visible even to the naked eye. These correlate with the patches of layer II multipolar neurons and their associated parvalbumin-containing interneurons (FIG. 8). In AD the multipolar neurons of entorhinal layer II are heavily invested by neurofibrillary tangles (FIG. 6) and contain no stainable rough endoplasmic reticulum or ribosomes. Their extensive destruction contributes greatly to the quantitative fact that entorhinal cortex suffers more pathology than does any other cortical area in AD. Curiously, these neurons are among some of the first affected in the cortex in AD,[8,30] and some pathology is observed in these neurons in normal aging.[112] Nevertheless, the essential point is that layer II neurons of the entorhinal cortex are the key elements of a modular pattern of organization, and they are destroyed or heavily damaged in all cases of AD.

A somewhat different pattern of modular change occurs in temporal area 35, the perirhinal cortex (FIG. 9). This proisocortical area forms the medial bank of the collateral sulcus in humans and adjoins the entorhinal cortex. Here neurofibrillary tangles form distinct columns usually four to five cell diameters in width that can be seen particularly in layers two and three. Pathologically free interspaces occur between the columns of neurofibrillary tangles. Braak and Braak[30] believe that the earliest age-related changes in the cortex occur here, and our observations support their assertion.

In the temporal isocortex a third type of modular change is suggested in AD. This is localized largely in layer three and consists of cell absent patches or domains

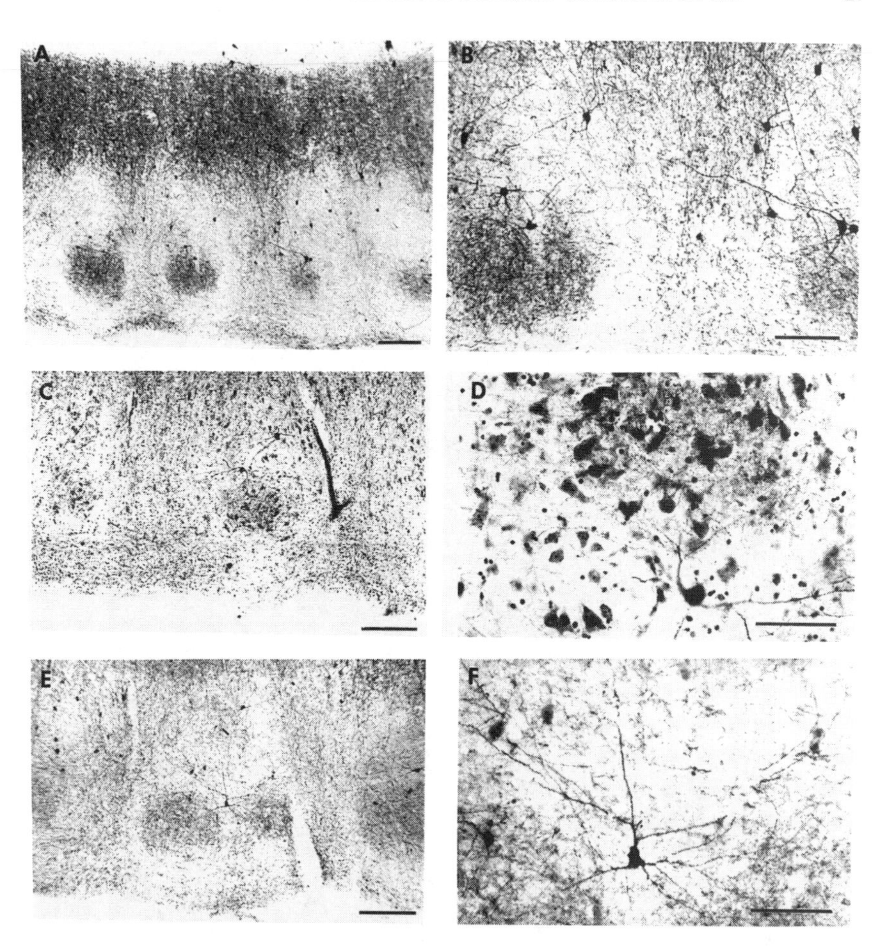

FIGURE 8. Photographs of parvalbumin immunostaining in the entorhinal cortex of normal cases. **B, D,** and **F** are higher magnifications of **A, C,** and **E,** respectively. **C** and **D** are counterstained for Nissl substance. Note the modular nature of this immunostaining in layer II, the vertical columns running from layer IV to layers I and II, and the dense band in layer IV (**A, C,** and **E**). In layer II, immunoreactive cell bodies are located deep to the cell islands and send their dendrites and axons into adjacent islands. Calibration bars: **A, C,** and **E:** 250 μm; **B, D,** and **F:** 100 μm.

that may be as wide as 8 to 12 cell diameters. Some of the neurons in these areas contain neurofibrillary tangles but others do not, suggesting that either they had tangles early in the disease process which had been removed already at death or they were affected and died of a disease process other than neurofibrillary tangle formation. These modular changes in layer III of the isocortex are of substantial interest because the pyramidal neurons of this layer in nonhuman primates give rise to the majority of corticocortical association connections, both of the intra- and interhemispheric

FIGURE 9. A column of neurofibrillary tangles stained with thioflavin S from the perirhinal cortex (Brodmann's area 35) in Alzheimer's disease. These occur as bands or strips in layers II and III of this cortex and set it apart from Brodmann's area 28 medially and 36 in the depths of the collateral fissure where different patterns of pathology are seen. Calibration bar: 50 μm.

variety. The cells of origin for individual subsets of these are grouped into large domains.[32]

In summary, a growing body of evidence suggests that AD pathology attacks at least some aspects of the modular organization of the cortex. If these are linked to the functional organization of the cortex, the result behaviorally could be quite devastating.

COMMENT AND CONCLUSION

From a neuroanatomical perspective, the topography of AD pathology is highly selective, affecting some types of cortex and sparing others. There is no foothold in current research for understanding this; why, for example, are the primary sensory and motor cortices spared and the limbic and multimodal association cortices highly affected? Alzheimer's disease pathology is also highly selective within affected areas, destroying some cortical layers, but sparing their neighbors. Again, there is no foothold in current research to understand this either. Clearly, larger cortical neurons acquire neurofibrillary tangles more so than do smaller ones, and accordingly, pyramidal neurons are more commonly targeted. It follows then that neurons that acquire neurofibrillary tangles have long axons of the corticocortical and corticofugal varieties which connect various parts of the cortex together and with subcortical structures along the neuraxis. However, it is probably inappropriate to view AD as a disease of corticocortical and corticofugal neural systems. Surely, in the affected association and limbic cortices, this is the case, but unaffected cortical areas also have corticocortical and corticofugal axons and they are spared. In fact, the largest neurons in the cortex, with the longest axons, are found in the motor cortices and neurofibrillary tangles are seldom seen there even at endstage AD after a long illness.

From a neural systems viewpoint, it is somewhat tempting to believe that multisynaptic pathways involving a series of interconnected neurons are targeted by pathology in AD. Those that eventually end in the entorhinal cortex and hippocampal formation lend themselves compellingly to this type of hypothesis. However, scrutiny of such a notion leaves gaps. For example, if AD pathology is patterned after a multisynaptic series of corticocortical connections, why aren't the primary areas affected by tangles in that they provide the first link in the series? In hippocampal formation, why aren't the granule cells of the dentate gyrus altered? They are bombarded by excitatory input from layer II entorhinal neurons in the intact brain, and these are some of the first cortical neurons affected in AD.

Clearly neuroanatomical treatments of the distribution of AD pathology have increased our appreciation of the selectivity of the lesion, and in some instances, this has correlated well with functional imaging results and the behavioral changes that accompany the disorder. Observations on the laminar distribution of pathology have also been important. The layers of origin for many cortical projections have been identified in nonhuman primate experimental studies allowing deductions about the neural systems compromised in AD. For example, it is clear that the hippocampal formation is nearly isolated or disconnected from the remainder of the cortex in AD by pathology in key neurons that mediate this linkage. However, this is due largely to the unique focal convergence of parallel neural systems onto the entorhinal cortex and the destruction of its output, the perforant pathway. A disconnection analysis of this sort has limitations for other neural systems.

Thus, it could be argued that we are at somewhat of an impasse to further understand how the neuroanatomy of cortical pathology in AD engineers dementia and a near total breakdown in cognition for its victims. With the information to date, we are also at an impasse for understanding why in many AD cases rather selective pathology in a brain with near normal weight is associated with dementia. As discussed in the section on modular organization of pathology in AD, a useful new direction may be emerging where the vertical organization and spatial patterns of cortical or pathological changes in AD are appreciated. These could be highly selective and disruptive from a functional point of view but not conspicuous from a neuropathological and/or neuroanatomical point of view. From a cognitive viewpoint they could bring the system down in a brain that otherwise appears relatively unaffected and thereby account for much of the variance observed in AD.

REFERENCES

1. Mutrux, S. 1947. Diagnostic différentiel histologique de la maladie d'Alzheimer et de la damence Sénile. Mschf. Psychiatr. Neurol. **113:** 100-107.
2. Brun, A. & L. Gustafson. 1976. Distribution of cerebral degeneration in Alzheimer's disease. Arch. Psychiatr. Nervenk. **223:** 15-33.
3. Brun, A. & E. Englund. 1981. Regional pattern of degeneration of Alzheimer's disease: Neuronal loss and histopathological grading. Histopathology **5:** 549-564.
4. Neary, D., J. S. Snowdon, D. M. A. Mann, D. M. Bowen, N. R. Sims, B. Northern, P. O. Yates & A. N. Davidson. 1986. Alzheimer's disease: A correlative study. J. Neurol. Neurosurg. Psychiatry **49:** 229-237.
5. Hof, P. R., C. Bouras, J. Constantinidis & J. H. Morrison. 1989. Balint's syndrome in Alzheimer's disease: Specific disruption of the occipitoparietal visual pathway. Brain Res. **493:** 368-375.
6. Hof, P. R., K. Cox & J. H. Morrison. 1990. Quantitative analysis of a vulnerable subset of pyramidal neurons in Alzheimer's disease. I. Superior frontal and inferior temporal cortex. J. Comp. Neurol. **301:** 44-54.
7. Hof, P. R. & J. H. Morrison. 1990. Quantitative analysis of a vulnerable subset of pyramidal neurons in Alzheimer's disease. II. Primary and secondary visual cortex. J. Comp. Neurol. **301:** 55-64.
8. Arnold, S. E., B. T. Hyman, J. Flory, A. R. Damasio & G. W. Van Hoesen. 1991. The topographical and neuroanatomical distribution of neurofibrillary tangles and neuritic plaques in the cerebral cortex of patients with Alzheimer's disease. Cereb. Cortex **1:** 103-116.
9. Arriagada, P. V., J. H. Growdon, E. T. H. Hedley-Whyte & B. T. Hyman. 1992. Neurofibrillary tangles but not senile plaques parallel duration and severity of Alzheimer's disease. Neurology (NY) **42:** 631-639.
10. Goodman, L. 1953. Alzheimer's disease: A clinico-pathological analysis of twenty-three cases with a theory on pathogenesis. J. Nerv. Ment. Dis. **117:** 97-130.
11. Hirano, A. & H. M. Zimmerman. 1962. Alzheimer's neurofibrillary changes. Arch. Neurol. **7:** 227-242.
12. Jamada, M. & P. Mehraein. 1968. Verteilungsmuster der senilin veranderungen im Gehirn. Die Beteiligung des limbischen Systems bei hernatropischen Prozessen des Senium and bei Morbus Alzheimer. Arch. Psychiatry Neurol. Sci. **211:** 308-324.
13. Ball, M. J. 1976. Neurofibrillary tangles and the pathogenesis of dementia: A quantitative study. Neuropath. Applied Neurobiol. **2:** 395-410.
14. Hooper, M. W. & F. S. Vogel. 1976. The limbic system in Alzheimer's disease. Am. J. Pathol. **85:** 1-13.
15. Ball, M. J. 1977. Neuronal loss, neurofibrillary tangles and granulovacuolar degeneration in the hippocampus with aging and dementia. Acta Neuropathol. **37:** 11-18.
16. Kemper, T. 1978. Senile dementia: A focal disease in the temporal lobe. *In* Senile Dementia: A Biomedical Approach. Nandy, K., Ed.: 105-113. Elsevier. New York.
17. Terry, R. D., A. Peck, R. DeTeresa, R. Schechter & D. S. Horoupian. 1981. Some morphometric aspects of the brain in senile dementia of the Alzheimer type. Ann. Neurol. **10:** 184-192.

18. Wilcock, G. K. & M. M. Esiri. 1982. Plaques, tangles and dementia, a qualitative study. J. Neurol. Sci. **56:** 343-356.
19. Wilcock, G. K. 1983. The temporal lobe in dementia of the Alzheimer's type. Gerontology **29:** 320-324.
20. Mann, D. M. A., P. O. Yates & B. Marcyniuk. 1985. Some morphometric observations on the cerebral cortex and hippocampus in presenile Alzheimer's disease, senile dementia of the Alzheimer's type and Down's syndrome in middle age. J. Neurol. Sci. **69:** 139-159.
21. Hyman, B. T., G. W. Van Hoesen, A. R. Damasio & C. L. Barnes. 1984. Alzheimer's disease: Cell-specific pathology isolates the hippocampal formation. Science **225:** 1168-1170.
22. Pearson, R. C. A., M. M. Esiri, R. W. Hjorns, G. K. Wilcock & T. P. S. Powell. 1985. Anatomical correlates of the distribution of pathological changes in the neocortex in Alzheimer's disease. Proc. Natl. Acad. Sci. USA **82:** 4531-4534.
23. Rogers, J. & J. H. Morrison. 1985. Quantitative morphology and regional and laminar distributions of the senile plaques in Alzheimer's disease. J. Neurosci. **5:** 2801-2808.
24. Saper, C. B., D. C. German & C. L. White, III. 1985. Neuronal pathology in the nucleus basalis and associated cell groups in senile dementia of the Alzheimer's type: Possible role in cell loss. Neurology **35:** 1089-1095.
25. Lewis, D. A., J. M. Campbell, R. D. Terry & J. H. Morrison. 1987. Laminar and regional distributions of neurofibrillary tangles and neuritic plaques in Alzheimer's disease: A quantitative study of visual and auditory cortices. J. Neurosci. **7:** 1799-1808.
26. Saper, C. B., B. H. Wainer & D. C. German. 1987. Axonal and transneuronal transport in the transmission of neurological disease: Potential role in system degenerations, including Alzheimer's disease. Neuroscience **23:** 389-398.
27. Moossy, J., G. S. Zubenko, J. Martines & G. R. Rao. 1988. Bilateral symmetry of morphologic lesions in Alzheimer's disease. Arch. Neurol. **45:** 251-254.
28. Papasozomenos, S. C. 1989. Tau protein immunoreactivity in dementia of the Alzheimer type. Lab. Invest. **60:** 123-137.
29. Pearson, R. C. A. & T. P. S. Powell. 1989. The neuroanatomy of Alzheimer's disease. Rev. Neurosci. **2:** 101-121.
30. Braak, H. & E. Braak. 1991. Neuropathological staging of Alzheimer's-related changes. Acta Neuropathol. (Berl.) **82:** 239-259.
31. Armstrong, R. A. 1993. Is the clustering of neurofibrillary tangles in Alzheimer's patients related to the cell of origin of specific cortico-cortical projections? Neurosci. Lett. **160:** 57-60.
32. Van Hoesen, G. W. & A. Solodkin. 1993. Some modular features of temporal cortex in humans as revealed by pathological changes in Alzheimer's disease. Cereb. Cortex **3:** 465-475.
33. Corsellis, J. A. N. 1976. Aging and the dementias. *In* Greenfield's Neuropathology. W. Blackwood & J. A. N. Corsellis, Eds.: 796-848. Edward Arnold. London.
34. Lamy, C., C. Duyckaerts, P. Delaere, Ch. Payan, J. Fermanian, V. Poulain & J. J. Hauw. 1989. Comparison of seven staining methods for senile plaques and neurofibrillary tangles in a prospective series of 15 elderly patients. Neuropathol. Appl. Neurobiol. **15:** 563-578.
35. Braak, H., E. Braak, I. Grundke-Iqbal & K. Iqbal. 1986. Occurrence of neuropil threads in the senile human brain and in Alzheimer's disease: A third location of paired helical filaments outside of neurofibrillary tangles and neuritic plaques. Neurosci. Lett. **65:** 351-355.
36. Tourtellotte, W. & G. W. Van Hoesen. 1991. The axonal origin of a subpopulation of dystrophic neurites in Alzheimer's disease. Neurosci. Lett. **129:** 11-16.
37. Su, J. H., B. J. Cummings & C. W. Cotman. 1993. Identification and distribution of axonal dystrophic neurites in Alzheimer's disease. Brain Res. **625:** 228-237.

38. MASLIAH, E., A. MILLER & R. D. TERRY. 1993. The synaptic organization of the neocortex in Alzheimer's disease. Med. Hypotheses **41:** 334-340.
39. TOMLINSON, B. E. & J. A. N. CORSELLIS. 1984. Aging and the dementias. *In* Greenfield's Neuropathology. J. Hume Adams, J. A. N. Corsellis & L. W. Duchen, Eds.: 951-1025. Wiley. New York.
40. MANN, D. M. A. 1985. The neuropathology of Alzheimer's disease: A review with pathogenetic, aetiological and therapeutic considerations. Mech. Aging Dev. **31:** 213-255.
41. CORSELLIS, J. A. N. 1970. The limbic areas in Alzheimer's disease and in other conditions associated with dementia. *In* Alzheimer's Disease and Related Conditions. G. E. W. Wolstenhome, M. O'Connor, Eds.: 37-45. Churchill. London.
42. VAN HOESEN, G. W. 1985. Neural systems of the non-human primate forebrain implicated in memory. *In* Memory Dysfunctions. S. Corkin, E. Gramzu & D. Olton, Eds.: Ann. N.Y. Acad. Sci. 97-112.
43. VAN HOESEN, G. W. & A. R. DAMASIO. 1987. Neural correlates of cognitive impairment in Alzheimer's disease. *In* Handbook of Physiology—The Nervous System. Vol V. Higher functions of the brain. Pt 1. V. B. Mountcastle, F. Plum, S. R. Geiger, Eds.: 871-898. American Physiological Society. Bethesda, MD.
44. VAN HOESEN, G. W., B. T. HYMAN & A. R. DAMASIO. 1991. Entorhinal cortex pathology in Alzheimer's disease. Hippocampus **1:** 1-18.
45. ALZHEIMER, A. 1907. Über eine eigenartige Erkrankung der Hirnrinde. Allg. Z. Psychiatr. **64:** 146-148.
46. BALL, M. J. 1978. Histopathology of Cellular Changes in Alzheimer's Disease. *In* Senile Dementia: A Biomedical Approach. K. Nandy, Ed.: 89-104. Elsevier. New York.
47. BALL, M. J. 1978. Topographic distribution of neurofibrillary tangles and granulovacuolar degeneration in hippocampal cortex of aging and demented patients: A quantitative study. Acta Neuropathol. **42:** 73-80.
48. BALL, M. J., M. FISMAN, V. HACHINSKI, W. BLUME, A. FOX, V. A. KRAL, A. J. KIRSHEN, H. FOX & H. MERSKEY. 1985. A new definition of Alzheimer's disease: A hippocampal dementia. Lancet **i:** 14-16.
49. BURGER, P. C. 1983. The limbic system in Alzheimer's disease. *In* Banbury Report. Biological Aspects of Alzheimer's Disease. R. Katzman, Ed. **15:** 37-44. Cold Spring Harbor Laboratory. Cold Spring Harbor, NY.
50. GIBSON, P. H. 1983. Form and distribution of senile plaques seen in silver impregnated sections in the brains of intellectually normal elderly people and people with Alzheimer-type dementia. Neuropathol. Applied Neurobiol. **9:** 379-389.
51. HERZOG, A. G. & T. L. KEMPER. 1980. Amygdaloid changes in aging and dementia. Arch. Neurol. **37:** 625-629.
52. KEMPER, T. L. 1983. Organization of the neuropathology of the amygdala in Alzheimer's disease. *In* R. Katzman, Ed. Banbury Report. Biological Aspects of Alzheimer's Disease. **15:** 31-35. Cold Spring Harbor Laboratory. Cold Spring Harbor, NY.
53. ESIRI, M. M. & G. K. WILCOCK. 1984. The olfactory bulbs in Alzheimer's disease. J. Neurol. Neurosurg. Psychiatry **47:** 56-60.
54. BRASHEAR, H. R., M. S. GODEC & J. CARLSON. 1988. The distribution of neuritic plaques and acetylcholinesterase staining in the amygdala in Alzheimer's disease. Neurology **38:** 1694-1699.
55. BRADY, D. R. & E. J. MUFSON. 1990. Amygdaloid pathology in Alzheimer's disease: Quantitative and qualitative analysis. Dementia **1:** 5-17.
56. KROMER-VOGT, L. J., B. T. HYMAN, G. W. VAN HOESEN & A. R. DAMASIO. 1990. Pathologic alterations in the amygdala in Alzheimer's disease. Neuroscience **37:** 377-385.
57. UNGER, J. W., L. W. LAPHAM, T. H. MCNEIL, T. A. ESKIN & R. W. HAMILL. 1991. The amygdala in Alzheimer's disease: Neuropathology and Alz-50 immunoreactivity. Neurobiol. Aging **12:** 389-399.

58. VAN HOESEN, G. W. 1982. The parahippocampal gyrus: New observations regarding its cortical connections in the monkey. Trends Neurosci. **5:** 345-350.
59. NOWAKOWSKI, R. S. & P. RAKIC. 1981. The site of origin and route and rate of migration of neurons to the hippocampal region of the rhesus monkey. J. Comp. Neurol. **196:** 129-154.
60. RAKIC, P. & R. S. NOWAKOWSKI. 1981. The time of origin of neurons in the hippocampal region of the rhesus monkey. J. Comp. Neurol. **196:** 99-128.
61. VAN HOESEN, G. W., B. T. HYMAN & A. R. DAMASIO. 1986. Cell-specific pathology in neural systems of the temporal lobe in Alzheimer's disease. *In* Progress in Brain Research. D. Swaab, Ed. **70:** 361-375. Elsevier. Amsterdam.
62. HYMAN, B. T., G. W. VAN HOESEN, L. J. KROMER & A. R. DAMASIO. 1985. The subicular cortices in Alzheimer's disease: Neuroanatomical relationships and the memory impairment. Soc. Neurosci. (Abstr.) **11:** 458.
63. SANIDES, F. 1962. Die Architektonik des menschlichen Stirhirns. Monogr. Neurol. Psychol. **98:** 1-201.
64. BRAAK, H. & E. BRAAK. 1985. On areas of transition between entorhinal allocortex and temporary isocortex in the human brain. Normal morphology and lamina-specific pathology in Alzheimer's disease. Acta Neuropathol. **68:** 325-332.
65. MORECRAFT, R. J. & G. W. VAN HOESEN. 1992. Cingulate input to the primary and supplementary motor cortices in the rhesus monkey. J. Comp. Neurol. **322:** 471-489.
66. MORECRAFT, R. J. & G. W. VAN HOESEN. 1993. Frontal granular cortex input to the cingulate (M_{III}) supplementary (M_{II}) and primary (M_I) motor cortices in the rhesus monkey. J. Comp. Neurol. **337:** 669-689.
67. CORWIN, J., M. SERBY, P. CONRAD & J. ROTROSEN. 1985. Olfactory recognition deficit in Alzheimer's and Parkinsonian dementias. IRCS Med. Sci. Libr. Compend. **13:** 260.
68. DOTY, R. L., P. F. REYES & T. GREGOR. 1987. Presence of both odor identification and detection deficits in Alzheimer's disease. Brain Res. Bull. **18:** 597-600.
69. WARNER, M. D., O. A. PEABODY, J. J. FLATTERY & J. R. TINKLENBERG. 1986. Olfactory deficits and Alzheimer's disease. Biol. Psychiatry **21:** 116-118.
70. JONES, E. G. & T. P. S. POWELL. 1970. An anatomical study of converging sensory pathways within the cerebral cortex of the monkey. Brain **93:** 793-820.
71. PANDYA, D. N. & H. G. J. M. KUYPERS. 1969. Corticocortical connections in the rhesus monkey. Brain Res. **13:** 13-36.
72. PANDYA, D. N. & B. SELTZER. 1982. Association areas of the cerebral cortex. Trends Neurosci. **5:** 386-390.
73. PANDYA, D. N. & E. H. YETERIAN. 1985. Architecture and connections of cortical association areas. *In* Cerebral Cortex, Vol 4. Association and Auditory Cortices. A. Peters & E. G. Jones, Eds.: 3-61. Plenum. New York.
74. DAMASIO, A. R. 1989. Multiregional retroactivation: A systems level model for some neural substrates of cognition. Cognition **33:** 25-62.
75. DAMASIO, A. R. 1989. The brain binds entities and events by multiregional activation from convergence zones. Neural. Comput. **1:** 123-132.
76. DAMASIO, A. R., H. DAMASIO, D. TRANEL & J. P. BRANDT. 1990. The neural regionalization of knowledge access devices: Preliminary evidence. *In* Symposia on Quantitative Biology, Vol 55. Cold Spring Harbor Laboratory. Cold Spring Harbor, NY.
77. DUYCKAERTS, C., J.-J. HAUW, F. BASTENAIRE, F. PIETTE, C. POULAIN, V. RAINSARD, F. JAVOY-AGID & P. BERTHAUX. 1986. Laminar distribution of neocortical senile plaques in senile dementia of the Alzheimer type. Acta Neuropathol. (Berl.) **70:** 249-256.
78. TERRY, R. D. & R. KATZMAN. 1983. Senile dementia of the Alzheimer type. Ann. Neurol. **10:** 184-192.
79. ROSENE, D. L. & G. W. VAN HOESEN. 1977. Hippocampal formation of the primate brain. *In* Cerebral Cortex, Vol 6. E. G. Jones & A. Peters, Eds.: 345-456. Plenum. New York.

80. Porrino, L. J., A. M. Crane & P. S. Goldman-Rakic. 1981. Direct and indirect pathways from the amygdala to the frontal lobe in the rhesus monkeys. J. Comp. Neurol. **198:** 121-136.
81. Kosel, K. C., G. W. Van Hoesen & D. L. Rosene. 1982. Nonhippocampal cortical projections from the entorhinal cortex in the rat and rhesus monkey. Brain Res. **244:** 201-213.
82. Amaral, D. G. & J. L. Price. 1984. Amygdalo-cortical projections in the monkey (*Macaca fascicularis*). J. Comp. Neurol. **230:** 465-496.
83. Van Hoesen, G. W., D. N. Pandya & N. Butters. 1972. Cortical afferents to the entorhinal cortex of the rhesus monkey. Science **175:** 1471-1473.
84. Van Hoesen, G. W. & D. N. Pandya. 1975. Some connections of the entorhinal (area 28) and perirhinal (area 35) cortices of the rhesus monkey. I. Temporal lobe afferents. Brain Res. **91:** 1-24.
85. Van Hoesen, G. W., D. N. Pandya & N. Butters. 1975. Some connections of the entorhinal (area 28) and perirhinal (area 35) cortices of the rhesus monkey. II. Frontal lobe afferents. Brain Res. **95:** 25-38.
86. Amaral, D. G. 1987. Memory: Anatomical organization of candidate brain regions. *In* Handbook of Physiology—The Nervous System, Vol. V. Higher Functions of the Brain, Pt I. V. B. Mountcastle, F. Plum & S. R. Geiger, Eds.: 211-294. American Physiological Society. Bethesda, MD.
87. Insausti, R., D. G. Amaral & W. M. Cowan. 1987. The entorhinal cortex of the monkey: II. Cortical afferents. J. Comp. Neurol. **264:** 356-395.
88. Amaral, D. G. & R. Insausti. 1990. Hippocampal formation. *In* The Human Nervous System. Paxinos G., Ed.: 711-755. Academic. New York.
89. Hyman, B. T., G. W. Van Hoesen & A. R. Damasio. 1987. Alzheimer's disease: Glutamate depletion in perforant pathway terminals. Ann. Neurol. **22:** 37-40.
90. Hyman, B. T., L. J. Kromer & G. W. Van Hoesen. 1988. A direct demonstration of the perforant pathway terminal zone in Alzheimer's disease using the monoclonal antibody Alz-50. Brain Res. **450:** 392-397.
91. Hyman, B. T., G. W. Van Hoesen & A. R. Damasio. 1990. Memory-related neural systems in Alzheimer's disease: An anatomic study. Neurology **40:** 1721-1730.
92. Hyman, B. T., G. W. Van Hoesen, L. J. Kromer & A. R. Damasio. 1986. Perforant pathway changes and the memory impairment of Alzheimer's disease. Ann. Neurol. **20:** 472-481.
93. Morrison, J. H., D. A. Lewis & M. J. Campbell. 1987. Distribution of neurofibrillary tangles and nonphosphorylated neurofilament protein-immunoreactive neurons in cerebral cortex: Implications for loss of corticocortical circuits in Alzheimer's disease. *In* Banbury Report, Vol 27, Molecular Neuropathology of Aging. P. Davies & C. E. Finch, Eds.: 109-124. Cold Spring Harbor Laboratory. Cold Spring Harbor, NY.
94. Mountcastle, V. B. 1957. Modality and topographic properties of single neurons of cat's somatic sensory cortex. J. Neurophysiol. **20:** 408-434.
95. Woolsey, T. A. & H. van der Loos. 1970. The structural organization of layer IV in the somatosensory region (SI) of mouse cerebral cortex. The description of a cortical barrel field composed of discrete cytoarchitectonic units. Brain Res. **17:** 205-242.
96. Hubel, D. H. & T. N. Wiesel. 1972. Laminar and columnar distribution of geniculocortical fibres in the macaque monkey. J. Comp. Neurol. **146:** 421-450.
97. Jones, E. G., H. Burton & R. Porter. 1975. Commissural and cortico-cortical 'columns' in the somatic sensory cortex of primates. Science **190:** 572-574.
98. Goldman, P. S. & W. J. H. Nauta. 1977. Columnar distribution of corticocortical fibres in the frontal association, limbic and motor cortex of the developing rhesus monkey. Brain Res. **122:** 393-414.
99. Hubel, D. H. & T. N. Wiesel. 1977. Functional architecture of macaque monkey visual cortex. Proc. R. Soc. Lond. (Biol.) **198:** 1-59.

100. Horton, J. C. & D. H. Hubel. 1981. Regular patchy distribution of cytochrome oxidase staining in primary visual cortex of macaque monkey. Nature **292:** 762.
101. Bugbee, N. M. & P. S. Goldman-Rakic. 1983. Columnar organization of corticocortical projections in squirrel and rhesus monkeys: Similarity of column width in species of differing cortical volume. J. Comp. Neurol. **220:** 355-364.
102. Goldman-Rakic, P. S. & M. L. Schwartz. 1982. Interdigitation of contralateral and ipsilateral columnar projections to frontal association cortex in primates. Science **216:** 755-757.
103. Goldman-Rakic, P. S. 1984. Modular organization of pre-frontal cortex. Trends Neurosci. **7:** 419.
104. DeFelipe, J., S. H. C. Hendry, T. Hashikawa, M. Molinari & E. G. Jones. 1990. A microcolumnar structure of monkey cerebral cortex revealed by immunocytochemical studies of double bouquet cell axons. Neuroscience **37:** 655-673.
105. Purves, D. & A. S. LaMantia. 1990. Construction of modular circuits in the mammalian brain. Cold Spring Harbor Symp. Quant. Biol. **55:** 362-368.
106. Hiorns, R. W., J. W. Neal, R. C. A. Pearson & T. P. S. Powell. 1991. Clustering of ipsilateral cortico-cortical projection neurons to area 7 in the rhesus monkey. Proc. R. Soc. Lond. (Biol.) **246:** 1-9.
107. Hevner, R. F. & M. T. T. Wong-Riley. 1992. Entorhinal cortex of the human, monkey and rat: Metabolic map as revealed by cytochrome oxidase. J. Comp. Neurol. **326:** 451-459.
108. Purves, D., D. R. Riddle & A. S. LaMantia. 1992. Iterated patterns of brain circuitry (or how the cortex gets its spots). Trends Neurosci. **15:** 362-368.
109. Purves, D. & A. LaMantia. 1993. Development of blobs in the visual cortex of Macaques. J. Comp. Neurol. **334:** 169-175.
110. Klinger, J. 1948. Die makroskopische Anatomie der Ammonsformation, Denkschrifter der Schweizerischen Naturforeschenden Gesellschaft, Band LXVIII, Abh. 1. Gebrüfrt zgtryx zsh/. Zürich.
111. Akiyama, H., T. Yamada, P. L. McGeer, T. Kawawata, I. Tooyama & T. Ishii. 1993. Columnar arrangement of β-amyloid protein deposits in the cerebral cortex of patients with Alzheimer's disease. Acta Neuropathol. **85:** 400-403.
112. Arriagada, P. & B. T. Hyman. 1990. Topographic distribution of Alzheimer neuronal changes in normal aging brains. J. Neuropathol. Exp. Neurol. **49:** 226.

Possible Role of Apoptosis in Alzheimer's Disease

CARL W. COTMAN, EDWARD R. WHITTEMORE, JOHN A. WATT, AILEEN J. ANDERSON, AND DERYK T. LOO

Irvine Research Unit in Brain Aging
University of California, Irvine
Irvine, California 92717-4550

The development and maintenance of adult tissues are achieved through interactions among several dynamically regulated mechanisms including cell proliferation, differentiation, and programmed cell death. In programmed cell death, cells are destroyed by a cell ''suicide'' program called apoptosis.[1,2] It is suggested that apoptosis provides a mechanism well suited to the elimination of ''unwanted'' cells. This mechanism is activated by response to a variety of highly specific signals including hormones, growth factors, viral infections, toxins, and certain types of injury.

It is increasingly clear that apoptosis is an important mechanism that is active throughout life. During development, apoptosis has a key role in cell selection. In adult tissues, apoptosis occurs during tissue regression (eg, in the liver following hyperplasia),[3] clonal deletion of thymocytes (eg, *in vivo* via self-directed antigens),[4] toxin damage (eg, dioxin and carbon tetrachloride diethylmitrosamine exposure),[4,5] and hormonal changes (eg, glucocorticoids in thymocytes).[6]

Apoptosis may also serve a protective role against disease and tissue damage.[5] As a cell death mechanism, apoptosis can selectively remove individual cells in tissue without damage to healthy cells. Old or damaged cells are preferentially eliminated, as in mitogen-induced hyperplasia in rat liver[7] or toxin-induced damage to proliferating cells in intestine.[8] Apoptosis also degrades DNA, thereby deleting faulty genetic messages that may have been altered by toxins or viruses.[9] Furthermore, dying cells are rapidly removed by phagocytosis which minimizes inflammatory reactions.[2] These features may serve as protective mechanisms in disease. Alternatively, apoptosis may participate in chronic disease.

APOPTOSIS AS A MECHANISM IN ALZHEIMER'S DISEASE

Recently, we suggested that apoptosis is one of the mechanisms leading to neuronal loss in Alzheimer's disease (AD). Furthermore, we suggested that β-amyloid (Aβ), a small peptide of approximately 42 amino acids, is one of the stimuli that can initiate the apoptotic program. When configured in a particular assembly state, Aβ appears to be interpreted by neurons as a signal to initiate the apoptotic program. In this chapter, we develop this idea in the context of recent findings. We describe recent data suggesting that Aβ organizes the onset of brain pathology as it accumulates.

Aβ causes neuronal processes to degenerate, thereby disrupting key brain circuits, places neurons at risk to a variety of insults, and eventually leads to programmed cell death. We discuss the possible mechanisms that may control these degenerative processes, particularly in relation to the regulation of intracellular calcium.

In the aging and AD brain Aβ accumulates in senile plaques and to lesser degree in neurofibrillary tangles. Senile plaques are infiltrated with dystrophic neurites and are colocalized with areas of cell loss and/or projection sites. The association of Aβ with neuritic pathology and neuronal loss suggests that this molecule may demonstrate bioactivity or, alternatively, it may be an inert byproduct of ongoing pathology.

Accordingly, several years ago, we began to investigate the possibility that Aβ may actively drive the course of brain dysfunction. Inasmuch as an accurate animal model of AD does not yet exist, investigation of the role of Aβ in AD requires an experimental approach in which neuronal conditions, Aβ concentrations, and molecular assembly state can be controlled and cellular responses quantified. If Aβ deposits initiate or contribute to neurodegeneration in AD as suggested by histologic examination of AD brain tissue, then Aβ should exhibit bioactivity in culture consistent with this proposed role. On the basis of this premise, we and others have used primary neuronal culture paradigms to examine the role of Aβ in AD.

β-AMYLOID CAN STIMULATE NEURONAL GROWTH OR LEAD TO NEURONAL DEATH

The results of numerous *in vitro* experiments suggest that Aβ is capable of either enhancing growth or promoting toxicity. These properties appear to be related to the aggregation state of Aβ. In brief, findings include:

- Acute application of Aβ in culture stimulates neuronal process outgrowth and enhances survival over short time intervals.[10,11] This effect prevails over short time periods and appears to be most consistent with soluble Aβ.
- Simulating the *in vivo* accumulation of Aβ observed in the AD brain by allowing cultured neurons to grow in the presence of Aβ aggregates results in the attraction of neurites; however, once attracted, these neurites become dystrophic and degenerate, appearing similar in morphology to the dystrophic neuronal fibers observed in AD.[12]
- Aβ is toxic to neurons.[11,13,14] It increases the susceptibility of neurons to excitotoxicity in both rodent[13] and human[15] neurons. These *in vitro* results have correctly predicted a similar outcome in the rodent brain.[16]
- Several investigators have observed reductions in glucose metabolism in the AD brain. We found that Aβ increases the vulnerability of neurons to injury when glucose levels are reduced.[17] Additionally, glucose deprivation induces increases in tau and ubiquitin immunoreactivity in cultured neurons similar to those observed in the AD brain.[18]
- Aβ can also directly induce neuronal degeneration in cultured neurons.[14,19] This appears to depend on the state of Aβ aggregation,[14,20,21] probably related to

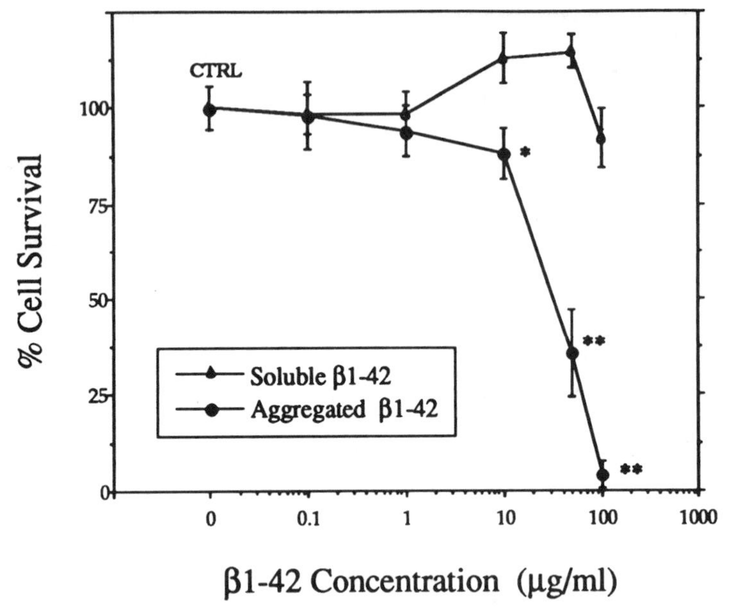

FIGURE 1. The effect of Aβ (β1-42) on primary neurons depends on peptide assembly state. (**A**) Soluble freshly prepared Aβ and (**B**) aged-assembled Aβ.[14]

β-pleated sheet structure.[22] That is, activation of neuronal degeneration appears to depend on the conversion of soluble Aβ peptide to a particular assembly state (Fig. 1).

Our finding that the assembly state of Aβ is a significant variable in producing neuronal degeneration allowed us and others to obtain consistent results across experiments and between laboratories. As a result, we noticed an unusual characteristic of Aβ-induced neurodegeneration. Previous work had demonstrated a strict correlation between lactate dehydrogenase (LDH) release and glutamate-mediated necrotic degeneration.[23] A clue that cell death initiated by Aβ peptides may be due to apoptosis rather than necrosis was the observation that LDH release measured 24 hours after treatment with Aβ did not correlate with the extent of morphologic degeneration observed at this time point, indicating that membrane integrity was still preserved. In addition, we noted that neurons degenerate asynchronously over 24 hours of exposure to Aβ and exhibited small, condensed, and irregularly shaped cell bodies.[24]

CHARACTERISTICS AND STAGES IN APOPTOSIS

Apoptosis was originally defined on the basis of morphologic and functional grounds.[1] Morphologic features still provide the most reliable markers and general characteristics of apoptosis. Apoptosis is characterized by cell surface protuberances (blebs), chromatin condensation, and nuclear shrinkage (pyknosis), followed by nuclear fragmentation (karyorrhexis) into multiple nuclear bodies. Cytoplasmic changes include polyribosome dispersal and cell shrinkage resulting from the blebbing process. Limited amounts of endoplasmic reticulum remain and mitochondria appear normal and unswollen.[2,25] Importantly, plasma membrane integrity is maintained until late in the course of apoptosis when secondary necrosis occurs. An additional mechanism in the apoptotic pathway is the activation of endogenous nucleases and subsequent DNA fragmentation into oligonucleosome-length fragments. In contrast, necrosis is characterized by dilation of the endoplasmic reticulum and mitochondria, and ultimately membrane disintegration with subsequent loss of cytoplasmic contents, leading to inflammation and further damage to surrounding healthy tissues. To further explore the possibilities of apoptosis, scanning electron microscopy (SEM) and transmission electron microscopy (TEM) were carried out to determine if Aβ-treated neurons also exhibit ultrastructural changes consistent with apoptosis.[24,26]

Neurons examined by SEM following 24-hour exposure to Aβ peptides exhibit severe surface blebbing and loss of neurites. Limited somal blebbing was observed as early as 4 hours following Aβ treatment, the earliest posttreatment time examined, and continued to increase in occurrence throughout the 24-hour experimental period. Blebbing occurred globally on affected cells and did not appear restricted to sites of direct amyloid contact (FIG. 2). The blebs were initially of small caliber but appeared larger with time. Ultimately, continued blebbing resulted in the progressive dissociation of the cell into multiple membrane-bound bodies. Transmission electron microscopic examination of treated hippocampal neurons revealed compact patches of condensed nuclear chromatin with no apparent plasma membrane disruption (FIG. 3). Nuclear morphology of untreated neurons was characterized by dispersed euchromatin dominating the nucleoplasm in conjunction with a prominent nucleolus. Limited condensed heterochromatin was typically observed at the margins of the nuclear envelope. The initial response to Aβ stimulation was the appearance of cytoplasmic vacuoles (FIG. 4). Closely following vacuole formation was the accumulation of heterochromatin at the periphery of the internal margin of the nuclear envelope. The chromatin patterns observed suggest a peripheral to central process of chromatin condensation, ultimately forming a single compact electron-dense body. The condensed nuclei subsequently underwent a series of membrane involutions that resulted in the fragmentation of the pyknotic nuclei into two or more membrane-bound bodies.

Concurrent with chromatin compaction, rapid dispersal of polyribosomal rosettes was observed, lending a fine granular and increasingly electron dense appearance to the cytoplasm. With continued compaction of the cytoplasm fewer organelles, most notably the microtubules, golgi complex, and smooth endoplasmic reticulum (ER), could be recognized, indicating continued degeneration as apoptosis proceeded. Ultimately, a reduced population of intact mitochondria, rough ER, and ribosomes was all that remained of the original compliment of cytoplasmic organelles. Thus, Aβ-induced ultrastructural changes in cultured neurons are consistent with apoptosis. We next sought biochemical evidence that Aβ-treated neurons degenerate through an apoptotic pathway.

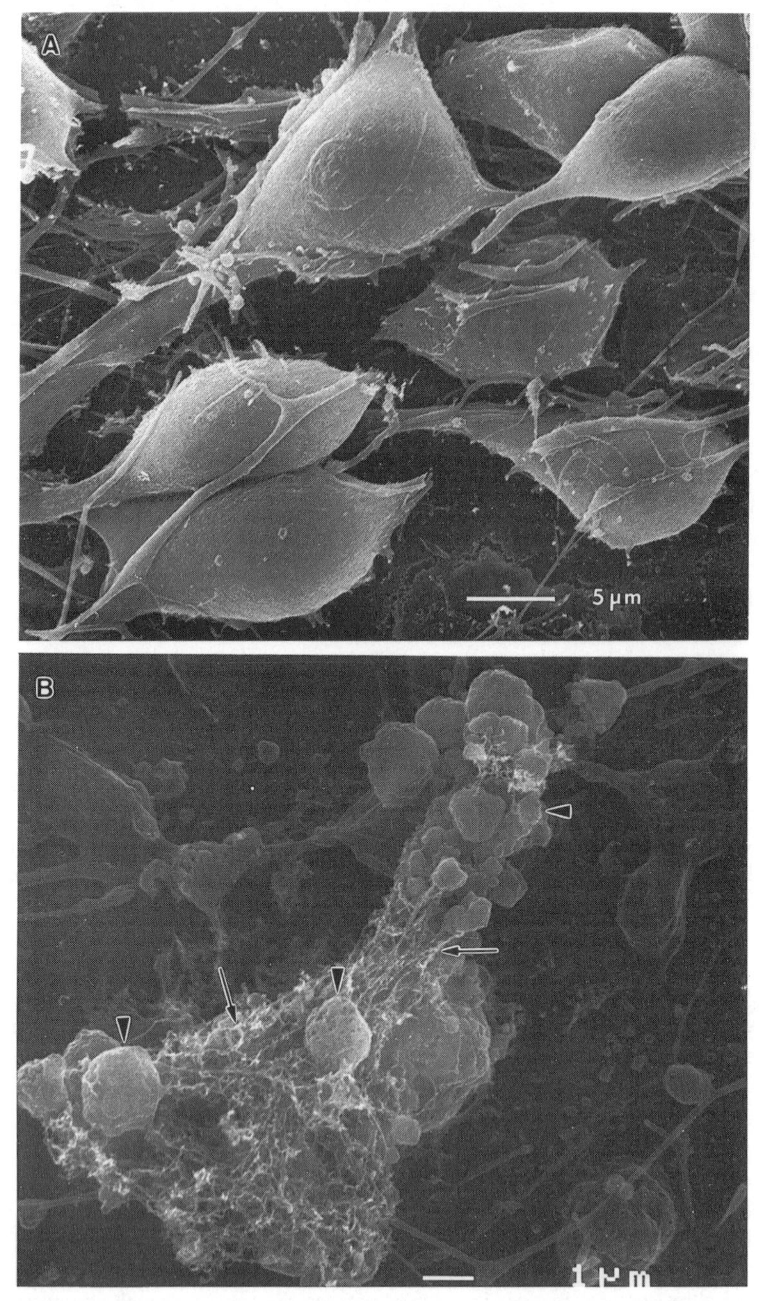

FIGURE 2. (**A**) Scanning electron micrograph of nontreated primary hippocampal neurons. (**B**) This Aβ-treated hippocampal neuron is enveloped in Aβ fibrils (*arrow*) and displays numerous apoptotic bodies (*arrowheads*).

DNA FRAGMENTATION

DNA fragmentation into oligonucleosome-length fragments is a discrete molecular event which occurs during apoptosis in many cells.[2,27] The process of fragmentation has been studied in detail in a variety of cell types using biochemical methods to monitor fragmentation among cell populations. This pattern is distinct from that seen with necrosis. Genomic DNA isolated from necrotic cells displays a smear following gel electrophoresis, indicative of random DNA cleavage.[2] In contrast, genomic DNA isolated from apoptotic cells displays a characteristic ''ladder'' pattern consisting of multiples of ~180 base pairs of DNA following fractionation by gel electrophoresis. This DNA cleavage is hypothesized to result from the production and/or activation of an endonuclease which cleaves DNA at exposed linker regions between histones.[28] One mechanism causing DNA fragmentation may occur through activation of a calcium-dependent endonuclease; such an endonuclease has been purified and characterized from thymocytes undergoing apoptosis.[28]

Although DNA fragmentation is not a necessary event in all types of apoptosis,[2,29,30] it does occur in Aβ-induced apoptosis. We hypothesize that Aβ-induced DNA fragmentation occurs after an induction period of several hours, follows nuclear condensation, and may be linked to immediate early genes (IEGs) and/or a rise in intracellular calcium. Following a 24-hour exposure to Aβ, DNA isolated from treated neurons exhibited a ladder of oligonucleosomal-length fragments of DNA (FIG. 4). Similar to young cells, mature cortical neurons exposed to Aβ also underwent endonucleolytic DNA cleavage. Similar findings for cortical cultures were recently reported by Forloni and colleagues[31] in 1993.

DNA fragmentation in other apoptosis paradigms has been correlated with the activation of endonucleases.[2,28] Aurintricarboxylic acid, an inhibitor of nucleases, was reported to suppress DNA fragmentation and block apoptosis in a variety of cell types.[30,32–35] Aurintricarboxylic acid suppressed Aβ-induced DNA fragmentation when added to the cultures simultaneously with Aβ peptides. DNA fragmentation was not a nonspecific consequence of neuronal injury, as neurons exposed to the calcium ionophore A23187 showed only random DNA degradation. Taken together, these morphologic and biochemical data suggest that specific stimuli, in this case Aβ, can induce primary neurons to undergo apoptosis.

POSSIBLE ROLE OF CALCIUM IN Aβ-MEDIATED APOPTOSIS

Calcium has been implicated as a critical second messenger in excitotoxicity and glutamate-mediated cell death in the presence of Aβ[36] (see also Mattson this volume). In addition, calcium has been linked to apoptosis in thymocytes, lymphocytes, and sympathetic neurons. In the case of thymocytes, an increase in calcium has been linked to the regulation of endonucleases involved in DNA fragmentation, cytoskeletal breakdown, intracellular kinases, and other signaling pathways associated with apoptosis.[27] Conversely, a decrease in intracellular calcium appears to regulate the apoptosis program in sympathetic neurons.[37,38] Thus, the involvement of calcium appears common to several cell types; however, the exact nature of the signal can be unique.

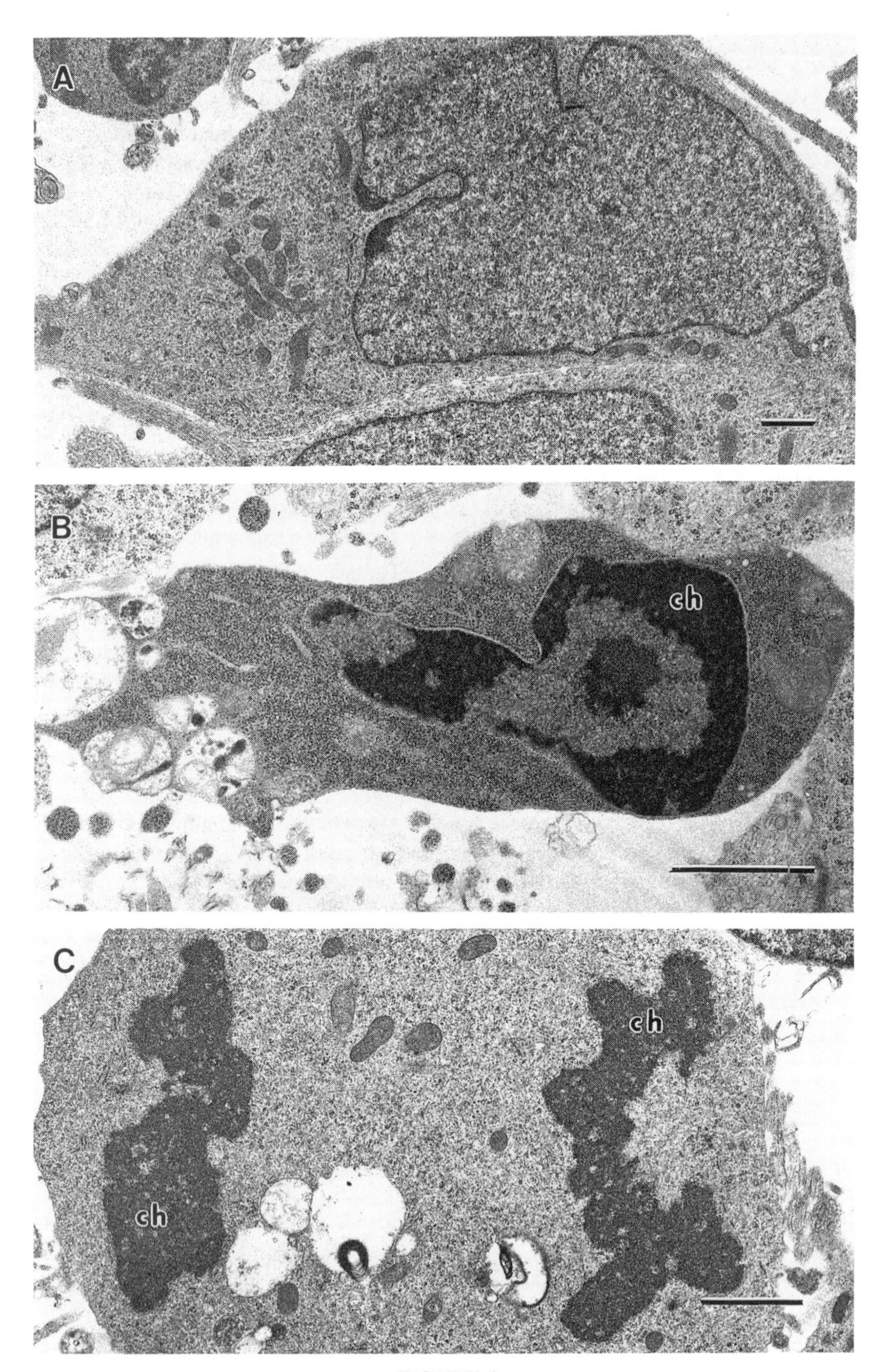

FIGURE 3.

If an increase in intracellular calcium is linked to Aβ-induced apoptosis, calcium-dependent mechanisms may participate in one or more aspects of the cell death program such as cytoskeletal breakdown or endonuclease activation. Alternatively, an increase in calcium may be a secondary event following actual cell death. It is necessary to evaluate the precise time course of calcium and demonstrate that stabilization of calcium can alter the course of the apoptotic program. We approached the evaluation of the role of calcium in apoptosis with studies using Fura-2 imaging[39] to examine the time course of Aβ-induced apoptosis as a function of possible alterations in intracellular free calcium. We also examined the effect of voltage-dependent calcium channel antagonists on Aβ-induced neuronal death to determine if direct regulation of calcium through these channels may serve to protect neurons.

Our initial results illustrate the extreme sensitivity of neurons subjected to Aβ. We examined the early changes in calcium employing measurement conditions similar to those already described.[36] Neurons were exposed to Aβ and then incubated with Fura-2 in a physiologic salt solution, 5 mM glucose buffered with Hepes. In this condition, calcium levels rose progressively over a period of 3-5 hours. No changes were observed in control cultures. Significantly, Aβ-treated neurons showed signs of necrotic degeneration within this brief period. That is, they appeared swollen and flattened with very thin balloon-like cytoplasmic boundaries. These morphologic characteristics were unexpected because similar morphologies were rarely, if ever, observed in our normal culture paradigm. We hypothesized that some aspect of the imaging protocol caused this change in response to Aβ. This could include media changes, introduction of Fura-2, or temperature reduction because these experiments, like experiments conducted in this and other laboratories, were carried out at room temperature.

The media used in the imaging protocol contained 5 mM glucose compared to 25 mM glucose in our standard DMEM-based culture medium. Inasmuch as neurons in the presence of Aβ were shown to be sensitive to reduced glucose,[17,18] we formulated a balanced buffer containing high glucose (25 mM) or low glucose (5 mM) and carried out Fura-2 measurements in these media. As shown in FIGURE 5, neurons imaged in the 25 mM glucose medium showed no significant increases in intracellular calcium during a 3-hour exposure to Aβ. In contrast, neurons in 5 mM glucose showed a significant rise in calcium during the 3-hour exposure to Aβ. These increases were variable in onset and some neurons maintained their calcium levels during this brief exposure.

Taken together, these data suggest that Aβ can make neurons more sensitive to changes in the extracellular environment as signaled by changes in their intracellular calcium levels. In the low glucose condition, cells showed morphologic features of

←

FIGURE 3. Transmission electron micrographs of cultured hippocampal neurons. (**A**) A neuron treated with the inactive scrambled sequence Aβ peptide. (**B** and **C**) Hippocampal neurons following treatment with the active Aβ peptide. Aβ-treated neurons show chromatin condensation (ch) and fragmentation, cytoplasmic vacuoles, and preservation of mitochondria and the plasma membrane.

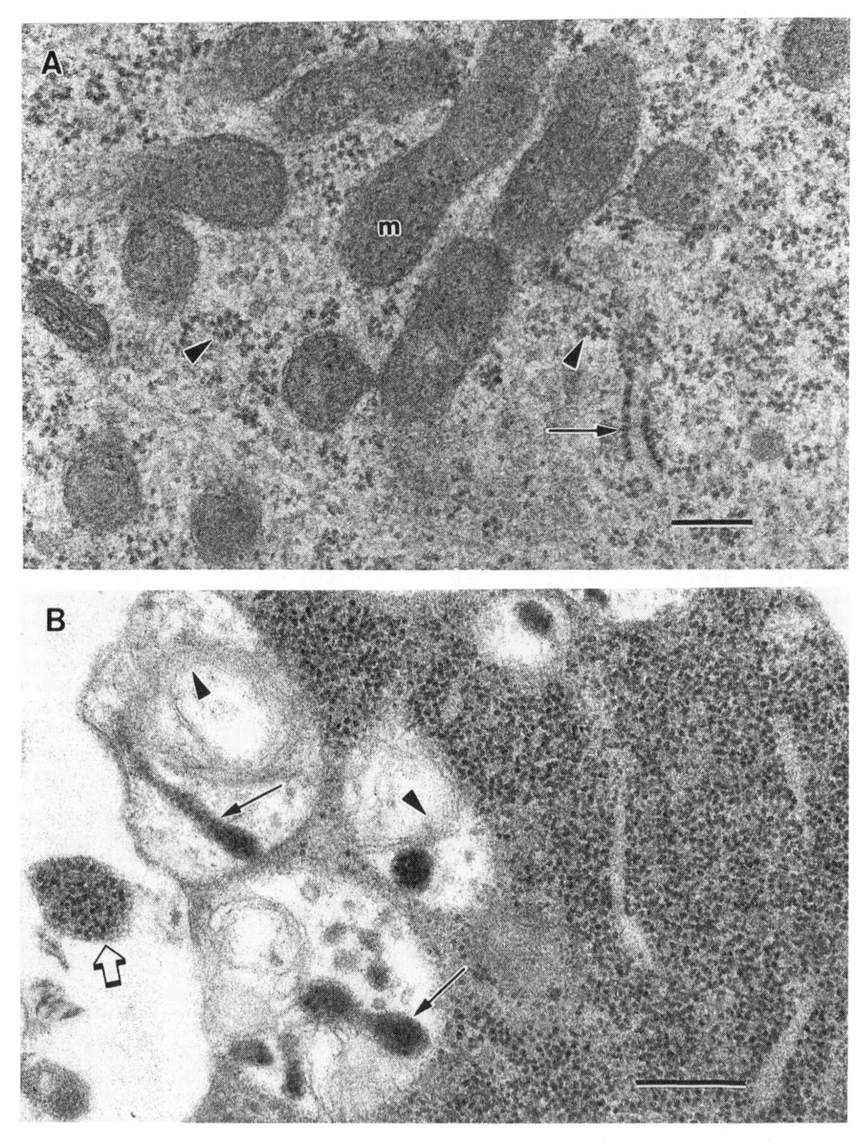

FIGURE 4. (**A**) At higher magnification, the cytoplasm of the control neuron shown in FIGURE 3A shows intact mitochondria (m), rough ER (*arrow*) and polyribosomal rosettes (*arrowheads*). (**B**) In contrast, the cytoplasm of the Aβ-treated neuron seen in FIGURE 3C contains vacuoles filled with fine filamentous material (*arrowheads*) and electron dense particulate matter (*arrows*). The polyribosomal rosettes have dispersed to monosomes, but the mitochondria remain indistinguishable from controls. Note the presence of the apoptotic body containing dispersed ribosomes and electron dense cytoplasm (*open arrow*).

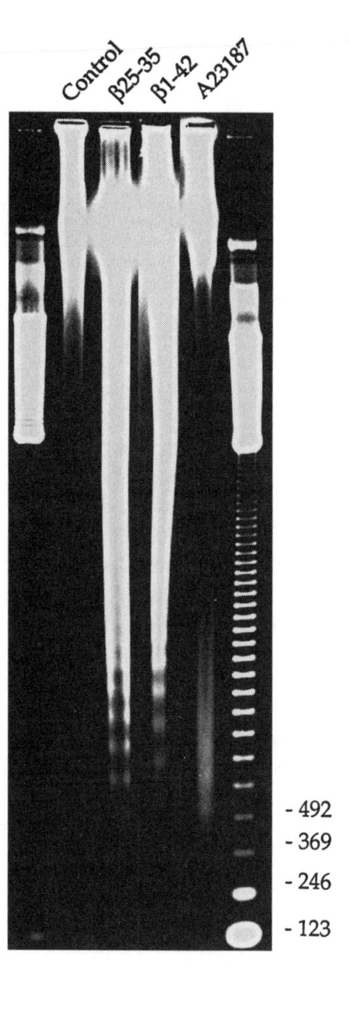

FIGURE 5. Aβ (β25-35 and β1-42) induce primary CNS neurons to undergo DNA cleavage into oligonucleosome-length fragments, characteristic of apoptosis. Cultures exposed to the calcium ionophore A23187 (1 μM) generate a diffuse smear of randomly degraded DNA, characteristic of necrosis.[24]

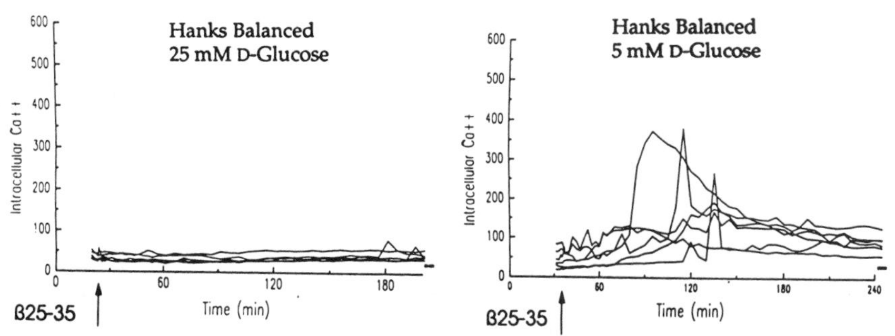

FIGURE 6. In the presence of high concentrations of glucose (25 mM) Aβ failed to induce significant increases in intracellular calcium in primary neurons. However, under conditions of low glucose (5 mM) Aβ induced a significant increase in intracellular calcium concentrations.

necrotic degeneration, whereas in the high glucose condition, degeneration appeared to be apoptotic. These data suggest that changes in energy state can affect the pathway of degeneration, even degeneration stemming from the same stimulus. Furthermore, the kinetics for regulation of intracellular Ca^{2+} depends on the degeneration pathway.

CALCIUM CHANNEL ANTAGONISTS CAN PROTECT AGAINST β-AMYLOID TOXICITY

To further evaluate the role of calcium in an apoptotic program, we examined the role of calcium channel antagonists on Aβ-induced cell death in culture. Two classes of Ca^{2+} entry blockers were examined for their ability to protect against Aβ neurotoxicity: glutamate receptor antagonists and voltage-sensitive Ca^{2+} channel blockers.[40] The presence of both NBQX, an AMPA kainate subtype blocker,[41] and MK-801, a blocker of the NMDA subclass of glutamate receptors, had no effect on Aβ neurotoxicity.

In comparison, the presence of the dihydropyridine Ca^{2+} channel blocker nimodipine (20 μM) attenuated injury by greater than 60% in 11 of 13 experiments. (The two other experiments showed between 30% and 50% protection.) Additionally, Co^{2+} (100 μM), a nonspecific blocker of voltage-sensitive Ca^{2+} channels, protected by greater than 60% in five of five experiments. The amount or appearance of aggregates was not noticeably affected by either nimodipine or Co^{2+}. On rare occasions some degeneration occurred to untreated (control) cultures; however, neither compound had deleterious effects on the cultures. Nimodipine and Co^{2+} showed mild protective effects; the protective effect of nimodipine against Aβ neurotoxicity was dose dependent between 1 and 20 μM.[40]

It is important to emphasize, however, that these data need to be interpreted with caution. Although the protective effect is clear, the precise target for nimodipine has not been proven. Indeed, the effective concentrations of the compound in preventing Aβ cell death are higher than expected for blocking voltage-dependent Ca^{2+} channels. Furthermore, the experimental *in vitro* paradigm shows that this compound can only delay neuronal death, because neurons will eventually degenerate in these cultures. Finally, the extrapolation to *in vivo* conditions needs to be established.

CONCLUSION

It has been clear for some time that during development, massive numbers of neurons, perhaps as many as one half, degenerate by a pathway of programmed cell death (apoptosis). It was only recently recognized that apoptosis may also contribute to neurodegenerative diseases such as Alzheimer's disease (AD). In AD, a peptide called Aβ accumulates in the brain. We and others showed that Aβ causes cultured neurons to degenerate, raising the intriguing possibility that Aβ may contribute to neurodegeneration in AD. Furthermore, we obtained evidence to support the notion that Aβ can cause neurons to die by an apoptotic pathway. That is, Aβ-treated neurons in culture display hallmarks of apoptosis including membrane blebbing, chromatin condensation, and DNA fragmentation into oligonucleosomal-length fragments. It is likely that calcium has a role in Aβ-mediated death. In particular, calcium channel antagonists can protect or at least delay Aβ-induced death. Taken together, these data suggest that Aβ may drive the course of brain dysfunction, place neurons at risk for insults, and, over time, initiate programmed cell death unless otherwise constrained. The mechanism by which Aβ induces toxicity will increase the understanding of the basic mechanism of neuronal death in AD and aid in the development of appropriate interventions.

REFERENCES

1. KERR, J. F., A. H. WYLLIE & A. R. CURRIE. 1972. Apoptosis: A basic biological phenomenon with wide-ranging implications in tissue kinetics. Br. J. Cancer **26:** 239-257.
2. WYLLIE, A. H. 1980. Cell death: The significance of apoptosis. Int. Rev. Cytol. **68:** 251-306.
3. BURSCH, W., B. DUSTERBERG & H. R. SCHULTE. 1986. Growth, regression and cell death in rat liver as related to tissue levels of the hepatomitogen cyproterone acetate. Arch. Toxicol. **59:** 221-227.
4. MCCONKEY, D. J., M. JONDAL & S. ORRENIUS. 1992. Cellular signaling in thymocyte apoptosis. Semin. Immunol. **4:** 371-377.
5. BURSCH, W., F. OBERHAMMER & R. SCHULTE-HERMANN. 1992. Cell death by apoptosis and its protective role against disease. TiPS **13:** 245-251.
6. MCCONKEY, D. J., P. NICOTERA, P. HARTZELL, G. BELLOMO, A. H. WYLLIE & S. ORRENIUS. 1989. Glucocorticoids activate a suicide process in thymocytes through an elevation of cytosolic calcium concentration. Arch. Biochem. Biophys. **269:** 365-370.
7. BURSCH, W., H. S. TAPER, B. LAUER & H. R. SCHULTE. 1985. Quantitative histological and histochemical studies on the occurrence and stages of controlled cell death (apoptosis) during regression of rat liver hyperplasia. Virchows Arch. B Cell Pathol. **50:** 153-166.
8. IJIRI, K. & C. S. POTTEN. 1983. Response of intestinal cells of differing topographical and hierarchical status to ten cytotoxic drugs and five sources of radiation. Br. J. Cancer **47:** 175-185.
9. ARENDS, M. J., R. G. MORRIS & A. H. WYLLIE. 1990. Apoptosis. The role of the endonuclease. Am. J. Pathol. **136:** 593-608.
10. WHITSON, J. S., D. J. SELKOE & C. W. COTMAN. 1989. Amyloid beta protein enhances the survival of hippocampal neurons *in vitro.* Science **243:** 1488-1490.
11. YANKNER, B. A., L. K. DUFFY & D. A. KIRSCHNER. 1990. Neurotrophic and neurotoxic effects of amyloid β-protein: Reversal by tachykinin neuropeptides. Science **250:** 279-282.
12. PIKE, C. J., B. J. CUMMINGS & C. W. COTMAN. 1992. β-amyloid induces neuritic dystrophy *in vitro:* Similarities with Alzheimer pathology. NeuroReport **3:** 769-772.
13. KOH, J. Y., L. L. YANG & C. W. COTMAN. 1990. β-amyloid protein increases the vulnerability of cultured cortical neurons to excitotoxic damage. Brain Res. **533:** 315-320.
14. PIKE, C. J., A. J. WALENCEWICZ, C. G. GLABE & C. W. COTMAN. 1991. Aggregation-related toxicity of synthetic β-amyloid protein in hippocampal cultures. Eur. J. Pharmacol. **207:** 367-368.
15. MATTSON, M. P., B. CHENG, D. DAVIS, K. BRYANT, I. LIEDERBURG & R. E. RYDEL. 1992. Beta-amyloid peptides destabilize calcium homeostasis and render human cortical neurons vulnerable to excitotoxicity. J. Neurosci. **12:** 376-389.
16. DORNAN, W. A., D. E. KANG, A. MCCAMPBELL & E. E. KANG. 1993. Bilateral injections of βA(25-35)+IBO into the hippocampus disrupts acquisition of spatial learning in the rat. NeuroReport **5:** 165-168.
17. COPANI, A., J. KOH & C. W. COTMAN. 1991. β-amyloid increases neuronal susceptibility to injury by glucose deprivation. NeuroReport **2:** 763-765.

18. CHENG, B. & M. P. MATTSON. 1992. Glucose deprivation elicits neurofibrillary tangle-like changes in hippocampal neurons: prevention by NGF and bFGF. Exp. Neurol. **117:** 114-123.
19. PIKE, C. J., A. J. WALENCEWICZ, C. G. GLABE & C. W. COTMAN. 1991. *In vitro* aging of β-amyloid protein causes peptide aggregation and neurotoxicity. Brain Res. **563:** 311-314.
20. PIKE, C. J., D. BURDICK, A. WALENCEWICZ, C. G. GLABE & C. W. COTMAN. 1993. Neurodegeneration induced by β-amyloid peptides *in vitro*: The role of peptide assembly state. J. Neurosci. **13:** 1676-1687.
21. PIKE, C. J. & C. W. COTMAN. 1993. Cultured GABA-immunoreactive neurons are resistant to toxicity induced by β-amyloid. Neuroscience **56:** 269-274.
22. PIKE, C. J., A. J. WALENCEWICZ-WASSERMAN, J. KOSMOSKI, D. H. CRIBBS, C. G. GLABE & C. W. COTMAN. "Structure-activity analyses of β-Amyloid peptides: Contributions of the β25-35 region to aggregation and neurotoxicity." J. Neurochem., in press.
23. KOH, J.-Y. & D. W. CHOI. 1987. Quantitative determination of glutamate mediated cortical neuronal injury in cell culture by lactate dehydrogenase afflux assay. I. Neurosci. Methods **20:** 83-90.
24. LOO, D. T., A. COPANI, C. J. PIKE, E. R. WHITTEMORE, A. J. WALENCEWICZ & C. W. COTMAN. 1993. Apoptosis is induced by β-amyloid in cultured central nervous system neurons. Proc. Natl. Acad. Sci. USA **90:** 7951-7955.
25. ARENDS, M. J. & A. H. WYLLIE. 1991. Apoptosis: Mechanisms and roles in pathology. Int. Rev. Exp. Pathol. **32:** 223-255.
26. WATT, J. A., C. J. PIKE, A. J. WALENCEWICZ-WASSERMAN & C. W. COTMAN. 1994. Ultrastructural analysis of β-amyloid-induced apoptosis in cultured hippocampal neurons. Brain Res., in press.
27. MCCABE, M. J., P. NICOTERA & S. ORRENIUS. 1992. Calcium-dependent cell death. Role of the endonuclease, protein kinase C, and chromatin conformation. Ann. N.Y. Acad. Sci. **663:** 269-278.
28. GAIDO, M. L. & J. A. CIDLOWSKI. 1991. Identification, purification, and characterization of a calcium-dependent endonuclease (NUC18) from apoptotic rat thymocytes. J. Biol. Chem. **266:** 18580-18585.
29. DUKE, R. C. & J. J. COHEN. 1986. IL-2 addiction: Withdrawal of growth factor activates a suicide program in dependent T cells. Lymphokine Res. **5:** 289-299.
30. PITTMAN, R. N., S. WANG, A. J. DIBENEDETTO & J. C. MILLS. 1993. A system for characterizing cellular and molecular events in programmed neuronal cell death. J. Neurosci. **13:** 3669-3680.
31. FORLONI, G., R. CHIESA, S. SMIROLDO, L. VERGA, M. SALMONA, F. TAGLIAVINI & N. ANGERETTI. 1993. Apoptosis mediated neurotoxicity induced by chronic application of beta amyloid fragment 25-35. NeuroReport **4:** 523-;526.
32. MCCONKEY, D. J., P. HARTZELL, P. NICOTERA & S. ORRENIUS. 1989. Calcium-activated DNA fragmentation kills immature thymocytes. FASEB J. **3:** 1843-1846.
33. BATISTATOU, A. & L. A. GREENE. 1991. Aurintricarboxylic acid rescues PC12 cells and sympathetic neurons from cell death caused by nerve growth factor deprivation: Correlation with suppression of endonuclease activity. J. Cell Biol. **115:** 461-471.
34. BATISTATOU, A. & L. A. GREENE. 1993. Internucleosomal DNA cleavage and neuronal cell survival/death. J. Cell. Biol. **122:** 523-532.
35. HELGASON, C. D., L. SHI, A. H. GREENBERG, Y. SHI, P. BROMLEY, T. G. COTTER, D. R. GREEN & R. C. BLEACKLEY. 1993. DNA fragmentation induced by cytotoxic T lymphocytes can result in target cell death. Exp. Cell Res. **206:** 302-310.
36. MATTSON, M. P. 1993. Calcium destabilizing and neurodegenerative effects of aggregated β-amyloid peptide are attenuated by basic FGF. Brain Res. **621:** 35-49.
37. KOIKE, T., D. P. MARTIN & E. J. JOHNSON. 1989. Role of Ca^{2+} channels in the ability of membrane depolarization to prevent neuronal death induced by trophic-factor depriva-

tion: Evidence that levels of internal Ca^{2+} determine nerve growth factor dependence of sympathetic ganglion cells. Proc. Natl. Acad. Sci. USA **86:** 6421-6425.
38. JOHNSON, E. M., T. KOIKE & J. FRANKLIN. 1992. A "calcium set-point hypothesis" of neuronal dependence on neurotrophic factor. Exp. Neurol. **115:** 163-166.
39. GRYNKIEWICZ, G., M. POENIE & R. Y. TSIEN. 1985. A new generation of Ca^{2+} indicators with greatly improved fluorescence properties. J. Biol. Chem. **260:** 3440-3450.
40. WEISS, J. H., C. J. PIKE & C. W. COTMAN. 1994. Ca^{2+} channel blockers attenuate β-amyloid peptide toxicity to cortical neurons in culture. J. Neurochem. **62:** 372-375.
41. SHEARDOWN, M. J., E. O. NIELSON, A. J. HANSEN, P. JACOBSEN & T. HONORE. 1990. 2,3-Dihydroxy-6-nitro-7-sulfamoyl-benzo-(F)quinoxaline: A neuroprotectant for cerebral ischemia. Science **247:** 571-574.

Calcium and Neuronal Injury in Alzheimer's Disease

Contributions of β-Amyloid Precursor Protein Mismetabolism, Free Radicals, and Metabolic Compromise[a]

MARK P. MATTSON [b]

Sanders-Brown Research Center on Aging
and
Department of Anatomy & Neurobiology
University of Kentucky
Lexington, Kentucky 40536-0230

BACKGROUND: INTRACELLULAR CALCIUM IN ADAPTIVE AND MALADAPTIVE CHANGES IN NEURONAL CYTOARCHITECTURE

Calcium is the premier second messenger mediating adaptive changes in neuroarchitecture in response to environmental signals such as neurotransmitters and neurotrophic factors.[1-3] Roles for intracellular free calcium levels ($[Ca^{2+}]_i$) in the regulation of the neuronal growth cone were demonstrated in studies of invertebrate[4] and mammalian[5] neurons. Calcium regulates structural plasticity in neurons by influencing the cytoskeleton and associated proteins in specific ways. For example, the microfilament-based motility of growth cones and the microtubule-based elongation of neurites are differentially sensitive to Ca^{2+}, a fact that allows for the control of neurite elongation rate, branching, and pathfinding by environmental signals such as neurotransmitters and growth factors.[3,6] Within an optimum range of $[Ca^{2+}]_i$, neuroarchitecture is regulated adaptively. However, when the $[Ca^{2+}]_i$ becomes too high, for too long a time-period, maladaptive degradation of cellular components occurs. Such a loss of $[Ca^{2+}]_i$ homeostasis is certainly involved in acute degenerative conditions such as stroke and severe epileptic seizures[7,8] and is very likely to play a role in the pathogenesis of more chronic insults including Alzheimer's disease (AD)[3,9] for reasons that follow.

In addition to their role in synaptic transmission in the mature nervous system, neurotransmitters regulate the outgrowth of neurites, synaptogenesis, and adaptive modifications in the structure of neural circuits.[6] Among neurotransmitters, glutamate

[a] Research from this laboratory was supported by the National Institutes of Health, the Alzheimer's Association, and the Metropolitan Life Foundation.

[b] Address for correspondence: Dr. Mark P. Mattson, 211 Sanders-Brown Building, University of Kentucky, Lexington, KY 40536-0230.

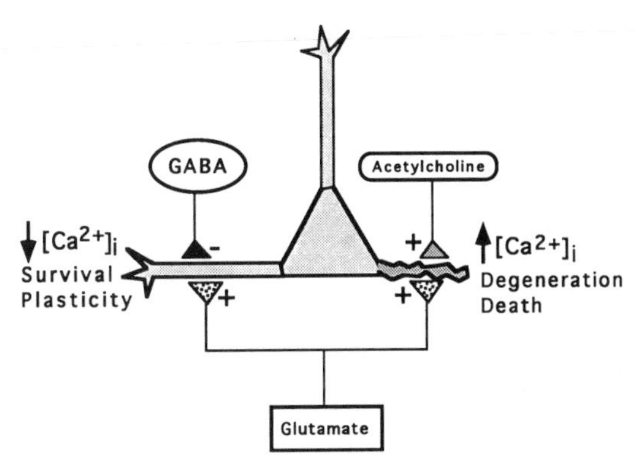

FIGURE 1. Interactive effects of excitatory and inhibitory neurotransmitters on neuronal plasticity and degeneration. In addition to excitatory input from glutamatergic neurons, pyramidal neurons in the hippocampus receive inhibitory inputs from GABAergic neurons and excitatory inputs from cholinergic neurons. The GABAergic and cholinergic inputs play important roles in modifying neuronal responses to glutamate. For example, activation of GABA receptors can protect neurons against glutamate toxicity, and activation of muscarinic receptors can potentiate glutamate toxicity.[14,15]

(the major excitatory transmitter in the mammalian brain) can be considered a key signal involved in developmental and synaptic plasticity.[10] Glutamate can stabilize the outgrowth of dendrites and promote synapse formation in developing hippocampal neurons,[5,11] and it plays a major role in the activity-dependent refinement of synaptic connections in the visual system.[12] Moreover, the mechanism of learning and memory in man (which involves the hippocampus) is critically dependent on glutamatergic transmission.[13] In addition to glutamate, other neurotransmitters modulate developmental and synaptic plasticity. For example, the dendrite outgrowth-inhibiting action of glutamate is antagonized by activation of GABA receptors[14] and potentiated by activation of muscarinic acetylcholine receptors[15] (FIG. 1). Similarly, activation of cholinergic and noradrenergic inputs to mature hippocampal neurons can modify long-term potentiation (LTP).[16]

Unfortunately, activation of glutamate signaling pathways is not always beneficial to neurons. Overactivation of glutamate receptors can cause a spectrum of maladaptive structural changes in neurons, ranging from neuritic pruning[5] to cell death.[17] Excitatory damage is mediated in large part by uncontrolled Ca^{2+} influx, resulting in sustained elevation of $[Ca^{2+}]_i$.[3,7,18] Activation of glutamate receptors can lead to Ca^{2+} influx through several portals including ligand-gated channels in the plasma membrane (NMDA and AMPA receptors), voltage-dependent calcium channels, and calcium release from internal stores resulting from activation of metabotropic receptors linked to inositol phospholipid hydrolysis (FIG. 2). Most neurons receive input from several different neurotransmitters, and data suggest that interactive actions can determine, in roughly an additive way, if degeneration occurs. For example, hippocampal pyramidal neurons receive excitatory glutamatergic and cholinergic inputs and inhibitory GABAergic input. A given level of excitation by glutamate that would not alone be destructive may be destructive when cholinergic inputs are also activated (FIG. 1).

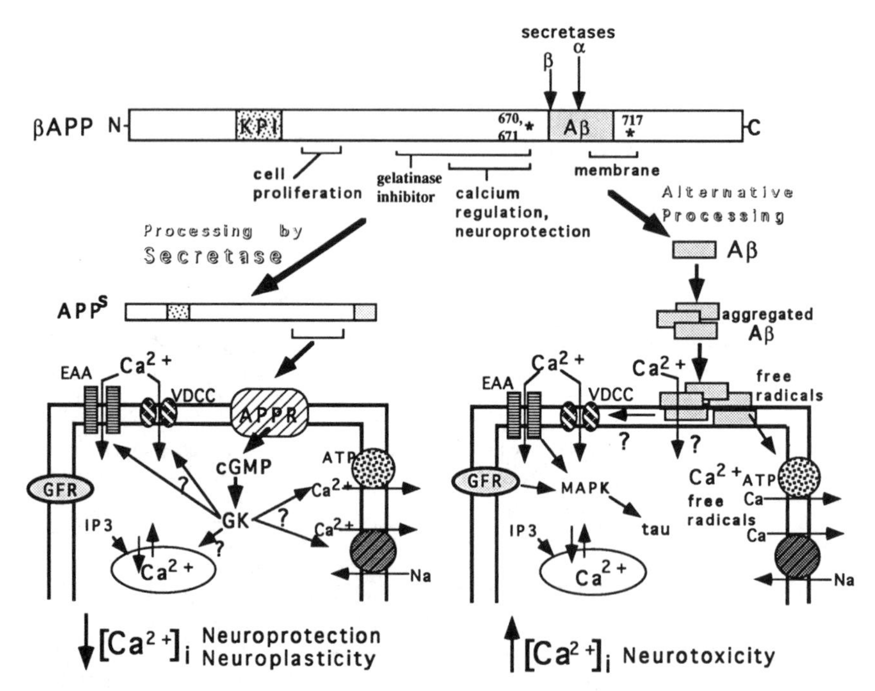

FIGURE 2. Effects of βAPP metabolites on neuronal calcium homeostasis. The structure of βAPP including biologically active regions, sites of proteolytic processing, and sites of known mutations in AD* is shown at the *top.* βAPP is a transmembrane glycoprotein with a large NH_2-terminal extracellular region and a shorter intracellular COOH-terminus. The Aβ domain lies partially within the membrane and partially in the extracellular environment. Two sites of secretory cleavage have been identified, one involving cleavage between amino acids 16 and 17 of Aβ (α-secretase) and the other involving cleavage at the NH_2-terminus of Aβ (β-secretase). Several functional domains in βAPP have been reported including: kunitz protease inhibitor (KPI) domains (APP751 and APP770); a domain that stimulates proliferation of non-neuronal cells; a gelatinase inhibitor region; and a region involved in regulation of $[Ca^{2+}]_i$ and neuroprotection. Several mutations in βAPP have been identified in families with inherited forms of AD including those involving codon 717 and codons 670-671. Processing of βAPP by secretase pathways liberates secreted forms of βAPP (APP^s). A domain common to all APP^ss can lower $[Ca^{2+}]_i$ and attenuate excitatory amino acid (EAA)-induced elevations of $[Ca^{2+}]_i$. The $[Ca^{2+}]_i$-lowering mechanism of APP^ss appears to be mediated by a specific cell surface receptor (APPR) linked to an increase in cyclic GMP levels and activation of cGMP kinase (GK). GK may lower $[Ca^{2+}]_i$ by reducing Ca^{2+} influx and/or enhancing Ca^{2+} extrusion. Alternative processing of βAPP (probably involving endocytosis and proteolysis in an acidic intracellular compartment) liberates Aβ from cells. Under certain environmental conditions (eg, altered pH or presence of free radicals) Aβ forms insoluble aggregates that can accumulate at the plasma membrane of neurons and disrupt $[Ca^{2+}]_i$ regulation. Generation of free radicals during the process of Aβ fragmentation and aggregation may damage $[Ca^{2+}]_i$-regulating proteins in the plasma membrane. Aβ may enhance Ca^{2+} influx and/or impair Ca^{2+} extrusion, leading to aberrant elevation of $[Ca^{2+}]_i$ and increased vulnerability to metabolic and excitotoxic insults. See text for discussion.

Conversely, activation of GABA receptors can raise the threshold for glutamate toxicity. These combinatorial effects of transmitters on neuronal survival appear to be related to their net effect on $[Ca^{2+}]_i$. Neurons that bear glutamate receptors are particularly vulnerable to excitotoxicity when energy supplies (glucose and oxygen) are low,[7,19] probably because removal of Ca^{2+} from the cytoplasm requires ATP. Indeed, manipulations that block activation of glutamate receptors or voltage-dependent Ca^{2+} channels can reduce ischemic brain injury. Conversely, conditions that compromise energy availability to neurons such as glucocorticoids[20] increase neuronal vulnerability to excitotoxicity.

Evidence for loss of neuronal $[Ca^{2+}]_i$ homeostasis and an excitotoxic contribution to neuronal degeneration in AD is considerable and rapidly increasing.[3,9,21,22] Neurons vulnerable in AD (eg, neurons of the hippocampus and entorhinal cortex) possess high levels of glutamate receptors, particularly *N*-methyl-D-aspartate (NMDA) receptors which are very efficient in passing Ca^{2+} into the cell.[23] In addition, neurons vulnerable in AD are also selectively vulnerable under conditions of metabolic compromise. For example, CA1 hippocampal neurons are vulnerable, and dentate granule cells are resistant in both stroke and AD. Neuronal degeneration in AD manifests as neurofibrillary tangles which are characterized by several structural and antigenic features including the accumulation of straight and paired-helical filaments and increased immunoreactivity with tau and ubiquitin antibodies.[24,25] Neurofibrillary tangle-like changes can be induced in neurons by glutamate and other $[Ca^{2+}]_i$-elevating conditions. For example, glutamate, calcium ionophores, and glucose deprivation increase neuronal immunoreactivity towards tau (Alz-50 and 5E2) and ubiquitin antibodies in cultured rat hippocampal and human cortical neurons.[26-29] Elevated $[Ca^{2+}]_i$ also caused altered electrophoretic mobility of tau and the accumulation of straight filaments in cultured human cortical neurons.[27] Chronic exposure of cultured human spinal cord neurons to glutamate induced paired helical filament-like structures.[30] Recent *in vivo* studies also implicate an excitotoxic mechanism in the genesis of neurofibrillary tangles. Thus, kainic acid injection into the hippocampus of rats elicited a rapid (hours) but transient accumulation of Alz-50 and 5E2 immunoreactivity in the somata of CA3 neurons.[31] Spectrin breakdown (a sign of activation of Ca^{2+}-dependent proteases) was also induced by kainate. The neuronal damage, appearance of tau immunoreactivity, and spectrin breakdown caused by kainic acid were all exacerbated by glucocorticoids[31] (FIG. 3). The latter observations are consistent with an excitotoxic mechanism of neuronal degeneration in AD inasmuch as glucocorticoids were previously shown to metabolically compromise neurons and increase their vulnerability to excitotoxicity,[20] and because evidence exists of altered glucocorticoid regulation[32] and reduced energy availability to neurons[33,34] in AD (see below).

There has been intense interest in identifying the kinase(s) responsible for the altered phosphorylation of tau in neurofibrillary tangles. It is clear that tau is a substrate for a variety of kinases including calcium/calmodulin-dependent protein kinase II,[35,36] protein kinase C,[37,38] cyclic AMP-dependent protein kinase,[36,39] and casein kinase II.[40] However, studies using tau antibodies specific for paired helical filament tau indicate that the specific phosphorylation pattern present in paired helical filament tau cannot be reproduced by the aforementioned kinases.[41] Recently, it was demonstrated that MAP kinase phosphorylates tau in a manner similar to that observed in paired helical filaments.[42,43] Phosphorylation of tau by MAP kinase results in a

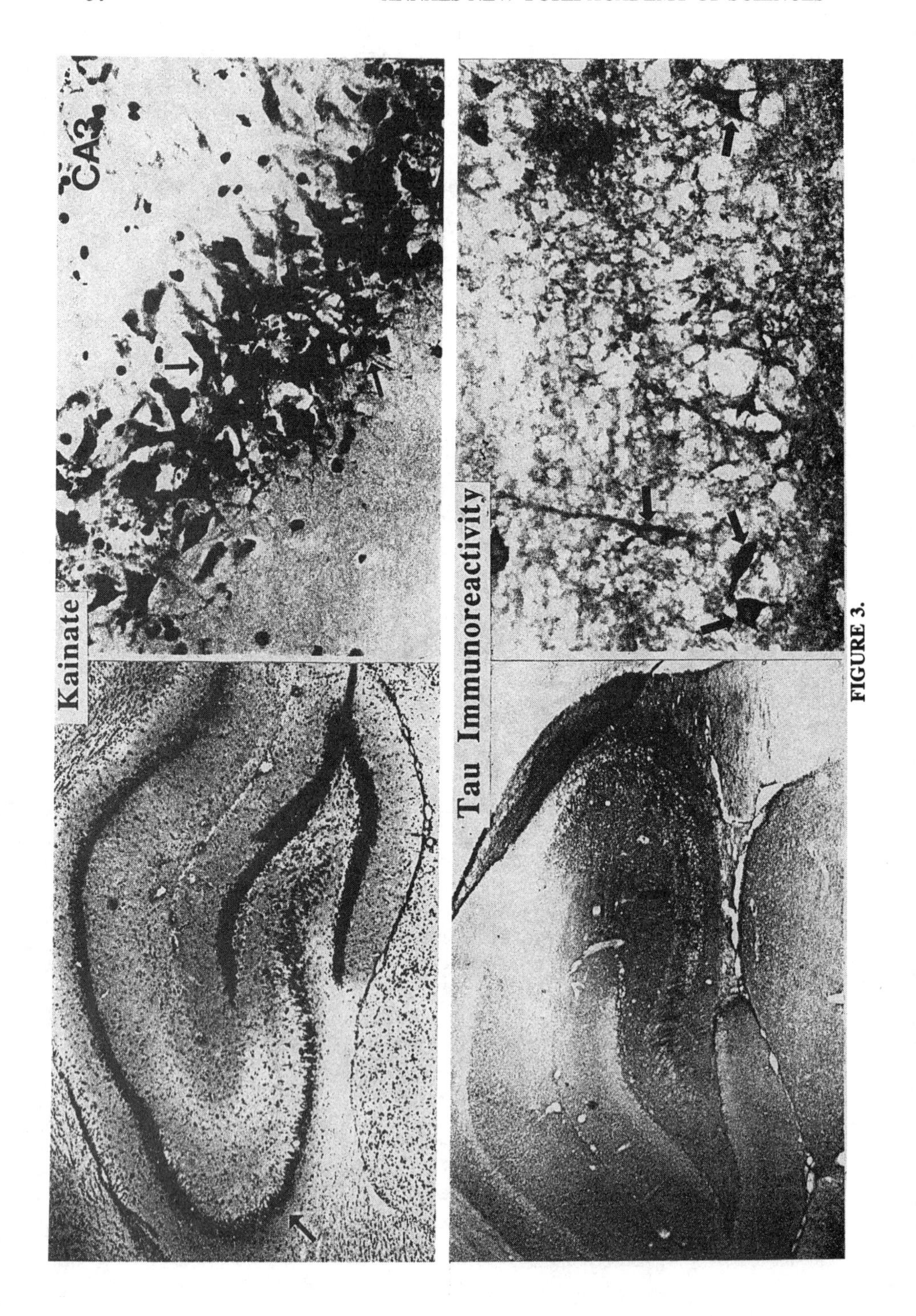

FIGURE 3.

10-fold reduction in the ability of tau to bind microtubules.[44] In addition, Lu *et al.*[45] recently showed that p44 MAP kinase can phosphorylate tau in cultured hippocampal neurons and that this reduces the ability of tau to promote microtubule assembly. Interestingly, stimulation of NMDA receptors in cell culture[46] and electroconvulsive shock *in vivo*[47] activate MAP kinases in neurons, and MAP kinase activation can be dependent upon calcium.[48] On the other hand, a variety of neurotrophic factors also activate MAP kinases.[49] I suggest that phosphorylation of tau by MAP kinases in AD may represent the activation of a decayed neuroprotective signal transduction mechanism that eventually fails (FIG. 2). The latter possibility leads to the interesting conclusion that in contrast to popular belief, alterations in tau are not causally involved in neuronal injury and death in AD but, rather, are a response to cell injury.

Additional evidence consistent with a causal role for excitatory amino acids in the pathogenesis of AD comes from analyses of cases of ingestion of known excitotoxins and studies of populations in the Western Pacific that have a very high incidence of dementing disorders with massive neurofibrillary degeneration in brain regions also vulnerable in AD. One example of excitotoxin ingestion is the case of domoic acid poisoning in Canada where several people consumed shellfish containing high levels of domoic acid. Many of these people developed memory loss.[50] The cause of the ALS-parkinsonism-dementia complex of Guam is not completely clear, but substantial evidence indicates involvement of an environmental excitotoxin.[51] A staple in the diet of these natives is flour made from the cycad seed which contains high levels of the excitotoxin cycasin and may contribute to neuronal degeneration in this disorder. A second hypothesis for the cause of the Guamian syndrome is a deficiency of calcium and magnesium in the diet which leads to hyperparathyroidism and increased calcium uptake by soft tissues including brain.[52]

Studies of $[Ca^{2+}]_i$ regulation in fibroblasts taken from patients with AD suggest that alterations in cellular calcium homeostatic mechanisms are not limited to the brain. For example, Peterson *et al.*[53] reported reduced $[Ca^{2+}]_i$ and altered morphology in fibroblasts from AD donors. Sisken and coworkers[54] provided evidence that fibroblasts from patients with AD exhibit altered $[Ca^{2+}]_i$ responses to serum and bradykinin, apparently due to a perturbed transduction mechanism for release of calcium from intracellular stores. Recently, it was shown that K^+ channels are altered in AD fibroblasts such that the cells do not show a $[Ca^{2+}]_i$ response to K^+ blockade by tetraethylammonium.[55] If a similar abnormality was present in neurons in AD, it could contribute to the aberrant elevation of $[Ca^{2+}]_i$ and cell injury.

FIGURE 3. Excitotoxic injury elicits neurofibrillary tangle-like antigenic changes in tau protein. **(Top)** Cresyl violet-stained coronal section of hippocampus from a rat injected with the excitotoxin kainate in region CA1 of the hippocampus. The rat was killed 3 hours after kainate injection. Note damage to neurons in region CA3 (magnification of a portion of CA3 is shown at *right*; note pyknotic cells). **(Bottom)** Tau immunoreactivity in a hippocampal section from the same stressed and kainate-injected rat shown in the *top panel*. The section was immunostained with antibody 5E2 which stains neurofibrillary tangles in AD. The low magnification micrograph shows the appearance of tau immunoreactivity in region CA3. At higher magnification (*right*) it can be seen that some neurons show intense immunoreactivity (*arrows*). The *arrow* in each low magnification micrograph shows the location of the region magnified in the *right panels.*

TABLE 1. Biologic Activities of β-Amyloid Precursor Protein Metabolites and Regulation of Their Production

Biologic Activities	References
Activities of Secreted Forms of β-Amyloid Precursor Protein	
APP^{a}s regulate cell proliferation in non-neuronal cells	77,78
APP^{s}s promote neuronal survival	84,86
APP^{s}s protect against excitotoxic/ischemia-like insults	86,92
APP^{s}s modulate glutamatergic transmission	86,87
APP^{s}s regulate neurite outgrowth	85,87
APP^{s}s regulate intracellular calcium levels	86,87
Activities of Amyloid β-Peptide	
Aβ promotes neurite outgrowth	160
Aβ is neurotoxic	109,114,115
Aβ renders neurons vulnerable to excitotoxicity	110,112,114
Aβ destabilizes neuronal calcium homeostasis	112,114,120
Aggregation of Aβ is related to toxic potency	113-115
Growth factors attenuate Aβ toxicity	114
Aβ forms calcium-conducting pores in membranes	125
Aβ generates free radical peptides which damage neurons	124, 156
Regulation of βAPP Metabolism	
βAPPs are axonally transported	88
βAPP expression is induced by injury (eg, ischemia and head trauma)	95-97
Acetylcholine promotes release of secreted βAPPs	89
Inositol phospholipid pathway promotes release of APP^{s}s	161
Electrical activity promotes secretase activity	90

ALTERNATIVE PROCESSING OF β-AMYLOID PRECURSOR PROTEIN

A long-standing problem with the calcium hypothesis of AD was that it was not clear how it related to another major hypothesis that designated the deposition of amyloid β-peptide (Aβ) as seminal to the pathogenesis of AD.[56] Recent data indicate that altered metabolism of βAPP contributes to loss of $[Ca^{2+}]_i$ homeostasis in AD.[22] Before discussing the βAPP-calcium link it is essential to consider what is known about the metabolism of βAPP (TABLE 1; FIG. 2). βAPP, a 695-770 amino acid glycoprotein, is the source of Aβ which accumulates as diffuse and aggregated plaques in the brains of AD victims (and to a lesser extent with normal aging). Aβ is a 40-42 amino acid segment of βAPP that lies partially in the extracellular environment and partially within the plasma membrane.[57] Several forms of βAPP exist including one form lacking (APP695) and two forms containing (APP751 and APP770) a kunitz protease inhibitor domain in the NH_2-terminal region.[58-60] Several alternative pathways of βAPP metabolism have been identified that give rise to different products that are secreted from cells. Two different enzymatic cleavages occur near the cell surface[61,62] that liberate secreted forms of βAPP (APP^{s}s) into the extracellular milieu; one cleavage is within the Aβ sequence (α-secretase) and another at the NH_2-terminus of Aβ (β-secretase). The cleavage of βAPP by α-secretase precludes release of intact Aβ, whereas the β-secretase cleavage leaves intact Aβ as part of a potentially amyloidogenic COOH-terminal fragment of βAPP remaining in the membrane. Although α-secretase has not been identified, a metalloproteinase inhibitor domain in the COOH-terminal region of APP^ss is a substrate for gelatinase A, a cell surface-associated enzyme that has α-secretase-like activity.[63] Alternative processing of full length βAPP or COOH-terminal fragments (apparently in an acidic intracellular compartment) can result in liberation of intact Aβ from cells (neurons, glia, and others) in low amounts.[64–68] Release of Aβ from cells occurs during normal metabolism of βAPP, and Aβ circulates at low levels in cerebrospinal fluid and blood.

It is believed that altered expression and/or processing of βAPP occurs in AD and related disorders such as Down's syndrome leading to increased liberation of Aβ from cells.[56] The gene for βAPP was mapped to chromosome 21, which was of obvious interest in Down's syndrome because this disorder is trisomy 21 and because essentially all persons with Down's syndrome develop Alzheimer pathology (accumulation of Aβ and neurofibrillary tangles). More recently it was shown that mutations in βAPP underlie at least some cases of inherited AD. Several different mutations in βAPP have been linked to AD including a conversion of valine to isoleucine, phenylalanine, or glycine at codon 717.[69–71] The 717 mutation lies 3 amino acids away from the Aβ sequence toward the carboxy terminus. A double mutation in βAPP (lysine-methionine to asparagine-leucine conversion at codons 670 and 671) just NH_2-terminal to Aβ was also recently reported in a family with AD.[72] It was shown that processing of βAPP is altered in cells expressing βAPP containing the 670-671 double mutation in a manner consistent with increased Aβ production.[73,74] In Down's syndrome the increased deposition of Aβ may result from altered βAPP processing due to overexpression of βAPP with the increased gene dosage. Alzheimer's disease in some families is linked to mutations in chromosomes other than 21 and therefore may not involve alterations in βAPP metabolism. However, increasing data suggest that some of the mutations on other chromosomes may also affect βAPP metabolism. For example, Tanzi and coworkers[75] recently identified a gene in the AD region of chromosome 19 that encodes an amyloid precursor-like protein (APLP1) which is structurally similar to βAPP but lacks the Aβ sequence. If similar mechanisms are involved in the processing of APLP1 and βAPPs, then it is possible that APLP1 mutations could indirectly affect βAPP processing. The genes affected in other AD families such as those mapping to chromosome 14[76] remain to be determined; however, gene products involved in βAPP metabolism, calcium regulation, glutamatergic transmission, or free radical metabolism seem likely (see below).

NORMAL FUNCTIONS OF β-AMYLOID PRECURSOR PROTEIN IN NEURONS: REGULATION OF $[Ca^{2+}]_i$?

The normal functions of βAPP are beginning to emerge (TABLE 1). The vast majority of evidence indicates that among the different metabolites of βAPP (APP^s, Aβ, and COOH-terminal fragments), APP^s are most likely to serve physiologic functions. Several biologic activities of APP^ss in non-neuronal cells have been reported. APP^{s}751 stimulated growth of fibroblasts in cell culture[77,78]. APP^{s}s also plays a role in blood coagulation as indicated by the identification of APP^{s}751 as protease nexin II[59,60] and as an inhibitor of coagulation factor XIa.[79] In addition, data indicate that APP^{s}s and/or membrane-associated βAPPs may play roles in cell substrate and cell-cell adhesion.[80,81] The structure and localization of βAPP suggest that it may serve as a receptor,[57] but the only evidence supporting this possibility is the report that βAPP binds a G-protein.[82] However, linkage of βAPP to a G-protein may as easily reflect a mechanism for routing βAPP in the cell because G-proteins play important roles in membrane trafficking.[83]

In neurons APP^{s}s are likely to function in developmental and synaptic plasticity.[22] βAPPs have been shown to affect the outgrowth and survival of developing neurons in culture.[84–87] βAPPs are axonally transported.[88] Stimulation of neurons by excitatory transmitters and electrical activity induces secretion of APP^{s}s[89,90] which is presumably released from presynaptic terminals. APP^{s}s can modify $[Ca^{2+}]_i$ responses to glutamate in hippocampal neurons,[87] suggesting a role in synaptic plasticity. In addition, APP^{s}s can protect neurons against excitotoxicity in rat hippocampal and human cortical cell cultures[86] and against ischemic brain damage *in vivo.*[91] Further evidence for an excitoprotective role for APP comes from the recent work of Schubert and Behl[92] who found that a clonal neuron-like cell line transfected with βAPP was more resistant to glutamate toxicity than was the parent cell line which did not express βAPP. A 17 amino acid peptide corresponding to a region adjacent to the kunitz inhibitor domain of APP751 reduced neurologic damage in rabbits subjected to spinal cord ischemia.[93] Interestingly, this same 17 amino acid peptide is responsible for the growth-regulating activity of APP^{s}s in fibroblasts.[94] Inasmuch as βAPP expression is increased in response to brain injury,[95–97] APP^{s}s may serve a neuroprotective function. In this way APP^{s}s are considered as trophic factors similar to classical neurotrophic factors such as NGF, bFGF, and IGFs (see below).

An understanding of the signal transduction mechanism(s) of APP^{s}s in neurons is emerging. APP^{s}695 and APP^{s}751 were recently shown to regulate $[Ca^{2+}]_i$ and to modify $[Ca^{2+}]_i$ responses to glutamate.[86,87] APP^{s}695 and APP^{s}751 rapidly reduced $[Ca^{2+}]_i$ in cultured rat hippocampal and human cortical neurons. This action of the βAPP^{s}s was observed with concentrations as low as 1-10 pM. The $[Ca^{2+}]_i$-lowering and neuroprotective actions of the APP^{s}s were blocked by antibodies that recognize a region (amino acids 444-592 or APP695) common to all forms of βAPP.[86] Taken together with the demonstration of high affinity binding sites for APP^{s}695 in cells,[98] these data suggest that the $[Ca^{2+}]_i$-lowering effect is mediated by specific cell surface receptors. Although the signal transduction mechanism leading to reduced $[Ca^{2+}]_i$ and protection against excitotoxicity is not completely clear, recent studies suggest the involvement of the cyclic GMP second messenger system. In cultured rat hippocampal neurons, APP^{s}s elevated cyclic GMP levels; membrane-permeant cyclic GMP analogs lowered $[Ca^{2+}]_i$; and cyclic GMP antagonists blocked the $[Ca^{2+}]_i$-lowering effect of APP^{s}s.[99] Activation of the cyclic GMP pathway by APP^{s}s is of interest because previous work in cerebellar slices indicated that cyclic GMP can protect neurons against excitotoxicity.[100]

The ability of APP^{s}s to attentuate $[Ca^{2+}]_i$ responses to glutamate and excitotoxicity suggests that APP^{s}s normally play roles in modulation of synaptic transmission and in neuroprotection. Inasmuch as glutamate plays an important role in LTP (a cellular correlate of learning and memory), one would predict that APP^{s}s modify LTP. Interestingly, an APP-like protein is involved in learning and memory in flies because flies with a mutation in this protein show behavioral deficits and because expression of human APP in the mutant flies restores the deficit.[101] In addition, infusion of APP antibodies into brains of rats impaired their learning and memory.[102] A role for APP^{s}s in neuroprotection is consistent with the observation that βAPP expression is increased in response to brain injury,[95-97] as is the expression of many different neurotrophic factors including FGF,[103] NGF,[104] and BDNF.[105] Moreover, βAPP expression is also upregulated in response to growth factors and cytokines,[106,107] further supporting a role for βAPPs in the brain's response to injury. On the basis of the data just described, my colleagues and I have proposed that aberrant metabolism of βAPP could result in neuronal degeneration by compromising a normal neuroprotective function of APP^{s}s[22,86] (FIG. 2).

AGGREGATED Aβ DISRUPTS CALCIUM HOMEOSTASIS AND CAN BE NEUROTOXIC

Since the original description of the histopathologic features of AD, it has been hypothesized that the accumulation of amyloid is causally involved in neuronal degeneration. Isolation of cerebrovascular amyloid and identification of the amino acid sequence of Aβ[108] allowed the synthesis of pure preparations of Aβ which were then administered to cultured neurons in a series of studies. Cumulative data from several laboratories indicate that, indeed, Aβ can be neurotoxic, under certain conditions. Aβ 1-40, 1-42, and 25-35 were toxic to cultured hippocampal and cortical neurons.[109-112] However, among those studies differences were observed in whether Aβ was directly or indirectly neurotoxic and in the time course of neuronal degeneration and neurotoxic potency of the peptides. The basis for the inconsistencies has been resolved in independent studies from several laboratories[113-115] which show that: (1) Aβ is neurotoxic when it is able to form aggregates and (2) different lots of synthetic Aβ have different aggregation properties. Aggregation properties of synthetic Aβ peptides, which have been studied in detail, are known to be influenced by factors including osmolarity, polarity, and pH of the solvent as well as peptide concentration.[116-118] Interestingly, data from Dyrks *et al.*[119] indicate that oxidation can induce Aβ aggregation and free radical scavengers can inhibit aggregation, an observation that I will revisit below in the context of changes in the brain associated with aging and AD. It should be noted that the relationship of Aβ aggregation and neurotoxicity in cell culture studies fits well with the fact that degenerated neurites surround compact (aggregated) plaques in AD, whereas no degeneration is observed in neurons associated with diffuse (unaggregated) amyloid.[25]

The mechanism of Aβ neurotoxicity apparently involves induction of free radical production in neurons,[156] disruption of $[Ca^{2+}]_i$-regulating mechanisms, resulting in aberrant elevations of $[Ca^{2+}]_i$ and increased sensitivity to excitatory stimuli.[22,114] Exposure of cultured human cerebrocortical or rat hippocampal neurons to Aβ results in progressive elevation of $[Ca^{2+}]_i$, the time course and magnitude of which depend upon the degree of Aβ aggregation at the time of exposure to the neurons.[114] That is, the $[Ca^{2+}]_i$ rises more rapidly in neurons exposed to preaggregated Aβ than in neurons exposed to unaggregated Aβ (which then aggregates with time in culture). In our experience, the time course of $[Ca^{2+}]_i$ elevation is on the order of hours to days, and we have not observed more rapid (minutes) increases in $[Ca^{2+}]_i$. The $[Ca^{2+}]_i$-elevating action of aggregated Aβ is observed at the single cell level such that neurons with large amounts of Aβ associated with their cell surface have a higher $[Ca^{2+}]_i$ than do neurons with little Aβ associated with their surface (FIG. 4). Importantly, neurons exposed to Aβ exhibit enhanced $[Ca^{2+}]_i$ responses to excitatory stimuli including glutamate and membrane depolarization with KCl.[112,114,120] This increased sensitivity to excitation is correlated with greatly increased vulnerability to excitotoxicity. Aβ also renders neurons more vulnerable to hypoglycemic injury.[121] Clearly, these data support a role for Aβ in an excitotoxic mechanism of neuronal degeneration in AD. Interestingly, the $[Ca^{2+}]_i$-destabilizing action of Aβ may not be limited to neurons. For example, Eckert *et al.*[122] showed that Aβ25-35 can enhance the mitogen-induced calcium response in lymphocytes. Moreover, Aβ caused a large calcium conductance in oocytes expressing both a glutamate receptor (NMDAR1 or GluR1) and the substance P receptor, but not in oocytes expressing only glutamate receptors or only substance P receptors.[123] It is not clear how the findings in oocytes relate to the mechanism of Aβ neurotoxicity because the oocyte responses were rapid and were observed with levels of Aβ (10-100 nM) well below those required for significant aggregation and neurotoxicity of the peptide.

The site of action of Aβ in disrupting neuronal $[Ca^{2+}]_i$ homeostasis appears to be the plasma membrane inasmuch as aggregates of Aβ accumulate at the plasma membrane and because Aβ toxicity is attenuated when Ca^{2+} influx is experimentally reduced.[114] Intuitively it seems likely that disruption of $[Ca^{2+}]_i$ homeostasis by aggregated Aβ results from nonspecific effects at the plasma membrane. For example, Aβ might perturb the regulation of Ca^{2+} channels or glutamate receptors resulting in increased Ca^{2+} influx, or Aβ may impair Ca^{2+} removal from the cell (ie, Ca^{2+} pump or Na^+/Ca^{2+} exchange). We recently showed that during the process of aggregation in aqueous solution Aβ fragments and generates free radical peptides.[124] Aβ induced accumulation of reactive oxygen species in neurons which was linked to loss of calcium homeostasis.[156] Moreover, Aβ (at neurotoxic concentrations) inactivated the cytoplasmic enzymes glutamine synthetase and creatine kinase[124] and compromised plasma membrane Ca^{2+}-ATPase activity (R. Mark and M. P. Mattson, unpublished data). Aβ could also conceivably disrupt Ca^{2+} regulation at the plasma membrane by forming pores in the membrane as suggested by the work of Arispe *et al.*[125] These data indicate that Aβ may damage $[Ca^{2+}]_i$-regulating proteins by a free radical-mediated process (FIG. 2).

Taken together, the studies of biologic activities and mechanisms of action of βAPP metabolites suggest that altered βAPP metabolism in aging and AD may shift processing to reduce the release of APP^ss and increase the production of Aβ. This shift would have two deleterious consequences for $[Ca^{2+}]_i$ regulation: a reduction in levels of $[Ca^{2+}]_i$-stabilizing neuroprotective APP^ss; and an increase in levels of the $[Ca^{2+}]_i$-destabilizing neurotoxic Aβ. It should also be considered that in some cases of AD altered processing of βAPP may not be required for pathologic deposits of Aβ to form. Because Aβ is normally released from brain cells, it may be that an alteration in the ability to sequester/remove Aβ from brain tissue occurs in AD.[126]

Another possibility is that alterations present in AD promote transformation of diffuse unaggregated Aβ into compact aggregated (neurotoxic) Aβ.[119]

NEUROTROPHIC FACTORS PROTECT NEURONS AGAINST LOSS OF CALCIUM HOMEOSTASIS AND FREE RADICALS

To this point I have focused on the roles of βAPP metabolites and excitatory amino acids in the pathogenesis of AD. An additional class of signaling molecules that is certainly relevant to AD is neurotrophic factors. During the last 5 years it has become clear that neurotrophic factors serve important functions that extend beyond their classical roles in promoting cell survival and neurite outgrowth during brain development. Most relevant to the pathogenesis of neurodegenerative conditions is the ability of neurotrophic factors to protect neurons from a variety of insults that involve Ca^{2+}- and free radical-mediated damage.[127] Cell culture and *in vivo* studies have demonstrated the ability of several neurotrophic factors to protect neurons in different brain regions against excitotoxic/ischemic insults. Examples include: basic fibroblast growth factor (bFGF) protects rat hippocampal neurons against glutamate toxicity *in vitro*[18] and against ischemic injury *in vivo.*[128] Nerve growth factor (NGF), bFGF, insulin-like growth factors (IGFs), brain-derived neurotrophic factor (BDNF), and neurotropin-3 (NT-3) can protect cultured rat hippocampal, rat septal, and human cortical neurons against glucose deprivation-induced damage.[129–131] NGF[132] and IGF-I[133] also protected brain neurons against ischemic damage *in vivo.* By comparing $[Ca^{2+}]_i$ in control glucose-deprived neurons and growth factor-treated glucose-deprived neurons we found that each of the growth factors just referred to prevents the large elevation of $[Ca^{2+}]_i$ normally caused by glucose deprivation. These growth factors do not prevent ATP depletion caused by glucose deprivation, indicating that they somehow enhance the ability of the neuron to maintain rest $[Ca^{2+}]_i$ in the face of reduced energy availability.[134]

The ways in which neurotrophic factors stabilize $[Ca^{2+}]_i$ and protect neurons against excitotoxic insults are beginning to be understood. Many neuronal growth factors bind to receptors that possess intrinsic tyrosine kinase activity.[49] Activation of these receptors can stimulate (or inhibit) an array of genes. A cascade of phosphorylation events involving activation of various kinases (eg, MAP kinases) seems to transduce the signal to the nucleus where primary response genes such as *fos* are induced.[135] It is not yet clear which of the gene products are responsible for trophic and neuroprotective actions of the different growth factors; however, recent findings suggest that growth factors can stabilize $[Ca^{2+}]_i$ by influencing specific $[Ca^{2+}]_i$-regulating proteins. For example, in hippocampal cell cultures bFGF and NT-3 increased the expression of the 28-kD calcium-binding protein calbindin.[136] Previous work showed that neurons expressing calbindin are superior in their ability to reduce $[Ca^{2+}]_i$ following a challenge and are relatively resistant to excitotoxicity.[137] Recently, we demonstrated that bFGF can suppress the expression of a 71-kDa NMDA receptor protein that mediates calcium influx and excitotoxicity in cultured hippocampal neurons.[138] The data are consistent with a role for growth factors in modulating glutamatergic transmission, a possibility reinforced by studies showing that NGF, bFGF, and EGF can modulate LTP.[139] Inasmuch as the expression of neurotrophic

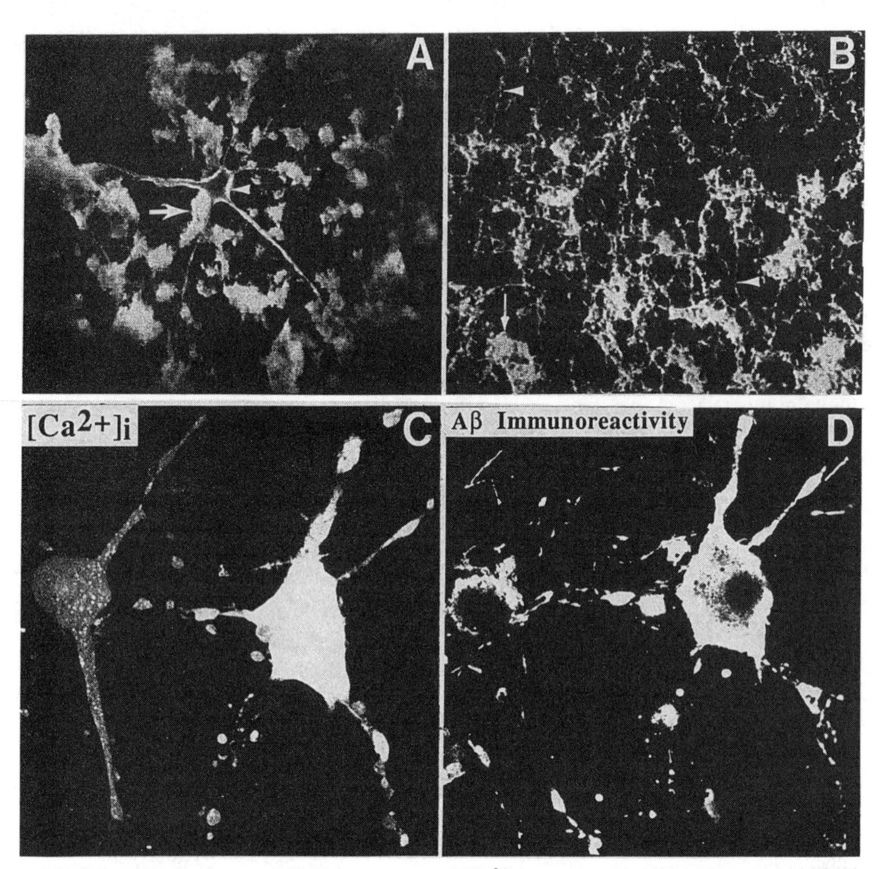

FIGURE 4. Aβ aggregates and disrupts neuronal $[Ca^{2+}]_i$ homeostasis. All images were acquired using a confocal laser scanning microscope with a 60X objective lens. (**A**) Aβ immunoreactivity in a rat hippocampal cell culture that had been exposed to 20 μM Aβ1-40 for 48 hours. Note large aggregates of Aβ immunoreactive material associated with the neuron (eg, *arrow*) and the culture surface. *Arrowhead* points to neuron cell body. (**B**) Aβ forms fibrillar aggregates. In this case Aβ was added to a culture dish lacking cells, allowed to incubate for 24 hours, and then immunostained with an antibody to Aβ. Loose (*arrowheads*) and compact (eg, *arrow*) fibrillar aggregates are observed. (**C** and **D**) Relationship between $[Ca^{2+}]_i$ and Aβ association with neurons. A hippocampal culture was exposed to 20 μM Aβ1-40 for 24 hours and then cells were loaded with the Ca^{2+} indicator dye fluo-3. An image of $[Ca^{2+}]_i$ was acquired (**C**) and then the cells were fixed and immunostained with Aβ antibody (**D**). Note that the neuron with the low $[Ca^{2+}]_i$ (neuron at *left*) had relatively little Aβ immunoreactivity associated with it relative to the neuron with the high $[Ca^{2+}]_i$ (neuron at *right*).

factors is modulated by electrical activity,[140,141] it seems likely that the ability of growth factors to stabilize $[Ca^{2+}]_i$ is a mechanistic basis for protection against the potentially devastating excitotoxicity that can result from overactivity in neural circuits.

In addition to stabilizing $[Ca^{2+}]_i$, growth factors can prevent or attenuate free radical damage by enhancing cellular systems designed to protect against radicals.

For example, NGF protected cultured PC12 cells against hydrogen peroxide.[142] NGF, bFGF, and IGFs attenuated iron-induced oxidative damage and death in cultured rat hippocampal neurons.[143] Growth factors increase the expression of free radical scavenging proteins including catalase, superoxide dismutase, and glutathione peroxidase.[144] Because conditions that increase $[Ca^{2+}]_i$ also tend to result in the generation of free radicals (eg, ischemia and activation of glutamate receptors), it makes sense that growth factors would elicit changes in neurons designed to protect against both $[Ca^{2+}]_i$ elevation and free radicals. Several studies indicate that increased levels of free radicals are present in the aging brain and in AD.[145,146] Moreover, evidence supports a major role for free radicals in the pathogenesis of other neurodegenerative disorders that may have similarities with AD in terms of mechanisms of neuronal injury. For example, amyotrophic lateral sclerosis which involves degeneration of motor neurons was recently linked to a mutation in the gene for superoxide dismutase, an enzyme that prevents accumulation of free radicals.[147] In addition, Parkinson's disease is believed to involve free radical-mediated damage to neurons in the substantia nigra.[148]

Relevant to the pathogenesis of AD are recent findings showing that neurotrophic factors can protect neurons against Aβ toxicity and neurofibrillary tangle-like antigenic changes. Cultured hippocampal neurons treated with bFGF exhibited increased resistance to Aβ neurotoxicity.[114] Basic FGF prevented the loss of calcium homeostasis normally caused by aggregated Aβ. Neurofibrillary tangle-like antigenic alterations in tau and ubiquitin induced by glucose deprivation were prevented by bFGF and NGF in cultured hippocampal neurons.[28] Again the neuroprotective mechanism involved stabilization of $[Ca^{2+}]_i$ by growth factors. Taken together with studies showing that NGF and bFGF can protect cholinergic septal neurons against axotomy-induced degeneration,[149,150] our data suggest that growth factors may counteract age- and disease-associated changes in the brain that tend to increase neuronal vulnerability to excitotoxicity (see below). Indeed, such a rationale has led to the initiation of clinical trials of NGF therapy in patients with AD.[151]

AGE-RELATED METABOLIC COMPROMISE MAY COMBINE WITH ALTERED βAPP METABOLISM TO DESTABILIZE NEURONAL CALCIUM HOMEOSTASIS

Since age is the primary risk factor for AD, any hypothesis concerning the pathophysiology of this disorder must take into account age-associated changes in the brain. Considerable data indicate that as we age several alterations occur that contribute to a relative metabolic compromise of neurons. Atherosclerosis in brain vessels would be expected to limit blood supply to the brain parenchyma (ie, ischemic conditions). The availability of glucose to neurons in AD is apparently reduced as a result of a reduction in glucose transport activity[34] and glucose utilization[33] in the brain. Increased levels of glucocorticoids that occur with age are believed to reduce glucose uptake into neurons and thereby increase their vulnerability to excitotoxicity.[20] In experimental animals glucocorticoids (and physiologic stress) can exacerbate excitotoxic damage and neurofibrillary tangle-like antigenic changes in hippocampal neurons[31,152] (FIG. 3). All of these findings suggest that neurons in the aging brain and particularly in AD receive a reduced energy supply. Again I note that neurons are particularly vulnerable to excitotoxicity and Aβ toxicity when energy levels are reduced (eg, hypoglycemia). Perhaps it is the age-associated compromise in energy availability to neurons that provides the environment in which the neurodegenerative tendency is expressed. For example, in amyotrophic lateral sclerosis neurons may be able to cope with increased levels of free radicals caused by a mutation in the superoxide dismutase gene during young adulthood, but not as the aging process compromises energy availability to neurons. In AD, a reasonable scenario can be envisioned in which Aβ deposits are particularly detrimental when energy availability to neurons is low, as may occur with aging.

Additional age-related changes in the brain that have been documented implicate loss of $[Ca^{2+}]_i$ homeostasis in the pathogensis of AD. For example, animal studies have shown a prolongation of the calcium spike in hippocampal neurons from aged rats[153] which may be exacerbated by glucocorticoids.[154] Such changes in calcium currents would be expected to contribute to a chronic elevation of $[Ca^{2+}]_i$. Dysregulation of the stress-responsive brain-hypothalamic-pituitary-adrenal system has been reported to occur with aging and in AD.[32,153] Data suggest a relative increase in glucocorticoid levels in aging and AD which may contribute to increased $[Ca^{2+}]_i$ by impairing glucose transport and thereby increasing vulnerability to excitotoxicity.[31,155]

In concert with perturbed energy metabolism, alterations in the status of systems that control free radical levels have been shown to occur in the brain with aging. Age-dependent reductions in levels of antioxidant enzymes, including catalase and superoxide dismutase, as well as increased levels of hydroxl radical in brain regions particularly vulnerable to aging and Alzheimer's disease have been reported.[146] Experimental studies have shown that free radical-trapping compounds can prevent and reverse age-associated changes in brain protein oxidation, enzyme activity, and memory functions.[145] Degenerative changes in the brain in Down's syndrome may also involve free radical-mediated processes. Indeed, the gene for superoxide dismutase is on the affected chromosome in this disorder (chromosome 21). Experimental evidence suggests that altered βAPP metabolism could exacerbate age-related changes in free radical metabolism. Indeed, APP^ss reduce and Aβ increases the vulnerability of cultured rat hippocampal and human cortical neurons to oxidative damage promoted by exposure to iron or hydrogen peroxide.[156] Increased free radicals with aging may also contribute to the formation of amyloid fibrils and Aβ aggregates. Dyrks *et al.*[119] have shown that free radicals promote and antioxidants suppress Aβ aggregation in cell-free systems. Thus, age-associated changes in brain metabolism may contribute to altered βAPP metabolism.

PREVENTIVE AND THERAPEUTIC CONSIDERATIONS

What are likely to be the most successful approaches to preventing and treating AD? Clearly, this question cannot be answered properly until the causes of this disorder are more fully understood and very early diagnosis (prior to neuronal degeneration) is possible. Indeed, it seems unlikely that AD will be treatable by means other than those that prevent or retard the progression of the underlying neuronal degeneration. Therefore, I would argue that approaches aimed at enhancing activity

TABLE 2. Factors That May Play a Role in Determining if Neuronal Calcium Homeostasis Is Lost or Maintained in Aging and Alzheimer's Disease

Factors	Mechanism of Action
Factors that destabilize $[Ca^{2+}]_i$:	
Excitatory amino acids	*Activation of receptor* Ca^{2+} channels and voltage-dependent Ca^{2+} channels
Ischemia (hypoglycemia, hypoxia)	Reduction of energy availability to neurons
Amyloid β-peptide	Disruption of Ca^{2+} regulation in plasma membrane
Glucocorticoids	Reduction of glucose transport/ decrease in energy supply
Factors that stabilize $[Ca^{2+}]_i$:	
Secreted forms of βAPP	Reduction of Ca^{2+} influx, enhancement of Ca^{2+} extrusion/buffering?
Growth factors (NGF, bFGF, IGFs, NT-3, BDNF)	Reduction of Ca^{2+} influx, enhancement of Ca^{2+} extrusion/buffering?
Inhibitory transmitters (GABA)	Reduction of membrane depolarization
Antioxidants (eg, vitamin E, PBN)	Reduction of free radical levels

in the remaining circuitry (eg, cholinesterase inhibitors) will not prove useful and may actually accelerate the progression of the degenerative (excitotoxic?) process[15] (FIG. 1). The data just described would argue that the most beneficial approaches may include manipulations of systems involved in: processing of βAPP; aggregation of Aβ; regulation of neuronal calcium homeostasis; and free radical metabolism (TABLE 2).

Inasmuch as increasing data indicate that the way in which cells process βAPP may have important consequences for deposition of Aβ, as well as the normal function of APP^ss, it is clearly of importance to gain a better understanding of the enzymes and mechanisms involved in βAPP processing. There would seem to be great potential in manipulating βAPP processing once key enzymes have been identified. For example, enhancing the activity or levels of α-secretase would be expected to reduce levels of Aβ and increase release of APP^ss. Because Aβ and APP^ss seem to be neurotoxic and neuroprotective, respectively, it seems likely that increasing α-secretase activity would be beneficial. Another approach would be to prevent aggregation of Aβ in that it is the aggregated form of Aβ that seems to be particularly detrimental to neurons. This might be accomplished by administration of antioxidants (eg, vitamin E) because free radicals appear to promote Aβ aggregation and are involved in the process of cell injury. A better understanding of the factors influencing Aβ aggregation may reveal other approaches to preventing aggregation.

Because accumulating data indicate that loss of $[Ca^{2+}]_i$ homeostasis may be a focal point for a variety of age- and genetic-related conditions contributing to different cases of AD, it would seem prudent to pursue preventive approaches aimed at stabilizing $[Ca^{2+}]_i$ in neurons. This will involve: exploring a large number of available compounds in the clinical setting; developing compounds with increased specificity; and expanding basic research into mechanisms of neuronal $[Ca^{2+}]_i$ regulation and the actions of neuroprotective growth factors. There is currently an array of compounds in clinical use for disorders other than AD that might suppress elevations of neuronal $[Ca^{2+}]_i$ and protect neurons against excitotoxic injury. They include: excitatory amino acid receptor antagonists; calcium channel blockers; anticonvulsants; and growth factors. Compounds in each of these classes has proven beneficial in animal studies of acute neurodegenerative disorders such as stroke and head trauma. With glutamate receptor antagonists and calcium channel blockers, however, it seems likely that long-term treatment may result in unwanted side effects. That is, virtually all neurons in the CNS, as well as non-neuronal cells, possess glutamate receptors and calcium channels. Therefore, function of many of these cells may be compromised by antagonists. On the other hand, the data of Disterhoft and coworkers[157] indicate that dihydropyridine calcium channel blockers can prevent age-associated deficits in learning and memory, suggesting their potential use in age-associated neurodegenerative disorders such as AD. Activation of excitatory amino acid receptors and voltage-dependent calcium channels is reduced by Mg^{2+}, and it is therefore reasonable to consider that administration of Mg^{2+} might reduce neuronal injury in AD. Anticonvulsants have a long clinical track record in chronic treatment of epilepsy, and their side effects are known. These compounds (eg, carbamazepine, phenytoin, and valproic acid) reduce membrane depolarization and so reduce calcium influx into neurons. Potassium channel agonists also reduce excitability and might be expected to reduce an excitotoxic form injury in AD.

Neuroprotective growth factors are of particular interest because: they were recently shown to be highly neuroprotective in various animal models of both acute and chronic neurodegenerative conditions (see above); they can stabilize both $[Ca^{2+}]_i$ homeostasis and free radicals; and they are normally present in the brain. Administration of neurotrophic factors seems to require central administration (into the ventricles or brain parenchyma) although other means of delivery may be possible.[158] Trials of NGF in AD are underway.[151] Another approach would be to use low molecular weight compounds that mimic the actions of neurotrophic factors. For example, recent findings indicate that very low concentrations of staurosporine and K-252 compounds can protect hippocampal, septal, and cortical neurons against hypoglycemic injury.[159] Interestingly, these compounds stabilize $[Ca^{2+}]_i$ levels and may act by a mechanism similar to neurotrophic factors because they can stimulate tyrosine phosphorylation of several different proteins.

As already indicated, free radicals could contribute to the pathogenesis of AD in several ways including direct damage to neurons and promotion of Aβ accumulation. A variety of compounds are available and many more are being developed that can reduce free radical levels. Examples include free radical scavengers such as vitamin E and ubiquinone and spin-trapping compounds such as *N*-tert-butyl-alpha-phenylnitrone. In light of the increasing evidence for the involvement of calcium dysregulation and free radicals in the pathogenesis of many neurodegenerative conditions (stroke, AD, Parkinson's disease, and amyotrophic lateral sclerosis), it seems wise to continue and expand research efforts in these areas.

SUMMARY

Alzheimer's disease (AD) is defined by degeneration of specific populations of neurons and the presence of insoluble aggregates of cytoskeletal proteins and amyloid β-peptide (Aβ) within affected brain regions. Alzheimer's disease does not appear to result from a single alteration, but in some cases of inherited AD a specific genetic defect can precipitate the disease. In this article, metabolic compromise, altered metabolism of the β-amyloid precursor protein (βAPP), and an excitotoxic form of neuronal injury are considered central to the pathogenesis AD. The hypothesis is forwarded that destabilization of neuronal Ca^{2+} homeostasis underlies neuronal degeneration and that multiple age-associated and/or genetic alterations contribute to the loss of Ca^{2+} homeostasis.

Recent studies showed that the secreted forms of βAPP (APP^ss) stabilize intracellular free calcium levels ($[Ca^{2+}]_i$) and protect neurons against excitotoxic insults. In contrast, Aβ which arises from alternative processing of βAPP forms free radical peptides and aggregates that destabilize $[Ca^{2+}]_i$ and make neurons vulnerable to metabolic insults. Increased expression (eg, Down's syndrome) or altered processing (eg, βAPP mutations) of βAPP may increase the Aβ/APP^s ratio. The death of neurons in AD most likely has an excitotoxic component because: the vulnerable neurons possess high levels of glutamate receptors; experimentally induced excitotoxicity shows several features similar to those of neurofibrillary tangles; and Aβ can destabilize $[Ca^{2+}]_i$ homeostasis and render neurons vulnerable to neurofibrillary degeneration. Selective vulnerability may result from cell type-specific differences in expression of proteins involved in regulating $[Ca^{2+}]_i$. In addition, many intercellular signals are involved in determining whether a neuron is able to maintain $[Ca^{2+}]_i$ within a range of concentrations conducive to cell survival and adaptive plasticity. In this regard, it was recently shown that several growth factors can stabilize $[Ca^{2+}]_i$ and protect neurons against excitotoxic injury and Aβ toxicity. Age-related changes in the brain (eg, ischemic conditions, reduced glucose uptake, and increased glucocorticoid levels) may compromise the mechanisms that normally regulate $[Ca^{2+}]_i$ adaptively.

ACKNOWLEDGMENTS

I appreciate the contributions of S. W. Barger, B. Cheng, Y. Goodman, R. Mark, J. Oeltgen, J. O'Keefe, and V. L. Smith-Swintosky to the original research from this laboratory. I also thank D. A. Butterfield, J. Carney, S. Christakos, I. Lieberburg, E. K. Michaelis, and R. E. Rydel for collaborative studies.

REFERENCES

1. KATER, S. B., M. P. MATTSON, C. S. COHAN & J. A. CONNOR. 1988. Calcium regulation of the neuronal growth cone. Trends Neurosci. **11:** 315-321.
2. KENNEDY, M. B. 1989. Regulation of synaptic transmission in the central nervous system: Long-term potentiation. Cell **59:** 777-787.
3. MATTSON, M. P. 1992. Calcium as sculptor and destroyer of neural circuitry. Exp. Gerontol. **27:** 29-49.

4. MATTSON, M. P. & S. B. KATER. 1987. Calcium regulation of neurite elongation and growth cone motility. J. Neurosci. **7:** 4034-4043.
5. MATTSON, M. P., P. DOU & S. B. KATER. 1988. Outgrowth-regulating actions of glutamate in isolated hippocampal pyramidal neurons. J. Neurosci. **8:** 2087-2100.
6. MATTSON, M. P. 1988. Neurotransmitters in the regulation of neuronal cytoarchitecture. Brain Res. Rev. **13:** 179-212.
7. SIESJO, B. K., F. BENGTSSON, W. GRAMPP & S. THEANDER. 1989. Calcium, excitotoxins, and neuronal death in the brain. Ann. N.Y. Acad. Sci. **568:** 234-251.
8. SCHARFMAN, H. E. & P. A. SCHWARTZKROIN. 1989. Protection of dentate hilar cells from prolonged stimulation by intracellular calcium chelation. Science **246:** 257-260.
9. KHACHATURIAN, Z. S. 1989. The role of calcium regulation in brain aging: Reexamination of a hypothesis. Aging **1:** 17-34.
10. MCDONALD, J. W. & M. V. JOHNSTON. 1990. Physiological and pathophysiological roles of excitatory amino acids during central nervous system development. Brain Res. Rev. **15:** 41-70.
11. MATTSON, M. P., R. E. LEE, M. E. ADAMS, P. B. GUTHRIE & S. B. KATER. 1988. Interactions between entorhinal axons and target hippocampal neurons: A role for glutamate in the development of hippocampal circuitry. Neuron **1:** 865-876.
12. CLINE, H. T. & M. CONSTANTINE-PATON. 1990. NMDA receptor agonist and antagonists alter retinal ganglion cell arbor structure in the developing frog retinotectal projection. J. Neurosci. **10:** 1197-1216.
13. COLLINGRIDGE, G. I. & T. V. P. BLISS. 1987. NMDA receptors: Their role in long-term potentiation. Trends Neurosci. **10:** 288-293.
14. MATTSON, M. P. & S. B. KATER. 1989. Excitatory and inhibitory neurotransmitters in the generation and degeneration of hippocampal neuroarchitecture. Brain Res. **478:** 337-348.
15. MATTSON, M. P. 1989. Acetylcholine potentiates glutamate-induced neurodegeneration in cultured hippocampal neurons. Brain Res. **497:** 402-406.
16. JOHNSTON, D., S. WILLIAMS, D. JAFFE & R. GRAY. 1992. NMDA-receptor-independent long-term potentiation. Annu. Rev. Physiol. **54:** 489-505.
17. ROTHMAN, S. M. & J. W. OLNEY. 1986. Glutamate and the pathophysiology of hypoxic-ischemic brain damage. Ann. Neurol. **19:** 105-111.
18. MATTSON, M. P., M. MURRAIN, P. B. GUTHRIE & S. B. KATER. 1989. Fibroblast growth factor and glutamate: Opposing roles in the generation and degeneration of hippocampal neuroarchitecture. J. Neurosci. **9:** 3728-3740.
19. NOVELLI, A., J. A. REILLY, P. G. LYSKA & R. C. HENNEBERRY. 1988. Glutamate becomes neurotoxic via the *N*-methyl-D-aspartate receptor when intracellular energy levels are reduced. Brain Res. **451:** 205-212.
20. SAPOLSKY, R. M. 1985. A mechanism for glucocorticoid toxicity in the hippocampus: Increased vulnerability to metabolic insults. J. Neurosci. **5:** 1228-1232.
21. GREENAMYRE, J. T. & A. B. YOUNG. 1989. Excitatory amino acids and Alzheimer's disease. Neurobiol. Aging **10:** 593-602.
22. MATTSON, M. P., S. W. BARGER, B. CHENG, I. LIEBERBURG, V. L. SMITH-SWINTOSKY & R. E. RYDEL. 1993. β-amyloid precursor protein metabolites and loss of neuronal calcium homeostasis in Alzheimer's disease. Trends Neurosci. **16:** 409-414.
23. MACDERMOTT, A. B. & N. DALE. 1987. Receptors, ion channels and synaptic potentials underlying the integrative actions of excitatory amino acids. Trends Neurosci. **10:** 280-284.
24. WISNIEWSKI, K., G. A. JERVIS, R. C. MORETZ & H. M. WISNIEWSKI. 1979. Alzheimer neurofibrillary tangles in diseases other than senile and presenile dementia. Ann. Neurol. **5:** 288-294.
25. SELKOE, D. J. 1989. Biochemistry of altered brain proteins in Alzheimer's disease. Ann. Rev. Neurosci. **12:** 463-490.

26. MATTSON, M. P. 1990. Antigenic changes similar to those seen in neurofibrillary tangles are elicited by glutamate and calcium influx in cultured hippocampal neurons. Neuron **4:** 105-117.
27. MATTSON, M. P., M. G. ENGLE & B. RYCHLIK. 1991. Effects of elevated intracellular calcium levels on the cytoskeleton and tau in cultured human cerebral cortical neurons. Mol. Chem. Neuropathol. **15:** 117-142.
28. CHENG, B. & M. P. MATTSON. 1992. Glucose deprivation elicits neurofibrillary tangle-like antigenic changes in hippocampal neurons: Prevention by NGF and bFGF. Exp. Neurol. **117:** 114-123.
29. SAUTIERE, P.-E., P. SINDOU, P. COURATIER, J. HUGON, A. WATTEZ & A. DELACOURTE. 1992. Tau antigenic changes induced by glutamate in rat primary culture model: A biochemical approach. Neurosci. Lett. **140:** 206-210.
30. DE BONI, U. & D. R. C. MCLACHLAN. 1985. Controlled induction of paired helical filaments of the Alzheimer type in cultured human neurons, by glutamate and aspartate. J. Neurol. Sci. **68:** 105-118.
31. ELLIOTT, E. M., M. P. MATTSON, P. VANDERKLISH, G. LYNCH, I. CHANG & R. M. SAPOLSKY. 1993. Corticosterone exacerbates kainate-induced alterations in hippocampal tau immunoreactivity and spectrin proteolysis in vivo. J. Neurochem. **61:** 57-67.
32. GREENWALD, B. S., A. A. MATHE, R. C. MOHS, M. I. LEVY, C. A. JOHNS & K. L. DAVIS. 1986. Cortisol and Alzheimer's disease. II. Dexamethasone suppression, dementia severity, and affective symptoms. Am. J. Psychiatry **143:** 442-446.
33. HOYER, S., K. OESTERREICH & O. WAGNER. 1988. Glucose metabolism as the site of the primary abnormality in early-onset dementia of Alzheimer type? J. Neurol. **235:** 143-148.
34. KALARIA, R. N. & S. I. HARIK. 1989. Reduced glucose transporter at the blood-brain barrier and in cerebral cortex in Alzheimer's disease. J. Neurochem. **53:** 1083-1088.
35. STEINER, B., E. M. MANDELKOW, J. BIERNAT, N. GUSTKE, H. E. MEYER, B. SCHMIDT, G. MIESKES, H. D. SOLING, D. DRECHSEL & M. W. KIRSCHNER. 1990. Phosphorylation of microtubule-associated protein tau: Identification of the site for Ca^{2+}-calmodulin dependent kinase and relationship with tau phosphorylation in Alzheimer tangles. EMBO J. **9:** 3539-3544.
36. JOHNSON, G. V. 1992. Differential phosphorylation of tau by cyclic AMP-dependent protein kinase and CA^{2+}/calmodulin-dependent protein kinase II. Metabolic and functional consequences. J. Neurochem. **59:** 2056-2062.
37. MATTSON, M. P. 1991. Evidence for the involvement of protein kinase C in neurodegenerative changes in cultured human cortical neurons. Exp. Neurol. **112:** 95-103.
38. CORREAS, I., J. CIAZ-NIDO & J. AVILA. 1992. Microtubule-associated protein tau is phosphorylated by protein kinase C on its tubulin binding domain. J. Biol. Chem. **267:** 15721-15728.
39. SCOTT, C. W., R. C. SPREEN, J. L. HERMAN, F. P. CHOW, M. D. DAVISON, J. YOUNG & C. B. CAPUTO. 1993. Phosphorylation of recombinant tau by cAMP-dependent protein kinase. Identification of phosphorylation sites and effect on microtubule assembly. J. Biol. Chem. **268:** 1166-1173.
40. MASLIAH, E., D. S. IIMOTO, M. MALLORY, T. ALBRIGHT, L. HANSEN & T. SAITOH. 1992. Casein kinase II alteration precedes tau accumulation in tangle formation. Am. J. Pathol. **140:** 263-268.
41. GOEDERT, M., M. G. SPILLANTINI, N. J. CAIRNS & R. A. CROWTHER. 1992. Tau proteins of Alzheimer paired helical filaments: Abnormal phosphorylation of all six brain isoforms. Neuron **8:** 159-168.
42. DREWES, G., B. LICHTENBERG-KRAAG, F. DORING, E. M. MANDELKOW, J. BIERNAT, J. GORIS, M. DOREE & E. MANDELKOW. 1992. Mitogen activated protein (MAP) kinase transforms tau protein into an Alzheimer-like state. EMBO J. **11:** 2131-2138.

43. GOEDERT, M., E. S. COHEN, R. JAKES & P. COHEN. 1992. p42 MAP kinase phosphorylation sites in microtubule-associate protein tau are dephosphorylated by protein phosphatase 2A. FEBS Lett. **312:** 95-99.
44. DRECHSEL, D. N., A. A. HYMAN, M. H. COBB & M. W. KIRSCHNER. 1992. Modulation of the dynamic instability of tubulin assembly by the microtubule-associated protein tau. Mol. Biol. Cell **3:** 1141-1154.
45. LU, Q., J. P. SORIA & J. G. WOOD. 1993. p44mpk MAP kinase induces Alzheimer type alterations in tau function and in primary hippocampal neurons. J. Neurosci. Res. **35:** 439-444.
46. BADING, H. & M. E. GREENBERG. 1991. Stimulation of tyrosine phosphorylation by NMDA receptor activation. Science **253:** 912-914.
47. BARABAN, J. M., R. S. FIORE, J. S. SANGHERA, H. B. PADDON & S. L. PELECH. 1993. Identification of p42 mitogen-activated protein kinase as a tyrosine kinase substrate activated by maximal electroconvulsive shock in hippocampus. J. Neurochem. **60:** 330-336.
48. CHAO, T.-S., K. L. BYRON, K. M. LEE, M. VILLEREAL & M. R. ROSNER. 1992. Activation of MAP kinase by calcium-dependent and calcium-independent pathways: Stimulation by thapsigargin and epidermal growth factor. J. Biol. Chem. **267:** 19876-19883.
49. SCHLESSINGER, J. & A. ULRICH. 1992. Growth factor signaling by receptor tyrosine kinases. Neuron **9:** 383-391.
50. ZATTORE, R. J. 1990. Memory loss following domoic acid intoxication from ingestion of toxic mussels. Can. Dis. Wkly. Rep. **16**(Suppl. 1E): 101-103.
51. SPENCER, P. S., P. B. NUNN, J. HUGON, A. C. LUDOLPH, S. M. ROSS, D. N. ROY & R. C. ROBERTSON. 1987. Guam amyotrophic lateral sclerosis-Parkinsonism-dementia linked to a plant excitant neurotoxin. Science **237:** 517-522.
52. GARRUTO, R. M. & Y. YASE. 1986. Neurodegenerative disorders of the Western Pacific: The search for mechanisms of pathogenesis. Trends Neurosci. **9:** 368-374.
53. PETERSON, C., R. R. RATAN, M. L. SHELANSKI & J. E. GOLDMAN. 1986. Cytosolic free calcium and cell spreading decrease in fibroblasts from aged and Alzheimer donors. Proc. Natl. Acad. Sci. USA **83:** 7999-8001.
54. MCCOY, K. R., R. D. MULLINS, T. G. NEWCOMB, G. M. NG, G. PAVLINKOVA, R. J. POLINSKY, L. E. NEE & J. E. SISKEN. 1993. Serum- and bradykinin-induced calcium transients in familial Alzheimer's fibroblasts. Neurobiol. Aging **14:** 447-455.
55. ETCHEBERRIGARAY, R., E. ITO, K. OEA, B. TOPEL-GREHL, G. E. GIBSON & D. L. ALKON. 1993. Potassium channel dysfunction in fibroblasts identifies patients with Alzheimer's disease. Proc. Natl. Acad. Sci. USA **90:** 8209-8213.
56. SELKOE, D. J. 1993. Physiological production of the β-amyloid protein and the mechanism of Alzheimer's disease. Trends Neurosci. **16:** 403-409.
57. KANG, J., H.-G. LEMAIRE, A. UNTERBECK, J. M. SALBAUM, C. L. MASTERS, K.-H. GRZESCKIK, G. MULTHAUP, K. BEYREUTHER & B. MULLER-HILL. 1987. The precursor of Alzheimer's disease amyloid A4 protein resembles a cell-surface receptor. Nature **325:** 733-736.
58. DYRKS, T., A. WEIDEMANN, G. MULTHAUP, J. M. SALBAUM, H.-G. LEMAIRE, J. KANG, B. MULLER-HILL, C. L. MASTERS & K. BEYREUTHER. 1988. Identification, transmembrane orientation and biogenesis of the amyloid A4 precursor of Alzheimer's disease. EMBO J. **7:** 949-957.
59. OLTERSDORF, T., L. C. FRITZ, D. B. SCHENK, I. LIEBERBURG, K. L. JOHNSON-WOOD, E. C. BEATTIE, P. J. WARD, R. W. BLACHER, H. F. DOVEY & S. SINHA. 1989. The secreted form of Alzheimer's amyloid precursor protein with the Kunitz domain is protease nexin II. Nature **347:** 144-147.
60. VAN NOSTRAND, W. E., S. L. WAGNER, M. SUZUKI, B. H. CHOI, J. S. FARROW, J. W. GEDDES & C. W. COTMAN. 1989. Protease nexin-II, a potent antichymotrypsin, shows identity to amyloid beta-protein precursor. Nature **341:** 546-549.

61. Esch, F. S., P. S. Keim, E. C. Beattie, R. W. Blacher, A. R. Culwell, T. Oltersdorf, D. McClure & P. J. Ward. 1990. Cleavage of amyloid β peptide during constitutive processing of its precursor. Science **248:** 1122-1124.
62. Seubert, P., T. Oltersdorf, M. G. Lee, R. Barbour, C. Blomquist, D. L. Davis, K. Bryant, L. C. Fritz, D. Galasko, L. J. Thal, I. Lieberburg & D. B. Schenk. 1993. Secretion of β-amyloid precursor protein cleaved at the amino terminus of the β-amyloid peptide. Nature **361:** 260-263.
63. Miyazaki, K., M. Hasegawa, K. Funahashi & M. Umeda. 1993. A metalloproteinase inhibitor domain in Alzheimer amyloid protein precursor. Nature **362:** 839-841.
64. Sisodia, S. S., E. H. Koo, K. Beyreuther & A. Unterbeck. 1990. Evidence that beta-amyloid protein in Alzheimer's disease is not derived by normal processing. Science **248:** 492-495.
65. Golde, T. E., S. Estus, L. H. Younkin, D. J. Selkoe & S. G. Younkin. 1992. Processing of the amyloid protein precursor to potentially amyloidogenic derivatives. Science **255:** 728-730.
66. Haass, C., M. G. Schlossmacher, A. Y. Hung, C. Vigo-Pelfrey, A. Mellon, B. Ostaszewski, I. Lieberburg, E. H. Koo, D. Schenk, D. B. Teplow & D. J. Selkoe. 1992. Amyloid β-peptide is produced by cultured cells during normal metabolism. Nature **359:** 322-325.
67. Seubert, P., C. Vigo-Pelfrey, F. Esch, M. Lee, H. Dovey, D. Davis, S. Sinha, M. Schlossmacher, J. Whaley, C. Swindlehurst, R. McCormack, R. Wolfert, D. J. Selkoe, I. Lieberburg & D. Schenk. 1992. Isolation and quantitation of soluble Alzheimer's β-peptide from biological fluids. Nature **359:** 325-327.
68. Shoji, M., T. E. Golde, J. Ghiso, T. T. Cheung, S. Estus, L. M. Shaffer, S.-D. Cai, D. M. McKay, R. Tintner, B. Frangione & S. G. Younkin. 1992. Production of the Alzheimer amyloid β protein by normal proteolytic processing. Science **258:** 126-129.
69. Chartier-Harlin, M. C., F. Crawford, H. Houlden, A. Warren, D. Hughes, L. Fidani, A. Goate, M. Rossor, P. Roques, J. Hardy & M. Mullan. 1991. Early-onset Alzheimer's disease caused by mutations at codon 717 of the β-amyloid precursor protein. Nature **353:** 844-846.
70. Goate, A., M. C. Chartier-Harlin, M. Mullan *et al.* 1991. Segregation of a missense mutation in the amyloid precursor protein gene with familial Alzheimer's disease. Nature **349:** 704-706.
71. Murrell, J., M. Farlow, B. Ghetti & M. D. Benson. 1991. A mutation in the amyloid precursor protein associated with hereditary Alzheimer's disease. Science **254:** 97-99.
72. Mullan, M., F. Crawford, K. Axelman *et al.* 1992. A pathogenic mutation of probable Alzheimer's disease in the APP gene at the N-terminus of β-amyloid. Nature Genet. **1:** 345-347.
73. Cai, X.-D., T. E. Golde & S. G. Younkin. 1993. Release of excess amyloid β protein from a mutant amyloid β protein precursor. Science **259:** 514-516.
74. Citron, M., T. Oltersdorf, C. Haass, L. McConlogue, A. Y. Hung, P. Seubert, C. Vigo-Pelfrey, I. Lieberburg & D. J. Selkoe. 1993. Mutation of the β-amyloid precursor protein in familial Alzheimer's disease causes increased β-protein production. Nature **360:** 672-674.
75. Wasco, W., J. D. Brook & R. E. Tanzi. 1993. The amyloid precursor-like protein (APLP) gene maps to the long arm of human chromosome 19. Genetics **15:** 237-239.
76. Schellenberg, G. D., T. D. Bird, E. M. Wijsman, H. T. Orr, L. Anderson, E. Nemens, J. A. White, L. Bonnycastle, J. L. Weber, M. E. Alonso, H. Potter, L. L. Heston & J. Martin. 1992. Genetic linkage evidence for a familial Alzheimer's disease locus on chromosome 14. Science **258:** 668-671.

77. SAITOH, T., M. SUNDSMO, J.-M. ROCH, N. KIMURA, G. COLE, D. SCHUBERT, T. OLTERSDORF & D. B. SCHENK. 1989. Secreted form of amyloid β protein precursor is involved in the growth regulation of fibroblasts. Cell **58:** 615-622.
78. SCHUBERT, D., G. COLE, T. SAITOH & T. OLTERSDORF. 1989. Amyloid beta protein precursor is a mitogen. Biochem. Biophys. Res. Commun. **162:** 83-88.
79. SMITH, R. P., D. A. HIGUCHI & G. J. BROZE JR. 1990. Platelet coagulation factor XIa-inhibitor, a form of Alzheimer amyloid precursor protein. Science **248:** 1126-1128.
80. SCHUBERT, D., L.-W. JIN, T. SAITOH & G. COLE. 1989. The regulation of amyloid β protein precursor secretion and its modulatory role in cell adhesion. Neuron **3:** 689-694.
81. BREEN, K. C., M. BRUCE & B. H. ANDERTON. 1991. Beta amyloid precursor protein mediates neuronal cell-cell and cell-surface adhesion. J. Neurosci. Res. **28:** 90-100.
82. NISHIMOTO, I., T. OKAMOTO, Y. MATSUURA, S. TAKAHASHI, T. OKAMOTO, Y. MURAYAMA & E. OGATA. 1993. Alzheimer amyloid protein precursor complexes with brain GTP-binding protein G_o. Nature **362:** 75-79.
83. BOMSEL, M. & K. MOSTOV. 1992. Role of heterotrimeric G proteins in membrane traffic. Mol. Biol. Cell **3:** 1317-1328.
84. ARAKI, W., N. KITAGUCH, Y. TOKUSHIMA, K. ISHII, H. ARATAKE, S. SHIMOHAMA, S. NAKAMURA & J. KIMURA. 1992. Trophic effect of β-amyloid precursor protein on cerebral cortical neurons in culture. Biochem. Biophys. Res. Commun. **181:** 265-271.
85. MILWARD, E. A., R. PAPADOOPOULOS, S. J. FULLER, R. D. MOIR, D. SMALL, K. BEYREUTHER & C. L. MASTERS. 1992. The amyloid protein precursor of Alzheimer's disease is a mediator of the effects of nerve growth factor on neurite outgrowth. Neuron **9:** 129-137.
86. MATTSON, M. P., B. CHENG, A. R. CULWELL, F. S. ESCH, I. LIEBERBURG & R. E. RYDEL. 1993. Evidence for excitoprotective and intraneuronal calcium-regulating roles for secreted forms of β-amyloid precursor protein. Neuron **10:** 243-254.
87. MATTSON, M. P. 1994. Secreted forms of β-amyloid precursor protein modulate dendrite outgrowth and calcium responses to glutamate in cultured embryonic hippocampal neurons. J. Neurobiol. **25:** 439-450.
88. KOO, E. H., S. S. SISODIA, D. A. ARCHER, L. J. MARTIN, A. WEIDEMANN, K. BEYREUTHER, C. L. MASTERS, P. FISHER & D. L. PRICE. 1990. Precursor of amyloid protein in Alzheimer's disease undergoes fast anterograde axonal transport. Proc. Natl. Acad. Sci. USA **87:** 1561-1565.
89. NITSCH, R. M., B. E. SLACK, R. J. WURTMAN & J. H. GROWDON. 1992. Release of Alzheimer amyloid precursor derivatives stimulated by the activation of muscarinic acetylcholine receptors. Science **258:** 304-307.
90. NITSCH, R. M., B. E. SLACK, J. H. GROWDON & R. J. WURTMAN. 1993. Release of amyloid b-protein precursor derivatives by electrical depolarization of rat hippocampal slices. Proc. Natl. Acad. Sci. USA **90:** 5191-5193.
91. SMITH-SWINTOSKY, V. L., L. C. PETTIGREW, S. D. CRADDOCK, R. E. RYDEL & M. P. MATTSON. 1994. Secreted forms of β-amyloid precursor protein protect against ischemic brain injury. J. Neurochem. **63:** 781-784.
92. SCHUBERT, D. & C. BEHL. 1993. The expression of amyloid beta protein precursor protects nerve cells from β-amyloid and glutamate toxicity and alters their interaction with the extracellular matrix. Brain Res. **629:** 275-282.
93. BOWES, M. P., T. SAITOH, J. A. ZIVIN, J.-M. ROCH & K. A. UEDA. 1992. A 17-mer peptide segment of the amyloid β/A4 protein precursor (APP) reduces neurologic damage in a rabbit spinal cord ischemia model. Soc. Neurosci. Abstr. **18:** 1437.
94. ROCH, J., I. P. SHAPIRO, M. P. SUNDSUMO, D. A. C. OTERO, L. M. REFOLO, N. K. ROBAKIS & T. SAITOH. 1992. Bacterial expression, purification, and functional mapping of the amyloid B/A4 protein precursor. J. Biol. Chem. **267:** 2214-2221.

95. SIMAN, R., J. P. CARD, R. B. NELSON & L. G. DAVIS. 1989. Expression of β-amyloid precursor protein in reactive astrocytes following neuronal damage. Neuron **3:** 275-285.
96. ABE, K., R. E. TANZI & K. KOGURE. 1991. Selective induction of Kunitz-type inhibitor domain-containing amyloid precursor protein mRNA after persistent focal ischemia in rat cerebral cortex. Neurosci. Lett. **125:** 172-174.
97. ROBERTS, G. W., S. M. GENTLEMAN, A. LYNCH & D. I. GRAHAM. 1991. βA4 amyloid protein deposition in brain after head injury. Lancet **338:** 1422-1423.
98. JOHNSON-WOOD, K. L., T. HENRIKSSON, P. SEUBERT, T. OLTERSDORF, I. LIEBERBURG & D. B. SCHENK. 1990. Identification of specific binding sites for ^{125}I secreted APP751 on intact fibroblast cells. Neurobiol. Aging **11:** 306.
99. BARGER, S. W., R. R. FISCUS, P. RUTH, F. HOFMANN & M. P. MATTSON. 1994. Modulation of glutamate responses by cGMP: Mechanism of action of secreted forms of the Alzheimer's β-amyloid precursor protein. J. Neurochem., in press.
100. GARTHWAITE, G. & J. GARTHWAITE. 1988. Cyclic GMP and cell death in rat cerebellar slices. Neuroscience **26:** 321-326.
101. LUO, L., T. TULLY & K. WHITE. 1992. Human amyloid precursor protein ameliorates behavioral deficit of flies deleted for Appl gene. Neuron **9:** 595-605.
102. HUBER, S., J. R. MARTIN, J. LOFLER & J.-L. MOREAU. 1993. Involvement of amyloid precursor proteins in memory formation in the rat: An indirect antibody approach. Brain Res. **603:** 348-352.
103. FINKLESTEIN, S. P., P. J. ASPOSTOLIDES, C. G. CADAY, J. PROSSER, M. G. PHILLIPS & M. KLAGSBRUN. 1988. Increased basic fibroblast growth factor (bFGF) immunoreactivity at the site of focal brain wounds. Brain Res. **460:** 253-259.
104. LOREZ, H., F. KELLER, G. RUESS & U. OTTEN. 1989. Nerve growth factor increases in adult brain after hypoxic injury. Neurosci. Lett. **98:** 339-344.
105. LINDVALL, O., P. ERNFORS, J. BENGZON, Z. KOLAIA, M.-L. SMITH, B. K. SIESJO & H. PERSSON. 1992. Differential regulation of mRNAs for nerve growth factor, brain-derived neurotrophic factor, and neurotrophin-3 in the adult rat brain following cerebral ischemia and hypoglycemic coma. Proc. Natl. Acad. Sci. USA **89:** 648-652.
106. REFOLO, L. M., S. R. J. SALTON, J. P. ANDERSON, P. MEHTA & N. K. ROBAKIS. 1989. Nerve and epidermal growth factors induce the release of the Alzheimer amyloid precursor from PC12 cell cultures. Biochem. Biophys. Res. Commun. **164:** 664-670.
107. QUON, D., R. CATALANO & B. CORDELL. 1990. Fibroblast growth factor induces beta-amyloid precursor mRNA in glial but not neuronal cultured cells. Biochem. Biophys. Res. Commun. **167:** 96-102.
108. GLENNER, G. G. & C. W. WONG. 1984. Alzheimer's disease: Initial report of the purification and characterization of a novel cerebrovascular amyloid protein. Biochem. Biophys. Res. Commun. **120:** 885-890.
109. YANKNER, B. A., L. K. DUFFY & D. A. KIRSCHNER. 1990. Neurotrophic and neurotoxic effects of amyloid β protein: Reversal by tachykinin neuropeptides. Science **250:** 279-282.
110. KOH, J.-Y., L. L. YANG & C. W. COTMAN. 1990. β-amyloid protein increases the vulnerability of cultured cortical neurons to excitotoxic damage. Brain Res. **533:** 315-320.
111. PIKE, C. J., A. J. WALENCEWICZ, C. G. GLABE & C. W. COTMAN. 1991. In vitro aging of β-amyloid protein causes peptide aggregation and neurotoxicity. Brain Res. **563:** 311-314.
112. MATTSON, M. P., B. CHENG, D. DAVIS, K. BRYANT, I. LIEBERBURG & R. E. RYDEL. 1992. β-amyloid peptides destabilize calcium homeostasis and render human cortical neurons vulnerable to excitotoxicity. J. Neurosci. **12:** 379-389.
113. BUSCIGLIO, J., A. LORENZO & B. A. YANKNER. 1992. Methodological variables in the assessment of beta amyloid neurotoxicity. Neurobiol. Aging **13:** 609-612.

114. MATTSON, M. P., K. J. TOMASELLI & R. E. RYDEL. 1993. Calcium-destabilizing and neurodegenerative effects of aggregated β-amyloid peptide are attenuated by basic FGF. Brain Res. **621:** 35-49.
115. PIKE, C. J., D. BURDICK, A. J. WALENCEWICZ, C. G. GLABE & C. W. COTMAN. 1993. Neurodegeneration induced by β-amyloid peptides in vitro: The role of peptide assembly state. J. Neurosci. **13:** 1676-1687.
116. BARROW, C. J. & M. G. ZAGORSKI. 1991. Solution structures of β peptide and its constituent fragments: Relation to amyloid deposition. Science **253:** 179-182.
117. HILBICH, C., B. KISTERS-WOIKE, J. REED, C. L. MASTERS & K. BEYREUTHER. 1991. Aggregation and secondary structure of synthetic amyloid βA4 peptides of Alzheimer's disease. J. Mol. Biol. **218:** 149-163.
118. BURDICK, D., B. SOREGHAN, M. KWON, J. KOSMOSKI, M. KNAUER, A. HENSCHEN, J. YATES, C. W. COTMAN & C. GLABE. 1992. Assembly and aggregation properties of synthetic Alzheimer's A4/β amyloid peptide analogs. J. Biol. Chem. **267:** 546-554.
119. DYRKS, T., E. DYRKS, T. HARTMANN, C. MASTERS & K. BEYREUTHER. 1992. Amyloidogenicity of βA4 and βA4-bearing amyloid protein precursor fragments by metal-catalyzed oxidation. J. Biol. Chem. **267:** 18210-18217.
120. HARTMANN, H., A. ECKERT & W. E. MULLER. 1993. β-amyloid protein amplifies calcium signalling in central neurons from the adult mouse. Biochem. Biophys. Res. Commun. **194:** 1216-1220.
121. COPANI, A., J.-Y. KOH & C. W. COTMAN. 1991. β-amyloid increases neuronal susceptibility to injury by glucose deprivation. NeuroReport **2:** 763-765.
122. ECKERT, A., H. HARTMANN & W. E. MULLER. 1993. β-amyloid protein enhances the mitogen-induced calcium response in circulating human lymphocytes. FEBS Lett. **330:** 49-52.
123. KIMURA, H. & D. SCHUBERT. 1993. Amyloid β-protein activates tachykinin receptors and inositol triphosphate accumulation by synergy with glutamate. Proc. Natl. Acad. Sci. USA **90:** 7508-7512.
124. HENSLEY, K., J. CARNEY, M. P. MATTSON, M. AKSENOVA, M. HARRIS, J. F. WU, R. FLOYD & D. A. BUTTERFIELD. 1994. A new model for β-amyloid aggregation and neurotoxicity based on the free radical generating capacity of the peptide. Proc. Natl. Acad. Sci. USA **91:** 3270-3274.
125. ARISPE, N., E. ROJAS & H. B. POLLARD. 1993. Alzheimer's disease amyloid β protein forms calcium channels in bilayer membranes: Blockade by tromethamine and aluminum. Proc. Natl. Acad. Sci. USA **90:** 567-571.
126. GOLDGABER, D., A. I. SCHWARZMAN, R. BHASIN, L. GREGORI, D. SCHMECHEL, A. M. SAUNDERS, A. D. ROSES & W. J. STRITTMATTER. 1993. Sequestration of amyloid β peptide. *In* Alzheimer's disease: Amyloid precursor proteins, signal transduction, and neuronal transplantation. R. M. Nitsch, J. G. Growdon, S. Corkin & R. J. Wurtman, eds.: 279-283. CBSMCT. Cambridge, MA.
127. MATTSON, M. P., B. CHENG & V. L. SMITH-SWINTOSKY. 1993. Neurotrophic factor mediated protection from excitotoxicity and disturbances in calcium and free radical metabolism. Semin. Neurosci. **5:** 295-307.
128. NOZAKI, K., S. P. FINKLESTEIN & M. F. BEAL. 1993. Basic fibroblast growth factor protects against hypoxia/ischemia and NMDA neurotoxicity in neonatal rats. J. Cereb. Blood Flow Metab. **13:** 221-228.
129. CHENG, B. & M. P. MATTSON. 1991. NGF and bFGF protect rat hippocampal and human cortical neurons against hypoglycemic damage by stabilizing calcium homeostasis. Neuron **7:** 1031-1041.
130. CHENG, B. & M. P. MATTSON. 1992. IGF-I and IGF-II protect cultured hippocampal and septal neurons against calcium-mediated hypoglycemic damage. J. Neurosci. **12:** 1558-1566.
131. CHENG, B. & M. P. MATTSON. 1994. NT-3 and BDNF protect hippocampal, septal and cortical neurons against metabolic compromise. Brain Res. **640:** 56-67.

132. SHIGENO, T., T. MIMA, K. TAKAKURA, D. I. GRAHAM, G. KATO, Y. HASHIMOTO & S. FURUKAWA. 1991. Amelioration of delayed neuronal death in the hippocampus by nerve growth factor. J. Neurosci. **11:** 2914-2919.
133. GLUCKMAN, P., N. KLEMPT, J. GUAN, C. MALLARD, E. SIRIMANNE, M. DRAGUNOW, M. KLEMPT, K. SINGH, C. WILLIAMS & K. NIKOLICS. 1992. A role for IGF-I in the rescue of CNS neurons following hypoxic-ischemic injury. Biochem. Biophys. Res. Commun. **182:** 593-599.
134. MATTSON, M. P., Y. ZHANG & S. BOSE. 1993. Growth factors prevent mitochondrial dysfunction, loss of calcium homeostasis and cell injury, but not ATP depletion in hippocampal neurons deprived of glucose. Exp. Neurol. **121:** 1-13.
135. HERSCHMAN, H. R. 1991. Primary response genes induced by growth factors and tumor promoters. Ann. Rev. Biochem. **60:** 281-319.
136. COLLAZO, D., H. TAKAHASHI & R. D. G. MCKAY. 1992. Cellular targets and trophic functions of neurotrophin-3 in the developing rat hippocampus. Neuron **9:** 643-656.
137. MATTSON, M. P., B. RYCHLIK, C. CHU & S. CHRISTAKOS. 1991. Evidence for calcium-reducing and excitoprotective roles for the calcium-binding protein calbindin-D28k in cultured hippocampal neurons. Neuron **6:** 41-51.
138. MATTSON, M. P., K. KUMAR, H. WANG, B. CHENG & E. K. MICHAELIS. 1993. Basic FGF regulates the expression of a functional 71 kDa NMDA receptor protein that mediates calcium influx and neurotoxicity in hippocampal neurons. J. Neurosci. **13:** 4575-4588.
139. SASTRY, B. R., S. S. CHIRWA, P. B. Y. MAY & H. MARETIC. 1988. Are nerve growth factors involved in long-term potentiation in the hippocampus and spatial memory? *In* Synaptic Plasticity and the Hippocampus. G. Buzaki & H. Haas, eds.: 101-105. Springer. Berlin.
140. BALLARIN, M., P. ERNFORS, N. LINDEFORS & H. PERSSON. 1988. Hippocampal damage and kainic acid injection induce a rapid increase in mRNA for BDNF and NGF in the rat brain. Exp. Neurol. **114:** 35-43.
141. GALL, C. & P. J. ISACKSON. 1989. Limbic seizures increase neuronal production of mRNA for nerve growth factor. Science **245:** 758-761.
142. JACKSON, G. R., L. APFELL, K. WERRBACH-PEREZ & J. R. PEREZ-POLO. 1990. Role of nerve growth factor in oxidant-antioxidant balance and neuronal injury. I. stimulation of hydrogen peroxide resistance. J. Neurosci. Res. **25:** 360-368.
143. ZHANG, Y., T. TATSUNO, J. M. CARNEY & M. P. MATTSON. 1993. Basic FGF, NGF, and IGFs protect hippocampal and cortical neurons against iron-induced degeneration. J. Cereb. Blood. Flow. Metab. **13:** 378-388.
144. NISTICO, G., M. R. CIRIOLO, D. RISKIN, M. IANNONE, A. DEMARTINO & G. ROTILIO. 1992. NGF restores decrease in catalase activity and increases superoxide dismutase and glutathione peroxidase activity in the brain of aged rats. Free Rad. Biol. Med. **12:** 177-181.
145. CARNEY, J. M., P. E. STARKE-REED, C. N. OLIVER, R. W. LANDUM, M. S. CHENG, J. F. WU & R. A. FLOYD. 1991. Reversal of age-related increase in brain protein oxidation, decrease in enzyme activity, and loss in temporal and spatial memory by chronic administration of the spin-trapping compound *N*-tert-butyl-alpha-phenylnitrone. Proc. Natl. Acad. Sci. USA **88:** 3633-3636.
146. SMITH, C. D., J. M. CARNEY, P. E. STARKE-REED, C. N. OLIVER, E. R. STADTMAN, R. A. FLOYD & W. R. MARKESBERY. 1991. Excess brain protein oxidation and enzyme dysfunction in normal aging and in Alzheimer's disease. Proc. Natl. Acad. Sci. USA **88:** 10540-15543.
147. ROSEN, D. R., T. SIDDIQUE, D. PATTERSON, D. A. FIGLEWICZ, P. SAPP *et al.* 1993. Mutations in Cu/Zn superoxide dismutase gene are associated with familial amyotrophic lateral sclerosis. Nature **362:** 59-62.

148. DEXTER, D. T., F. R. WELLS, A. LEES, F. JAVOY-AGID, Y. AGID, P. JENNER & C. D. MARSDEN. 1989. Increased nigral iron content and alterations in other metal ions occurring in brain in Parkinson's disease. J. Neurochem. **52:** 1830-1836.
149. ANDERSON, K. J., D. DAM, S. LEE & C. W. COTMAN. 1988. Basic fibroblast growth factor prevents death of lesioned cholinergic neurons in vivo. Nature **332:** 360-361.
150. HEFTI, F., J. HARTIKKA & B. KNUSEL. 1989. Function of neurotrophic factors in the adult and aging brain and their possible use in the treatment of neurodegenerative diseases. Neurobiol. Aging **10:** 515-533.
151. OLSON, L., A. NORDBERG, H. VONHOLST, L. BACKMAN, T. EBENDAL, I. ALAFUZOFF, K. AMBERLA, P. HARWIVIG, A. HERLITZ, A. LILJA *et al.* 1992. Nerve growth factor affects C-11-nicotine binding, blood flow, EEG, and verbal episodic memory in an Alzheimer patient: Case report. J. Neurol. Trans. (Parkinsons) **4:** 79-84.
152. STEIN-BEHRENS, B., M. P. MATTSON, I. CHANG, M. YEH & R. M. SAPOLSKY. 1993. Stress exacerbates neuron loss and cytoskeletal pathology in the hippocampus. J. Neurosci. **14:** 5373-5380.
153. LANDFIELD, P. W. 1987. Increased calcium current hypothesis of brain aging. Neurobiol. Aging **8:** 346-347.
154. KERR, D., L. CAMPBELL, S. HAO & P. LANDFIELD. 1989. Corticosteroid modulation of hippocampal potentials: Increased effect with aging. Science **245:** 1505-1509.
155. ELLIOTT, E. M. & R. M. SAPOLSKY. 1993. Corticosterone impairs hippocampal neuronal calcium regulation—possible mediating mechanisms. Brain Res. **602:** 84-90.
156. GOODMAN, Y. & M. P. MATTSON. 1994. Secreted forms of β-amyloid precursor protein protect hippocampal neurons against amyloid β-peptide induced oxidative injury. Exp. Neurol. **128:** 1-12.
157. DISTERHOFT, J. F., J. R. MOYER & L. T. THOMPSON. 1994. The calcium rationale in Alzheimer's disease. Ann. N.Y. Acad Sci., this volume.
158. FRIDEN, P. M., L. R. WALU, P. WATSON, S. R. DOCTROW, J. W. KOZARICH, C. BACKMAN, H. BERGMAN, B. HOFFER, F. BLOOM & A.-C. GRANHOLM. 1993. Blood-brain barrier penetration and in vivo activity of an NGF conjugate. Science **259:** 373-377.
159. CHENG, B., S. W. BARGER & M. P. MATTSON. 1994. Staurosporine, K-252a and K-252b stabilize calcium homeostasis and promote survival of CNS neurons in the absence of glucose. J. Neurochem. **62:** 1319-1329.
160. WHITSON, J. S., C. G. GLABE, E. SHITANI, A. ABCAR & C. W. COTMAN. 1990. β-amyloid protein promotes neuritic branching in hippocampal cultures. Neurosci. Lett. 110: 319-324.
161. BUXBAUM, J. D., S. E. GANDY, P. CICCHETTI, M. E. EHRLICH, A. J. CZERNIK, R. P. FRACASSO, T. V. RAMABHADRAN, A. J. UNERBECK & P. GREENGARD. 1990. Processing of Alzheimer β/A4 amyloid precursor protein: Modulation by agents that regulate protein phosphorylation. Proc. Natl. Acad. Sci. USA **87:** 6003-6006.

Calcium-Activated Neutral Proteinase (Calpain) System in Aging and Alzheimer's Disease[a]

R. A. NIXON,[b,c] K.-I. SAITO,[d] F. GRYNSPAN,
W. R. GRIFFIN, S. KATAYAMA,[d] T. HONDA,[d]
P. S. MOHAN, T. B. SHEA, AND M. BEERMANN

[b]Laboratories for Molecular Neuroscience
Mailman Research Center
McLean Hospital
and
Consolidated Departments of Psychiatry and [b]Program in Neuroscience
Harvard Medical School
Belmont, Massachusetts 02178

Calcium initiates major changes in cell architecture and function during development and adult plasticity, but it may trigger regressive events, including irreversible degeneration, when levels exceed a certain threshold.[1–5] The messages conveyed by shifting calcium levels are executed through an array of calcium-dependent enzymes. Perhaps better than any other calcium effector, the calcium-activated neutral proteases (CANPs or calpains) illustrate how constructive and destructive potentials exist within the same enzyme. At physiologic calcium levels, calpains act at the membrane-cytoskeleton interface to influence important signal pathways that control diverse behaviors of intracellular proteins and organelles.[6–8] At abnormally high calcium levels, however, these proteases are a potent destructive force, capable of cleaving more than half of the cell's protein pools in 1 hour.[9] How well neurons regulate this enormous proteolytic potential is a key determinant of cell survival in many degenerative disease states involving abnormal calcium homeostasis. We proposed in a previous review[8] that calpains may be the principal effector of calcium-induced neurodegeneration in Alzheimer's disease and, in addition, through their diverse actions on the cytoskeleton, membrane skeleton, and membrane dynamics, they may contribute in specific ways to neurofibrillary lesions, altered membrane protein trafficking, and synaptic dysfunction in Alzheimer's disease. Although experimental support for this earlier hypothesis

[a]The work from our laboratories was supported in part by the National Institute on Aging (AG10916), the Mitsubishi Kasei Corporation, and the Anna and Seymour Gitenstein Foundation.

[c]Address for correspondence: Dr. Ralph A. Nixon, Laboratories for Molecular Neuroscience, Mailman Research Center, McLean Hospital, 115 Mill Street, Belmont, MA 02178.

[d]Current address: Pharmaceutical Laboratories, Research Center, Mitsubishi Kasei Corporation, Yokohama, Japan.

came largely from studies of non-neural tissues, recent advances in understanding the calpain system in normal and pathologic states, including Alzheimer's disease, are beginning to strengthen the conceptual, if not experimental, framework for these ideas. Here, we will consider evidence that activation of the calpain system may be involved in some degenerative phenomena of normal aging and is a critical step along the final common pathway to degeneration in certain neuropathologic states, including Alzheimer's disease. Moreover, in view of the suspected influences of calpains on the membrane and cytoskeleton and amplification of their effects through specific protein kinases, persistent low level activation of calpain-mediated proteolysis might be expected to have slowly progressive effects relevant to amyloid precursor protein processing and cytoskeletal pathology.

THE CALPAIN-CALPASTATIN SYSTEM

Calpains are cysteine proteinases that are optimally active within the neutral pH range and require calcium for activity. Their properties were extensively reviewed,[6-8,10-14] and only the most pertinent aspects are briefly mentioned here. Two main types of calpains encoded by distinct genes are distinguished by their calcium requirement *in vitro*. Calpain I, or μcalpain (μCANP), is optimally active at calcium concentrations in the low micromolar range. Calpain II, or mcalpain (mCANP), in its purified state requires nearly millimolar calcium levels for full activity against conventional substrates *in vitro,* but *in vivo* various intracellular factors likely allow it to act at physiologic calcium levels.[15-18] The isoenzymes are composed of different large subunits (80-kD subunit) and identical small subunits (30-kD subunit). Because the catalytic domain structure is conserved within the cysteine protease family,[19] currently available synthetic protease inhibitors also inhibit other cellular cysteine proteases.

Most calpain in cells exists in a latent form and is activated in the presence of calcium ions by intramolecular cleavage of a 9-kD domain from the small subunit[20-22] and a short sequence from the NH_2-terminus of calpain I[22,23] and possibly calpain II.[24,25] The calcium levels required for autolysis and activity towards other proteins are lowered when the protease interacts with certain protein substrates,[26-30] acidic membrane lipids,[31-33] especially phosphoinositides,[15] or other factors,[34,35] including possible endogenous protein activators.[36] A specific endogenous protein inhibitor, called calpastatin, also modulates calpain activity. Calpastatins ranging in size from 70-140 kD in different tissues are derived from a single calpastatin gene by differential mRNA splicing[37] and posttranslational modification by phosphorylation[38,39] and proteolysis.[40] The calpastatin subunit in most tissues, including brain, has four repetitive inhibitory domains.[41] Cleavage of these subunits in the process of inhibiting calpain releases smaller inhibitory domains that may increase inhibitory efficacy (unpublished data). An important, still unanswered question is whether calpastatin tonically regulates calpain activation or simply serves to limit the action of calpains, once activated.[12]

A clue to the functions of calpains is their relatively restricted specificity. Calpains cleave at highly selective sites within a substrate rather than breaking down proteins into small fragments or amino acids. One intriguing example of this restricted specificity is the affinity of calpains for regions of certain enzymes between a regulatory

TABLE 1. Enzyme Activities Modulated by the Limited Proteolytic Action of Calpains

Enzyme
Calcium dependent
Protein kinase C[42-46a]
Ca^{2+} and calmodulin-dependent protein kinase[47a]
Calcineurin[48a]
Calcium-dependent cyclic nucleotide phosphodiesterase[49]
Calcium ATPase[50a]
Calcium independent
Tyrosine hydroxylase[51a]
CAMP-dependent protein kinase[52]
Phosphorylase kinase[53]
Glycogen synthase[54]

[a] Enzymes containing a regulatory domain removed by calpain cleavage.

domain and the catalytic domain. Cleavage within this region renders the enzyme cofactor independent and also increases its activity in most cases. A number of calcium-dependent kinases and phosphatases are regulated this way (TABLE 1), the best studied example being protein kinase C (PKC), which, like calpains, is translocated to membranes and activated in response to a calcium flux. As part of this activation process, calpain releases a 50-kD catalytic domain of PKC into the cytoplasm,[9,69,73,118,119] designated PKM, which no longer requires calcium or phospholipid for activity[55,56] and has a different substrate specificity from that of PKC.[57] Although the effect of PKM may be short-lived under normal conditions,[58,59] its ability to diffuse into the cytosol may allow the modified kinase to act transiently on a new group of substrates. Thus, by converting PKC to PKM, calpains could possibly trigger shifts in phosphorylation events from the membrane to locations within the cytoplasm and cytoskeleton. Modulation of protein kinases is one example of how calpains may amplify their effects on cellular function.

CALPAINS AND AGING

Changes in the rate or efficiency of proteolysis have figured prominently in many hypotheses on the mechanism of aging, and the calpain system continues to gain attention as a likely mediating factor in the aging process. Lynch and colleagues[60] first proposed that the activity of the calpain system may be related to rates of aging after they noted a strong inverse correlation between soluble levels of calpain activity in brain extracts and the life span across seven orders of mammals. Members of the Chiroptera order, which have exceptional maximum life spans that are 5 times longer than those of other mammals of comparable size, also exhibited exceptional (five- to sevenfold lower) calpain activities.[61] Although the relationship of these *in vitro* measurements to *in vivo* calpain activity has not been established, the findings are in accord with other studies of calpain in aging-related phenomena and with much of the data linking abnormal calcium homeostasis to brain aging reviewed in this

TABLE 2. Calcium-Induced Degenerative States Mediated in Part by Calpain Activity

In Vitro
Wallerian degeneration[74–76]
Organophosphorus neuropathy[77]
Spinal cord compression[78]
Excitotoxin-induced injury–hippocampus[79–81]
Hypoxia-induced injury–hippocampus[82–84]
Capsaicin-induced injury–dorsal root ganglion[85]
In Vivo
Ischemia/hypoxia-induced injury[82,86]
Excitotoxin-induced injury[80]

Volume. A particularly noteworthy example is the process of cataract formation, a well-established degenerative phenomenon of aging, and a response of the lens to specific insults including excess sugar, UV irradiation, trauma, and toxic chemicals. Cataract formation seems to result from the specific proteolytic cleavage of β-crystallin leading to its denaturation.[62] and α-crystallin which normally acts as a chaperone to reduce β-crystallin denaturation.[63,64] Calpain II, a major protease activity in lens,[65] is now believed to be responsible for both proteolytic events in experimental *in vivo* models of cataract formation induced by selenite, calcium ionophore A23187, galactose, xylose, or diamide.[66–68]

Alterations of calpain-mediated proteolysis are also evident in other tissues during normal aging. The breakdown of spectrin, which is often used as an index of calpain activity, was higher in fibroblasts from aged individuals than in those from young adults, when the cells were analyzed under resting conditions or after stimulated calcium influx.[69] The ability of calpain I to degrade the membrane protein, band 3, was increased in erythrocyte ghosts from aged individuals, implying a change in the susceptibility of the substrate to calpain.[70] Unlike many enzymes during aging,[71] total calpain activity increases in various tissues of older rabbits,[72] and calpain II levels measured *in vitro* activity in partially purified extracts increase in the aging brain.[73]

CALPAINS AND NEURODEGENERATION

Calpains have been implicated as the incisive factor in the degenerative process in a growing array of neuropathologic states triggered by increased intracellular calcium levels. In the examples listed in TABLE 2, the final common pathway to degeneration involves calcium entry into cells. Calpain activation in most cases was demonstrated by identifying the selective breakdown of preferred substrates of calpain, MAP2 and fodrin, and the importance of these events to the degenerative process was evident from the significant ability of calpain inhibitors to substantially protect against cell injury and to preserve neuronal function (TABLE 2). Calcium influx through voltage-gated channels (organophosphorus neurotoxicity)[77] and receptor-linked channels of the glutamate-responsive and glutamate-independent types

(NMDA[79–81] and capsaicin toxicity,[85] respectively) have each been shown to trigger excessive activation of the calpain system. Not all calcium-induced degenerative states, however, necessarily depend on calpain activation. Calcium-dependent excitotoxicity in cerebellar granule neuron culture, unlike that in the hippocampus and cortex, seems to be unaffected by calpain inhibitors. In this regard, it is interesting that in Alzheimer's disease, where the calpain system may be abnormally activated,[89] cerebellar granule cells are also resistant to degeneration, whereas pyramidal cells are highly vulnerable. Different calcium-dependent degenerative mechanisms may therefore exist in neurons of varying type or stage of development and have importance as determinants of neuronal vulnerability in disease states.

The potential clinical relevance of calpain's role in neurodegeneration is suggested by the *in vivo* efficacy of leupeptin in limiting neural damage in ischemia,[82,86] excitotoxicity,[80] and vasospasm[90] and in promoting peripheral nerve regeneration after axonal injury.[91,92]

ALZHEIMER'S DISEASE AND OTHER DEMENTIAS

Calpain I

Regressive neuronal changes and the loss of neurons are prominent neuropathologic features of Alzheimer's disease.[93–95] Synapse loss in particular, which is extensive in vulnerable brain regions, shows the strongest correlation of any neuropathologic marker with the severity of clinical dementia.[93] In view of calpain's known actions, its enrichment in synaptic endings[96] and other sensitive neuronal compartments,[97,98] and its suspected alteration in aging and other degenerative states, we considered the calpain system to be an attractive candidate for mediating synaptic loss and cell death in Alzheimer's disease and have been investigating its activity in the brains and non-neural tissues of patients with Alzheimer's disease.[89,108,125]

In vivo calpain activity is difficult to evaluate directly. Only a small proportion of the total enzyme pool may be active at any given time, and total levels of the protease, as measured by immunochemical or enzymatic assay, are therefore relatively insensitive to changes in calpain system function. In one approach to the assessment of active calpain levels in cells, we took advantage of the observation that calpain I activation requires autolysis of 26 amino acids from the amino terminus, which generates a 76-kD "activated" isoform distinct from the 80-kD native isoform on electrophoresis gels[22–25] (FIG. 1). One or more autolytic intermediates may also be formed in the process. Because hydrolysis of protein substrates *in vivo* at physiologic calcium levels requires calpain I autolysis,[99] the ratio of the activated isoform(s) of the enzyme to the precursor isoform in tissues is a potential index of calpain I function *in vivo*. Supporting this assumption are observations that this ratio increases in cultured human neuroblastoma cells during retinoic acid-induced differentiation and after treatment with phorbol esters or calcium ionophores, and the increase is accompanied by accelerated loss of calpain substrates such as fodrin and MAP II, but not loss of poor calpain substrates (Grynspan, Shea, Griffin, and Nixon, unpublished results). The generation of activated isoforms of calpain I also occurs in relation to changes in platelet shape, which are calpain-mediated.[100] The level of activated calpain isoforms,

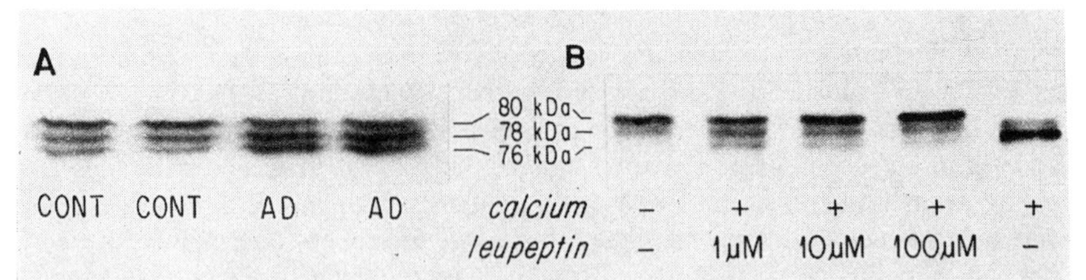

FIGURE 1. **(A)** Immunoblot analysis. Calpain I in prefrontal cortex from control subjects (CONT) (*lanes 1 and 2*) and patients with Alzheimer's disease (AD) (*lanes 3 and 4*) demonstrating the three major isoforms of calpain I. **(B)** Autolysis of calpain I from human erythrocytes *in vitro*. Calpain I (*lane 1*) was incubated in the presence of 50 μM free calcium and a calpain substrate, human erythrocyte spectrin, for 30 minutes at 30°C in the presence of varying concentrations of the cysteine protease inhibitor leupeptin. Hydrolysis of spectrin is accompanied by the formation of the 76-kD isoform. Enzyme activity was maintained in the absence of the 80-kD isoform and remains proportional to the amount of the 76-kD isoform.[89] Reprinted, with permission, from Saito *et al.*[89]

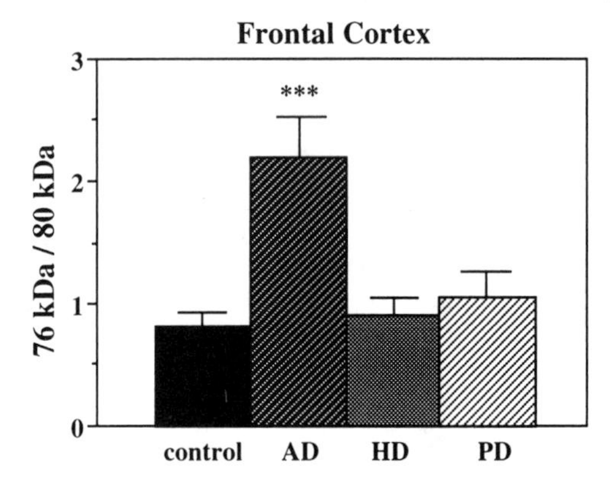

FIGURE 2. The ratio of the 76-kD (activated) calpain I isoform to the 80-kD (precursor) isoform, termed the activation index, measured in prefrontal cortex from control subjects (CONT) and patients with Alzheimer (AD), Huntington (HD), and Parkinson's disease (PD) measured by immunoassay. Cases were matched with respect to age and postmortem interval,[89] and the values depicted represent the mean ± SEM of 18, 22, 12, and 14 cases, respectively.*** $p < 0.001$. Figure adapted from Saito *et al.*[89]

however, is not a direct measure of calpain activity because various factors may modulate the activity of the activated protease, including most notably calpastatin.[26–40] Activated calpain may also be generated artifactually during preparation and analysis of cells or during the postmortem interval before analysis. Despite these limitations, activated calpain I is a sensitive and specific measure of one aspect of calpain I function, and shifts in the ratio of activated to precursor isoforms in disease states and experimental conditions provide potentially useful information about changes in calpain I activity *in vivo*.

Using an immunoassay to quantitate calpain I isoforms, we found that the relative activation of calpain I was greatly increased in the brains of patients with Alzheimer's disease. The ratio of activated (76 kD) to precursor (80 kD) isoforms of calpain I, the "activation index," was elevated threefold in samples of postmortem prefrontal cortex from 22 patients with Alzheimer's disease compared to those from 16 neurologically normal individuals, 18 patients with Huntington's disease, and 14 patients with Parkinson's disease (FIG. 2). In agreement with earlier preliminary studies[101–103] showing no difference in total *in vitro* activity in the brain in Alzheimer's disease, we found the sum of the three isoforms in each group to be unchanged. Although it is possible that measured levels of activated calpain were higher than those existing *in vivo* because of some autolysis during the postmortem interval, the abnormally high activation index in brains in Alzheimer's disease could not be explained by postmortem variables because the brains in each test group were matched with respect to postmortem interval, and the ratio of the 76-kD isoforms to the 80-kD calpain showed no relation to the length of the postmortem interval between 4 and 27 hours.

Although calpain activation would be expected in degenerating neurons, several lines of evidence suggest that abnormal calpain I activation in the brain in Alzheimer's disease is not simply a consequence of end-stage neuronal degeneration. The cerebellum and putamen, two brain regions where few neurons have advanced to a stage of overt degeneration in Alzheimer's disease, also exhibited abnormally increased calpain I activation, but to a lesser extent (45-60% above normal) than in the neocortex. Moreover, the threefold increase in calpain I activation in the prefrontal cortex of the brain in Alzheimer's disease greatly exceeded the change expected from the small proportion of neocortical neurons exhibiting the chromatolytic changes of end-stage degeneration. By comparison, activation of calpain I was increased only 50% in samples of putamen from patients with Huntington's disease where neuronal cell loss is considerable. These observations suggest that calpain activation may reflect widespread metabolic alteration that precedes as well as contributes to neuronal cell death. This view is consistent with evidence for higher calpain activity in erythrocytes of male patients with Alzheimer's disease,[104] and increased degradation of spectrin[105] and diminished content of protein kinase C in fibroblasts[106] from patients with Alzheimer's disease because both proteins are preferred substrates to calpain. Brain spectrin breakdown is also increased in cortex and hippocampus of the Alzheimer brain.[106]

Calpastatin

Additional observations have raised the possibility that the differential vulnerability of neuronal populations to increased calpain I activation may partly depend on the cell's ability to maintain adequate levels of calpastatin, its endogenous protease inhibitor. Two affinity-purified antibodies raised against purified human brain calpastatin strongly labeled pyramidal neurons and their long dendritic arborizations in human neocortex. In the brains of normal individuals, calpastatin staining was uniformly strong in all cortical layers, but in the brains of patients with Alzheimer's disease, a prominent decrease in calpastatin immunoreactivity was observed in cortical layers II-V. Preliminary computer-assisted densitometric analyses on 12 Alzheimer and 10 control brains immunostained under identical conditions confirmed statistically significant decreases in calpastatin levels in layers II, III, and V and relatively preserved immunoreactivity in layer VI of the brain in Alzheimer's disease.[108] These findings suggest that depletion of calpastatin may be one critical factor that determines if a given level of calpain system activation would lead to degenerative changes. Because calpastatin may be cleaved as it inhibits calpain,[39,40] depletion of calpastatin could reflect failure to adequately replace calpastatin lost as a result of persistently elevated calpain activity.

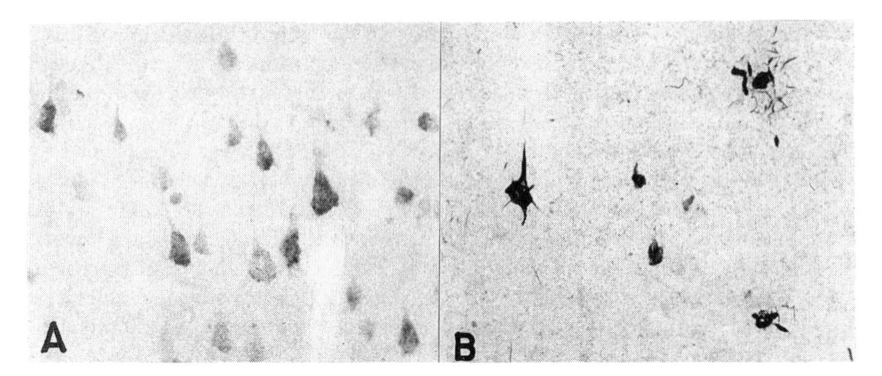

FIGURE 3. An affinity-purified polyclonal antiserum to calpain II decorates pyramidal neurons of layer V in the prefrontal cortex of a normal control brain. In the same brain region of patients with Alzheimer's disease, the number of immunoreactive neurons was decreased and neurofibrillary tangles, dystrophic neurites, and neurophil threads were intensely immunostained.

Calpain II and Neurofibrillary Pathology

Changes in calpain II functional activity *in vivo* are more difficult to evaluate because autolytic activation, if brain calpain II is in fact activated this way, does not generate additional distinct, recognizable isoforms. As observed for calpain I, total levels of calpain II were unchanged in all regions of Alzheimer's brain studied.[89] In addition, total calpain II activity, which is more stable *in vitro* than is that of calpain I, was not altered in Alzheimer's disease.[89] Because of the widespread distribution of calpain II among neurons and glial cells, however, changes in its content may be masked by neuronal cell loss, astrogliosis, or other factors. In this regard, immunocytochemical studies of the brain in Alzheimer's disease suggest that calpain II levels may be increased in neurons exhibiting degenerative changes. A polyclonal antibody against calpain II, which also recognized calpain I, immunostained large pyramidal neurons in neocortex.[109] In the Alzheimer brain, loss of many of these cells was noted, but a high percentage of neurons undergoing neurofibrillary tangle formation displayed calpain immunoreactivity.[110] We recently observed that polyclonal antibodies specific for calpain II decorate neurofibrillary tangles (NFTs) in the Alzheimer brain (FIG. 3) and levels of calpain II may be increased in some neurons within vulnerable populations (Grynspan, Katayama, and Nixon, unpublished data). The association of calpain II with NFTs is interesting in that protein constituents of NFT—tau protein and neurofilament proteins—are normally good substrates for calpain but become more resistant to calpain-mediated hydrolysis when extensively phosphorylated.[26,111] It may be reasonable to propose that abnormal accumulation of tau and other proteins in NFTs may result in part from a relative resistance of these substrates to normal calpain hydrolysis because of their abnormal state of phosphorylation within the tangle. Binding of aluminum to phosphorylated domains of neurofilament proteins was also shown to inhibit their proteolysis by calpains[112,113] as well as their dephosphorylation. Increased concentration of aluminum associated with NFTs[114] may similarly inhibit the actions of calpain on constituent proteins. Collectively these effects may explain the observation that certain relatively good calpain substrates accumulate in degenerating neurons in Alzheimer's disease in the presence of apparently increased activation of the calpain system.

Calpains, Vesicular Trafficking, and APP Processing

In view of the suspected biologic roles of calpains, other consequences might be expected from the direct actions of the protease and amplification of its effects through protein kinase pathways. During an influx of calcium into certain cells, calpain translocates to the membrane where the rate of activation is accelerated.[122] The fact that many of the best calpain substrates are membrane skeleton proteins and cytoplasmic domains of various integral membrane proteins[8,10,12] supports other evidence that calpains may regulate central events related to membrane and vesicular trafficking. Calpain-mediated proteolysis of certain of these proteins has been implicated in the fusion of membranes during erythrocyte differentiation[116,117] or in shape changes in platelets[100] or erythrocytes[115,119] in response to calcium. Similar events are proposed for the fusion of membranes during myogenesis[116] and during membrane resealing after neurite transection.[118] Finally, calpains have been localized in substantial concentration on clathrin-coated vesicles.[130]

These considerations are relevant to the cell and molecular pathology of Alzheimer's disease, in which abnormalities of the endosomal-lysosomal system are an early marker of affected neurons[120,121] and may be related to alterations of β-APP processing and production of β-amyloid in Alzheimer's disease. A link between calpains, protein kinase C, and β-APP metabolism is suggested by the observation that phorbol esters, which activate a signal transduction cascade involving calpain and protein kinase C, induce a phosphorylation dependent increase in the secretion of APP derivatives in PC12 cells.[122] The stimulated secretion is blocked by inhibitors of protein kinase C.[123] In many cell types, activation of calpain is considered the mechanism by which PKC becomes downregulated in response to persistent calcium influx.[124] A diminished content of PKC was demonstrated in fibroblasts of patients with Alzheimer's disease[106] which were also shown in other studies to display evidence of calpain activation.[105] Together these results allow the prediction that β-APP secretion in the brain in Alzheimer's disease should be decreased in relation to the extent of PKC downregulation and calpain activation. In this regard, we observed that the degree of abnormal calpain I activation found in the neocortex in Alzheimer's disease shows a strong inverse correlation ($r = 0.87$; $p < 0.001$) with levels of soluble β-APP, which have been identified as the secreted (protease nexin II) domain of β-APP.[125] Whether these two events are interrelated or are independent consequences of another event such as calcium influx remains to be experimentally tested. Diminished β-APP secretion, in turn, may impact negatively on calcium homeostasis because soluble β-APP has been shown to reduce basal intracellular calcium levels in cultured neurons and prevent the rise in calcium induced by hypoglycemia, thereby protecting against cell damage.[126] Moreover, diminished β-APP secretion is also associated with increased secretion of alternately processed forms,[127] including β-amyloid[128] peptides, and with alternative β-APP processing through endosomal/lysosomal routes.[127,130]

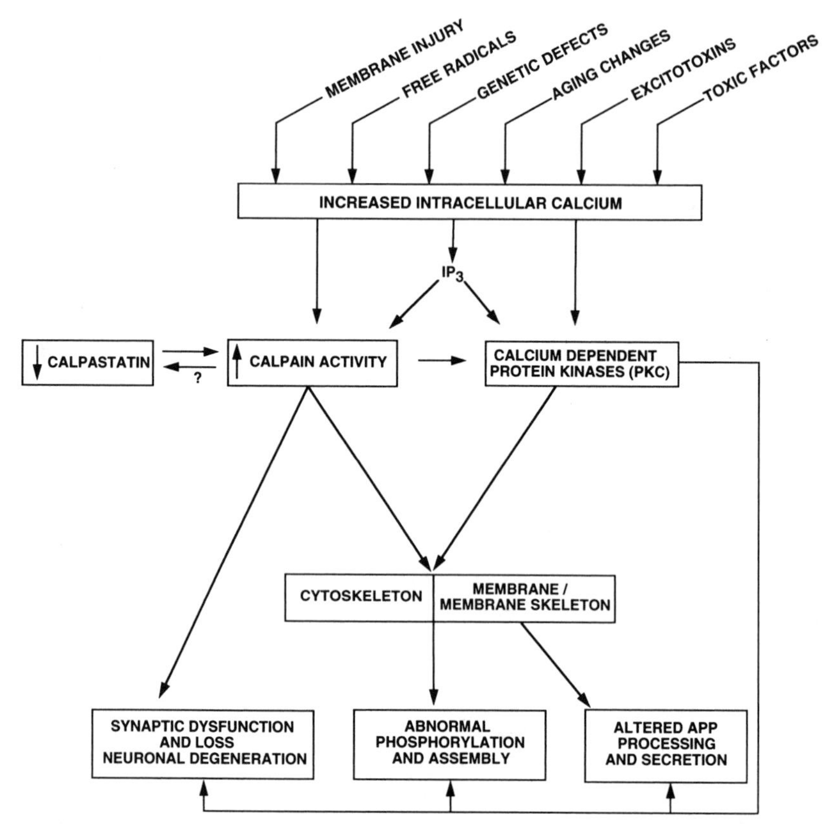

FIGURE 4. Speculative model relating abnormal activation of the calpain system to the neuropathologic lesions of Alzheimer's disease based on suspected physiologic actions of calpains. Further details are discussed in the text.

In summary, a growing number of findings point to excessive activation of the calpain system in Alzheimer's disease. Because calpains are direct effectors of calcium signals, these alterations may be a valuable window into possible changes in calcium homeostasis in Alzheimer's disease and aging. Among the many calcium-dependent enzymes, calpains may have a central role in calcium-induced cellular injury as the most proximate enzyme capable of leading to irreversible structural damage (FIG. 4). Moreover, dysregulation of calcium-dependent protein kinases and protein phosphatases and the direct or indirect effects on membrane and membrane-skeleton proteins are potential factors in the development of neurofibrillary pathology and altered β-APP processing (FIG. 4). Blocking calpain activation has the potential of slowing the process by which cells degenerate regardless of the primary etiologic factors that trigger changes in calcium homeostasis.

SUMMARY

Calpains (CANPs) are a family of calcium-dependent cysteine proteases under complex cellular regulation. By making selective limited proteolytic cleavages, they activate or alter the regulation of certain enzymes, including key protein kinases and phosphatases, and induce specific cytoskeletal rearrangements, accounting for their suspected involvement in intracellular signaling, vesicular trafficking, and structural stabilization.

Calpain activity has been implicated in various aging phenomena, including cataract formation and erythrocyte senescence. Abnormal activation of the large stores of latent calpain in neurons induces cell injury and is believed to underlie neurodegeneration in excitotoxicity, Wallerian degeneration, and certain other neuropathologic states involving abnormal calcium influx. In Alzheimer's disease, we found the ratio of activated calpain I to its latent precursor isoform in neocortex to be threefold higher than that in normal individuals and those with Huntington's or Parkinson's disease. Immunoreactivity toward calpastatin, the endogenous inhibitor of calpain, was also markedly reduced in layers II-V of the neocortex in Alzheimer's disease. The excessive calpain system activation suggested by these findings represents a potential molecular basis for synaptic loss and neuronal cell death in the brain in Alzheimer's disease given the known destructive actions of calpain I and its preferential neuronal and synaptic localization. In surviving cells, persistent calpain activation may also contribute to neurofibrillary pathology and abnormal amyloid precursor protein trafficking/processing through its known actions on protein kinases and the membrane skeleton.

The degree of abnormal calpain activation in the brain in Alzheimer's disease strongly correlated with the extent of decline in levels of secreted amyloid precursor protein in brain. Cytoskeletal proteins that are normally good calpain substrates become relatively calpain resistant when they are hyperphosphorylated, which may contribute to their accumulation in neurofibrillary tangles. As a major effector of calcium signals, calpain activity may mirror disturbances in calcium homeostasis and mediate important pathologic consequences of such disturbances.

ACKNOWLEDGMENTS

We thank Johanne Khan for typing and for assistance in preparing the manuscript.

REFERENCES

1. KHACHATURIAN, Z. S. 1989. Aging **1:** 17-34.
2. KHACHATURIAN, Z. S. 1993. Neurosci. Res. Comm. **13**(S): S3-S6.
3. MATTSON, M. P. 1992. **27:** 29-49.
4. KATER, S. B., M. P. MATTSON & P. B. GUTHRIE. 1989. Ann. N.Y. Acad. Sci. **568:** 252-261.
5. MILLS, L. R. & S. B. KATER. 1990. Neuron **2:** 149-163.
6. SUZUKI, K. & S. OHNO. 1990. Cell Struct. Funct. **15:** 1-6.
7. MURACHI, T. 1989. Biochem. Int. **18:** 263-294.
8. NIXON, R. A. 1989. Ann. N.Y. Acad. Sci. **568:** 198-206.

9. Nixon, R. A., R. Quackenbush & A. Vitto. 1986. J. Neurosci. **6:** 1252-1263.
10. Johnson, R. 1990. Int. J. Biochem. **22:** 811-822.
11. Mellgren, R. L. & T. Murachi, Eds. 1990. Intracellular Calcium-Dependent Proteolysis. CRC Press. Boca Raton, FL.
12. Goll, D. E., V. F. Thompson, R. G. Taylor & T. Zalewska. 1992. BioEssays **14:** 549-556.
13. Zimmerman, U.-J. P. & W. W. Schlaepfer. 1984. Prog. Neurobiol. **23:** 63-78.
14. Suzuki, K., S. Imajoh, Y. Emori, H. Kawasaki, Y. Minami & S. Ohno. 1988. Adv. Enzyme Reg. **27:** 153-169.
15. Saido, T. C., M. Shibata, T. Takenawa, H. Murofushi & K. Suzuki. 1992. J. Biol. Chem. **267:** 24585-24590.
16. Mellgren, R. L. & C. B. Rozanov. 1990. Biochem. Biophys. Res. Commun. **168:** 589-595.
17. Chakrabarti, A. K., S. Dasgupta, N. L. Banik & E. L. Hogan. 1990. Biochim. Biophys. Acta **1038:** 195-198.
18. Salamino, F., R. DeTullio, P. Mengotti, P. L. Viotti, E. Melloni & S. Pontremoli. 1993. Biochem. J. **15:** 191-197.
19. Suzuki, K. 1990. *In* Intracellular Calcium-Dependent Proteolysis. R. L. Mellgren & T. Murachi, Eds.: 25-35. CRC Press. Boca Raton, FL.
20. Hathaway, D. R., D. K. Werth & J. R. Haeberle. 1982. J. Biol. Chem. **257:** 9072-9077.
21. Vitto, A. & R. A. Nixon. 1986. J. Neurochem. **47:** 1039-1051.
22. Suzuki, K., S. Tsuji, S. Ishiura, Y. Kimura, S. Kubota & K. Imahori. 1981. J. Biochem. **90:** 1787-1793.
23. Suzuki, K., S. Tsuji, S. Kubota, Y. Kimura & K. Imahori. 1981. J. Biochem. **90:** 275-278.
24. Inomata, M., M. Hayashi, M. Nakamura, K. Imahori & S. Kawashima. 1985. J. Biochem. **98:** 407-416.
25. Croall, D. E., C. A. Slaughter, H. S. Wortham, C. M. Skelly, L. DeOgny & C. R. Moomaw. 1992. Biochim. Biophys. Acta **1121:** 47-53.
26. Goldstein, M. E., N. H. Sternberger & L. A. Sternberger. 1987. J. Neuroimmunol. **14:** 149-169.
27. Nixon, R. A., R. Quackenbush & A. Vitto. 1986. J. Neurosci. **6:** 1252-1263.
28. Harris, A. S. & J. S. Morrow. 1988a. J. Neurosci. **8:** 2640-2651.
29. Harris, A. S. & J. S. Morrow. 1988b. J. Cell Biol. **107:** 25a.
30. Seubert, P., M. Baudry, S. Dudek & G. Lynch. 1987. Synapse **1:** 20-24.
31. Coolican, S. A. & D. R. Hathaway. 1984. J. Biol. Chem. **259:** 11627-11630.
32. Imajoh, S., H. Kawasaki & K. Suzuki. 1986. J. Biochem. **99:** 1281-1284.
33. Melloni, E., S. Pontremoli, M. Michetti, O. Sacco, B. Sparatore & F. Salamino. 1985. Proc. Natl. Acad. Sci. USA **82:** 6435-6439.
34. Takeyama, Y., H. Nakanishi, Y. Uratsuji, A. Kishimoto & Y. Nishizuka. 1986. FEBS Lett. **194:** 110-114.
35. Pontremoli, S., E. Melloni, M. Michetti, F. Salamino, B. Sparatore & B. L. Horecker. 1988. Proc. Natl. Acad. Sci. USA **85:** 1740-1743.
36. Hathaway, D. R. & P. McClelland. 1990. *In* Intracellular Calcium-Dependent Proteolysis. R. L. Mellgren & T. Murachi, Eds.: 91-102. CRC Press. Boca Raton, FL.
37. Lee, W. J., H. Ma, E. Takano, H. Q. Yang, M. Hatanaka & M. Maki. 1992. J. Biol. Chem. **267:** 8437-8442.
38. Mellgren, R. L., R. D. Lane, M. G. Hegazy & E. M. Reimann. 1985. Adv. Prot. Phosphatases **1:** 259-275.
39. Adachi, Y., A. Ishida-Takahashi, C. Takahashi, E. Takano, T. Murachi & M. Hatanaka. 1991. J. Biol. Chem. **266:** 3968-3972.
40. Nakamura, M., M. Inomata, S. Imajoh, K. Suzuki & S. Kawashima. 1989. Biochemistry **28:** 449-455.

41. NAKAMURA, M., M. INOMATA, S. IMAJOH, K. SUZUKI & S. KAWASHIMA. 1989. Biochemistry **28:** 449-455.
42. PONTREMOLI, S., E. MELLONI, G. DAMIANI, F. SALAMINO, B. SPARATORE, M. MICHETTI & B. L. HORECKER. 1988. J. Biol. Chem. **263:** 1915-1919.
43. TAPLEY, P. M. & A. W. MURRAY. 1984. Biochem. Biophys. Res. Commun. **18:** 835-841.
44. PONTREMOLI, S., E. MELLONI, M. MICHETTI, O. SACCO, F. SALAMINO, B. SPARATORE & B. L. HORECKER. 1986. J. Biol. Chem. **261:** 8309-8313.
45. MELLONI, E., S. PONTREMOLI, M. MICHETTI, O. SACCO, B. SPARATORE & B. L. HORECKER. 1986. J. Biol. Chem. **261:** 4101-4105.
46. INOUE, M., A. KISHIMOTO, Y. TAKAI & Y. NISHIZUKA. 1977. J. Biol. Chem. **252:** 7610-7616.
47. KENNEDY, M. B. 1988. Nature **335:** 770-772.
48. TALLANT, E. A., L. M. BRUMLEY & R. W. WALLACE. 1988. Biochemistry **272:** 205-2211.
49. ITO, M., T. TANAKA, K. NUNOKI, H. HIDAKA & K. SUZUKI. 1987. Biochem. Biophys. Res. Commun. **145:** 1321-1328.
50. WANG, K. K. W., B. D. ROUFOGALIS & A. VILLALOBO. 1988. Arch. Biochem. Biophys. **267:** 317-XXX.
51. KIUCHI, K., K. KIUCHI, K. TITANI, K. FUJITA, K. SUZUKI & T. NAGATSU. 1991. Biochemistry **30:** 10416-10410.
52. MULLER, U. & H. C. SPATZ. 1989. J. Neurogenet. **6:** 95-114.
53. HUSTON, R. B. & E. G. KREBS. 1968. Biochemistry **7:** 2116-2122.
54. BELOCOPITOW, E., M. M. APPLEMAN & H. N. TORRES. 1965. J. Biol. Chem. **240:** 3473-3478.
55. KISHIMOTO, A., K. MIKAWA, K. HASHIMOTO, I. YASUDA, S.-I. TANAKA, M. TOMINAGA, T. JURODE & Y. NISHIZUKA. 1989. J. Biol. Chem. **264:** 4088-4092.
56. NISHIZUKA, Y. 1986. Science **233:** 305-312.
57. PONTREMOLI, S., E. MELLONI, M. MICHETTI, B. SPARATORE, F. SALAMINO, O. SACCO & B. HORECKER. 19878. Proc. Natl. Acad. Sci. USA **84:** 3604-3608.
58. MURRAY, A. W., A. FOURNIER & S. J. HARDY. 1987. TIBS **12:** 53-54.
59. PONTREMOLI, S., M. MICHETTI, E. MELLONI, B. SPARATORE, F. SALAMINO & B. L. HORECKER. 1990. Proc. Natl. Acad. Sci. USA **87:** 3705-3707.
60. LYNCH, G., J. LARSON & M. BAUDRY. 1986. *In* Treatment Development Strategies for Alzheimer's Disease. T. Krook, R. T. Bartus, S. Ferris & S. Gershon, Eds.: 119-149. Mark Powley Associates, Inc. Madison, CT.
61. BAUDRY, M., R. DUBRIN, L. BEASLEY, M. LEON & G. LYNCH. 1986. Neurobiol. Aging **7:** 255-258.
62. DAVID, L. L., J. W. WRIGHT & T. R. SHEARER. 1992. Biochim. Biophys. Acta **1139:** 210-216.
63. YOSHIDA, H., T. MURACHI & I. TSUKAHARA. 1984. Biochim. Biophys. Acta **798:** 252-259.
64. KELLEY, M. J., L. L. DAVID, N. IWASAKI, J. WRIGHT & T. R. SHEARER. 1993. J. Biol. Chem. **268:** 18844-18849.
65. VARNUM, M. D., L. L. DAVID & T. SHEARER. 1989. Exp. Eye Res. **49:** 1053-1065.
66. SHEARER, T. R., AZUMA, M., L. L. DAVID & T. MURACHI. 1991. Invest. Ophthalmol. & Visual Sci. **32:** 533-540.
67. AZUMA, M., T. R. SHEARER, T. MATSUMOTO, L. L. DAVID & T. MURACHI. 1990. Exp. Eye Res. **51:** 393-401.
68. AZUMA, M. & T. R. SHEARER. 1992. FEBS Lett. **307:** 313-317.
69. PETERSON, C. & J. E. GOLDMAN. 1986. Proc. Natl. Acad. Sci. USA **83:** 2758-2762.
70. SCHWARZ-BEN MEIR, N., T. GLASER & N. S. KOSOWER. 1991. Biochem. J. **275:** 47-52.
71. FLORINI, J. 1980. Handbook of Biochemistry in Aging. CRC Press. Boca Raton, FL.

72. BLOMGREN, K., E. NILSSON & J.-O. KARLSSON. 1989. Comp. Biochem. Physiol. **93B:** 403-407.
73. KENESSEY, A., M. BANAY-SCHWARTZ, T. DEGUZMAN & A. LAJTHA. 1990. Neurochem. Res. **15:** 243-249.
74. GILBERT, D. S., B. J. NEWBY & B. H. ANDERTON. 1975. Nature **256:** 586-589.
75. SCHLAEPFER, W. W., C. LEE, J. Q. TROJANOWSKI & V. M.-Y. LEE. 1984. J. Neurochem. **43:** 857-864.
76. SCHLAEPFER, W. W., C. LEE, V. M.-Y. LEE & U.-J. P. ZIMMERMAN. 1985. J. Neurochem. **44:** 502-509.
77. EL FAWAL, H. A. N. & M. F. ERLICH. 1993. Ann. N.Y. Acad. Sci. **679:** 325-335.
78. BANIK, N. L., E. L. HOGAN, J. M. POWERS & K. P. SMITH. 1986. J. Neurol. Sci. **73:** 245-256.
79. SIMAN, R., J. C. NOSZEK & C. KEGERISE. 1989. J. Neurosci. **9:** 1579-1590.
80. SIMAN, R. & J. C. NOSZEK. 1989. Neuron **1:** 279-287.
81. SIMAN, R. 1990. *In* Neurotoxicity of Excitatory Amino Acids. A. Guidotti, Ed.: 145-161. Raven Press. New York.
82. RAMI, A. & J. KRIEGLSTEIN. 1993. Brain Res. **609:** 67-70.
83. ARAI, A., M. KESSLER, K. LEE & G. LYNCH. 1990. Brain Res. **532:** 63-68.
84. SEUBERT, P., LARSON, J., M. W. OLIVER, M. W. JUNG, BAUDRY & G. LYNCH. 1988. Brain Res. **460:** 189-194.
85. GIBBONS, S. J., J. R. BRORSON, D. BLEAKMAN, P. S. CHARD & R. J. MILLER. 1993. Ann. N.Y. Acad. Sci. **679:** 22-33.
86. LEE, K. S., S. FRANK, P. VANDERKLISH, A. ARAI & G. LYNCH. 1991. Proc. Natl. Acad. Sci. USA **88:** 7233-7237.
87. DISTASI, A. M. M., V. GALLO, M. CECCARINI & T. C. PETRUCCI. 1991. Neuron **6:** 445-454.
88. MANEV, H., M. FAVARON, R. SIMAN, A. GUIDOTTI & E. COSTA. 1991. J. Neurochem. **57:** 1288-1295.
89. SAITO, K.-I., J. S. ELCE, J. E. HAMOS & R. A. NIXON. 1993. Proc. Natl. Acad. Sci. USA **90:** 2628-2632.
90. MINAMI, N., E. TANI, Y. MAEDA, I. YAMAURA & M. FUKAMI. 1992. J. Neurosurg. **76:** 111-118.
91. HURST, L. D., M. A. BADALAMENTE, J. ELLSTEIN & A. STRACHER. 1984. J. Hand Surg. **9:** 564-572.
92. BADALAMENTE, M. A., L. C. HURST & A. STRACHER. 1989. Proc. Natl. Acad. Sci. USA **86:** 5983-5987.
93. TERRY, R. D., E. MASLIAH, D. P. SALMON, N. BUTTERS, R. DETERESA, R. HILL, L. A. HANSEN & R. KATZAN. 1991. Ann. Neurol. **30:** 572-580.
94. HAMOS, J. E., L. J. DEGENNARO & D. A. DRACHMAN. 1989. Neurology **39:** 355-361.
95. DEKOSKY, S. T. & S. W. SCHEFF. 1990. Ann. Neurol. **27:** 457-464.
96. PERLMUTTER, L. S., R. SIMAN, C. GALL, P. SEUBERT, M. BAUDRY & G. LYNCH. 1988. Synapse **2:** 79-88.
97. HAMAKUBU, T., R. KANNAGI, T. MURACHI & A. MATUS. 1986. J. Neurosci. **6:** 3103-3111.
98. FUKUDA, T., E. ADACHI, S. KAWASHIMA, I. YOSHIYA, & P. H. HASHIMOTO. 1990. J. Comp. Neurol. **302:** 100-109.
99. HIYASHI, M., M. INOMATA, Y. SAITO, H. ITO & S. KAWASHIMA. 1991. Biochim. Biophys. Acta **1094:** 249-256.
100. SAIDO, T. C., H. SUZUKI, H. YAMAZAKI, K. TANOUE & K. SUZUKI. 1993. J. Biol. Chem. **268:** 7422-7426.
101. KAWASHIMA, S., Y. IHARA & M. INOMATA. 1989. Biomed. Res. **10:** 17-23.
102. MANTLE, D. & E. K. PERRY. 1991. J. Neurol. Sci. **102:** 220-224.

103. NILSSON, E., I. ALAFUZOFF, K. BLENNOW, K. BLOMGREN, C. M. HALL, I. JANSON, I. KARLSSON, A. WALLIN, C. G. GOTTFRIES & J.-O. KARLSSON. 1990. Neurobiol. Aging **11:** 425-431.
104. KARLSSON, J.-O., K. BLENNOW, B. HOLMBERG, I. JANSON, I. KARLSSON, E. NILSSON, A. WALLIN & C. G. GOTTFRIES. 1992. Dementia **3:** 200-204.
105. PETERSON, C., P. VANDERKLISH, P. SEUBERT, C. COTMAN & G. LYNCH. 1991. Neurosci. Lett. **121:** 239-243.
106. SAITOH, T., E. MASLIAH, L. W. JIN, G. M. COLE, T. WIELOCH & I. P. SHAPIRO. 1991. Lab. Invest. **64:** 596-616.
107. MASLIAH, E., D. S. IIMOTO, T. SAITOH, L. A. HANSEN & R. D. TERRY. 1990. Brain Res. **531:** 36-44.
108. NIXON, R. A., K.-I. SAITO, A. M. CATALDO, J. HAMOS, D. HAMILTON, T. HONDA & A. POPE. 1992. Soc. Neurosci. Abstr. **18:** 198.
109. IWAMOTO, N. & P. C. EMSON. 1991. Neurosci. Lett. **128:** 81-84.
110. IWAMOTO, N., W. THANGNIPON, C. CRAWFORD & P. C. EMSON. 1991. Brain Res. **561:** 177-180.
111. LITERSKY, J. M. & G. V. W. JOHNSON. 1992. J. Biol. Chem. **267:** 1563-1568.
112. NIXON, R. A., J. F. CLARKE, K. B. LOGVINENKO, M. K. H. TAN, M. HOULT & F. GRYNSPAN. 1990. J. Neurochem. **55:** 1950-1959.
113. SHEA, T. B., M. L. BEERMANN & R. A. NIXON. 1992. J. Neurochem. **58:** 542-547.
114. GOOD, P. F., D. P. PERL, L. M. BIERER & J. SCHMEIDLER. 1992. Ann. Neurol. **31:** 286-292.
115. WHATMORE, J. L., E. K. Y. TANG & J. A. HICKMAN. 1992. Exp. Cell Res. **200:** 316-325.
116. J. E. SCHOLLMEYER. 1986. Exp. Cell Res. **162:** 411-422.
117. THOMAS, P., A. R. LIMBRICK & D. ALLAN. 1983. Biochem. Biophys. Acta **730:** 351-358.
118. XIE, X.-YI & J. N. BARRETT. 1991. J. Neurosci. **11:** 3257-3267.
119. GLASER, T. & N. S. KOSOWER. 1986. FEBS Lett. **206:** 115-120.
120. NIXON, R. A., A. M. CATALDO, P. A. PASKEVICH, D. J. HAMILTON, T. R. WHEELOCK & L. KANALEY-ANDREWS. 1992. Ann. N.Y. Acad. Sci. **674:** 65-88.
121. CATALDO, A. M., C. Y. THAYER, E. D. BIRD, T. R. WHEELOCK & R. A. NIXON. 1990. Brain Res. **513:** 181-192.
122. CAPORASO, G. L., S. E. GANDY, J. D. BUXBAUM, T. V. RAMABHADRAN & P. GREENGARD. 1992. Proc. Natl. Acad. Sci. USA **89:** 3055-3059.
123. SLACK, B. E., R. M. NITSCH, E. LIVNEH, G. M. KUNZ, JR., J. BREU, H. ELDAR & R. J. WURTMAN. 1993. J. Biol. Chem. **5:** 21097-21101.
124. ADACHI, Y., T. MURACHI, M. MAKE, K. ISHII & M. HATANAKA. 1990. Biomed. Res. **11:** 313-317.
125. HONDA, T., J. HAMOS & R. A. NIXON. 1992. Soc. Neurosci. Abstr. **18:** 733.
126. MATTSON, M. P., B. CHENG, A. R. CULWELL, F. S. ESCH, I. LIEBERBURG & R. E. RYDEL. 1993. Neuron **10:** 243-254.
127. GANDY, S. & P. GREENGARD. 1992. Trends Pharmacol. Sci. **23:** 108-113.
128. HAASS, C., M. G. SCHLOSSMACHER, A. Y. HUNG, C. VIGO-PELFREY, A. MELLON, B. L. OSTASZEWSKI, I. LIEBERBURG, E. H. KOO, D. SCHENK, D. B. TEPLOW & D. J. SELKOE. 1992. Nature **359:** 322-325.
129. ESTUS, S., T. E. GOLDE, T. KUNISHITA, D. BLADES, D. LOWERY, M. EISEN, M. USIAK, X. QU, T. TABIRA, B. D. GREENBERG & S. G. YOUNKIN. 1992. Science **255:** 726-728.
130. NAKAMURA, M., M. MORI, Y. MORISHITA, S. MORI & S. KAWASHIMA. 1992. Exp. Cell Res. **200:** 513-522.

Phosphorylation of Neuronal Cytoskeletal Proteins in Alzheimer's Disease and Lewy Body Dementias[a]

JOHN Q. TROJANOWSKI[b] AND VIRGINIA M.-Y. LEE

Department of Pathology and Laboratory Medicine
Division of Anatomic Pathology
The University of Pennsylvania School of Medicine
Philadelphia, Pennsylvania 19104

Accumulating data from various lines of research over the last decade support the notion that Alzheimer's disease (AD) is a heterogeneous disorder. For example, recent studies of the molecular genetics of familial (FAD) and sporadic AD implicate genes on three different chromosomes (ie, 14, 19, and 21) in the etiology of AD, and the apolipoprotein E (ApoE) 4 allele is a risk factor for both FAD and sporadic AD (FIG. 1).[1-5] Furthermore, mutations in the β-amyloid peptide (Aβ) or the Aβ precursor proteins (APPs) have been detected in a few families with FAD.[1-3,5] The evolving notion that AD represents a group of phenotypically similar neurodegenerative conditions with different initiating events has important implications for the treatment of AD. For example, this heterogeneity could make it necessary to "solve" the riddle of AD multiple times in order to prevent or block the progression of each variant of this late life dementia. Thus, as in cancer or infectious diseases, different strategies may be required for the treatment of etiologically distinct forms of AD. Alternatively, since a late life dementia is defined operationally as AD if postmortem examination reveals abundant telencephalic amyloid-rich senile plaques (SPs) and neurofibrillary tangles (NFTs),[6] these AD lesions may represent the "final common pathway" upon which different initiating events converge. Hence, the formation of plaques and tangles may be the point at which different disease mechanisms converge and then lead to the dysfunction and death of CNS neurons as well as to the dementia of AD. The convergence of different initiating events on the plaque and the tangle in the AD brain may be critical for the treatment of AD because these lesions could be pursued as therapeutic targets common to all forms of AD (FIG. 1).

Although SPs and NFTs (as well as related neurofibrillary lesions) are not restricted to AD, a close correlation exists between the burden of NFTs and the dementia in AD.[7-9] Furthermore, whereas neuronal cytoskeletal proteins known as τ are the building blocks of NFTs and related neurofibrillary lesions, other neuronal cytoskeletal proteins also accumulate as intraneuronal inclusions in AD brains.[10-18] For example,

[a] This work was supported by grants from the National Institutes of Health and the Dana Foundation.

[b] Address for correspondence: Dr. J. Q. Trojanowski, Department of Pathology and Laboratory Medicine, University of Pennsylvania School of Medicine, HUP, Maloney Bldg., Room A009, Philadelphia, PA 19104-4283.

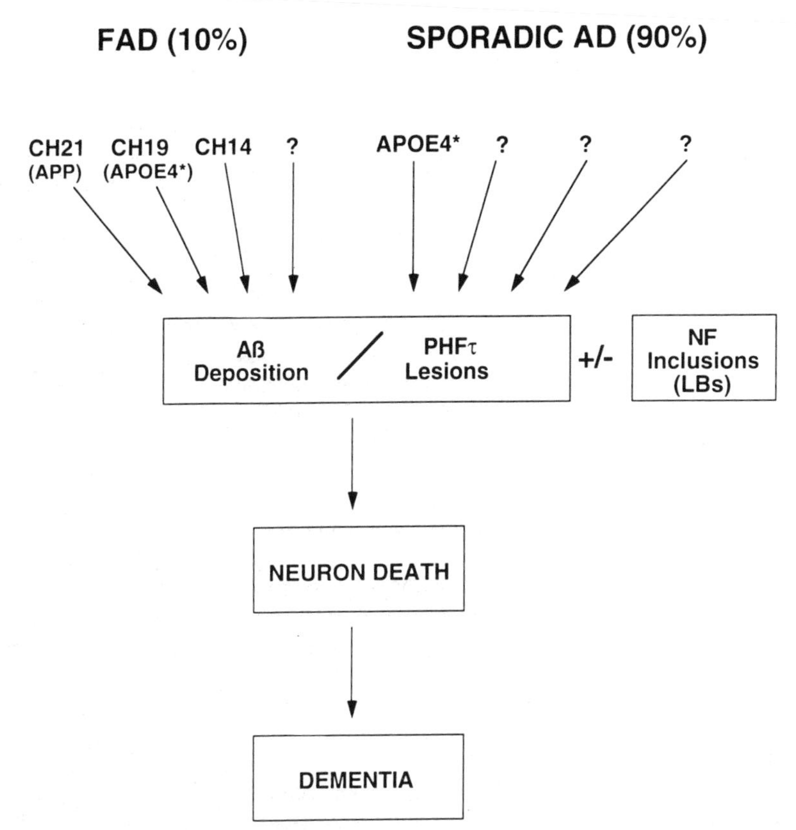

FIGURE 1. A summary of risk factors or etiologies of Alzheimer's disease (AD) that converge to generate characteristic AD brain lesions (amyloid deposits and neurofibrillary lesions) as well as lesions (LBs) found only in some AD cases. Amyloid and neurofibrillary lesions may be a final common pathway leading to neuron death and dementia in AD. The sequence of events shown here is hypothetical, but familial AD (FAD) accounts for about 10% of cases and is linked to chromosome 21 (CH21), 19 (CH19), and 14 (CH14). The APO E4 allele (APOE4*) on chromosome 19 is a risk factor for FAD and sporadic AD.

neurofilament-rich cortical Lewy bodies (LBs) similar to those found in Parkinson's disease also are detected in the brains of some patients with AD, and the presence of numerous cortical LBs, SPs, and NFTs defines a subtype of AD referred to as the LB variant of AD.[15,18] Alternatively, the presence of abundant cortical LBs in the brain of a demented patient without SPs and NFTs is recognized as a clinicopathologic entity known as diffuse LB disease.[15,18] Thus, accumulating evidence suggests that alterations in the normal metabolism and posttranslational processing (eg, phosphorylation) of τ and neurofilament proteins could lead to the formation of neurofibrillary lesions and LBs, respectively. Here we review recent insights into mechanisms that might account for the disruption of the normal metabolism and posttranslational processing of τ and neurofilament proteins in neurodegenerative diseases of the elderly. Additionally, we also discuss how neurons might become dysfunctional and degenerate as a result of the abnormal metabolism of τ and neurofilament proteins in the CNS.

PATHOLOGIC MARKERS OF ALZHEIMER'S DISEASE

Straight and paired helical filaments (PHFs) as well as amorphous material aggregate together in neuronal perikarya to form NFTs. Although Aβ fibrils and PHFs share many of the properties of other amyloids, Aβ amyloid fibrils and PHFs are composed of different proteins. Specifically, Aβ amyloid fibrils are formed from 39-43 amino acid long peptides cleaved from one or more APPs,[1-3,5] whereas PHFs are composed of hyperphosphorylated forms of normal adult CNS τ proteins referred to as A68 or PHFτ.[10-14,16-18] Despite intense research efforts into the pathologic significance of Aβ amyloid deposits since the discovery of the Aβ peptide a decade ago, the role of Aβ in the pathogenesis of AD remains enigmatic, and it is an unresolved puzzle why a plethora of Aβ deposits can be found in the brains of elderly individuals who show no antemortem evidence of cognitive deficits.[8] Furthermore, neither the function of the Aβ peptides that are constitutively secreted by normal human brain cells,[19-21] nor that of any of the three major species of APPs have been defined clearly.[1-3,5] On the other hand, considerable information is available concerning the basic biology and functions of τ proteins, a group of low molecular weight microtubule-associated proteins that bind to microtubules (MTs) and stabilize MTs in the polymerized state.[10-14,16-18] Normal adult human brain τ consists of six isoforms each of which contains either three or four consecutive MT binding motifs (ie, imperfect repeats of 31 or 32 amino acids) with or without inserts in the amino-terminal third of τ that are 29 or 58 amino acids in length. The function of these amino terminal inserts is unknown. Hence, the largest τ isoform is 441 amino acids long (with 4 tandemly repeated MT binding motifs and a 58 amino acid long amino terminal insert), whereas the smallest τ isoform is 352 amino acids long (with three consecutive MT binding repeats and no amino terminal insert).[10,22] The expression of these alternatively spliced τ proteins is developmentally regulated in the human CNS. Specifically, the shortest τ isoform (ie, the so-called "fetal" form of τ) is expressed early in the developing human nervous system, while this isoform plus the five other alternatively spliced CNS τ proteins are expressed in the adult human brain.[10,22] Presumably, the developmental regulation of the number of CNS τ isoforms and their phosphorylation state (see below) permits greater flexibility in establishing the appropriate equilibrium between polymerized MTs and depolymerized MTs in the immature, adolescent, and mature CNS.

HYPERPHOSPHORYLATED CNS τ PROTEINS ARE THE BUILDING BLOCKS OF PAIRED HELICAL FILAMENTS IN NEUROFIBRILLARY TANGLES, NEUROPIL THREADS, AND DYSTROPHIC NEURITES

Although PHFs are abundant in neuronal perikarya where they are the major structural elements of NFTs, PHFs also accumulate as neuropil threads (NTs) dispersed throughout the CNS gray matter and in olfactory epithelium as well as in dystrophic neurites intermingled with Aβ fibrils in neuritic amyloid plaques.[10–14,16–18,23–37] Initial reports suggesting that τ was the major subunit of PHFs in AD NFTs[38–43] were supported by the biochemical isolation of τ fragments from purified PHFs.[44–46] Taken together, these studies provided compelling evidence that the PHFs in AD neurofibrillary lesions were formed from full-length polymers of τ. This notion was firmly established when SDS-soluble PHFs were purified from AD brains and shown to consist of CNS τ proteins that were abnormally phosphorylated at specific residues (eg at Ser^{396}).[47] These findings were rapidly confirmed,[48–55] and it was shown that all six human CNS τ isoforms contribute to the formation of AD PHFs.[56] Furthermore, *in vitro* studies then showed that recombinant τ fragments comprised of nonphosphorylated MT binding repeats were able to form PHF-like structures.[57,58] Although all alternatively spliced CNS τ isoforms may be converted into PHFτ during the progression of AD, a higher molecular weight τ isoform of the peripheral nervous system (so-called "big" τ) that contains an additional amino terminal insert[59,60] has not been found in PHFτ.

On the basis of the studies reviewed, we concluded[47] that the generation of PHFτ results in whole or in part from the aberrant hyperphosphorylation of normal adult brain τ (FIG. 2). Accordingly, we and others have sought to identify the sites of aberrant phosphorylation that distinguish PHFτ from normal adult CNS τ.[47,48,51,61–63] For example, two Ser residues that are abnormally phosphorylated in PHFτ relative to adult human CNS τ (ie, Ser^{202} and Ser^{396}) were identified using immunologic methods and synthetic phosphopeptides as well as recombinant human τ subjected to site-directed mutagenesis and *in vitro* phosphorylation.[47,48,51] These studies showed that the abnormally phosphorylated Ser^{202} and Ser^{396} (using the numbering system of the longest human CNS τ isoform[22]) in PHFτ reside within the epitopes recognized by the well-characterized PHFτ-specific monoclonal antibodies known as AT8 and PHF1. Notably, the epitope detected by AT8, which recognizes phosphorylated Ser^{202} in PHFτ, is nearly complementary to the epitope detected by the Tau1 monoclonal antibody, which recognizes dephosphorylated PHFτ and normal adult CNS τ, but not native PHFτ.[51] Furthermore, PHF1 recognizes phosphorylated Ser^{396} in PHFτ just like the T3P antibody,[47,48] but PHF1 also appears to recognize phosphorylated Ser^{404} in PHFτ (Lee *et al.,* unpublished data). Mass spectrometry also has been used to identify additional abnormal phosphorylation sites in PHFτ,[63] and these include Thr^{181}, Thr^{231}, Ser^{235}, and Ser^{262}. However, Ser^{262} is phosphorylated in a fraction of PHFτ, and it may represent a secondary or less constant site of abnormal phosphorylation in PHFτ. Thus, all the major sites of abnormal phosphorylation in PHFτ are Ser/Pro or Thr/Pro sites. Inasmuch as there are 17 Ser/Pro or Thr/Pro sites in the largest human brain τ isoform and 14 of these sites are shared by all 6 CNS τ proteins, it is important now to determine which of these sites are aberrantly phosphorylated in PHFτ (relative to fetal and adult human CNS τ) and which sites interfere with the normal function of τ when they are aberrantly phosphorylated. Given the proximity of Ser^{396} and Ser^{404} to the MT binding domain just carboxy terminal to the last MT binding repeat, as well as the inverse relationship between the extent to which τ is phosphorylated and the ability of τ to bind MTs, it is not surprising that the aberrant phosphorylation of Ser^{396} in PHFτ relative to normal adult human CNS τ has been

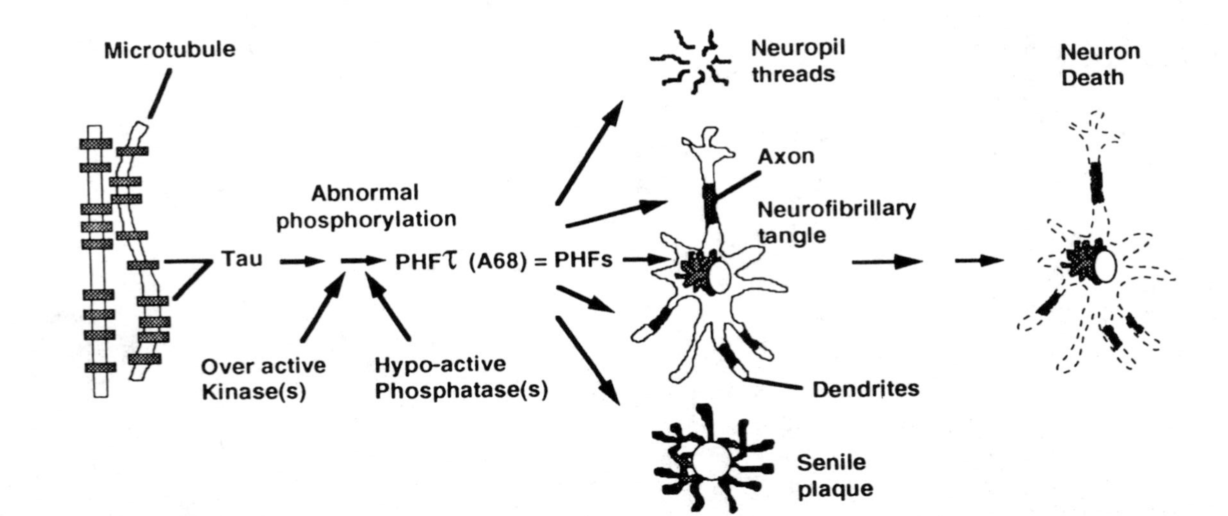

FIGURE 2. A hypothetical mechanism for the conversion of normal brain τ (rectangles overlying two MTs on the left) into PHFτ (A68). PHFτ is generated in neuronal perikarya and their processes as a consequence of overactivity or inappropriate activation of kinase(s) and/or hypoactivity or loss of phosphatase(s) that regulates the phosphorylation state of CNS τ. PHFτ then accumulates in neuronal processes as NTs (*top*), in neuronal perikaryal as NFTs (*middle*), and in plaque-associated dystrophic neurites (*bottom*). PHFτ loses the ability to bind MTs, and this could lead to the depolymerization of MTs, disruption of axonal transport, and dysfunction and/or degeneration of neurons in AD. The accumulation of PHFs in neurons could exacerbate this process by sequestering normal τ and by physically blocking the transport of proteins and organelles in neuronal perikarya, axons, and dendrites. The death of the neuron (*on the right*) is the ultimate outcome of all of these events.

implicated in the loss of the ability of PHFτ to bind to MTs.[48] Nonetheless, this loss of function also could result from abnormal hyperphosphorylation of other τ residues.

Comparisons of PHFτ with adult and fetal CNS τ proteins are important because several reports have demonstrated that PHFτ recapitulates the phosphorylation state of fetal CNS τ.[48,51,64,65] For example, Ser^{202} and Ser^{396} are phosphorylated in PHFτ and in the smallest τ isoform when it is expressed in the fetal CNS, but not in any of the six τ isoforms in the normal adult CNS including the smallest τ protein that is present in both the fetal and the adult brain.[48,51] Despite the fact that fetal CNS τ is phosphorylated at Ser^{202} and Ser^{396} as is PHFτ, fetal τ is capable of binding to MTs, whereas PHFτ completely loses the ability to bind MTs.[48] However, this loss of function is reversible because dephosphorylation of PHFτ restores the ability of PHFτ to bind MTs. In addition, the normal programmed death of large numbers of neurons in the developing CNS occurs without any of the neurofibrillary lesions that are found in the brain in AD.[65–67] Taken together, these observations suggest that the accumulation of neurofibrillary lesions in AD may be due to the aberrant reactivation of fetal protein kinases and the inactivation of protein phosphatases in the AD brain that normally determine the phosphorylation state of fetal τ in the developing brain. To understand this issue better, it is important to understand how the phosphorylation state of τ is regulated in fetal and adult human brain.

By using analytic strategies similar to those described above,[48,51,63] all of the abnormal phosphorylation sites that distinguish PHFτ from normal τ in the fetal and adult CNS can be identified. The accomplishment of this objective is important because it will set the stage for identifying the kinases and phosphatases that are involved in the generation of PHFτ in the AD brain (FIG. 2). Indeed, the availability of detailed information on the unique properties of PHFτ has facilitated efforts to identify several candidate kinases that might be involved in the conversion of normal τ into PHFτ. Some of these candidate kinases include mitogen-activated protein (MAP) kinases (also known as extracellular signal-regulated kinases or ERKs), glycogen synthase kinase-3, proline-directed protein kinase, calcium/calmodulin-dependent protein kinase (CaMK II), and cyclic-AMP-dependent protein kinase (also known as PKA).[62,68–75] Although there is less information on protein phosphatases that might be involved in the generation of aberrantly phosphorylated PHFτ in the AD brain, preliminary evidence has implicated protein phosphatase 2A1 (PP2A1) and protein phosphatase 2B (PP2B or calcineurin) in the conversion of normal τ into PHFτ.[68,76] However, several different kinases and phosphatases may be involved simultaneously or sequentially in the progressive conversion of normal τ into PHFτ, and it may be necessary to elucidate the contributions of all of these enzymes to the generation of PHFτ before the role of neurofibrillary lesions in the dementia of AD can be fully understood.

BIOLOGIC SIGNIFICANCE OF THE ACCUMULATION OF PHFτ IN THE BRAIN

Despite gaps in our understanding of the detailed pathobiology of AD neurofibrillary lesions, the available information suggests that the conversion of normal τ into PHFτ may have deleterious effects on neurons during the progression of AD. For example, it is well known that the accumulation of PHFτ in AD cortex[49,53,77] correlates with the abundance of NFTs as well as with diminished levels of normal MT binding competent τ in the CNS.[49] Furthermore, because hyperphosphorylated PHFτ is unable to bind to MTs,[48] we proposed[14,15,17,18] that the conversion of normal τ into PHFτ could lower the levels of MT binding τ, destabilize MTs,[78] disrupt axonal transport, and lead to the "dying back" of axons in AD. This could in turn result in the degeneration or loss of corticocortical connections, impaired synaptic transmission, and, finally, the emergence of cognitive impairments (FIG. 2).

Furthermore, the release of PHFτ from dying neurons or dystrophic neurites could result in interactions between PHFτ and the soluble Aβ normally secreted by neurons or other CNS cells.[19–21] In fact, some experimental data suggest that the release of PHFτ into the extracellular space could contribute to the formation of SPs. For example, injections of human PHFτ into the rodent brain induce co-deposits of Aβ at the injection site.[79] Notably, SPs are complex amyloid deposits that disrupt the surrounding neuropil and are invariably associated with PHFτ and other components.[17,18,23,80–82] This is in sharp contrast to the pre-amyloid plaques that are comprised almost exclusively of Aβ and do not disrupt synaptic profiles or elicit a gliotic response as do SPs.[17,82] Thus, it is interesting that other proteins associated with NFTs and SPs in the AD brain (i.e., ubiquitin and α_1-antichymotrypsin[12,17,18,82]) also were co-deposited in the PHF τ injection sites of the rodent brain.[79] However, similar effects were not produced by cerebral injections of other proteins including: normal adult and fetal human τ, dephosphorylated PHFτ, high molecular weight NF proteins, or α_1-antichymotrypsin. These studies imply that mechanisms leading to SP formation could involve interactions between PHFτ and Aβ. Thus, we speculate that interactions between secreted, soluble Aβ and other molecules (including PHFτ and other SP components) in the extracellular space could serve as a nidus for the induction of Aβ fibril formation and the generation of complex neuritic plaques. FIGURE 3 depicts a hypothetical scenario summarizing how such events might occur in the AD brain. Although the details of this hypothetical scheme remain to be elucidated, it appears increasingly likely that the conversion of soluble Aβ into insoluble amyloid fibrils in the extracellular space leading to the formation of neuritic amyloid plaques is a multistep process that may involve a number of different cofactors.[17,18,83–86]

ORIGIN OF LEWY BODIES FROM NEUROFILAMENT PROTEINS IN ALZHEIMER'S DISEASE AND LEWY BODY DEMENTIAS

Although most commonly associated with idiopathic Parkinson's disease, LBs are neither restricted to Parkinson's disease alone, nor confined only to monoaminergic neurons in the brains of patients with Parkinson's disease. In fact, LBs increasingly are recognized in the brains of patients with other neurodegenerative diseases, and especially in the neocortex of patients with a late life dementia that is clinically similar to AD.[15–17,87–93] The magnitude of the burden of cortical LBs in demented patients has been appreciated only recently after the introduction of immunohistochemical methods to detect LBs *in situ* using antibodies to ubiquitin or neurofilament proteins. Indeed, the detection of numerous cortical LBs in selected regions of the

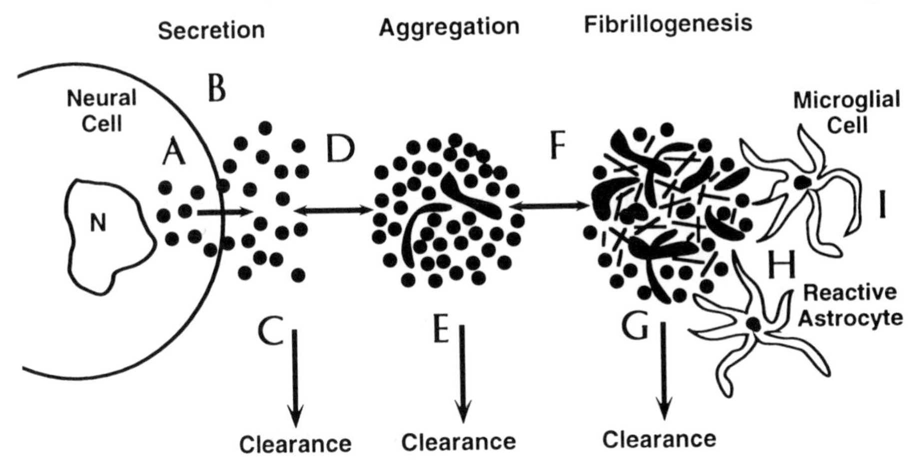

FIGURE 3. A hypothetical schematic showing a series of events leading to the formation of SPs in AD. Soluble Aβ (*filled circles*) is produced in neural cells (eg, neurons) in the brain (**A;** N = nucleus) and secreted into the extracellular space (**B**) from whence it is cleared in normal younger individuals (**C**). With advancing age, soluble Aβ aggregates (**D**) into pre-amyloid plaques (*closely packed filled circles to the right of* **D**). Some aggregated Aβ also may be cleared from these plaques (**E**). In AD, soluble Aβ may come into contact with PHFτ (*filled curvilinear profiles*) released from degenerating neurons as well as with other factors (eg, proteoglycans) in the extracellular space (*filled curvilinear profiles*). As a consequence, soluble Aβ may be induced to form fibrillar Aβ (**F**), leading to the formation of SPs (*to the right of* **F**) that accumulate many components including soluble Aβ (*filled circles*), PHFτ (*curvilinear profiles*), fibrillar Aβ (*short linear profiles*), other fibrillogenic factors (*curvilinear profiles*), microglial cells (**I**) and reactive astrocytes (**H**). Nonfibrillar (soluble) Aβ (and possibly fibrillar Aβ) might be cleared from SPs (**G**), but the continued presence of factors that induce Aβ fibrillogenesis and insolubility favors the persistence of SPs in the AD brain.

postmortem brain with these antibodies defines a late life dementia as diffuse Lewy body disease when no other diagnostic brain lesions are present. Alternatively, the detection of cortical LBs in conjunction with AD pathology in the postmortem brain of a demented individual is referred to as the LB variant of AD.

The morphology of LBs in Parkinson's disease is well established at the light and electron microscopic levels, especially in substantia nigra. These intraneuronal inclusions correspond to spherical accumulations of filaments arranged in a loose, radiating array in the periphery or "corona" of the LB, whereas a matted meshwork of filaments is found in the center or "core" of the LB.[15,18,92,93] The corona of LBs in substantia nigra is striking because filaments 7-20 nm in diameter radiate through the corona from the core region, and the more clearly discernible filaments resemble neurofilaments. However, these filaments, as well as those in the core of the LB, lack the projecting "side arms" typical of normal neurofilaments in axons.[94] Furthermore, amorphous material and membranous profiles are admixed with these neurofilament-like structures (especially in the LB core), and it is uncertain if this less well-defined material represents nonspecifically trapped elements or material that contributes in some mechanistic way to the failure of these filaments or their individual subunits to undergo axonal transport and to aggregate instead in the neuronal perikaryon as LBs. Although cortical LBs have not been studied as exhaustively, these LBs have a less stereotyped light and electron microscopic appearance compared to that of substantia nigra LBs. For example, cortical LBs usually lack a definable central core and a distinct peripheral "halo" or corona that are typical of LBs in the substantia nigra.

By contrast with the morphologic properties of LBs, the protein composition of cortical and subcortical LBs has proven difficult to define at the biochemical level because of the lack of reliable methods for routine purification of LBs in bulk from postmortem brains. Furthermore, the density of LBs in a given volume of brain is less than that of NFTs or SPs in AD, and diffuse Lewy body disease is less common overall than is AD. However, recent efforts to isolate and characterize cortical LBs appear promising,[95,96] and these preliminary reports confirm findings from earlier *in situ* immunohistochemical studies that implicated neurofilament subunits as the major proteins in cortical and nigral LBs.[15,16,90,94,95–101] Given the difficulties inherent in the bulk isolation of LBs for biochemical and immunochemical analysis, it is not surprising that immunohistochemical studies have contributed most of the significant details on the composition of LBs in diffuse Lewy body disease, the LB variant of AD as well as in Parkinson's disease. Indeed, neurofilament protein and ubiquitin immunoreactivity have been demonstrated consistently in LBs by different immunohistochemical procedures in reports published from many laboratories over the last decade, and these reports have established that cortical and subcortical LBs contain each of the neurofilament triplet proteins.[15,18] These neurofilament subunits include the low (NF-L), middle (NF-M), and high (NF-H) molecular weight neurofilament proteins that form the heteropolymeric, 10-nm diameter, intermediate filaments of neurons. Neurofilaments are thought to have important functions in axons as they are one of the most prominent cytoskeletal organelles in large diameter axons.[16,18,102] NF-H and NF-M are heavily phosphorylated relative to NF-L, and many antibodies raised to NF proteins recognize specific phosphorylation-dependent epitopes in NF-H and NF-M.[18] Thus, it was surprising that LBs were labeled by many antibodies specific for heavily phosphorylated epitopes located in the carboxy terminal extensions or "tail" domains of NF-H and NF-M because phosphorylation of these epitopes normally occurs in axons.[94,101,102] Findings from these epitope mapping analyses and earlier immunohistochemical studies of LBs suggest that NF-H and NF-M are abnormally phosphorylated in perikaryal LBs inasmuch as NF-H and NF-M normally become heavily phosphorylated only in the axon.[102] However, recent studies have identified dystrophic processes in the CA2/3 region of the hippocampus of patients with diffuse Lewy body disease that are immunologically similar to perikaryal LBs.[103] Hence, it is possible that these lesions arise as a consequence of pathologic mechanisms in axons that are similar to those that lead to the formation of LBs in the perikaryon.[103]

Thus, as with PHFτ in neurofibrillary lesions in AD, abnormal or aberrant hyperphosphorylation of the normal neuronal cytoskeletal proteins NF-H and NF-M in neuronal cell bodies has been implicated in the formation of LBs which serve as pathologic signatures of diseases such as Parkinson's and diffuse Lewy body disease. Although practically no information is available on the putative kinases and phosphatases that may regulate abnormal phosphorylation of NF proteins in LBs, calcium/calmodulin-dependent protein kinase (CaMK II) immunoreactivity has been detected in LBs.[104] On the other hand, the phosphorylation state of NF-H and NF-M in LBs may be a late event that occurs after LBs are formed because the expression of highly phosphorylated NF proteins in perikarya is a common response by neurons to a variety of injuries regardless of the cause. Alternatively, LB formation may be due to faulty targeting of most neurofilaments into perikarya (instead of axons) coupled with impaired or altered proteolysis of neurofilaments before or after they aggregate into LBs. The abundant ubiquitin immunoreactivity in LBs[15,18] suggests that neurofilaments in these inclusions indeed are targeted for proteolysis, but efforts by affected neurons to eliminate LBs by proteolytic degradation are ineffective for reasons that remain unknown. Although proteases involved in the normal turnover of neurofilaments have been studied,[102] there is little information on the proteases that prevent the incorporation of neurofilaments into LBs in most normal elderly individuals. Finally, it should be noted that the accumulation of neurofilaments and neurofilament proteins in perikaryal LBs is associated with several other cytoskeletal and noncytoskeletal proteins that could play a role in LB formation.[15,18] However, the contribution of these other components to the formation of LBs is currently unknown.[15,16,18]

BIOLOGIC SIGNIFICANCE OF THE ACCUMULATION OF LEWY BODIES IN THE BRAIN

Information on the possible biologic consequences of LB formation in neurons is incomplete, and none of the conventional animal models of Parkinson's disease has proven amenable for studies of the pathobiology of LBs. However, the recent demonstration that mRNAs encoding NF-H and NF-L subunits are reduced in LB-containing neurons may signify that the formation of LBs from normal or partially degraded neurofilament proteins could alter neurofilament synthesis and the translocation of assembled neurofilaments into axons.[105] Therefore, we proposed[18,105] that axons of LB containing nigral and cortical neurons could undergo a "dying back" process from diminished levels of neurofilaments. Accordingly, this could "disconnect" the substantia nigra from the basal ganglia or one cortical region from another. In this manner, the formation of LBs and their accumulation in subcortical and cortical neurons could have deleterious long-term effects on neuronal function and survival (FIG. 4). Although this mechanistic scenario is speculative, portions of it appear plausible now in light of recently described data on transgenic mice that were shown to overexpress NF-L or NF-H proteins and to develop intraneuronal accumulations of neurofilaments due to altered neurofilament metabolism.[106,107] As a result of these perturbations, tangled masses of neurofilaments accumulated in the perikarya of spinal cord motor neurons. These neurons and their axons then degenerated which led to muscle atrophy, motor weakness, and the death of some of the transgenic mice. These transgenic mice were regarded as animal models of motor neuron disease because of similarities between the clinical and pathological phenotype of mice and that of human motor neuron diseases (ie, amyotrophic lateral sclerosis) including perikaryal and axonal accumulations of phosphorylated neurofilaments.[108–110] However, the findings in these two studies also are compatible with the notion that LB formation in cortical and subcortical neurons could have deleterious consequences on the affected populations of neurons in human LB disorders (FIG. 4).

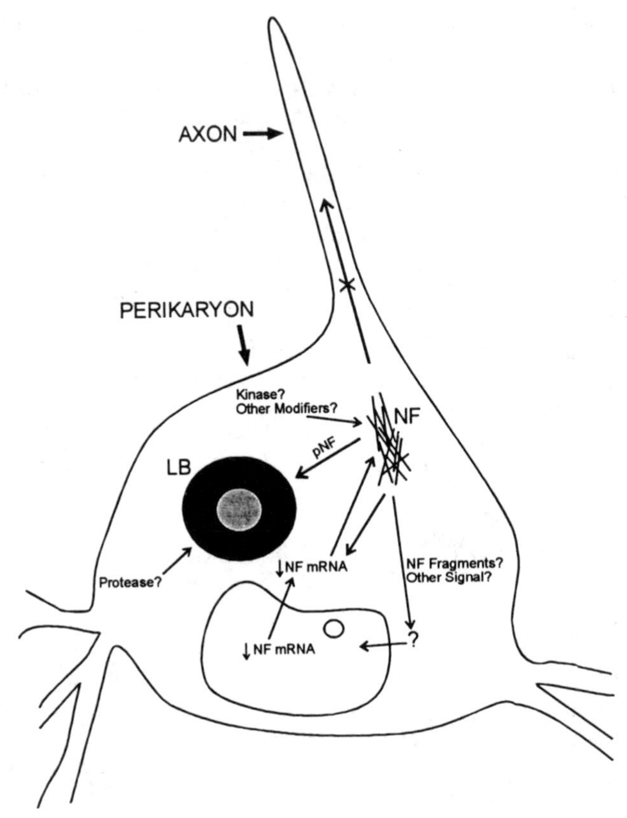

FIGURE 4. A hypothetical series of events leading to the formation of LBs in cortical and subcortical neurons as well as potential deleterious consequences of LB formation in neurons. Neurofilaments (NF) may aggregate into perikaryal LBs due to aberrant phosphorylation of neurofilaments or neurofilament subunits (pNF) in the perikaryon where proteases fail to degrade these inclusions completely despite the ubiquitination of LBs. Proteolytic neurofilament fragments in perikarya or other signals may downregulate neurofilament mRNA levels, leading to reduced neurofilament subunit synthesis. This and the diversion of neurofilaments into LBs may deplete the axon of neurofilaments, leading to the "dying" back of axons, degeneration of affected neurons, and functional impairments in LB disorders.

CONCLUDING REMARKS

Disruption of the neuronal cytoskeleton is a common feature of chronic dementias of the elderly including AD, diffuse Lewy body disease, and the LB variant of AD. Furthermore, the abnormal accumulation of neuronal cytoskeletal proteins in these disorders (eg, NFTs and LBs) may be due in part to similar mechanisms, that is, the abnormal phosphorylation of neurofilament and τ proteins. Although the disruption of the neuronal cytoskeleton and the incorporation of abnormally phosphorylated CNS τ and neurofilament proteins into pathologic inclusions provide markers for the disease state, these events also could compromise the function and viability of neurons, thereby contributing to the emergence of dementia in AD, the LB variant of AD, and diffuse Lewy body disease. Thus, we anticipate that advances in understanding the mechanisms whereby neurofibrillary lesions and LBs are generated from normal adult τ or neurofilament proteins, respectively, will provide further insights into how these lesions contribute to neuronal dysfunction and degeneration. Such insights will set the stage for developing strategies for the therapy of AD, diffuse Lewy body disease, the LB variant of AD as well as Parkinson's disease.

ACKNOWLEDGMENTS

We thank our colleagues in the Departments of Pathology and Laboratory Medicine, Neurology, Psychiatry, and the Penn Alzheimer Center for their many contributions to the work reviewed here.

REFERENCES

1. Hardy, J. 1993. Genetic mistakes point the way for Alzheimer's disease. J. NIH Res. **5:** 46-49.
2. Mullan, M. & F. Crawford. 1993. Genetic and molecular advances in Alzheimer's disease. Trends Neurosci. **16:** 398-403.
3. Price, D. L., D. R. Borchelt & S. S. Sisodia. 1993. Alzheimer disease and the prion disorders amyloid β-protein and prion protein amyloidoses. Proc. Natl. Acad. Sci. USA **90:** 6381-6384.
4. Saunders, A. M., W. J. Strittmatter, D. Schmechel, P. H. St. George-Hyslop, M. A. Pericak-Vance, S. H. Joo, B. L. Rosi, J. F. Gusella, D. R. Crapper-MacLachlan, M. J. Alberts, C. Hulette, B. Crain, D. Goldgaber & A. D. Rosers. 1993. Association of Apolipoprotein E Allele ϵ4 with late-onset familal and sporadic Alzheimer's disease. Neurology **43:** 1467-1472.
5. Selkoe, D. J. 1993. Physiological production of the β-amyloid protein and the mechanisms of Alzheimer's disease. Trends Neurosci. **16:** 403-409.
6. Khachaturian, Z. S. 1985. Diagnosis of Alzheimer's disease. Arch. Neurol. **42:** 1097-1106.
7. Arriagada, P. A., J. H. Growdon, E. T. Hedley-White & B. T. Hyman. 1992. Neurofibrillary tangles but not senile plaques parallel duration and severity of Alzheimer's disease. Neurology **42:** 631-639.
8. Dickson, D. W., H. A. Crystal, L. A. Mattiace, D. M. Masur, A. D. Blau, P. Davies, S.-H. Yen & M. Aronson. 1991. Identification of normal and pathological aging in prospectively studied nondemented elderly humans. Neurobiol. Aging **13:** 179-189.
9. McKee, A. C., K. S. Kosik & N. W. Kowall. 1991. Neuritic pathology and dementia in Alzheimer's disease. Ann. Neurol. **30:** 156-165.
10. Goedert, M. 1993. Tau protein and the neurofibrillary pathology of Alzheimer's disease. Trends Neurosci. **16:** 460-465.
11. Iqbal, K. & I. Grundke-Iqbal. 1991. Ubiquitination and abnormal phosphorylation of PHFs in Alzheimer's disease. Mol. Neurobiol. **5:** 399-410.
12. Kosik, K. S. 1992. Alzheimer's disease: A cell biological perspective. Science **256:** 780-783.
13. Lee, V. M.-Y. 1991. Unraveling the mystery of the paired helical filaments of Alzheimer's disease. J. NIH Res. **3:** 54-56.
14. Lee, V. M.-Y. & J. Q. Trojanowski. 1992. The disordered neuronal cytoskeleton in Alzheimer's disease. Curr. Opin. Neurobiol. **2:** 653-656.

15. POLLANEN, M. S., D. W. DICKSON & C. BERGERON. 1993. Pathology and biology of the Lewy Body. J. Neuropathol. Exp. Neurol. **52:** 183-191.
16. TROJANOWSKI, J. Q., M. L. SCHMIDT, L. OTVOS, JR., H. ARAI, W. D. HILL & V. M.-Y. LEE. 1990. Vulnerability of the neuronal cytoskeleton in Alzheimer's disease: Widespread involvement of all three major filament systems. Ann. Rev. Gerontol. Geriat. **10:** 167-182.
17. TROJANOWSKI, J. Q., M. L. SCHMIDT, R.-W. SHIN, G. T. BRAMBLETT, M. GOEDERT & V. M.-Y. LEE. 1994. PHFτ (A68): From pathological marker to potential mediator of neuronal dysfunction and degeneration in Alzheimer's disease. Clin. Neurosci. **3:** 45-54.
18. TROJANOWSKI, J. Q., M. L. SCHMIDT, R.-W. SHIN, G. T. BRAMBLETT, D. RAO & V. M.-Y. LEE. 1993. Altered tau and neurofilament proteins in neurodegenerative diseases: Diagnostic implications for Alzheimer's disease and Lewy body dementias. Brain Pathol. **3:** 45-54.
19. SEUBERT, P., C. VIGO-PELFREY, F. ESCH, M. LEE, H. DOVEY, D. DAVIS, S. SINHA, M. SCHLOSSMACHER, J. WHALEY, C. SWINDLEHURST, R. MCCORMACK, R. WOLFERT, D. SELKOE, I. LIEBERBURG & D. SCHENK. 1992. Isolation and quantification of soluble Alzheimer's β-peptide from biological fluids. Nature **359:** 325-327.
20. SHOJI, M., T. E. GOLDE, J. GHISO, T. T. CHEUNG, S. ESTUS, L. M. SHAFFER, X.-D. CAI, D. M. MCKAY, R. TINTNER, B. FRANGIONE & S. G. YOUNKIN. 1992. Production of the Alzheimer amyloid β protein by normal proteolytic processing. Science **258:** 126-129.
21. WERTKIN, A. M., R. S. TURNER, S. J. PLEASURE, T. E. GOLDE, S. G. YOUNKIN, J. Q. TROJANOWSKI & V. M.-Y. LEE. 1993. Human neurons derived from a teratocarcinoma cell line express solely the 695-amino acid precursor protein and produce intracellular β-amyloid or A4 peptides. Proc. Natl. Acad. Sci. USA **90:** 9513-9517.
22. GOERDERT, M., M. G. SPILLANTINI, R. JAKES, D. RUTHERFORD & R. A. CROWTHER. 1989. Multiple isoforms of human microtubule-associated protein tau: Sequences and localization in neurofibrillary tangles of Alzheimer's disease. Neuron **3:** 519-526.
23. ARAI, H., V. M.-Y. LEE, L. OTVOS, B. GREENBERG, D. E. LOWERY, S. K. SHARMA, M. L. SCHMIDT & J. Q. TROJANOWSKI. 1990. Defined neurofilament, τ, and β-amyloid precursor protein epitopes distinguish Alzheimer from non-Alzheimer senile plaques. Proc. Natl. Acad. Sci. USA **87:** 2249-2253.
24. ARAI, H., M. L. SCHMIDT, V. M.-Y. LEE, H. I. HURTIG, B. D. GREENBERG, C. ADLER & J. Q. TROJANOWSKI. 1992. Epitope analysis of senile plaque components in the hippocampus of patients with Parkinson's disease. Neurology **42:** 1315-1322.
25. BRAAK, H. & E. BRAAK. 1988. Neuropil threads occur in dendrites of tangle-bearing nerve cells. Neuropathol. Appl. Neurobiol. **14:** 39-44.
26. BRAAK, H., E. BRAAK, I. IQBAL-GRUNDKE & K. IQBAL. 1986. Occurrence of neuropil threads in the senile human brain and in Alzheimer's disease: A third location of paired helical filaments outside of neurofibrillary tangles and neuritic plaques. Neurosci. Lett. **66:** 212-222.
27. IHARA, Y. 1988. Massive somatodendritic sprouting of cortical neurons in Alzheimer's disease. Brain. Res. **459:** 138-144.
28. LEE, J. H., M. GOEDERT, W. D. HILL, V. M.-Y. LEE & J. Q. TROJANOWSKI. 1993. Tau proteins are abnormally expressed in olfactory epithelium of Alzheimer's disease and developmentally regulated in fetal spinal cord. Exp. Neurol. **121:** 93-105.
29. MASLIAH, E., M. ELLISMAN, B. CARRAGHER, M. MALLORY, S. YOUNG, L. HANSEN, R. DETERESA & R. D. TERRY. 1992. Three dimensional analysis of the relationship between synaptic pathology and neuropil threads in Alzheimer's disease. J. Neuropathol. Exp. Neurol. **51:** 404-414.
30. PERRY, G., M. KAWAI, M. TABATON, M. ONORATO, P. MULVILL, P. RICHEY, A. MORANDI, J. A. CONNOLLY & P. GAMBETTI. 1990. Neuropil threads of Alzheimer's disease show a marked alteration of the normal cytoskeleton. J. Neurosci. **11:** 1748-1755.

31. SCHMIDT, M. L., V. M.-Y. LEE & J. Q. TROJANOWSKI. 1990. Relative abundance of tau and neurofilament epitopes in hippocampal neurofibrillary tangles. Am. J. Pathol. **136:** 1069-1075.
32. SCHMIDT, M. L., V. M.-Y. LEE & J. Q. TROJANOWSKI. 1991. Comparative epitope analysis of neuronal cytoskeletal proteins in Alzheimer's disease senile plaque neurites and neuropil threads. Lab. Invest. **64:** 352-357.
33. SCHMIDT, M. L., J. MURRAY & J. Q. TROJANOWSKI. 1993. Continuity of neuropil threads with tangle-bearing and tangle-free neurons in Alzheimer's disease cortex. A confocal laser scanning microscopy study. Mol. Chem. Neuropathol. **18:** 299-312.
34. SHIN, R.-W., K. OGOMORI, T. KITAMOTO & J. TATEISHI. 1989. Increased tau accumulation in senile plaques as a hallmark in Alzheimer's disease. Am. J. Pathol. **134:** 1365-1371.
35. TABATON, M., S. CAMMARATA, G. L. MANCARDI, G. CORDONE, G. PERRY & C. LOEB. 1991. Abnormal tau-reactive filaments in olfactory mucosa in biopsy specimens of patients with probable Alzheimer's disease. Neurology **41:** 391-394.
36. TALAMO, B. R., R. RUDELL, K. S. KOSIK, V. M.-Y. LEE, S. NEFF, L. ADELMAN & J. S. KAUER. 1989. Pathological changes in olfactory neurons in patients with Alzheimer's disease. Nature **337:** 736-739.
37. TROJANOWSKI, J. Q., P. D. NEWMAN, W. D. HILL & V. M.-Y. LEE. 1991. Human olfactory epithelium in normal aging, Alzheimer's disease and other neurodegenerative disorders. J. Comp. Neurol. **310:** 365-376.
38. BRION, J. P., H. PASSARIER, J. NUNEZ & J. FLAMENT-DURAND. 1985. Immunologic determinants of tau protein are present in neurofibrillary tangles of Alzheimer's disease. Arch. Biol. **95:** 229-235.
39. DELACOURTE, A. & A. DEFOSSEZ. 1986. Alzheimer's disease: Tau proteins, the promoting factors of microtubule assembly are major components of paired helical filaments. J. Neurol. Sci. **76:** 173-186.
40. GRUNDKE-IQBAL, I., K. IQBAL, Y.-C. TUNG, M. QUINLAN, H. M. WISNIEWSKI & L. I. BINDER. 1986. Abnormal phosphorylation of the microtubule-associated protein tau in Alzheimer cytoskeletal pathology. Proc. Natl. Acad. Sci. USA **83:** 4913-4917.
41. KOSIK, K., C. L. JOACHIM & D. J. SELKOE. 1986. Microtubule-associated protein tau is a major antigenic component of paired helical filaments in Alzheimer's disease. Proc. Natl. Acad. Sci. USA **83:** 4044-4048.
42. KOSIK, K. S., L. D. ORECCHIO, L. BINDER, J. Q. TROJANOWSKI, V. M.-Y. LEE & G. LEE. 1988. Epitopes that span the tau molecule are shared with paired helical filaments. Neuron **1:** 817-825.
43. WOOD, J. G., S. S. MIRRA, N. J. POLLOCK & L. I. BINDER. 1986. Neurofibrillary tangles of Alzheimer disease share antigenic determinants with the axonal microtubule-associated protein tau. Proc. Natl. Acad. Sci. USA **83:** 4040-4043.
44. GOEDERT, M., C. M. WISCHIK, R. A. CROWTHER, J. E. WALKER & A. KLUG. 1988. Cloning and sequencing of the cDNA encoding a core protein of the paired helical filament of Alzheimer disease: Identification as the microtubule-associated protein tau. Proc. Natl. Acad. Sci. USA **85:** 4051-4055.
45. KONDO, J., T. HONDA, H. MORI, Y. HAMADA, R. MIURA, M. OGAWARA & Y. IHARA. 1988. The carboxyl third of tau is tightly bound to paired helical filaments. Neuron **1:** 827-834.
46. WISCHIK, C. M., M. NOVAK, H. C. THOGERSEN, P. C. EDWARDS, M. J. RUNSWICK, R. JAKES, J. WALKER, C. MILSTEIN, M. ROTH & A. KLUG. 1988. Isolation of a fragment of tau derived from the core of the paired helical filament of Alzheimer disease. Proc. Natl. Acad. Sci. USA **85:** 4506-4510.
47. LEE, V. M.-Y., B. J. BALIN, L. OTVOS, JR. & J. Q. TROJANOWSKI. 1991. A68: A major subunit of paired helical filaments and derivatized forms of normal tau. Science **251:** 675-678.

48. Bramblett, G. T., M. Goedert, R. Jakes, S. E. Merrick, J. Q. Trojanowski & V. M.-Y. Lee. 1993. Abnormal tau phosphorylation at Ser^{396} in Alzheimer's disease recapitulates development and contributes to reduced microtubule binding. Neuron **10:** 1089-1099.
49. Bramblett, G. T., J. Q. Trojanowski & V. M.-Y. Lee. 1992. Regions with abundant neurofibrillary pathology in human brain exhibit a selective reduction in levels of binding-competent τ and the accumulation of abnormal τ-isoforms (A68 proteins). Lab. Invest. **66:** 212-222.
50. Brion, J. P., D. P. Hanger, A. M. Couck & B. Anderton. 1991. A68 proteins in Alzheimer's disease are composed of several tau isoforms in a phosphorylated state which affects their electrophoretic mobilities. Biochem. J. **279:** 831-836.
51. Goedert, M., R. Jakes, R. A. Crowther, J. Six, U. Luebke, M. Vandermeeren, P. Cras, J. Q. Trojanowski & V. M.-Y. Lee. 1993. The abnormal phosphorylation of tau protein at $serine^{202}$ in Alzheimer's disease recapitulates phosphorylation during development. Proc. Natl. Acad. Sci. USA **90:** 5066-5070.
52. Greenberg, S. G., P. Davies, J. D. Scheim & L. I. Binder. 1992. Hydrofluoric acid-treated τ PHF proteins display the same biochemical properties as normal tau. J. Biol. Chem. **267:** 564-569.
53. Khatoon, S., I. Grundke-Iqbal & K. Iqbal. 1992. Brain levels of microtubule-associated protein τ are elevated in Alzheimer's disease. A radioimmuno-slot-blot assay for nanograms of the protein. J. Neurochem. **59:** 750-753.
54. Ksiezak-Reding, H. & S.-H. Yen. 1991. Structural stability of paired helical filaments requires microtubule-binding domains of tau: A model for self-association. Neuron **6:** 717-728.
55. Shin, R.-W., T. Iwaki, T. Kitamoto, Y. Sato & J. Tateishi. 1992. Massive accumulation of modified tau and severe depletion of normal tau characterize the cerebral cortex and white matter of Alzheimer's disease. Am. J. Pathol. **140:** 937-945.
56. Goedert, M., M. G. Spillantini, N. J. Cairns & R. A. Crowther. 1992. Tau proteins of Alzheimer paired helical filaments: Abnormal phosphorylation of all six brain isoforms. Neuron **8:** 159-168.
57. Crowther, R. A., O. F. Olesen, R. Jakes & M. Goedert. 1992. The microtubule binding repeats of tau protein assemble into filaments like those found in Alzheimer's disease. FEBS Lett. **309:** 199-202.
58. Wille, H., G. Drewes, J. Biernat, E. M. Mandelkow & E. Mandelkow. 1992. Alzheimer-like paired helical filaments and antiparallel dimers formed from microtubule-associated tau *in vitro.* J. Cell Biol. **118:** 573-584.
59. Couchie, D., C. Mavilia, I. S. Georgieff, R. K. H. Liem, M. L. Shelanski & J. Nunez. 1992. Primary structure of high molecular weight tau present in the peripheral nervous system. Proc. Natl. Acad. Sci. USA **89:** 4378-4381.
60. Goedert, M., M. G. Spillantini & R. A. Crowther. 1992. Cloning of a big tau microtubule-associated protein characteristic of the peripheral nervous system. Proc. Natl. Acad. Sci. USA **89:** 1983-1987.
61. Biernat, J., E. M. Mandelkow, C. Schroeter, B. Lichtenberg-Kraag, B. Steiner, B. Berling, H. Meyer, M. Mercken, A. Vandermeeren, M. Goedert & E. Mandelkow. 1991. The switch of tau protein to an Alzheimer-like state induces the phosphorylation of two serine-proline motifs upstream of the microtubule binding region. EMBO J. **11:** 1593-1597.
62. Drewes, G., B. Lichtenberg-Kraag, F. Doering, E. M. Mandelkow, J. Biernat, J. Goris, M. Doree & E. Mandelkow. 1992. Mitogen-activated protein (MAP) kinase transforms tau protein into an Alzheimer-like state. EMBO J. **6:** 2131-2138.
63. Hasagewa, M., M. Morishima-Kawashima, K. Takio, M. Suzuki, K. Litani & Y. Ihara. 1992. Protein sequence and mass spectrometric analyses of tau in the Alzheimer's disease brain. J. Biol. Chem. **267:** 17047-17054.

64. Kanemaru, K., K. Takio, R. Miura, K. Titani & Y. Ihara. 1992. Fetal-type phosphorylation of the tau in paired helical filaments. J. Neurochem. **58:** 1667-1675.
65. Lee, J. H., M. Goedert, W. D. Hill, V. M.-Y. Lee & J. Q. Trojanowski. 1993. Tau proteins are abnormally expressed in olfactory epithelium of Alzheimer's disease and developmentally regulated in fetal spinal cord. Exper. Neurol. **121:** 93-105.
66. Tohyama, T., V. M.-Y. Lee, L. B. Rorke & J. Q. Trojanowski. 1991. Molecular milestones that signal axonal maturation and the commitment of human spinal cord precursor cells to the neuronal or glial phenotype in development. J. Comp. Neurol. **310:** 285-299.
67. Yachnis, A. T., L. B. Rorke, V. M.-Y. Lee & J. Q. Trojanowski. 1993. Expression of neuronal and glial polypeptides during histogenesis of the human cerebellar cortex including observations on the dentate nucleus. J. Comp. Neurol. **334:** 356-369.
68. Goedert, M., E. S. Cohen, R. Jakes & P. Cohen. 1992. p42 MAP kinase phosphorylation sites in microtubule-associated protein tau are dephosphorylated by protein phosphatase 2A1. FEBS Lett. **312:** 95-99.
69. Hanger, D. P., K. Hughes, J. R. Woodgett, J.-P. Brion & B. H. Anderton. 1992. Glycogen synthase kinase-3 induces Alzheimer's disease-like phosphorylation of tau: Generation of paired helical filament epitopes and neuronal localisation of the kinase. Neurosci. Lett. **147:** 58-62.
70. Ledesma, M. D., L. Correas, J. Avila & J. Diaz-Nido. 1992. Implication of brain cdc2 and MAP2 kinases in the phosphorylation of tau protein in Alzheimer's disease. FEBS Lett. **308:** 218-224.
71. Mandelkow, E. M., G. Drewes, J. Biernat, N. Gustke, J. Van Lint, J. L. Vandenhaude & E. Mandelkow. 1992. Glycogen synthase-3 and the Alzheimer-like state of microtubule-associated protein tau. FEBS Lett. **394:** 315-327.
72. Robertson, J., T. L. F. Loviny, M. Goedert, R. Jakes, K. J. Murray, B. H. Anderton & D. P. Hanger. 1993. Phosphorylation of tau by cyclic-AMP-dependent protein kinase. Dementia **4:** 256-263.
73. Steiner, B., E. M. Mandelkow, J. Biernat, N. Gustke, H. E. Meyer, B. Schmidt, G. Mieskes, H. D. Soling, D. Drechsel, M. W. Kirschner, M. Goedert & E. Mandelkow. 1990. Phosphorylation of microtubule-associated protein tau. Identification of the site for Ca^{2+}-calmodulin dependent kinase and relationship with tau phosphorylation in Alzheimer's tangles. EMBO J. **9:** 3539-3544.
74. Trojanowski, J. Q., M. Mawal-Dewan, M. L. Schmidt, J. Martin & V. M.-Y. Lee. 1993. Localization of the mitogen activated protein kinase ERK2 in Alzheimer's disease neurofibrillary tangles and senile plaque neurites. Brain Res. **618:** 333-337.
75. Vulliet, R., S. M. Halloran, R. K. Braun, A. J. Smith & G. Lee. 1992. Proline-directed phosphorylation of human tau protein. J. Biol. Chem. **267:** 22570-22574.
76. Harris, K. A., G. A. Oyler, G. M. Doolittle, I. Vincent, R. A. W. Lehman, R. L. Kincaid & M. L. Billingsley. 1993. Okadaic acid induces hyperphosphorylated forms of tau protein in human brain slices. Ann. Neurol. **33:** 77-87.
77. Ghanbari, H. A., B. E. Miller, H. J. Haigler, M. Arato, G. Bissette, P. Davies, C. B. Nemeroff, E. K. Perry, R. Perry, R. Ravid, D. F. Swaab, W. O. Whetsell & F. P. Zemlan. 1990. Biochemical assay of Alzheimer's disease-associated protein(s) in human brain tissue: A clinical study. JAMA **263:** 2907-2910.
78. Drechsel, D. N., A. A. Hyman, M. H. Cobb & M. Kirschner. 1992. Modulation of the dynamic instability of tubulin assembly by the microtubule-associated protein tau. Mol. Biol. Cell **3:** 1147-1154.
79. Shin, R.-W., G. T. Bramblett, V. M.-Y. Lee & J. Q. Trojanowski. 1993. Alzheimer disease A68 proteins injected into rat brain induce co-deposits of β-amyloid, ubiquitin and α1-antichymotrypsin. Proc. Natl. Acad. Sci. USA **90:** 6825-6828.
80. Li, Y.-T., D. Woodruff-Pak & J. Q. Trojanowski. 1994. Amyloid plaques in cerebellar cortex and the integrity of Purkinje cell dendrites. Neurobiol. Aging **15:** 1-9.

81. Standaert, D. G., V. M.-Y. Lee, B. D. Greenberg, D. E. Lowery & J. Q. Trojanowski. 1991. Molecular features of hypothalamic plaques in Alzheimer's disease. Am. J. Pathol. **139:** 681-691.
82. Masliah, E., A. Miller & R. D. Terry. 1993. The synaptic organization of the neocortex in Alzheimer's disease. Med. Hypotheses **41:** 334-340.
83. Greenberg, B. D., F. J. Keszdy & R. Kisilevsky. 1991. Amyloidosis as a therapeutic target in Alzheimer's disease. Ann. Rep. Med. Chem. **26:** 229-238.
84. Jarrett, J. T. & P. T. Lansbury, Jr. 1993. Seeding "one-dimensional crystallization" of amyloid: A pathogenic mechanism in Alzheimer's disease and scrapie. Cell **73:** 1055-1058.
85. Perry, G. 1993. Neuritic plaques in Alzheimer disease originate from neurofibrillary tangles. Med. Hypotheses **40:** 257-258.
86. Snow, A. D. & T. N. Wight. 1989. Proteoglycans in the pathogenesis of Alzheimer's disease and other amyloidoses. Neurobiol. Aging **10:** 481-497.
87. Hansen, L., D. Salmon, D. Galasko, E. Masliah, R. Katzman, R. DeTeresa, L. Thal, M. M. Pay, R. Hofstetter, M. Klauber, V. Rice, N. Butters & M. Alford. 1990. The Lewy body variant of Alzheimer's disease: A clinical and pathologic entity. Neurology **40:** 1-8.
88. Ince, P., D. Irving, F. MacArthur & R. H. Perry. 1991. Quantitative neuropathological study of Alzheimer-type pathology in the hippocampus: Comparison of senile dementia of Alzheimer type, senile dementia of Lewy body type, Parkinson's disease and non-demented elderly control patients. J. Neurol. Sci. **106:** 142-152.
89. Dickson, D. W., P. Davies, R. Mayeux, H. Crystal, D. S. Horoupian, A. Thompson & J. E. Goldman. 1987. Diffuse Lewy body disease: Neuropathological and biochemical studies of six patients. Acta Neuropathol. **75:** 8-15.
90. Dickson, D. W., H. Crystal, L. A. Mattiace, Y. Kess, A. Schwager, H. Ksiezak-Reding, P. Davies & S.-H. Yen. 1989. Diffuse Lewy body disease: Light and electron microscopic immunocytochemistry of senile plaques. Acta Neuropathol. **78:** 572-584.
91. Gibb, W. R. G., M. M. Esiri & A. J. Lees. 1985. Clinical and pathological features of diffuse cortical Lewy body disease (Lewy body dementia). Brain **110:** 1131-1153.
92. Perry, R. H., D. Irving, G. Blessed, A. Fairbairn & E. K. Perry. 1990. Senile dementia of Lewy body type: A clinically and neuropathologically distinct form of Lewy body dementia in the elderly. J. Neurol. Sci. **95:** 119-139.
93. Gibb, W. R. G. & A. J. Lees. 1989. The significance of the Lewy Body in the diagnosis of idiopathic Parkinson's disease. Neuropathol. Appl. Neurobiol. **15:** 27-44.
94. Hill, W. D., Jr., V. M.-Y. Lee, H. I. Hurtig, J. M. Murray & J. Q. Trojanowski. 1991. Epitopes located in spatially separate domains of each neurofilament subunit are present in Parkinson's disease Lewy bodies. J. Comp. Neurol. **109:** 150-160.
95. Pollanen, M. S., C. Bergeron & L. Weyer. 1992. Detergent-insoluble cortical Lewy body fibrils share epitopes with neurofilament and tau. J. Neurochem. **58:** 1953-1956.
96. Pollanen, M. S., C. Bergeron & L. Weyer. 1993. Deposition of detergent resistant neurofilaments into Lewy body fibrils. Brain Res. **603:** 121-124.
97. Goldman, J. E., S.-H. Yen, F.-C. Chiu & N. S. Peress. 1983. Lewy bodies of Parkinson's disease contain neurofilament antigens. Science **221:** 1082-1084.
98. Bancher, C., H. Lassman, H. Budka, K. Jellinger, I. Grundke-Iqbal, K. Iqbal, G. Wiche, F. Seitelberger & H. M. Wisniewski. 1989. An antigenic profile of Lewy bodies: Immunocytochemical indication for protein phosphorylation and ubiquitination. J. Neuropathol. Exp. Neurol. **48:** 81-93.
99. Galloway, P. G., I. Grundke-Iqbal, K. Iqbal & G. Perry. 1988. Lewy bodies contain epitopes both shared and distinct from Alzheimer neurofibrillary tangles. J. Neuropathol. Exp. Neurol. **47:** 654-663.
100. Pappolla, M. A. 1986. Lewy bodies of Parkinson's disease. Arch. Pathol. Lab. Med. **110:** 1160-1163.

101. SCHMIDT, M. L., J. M. MURRAY, V. M.-Y. LEE, D. W. HILL, A. WERTKIN & J. Q. TROJANOWSKI. 1991. Epitope map of neurofilament protein domains in cortical and peripheral nervous system Lewy bodies. Am. J. Pathol. **139:** 681-691.
102. NIXON, R. A. 1993. The regulation of neurofilament protein dynamics by phosphorylation: Clues to neurofibrillary pathobiology. Brain Pathol. **3:** 29-38.
103. DICKSON, D. W., M. L. SCHMIDT, V. M.-Y. LEE, M.-L. ZHAO, S.-H. YEN & J. Q. TROJANOWSKI. 1994. Immunoreactivity profile of hippocampal CA2/3 neurites in diffuse Lewy body disease. Acta Neuropathol. **87:** 269-276.
104. IWATSUBO, T., I. NAKANO, K. FUKUNAGA & E. MIYAMOTO. 1991. Ca^{2+}/calmodulin-dependent protein kinase II immunoreactivity in Lewy bodies. Acta Neuropathol. **82:** 159-163.
105. HILL, W. D., J. A. COHEN, M. ARAI & J. Q. TROJANOWSKI. 1993. Neurofilament protein mRNA is reduced in Parkinson's disease substantia nigra pars compacta neurons. J. Comp. Neurol. **329:** 328-336.
106. COTE, F., J.-F. COLLARD & J.-P. JULIEN. 1993. Progressive neuronopathy in transgenic mice expressing the human neurofilament heavy gene: A mouse model of amyotrophic lateral sclerosis. Cell **73:** 35-46.
107. XU, Z., L. C. CORK, J. W. GRIFFIN & D. W. CLEVELAND. 1993. Increased expression of neurofilament subunit NF-L produces morphological alterations that resemble the pathology of human motor neuron disease. Cell **73:** 23-33.
108. ITOH, T., G. SOBUE, E. KEN, T. MITSUMA, A. TAKAHASHI & J. Q. TROJANOWSKI. 1992. Phosphorylated high molecular weight neurofilament protein in the peripheral motor, sensory and sympathetic neuronal perikarya: System-dependent normal variations and changes in amyotrophic lateral sclerosis and multiple system atrophy. Acta Neuropathol. **83:** 240-245.
109. SCHMIDT, M. L., M. J. CARDEN, V. M.-Y. LEE & J. Q. TROJANOWSKI. 1987. Phosphate dependent and independent neurofilament epitopes in the axonal swellings of patients with motor neuron disease and controls. Lab. Invest. **56:** 282-294.
110. SOBUE, G., Y. HASHIZUMA, T. YASUDA, E. MUKAI, T. KUMAGAI, T. MITSUMA & J. Q. TROJANOWSKI. 1990. Phosphorylated high molecular weight neurofilament protein in lower motor neurons in amyotrophic lateral sclerosis and other neurodegenerative diseases involving ventral horn cells. Acta Neuropathol. **79:** 402-408.

Alzheimer's Disease: Membrane-Associated Metabolic Changes[a]

R. J. McCLURE,[b] J. N. KANFER,[e]
K. PANCHALINGAM,[b] W. E. KLUNK,[b] AND
J. W. PETTEGREW[b–d,f]

Departments of [b]Psychiatry, [c]Neurology, and
[d]Health Services Administration
University of Pittsburgh
Pittsburgh, Pennsylvania 15213

[e]Department of Biochemistry and Molecular Biology
University of Manitoba
Winnipeg, Manitoba, Canada R3E OW3

The neuropathological certification of Alzheimer's disease (AD) is based upon a greater density of senile plaques and neurofibrillary tangles in the cerebral cortex of the brain in AD compared to appropriate controls.[1] However, these pathological observations are neither specific to AD nor correlate with the clinical severity of the disease. It has been suggested that reduction in cortical synaptic densities correlates with cognitive deficits.[2,3] The most characteristic neurochemical hallmark of AD is reduced acetylcholine and choline acetyltransferase, the enzyme responsible for acetylcholine biosynthesis.[4,5] Extensive studies of the role of the acetylcholine deficit in the pathophysiology of AD and, more recently, the role of β amyloid, the main component of senile plaques, have been performed. However, the fundamental molecular defect responsible for AD has not been clearly defined.

A diversity of changes have been reported in the brain in AD including changes in membrane structure, lipid content or composition, neurotransmitter receptors, and enzyme activities associated with different subcellular organelles, ion pumps, and transporters. The complexity and diversity of these changes have hindered the development of a unifying theory for the pathophysiology of AD. Alterations of lipids, the only structural membrane component common to cells or subcellular organelles, may represent the basis for the myriad of changes, including the reduction of energetics, seen in AD. This review will focus on changes in brain tissue in AD with the view that membrane alterations may underlie the abundance of reported, unrelated dysfunctions in AD.

Brain plasma membranes appear to be the most vulnerable of the various cellular organelles to normal aging,[6] and brain aging can be characterized as a gradual decline

[a]This work was supported in part by National Institutes of Health grants AG08371, AG08974, AG50133, and AG9017.

[f]Address for correspondence: Jay W. Pettegrew, MD, Room A741 Crabtree Hall, 130 DeSoto Street, Pittsburgh, PA 15213.

in the ability to maintain normal membrane function. *In vivo* ^{31}P magnetic resonance spectroscopy (MRS) studies of normal volunteers reveal decreasing levels of phosphomonoesters (PME) and increasing levels of phosphodiesters (PDE) in individuals over the age of 50 years.[7] Alterations in phosphate esters reflect alterations in the metabolism of membrane phospholipids. These findings probably correlate with the loss of neuritic processes during normal aging. Similar changes have been observed by *in vitro* ^{31}P MRS techniques in aging rats.[8]

In vitro ^{31}P MRS studies of the brain in AD demonstrate elevated PME levels early in the course of the disease followed by elevations in the levels of PDE in later stages of the disease.[9–13] These changes in phospholipid metabolites indicate an alteration in phospholipid metabolism in AD that is distinct from that in normal aging. Altered membrane properties and function may be implicated in the pathophysiology of AD.

Changes in mitochondrial energy metabolism evidenced by changes in metabolic levels of phosphocreatine (PCr), adenosine diphosphate (ADP), and adenosine triphosphate (ATP) detected by *in vivo* ^{31}P MRS[14] also may play an important role in the pathophysiology of AD.[15] Energy-deprived neurons are much more susceptible to neurotoxic stress.[16] Alterations in mitochondrial membranes could lead to changes in high-energy phosphate metabolism and eventually to synaptic degeneration. We suggest that alterations in membrane metabolism, structure, and function contribute to the pathophysiology of AD. In this report, information is reviewed concerning membrane and energy-associated alterations in AD.

ALTERATIONS IN ENERGY AND MEMBRANE PHOSPHOLIPID METABOLITES IN ALZHEIMER'S DISEASE

^{31}P magnetic resonance spectroscopy is an excellent technique to evaluate the status of energy and phospholipid metabolism in brain. Magnetic resonance spectroscopy studies of extracts of autopsy tissue offer the advantage of quantitating a wide range of metabolites with a single procedure without the need for special derivatizations which may increase experimental variability. Phosphocreatine, ATP, and inorganic orthophosphate (Pi) levels are a reflection of energy metabolism, whereas PME, PDE, and phospholipid content are a reflection of phospholipid metabolism. ^{31}P magnetic resonance spectroscopy studies of brain tissue extracts provide an excellent technique to determine the concentration of these metabolites and to assess the status of energy and phospholipid metabolism in AD. The *in vitro* MRS analyses discussed in this report were performed on either hydrophilic compounds present in perchloric acid extracts[12] or hydrophobic compounds present in organic solvent extracts[17] of human autopsy brain tissues. The results have generally been expressed in mole percent of individual compounds. A systematic comparison of experimental data expressed as mole percent and μmole per gram based on an internal standard yields essentially identical results (FIG. 1).[18] *In vivo* ^{31}P MRS studies provide a noninvasive and safe technique to detect and monitor changes in energy and membrane metabolism in longitudinal clinical studies of the progression of AD.

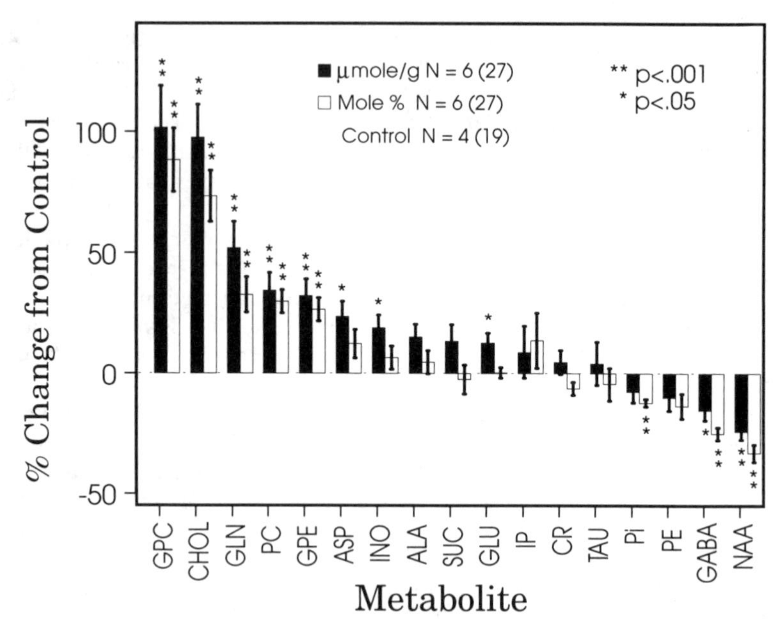

FIGURE 1. Comparison of the % change of metabolites in Alzheimer's brain perchloric acid extracts with that of age-matched control brain expressed in units of mole % (*open bars*) and µmole/g (*solid bars*). Abbreviations: ALA = alanine; ASP = aspartate; CHOL = choline; CR = creatine; GABA = γ-aminobutyric acid; GLN = glutamine; GLU = glutamate; GPC = glycerophosphocholine; GPE = glycerophosphoethanolamine; INO = myoinositol; IP = phosphoinositol; NAA = *N*-acetyl-L-aspartate; PC = phosphocholine; PE = phosphoethanolamine; Pi = orthophosphate; SUC = succinate; and TAU = taurine. (Adapted for use, with kind permission from Klunk *et al.*,[18] Elsevier Science Ltd, The Boulevard, Langford Lane, Kidlington 0X5 1GB, UK.)

^{31}P Magnetic Resonance Spectroscopy of Hydrophilic Compounds in Alzheimer's Disease Autopsy Brain

Phosphomonoesters

Increased amounts of PME and PDE and decreased Pi were found in AD samples compared to controls.[9–13,19] It was subsequently reported that the mole percent of PME was negatively correlated and the mole percent of PDEs was positively correlated with the number of senile plaques, but not neurofibrillary tangles[20] (FIG. 2). This increase in acknowledged phospholipid precursors (PME) and catabolites (PDE) may be an indication of accelerated membrane phospholipid metabolism in AD brain. The spectra for these earlier studies were acquired with a 4.7-Tesla, 200-MHz instrument which cannot completely resolve the individual PMEs and PDEs present in the samples. The availability of a 11.7-Tesla, 500-MHz instrument provides the capability to analyze the individual components of the PME region of the spectrum.[18] The results of an *in vitro* ^{31}P and ^{1}H MRS study[18] of PCA extracts of AD and control

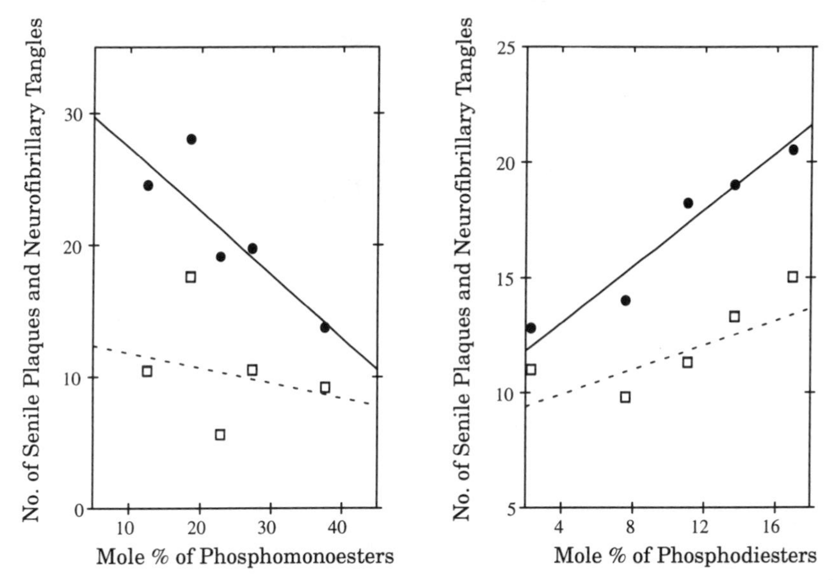

FIGURE 2. Plots of mole percentage of phosphomonoesters and phosphodiesters versus quantitative counts of senile plaques (*circles*) and neurofibrillary tangles (*squares*). (Reprinted, with permission from the American Medical Association, from Pettegrew *et al.*[20])

brain at 11.7 Tesla are shown in FIGURE 1. The PMEs present are phosphocholine (PC), phosphoethanolamine (PE), and inositol-1-phosphate (IP) with a statistically significant elevation in the levels of PC in the brain in AD compared to controls (FIG. 1). This analysis also shows slightly decreased PE and IP levels in the brain in AD. The net effect, however, is an increased level of PME. Reduced levels of PE in AD brain have previously been reported.[21,22] It is generally accepted that PC and PE are intermediates in PtdC and PtdE biosynthesis, respectively, as shown in FIGURE 3. PC, PE, and IP also are catabolic products of phospholipids (FIG. 3). Given the elevation of PC and the decrease in PE (FIG. 1) and the role that these intermediates play in lipid biosynthesis, PtdC biosynthesis would be expected to be maintained at a steady-state level in the brain in AD (FIG. 3 [2]), but PtdE biosynthesis would be expected to decrease in AD brain (FIG. 3). Although increased biosynthesis of PtdC should elevate or maintain PtdC levels in the brain in AD, catabolic pathways depleting PtdC (FIG. 3 [1 and 2]) probably predominate because phospholipid levels are reported to be decreased in AD.[23]

Alterations in phospholipid content of the brain in AD are complex. Lipid analyses were conducted on four gray matter regions and one white matter region of age-matched controls and type I AD (early onset, mean age of disease onset before 65 years of age) and type II AD (late onset, mean age of onset in the late 70s and 80s). Depending on the particular brain region, there was a 13–26% phospholipid reduction in the gray matter region of type I AD brain samples but none in type II AD brain samples. The reverse was observed for white matter analysis in which there was a 15% phospholipid reduction in type II AD brain samples but no differences in type

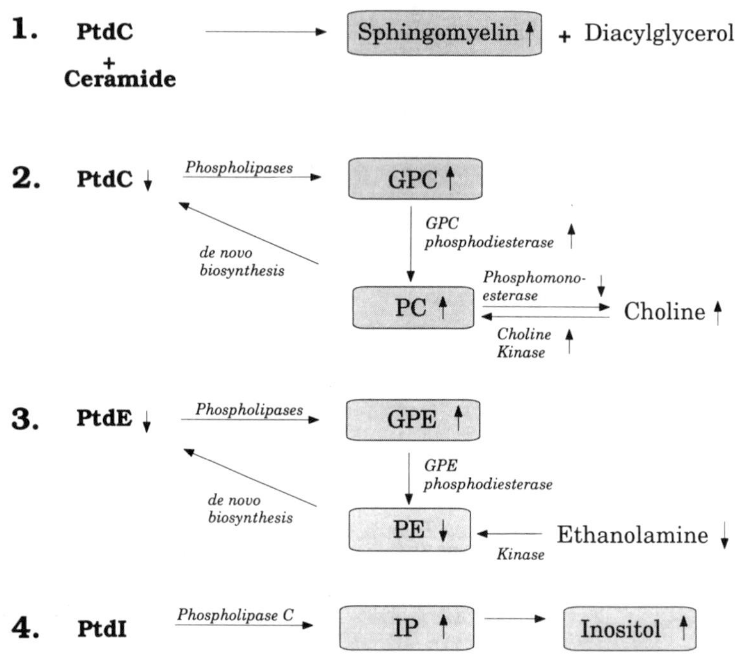

FIGURE 3. Metabolism of phospholipids. The highlighted metabolites were measured by magnetic resonance spectroscopy. (1) A pathway that would increase sphingomyelin at the expense of PtdC. (2) A pathway resulting in elevated GPC and PC levels. The *de novo* biosynthesis of PtdC from PC apparently does not compensate for the increased degradation of PtdC to GPC. (3) A pathway showing the degradation of PtdE to form GPE and PE. (4) The breakdown pathway of PtdI to yield IP and finally inositol. Elevations of choline kinase and GPC phosphodiesterase and decreased phosphomonoesterase activity result in elevated PC. Abbreviations: GPC = glycerophosphocholine; GPE = glycerophosphoethanolamine; IP = inositol phosphate; PC = phosphocholine; PE = phosphoethanolamine; PtdC = phosphatidylcholine; PtdE = phosphatidylethanolamine; and PtdI = phosphatidylinositol.

I AD brain samples.[24] By examining the differences in ganglioside content and cholesterol/phospholipid ratios of gray and white matter in these samples, the authors suggest that synapse loss predominates in AD type I and demyelination is responsible for lipid changes in AD type II.[24]

Levels of myoinositol were elevated (FIG. 1) in the brain in AD.[18] Elevations of myoinositol and IP are commonly regarded as arising from increased phospholipase C (PLC) hydrolysis of phosphatidyl inositol (PtdI) (FIG. 3 [4]). In addition, increased brain inositol in AD patients was observed with *in vivo* ^{1}H MRS.[25] An increase in IP and inositol in AD may indicate an accelerated breakdown of PtdI in AD. The release of calcium from its intracellular stores is stimulated by inositol 1,4,5 triphosphate (IP_3), one of the hydrolytic products of PLC activity. It is suggested that in AD, disruption of ATP homeostasis would compromise intracellular Ca^{2+} homeostasis through reduced functioning of Ca^+-ATPase pumps. This circumstance would be aggravated by increased release of Ca^{2+} from intracellular stores which are principally located in the endoplasmic reticulum. Increased cytosolic-free Ca^{2+} is seen in lymphocytes[26] from AD patients, and decreased neuronal calbindin mRNA[27] and reduced numbers of calbindin immunostained neurons are seen in the nucleus basalis of Meynert of the brain in AD.[28]

Phosphodiesters

Statistically significant elevations are seen for PDEs glycerophosphorylcholine (GPC) and glycerophosphoethanolamine (GPE) in the AD brain versus controls[18] (FIG. 1). These ^{31}P MRS findings are supported by other studies using a combination of chromatographic techniques and enzymatic assays that also report elevated PDE levels in AD versus controls.[29-31] The increase of GPC of 0.79 μmol/g is twice as great as the increase of GPE of 0.38 μmol/g. PDEs are generally regarded as catabolic breakdown products of phospholipids. The increased GPC (FIG. 3 [2]) may reflect an accelerated breakdown of membrane PtdC in AD.[32] Increased levels of GPE and decreased levels of PE suggest that attempts to maintain steady-state levels of PtdE (FIG. 3 [3]) by the AD brain are unsuccessful. Abnormal phospholipid brain metabolism could be an early ''molecular trigger'' in AD.

^{31}P Magnetic Resonance Spectroscopy of Membrane Phospholipids in the AD Brain at Autopsy

A representative ^{31}P MRS spectrum of human brain lipid extract, presented in FIGURE 4, shows prominent peaks for phosphatidylcholine (PtdC), phosphatidylethanolamine (PtdE), phosphatidylserine (PtdS), and sphingomyelin, with small amounts of PtdI, cardiolipin, and phosphatidic acid (PtdA). A comparison of AD brain samples with low or high senile plaque counts to control samples shows a consistent elevation of cardiolipin and sphingomyelin and a decreased phosphatidyl choline plasmalogen content (Panchalingam and Pettegrew, unpublished results; FIG. 5). The increased sphingomyelin content may reflect an attempt to compensate for the reduction in PtdC (FIG. 3 [1]).

The marked increase in cardiolipin in the brain in AD may be associated with possible mitochondrial alterations. Although small amounts of cardiolipin may be present in the endoplasmic reticulum, this phospholipid is regarded as a mitochondrial marker and is localized predominantly in the inner mitochondrial membrane.[33] Animal studies, such as experimental ischemia in rats resulting in reduced cardiolipin and mitochondrial respiratory enzyme activity[34,35] and the reduction of cardiolipin content with muscle denervation,[36] indicate that an increase in cardiolipin levels reflects increased mitochondrial content. The increased cardiolipin reported in AD may reflect increased levels of mitochondria. Also, ubiquinone, a neutral lipid that serves as a carrier in the electron-transport chain and is primarily concentrated in the mitochondria, is significantly more abundant in the brain in AD than in controls.[37] This is in contrast to the reduction in ubiquinone levels in normal aging. However, a report indicates that there is no increase in the number of mitochondria in the cerebral cortex of the brain in AD,[38] and ^{31}P MRS *in vivo* studies of AD patients to be discussed

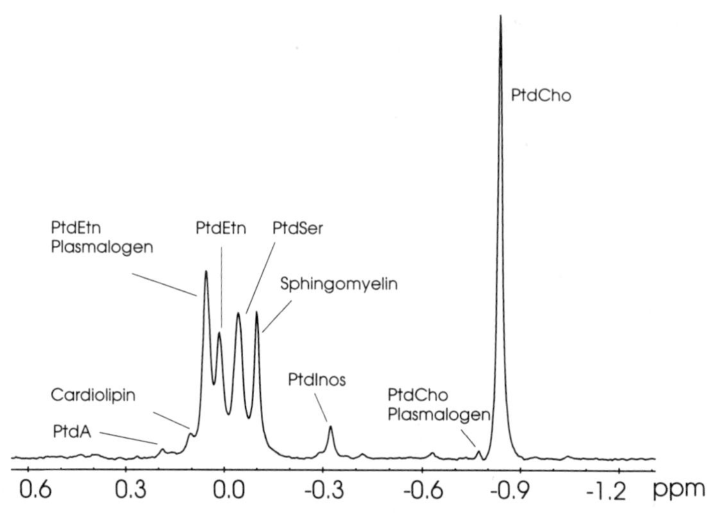

FIGURE 4. ^{31}P magnetic resonance spectroscopy spectrum of a Folch extract of human brain. Spectrum was taken with an 11.7-Tesla, 500-MHz Bruker at room temperature. ^{31}P MRS spectrum at 11.7 Tesla of a Folch extract of human brain. Abbreviations: PtdA = phosphatidic acid; PtdCho = phosphatidylcholine; PtdEtn = phosphatidylethanolamine; PtdInos = phosphatidylinositol; and PtdSer = phosphatidylserine.

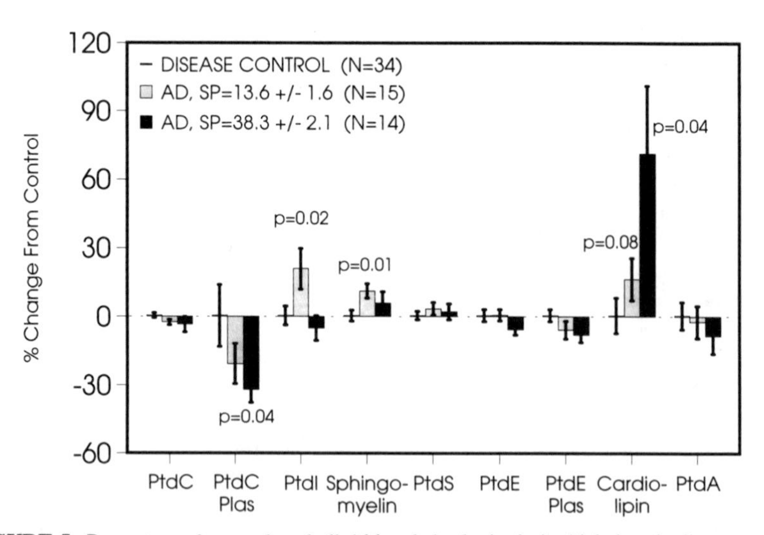

FIGURE 5. Percentage change phospholipid levels in the brain in Alzheimer's disease versus controls. Mole percent phospholipids obtained from ^{31}P MRS of a Folch extract of human brain. Abbreviations: PtdA = phosphatidic acid; PtdCho = phosphatidylcholine; PtdEtn = phosphatidylethanolamine; PtdInos = phosphatidylinositol; and PtdSer = phosphatidylserine.

do not indicate an increase in mitochondrial function. Therefore, the significance of increased levels of cardiolipin and ubiquinone in the brain in AD is not clear and may represent an increase in nonfunctional mitochondria in localized brain areas.

In Vivo ^{31}P *Magnetic Resonance Spectroscopy*

Phospholipid Metabolites

The alterations in phospholipid metabolism detected in the *in vitro* ^{31}P MRS examination of postmortem AD tissue are also detectable by *in vivo* ^{31}P MRS.[14,39] An *in vivo* ^{31}P MRS investigation of the frontal and temporoparietal regions of patients with AD, patients with multiple subcortical cerebral infarction (MSID), and age-matched controls found significant elevations of PME levels in the temporoparietal region of the brain in AD compared to either controls or patients with MSID.[39] The resonances of individual PMEs are not resolved in the *in vivo* spectra. A recent ^{31}P *in vivo* MRS longitudinal study of the prefrontal cortex of five mildly demented and 9 moderately demented patients with AD and 21 controls demonstrates that PME levels are significantly elevated in AD patients clinically characterized as mildly demented based upon their Mattis scores.[14] As the mildly demented AD patients became moderately demented, the PME levels decreased to levels obtained for controls. The PME levels significantly correlate with the Mattis score, which suggests that PME levels are highest at the onset of dementia. Other *in vivo* ^{31}P MRS studies of AD brain regions from whole axial cross-sectional slices through the corpus callosal white matter and lateral cerebral ventricles found no significant differences in the concentrations of phospholipid membrane metabolites.[40,41] This inconsistency probably arises from differences in technique and the regions of the AD examined, that is, a smaller volume of predominantly gray matter in the prefrontal cortex[14] versus a cross-sectional axial slice through a larger volume of predominantly white matter and cerebrospinal fluid.[40,41] Patients with AD and the controls also were not age matched in the ^{31}P MRS investigation of the axial slice.[40] Although PME levels were elevated by the same relative amount in the axial slice, a higher variance of the experimental data prevented this increase from reaching statistical significance. The elevations of GPE and GPC observed in the *in vitro* autopsy brain studies are not observed in the *in vivo* ^{31}P MRS studies. Strong contributions from long correlation time components present in the *in vivo* PDE peak arise from phospholipids present in membrane structures. These long correlation time components may mask changes in the levels of the low molecular weight, less abundant, water-soluble PDEs that have short correlation times. Therefore, it is not possible to distinguish between signals from PDEs and those from phospholipids by *in vivo* ^{31}P MRS.

High-Energy Metabolites

In vivo ^{31}P MRS offers a unique opportunity to measure energy metabolism in the brain. Characteristic high-energy metabolites cannot be measured in autopsy brain tissues because they are rapidly degraded postmortem. An *in vivo* ^{31}P MRS study of the frontal and temporoparietal regions of patients with AD showed a significant decrease in the PCr/Pi ratio as compared to those of MSID patients and controls.[39] A more recent longitudinal *in vivo* ^{31}P MRS study of the prefrontal cortex of patients with AD shows that PCr levels are decreased in mildly demented AD patients.[14] The PCr levels correlated significantly with the Mattis score for AD patients, showing that levels of PCr increased as the dementia worsened. The decrease in PCr in mildly demented patients with AD probably reflects a greater demand to replenish ATP levels in an effort to maintain this crucial energy source.[42,43] A significant decrease in ADP levels also is indicated in mildly demented patients with AD; levels of ATP showed no change in AD brain compared to that in controls. V/V_{max}, an expression of the oxidative metabolic rate for mitochondria calculated from the Pi/PCr ratio, decreases as dementia worsens. These results taken together suggest that both of the immediate precursors of ATP are diminished early in the brain in AD which is under oxidative stress as indicated by the elevated V/V_{max} levels. Given the high metabolic rate of the brain, this decrease in energy availability may directly lead to neuronal dysfunction.

Reduced energy metabolism in the brain in AD demonstrated by *in vivo* ^{31}P MRS findings is supported by other studies. Significant reductions of cytochrome oxidase (Complex 4) activity from 17-26% were found in the temporal and frontal cortices of AD samples but not in other brain regions.[44] The ratio of oxygen uptake was significantly reduced[45] in homogenates of AD brain in the presence or absence of ADP. The results of these studies are strong indications of reduced energy metabolism in AD brain. This is in contrast to the significant increase in cardiolipin, a putative mitochondrial marker, found in the brain in AD. One interpretation is that the apparent increase in mitochondria may reflect a futile, nonfunctional compensatory attempt by the brain in AD to maintain adequate oxidative phosphorylation to sustain its energy requirements.

Clinical studies provide additional evidence of a possible role of energetic stress in AD. A study of autopsy brain tissue of patients with coronary artery disease showed an increased number of senile plaques.[46] Also, women with a history of myocardial infarction have a fivefold increase in the incidence of AD.[47] These studies suggest that brain oxidative stress may contribute to the pathophysiology of AD.

CHANGES IN MEMBRANE STRUCTURE IN ALZHEIMER'S DISEASE

Small Angle X-ray Diffraction Studies

Small angle x-ray diffraction analysis of lipid extracts of membranes isolated from cortical gray matter of the superior temporal gyrus of patients with AD shows significant structural changes compared to those in controls. An average 4 Å reduction in the lipid bilayer width in AD brain samples was observed.[48] Analysis of the lipid composition of these samples revealed a 30% decrease in the cholesterol:phospholipid (C:PL) mole ratio in AD compared to controls. Increasing the cholesterol content of AD samples to the ratio observed in controls restored the membrane bilayer width and electron density profile to normal.[48] This indicates that the cholesterol deficit plays a major role in AD lipid membrane structure perturbation. There are conflicting reports describing both increases[49] and decreases[50] in total cholesterol in normal aging, depending on the brain regions.

In addition to the effect that the C:PL mole ratio has on bilayer width, it is also an important determinant of membrane fluidity. Even modest changes in neural plasma membrane cholesterol content can substantially modulate the activity of certain neurotransmitter receptors, including serotonin and opiate receptors.[51,52]

Fluorescence Studies of Brain Tissue in Alzheimer's Disease

Fluorescence studies of brain membrane samples from patients with AD using the 1,6-diphenyl-1,3,5-hexatriene (DPH) fluorophore did not show significant differences from controls,[53,54] indicating that the dynamics of the hydrophobic core of these membranes were not altered. However, fluorescence studies using the 1-[4-(trimethylamino)-phenyl]-6-diphenyl-1,3,5-hexatriene derivative of DPH indicate decreased motion at the lipid-aqueous interface of AD hippocampal brain membranes compared to that of controls.[54]

PHOSPHOLIPID ENZYMATIC ALTERATIONS IN ALZHEIMER'S DISEASE

The origin of the alteration in phospholipid metabolic levels in AD is not known. In this section, reported alterations in enzymatic activities or concentrations of enzymes involved in phospholipid metabolism in AD will be addressed. Phospholipase catalyzed hydrolysis of PtdC and phosphatidylinositol diphosphate (PIP_2) yields products that are involved in cell signaling through intricate modulation of Ca^{2+} concentration and kinase activities to evoke cellular responses.[55]

Phospholipase A_2

Phospholipase A_2, which hydrolyzes phospholipids to produce free fatty acids and lysophospholipids, shows similar activity in AD and control brain homogenates.[56]

Phospholipase C

Phospholipase C is an integral membrane protein that is part of the receptor-G-protein signal transduction system. The dystrophic neurites in senile plaques and the intraneural and extraneural neurofibrillary tangles of AD brain were immunostained with an antiphospholipase C δ antibody. Western blots of paired helical filaments isolated from AD brain also immunostained with this antibody.[57] Although this indicates that phospholipase C δ is associated with AD degradative processes, PI-specific phospholipase activity is reported to be unaltered in AD.[58]

Phospholipase D

Synaptosomal phospholipase D cleaves synaptosomal P+dC, releasing choline that can be used by synaptosomal ChAT to produce acetylcholine.[59] Activities of phospholipase D are decreased in AD autopsy brain homogenates as compared to those of controls.[60] This suggests that the release of choline from PtdC for acetylcholine formation is impaired in AD brain, possibly contributing to the characteristic cholinergic deficit.

Phosphodiesterase

The phosphomonoester PC represents another potential choline reservoir in the brain. The release of inorganic phosphate from PC was reduced with AD homogenates in the temporal but not the occipital brain regions compared to that in controls. No reductions in acid, neutral, or alkaline phosphatase activities were found in these samples.[61] Phosphodiesterase activity is increased in AD brain compared to that in controls. There was a 37% increase in the ability to hydrolyze GPC to produce PC in the brain in AD compared to that of controls. This activity inversely correlated with the number of neurofibrillary tangles. A very profound reduction was observed in choline kinase activity in the AD disease group.[56] Other investigators have observed that this GPC phosphodiesterase activity is similar in both AD and control brains.[30] However, a great difference exists in the absolute measured activities between these two studies that may be due to differences in the assay procedures.

Monoglyceride Lipase

Monoglyceride lipase activity was increased two- to fivefold in plasma and synaptosomal membrane fractions from the hippocampus and nucleus basalis of the brain in AD compared to that in controls. The diacylglycerol lipase activity of the nucleus basalis membranes prepared from the AD brain was fivefold greater than that of control brains. Soluble lysophospholipase activity was increased threefold in the hippocampal and nucleus basalis of the brain in AD.[57] These results suggest that lipid catabolism in AD brain membranes is increased compared to that in controls.

Phosphatidylinositol Kinase Activity

Phosphatidylinositol is converted to phosphatidylinositol monophosphate and PIP_2 by phosphatidylinositol kinase and phosphatidylinositol phosphate kinase, respectively. PIP_2 is the substrate for phospholipase C hydrolysis. Studies of phosphatidylinositol kinase and phosphatidylinositol phosphate kinase activities of postmortem brain tissues showed a significant reduction in phosphatidylinositol phosphate activity but no reduction in phosphatidylinositol activity in AD versus controls.[62] This may lead to inhibition of second messengers PIP_2 and diacylglycerol and may significantly alter the receptor-G protein-phospholipase C signal transduction.

ALTERATION OF MEMBRANE PROTEINS IN ALZHEIMER'S DISEASE

The biochemical transmission of an extracellular signal to evoke an intracellular response is an essential condition for inter- and intraneuronal communications. Neurotransmitters play a major role in this process by interacting with their transmembrane receptors to generate intracellular responses through either G protein-coupled receptors or ligand-gated ion channels. Glutamatergic and GABA-ergic transmembrane receptors are examples of the ligand-gated ion channels regulating Ca^{2+} and Cl^- ion flux, respectively, across the plasma membrane. Cholinergic, histaminergic, serotoninergic, norandrenergic, and dopaminergic transmembrane receptors are examples of the G protein-coupled systems that generate second messengers that in turn modulate the activities of protein kinases and the intracellular free Ca^{2+} concentration.[63] All the major protein components of the G protein-receptor-coupled systems reside in the cell membrane. Mechanisms of signal transduction in these systems probably involve conformational changes of proteins associated with each step in the system that ultimately produces second messengers. Because these conformational changes take place within the membrane matrix, changes in the physical or chemical properties of the membrane bilayer may alter the transduction of the extracellular neurotransmitter signal, causing inappropriate intracellular responses. The neurotransmitter receptors and their transmission systems in AD were recently reviewed by Greenamyre and Maragos.[64] None of these systems shows consistent, definitive changes associated with AD. Membrane structural and dynamic changes that alter membrane function, such as receptors, ion channels, antigens, enzymes, and components of the signaling systems, are involved in the pathogenesis of AD.

REFERENCES

1. Tomlinson, B. E. & J. A. N. Corsellis. 1984. Aging and the dementias. *In* Greenfield's Neuropathology. J. H. Adams, J. A. N. Corsellis & L. W. Duchen, Eds.: 951-1025. Wiley. New York.
2. DeKosky, S. T. & S. W. Scheff. 1990. Synaptic loss in frontal cortex biopsies in Alzheimer's disease: Correlation with cognitive severity. Ann. Neurol. **27:** 457-464.
3. Terry, R. D., E. Masliah, D. P. Salmon, N. Butters, R. DeTeresa, R. Hill, L. A. Hansen & R. Katzman. 1991. Physical basis of cognitive alterations in Alzheimer's disease: Synapse loss is the major correlate of cognitive impairment. Ann. Neurol. **30:** 572-580.
4. Marchbanks, R. M. 1982. Biochemistry of Alzheimer's disease. J. Neurochem. **39:** 9-15.
5. Collerton, D. 1986. Cholinergic function and intellectual decline in Alzheimer's disease. Neuroscience **19:** 1-28.
6. Bosman, G. J., I. G. Bartholomeus & W. J. de Grip. 1991. Alzheimer's disease and cellular aging: Membrane-related events as clues to primary mechanisms. Gerontology **37:** 95-112.
7. Panchalingam, K., J. W. Pettegrew, S. Strychor & M. Tretta. 1990. Effect of normal aging on membrane phospholipid metabolism by 31P in vivo NMR spectroscopy (abstract). Soc. Neurosci. (Abstr.) **16:** 843.
8. Pettegrew, J. W., K. Panchalingam, G. Withers, D. McKeag & S. Strychor. 1990. Changes in brain energy and phospholipid metabolism during development and aging in the Fischer 344 rat. J. Neuropathol. Exp. Neurol. **49:** 237-249.

9. PETTEGREW, J. W., N. J. MINSHEW, M. M. COHEN, S. J. KOPP & T. GLONEK. 1984. P-31 NMR changes in Alzheimer's and Huntington's disease brain (Abstr.). Neurology **34(suppl 1):** 281.
10. BARANY, M., Y. -C. CHANG, C. ARUS, T. RUSTAN & W. H. FREY. 1985. Increased glycerol-3-phosphorylcholine in post-mortem Alzheimer's brain [letter]. Lancet **1:** 517.
11. PETTEGREW, J. W., S. J. KOPP, N. J. MINSHEW, T. GLONEK, J. M. FELIKSIK, J. P. TOW & M. M. COHEN. 1987. ^{31}P nuclear magnetic resonance studies of phosphoglyceride metabolism in developing and degenerating brain: Preliminary observations. J. Neuropathol. Exp. Neurol. **46:** 419-430.
12. PETTEGREW, J. W., J. MOOSSY, G. WITHERS, D. MCKEAG & K. PANCHALINGAM. 1988. ^{31}P Nuclear magnetic resonance study of the brain in Alzheimer's disease. J. Neuropathol. Exp. Neurol. **47:** 235-248.
13. MIATTO, O., R. G. GONZALEZ, F. BUONANNO & J. GROWDON. 1986. *In vitro* ^{31}P NMR spectroscopy detects altered phospholipid metabolism in Alzheimer's disease. Can. J. Neurol. Sci. **13(suppl):** 535-539.
14. PETTEGREW, J. W., K. PANCHALINGAM, W. E. KLUNK, R. J. MCCLURE & L. R. MUENZ. 1994. Alterations of cerebral metabolism in probable Alzheimer's disease: A preliminary study. Neurobiol. Aging **15:** 117-132.
15. BEAL, M. F. 1992. Does impairment of energy metabolism result in excitotoxic neuronal death in neurodegenerative illnesses? Ann. Neurol. **31:** 119-130.
16. HENNEBERRY, R. C. 1989. The role of neuronal energy in the neurotoxicity of excitatory amino acids. Neurobiol. Aging **10:** 611-613.
17. MENESES, P. & T. GLONEK. 1988. High resolution ^{31}P NMR of extracted phospholipids. J. Lipid Res. **29:** 679-690.
18. KLUNK, W. E., C. J. XU, K. PANCHALINGAM, R. J. MCCLURE & J. W. PETTEGREW. 1994. Analysis of magnetic resonance spectra by mole percent: Comparison to absolute units. Neurobiol. Aging **15:** 133-140.
19. SMITH, C. D., L. G. GALLENSTEIN, W. J. LAYTON, R. J. KRYSCIO & W. R. MARKESABERY. 1993. ^{31}P magnetic resonance spectroscopy in Alzheimer's and Pick's disease. Neurobiol. Aging **14:** 85-92.
20. PETTEGREW, J. W., K. PANCHALINGAM, J. MOOSSY, J. MARTINEZ, G. RAO & F. BOLLER. 1988. Correlation of phosphorus-31 magnetic resonance spectroscopy and morphologic findings in Alzheimer's disease. Arch. Neurol. **45:** 1093-1096.
21. PERRY, T. L., V. W. YONG, C. BERGERON, S. HANSEN & K. JONES. 1987. Amino acids, glutathione, and glutathione transferase activity in the brains of patients with Alzheimer's disease. Ann. Neurol. **21:** 331-336.
22. ELLISON, D. W., M. F. BEAL & J. B. MARTIN. 1987. Phosphoethanolamine and ethanolamine are decreased in Alzheimer's disease and Huntington's disease. Brain Res. **417:** 389-392.
23. KIENZL, E., L. PUCHINGER, K. JELLINGER, H. STACHELBERGER & K. VARMUZA. 1993. Studies of phospholipid composition in Alzheimer's disease brain. Neurodegen. **2:** 101-109.
24. SVENNERHOLM, L. & C.-G. GOTTFRIES. 1994. Membrane lipids, selectively diminished in Alzheimer brains, suggest synapse loss as a primary event in early-onset form (Type I) and demyelination in late-onset form (type II). J. Neurochem. **62:** 1039-1047.
25. MILLER, B. L., R. A. MOATS, T. SHONK, T. ERNST, S. WOOLLEY & B. D. ROSS. 1993. Alzheimer's disease: Depiction of increased cerebral myo-inositol with proton MR spectroscopy. Radiology **187:** 433-437.
26. ADUNSKY, A., D. BARAM, M. HERSHKOWITZ & Y. A. MEKORI. 1991. Increased cytosolic free calcium in lymphocytes of Alzheimer patients. J. Neuroimmunol. **33:** 167-172.
27. IACOPINO, A. M. & S. CHRISTAKOS. 1990. Specific reduction of calcium-binding protein (28 Kda calbindin-D) gene expression in aging and neurodegenerated diseases. Proc. Natl. Acad. Sci. USA **87:** 4078-4082.

28. SUTHERLAND, M. K., L. WONG, M. J. SOMERVILLE, Y. K. YOONG, C. BERGERON, M. PARMENTUR & D. R. MALACHLAN. 1993. Reduction of calbindin-28K mRNA in Alzheimer's as compared to Huntington hippocampus. Brain Res. Mol. Brain Res. **18:** 32-42.
29. BLUSZTAJN, J. K., I. L. GONZALEZ-COVIELLA, M. LOGUE, J. H. GROWDON & R. J. WURTMAN. 1990. Levels of phospholipid catabolic intermediates, glycerophosphocholine and glycerophosphoethanolamine, are elevated in Alzheimer's disease but not in Down's syndrome patients. Brain Res. **536:** 240-244.
30. NITSCH, R. M., J. K. BLUSZTAJN, A. G. PITTAS, B. E. SLACK, J. H. GROWDON & R. J. WURTMAN. 1992. Evidence for a membrane defect in Alzheimer disease brain. Proc. Natl. Acad. Sci. USA **89:** 1671-1675.
31. NITSCH, R., A. PITTAS, J. K. BLUSZTAJN, B. E. SLACK, J. H. GROWDON & R. J. WURTMAN. 1991. Alterations of phospholipid metabolites in postmortem brain from patients with Alzheimer's disease. Ann. N.Y. Acad. Sci. **640:** 110-113.
32. PETTEGREW, J. W., W. E. KLUNK, R. J. MCCLURE, K. PANCHALINGAM & S. STRYCHOR. 1991. Alzheimer's disease: New treatment strategies. *In* Phosphomonoesters, Phospholipids and High-Energy Phosphates in Alzheimer's Disease: Alterations and Physiological Significance. Z. S. Khachaturian & J. P. Blass, Eds.: 193-212. Marcel Dekker Inc. New York.
33. CULLIS, P. R. & M. J. HOPE. 1991. Physical properties and functional roles of lipids in membranes. *In* Biochemistry of Lipids, Lipoproteins and Membranes. D. E. Vance & J. Vance, Eds.: 1-42. Elsevier Press. New York.
34. NAKAHARA, I., H. KIKUCHI, W. TAKI, S. NISHI, M. KITO, Y. YONEKAWA, Y. GOTO & N. OGOTA. 1992. Changes in major phospholipids of mitochondria during postischemic reperfusion in rat brain. J. Neurosurg. **76:** 244-250.
35. NAKAHARA, I., H. KIKUCHI, W. TAKI, S. NISHI, M. KITO, Y. YONEKAWA, Y. GOTO & N. OGATA. 1991. Degradation of mitochondrial phospholipids during experimental cerebral ischemia in rats. J. Neurochem. **57:** 839-844.
36. WICKS, K. L. & D. A. HOOD. 1991. Mitochondrial adaptations in denervated muscle: Relationship to muscle performance. Am. J. Physiol. **260:** C841-850.
37. SODERBERG, M., C. EDLUND, I. ALAFUZOFF, K. KRISTENSSON & G. DALLNER. 1992. Lipid composition in different regions of the brain in Alzheimer's disease/senile dementia of the Alzheimer's type. J. Neurochem. **59:** 1646-1653.
38. SUMPTER, P. Q., D. M. MANN, C. A. DAVIES, P. O. YATES, J. S. SNOWDEN & D. NEARY. 1986. An ultrastructural analysis of the effects of accumulation of neurofibrillary tangle in pyramidal neurons of the cerebral cortex in Alzheimer's disease. Neuropathol. Appl. Neurobiol. **12:** 305-319.
39. BROWN, G. G., S. R. LEVINE, J. M. GORELL, J. W. PETTEGREW, J. W. GDOWSKI, J. A. BUERI, J. A. HELPERN & K. M. WELCH. 1989. *In vivo* ^{31}P NMR profiles of Alzheimer's disease and multiple subcortical infarct dementia. Neurology **39:** 1423-1427.
40. BOTTOMLEY, P. A., J. P. COUSINS, D. L. PENDREY, W. A. WAGLE, C. J. HARDY, F. A. EAMES, R. J. MCCAFFREY & D. A. THOMPSON. 1992. Alzheimer dementia: Quantification of energy metabolism and mobile phosphoesters with P-31 NMR spectroscopy. Radiology **183:** 695-699.
41. MURPHY, D. G. M., P. A. BOTTOMLEY, J. SALERNO, W. WILLIAMS, M. B. SCHAPIRO, S. I. RAPOPORT, J. ALGER & B. HORWITZ. 1992. *In vivo* brain glucose and phosphorus metabolism in Alzheimer's disease. Soc. Neurosci. (Abstr.) **18:** 567.
42. CHANCE, B. 1989. Metabolic heterogeneities in rapidly metabolizing tissues. J. Appl. Cardiol. **4:** 207-221.
43. CHANCE, B. & B. SCHOENER. 1966. High and low energy states of cytochromes. J. Biol. Chem. **241:** 4577-4588.
44. KISH, S., C. BERGERON, A. RAJPUT, S. DOZIC, F. MASTROGRACOSNO, L. J. CHONG, J. M. WILSON, L. M. DISTEFANO & J. N. NOBREGIA. 1992. Brain cytochrome oxidase in Alzheimer's disease. J. Neurochem. **59:** 776-779.

45. SIMS, N. R., J. M. FINEGAN, J. P. BLASS, D. M. BOWEN & D. NEARY. 1987. Mitochondrial function in brain tissue in primary degenerative dementia. Brain Res. **436:** 30-38.
46. SPARKS, D. L., J. C. HUNSAKER, III, S. W. SCHEFF, R. J. KRYSCIO, J. L. HENSON & W. R. MARKESBERY. 1990. Cortical senile plaques in coronary artery disease, aging and Alzheimer's disease. Neurobiol. Aging **11:** 601-607.
47. ARONSON, M. K., W. L. OOI, H. MORGENSTERN, M. S. HAFNER, D. MASUR & H. CRYSTAL. 1990. Women myocardial infarction and dementia in the very old. Neurology **40:** 1102-1106.
48. MASON, R. P., W. J. SHOEMAKER, L. SHAJENKO, T. E. CHAMBERS & G. HERBETTE. 1992. Evidence for changes in the Alzheimer's disease brain cortical membrane structure mediated by cholesterol. Neurobiol. Aging **13:** 413-419.
49. ROUSER, G., G. KITCHENSKY, A. YAMAMOTO & C. F. BAXTER. 1972. Lipids in the nervous system of different species as a function of age. Adv. Lipid. Res. **10:** 261-360.
50. SODERBERG, M., K. EDLUND, K. KRISTENSSON & G. DALLNER. 1990. Lipid compositions of different regions of the human brain during aging. J. Neurochem. **54:** 415-423.
51. HERON, D. S., M. HERSHKOWITZ, M. SHINITZKY & D. SAMUEL. 1980. Lipid fluidity markedly modulates the binding of serotonin to mouse brain membranes. Proc. Natl. Acad. Sci. USA **77:** 7467-7563.
52. HERON, D. S., M. HERSHKOWITZ, M. SHINITZKY & D. SAMUEL. 1980. The lipid fluidity of synaptic membranes and the binding of serotonin and opiate ligands. *In* Neurotransmitters and Their Receptors. U. Z. Littauer, Y. Duadi, I. Silman, V. I. Teichberg & Z. Vogel, Eds.: 125-138. John Wiley. New York.
53. HAJIMOHAMMADREZA, I. & M. BRAMMER. 1992. Brain membrane fluidity and lipid peroxidation in Alzheimer's disease. Neurosci. Lett. **112:** 333-337.
54. ZUBENKO, G. S. 1986. Hippocampal membrane alterations in Alzheimer's disease. Brain Res. **385:** 115-121.
55. NISHIZUKA, Y. 1992. Intracellular signaling by hydrolysis of phospholipids and activation of phospholipids and activation of protein kinase C. Science **258:** 607-614.
56. KANFER, J. N., J. W. PETTEGREW, J. MOOSSY & D. G. MCCARTNEY. 1993. Alterations of selected enzymes of phospholipid metabolism in Alzheimer's disease brain tissue as compared to non-Alzheimer's disease controls. Neurochem. Res. **18:** 331-334.
57. SHIMOHAMA, S., Y. HOMMA, T. SUENAGA, S. FUJIMOTO, T. TANIGUCHI, W. ARAKI, Y. YAMAOKA, T. TAKENAWA & J. KIMURA. 1991. Aberrant accumulation of phospholipase C-delta in Alzheimer's brains. Am. J. Pathol. **139:** 737-742.
58. SHIMOHAMA, S., S. FUJIMOTO, T. TANIGUCHI & J. KIMURA. 1992. Phosphatidylinositol-specific phospholipase C activity in the postmortem human brain: No alteration in Alzheimer's disease. Brain Res. **579:** 347-349.
59. HATTORI, H. & J. N. KANFER. 1985. Synaptosomal phospholipase D potential role in providing choline for acetylcholine synthesis. J. Neurochem. **45:** 1578-1584.
60. KANFER, J. N., H. HATTORI & D. ORIHEL. 1986. Reduced phospholipase D activity in brain tissue samples from Alzheimer's disease patients. Ann. Neurol. **20:** 265-267.
61. KANFER, J. N. & D. G. MCCARTNEY. 1986. Reduced phosphorylcholine hydrolysis by homogenates of temporal regions of Alzheimer's brain. Biochem. Biophys. Res. Commun. **139:** 315-319.
62. JOLLES, J., J. BOTHMER, M. MARKERINK & R. RAVID. 1992. Phosphatidylinositol kinase is reduced in Alzheimer's disease. J. Neurochem. **58:** 2326-2329.
63. FOWLER, C. J., C. O'NEIL, A. GARLINK & R. F. COWBURN. 1990. Alzheimer's disease: Is there a problem beyond recognition? Trends Pharmacol. Sci. **11:** 183-184.
64. GREENAMYRE, T. & W. F. MARAGOS. 1993. Neurotransmitter receptors in Alzheimer disease. Cerebrovasc. & Brain Metab. Rev. **5:** 61-94.

Probing Membrane Bilayer Interactions of 1,4-Dihydropyridine Calcium Channel Blockers

Implications for Aging and Alzheimer's Disease

R. PRESTON MASON[a]

Neurosciences Research Center
Medical College of Pennsylvania, Allegheny Campus
Pittsburgh, Pennsylvania 15212-4772

BACKGROUND

Use of X-ray Diffraction to Examine Membrane Lipid Bilayer Structure. The use of x-ray diffraction to solve the structure of biological molecules and macromolecules at atomic resolution is well established. The protein and nucleic acid samples used for these studies are generally in the form of a highly ordered, crystalline solid. There are important instances, however, of applying x-ray diffraction techniques to gain molecular structure information from *non*crystalline samples that have a regular, periodic molecular arrangement. For example, an x-ray "diffraction pattern" from deoxynucleic acid (DNA) fibers in the β-form appears as a "helical cross" with reflections of 34 Å periodicity, corresponding to 10 nucleotide residues in each complete turn of the double helix, and reflections of 3.4 Å periodicity, which correspond to the average molecular distance between adjacent nucleotides. This information, obtained from Franklin and Wilkins, assisted Watson and Crick in proposing the double-helix model for DNA in 1953.

The noncrystalline membrane lipid bilayer has also been studied using small angle x-ray diffraction. Analogous to a liquid crystalline display, in which molecules are aligned by an electric field, the various lipid molecules (eg, phospholipid and unesterified cholesterol) that constitute the membrane are oriented in a bilayer by their amphipathic chemical properties. Specifically, the hydrophilic headgroup is in contact with the water phase, whereas the hydrophobic acyl chains point away from the water phase and interact with neighboring chains. This defined orientation is periodic when the membrane lipid bilayers are layered or "stacked" to form multibilayers. Membrane multibilayers can be formed experimentally from model and biological preparations; these structures coherently scatter x-rays in a "pattern" that is characteristic of the sample's molecular properties (i.e., membrane width, intermolec-

[a] Address for correspondence: Neurosciences Research Center, Medical College of Pennsylvania, 320 East North Avenue, Pittsburgh, PA 15212-4772.

ular acyl chain packing distance, and electron density distribution). This technique has been used to study the molecular structure of cerebral cortical synaptoneurosomal membranes in the presence and absence of an imidazobenzodiazepine analog.[1] In addition, we used this approach to examine changes in the structure of reconstituted lipid bilayers from Alzheimer's disease neural membranes relative to age-matched controls.[2]

The Membrane Mediates Interactions of Ca^{2+} Channel Blockers with Their Receptors. 1,4-Dihydropyridine (DHP) Ca^{2+} channel blockers modulate the transmembrane influx of Ca^{2+} in certain contractile and neuronal cells by specifically binding to voltage-sensitive Ca^{2+} channels found in the plasma membranes of these cells. Although these compounds have broad use in the treatment of cardiovascular disease, recent reports indicate that certain DHPs, including nimodipine and amlodipine, have the ability to reduce behavioral and sensorimotor deficits associated with aging in animal models.[3-5]

The plasma membrane forms a complex chemical environment for protein molecules which mediate various functions of the cell, including a subset which function as receptors to which ligands or drugs bind to initiate their effects on cells. Certain hydrophobic and amphiphilic drugs, including DHP Ca^{2+} channel blockers, appear to partition into the membrane prior to protein receptor binding.[6-11] A recent report from Triggle's laboratory[10] indicates that the active 1,4-DHP moiety must partition to a specific depth in the membrane hydrocarbon core in order to achieve successful receptor recognition and binding. Using various DHP probes, it was postulated that the binding site on the Ca^{2+} channel is deep in the hydrocarbon core, 11 Å from the membrane surface. These findings suggest that partitioning of a DHP molecule to a particular depth in the membrane represents a critical component of its receptor binding mechanism.

Membrane Cholesterol Changes Affect Structure/Function Relationships. A primary site for cholesterol (unesterified) is the cell plasma membrane, accounting for 30-40 mol% of the total lipid content of neural plasma membranes isolated from cortical and cerebellar gray matter of aged human subjects.[2] The ratio of cholesterol to phospholipid (C:P) in membranes correlates directly with membrane microviscosity; the cholesterol molecule incorporates into the membrane hydrocarbon core and reduces freedom of motion in lipid hydrocarbon chains near the glycerol backbone.[12-14] Changes in the membrane C:P ratio have significant pharmacological effects, including alterations in the specific binding activity of serotonin, opiate, and β-adrenergic analogs in mouse brain membranes.[15,16] The mechanism by which cholesterol modulates membrane protein activity is not understood; however, *altered membrane biophysical properties due to cholesterol changes may underlie perturbations in protein structure/function relationships.*

Changes in Neural Membrane Cholesterol Composition in Alzheimer's Disease. Alzheimer's disease (AD) is a neurodegenerative disorder characterized by progressive loss of higher intellectual function in the absence of focal neurological defects. Recent studies demonstrated that genetic variability in the apolipoprotein E (apoE) gene on chromosome 19 is an important risk factor in late-onset (>60 years of age) AD.[17] The apoE particle is synthesized by several organs and is critical for cholesterol transport to cells of the brain. In addition, it was demonstrated that apoE isoforms, particularly apoE4, bind to and form very stable complexes with synthetic β/A4 peptide. Consistent with this finding, apoE is localized in senile plaques, perivascular amyloid deposits, and neurofibrillary tangles, the hallmark lesions of AD. By forming complexes with β/A4, the transport function of apoE may be compromised, resulting in altered distribution of cholesterol in certain brain cells.[18]

This article reviews significant changes in the lipid composition of neural membranes isolated from affected AD cortical tissue relative to age-matched controls and subjects with Parkinson's disease (PD). The effect of altered membrane cholesterol content on equilibrium membrane-based partition coefficients ($K_{P[mem]}$) of various Ca^{2+} channel blockers is also presented. Small angle x-ray diffraction results are used to assess the molecular membrane location of the charged DHP amlodipine in reconstituted brain lipid bilayers. Collectively, these findings indicate basic alterations in the composition of neural membranes in AD which may, in turn, modulate interactions of Ca^{2+} channel blockers which target specific sites in the membrane.[19]

METHODS AND MATERIALS

All chemicals used were reagent grade or better and made up in ultra-pure deionized water. Bovine brain phosphatidylcholine (BBPC), dioleoyl phosphatidylcholine (DOPC), and cholesterol were purchased from Avanti Polar Lipids, Inc. (Alabaster, Alabama). Thin layer chromatography was used to ascertain purity and the presence of degradative products in lipid and cholesterol suspensions. Lipids were stored in a desiccator at $^-20$°C. The fatty acid composition of BBPC included 16:0 (30%), 18:1 (30%), 18:0 (14%), 18:2 (9%), 20:4 (6%), and 22:6 (3%) as determined by gas-liquid chromatographic analysis. The overall ratio of saturated to unsaturated fatty acids was 0.8:1. Amlodipine besylate used in x-ray diffraction experiments was provided as a gift from Pfizer Central Research (Groton, Connecticut) and was dissolved in ethanol before use. Labeled [^{3}H]amlodipine maleate dissolved in ethanol was provided by Pfizer via Amersham International (UK). [^{3}H]Isradipine, [^{3}H]nimodipine, [^{3}H]cholesterol, and [^{14}C]DOPC in ethanol were obtained from DuPont New England Nuclear Research Products (Boston, Massachusetts). All materials were stored at $^-20$°C and protected from light.

For neural membrane composition studies, samples of superior temporal gyrus and cerebellum were snap-frozen in liquid nitrogen ($^-70$°C) and coded by the staff at the Institute for Biogerontology Research (Sun City, Arizona). The eight controls had no history of dementia before death. The diagnoses of the nine patients with AD were confirmed at autopsy. The sex ratio (AD: 5 males, 4 females; PD: 5 males, 0 females; controls: 5 males, 4 females) was similar for the AD and control groups. The postmortem delay (AD: 3.20 ± 3.01 vs PD: 3.25 ± 1.30 vs controls: 2.28 ± 0.84 hours) was not significantly different among the 22 patients. For comparative studies, 6 AD, 5 PD, and 6 controls had similar ages at the time of death (AD: 74 ± 6 vs PD: 76 ± 5 vs controls: 74 ± 4 years, mean ± SD). Stained sections taken from both temporal gyrus and cerebellum were rated blindly by an experienced neuropathologist for extent of cell loss and gliosis.

Isolation of Brain Membranes. Gray matter from the superior temporal gyrus and cerebellum was dissected from frozen brain tissue. The material was then weighed and homogenized in 30 volumes of 150 mM Tris-HCl of ice cold buffer (pH 7.4) and homogenized for 30 seconds using a Tekman homogenizer. The homogenate was subjected to centrifugation at 20,000 × *g* for 25 minutes at 4°C. The supernatant was discarded and the final pellet was resuspended in 3 volumes of Tris buffer before lipid extraction.

Membrane protein was determined using bovine serum albumin as a standard.[20] The phospholipid phosphorus content of the samples was determined by previously described methods.[21,22] Specifically, 75 μl of 72% perchloric acid was added to 1-40 nmol of phospholipid phosphorus in borosilicate tubes and hydrolyzed for 1 hour at 180°C in a sand bath. The following reagents were prepared for the phosphomolybdate-complex color development: solution A, 1.57% ascorbate; solution B, 5.04 g ammonium molybdate, 33 ml concentrated H_2SO_4, and 84 ml distilled deionized water; solution C, 21.0 ml solution A and 2.0 ml solution B. Solution C (0.9 ml) is added to the cooled perchloric acid digest, color developed at 80°C for 15 minutes, and absorbance determined at 820 nm. A colorimetric assay was used to measure total cholesterol in native membranes by a modification of the cholesterol oxidase method.[23,24] Cholesterol was extracted from native membranes using a chloroform:methanol:water system.[25] The efficiency of cholesterol extraction was quantitated with [^{3}H]cholesterol (specific activity 93.8 Ci/mmol, NEN, Boston, Massachusetts) and determined to be 90%.

Preparation of Membrane Multilayer X-ray Diffraction Samples. Oriented membrane multilayer samples of bovine brain phosphatidylcholine/cholesterol at a 1/0.6 mole ratio were prepared as follows. Phospholipid and cholesterol dissolved in chloroform were dried down with a stream of N_2 gas to a thin film on the sides and bottom of a glass 13 × 100 mm test tube while vortexing. Residual solvent was removed under vacuum overnight. A specified volume of buffer (0.5 mM HEPES, 2.0 mM NaCl, pH 7.3) was added to the dried lipids while vortexing, yielding a final phospholipid concentration of 5 mg/ml. Multilamellar vesicles (MLV) were formed by vortexing the buffer and lipids for three minutes at ambient temperature.[26] Multilayer samples for small angle x-ray scattering were prepared by centrifuging the vesicles in an SW-28 rotor (Beckman Instruments, Inc., Fullerton, California) at 35,000 × g for 1 hour at 5°C in Lucite sedimentation cells, each containing an aluminum foil substrate.[27] For drug-containing samples, amlodipine besylate was added to the vesicle suspension at a 1:50 drug:phospholipid mole ratio before centrifugation. The final mass concentration of amlodipine in the total lipid sample was approximately 1%. On completion of the spin, supernatants were removed and each sample was mounted on a curved glass support. The samples equilibrated overnight in a glass vial containing a saturated salt solution which served to define a specific relative humidity of 95% at 5°C ($ZnSO_4$). Oriented membrane samples were then placed in sealed brass canisters containing aluminum foil windows in which temperature and relative humidity were controlled.

X-ray Diffraction Data Collection and Reduction. Small angle x-ray scattering was carried out by aligning the samples at near-grazing incidence with respect to the x-ray beam. The radiation source was a collimated, monochromatic x-ray beam (CuK_α, λ = 1.54 Å) from an Elliot GX-18 rotating anode microfocus generator (Enraf Nonius, Bohemia, New York) operated at 40 kV and 30 mA in the Biomolecular Structure Analysis Center at the University of Connecticut Health Center (Farmington, Connecticut). A helium tunnel was positioned between the sample and detector to reduce scattering from the air. The fixed geometry beamline used a single Franks mirror providing nickel-filtered radiation ($K_{\alpha 1}$ and $K_{\alpha 2}$ unresolved) at the detection plane. The beam height at the sample was ~1 mm.

Bragg's diffraction from the oriented membrane multilayer samples was recorded on a one-dimensional position-sensitive electronic detector (Innovative Technologies, Inc., Newburyport, Massachusetts). In addition to direct calibration of the detector system, lead stearate was used to verify the calibration. The sample-to-detector distance used in these experiments was 95 mm. Each individual lamellar diffraction peak was background-corrected using a linear subtraction routine that averaged the noise. The intensity functions were corrected by a factor of $s = 2 \sin\Theta/\lambda$, the Lorentz correction, in which λ is the wavelength of radiation (1.54 Å) and Θ is the Bragg angle equal to one half the angle between the incident beam and the scattered beam. A swelling analysis was used to assign unambiguous phases to the experimental structure factors.[28] Control and test electron density profiles were compared only after an identical number of structure factors was used in the Fourier synthesis.

Drug/Membrane Partition Coefficients. The equilibrium membrane partition coefficient ($K_{P[mem]}$) is a unitless measurement which represents the mass distribution of drug in the membrane versus aqueous buffer. This value is an expression of the drug's lipophilicity or affinity for the membrane lipid bilayer. Dihydropyridine partition coefficient experiments using DOPC multilamellar vesicles (MLV) of varying cholesterol content were carried out as previously described in detail.[19] Briefly, specific quantities of cholesterol and phospholipid in C:P mole ratios of 0:1, 0.3:1, and 0.6:1 were mixed and dried down in a glass test tube, while vortexing, to a thin film under a stream of N_2 gas. Tris buffer (150 mM NaCl and 10 mM Tris HCl, pH 7.0) was then added to yield a lipid concentration of 2 mg/ml, and the solution was vortexed at room temperature (above the thermal phase transition temperature) to form MLV. Reaction mixtures contained radiolabeled DHP analogs (5×10^{-10} M) with or without DOPC (20 μg/ml) and were incubated at ambient temperature for 30 minutes before filtration. The final concentration of ethanol in the reaction mixtures was <0.001%.

Rapid separation of membrane-bound versus free drug was accomplished using rapid filtration techniques. All solutions were filtered through Whatman GF/C glass fiber filters on a Brandel M-48 cell harvester (Brandel, Gaithersburg, Maryland). Using vesicles (20 μg/ml) radiolabeled with either [^{3}H]cholesterol or [^{14}C]DOPC, we were able to accurately and reproducibly measure the amount of membrane lipid retained on the filters during the filtration process (80%).

Reaction mixtures of 5 ml total contained either drug and membranes or drug alone. The filters for these experiments were not washed or dried, but were counted immediately for radioactivity in Pico-Fluor 15 (Packard, Meriden, Connecticut). All filters were counted for radioactivity (Tracor Analytic Delta 300/6891 Liquid Scintillation System). The total nonspecific binding of drug to membranes was corrected for any nonspecific binding of drug to filter paper. The count rate (CPM, counts per minute) for drug in the filtrate (ie, "free drug") was calculated by subtracting the CPM for drug bound to membrane and filter from the CPM for total drug added to the reaction mixtures (assessed by dispensing the same amount of drug directly into scintillation vials). The amount (g) of lipid bound to filters was calculated on the basis of the experimentally determined percentage recovery of lipid on the filters, as just described. Membrane partition coefficients were calculated for each trial using the following equation:

TABLE 1. C:P Mole Ratio in the Brain in Alzheimer's Disease and in Controls (Mean ± SD)[a,b]

Brain Area	Control Brain (n = 6)	Brain in Alzheimer's Disease (n = 6)
Superior temporal gyrus	0.66 ± 0.05[c]	0.46 ± 0.08[d]
Cerebellum	0.45 ± 0.08	0.50 ± 0.08

NOTE: c vs. d, p <0.01, Wilcoxon two-sample rank test (two-tailed).

[a] Patient ages at time of death were 74 ± 4 years (control) and 74 ± 6 years (AD).

[b] Table reproduced with permission from reference 2.

$$K_{P[mem]} = \frac{\text{(CPM drug bound to membrane/grams lipid)}}{\text{(CPM free drug/grams buffer)}}$$

Statistical analyses were used to evaluate nonspecific equilibrium binding parameters for the DHPs. Three trials consisting of 12 samples (n = 36) were carried out for experiments with [^{3}H]nimodipine and [^{3}H]isradipine, whereas two trials (n = 24) were carried out for amlodipine. For each set of experiments, $K_{P[mem]}$ values were obtained and standard error of the mean was calculated.

RESULTS

Analysis of Neural Membrane Protein, Cholesterol, and Phospholipid Content. The protein, phospholipid, and cholesterol content of human neural plasma membranes from the gray matter of superior temporal gyrus and cerebellum were measured and compared. In TABLE 1, the results of this analysis demonstrated a significant 30% decrease ($p < 0.01$, Wilcoxon two-sample rank test two-tailed), in the C:P mole ratio of AD temporal gyrus when compared with control samples. By contrast, the C:P mole ratio in the cerebellum did not change significantly. Moreover, the mass ratio of protein to phospholipid was not significantly different in either the temporal gyrus or the cerebellum of AD and control tissue. The measured C:P mole ratio value of fresh AD temporal gyrus was not significantly different from that obtained from the same tissue source after freezing. As a further control, tissue samples from five age-matched, PD brains were analyzed. The average C:P mole ratios from temporal gyrus (0.57 ± 0.10) and cerebellum (0.48 ± 0.06) of the PD brains were not significantly different from those of the controls.

Correlation between Age and Membrane Cholesterol Content. A significant correlation was noted between age and C:P mole ratio in the temporal gyrus but not in the cerebellum of both AD and control samples. The average age of the eight controls and nine patients with AD was 75 ± 9 years. A Pearson correlation coefficient (r) of 0.72 ($p < 0.025$) and 0.78 ($p < 0.01$) was calculated for the eight controls and 9 AD samples, respectively. The slope of the change (age vs C:P mole ratio) was 0.01

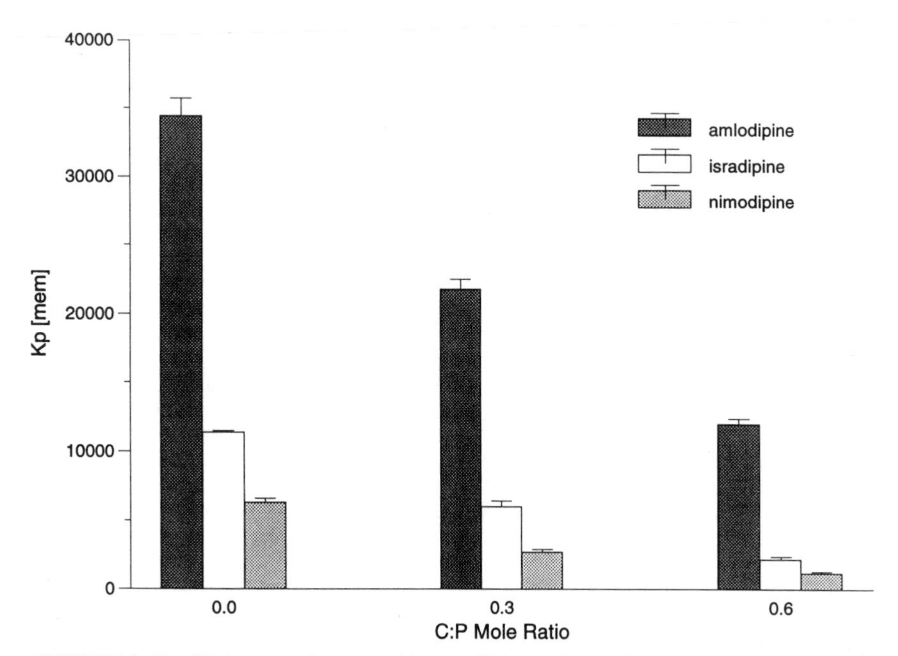

FIGURE 1. Equilibrium membrane partition coefficients ($K_{P[mem]}$) for 1,4-dihydropyridine Ca^{2+} channel blockers in liposomes of varying cholesterol content. [^{3}H]Amlodipine, [^{3}H]isradipine, and [^{3}H]nimodipine were used in these studies at a final concentration of 5×10^{-10} M. Liposomes were composed of dioleoylphosphatidylcholine at a fixed concentration of 20 μg/ml. The reaction was carried out at 21°C, pH 7.0, and terminated after 30 minutes incubation. Data were reproduced in part from reference 19.

for both control and AD samples. In these same samples, the mass ratio of phospholipid to protein was similar for controls (0.70 ± 0.05), patients with AD (0.69 ± 0.05), and patients with PD (0.78 ± 0.06, $n = 5$) and did not correlate significantly with age. By contrast, the r values were –0.27 (ns) and –0.07 (ns) in the cerebellum of control and AD samples, respectively.

Effect of Membrane Cholesterol Content on Dihydropyridine Partition Coefficients. FIGURE 1 summarizes the effect of membrane cholesterol content on the equilibrium membrane partition coefficient ($K_{P[mem]}$) values of several DHPs including [^{3}H]amlodipine, [^{3}H]isradipine, and [^{3}H]nimodipine. These data show a strong negative correlation between the C:P ratio of the liposomes and DHP $K_{P[mem]}$ values. For example, an increase in liposome cholesterol to phospholipid mole ratio from 0:1 to 0.6:1 resulted in an 81% decrease in $K_{P[mem]}$ values for isradipine and nimodipine. At a 0.6:1 C:P mole ratio, the $K_{P[mem]}$ for amlodipine was 5-fold and 10-fold greater than that of isradipine and nimodipine, respectively. For these experiments, the amount of phospholipid (20 μg/ml) was kept constant.

Molecular Interaction of a Charged Dihydropyridine Ca^{2+} Channel Blocker with Brain Phospholipid/Cholesterol Lipid Bilayers. To examine the structure of a brain lipid bilayer in the presence of a DHP Ca^{2+} blocker, small angle x-ray diffraction

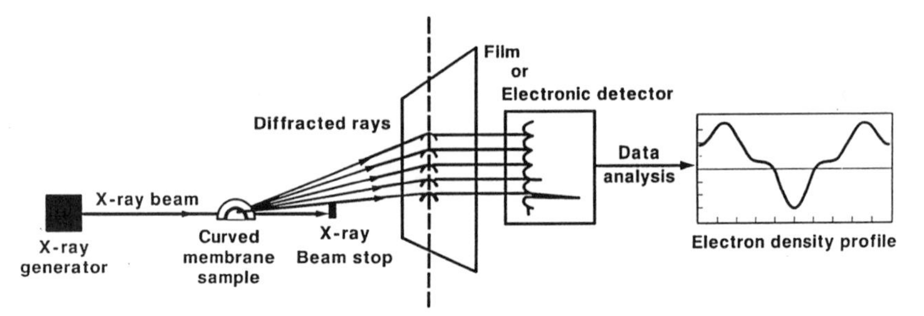

FIGURE 2. Summary of the small angle x-ray diffraction method. Membrane diffraction experiments used monochromatic radiation (λ = 1.54 Å) from a high-brilliance rotating anode x-ray generator source. The oriented brain lipid multilayers were placed on a curved glass mount so that the plane of the membranes was oriented around an axis perpendicular to the incoming focused x-ray beam at discrete angles. Coherent scattering from the sample was collected on either film or a one-dimensional position-sensitive electronic detector. The diffraction order intensities were then integrated and used to generate a one-dimensional electron density "profile" of the brain lipid bilayer.

was applied to oriented membrane multibilayers composed of bovine brain phosphatidylcholine and cholesterol at a 1:0.6 mole ratio (FIG. 2). The DHP amlodipine was used in this study as it has high affinity for the membrane and an electron-dense chlorine substituent as part of its chemical structure. X-ray scattering from the control samples produced six strong, reproducible diffraction orders (FIG. 3) with a unit cell periodicity or d-space (distance from the center of one lipid bilayer to the next, including water) of 57.2 Å at 5°C. Fourier analysis of the diffraction data produced a one-dimensional electron density "profile" of the membrane (FIG. 4) at 10 Å resolution. Following the addition of amlodipine at a 1:45 drug:phospholipid mole ratio, significant changes in membrane structure were observed, yielding information on the drug's location. Specifically, there was a broad increase in electron density ± 8-21 Å from the center of the lipid bilayer and an outward displacement of the phospholipid headgroups by ~1 Å relative to the control sample. The d-space of the amlodipine-containing sample was 58.7 Å.

DISCUSSION

Small angle x-ray diffraction approaches were used to directly characterize the structure of brain lipid bilayers in the absence and presence of the 1,4-DHP Ca^{2+} channel blocker amlodipine. One-dimensional electron density profiles generated from the diffraction data indicate that amlodipine occupies a discrete, time-averaged location in the membrane hydrocarbon core adjacent to the phospholipid headgroups. Relative to control samples, the addition of amlodipine produced a broad increase in hydrocarbon core electron density consistent with the long axis of the molecule lying parallel to the phospholipid acyl chains (FIG. 4). This increase in electron density is attributed primarily to the conjugated ring structure and electron-dense

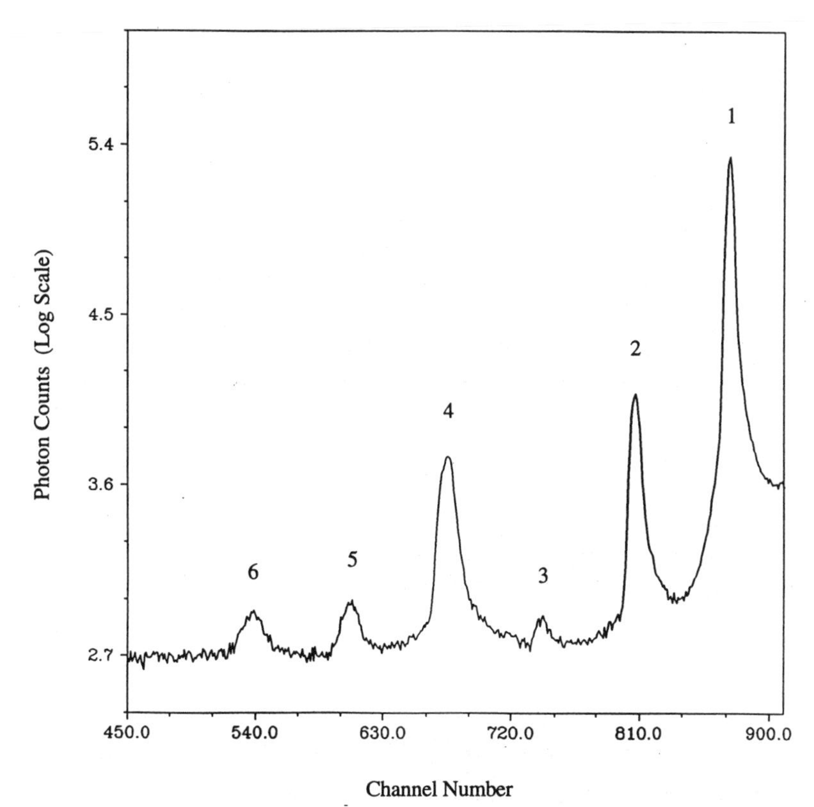

FIGURE 3. X-ray diffraction pattern from oriented brain lipid bilayer membranes. Six strong diffraction orders were collected with a one-dimensional position-sensitive electronic detector after 1 hour of exposure to x-rays. The experimental conditions were 95% relative humidity and 5°C. The sample-to-detector distance was 95 mm.

chlorine substituent of amlodipine. The insertion of amlodipine into the membrane at this location may also increase the local packing density of the phospholipid acyl chains, also resulting in an apparent increase in electron density. Finally, the addition of amlodipine caused an outward displacement of electron density peaks corresponding to the phosphate headgroups, further evidence for the effect of amlodipine on phospholipid conformation.

The location of amlodipine in the membrane hydrocarbon core is similar to that of unesterified cholesterol, a primary constituent of neural plasma membranes in aged and AD tissue (TABLE 1). In the membrane hydrocarbon core, cholesterol favors hydrogen-bond formations with the 3β-hydroxyl group and has hydrophobic steric contact with the condensed steroid nucleus.[29] Cholesterol intercalates into the membrane hydrocarbon core where it reduces freedom of motion in lipid hydrocarbon chains adjacent to the glycerol backbone.[14] An increase in membrane cholesterol content would be expected to reduce the available free volume available for DHP

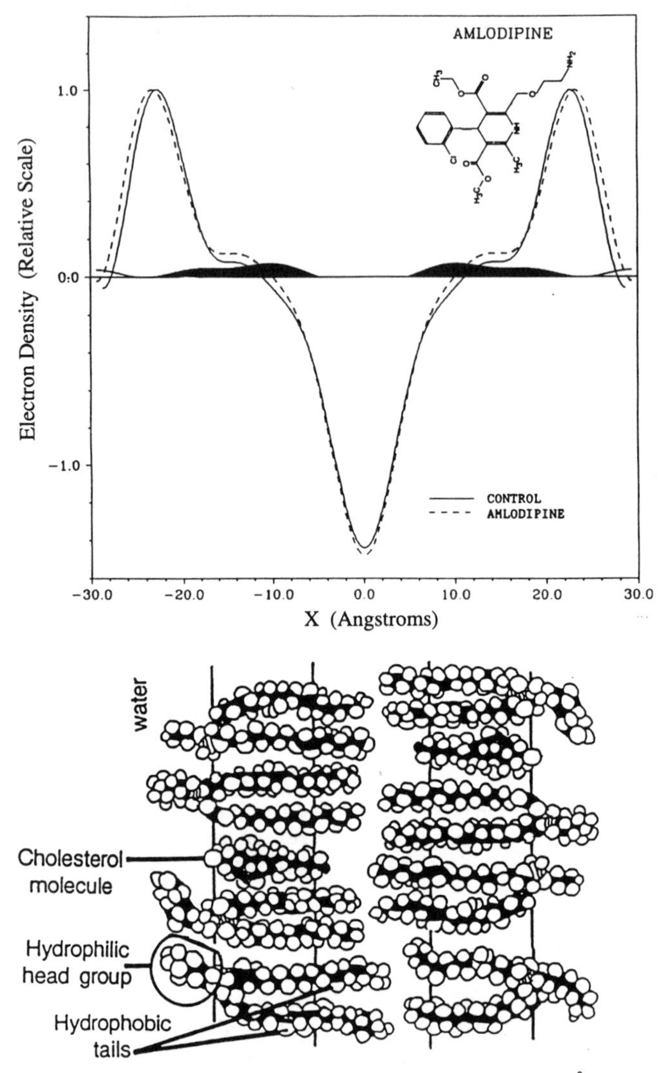

FIGURE 4. Superimposed one-dimensional electron density profiles (10 Å resolution) for a centrosymmetric brain lipid bilayer in the absence (*solid line*) and presence (*dashed line*) of amlodipine at a 1:45 drug/phospholipid mole ratio (**upper panel**). The two maxima of electron density correspond to the electron dense phosphate headgroups, whereas the minimum of electron density in the center of the figure corresponds to the membrane bilayer center. The addition of amlodipine produced positive differences in hydrocarbon core electron density as indicated by the *shaded areas*. Profiles were correlated with a space-filled model of a lipid bilayer (**lower panel**). Experiments were carried out at 95% relative humidity and 5°C. Membrane widths, including water, were 57.2 Å and 58.7 Å for the controls and amlodipine-containing samples, respectively.

partitioning into this region of the membrane hydrocarbon core.[30] Indeed, a strong and differential inverse relationship was observed between the membrane C:P ratio and the equilibrium membrane partition coefficients ($K_{P[mem]}$) for these compounds (FIG. 1). This was also observed for the Ca^{2+} channel blockers diltiazem and verapamil and has been reproduced in reconstituted cardiac liposome preparations.[19]

Interestingly, amlodipine had high membrane affinity ($K_{P[mem]} > 10^4$) although it is charged ($pK_a = 9.02$) and readily soluble in water when compared with neutral DHP analogs. The explanation for amlodipine's strong membrane interactions may be related to its strong amphiphilic properties. The chemistry of amlodipine includes a charged 2-aminoethoxymethyl substituent at the 2-position of the dihydropyridine ring which, in the crystal structure, extends away from the rest of the molecule.[31] This charged "extension" may facilitate ionic bonding with cationic oxygen in the phospholipid headgroups while the remainder of the molecule is energetically favored to interact with fatty acyl chains in the hydrocarbon core (FIG. 5). The charged anchor of amlodipine may also underlie its slow dissociation from the membrane[30] and receptor site.[11] Indeed, under pH conditions in which amlodipine is primarily neutral, its dissociation properties are rapid and similar to those of a neutral DHP.[11]

The high nonspecific membrane affinity and well-defined equilibrium location of amlodipine in the membrane hydrocarbon core are consistent with an intrabilayer protein receptor site which is accessed from the surrounding lipid environment.[7,8,32] In further support of this model, Triggle's laboratory demonstrated that DHP binding to its receptor only occurs when the active portion of the DHP molecule is able to partition to a particular depth in the membrane hydrocarbon core.[10] Using probes that consist of a 1,4-DHP moiety linked to a permanently charged group via polymethylene segments of varying length, it appeared that the DHP binding site on the Ca^{2+} channel is in the hydrocarbon core, approximately 11 Å from the external plasma membrane surface.[10] Although it is difficult to assign a particular depth to the "active" portion of amlodipine (eg, ester oxygen of the 1,4-DHP moiety) from the x-ray diffraction data, it is clear that the molecule alters the membrane hydrocarbon core in a region consistent with an intrabilayer site. Moreover, in a liquid crystalline membrane bilayer, the location of amlodipine is not static but distributed over a limited region (FIG. 4).

As a primary constituent of neural plasma membranes, cholesterol plays a significant role in the pharmacology of synaptic membranes.[15,16] Our laboratory reported a significant reduction in neural membrane cholesterol:phospholipid mole ratios of AD samples (TABLE 1, ref. 2). The underlying cause of this change in neural membrane C:P mole ratios is not understood, but it may be related in part to a perturbation in cholesterol metabolism. Indeed, interest in cholesterol distribution in the AD brain has intensified as a result of the finding that genetic variability in the apoE gene on chromosome 19 is an important risk factor in late-onset AD.[17] A compromise in the ability of the apoE particle to properly transport cholesterol to various cells and subcellular compartments in the brain (eg, plasma membrane) as a result of binding to βA4[18] may contribute to these observed changes in membrane composition.

SUMMARY

The results of this study demonstrate that the equilibrium nonspecific binding of DHP Ca^{2+} channel blockers to the membrane bilayer is highly dependent on cholesterol

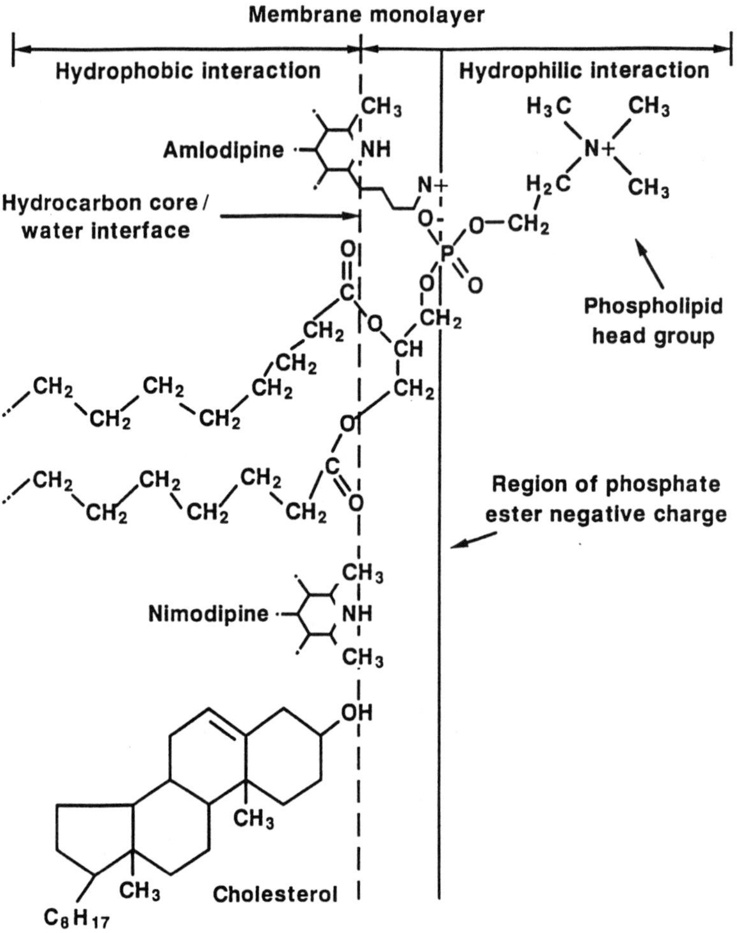

FIGURE 5. Schematic model of drug/membrane interactions based on current and previous[31,33] x-ray and neutron diffraction results. The location of amlodipine near the hydrocarbon core/water interface allows for the van der Waals interactions with phospholipid acyl chains and for ionic bonding between the drug's charged ethanolamine side group and the anionic oxygen of the phosphate moiety. These interactions may account for amlodipine's high $K_{P[mem]}$ values. Dihydropyridine and cholesterol occupy a similar location in the membrane hydrocarbon core (molecules are not drawn to scale). By ordering the phospholipid acyl chains adjacent to the glycerol backbone, cholesterol would reduce the available volume for DHP partitioning into this region of the membrane. Figure reproduced with permission from reference 19.

content. The molecular explanation for this observation appears to be related to the fact that cholesterol and DHPs occupy a similar molecular location in the membrane hydrocarbon core (FIG. 4). The membrane location of amlodipine may also be critical for subsequent receptor recognition and binding to voltage-sensitive Ca^{2+} channels in peripheral and CNS tissue. Finally, changes in the cholesterol content of neural plasma membranes isolated from diseased cortical regions of subjects with AD were reported and may be indicative of a general defect in lipid metabolism. Further studies are underway to characterize in greater detail possible changes in cholesterol content with aging and AD. The implication of these changes for structure/function relationships in the membrane is also being explored.

ACKNOWLEDGMENTS

The author would like to acknowledge the excellent editorial assistance of Pamela Mason. The diffraction experiments for this study were carried out in the Biomolecular Structure Analysis Center of the University of Connecticut Health Center.

REFERENCES

1. Moring, J., W. J. Shoemaker, V. Skita, R. P. Mason, H. C. Hayden, R. M. Salomon & L. G. Herbette. 1990. Rat cerebral cortical synaptoneurosomal membranes: Structure and interactions with imidazobenzodiazepine and 1,4-dihydropyridine calcium channel drugs. Biophys. J. **58:** 513-531.
2. Mason, R. P., W. J. Shoemaker, L. Shajenko, T. E. Chambers & L. G. Herbette. 1992. Evidence for changes in the Alzheimer's disease brain cortical membrane structure mediated by cholesterol. Neurobiol. Aging **13:** 413-419.
3. Deyo, R. A., K. T. Straube & J. F. Disterhoft. 1989. Nimodipine facilitates associative learning in aging rabbits. Science **243:** 809-811.
4. Schuurman, T., E. Horvath, D. G. Spencer, Jr. & J. Traver. 1987. Nimodipine and motor deficits in the aged rat. Neurosci. Res. Commun. **1:** 9-15.
5. Quartermain, D., A. Hawxhurst & B. Ermita. 1993. Effect of the calcium channel blocker amlodipine on memory in mice. Behav. Neural Biol. **60:** 211-219.
6. Kokubun, S. & H. Reuter. 1984. Dihydropyridine derivatives prolong the open state of Ca^{++} channels in cultured cardiac cells. Proc. Natl. Acad. Sci. USA **81:** 4824-4827.
7. Herbette, L. G., Y. M. H. Vant Erve & D. G. Rhodes. 1989. Interaction of 1,4-dihydropyridine Ca^{2+} channel antagonists with biological membranes: Lipid bilayer partitioning could occur before drug binding to receptors. J. Mol. Cell. Cardiol. **21:** 187-201.
8. Rhodes, D. G., J. G. Sarmiento & L. G. Herbette. 1985. Kinetics of binding of membrane-active drugs to receptor sites: Diffusion-limited rates for a membrane bilayer approach of 1,4-dihydropyridine Ca^{2+} channel antagonists to their active site. Mol. Pharmacol. **27:** 612-623.
9. Mason, R. P., G. E. Gonye, D. W. Chester & L. G. Herbette. 1989. Partitioning and location of Bay K 8644, 1,4-dihydropyridine Ca^{2+} channel agonist, in model and biological lipid membranes. Biophys. J. **55:** 769-778.
10. Baindur, N., A. Rutledge & D. J. Triggle. 1993. A homologous series of permanently charged 1,4-dihydropyridines: Novel probes designed to localize drug binding sites on ion channels. J. Med. Chem. **36:** 3743-3745.
11. Kass, R. S. & J. P. Arena. 1989. Influence of pH on calcium channel block by amlodipine, a charged dihydropyridine compound: Implications for location of dihydropyridine receptor. J. Gen. Physiol. **93:** 1109-1127.
12. Franks, N. P. 1976. Structural analysis of hydrated egg lecithin and cholesterol bilayers. J. Mol. Biol. **100:** 345-358.
13. McIntosh, T. J. 1978. The effect of cholesterol on the structure of phosphatidylcholine bilayers. Biophys. Biochim. Acta **513:** 43-58.

14. STOCKTON, G. W. & I. C. P. SMITH. 1976. A deuterium nuclear magnetic resonance study of the condensing effect of cholesterol on egg phosphatidylcholine bilayer membranes. Chem. Phys. Lipids **17:** 251-263.
15. HERON, D. S., M. SHINITZKY, M. HERSHKOWITZ & D. SAMUEL. 1980. Lipid fluidity markedly modulates the binding of serotonin to mouse brain membranes. Proc. Natl. Acad. Sci. USA **77:** 7463-7467.
16. HERSHKOWITZ, M., D. HERON, D. SAMUEL & M. SHINITZKY. 1982. The modulation of protein phosphorylation and receptor binding in synaptic membranes by changes in lipid fluidity: Implications for aging. Prog. Brain Res. **56:** 419-434.
17. CORDER, E. H., A. M. SAUNDERS, W. J. STRITTMATTER, D. E. SCHMECHEL, P. C. GASKELL, G. W. SMALL, A. D. ROSES, J. L. HAINES & M. A. PERICAK-VANCE. 1993. Gene dose of apolipoprotein E type 4 allele and the risk of Alzheimer's disease in late onset families. Science **261:** 921-923.
18. STRITTMATTER, W. J., A. M. SAUNDERS, D. SCHMECHEL, M. PERICAK-VANCE, J. ENGHILD, G. S. SALVESEN & A. D. ROSES. 1993. Apolipoprotein E: High-avidity binding to β-amyloid and increased frequency of type 4 allele in late-onset familial Alzheimer disease. Proc. Natl. Acad. Sci. USA **90:** 1977-1981.
19. MASON, R. P., D. M. MOISEY & L. SHAJENKO. 1992. Cholesterol alters the binding of Ca^{2+} channel blockers to the membrane lipid bilayer. Mol. Pharmacol. **41:** 315-321.
20. LOWRY, O. H., N. J. ROSEBROUGH, A. L. FARR & R. J. RANDALL. 1951. Protein determination with the phenol reagent. J. Biol. Chem. **193:** 265-275.
21. CHEN, P. S., T. Y. TORIBARA & H. WARNER. 1956. Microdetermination of phosphorus. Anal. Chem. **28:** 1756-1758.
22. CHESTER, D. W., L. G. HERBETTE, R. P. MASON, A. F. JOSLYN, D. J. TRIGGLE & D. E. KOPPEL. 1987. Diffusion of dihydropyridine calcium channel antagonists in cardiac sarcolemmal lipid multilayers. Biophys. J. **52:** 1021-1030.
23. HEIDER, J. G. & R. L. BOYETTE. 1978. The picomole determination of free and total cholesterol in cells in culture. J. Lipid Res. **19:** 514-518.
24. CHESTER, D. W., M. E. TOURTELOTTE, D. L. MELCHIOR & A. H. ROMANO. 1986. The influence of fatty acid modulation of bilayer physical state on cellular and membrane structure and function. Biochim. Biophys. Acta **860:** 383-398.
25. FOLCH, J., M. LEES & G. H. SLOANE STANLEY. 1957. A simple method for the isolation and purification of total lipids from animal tissues. J. Biol. Chem. **226:** 497-509.
26. BANGHAM, A. D., M. M. STANDISH & J. C. WATKINS. 1965. Diffusion of univalent ions across the lamellae of swollen phospholipids. J. Mol. Biol. **13:** 238-252.
27. BLASIE, J. K., C. R. WORTHINGTON & M. M. DEWEY. 1969. Molecular localization of frog retinal receptor photopigment by electron microscopy and low-angle x-ray diffraction. J. Mol. Biol. **39:** 407-416.
28. MOODY, M. F. 1963. X-ray diffraction pattern of nerve myelin: A method for determining the phases. Science **142:** 1173-1174.
29. HUANG, C. 1977. A structural model for the cholesterol-phosphatidylcholine complexes in bilayer membranes. Lipids **12:** 348-356.
30. STRAUME, M. & B. J. LITMAN. 1987. Influence of cholesterol on equilibrium and dynamic bilayer structure of unsaturated acyl chain phosphatidylcholine vesicles as determined from higher order analysis of fluorescence anisotropy decay. Biochemistry **26:** 5121-5126.
31. MASON, R. P., S. CAMPBELL, S. WANG & L. G. HERBETTE. 1989. A comparison of bilayer location and binding for the charged 1,4-dihydropyridine Ca^{2+} channel antagonist

amlodipine with uncharged drugs of this class in cardiac and model membranes. Mol. Pharmacol. **36:** 634-640.

32. MASON, R. P. & D. W. CHESTER. 1989. Diffusional dynamics of an active rhodamine-labeled 1,4-dihydropyridine in sarcolemmal lipid multibilayers. Biophys. J. **56:** 1193-1201.
33. HERBETTE, L. G., D. W. CHESTER & D. G. RHODES. 1986. Structural analysis of drug molecules in biological membranes. Biophys. J. **49:** 91-94.

Calcium-Mediated Processes in Neuronal Degeneration[a]

BO K. SIESJÖ [b]

Laboratory for Experimental Brain Research
University of Lund
Experimental Research Center
University Hospital
S-221 85 Lund, Sweden

Calcium plays a ubiquitous role in the transmittal of information from the external environment to the cell interior where the ion, acting as a second messenger, transforms external cues into metabolic responses that serve to alter membrane function and gene expression.[1–5] The source of the messenger calcium is either the pool of extracellular Ca^{2+} ions or calcium sequestered in internal pools, such as the endoplasmic reticulum (ER), or so-called calciosomes.[6–12] In the first instance, Ca^{2+} enters by voltage-sensitive calcium channels (VSCCs) or by agonist-operated ones (AOCCs). In the second instance, calcium is either released from the ER by inositoltrisphosphate (IP_3), formed during breakdown of phosphatidyl inositol bisphosphate (PIP_2) by the action of a phospholipase C, or from calciosomes by a rise in Ca^{2+}_i (calcium-triggered calcium release).

In this stimulus-response-metabolism coupling, glutamate and related excitatory amino acids (EAAs) play an important role. Thus, apart from being the most important agonist at receptors linked to cation channels, glutamate also activates metabotropic receptors coupled to phospholipase C.[13–16] Clearly, an important link exists between excitatory amino acids, particularly glutamate, and cell calcium metabolism.

Transients arising as a result of influx or release of calcium are attenuated by intracellular binding and sequestration and terminated by extrusion of calcium from the cells.[7–9,17] In this way, a delicate balance exists between influx/release and sequestration/efflux in which Ca^{2+}_i is kept within physiologic limits, yet fulfilling its role as a trigger of cellular responses. Some of these responses, like the Ca^{2+}-triggered release of transmitter from nerve endings, may be associated with marked increases in Ca^{2+}- in cellular subdomains.[18,19] Clearly though, if either influx/release is enhanced, or sequestration/efflux is inhibited, overall Ca^{2+}_i can rise to unphysiologic levels.

By employing calcium as a second messenger, cells carry the seeds of their own destruction. This is because calcium, when accumulating intracellularly in pathologic amounts, can trigger reactions that are potentially destructive. Calcium has been

[a] Studies from the authors' own laboratory were supported by continuous grants from the U.S. Public Health Service via the National Institutes of Health and by the Swedish Medical Research Council.

[b] Address for reprints: Bo K. Siesjö, University of Lund, Laboratory for Experimental Brain Research, Experimental Research Center, University Hospital, S-221 85 Lund, Sweden.

proposed to mediate cell death in a variety of tissues in a variety of disease conditions.[20] Neurons may be particularly prone to incur damage because of loss of their calcium homeostasis in that they contain a variety of VSCCs and AOCCs. It was originally assumed that excessive calcium influx by VSCCs is responsible for selective neuronal damage in ischemia, hypoglycemic coma, and status epilepticus[21] and subsequently that cell death in aging and Alzheimer's disease is calcium-mediated.[22] Later work emphasized the pathogenetic role of enhanced release and/or reduced reuptake of glutamate and related EAAs.[23-26] Inasmuch as most of the toxicity of EAAs seems to be exerted by influx/release of Ca^{2+}, the excitotoxic hypothesis has given novel insights into the mechanisms whereby calcium can mediate cell death,[27-29] mechanisms that may also be relevant to the premature death of neurons in aging and Alzheimer's disease.[26,30]

CELL CALCIUM METABOLISM

To discuss how the loss of cell calcium homeostasis can lead to cell death, I will briefly review pathways of calcium influx as well as mechanisms of binding, sequestration, and extrusion, making use of two schematic diagrams modified from previous articles.[31,32] I will then consider cell calcium metabolism in the energy-perturbed cell, describing the major mechanisms of calcium-related damage. In this discussion, I will explore how a pathologic rise in Ca^{2+}_i can trigger enhanced production of free radicals and will consider the interaction between Ca^{2+} and H^+.

Routes of Calcium Entry

As shown in FIGURE 1, calcium enters cells by two major routes, VSCCs and AOCCs. VSCCs from different tissues have been functionally and pharmacologically characterized; more recently, several channels have also been cloned and their molecular structures determined.[33-36] A large variety of such channels exist, being localized to different parts of neurons and glial cells. In the present context, we will consider four major types of VSCCs which have been named T (for transient), L (long-lasting), N (neuronal), and P (which stands for Purkinje cells in which the channel was first described). These channels differ with respect to threshold characteristics, duration of channel opening, and modulation by blockers. The figure makes the simplifying assumption that the N and P types of VSCCs abound at presynaptic endings and that they, thereby, modulate the release of transmitters which occurs in response to presynaptic influx of Na^+ and depolarization. The L and T types of VSCCs may be mainly localized to postsynaptic membranes.

Interest in AOCCs has surged in recent years, particularly in AOCCs gated by glutamate receptors.[13-16] Although additional subtypes exist, interest is now focused on two "ionotropic" receptors, selectively activated by α-amino-hydroxy-5-methyl-4-isoxazole-propionic acid (AMPA) and *N*-methyl-D-aspartate (NMDA), and on a metabotropic, quisqualate-preferring receptor, selectively activated by (+)-amino-1,3-cyclopentane-*trans*-dicarboxylic acid (*trans*-ACPD) (FIG. 2). The first two ionotropic receptors gate ion channels, whereas the third (metabotropic) is coupled to phospholi-

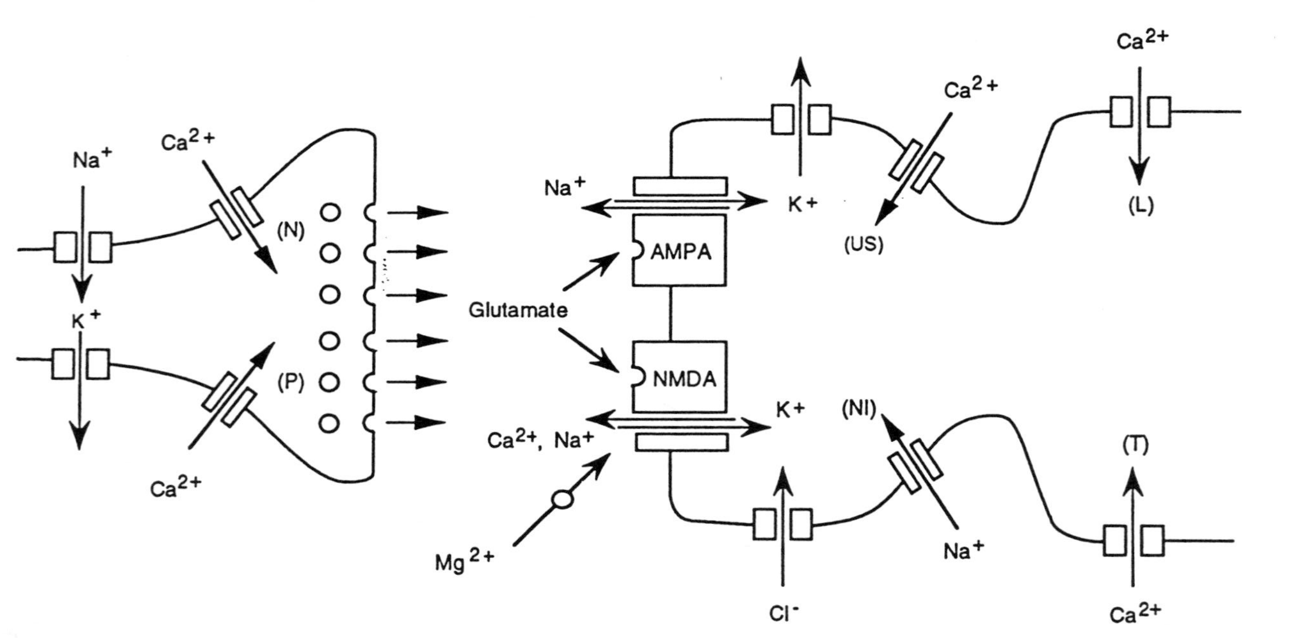

FIGURE 1. Schematic diagram illustrating pre- and postsynaptic ion channels, with emphasis on voltage-sensitive (VSCC) and agonist-operated (AOCC) calcium channels. Presynaptically, the VSCCs involved in transmitter release are assumed to be of the N and P types, whereas the L and T types are assumed to be localized to dendrites. It is further assumed that unspecific channels exist which are permeable to calcium (US). Release of the excitatory transmitter glutamate (Glu) is shown to activate two types of receptors selectively sensitive to amino-3-hydroxy-5-methyl-4-isoazole proprionic acid (AMPA) and to *N*-methyl-D-aspartate (NMDA), respectively. The AMPA receptor gates a channel that is permeable to monovalent cations (Na^+, K^+, and H^+), whereas the NMDA-gated channel also is permeable to Ca^{2+}. Normally, this channel is blocked by Mg^{2+}, but the block is voltage-dependent. By leading to depolarization, AMPA receptor activation and Na^+ influx relieve the block, allowing Ca^{2+} to enter. Depolarization also occurs by entry of Na^+ via voltage-sensitive Na^+ channels, some of which are slowly inactivating (IA). Depolarization also allows Ca^{2+} to enter via VSCC of the L, T, P, and N types. Inhibition is assumed to be mediated by activation of K^+ and Cl^- conductances localized to both pre- and postsynaptic membranes. Modified from ref. 119.

pase C (PLC), and its activation therefore leads to breakdown of phosphatidyl inositol bisphosphate (PIP_2) to inositol trisphosphate (IP_3) and diacylglyceride (DAG).[6,10,37]

It is now known that glutamate activation of ionotropic and metabotropic receptors leads to further reactions and also that a coupling exists between IP_3-triggered release of calcium and the influx of calcium via membrane channels.[11,12,38] For example, calcium released from the ER can trigger further release from intracellular stores and also elicit calcium oscillations. Furthermore, activation of PLC does trigger release not only of calcium from the ER and other intracellular stores, but also of calcium influx via membrane channels, which are in close contiguity with the ER. The agonist responsible for activation of such channels may be IP_4 which is formed when a kinase phosphorylates IP_3. The nature of the channels is not known. Clearly, stimulation of glutamate receptors elicits a cascade of reactions encompassing calcium influx through several different types of membrane channels, some of which are modulated by intracellular messengers.

The schematic diagram in FIGURE 1 helps to define the interrelationship between VSCCs and AOCCs. Crucial to an understanding of factors governing calcium influx are the properties of the glutamate receptors. The AMPA receptor gates a channel that is unselectively permeable to monovalent cations (Na^+, K^+, and H^+). Because of these properties, receptor activation will lead to Na^+ influx and thereby to membrane depolarization. The other ionotropic receptor, that one selectively activated by NMDA, gates a channel that is permeable to Ca^{2+}. An important property of this channel is that when open, it is blocked by Mg^{2+} in physiologic concentrations. However, this block is voltage-dependent, and when the membrane depolarizes, the block is relieved. This means that AMPA receptor activation, with an ensuing depolarization, allows calcium influx via the NMDA-gated cation channel. Furthermore, depolarization also allows calcium influx via VSCCs. At least under some circumstances, when its subunit composition changes, the channel gated by AMPA receptors may become permeable to Ca^{2+}.[39]

Two further points should be made. First, presynaptic influx of calcium and release of transmitters are modulated by drugs that act on the Ca^{2+} channels involved (presumably mainly N and P) as well as on the Na^+ channels that mediate depolarization; in addition, K^+ channels exist which, by allowing efflux of K^+, hyperpolarize membranes, thereby decreasing transmitter release.[40–43] Some of these conductances are activated by a rise in Ca^{2+}_i, others by a fall in ATP concentration or in the ATP/ADP ratio. Activation of presynaptic receptors coupled to Cl^- channels would subserve the same function, that is, one of inhibition of transmitter release. Second, the leading cause of calcium influx across postsynaptic membranes is Na^+ influx and depolarization. Na^+ influx can occur via the AMPA receptor-gated cation channel or via voltage-dependent channels, some of which are slowly inactivating (or noninactivating).[44] Depolarization can also occur by activation of the T type of VSCC which is a low threshold one, opening at very negative membrane potentials. The channel supports pacemaker activity and is responsible for "fast excitation," predominantly at postsynaptic sites.

Clearly, calcium influx in response to transmitter release depends on a host of factors and occurs via multiple channels, some of which may be unspecific (US), that is, not normally included in the classification of VSCCs and AOCCs.

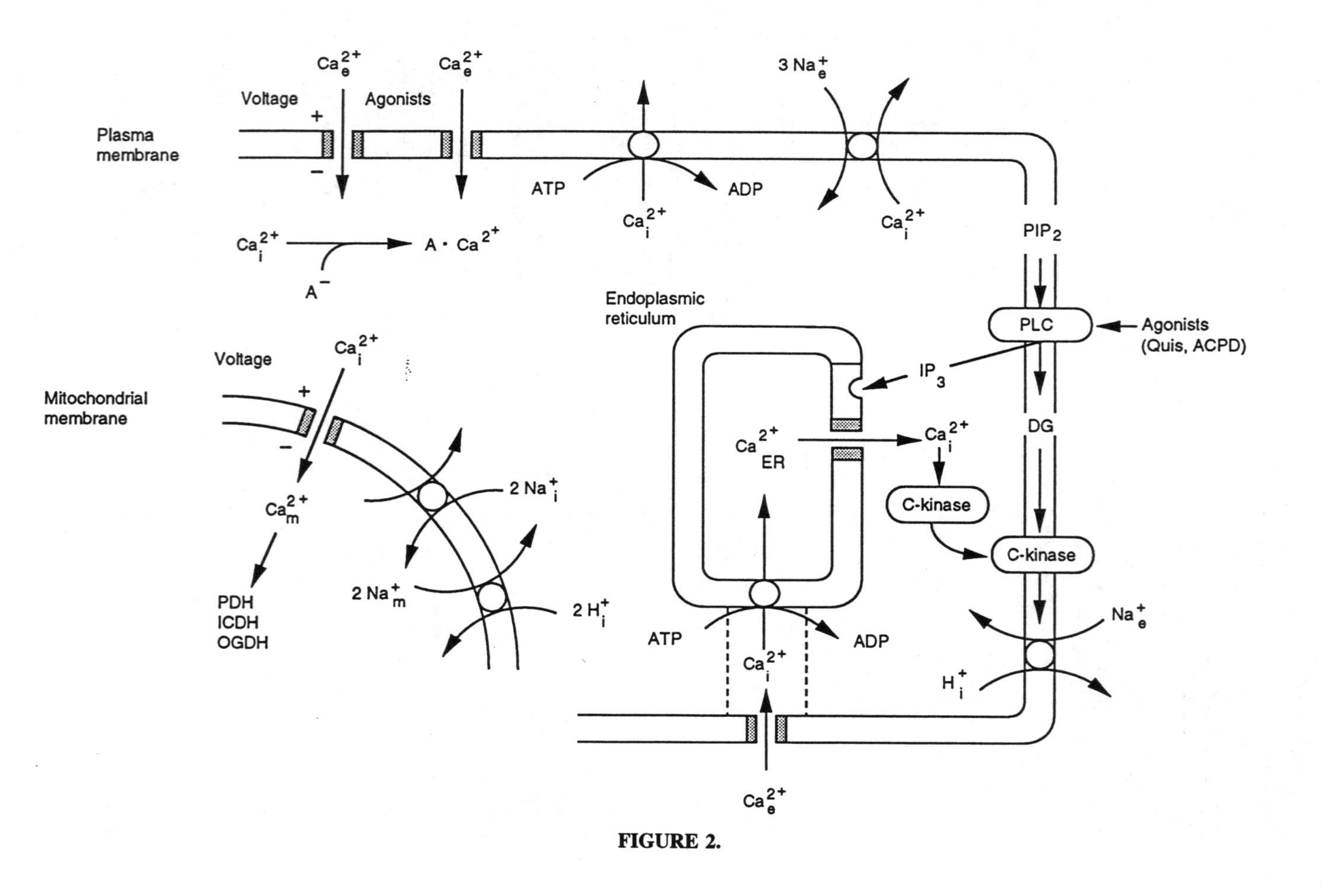
Plasma membrane
Voltage
+
−
Ca2+e
Agonists
Ca2+e
ATP
ADP
Ca2+i
3 Na+e
Ca2+i
Ca2+i
A · Ca2+
A−
PIP2
PLC
Agonists (Quis, ACPD)
IP3
DG
Endoplasmic reticulum
Ca2+ER
Ca2+i
C-kinase
C-kinase
Na+e
H+i
ATP
ADP
Ca2+i
Ca2+e
Mitochondrial membrane
Voltage
Ca2+i
Ca2+m
PDH
ICDH
OGDH
2 Na+i
2 Na+m
2 H+i

FIGURE 2.

Binding, Sequestration, and Extrusion

In neurons, activation of surface receptors and/or depolarization gives rise to an appreciable influx of Ca^{2+}. Under certain nonpathologic circumstances, such as spreading depression (SD), almost all extracellular calcium is translocated into intracellular fluids. With an extracellular calcium concentration (Ca^{2+}_e) of 1.3 mM and a 20% extracellular fluid volume, this gives a calcium load of about 250 $\mu M \cdot kg^{-1}$ of intracellular fluid. In energy-competent cells, virtually all of the calcium entering cells is bound or sequestered, and subsequently it is extruded from the cell by mechanisms that are outlined in FIGURE 2. Binding occurs to phospholipids and proteins. Some of the latter, like calcibindin and parvalbumin, serve as calcium buffers, while others, such as calmodulin, are effector proteins.[8,17] The calcium-binding capacity of some of these proteins is reduced by acidosis, suggesting competition between Ca^{2+} and H^+ for common binding sites.[45,46]

Both the extrusion of calcium from the cell and its intracellular sequestration are energy-requiring events.[7,9] Extrusion occurs by a high affinity-low capacity, calmodulin-dependent ATPase and by a low affinity-high capacity 3 Na^+/Ca^{2+}- exchanger. The latter is electrogenic, and the exchange is thus driven by both the Na^+ gradient and the membrane potential. Inasmuch as the exchanger is electrogenic, it can also catalyze calcium uptake in exchange for Na^+ efflux, if the conditions deviate from the physiologic ones. This is believed to occur if the membrane depolarizes and/or the Na^+ gradient is reduced (for literature, see ref. 47). It is not unlikely that partial collapse of the Na^+ gradient, with a rise in Na^+_i, can cause an increase in Ca^{2+}_i by enhancing Ca^{2+}/Na^+ exchange across both the mitochondrial and the plasma membranes.

Calcium transport across the membranes of intracellular organelles, like the ER, also occurs at the expense of ATP energy. This means that at any given moment, Ca^{2+}_i is a function of the balance between pump activity and leak pathways across plasma and intracellular membranes. A rise in Ca^{2+}_i may result either from a decrease in pump activity, as occurs when cellular energy production fails, or from an increase

FIGURE 2. Schematic diagram of cellular calcium homeostasis. Cellular calcium entry occurs through voltage-sensitive and agonist-operated calcium channels (*top left*). In the normal situation, almost all intracellular calcium is bound or sequestered. Numerous high- and low-molecular-weight compounds with negative sites (A^-) are present to bind Ca^{2+}. Calcium can be sequestered into the mitochondria (*middle left*) and by the endoplasmic reticulum (ER; *middle right*). Intramitochondrial calcium (Ca^{2+}_m) activates enzymes such as the pyruvate, isocitrate, and oxoglutarate dehydrogenases (PDH, ICDH, and OGDH, respectively). Agonist activation of receptors coupled to membrane-bound phospholipase C (PLC; right upper) produces inositol trisphosphate (IP_3), which releases calcium from the ER. Such agonists encompass glutamate, quisqualate, and ACPD. The calcium released from the ER may be replaced by external calcium by entry through a separate channel (*bottom*). Extrusion of calcium from the cell normally occurs through calcium-activated ATPase (*top middle*) and through $3Na^+$-Ca^{2+} exchange (*top right*). Activation and membrane translocation of PKC is conventionally thought to stimulate Na^+- H^+ exchange as well as membrane channels or receptors, such as that activated by AMPA. Reproduced with permission from ref. 119.

in the calcium leak.[31,32,48] As will be discussed, ischemia and hypoglycemia produce a rise in Ca^{2+}_i both by interfering with calcium pumps and by enhancing leak pathways.

Mitochondria play a special role in the regulation of Ca^{2+}_i. Under normal circumstances, that is, low or moderately raised values for Ca^{2+}_i, translocation of Ca^{2+} between cytoplasma and mitochondria probably serves to regulate intramitochondrial calcium concentration and thereby the activity of intramitochondrial dehydrogenases.[3] However, when Ca^{2+}_i rises above the "set point," mitochondria serve as an important site for calcium sequestration.[49] This is probably a major reason that cells can tolerate large calcium loads, with only moderate rises in Ca^{2+}_i. Clearly though, when the buffering capacity of mitochondria decreases, as occurs in ischemia/hypoxia, Ca^{2+}_i can rise appreciably.

MECHANISMS OF CALCIUM-RELATED CELL DAMAGE

An excessive rise in Ca^{2+}_i threatens cells in several ways. As recognized decades ago, tissues such as the heart experience ischemia with recirculation because mitochondria become overloaded with calcium. This occurs because when regaining their membrane potentials, mitochondria can take up massive amounts of calcium if Ca^{2+}_i remains elevated because of continued influx through "leaky" plasma membranes. This is less likely to occur in brain tissues because the low permeability of the blood-brain and the blood-CSF barriers restricts the availability of calcium. Thus, in the short perspective, the source of the rise in Ca^{2+}_i is the calcium contained in extracellular fluid (about 0.25 mM kg^{-1} of intracellular water), and it is only during sustained, incomplete ischemia that a substantial increase in total calcium content of brain tissues occurs. In many circumstances, therefore, the detrimental effects of a raised calcium concentration should be exerted on targets other than the mitochondria.

FIGURE 3 summarizes events triggered by a rise in Ca^{2+}_i. Most of these events reflect the action of calcium on enzymes, notably lipases, protein kinases (or phosphatases), proteases, and endonucleases.[48,50] The effect may be degradation beyond repair or changes in membrane function which allow only limited cell survival. Such changes encompass alterations in the function of receptors and ion channels which are secondary to changes in protein phosphorylation, inhibition of protein synthesis, or altered gene expression.

Lipolysis

Phospholipids are hydrolyzed by the action of phospholipases, yielding free fatty acids and other compounds, many of which have important biologic activities. It is likely that such compounds modulate the outcome of ischemic or hypoglycemic insults.[26,31,32,51] Inasmuch as resynthesis of the parent compounds requires energy in the form of ATP (or CTP), concentrations of some of the biologically active degradation products of phospholipids represent the balance between the rate of hydrolysis, which is Ca^{2+}-dependent, and the availability of ATP (and CTP).

Several phospholipids are broken down by phospholipases of the A_2 type (PLA_2) to lysophospholipids and free fatty acids including arachidonic acid. Both types of primary degradation products can act as detergents and ionophores, thereby leading

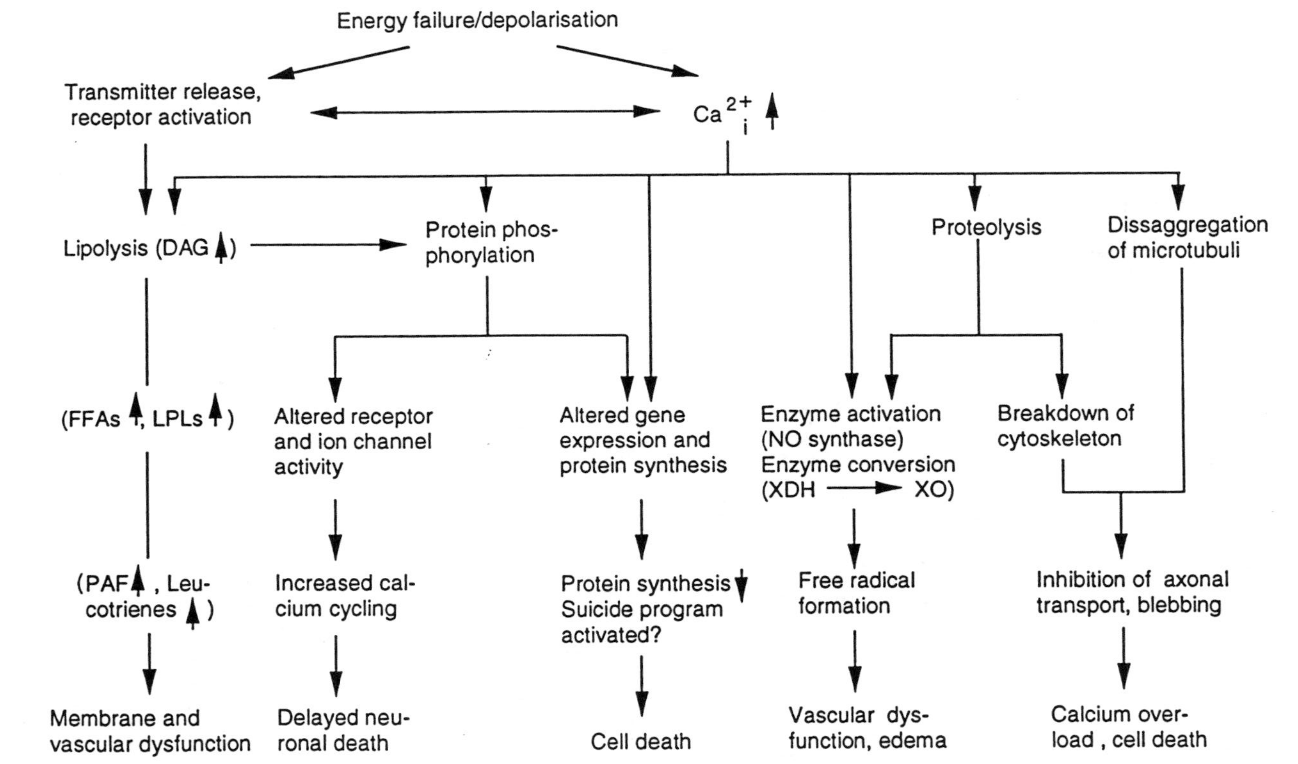

FIGURE 3. Diagram illustrating the primary effects of depolarization/receptor activation and the secondary effects of increased intracellular calcium concentration (Ca^{2+}_i). The diagram suggests that the major adverse effects of a massive rise in Ca^{2+}_i involve lipolysis, altered protein phosphorylation, proteolysis, and disassembly of microtubuli. Lipolysis encompasses production of diacylglycerides (DAG), lysophospholipids (LPL), free fatty acids (FFA), and platelet-activating factor (PAF). An altered protein phosphorylation is envisaged to influence protein synthesis and lead to altered gene transcription, possibly with activation of a latent suicide program. A rise in Ca^{2+}_i also activates enzymes (eg, nitric oxide synthase) that lead to the production of free radicals; calcium-activated proteolysis could act similarly by converting xanthine dehydrogenase (XDH) to xanthine oxidase (XO). Reproduced with permission from ref. 119.

to membrane dysfunction. An additional pathway of phospholipid breakdown by PLA_2 is the one leading to the formation of platelet-activating factor (PAF) from 1-alkyl-2-acyl-sn-glycerophosphocholine, a minor component of the phosphatidylcholine fraction of many cell membranes.[52,53] Although the primary products are AA and lyso-PAF, a biologically inactive compound, PAF is formed in an additional reaction catalyzed by an acetyltransferase. Platelet-activating factor is proinflammatory and vasoactive, promoting hemostasis by causing platelet aggregation and adherence of leukocytes to endothelial cells. Whenever formed, PAF can also activate PLA_2, thereby triggering hydrolysis of other phospholipids with the additional formation of free fatty acids, including arachidonic acid.

Both PLA_2 and PLC may be activated as a result of stimulation of surface receptors.[6,10,37,54] Receptor coupling to PLC causes hydrolysis of PIP_2 to DAG and IP_3, both acting as second messengers. A major effect of the DAGs formed is to activate protein kinase C (PKC). This enzyme is partly localized in the cytoplasm, and to be activated by DAG, which acts in the plane of the membrane, the enzyme must be translocated from cytoplasm to membrane.[6,37,55] Because such translocation is triggered by calcium, a rise in Ca^{2+}_i, of whatever cause, will activate PKC provided that DAG is formed. It has been postulated that a massive rise in Ca^{2+}_i can cause sustained translocation and activation of PKC and that such activation, like the activation of other protein kinases, may alter membrane function by phosphorylating receptors, ion channels, or translocases.[56,57] Although a resulting upregulation of synaptic efficacy may explain physiologic events such as long-term potentiation,[58] it is questionable if the mechanisms are the same after transient ischemic insults (see below).

Arachidonic acid is a key compound in the series of events that are triggered by phospholipase activation. This is because arachidonic acid is the substrate of the cyclooxygenase and lipoxygenase enzymes and thus the precursor of prostaglandins and leukotrienes. The role played by these oxidation products of arachidonic acid in the pathogenesis of ischemic/hypoxic brain damage is not accurately known, but a vascular effect has been suggested[59,60]; however, see also ref. 61. It is possible that leukotrienes, working in conjunction with PAF and free radicals, have a proinflammatory effect, promoting microvascular dysfunction and edema.[52,62]

Proteolysis and Disassembly of Microtubuli

Proteolytic damage to cytoskeletal components as well as disassembly of microtubuli is a normally occurring calcium-dependent event that subserves functions underlying plasticity and memory function.[63,64] The enzymes catalyzing proteolysis encompass the calpains which exist in two major forms; soluble and membrane-bound. Because the membrane-bound forms are activated by lower calcium concentrations, translocation of the enzyme to plasma and ER membranes is probably a requirement for normal activity. At very high calcium concentrations, though, both forms of the enzyme can become activated.

Pathophysiologic events triggered by excessive activation of proteases (or excessive disassembly of microtubuli) probably lead to such extensive cleavage of the cytoskeleton that intracellular communication via axoplasmic transport comes to a halt, and severage of the proteins that normally anchor the plasma membrane to the cytoskeleton may cause "blebbing" of the plasma membrane, an alteration that is considered a cause of abnormal membrane permeability, notably to calcium.[50] The importance of proteolytic events in the pathogenesis of neuronal necrosis following transient ischemia is supported by results demonstrating amelioration of the damage by calpain inhibitors.[65] However, there is as yet no data demonstrating blebbing of plasma membrane in brain cells. In theory, such a lesion is not unlikely to occur in cells with a high density of glutamatergic receptors. This is because such cells will incur a massive influx of calcium, activating the calpains, but will also be subjected to the physical strains imparted by the massive influx of Na^+ and Cl^-.

Protein Phosphorylation, Gene Expression, and Protein Synthesis

Protein synthesis, which is an energy-demanding task, is depressed or arrested when cellular ATP concentrations fall, such as occurs in ischemia, hypoxia, and hypoglycemia; in fact, even moderate perturbation of the cellular energy state occurring in status epilepticus is accompanied by depressed protein synthesis.[66,67] During recirculation following a transient ischemic insult, overall protein synthesis recovers very slowly (resistant areas) or not at all (vulnerable areas), suggesting long-lasting (or lasting) inhibition of initiation of protein synthesis, probably at the level of the eukaryotic initiation factor 2α.[66–69] However, new mRNA transcripts are expressed in metabolically perturbed tissue, and new proteins are preferentially synthesized. New mRNA transcripts of the immediate early genes, such as c-*fos* and c-*jun*, are translated into the protein components of the API complex which, by binding DNA promotor regions, triggers further transcription of other genes.[2] The sequence of events encompasses transcription of mRNAs for several growth factors.[70] The new proteins that are thus synthesized include heat shock and stress proteins which are highly conserved from prokaryotes to man. Probably synthesis of such proteins represents a homeostatic mechanism that helps cells survive damage.

In this sequence of events, calcium probably plays a pivotal role because a rise in Ca^{2+}_i may by itself serve as a trigger for the transcription of new genes.[2,4] Furthermore, apart from activating PKC I, II, and III, calcium also activates calmodulin-dependent protein kinase II (CAM-PK II). As already mentioned, these kinases may phosphorylate proteins in membrane receptors and ion channels, thereby altering membrane function. However, together with a series of phosphatases these and other kinases (such as MAPK [mitogen-activated protein kinase] and tyrosine kinases) form an intricate signal transduction chain that regulates not only gene expression, but also protein synthesis.[69,71] It is possible, therefore, that cell dysfunction due to a loss of cell calcium homeostasis is mediated by sustained alteration of protein kinases and/or phosphatases. Conceivably, sustained inhibition of protein synthesis is detrimental because it prevents the translation of proteins that are required for the survival of cells that have been damaged by the preceding insult. By the same token, failure to synthesize such proteins could be lethal to cells that are marginally deprived of oxygen or glucose.

Free Radicals

FIGURE 3 suggests a link between a rise in Ca^{2+}_i, and enhanced production of free radicals. Two major sources of oxygen radicals ($\cdot O_2$- and $\cdot OH$) and H_2O_2 have been suggested. One, which is triggered by the $Ca^{2\pm}$-mediated activation of PLA_2 and PLC, involves the production of free radicals mainly ($\cdot O_2$-) as metabolic intermediates when arachidonic acid is metabolized by cyclooxygenase and lipoxygenase. The other pathway, in which hypoxanthine and xanthine are metabolized to uric acid by xanthine oxidase leads to the formation of $\cdot O_2$- and H_2O_2. According to a now classic concept, reperfusion injury results when xanthine dehydrogenase is converted to xanthine oxidase by limited proteolysis, a calcium-dependent event.[72] Such conversion may not occur in the brain.[73] Thus, the major producers of $\cdot O_2$- may be preexisting xanthine oxidase, oxidative breakdown of arachidonic acid, and mitochondrial electron transport.[74]

It is widely believed that the initial oxidizing species is $\cdot OH$, the hydroxyl radical.[74] A major source of $\cdot OH$ is the iron-catalyzed Haber-Weiss reaction, that is, the reaction between $\cdot O_2$- and H_2O_2 in the presence of "delocalized" Fe^{2+}/Fe^{3+}. Such delocalization, that is, the release of Fe^{2+}/Fe^{3+} from protein bindings, may occur in response to acidosis.[75] An alternative source of $\cdot OH$ has been suggested in that, at least *in vitro*, $\cdot O_2$- and nitric oxide can react to yield peroxynitrate, which then decomposes to $\cdot OH$ in a pH-dependent reaction.[76]

Microvessels may be the primary target of free radical damage, in part because they contain a high concentration of xanthine oxidase/xanthine dehydrogenase, but also because they are exposed to high O_2 tensions and to circulating thrombocytes and leukocytes. If the latter attach to endothelial cells and become activated, they can initiate inflammatory reactions that lead to microvascular dysfunction.[77] For further literature, see refs. 31, 32, and 52.

Interaction between Ca^{2+} and H^+

Low extracellular pH is known to affect both calcium influx and calcium extrusion in muscle tissues,[20] and as a decrease in pH_i can also release bound or sequestered calcium,[45,46] the effect of acidosis or alkalosis on Ca^{2+}_i is difficult to predict.

Recent results demonstrate that lowering the extracellular pH (pH_e) decreases NMDA receptor-gated currents, reduces $^{45}Ca^{2+}$ uptake by neurons in culture during glutamate exposure or anoxia, and ameliorates neuronal damage *in vitro* due to agonist exposure or lack of oxygen.[78,79,80] It is tempting to assume, therefore, that acidosis protects cells from anoxia by reducing calcium influx through NMDA-gated channels.[81] In support of this contention are results showing that acidosis limits the rise in Ca^{2+}_i during glutamate exposure or K^+-induced depolarization.[82]

The question arises if these *in vitro* results are applicable to the *in vivo* situation. Available results are conflicting. In conformation of the *in vitro* results, decreases in pH_e (and pH_i) reduced Ca^{2+} influx into cells during ischemia, as reflected in the rate of fall in Ca^{2+}_e.[83] FIGURE 4 shows results obtained when intraischemic acidosis was varied by alterations in preischemic plasma glucose concentration. In hypoglycemic animals ($pH_e \approx 7.1$) the rate of fall in Ca^{2+}_e was rapid, 90% of the reduction occurring within 5 seconds. In normoglycemic subjects ($pH_e \approx 6.8$) it was clearly slower, and

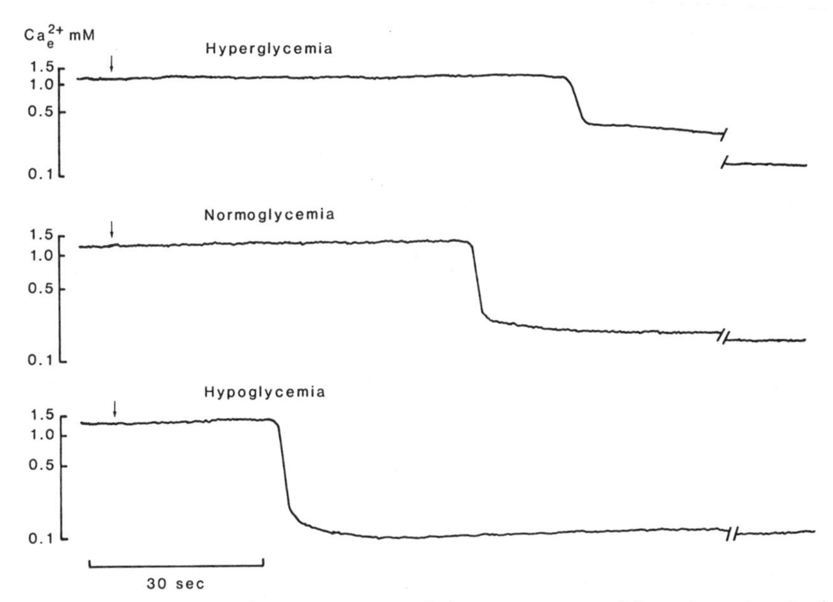

FIGURE 4. Changes in Ca^{2+}_e in the rat cortex in hyper-, normo-, and hypoglycemic animals following cardiac arrest. The rate of fall in Ca^{2+}_e was clearly influenced by the preischemic plasma glucose level. The maximal slope of fall in Ca^{2+}_e in hyperglycemic animals was markedly slower than that in hypo- or normoglycemic animals. The fast decrease was followed by a slow, gradual fall in Ca^{2+}_e. *Arrows* indicate the onset of ischemia. *Double bars* indicate a break in the traces between 2 and 6.8 minutes (hyperglycemic group) and between 2 and 4.8 minutes (hypo- and normoglycemic group). Reproduced with permission from ref. 83.

in hyperglycemic ones ($pH_e \approx 6.4$) not only was it additionally slowed down, but also it occurred in two distinct phases, the transition occurring at Ca^{2+}_e values of about 0.6 mM. The second phase was very slow, a stable value of about 0.1 mM being obtained first after 5-7 minutes. This marked delay in calcium influx was obviously due to acidosis because it could be reproduced by excessive hypercapnia in normoglycemic animals. The effect of acidosis could also be duplicated by MK-801, suggesting that low pH reduces calcium influx by blocking NMDA-gated ion channels. Clearly, acidosis and MK-801 only *reduce* the rate of uptake of Ca^{2+} during ischemia.

The controversy revolves around the effects of acidosis on ischemic brain damage. Thus, whereas acidosis *in vitro* protects neurons against glutamate-induced or anoxic damage,[81] hyperglycemia *in vivo* invariably aggravates damage due to dense transient ischemia.[84,85] At least in part, such exaggeration should be due to enhanced acidosis because hypercapnia also aggravates damage when induced in normoglycemic animals.[86] Conceivably, the beneficial effect that acidosis may have by reducing the rate of calcium influx into cells possessing a high density of NMDA receptors is outweighed by the effects of acidosis on other cellular targets such as endothelial cells or astrocytes. In this context, it is of interest that following a long period of transient ischemia, hyperglycemic animals failed to regulate Ca^{2+}_i as did normoglycemic animals.[87]

Glutamate Exposure, Intracellular Calcium, and Cell Death in Vitro

Many reports attest to the fact that transient exposure of neurons in culture (or slices) to glutamate and related EAAs leads to rapid or delayed cell death and that such devitalization of cells is coupled to influx of Ca^{2+} from the medium. (For recent articles and further literature, see refs. 88, 89, 92 and 93.) There are two major controversial issues in this field. First, it has been proposed that neurons *in vitro* may be particularly susceptible to excitotoxic cell death because they lack the trophic support that cells have *in vivo*.[90,91] If this is true, excitotoxic, calcium-mediated cell death in cultured cells may not replicate conditions existing *in vivo*. Second, it has been debated whether excitotoxic, calcium-induced damage is related to the ensuing rise in Ca^{2+}_i.

Additional information now exists on the second issue. Thus, although it was proposed that excitotoxic damage does not correlate with the rise in Ca^{2+}_i, subsequent results suggest that neuronal cell death correlates with the influx of calcium.[92,93] The results of Tymianski *et al.* (1993)[93] are particularly relevant to this question. These investigators noted that although both exposure to glutamate and K^+-induced depolarization increased Ca^{2+}_i, only the former was toxic. They could also demonstrate that independent of the Ca^{2+}_i recorded, the site of Ca^{2+} influx correlated with cell death. Notably, influx of Ca^{2+} through NMDA receptor-gated channels was potentially more detrimental than was influx through other channels. Thus, if the site of calcium entry is taken into account, the severity of the Ca^{2+} load correlates with the damage incurred.

CALCIUM-RELATED DAMAGE *IN VIVO*

In vivo, the coupling among release of EAA, influx of calcium, and the rise in Ca^{2+}_i and cell damage is much less obvious than *in vitro*. It has been established that both ischemia and hypoglycemia lead to increased release/diminished reuptake of EAAs; however, it is much less clear that the cell damage incurred in these conditions is "excitotoxic." For example, although it has been proposed that interruption of the glutamatergic input to selectively vulnerable brain regions ameliorates ischemic and hypoglycemic brain damage (eg, ref. 94), contradictory results exist.[95] Furthermore, available results make it clear that although both NMDA and AMPA receptor antagonists ameliorate brain damage due to focal ischemia and hypoglycemia, only AMPA receptor antagonists are efficacious in global/forebrain ischemia.[32,96–98] In view of these contradictory results, it is justified to scrutinize calcium metabolism in four different conditions. These are illustrated in FIGURE 5. In this figure, unbroken lines denote recorded changes in Ca^{2+}_e and interrupted lines the assumed values for Ca^{2+}_i. (For literature, see refs. 31, 32, and 48.)

Spreading Depression

Spreading depression, elicited by topical application of K^+ or glutamate, by electrical stimulation, or by a stab wound, is accompanied by cellular influx of Ca^{2+},

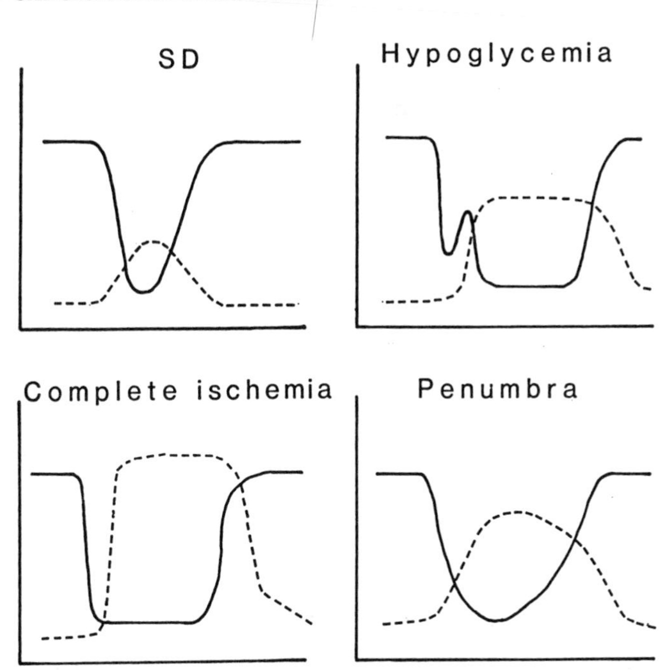

FIGURE 5. Changes in measured extracellular Ca^{2+} concentration (*unbroken lines*) and assumed intracellular calcium concentration (*interrupted lines*) during spreading depression (SD), hypoglycemia, complete global ischemia, and focal ischemia (penumbra zone). Modified from refs. 31 and 32.

Na^+, and Cl^-, and efflux of K^+ (eg, ref. 99). Thus, whenever elicited, SD leads to a calcium transient (overload). Yet, when repeated over a period of 4-5 hours, SD fails to induce permanent neuronal damage.[100] This suggests that in tissues with normal energy metabolism, calcium binding, sequestration, and extrusion are efficient enough to maintain Ca^{2+}_i below values that elicit harmful effects. The situation may be different in energy-compromised tissues in which SD gives rise to prolonged Ca^{2+} transients.[101] However, it remains to be seen if an encrouchment of energy production per se is sufficient to cause damage as a result of depolarizations of the SD type.

Hypoglycemic Coma

In SD, the calcium transient is brief, and it normally occurs in tissue that is uncompromised by energy failure. In hypoglycemic coma, the ionic transient occurs in tissue that is energetically compromised.[102,103] Thus, even though there may be initial, partial recovery, following the initial depolarization, Ca^{2+}_e then falls to very low levels which are sustained until the transient is terminated by glucose infusion. Although hypoglycemic coma leads to cellular uptake of calcium and to a rise in Ca^{2+}_i,[104] the actual values obtained are not known. We speculate that hypoglycemic coma leads to less damage than ischemia (for comparable insult durations) because 20-30% of the ATP stores are preserved.[105] Accordingly, the rise in Ca^{2+}_i should be less marked than in dense ischemia. This contention is supported by the fact that a hypoglycemic calcium transient of 10 minutes' duration does not normally lead to neuronal cell damage.[106]

Ischemia

It is profitable to subdivide ischemic conditions into two major ones: (1) forebrain or global ischemia of a duration that makes long-term recovery possible, and (2) focal ischemia due to middle cerebral artery occlusion.[23] The characteristic features of the former are that ischemia is dense and transient and that damage occurs after a delay whose length varies among vulnerable neuronal populations. In contrast, focal ischemia is usually more long-lasting (or permanent); furthermore, the ischemia affects only a part of the brain, more specifically an area whose periphery receives some blood supply from collateral vessels. Usually, there is therefore a densely ischemic core and a less densely ischemic perifocal, penumbra zone. The importance of this is that cells in the penumbra are potentially viable for one or several hours and can be salvaged by pharmacologic treatment.

These basic differences in the characteristics of ischemic damage have important implications for calcium-related pathology. In global/forebrain ischemia, cells are exposed to a calcium transient, from which they may recover; alternatively, this calcium transient could be the cause of their delayed death. Several mechanisms may be envisaged which lead to this end result. A likely cause is a disturbance of the signal transduction chain which starts with glutamate and growth factor receptors and which leads to transcription of mRNA and translation of DNA via a web of reactions encompassing a series of intracellular messengers and several types of kinases and phosphatases.[68,107,108] It is likely that calcium is involved in both the initial transient and the ultimate cell death. Thus, cell permeability to calcium may increase by phosphorylation/dephosphorylation of receptors or ion channels; alternatively, the defect is in the balance between intracellular release and resequestration of calcium.[109] It is currently under debate if delayed neuronal necrosis represents one form of programmed cell death, encompassing expression of a genetic suicide program, and if failure to transcribe mRNAs and translate DNAs for trophic factors is what starts the sequence of events leading to cell death (see ref. 110a and 110b for opposing views).

Mechanisms of cell death may be different in the penumbra zone of a stroke lesion. Thus, this tissue is energetically compromised, and cells are gradually recruited in the infarction process. Irregular K^+ transients occur in the penumbra zone[111], and because these are accompanied by cellar uptake of Ca^{2+},[112] it does not seem unlikely that repeated calcium transients in energetically compromised tissue leads to cumulative damage.[31,32] This notion is supported by results showing that a lowering of either plasma glucose content or CBF leads to progressive increases in the duration of calcium transients accompanying SD (ref. 101 and unpublished results). However, it remains to be seen if, and under what conditions, such transients lead to cell damage. Possibly, calcium transients are only deleterious if they work in conjunction with other adverse factors, such as damage to microvessels and glial cells.

Synthesis and Prospectus

Spreading depression, even if repeatedly elicited in partly energy-compromised tissue, does not cause brain damage. As stated, this may be because the Ca^{2+} transient, if occurring in energy-competent tissue, does not lead to a rise in Ca^{2+} of sufficient magnitude and duration to trigger cell damage. In hypoglycemic coma, the rise in Ca^{2+}_i is often long enough to trigger irreversible damage; however, for equal insult duration, hypoglycemic coma yields less damage than does dense ischemia.[102] This may be explained in one of several ways: (1) hypoglycemic coma yields less damage because it only reduces ATP content to 20-30% of control, (2) it is not accompanied by acidosis, (3) it does not lead to a reduction/oxidation change as observed in ischemia, or (4) it is not accompanied by the type of microvascular damage that is observed in underperfused tissues or those subjected to ischemia-reperfusion.

Tentatively, this suggests that ischemia leads to worse damage because it is accompanied by enhanced production of free radicals. Such damage may affect lipid components of tissue or proteins.[74,113,114] In gerbils, transient ischemia is accompanied by oxidation of proteins,[114] but in the rat such oxidation has not been observed.[115] Nonetheless, free radical scavengers ameliorate the damage incurred by relatively brief, transient ischemia, suggesting that free radicals contribute to neuronal necrosis in a setting in which little damage occurs to microvessels.[116,117]

These results lead to the tentative conclusion that calcium transients are particularly detrimental in a setting in which either acidosis or a redox change triggers the production of free radicals. This may be the case not only in ischemia-reperfusion, but also in aging and Alzheimer's disease.[118] We can thus envisage a cascade of potentially harmful reactions, which start with the release of excitatory amino acids and their interaction with postsynaptic receptors, continues with the influx of calcium and an unphysiologic rise in Ca^{2+}_i, and ends in calcium-triggered events, including the enhancement of reactions leading to the production of destructive, free radical species. Obviously, this cascade of reactions may cause damage not only in acute disease, such as in ischemia-reperfusion, but also in diseases leading to gradual destruction of neurons, such as aging and Alzheimer's disease.

ACKNOWLEDGMENT

The secretarial work of Katarina Månson is gratefully acknowledged.

REFERENCES

1. KENNEDY, M. 1989. Regulation of neuronal function by calcium. Trends Neurosci. **12:** 417-424.
2. MORGAN, J. I. & T. CURRAN. 1989. Stimulus-transcription coupling in neurons: Role of cellular immediate-early genes. TINS. **12:** 459.
3. MCCORMACK, J. G., A. P. HALESTRAP & R. M. DENTON. 1990. Role of calcium ions in regulation of mammalian intramitochondrial metabolism. Physiol. Rev. **70:** 391-425.
4. GINTY, D. D., H. BADING & M. E. GREENBERG. 1992. Trans-synaptic regulation of gene expression. Curr. Opin. Neurobiol. **2:** 312-316.

5. BADING, H., D. GINTY & M. GREENBERG. 1993. Regulation of gene expression in hippocampal neurons by distinct calcium signalling pathways. Science **260:** 181-186.
6. NISHIZUKA, Y. 1986. Studies and perspectives of protein kinase C. Science **233:** 305-312.
7. CARAFOLI, E. 1987. Intracellular calcium homeostasis. Ann. Rev. Biochem. **56:** 395-433.
8. MCBURNEY, R. & I. NEERING. 1987. Neuronal calcium homeostasis. Trends Neurosci. **10:** 164-169.
9. BLAUSTEIN, M. 1988. Calcium transport and buffering in neurons. Trends Neurosci. **11:** 438-443.
10. BERRIDGE, M. J. & R. F. IRVINE. 1989. Inositol phosphates and cell signalling. Nature **341:** 197-205.
11. IRVINE, R. 1990. "Quantal" Ca^{2+} release and the control of Ca^{2+} entry by inositol phosphates—a possible mechanism. FEBS Lett. **263:** 5-9.
12. BERRIDGE, M. 1991. Calcium oscillations. J. Biol. Chem. **265:** 9583-9586.
13. MAYER, M. & G. WESTBROOK. 1987. The physiology of excitatory amino acids in the vertebrate central nervous system. Prog. Neurobiol. **28:** 197-276.
14. LODGE, D. & G. L. COLLINGRIDGE, Eds. 1990. The pharmacology of excitatory amino acids. Trends Pharmacol. Sci. 1-89.
15. SLADECZEK, F., M. RÉCASENS & J. BOCKAERT. 1988. A new mechanism for glutamate receptor action: Phosphoinositide hydrolysis. TINS. **11:** 545-549.
16. BASKYS, A. 1992. Metabotropic receptors and slow excitatory actions of glutamate agonists in the hippocampus. Trends Neurosci. **15:** 92-96.
17. BAIMBRIDGE, K., M. CELIO & J. ROGERS. 1992. Calcium-binding proteins in the nervous system. Trends Neurosci. **15:** 303-308.
18. AUGUSTINE, G. J., E. M. ADLER & M. P. CHARLTON. 1991. The calcium signal for transmitter secretion from presynaptic nerve terminals. Ann. N.Y. Acad. Sci. **635:** 365-381.
19. LLINÁS, R., M. SUGIMORI & R. SILVER. 1992. Microdomains of high calcium concentration in a presynaptic terminal. Science **256:** 677-679.
20. SIESJÖ, B. K. 1988. Historical overview. Calcium, ischemia, and death of brain cells. Ann. N.Y. Acad. Sci. **522:** 638-661.
21. SIESJÖ, B. K. 1981. Cell damage in the brain: A speculative synthesis. J. Cereb. Blood Flow Metab. **1:** 155-185.
22. KHACHATURIAN, Z. S. 1984. Towards theories of brain aging. *In* Handbook of Studies on Psychiatry and Old Age. D. S. Kay & G. W. Burrows, Eds. Vol. 7-30. Elsevier. Amsterdam.
23. MELDRUM, B. 1985. Possible therapeutic applications of antagonists of excitatory amino acid neurotransmitters. Clin. Sci. **68:** 113-122.
24. ROTHMAN, S. M. & J. W. OLNEY. 1987. Excitotoxicity and the NMDA receptor. TINS. **10:** 299-302.
25. WIELOCH, T. 1985. Hypoglycemia-induced neuronal damage prevented by an *N*-methyl-D-aspartate antagonist. Science **230:** 681-683.
26. FAROOQUI, A. A. & L. A. HORROCKS. 1991. Excitatory amino acid receptors, neural membrane phospholipid metabolism and neurological disorders. Brain Res. **16:** 171-191.
27. CHOI, D. W. 1988. Glutamate neurotoxicity and diseases of the nervous system. Neuron **1:** 623-634.
28. CHOI, D. W. & S. M. ROTHMAN. 1990. The role of glutamate neurotoxicity in hypoxic/ischemic neuronal death. Ann. Rev. Neurosci. **13:** 171-182.
29. MATTSON, M. 1992. Calcium as sculptor and destroyer of neural circuitry. Exp. Gerontol. **27:** 29-49.
30. KHACHATURIAN, Z. S., C. W. COTMAN & J. W. PETTEGREW. 1989. Calcium, membranes, aging, and Alzheimer's disease. Ann. N.Y. Acad. Sci. Vol. **568.**

31. SIESJÖ, B. K. 1992. Pathophysiology and treatment of focal cerebral ischemia. I. Pathophysiology. J. Neurosurg. **77:** 169-184.
32. SIESJÖ, B. K. 1992. Pathophysiology and treatment of focal cerebral ischemia. II. Mechanisms of damage and treatment. J. Neurosurg. **77:** 337-354.
33. TSIEN, R., P. ELLINOR & W. HORNE. 1991. Molecular diversity of voltage-dependent Ca^{2+} channels. Trends Pharmacol. Sci. **12:** 349-354.
34. MILLER, R. J. 1992. Voltage sensitive Ca^{2+} channels. J. Biol. Chem. **267:** 1403-1406.
35. SNUTCH, T. P. & P. B. REINER. 1992. Ca^{2+} channels: diversity of form and function. Curr. Opin. Neurobiol. **2:** 247-253.
36. SPEDDING, M. & R. PAOLETTI. 1992. Classification of calcium channels and the sites of action of drugs modifying channel function. Pharmacol. Rev. **44:** 363-376.
37. SHINOMURA, T., Y. ASAOKA, M. OKA, K. YOSHIDA & Y. NISHIZUKA. 1991. Synergistic action of diacylglycerol and unsaturated fatty acid for protein kinase C activation: Its possible implications. Proc. Natl. Acad. Sci. USA **88:** 5149-5153.
38. TSIEN, R. W. & R. Y. TSIEN. 1990. Calcium channels, stores, and oscillations. Ann. Rev. Cell Biol. **6:** 715-760.
39. HOLLMANN, M., M. HARTLEY & S. HEINEMANN. 1991. Ca^{2+} permeability of KA-AMPA-gated glutamate receptor channels depends on subunit composition. Science **252:** 851-853.
40. ROBITAILLE, R. & M. CHARLTON. 1992. Presynaptic calcium signals and transmitter release are modulated by calcium-activated potassium channels. J. Neurosci. **12:** 297-305.
41. GOLA, M. & M. CREST. 1993. Colocalization of active KCa channels and Ca^{2+} channels with Ca^{2+} domains in Helix neurons. Neuron **10:** 689-699.
42. BEN-ARI, Y. 1990. Hippocampal potassium ATP channels and anoxia: Presynaptic, postsynaptic or both? TINS. **13:** 409-410.
43. ASHCROFT, F. M. 1988. Adenosine 5′-triphosphate-sensitive potassium channels. Ann. Rev. Neurosci. **11:** 97-118.
44. TAYLOR, C. P. 1993. Na^{+} currents that fail to inactivate. Trends Neurosci. **16:** 455-459.
45. ABERCROMBIE, R. F. & C. E. HART. 1986. Calcium and proton buffering and diffusion in isolated cytoplasm from myxicola axons. Am. J. Physiol. **250:** C391-C405.
46. BAKER, P. F. & J. A. UMBACH. 1987. Calcium buffering in axons and axoplasm of loligo. J. Physiol. **383:** 369-394.
47. SIESJÖ, B. K., F. BENGTSSON, W. GRAMPP & S. THEANDER. 1989. Calcium, excitotoxins, and neuronal death in the brain. Ann. N.Y. Acad. Sci. **568:** 234-251.
48. SIESJÖ, B. 1991. The role of calcium in cell death. *In* Neurodegenerative disorders: Mechanisms and prospects for therapy. D. Price, A. Aguayo & H. Thoenen, Eds. Vol. 35-59. John Wiley & Sons Ltd. Chichester.
49. NICHOLLS, D. G. 1985. A role for the mitochondrion in the protection of cells against calcium overload? Prog. Brain Res. **63:** 97-106.
50. ORRENIUS, S., D. MCCONKEY, D. JONES & P. NICOTERA. 1988. Ca^{2+}-activated mechanisms in toxicity and programmed cell death. ISI Atlas Sci.: Pharmacology **2:** 319-324.
51. BAZAN, N. 1989. Arachidonic acid in the modulation of excitable membrane function and at the onset of brain damage. Ann. N.Y. Acad. Sci. **559:** 1-16.
52. LINDSBERG, P., J. HALLENBECK & G. FEURESTEIN. 1991. Platelet-activating factor in stroke and brain injury. Ann. Neurol. **30:** 117-129.
53. BAZAN, N. G., S. P. SQUINTO, P. BRAQUET, P. T. & M. V. L. 1991. Platelet-activating factor and polyunsaturated fatty acids in cerebral ischemia or convulsions: intracellular PAF-binding sites and activation of a FOS/JUN/AP-1 transcriptional signaling system. Lipids **26:** 1236-1242.
54. AXELROD, J., R. M. BURCH & C. L. JELSEMA. 1988. Receptor-mediated activation of phospholipase A_2 via GTP-binding proteins: Arachidonic acid and its metabolites as second messengers. TINS. **11:** 117-123.

55. Huang, K.-P. 1989. The mechanism of protein kinase C activation. TINS. **12:** 425-432.
56. Alkon, D. & H. Rasmussen. 1988. A spatial-temporal model of cell activation. Science **239:** 998-1056.
57. Manev, H., M. Favaron, A. Guidotti & E. Costa. 1989. Delayed increase of Ca^{2+} influx elicited by glutamate: Role in neuronal death. Mol. Pharmacol. **36:** 106-112.
58. Malenka, R. C., J. A. Kauer, D. J. Perkel & R. A. Nicoll. 1989. The impact of postsynaptic calcium on synaptic transmission: Its role in long-term potentiation. Trends Neurosci. **12:** 444-450.
59. Kochanek, P., A. Dutka, K. Kumaroo & J. Hallenbeck. 1988. Effects of prostacyclin, indomethacin, and heparin on cerebral blood flow and platelet adhesion after multifocal ischemia of canine brain. Stroke **19:** 693-699.
60. Pettigrew, L., J. Grotta, H. Rhoades & K. Wu. 1989. Effect of thromboxane synthase inhibition on eicosanoid levels and blood flow in ischemic rat brain. Stroke **20:** 627-632.
61. Nakagomi, T., T. Sasaki, T. Kirino, A. Tamura, M. Noguchi, I. Saito & K. Takakura. 1989. Effect of cyclooxygenase and lipoxygenase inhibitors on delayed neuronal death in the gerbil hippocampus. Stroke **20:** 925-929.
62. Minamisawa, H., A. Terashi, Y. Katayama, Y. Kanda, J. Shimizu, T. Shiratori, K. Inamura, H. Kaseki & Y. Yoshino. 1988. Brain eicosanoid levels in spontaneously hypertensive rats after ischemia with reperfusion: Leukotriene C_4 as a possible cause of cerebral edema. Stroke **19:** 372-377.
63. Siman, R. & J. Noszek. 1988. Excitatory amino acids activate calpain I and induce structural protein breakdown *in vivo*. Neuron **1:** 279-287.
64. Nixon, R. A. 1989. Calcium-activated neutral proteinases as regulators of cellular function. Ann. N.Y. Acad. Sci. **568:** 198-208.
65. Lee, K., S. Frank, P. Vanderklish, A. Arai & G. Lynch. 1991. Inhibition of proteolysis protects hippocampal neurons from ischemia. Proc. Natl. Acad. Sci. USA **88:** 7233-7237.
66. Nowak, T. S. 1990. Protein synthesis and the heat shock/stress response after ischemia. Cerebrovasc. Brain Metab. Rev. **2:** 345-366.
67. Hossmann, K.-A. 1993. Disturbances of cerebral protein synthesis and ischemic cell death. *In* Neurobiology of Ischemic Brain Damage. K. Kogure, K.-A. Hossmann & B. K. Siesjö, Eds. Progress in Brain Research. Vol. **96:** 161-177. Elsevier Science Publishers. B.V. Amsterdam.
68. Wieloch, T., K. Bergstedt & B. R. Hu. 1993. Protein phosphorylation and the regulation of mRNA translation following cerebral ischemia. *In* Neurobiology of Ischemic Brain Damage. K. Kogure, K.-A. Hossmann & B. K. Siesjö, Eds. Progress in Brain Research, Vol. 96: 179-191. Elsevier Science. Amsterdam.
69. Hu, B.-R., Y.-B. Ou Yang & T. Wieloch. 1993. Depression of neuronal protein synthesis initiation by tyrosine kinase inhibitors. J. Neurochem. **61:** 1789-1794.
70. Lindvall, O., P. Ernfors, J. Bengtzon, Z. Kokaia, M.-L. Smith, B. K. Siesjö & H. Persson. 1992. Differential regulation of mRNAs for nerve growth factor, brain-derived neurotrophic factor and neurotrophin-3 in the adult rat brain following cerebral ischemia and hypoglycemic coma. Proc. Natl. Acad. Sci. USA **89:** 648-652.
71. Hu, B.-R. & T. Wieloch. 1993. Changes in tyrosine phosphorylation in neocortex following transient cerebral ischaemia. NeuroReport **4:** 219-222.
72. McCord, J. 1985. Oxygen-derived free radicals in postischemic tissue injury. N. Engl. J. Med. **312:** 159-163.
73. Mink, R., A. Dukta, K. Kumaroo & J. Hallenbeck. 1990. No conversion of xanthine dehydrogenase to oxidase in canine cerebral ischemia. Am. J. Physiol. (Heart Circ. Physiol. **28):** H1655-H1659.
74. Halliwell, B. 1992. Reactive oxygen species and the central nervous system. J. Neurochem. **59:** 1609-1623.

75. SIESJÖ, B. K., G. BENDEK, T. KOIDE, E. WESTERBERG & T. WIELOCH. 1985. Influence of acidosis on lipid peroxidation in brain tissues in vitro. J. Cereb. Blood Flow Metab. **5:** 253-258.
76. BECKMAN, J., T. BECKMAN, J. CHEN, P. MARSHALL & B. FREEMAN. 1990. Apparent hydroxyl radical production by peroxynitrite: Implications for endothelial injury from nitric oxide and superoxide. Proc. Natl. Acad. Sci. USA **87:** 1620-1624.
77. HARLAN, J. M. 1987. Neutrophil-mediated vascular injury. Acta Med. Scand. **715:** 123-129.
78. GIFFARD, R., H. MONYER, C. CHRISTINE & D. CHOI. 1990. Acidosis reduces NMDA receptor activation, glutamate neurotoxicity, and oxygen-glucose deprivation neuronal injury in cortical cultures. Brain Res. **506:** 339-342.
79. TANG, C.-M., M. DICHTER & M. MORAD. 1990. Modulation of the *N*-methyl-D-aspartate channel by extracellular H^+. Proc. Natl. Acad. Sci. **87:** 6445-6449.
80. TOMBAUGH, G. C. & R. M. SAPOLSKY. 1990. Mild acidosis protects hippocampal neurons from injury induced by oxygen and glucose deprivation. Brain Res. **506:** 343-345.
81. TOMBAUGH, G. C. & R. M. SAPOLSKY. 1993. Evolving concepts about the role of acidosis in ischemic neuropathology. J. Neurochem. **61:** 793-803.
82. OU YANG, Y.-B., T. KRISTIÁN, P. MELLERGÅRD & B. SIESJÖ. 1993. The influence of pH on glutamate- and depolarisation-induced increases of intracellular calcium concentration in cortical neurons in primary culture. Brain Res., in press.
83. KRISTIÁN, T., K. KATSURA, G. GIDÖ & B. K. SIESJÖ. 1993. The influence of pH on cellular calcium influx during ischemia. Brain Res., in press.
84. SIESJÖ, B. K. 1988. Acidosis and ischemic brain damage. Neurochem. Pathol. **9:** 31-88.
85. SIESJÖ, B. K., K. KATSURA, P. MELLERGÅRD, A. EKHOLM, J. LUNDGREN & M.-L. SMITH. 1993. Acidosis-related brain damage. *In* Neurobiology of Ischemic Brain Damage. K. Kogure, K.-A. Hossmann & B. K. Siesjö, Eds. Vol. Prog. Brain Res. **96:** 23-48. Elsevier Science. Amsterdam.
86. KATSURA, K., T. KRISTIÁN, M.-L. SMITH & B. SIESJÖ. 1993. Acidosis induced by hypercapnia exaggerates ischemic brain damage. J. Cereb. Blood Flow Metab., in press:
87. ARAKI, N., J. H. GREENBERG, J. T. SLADKY, D. UEMATSU, A. KARP & M. REIVICH. 1992. The effect of hyperglycemia on intracellular calcium in stroke. J. Cereb. Blood Flow Metab. **12:** 469-476.
88. RANDALL, R. D. & S. A. THAYER. 1992. Glutamate-induced calcium transient triggers delayed calcium overload and neurotoxicity in rat hippocampal neurons. J. Neurosci. **12:** 1882-1895.
89. DUBINSKY, J. M. 1993. Intracellular calcium levels during the period of delayed excitotoxicity. J. Neurosci. **13:** 623-631.
90. DUX, E., U. OSCHLIES, C. WIESSNER & K.-A. HOSSMANN. 1992. Glutamate-induced ribosomal disaggregation and ultrastructural changes in rat cortical neuronal culture: Protective effect of horse serum. Neurosci. Lett. **141:** 173-176.
91. MATTSON, M. & B. RYCHLIK. 1990. Glia protect hippocampal neurons against excitatory amino acid-induced degeneration: Involvement of fibroblast growth factor. Int. J. Dev. Neurosci. **8:** 399-415.
92. HARTLEY, D. M., M. C. KURTH, L. BJERKNESS, J. H. WEISS & D. W. CHOI. 1993. Glutamate receptor-induced $^{45}Ca^{2+}$ accumulation in cortical cell culture correlates with subsequent neuronal degeneration. J. Neurosci. **13:** 1993-2000.
93. TYMIANSKI, M., M. P. CHARLTON, P. L. CARLEN & C. H. TATOR. 1993. Source specificity of early calcium neurotoxicity in cultured embryonic spinal neuron. J. Neurosci. **13:** 2085-2104.
94. WIELOCH, T. 1985. Neurochemical correlates to selective neuronal vulnerability. Prog. Brain. Res. **63:** 69-85.
95. BUCHAN, A. M. & W. A. PULSINELLI. 1990. Septo-hippocampal deafferentation protects CA1 neurons against ischemic injury. Brain Res. **512:** 7-14.

96. DIEMER, N. H., F. F. JOHANSEN & M. B. JÖRGENSEN. 1990. *N*-methyl-D-aspartate and non-*N*-methyl-D-aspartate antagonists in global cerebral ischemia. Stroke **21:** 39-42.
97. NELLGÅRD, B. & T. WIELOCH. 1992. Postischemic blockade of AMPA but no NMDA receptors mitigates neuronal damage in the rat brain following transient severe cerebral ischemia. J. Cereb. Blood Flow Metab. **12:** 2-11.
98. BUCHAN, A. M., H. LI, C. CHO & W. A. PULSINELLI. 1991. Blockade of the AMPA receptor prevents CA1 hippocampal injury following severe but transient forebrain ischemia in the adult rat. Neurosci. Lett. **132:** 255-258.
99. HANSEN, A. J. 1985. Effects of anoxia on ion distribution in the brain. Physiol. Rev. **65:** 101-148.
100. NEDERGAARD, M. & A. J. HANSEN. 1988. Spreading depression is not associated with neuronal injury in the normal brain. Brain Res. **449:** 395-398.
101. GIDÖ, G., K. KATSURA, T. KRISTIÁN & B. K. SIESJÖ. 1993. Influence of plasma glucose concentration on rat brain extracellular calcium transients during spreading depression. J. Cereb. Blood Flow Metab. **13:** 179-182.
102. AUER, R. N. & B. K. SIESJÖ. 1988. Biological differences between ischemia, hypoglycemia and epilepsy. Ann. Neurol. **24:** 699-707.
103. SIESJÖ, B. K. 1988. Hypoglycemia, brain metabolism, and brain damage. Diabetes/Metabolism Rev. **4:** 113-144.
104. UEMATSU, D., J. GREENBERG, M. REIVICH & A. KARP. 1989. Cytosolic free calcium, NAD/NADH redox state and hemodynamic changes in the cat cortex during severe hypoglycemia. J. Cereb. Blood Flow Metab. **9:** 149-155.
105. CHAPMAN, A., E. WESTERBERG & B. K. SIESJÖ. 1981. The metabolism of purine and pyrimidine nucleotides in rat cortex during insulin-induced hypoglycemia and recovery. J. Neurochem. **36:** 179-189.
106. GIDÖ, G., T. KRISTIÁN, K. KATSURA & B. K. SIESJÖ. 1993. The influence of repeated spreading depression-induced calcium transients on neuronal viability in moderately hypoglycemic rats. Exp. Brain Res., in press.
107. HU, B. & T. WIELOCH. 1993. Stress induced inhibition of protein synthesis initiation. Modulation of initiation factor 2 and guanine exchange factor activities following transient cerebral ischemia in the rat. J. Neurosci. **13:** 1830-1838.
108. HU, B.-R. & T. WIELOCH. 1993. Casein kinase II activity in the postischemic rat brain increases in brain regions resistant to ischemia and decreases in vulnerable areas. J. Neurochem. **60:** 1722-1728.
109. KIRINO, T., H. ROBINSON, A. MIWA, A. TAMURA & N. KAWAI. 1992. Disturbance of membrane function preceding ischemic delayed neuronal death in the gerbil hippocampus. J. Cereb. Blood Flow Metab. **12:** 408-417.
110. (a) DESPHANDE, J., K. BERGSTEDT, T. LINDÉN, H. KALIMO & T. WIELOCH. 1992. Ultrastructural changes in the hippocampal CA1 region following transient cerebral ischemia: Evidence against programmed cell death. Exp. Brain Res. **88:** 91-105.
(b) MACMANUS, J. P., A. M. BUCHAU, J. E. HILL, J. RASQUINHA & E. PRESTON. 1993. Global ischemia can cause DNA fragmentation indicative of apoptosis in rat brain. Neurosci. Lett. **164:** 89-92.
111. NEDERGAARD, M. & A. J. HANSEN. 1993. Characterization of cortical depolarizations evoked in focal cerebral ischemia. J. Cereb. Blood Flow Metab. **13:** 568-574.
112. GILL, R., P. ANDINÉ, L. HILLERED, L. PERSSON & H. HAGBERG. 1992. The effect of MK-801 on cortical spreading depression in the penumbral zone following focal ischemia in the rat. J. Cereb. Blood Flow Metab. **12:** 371-379.
113. CARNEY, J. M. & R. A. FLOYD. 1991. Protection against oxidative damage to CNS by a-phenyl-tert-butyl nitrone (PBN) and other spin-trapping agents: A novel series of nonlipid free radical scavengers. J. Mol. Neurosci. **3:** 47-57.
114. OLIVER, C., P. STARKE-REED, E. STADTMAN, G. LIU, J. CARNEY & R. FLOYD. 1990. Oxidative damage to brain proteins, loss of glutamine synthetase activity, and produc-

tion of free radicals during ischemia/reperfusion-induced injury to gerbil brain. Proc. Natl. Acad. Sci. **87:** 5144-5147.

115. Folbergrová, J., Y. Kiyota, K. Pahlmark, H. Memezawa, M.-L. Smith & B. K. Siesjö. 1993. Does ischemia with reperfusion lead to oxidative damage to proteins in the brain? J. Cereb. Blood Flow Metab. **13:** 145-152.
116. Cao, W., J. Carney, A. Duchon, R. Floyd & M. Chevion. 1988. Oxygen free radical involvement in ischemia and reperfusion injury to brain. Neurosci. Lett. **88:** 233-238.
117. Pahlmark, K., J. Folbergrová, M.-L. Smith & B. K. Siesjö. 1993. Effects of dimethylthiourea on selective neuronal vulnerability in forebrain ischemia in rats. Stroke **24:** 731-737.
118. Smith, C., J. Carney, P. Starke-Reed, C. Oliver, E. Stadtman, R. Floyd & W. Markesbery. 1988. Excess brain protein oxidation and enzyme dysfunction in normal aging and in Alzheimer disease. Proc. Natl. Acad. Sci. USA **88:** 10540-10543.
119. Siesjö, B. K. 1993. Basic mechanisms of traumatic brain damage. Ann. Emerg. Med. **22:** 959-969.

Calcium and Excitotoxic Neuronal Injury

DENNIS W. CHOI [a]

Center for the Study of Nervous System Injury
and
Department of Neurology
Washington University School of Medicine
St. Louis, Missouri 63110

Excitotoxic injury mediated by glutamate or related compounds contributes to central neuronal death in several human pathological conditions, such as after food ingestion, hypoxia-ischemia, trauma, or prolonged seizures.[1-4] Whether excitotoxicity also contributes to neuronal loss in certain neurodegenerative disorders, such as Alzheimer's disease, Huntington's disease, or motor neuron disease, is unknown, but is the subject of current inquiry. In any case, the mission of gaining insight into the basis of Alzheimer's disease is served by elucidating the mechanisms underlying excitotoxic neuronal death, because excitotoxic death may converge, that is, share common aspects with other forms of neuronal death. Thus, even if excitotoxicity has no direct involvement in Alzheimer's disease, its elucidation may lead to useful insights into the nature of other pathological neuronal deaths and aid the development of possible therapeutic approaches.

I will outline here some current ideas about the role of calcium overload in excitotoxic death derived largely from work on neurons *in vitro* carried out in several laboratories.

MECHANISMS UNDERLYING GLUTAMATE NEUROTOXICITY ON CULTURED CORTICAL NEURONS

Glutamate neurotoxicity can be hypothetically modeled as a three-stage process analogous to current models of long-term potentiation: induction, amplification, and expression.[5,6] First, extracellular glutamate initiates death induction by activating neuronal membrane receptors and triggering a set of defined intracellular derangements, particularly intracellular Ca^{2+} overload. Second, modulatory events amplify these derangements, increasing their intensity and recruiting additional neurons into the injury process. Finally, death is expressed when these derangements set in motion the final cascades directly responsible for neuronal disintegration.

[a] Address for correspondence: Dennis W. Choi, Department of Neurology, Box 8111, Washington University School of Medicine, St. Louis, MO 63110.

Induction

The induction of glutamate neurotoxicity consists of glutamate receptor activation and the intracellular derangements that follow immediately. These initial derangements serve as triggers for subsequent lethal events, but are not themselves truly irreversible; neurons can be rescued following full induction.[7] Induction is most simply accomplished by receptor overstimulation, but normal physiological levels of receptor activation may become neurotoxic if neuronal energy levels are compromised.[8]

Glutamate activates three major families of ionophore-linked receptors classified by their preferred agonists: *N*-methyl-D-aspartate (NMDA), kainate, and alpha-amino-3-hydroxy-5-methyl-4-isoxazolepropionic acid (AMPA).[9] Multiple functional receptor subunits from each family have been cloned.[10–12] The channels gated by all three receptor subtypes are permeable to both Na^+ and K^+. Channels gated by NMDA receptors, but only a minority subset of channels gated by AMPA or kainate receptors (see below), additionally possess high permeability to Ca^{2+}.[13] Glutamate also activates a family of metabotropic receptors that activate second messenger systems rather than directly gating ion channels.[14,15]

If glutamate exposure is intense, widespread cortical neuronal death can be triggered by exposure times as short as 2-3 minutes,[16] a phenomenon my colleagues and I call rapidly triggered excitotoxicity. Two components of injury are distinguishable[17]: (1) an acute component, marked by immediate neuronal swelling and dependent on the presence of extracellular Na^+ and Cl^- (see also refs. 18 and 19); and (2) a delayed component marked by neuronal disintegration occurring over a period of hours after exposure, dependent on the presence of extracellular Ca^{2+}. The first component probably reflects the influx of extracellular Na^+, accompanied passively by the influx of Cl^- and water, resulting in cell volume expansion. The second component depends on the presence of extracellular Ca^{2+} and is likely triggered by excessive Ca^{2+} influx. Although either the acute Na^+-dependent component or the delayed Ca^{2+}-dependent component of glutamate neurotoxicity can alone produce irreversible neuronal injury, the latter component predominates at lower levels of glutamate exposure.

Experiments with glutamate antagonists suggest that both AMPA/kainate (these two receptor types are not easily distinguished with present antagonists) and NMDA type glutamate receptors contribute to acute neuronal swelling, but that most delayed disintegration requires NMDA receptor activation. Death after brief intense glutamate exposure can be almost completely blocked by selective blockade of NMDA receptors,[20] but selective blockade of AMPA/kainate receptors has only a small effect on late neuronal death.[21] Only when both NMDA and AMPA/kainate receptors are blocked is acute glutamate-induced neuronal swelling eliminated.

Although even 15-30 minutes of exposure to high concentrations of AMPA or kainate induce relatively little neuronal death, most cortical neurons die after exposure times of several hours.[20,22] With 24-hour exposure, 10 μM concentrations of either kainate or AMPA are highly lethal[23]; my colleagues and I call this "slowly triggered excitotoxicity" to emphasize its longer receptor activation requirement in comparison to that of rapidly triggered, NMDA receptor-mediated toxicity.

Abnormal entry of extracellular Ca^{2+} may be the primary factor responsible for the induction of both rapidly triggered and slowly triggered excitotoxicity. The dependence of rapidly triggered toxicity upon extracellular Ca^{2+} and NMDA receptor activation is consistent with the idea that it is initiated by excessive Ca^{2+} influx through the Ca^{2+}-permeable NMDA receptor-gated channel. Slowly triggered, AMPA/kainate receptor-mediated excitotoxicity may also be initiated by excessive Ca^{2+} influx. Inasmuch as most channels gated by AMPA or kainate receptors have limited Ca^{2+} permeability, AMPA or kainate-induced Ca^{2+} influx may occur largely through indirect routes, for example, through voltage-gated Ca^{2+} channels, reverse operation of the Na^{+}-Ca^{2+} exchanger, or membrane stretch-activated conductances.[24] Although kainate and AMPA can induce striking increases in intracellular free Ca^{2+} in many central neurons[25] including our cortical neurons (ref. 26 and unpublished results), its ability to produce bulk movement of extracellular $^{45}Ca^{2+}$ into cortical neurons is severalfold slower than that of NMDA.[16,27] These observation sets do not conflict, as substantial increases in cytosolic free Ca^{2+} might occur with relatively small amounts of net Ca^{2+} influx. The extant correlation between the rate of bulk Ca^{2+} influx associated with exposure to NMDA versus AMPA or kainate and the rate of lethal injury induction supports the idea that the critical difference between rapidly triggered and slowly triggered excitotoxicity on cortical neurons may be this rate of net Ca^{2+} influx. When enough total Ca^{2+} enters to overwhelm cell homeostatic mechanisms, a final rise in intracellular free Ca^{2+} may occur,[28] triggering destructive events.

An interesting test of the Ca^{2+} influx/overload hypothesis was recently provided by the finding that a minority subset of AMPA receptors gate channels permeable to Ca^{2+},[29] likely reflecting a molecular composition lacking an edited GluR2/GluRB subunit.[30,31] These receptors can be identified histochemically by staining for kainate-activated Co^{2+} uptake, because Co^{2+} permeates through these AMPA or kainate receptor-gated Ca^{2+} channels but not through voltage-gated Ca^{2+} channels.[32]

We found that kainate-activated Co^{2+} uptake labels approximately 13-15% of the neuronal population in cortical cultures, and in fact this neuronal subpopulation was especially vulnerable to damage induced by short exposure to kainate but not NMDA.[26,33] Cultured cerebellar Co^{2+} uptake-positive neurons were also found to exhibit enhanced vulnerability to kainate.[34] Nearly half of the Co^{2+} uptake-positive cells in our cortical cultures failed to immunostain with antibody directed against AMPA subunits GluR2/3, consistent with the idea that at least this group lacked expression of GluR2 (GluRB).[35,36] Interestingly, many neurons in this group expressed immunostaining for GluR1 (GluRA), GluR4 (GluRD), and glutamate acid decarboxylase, each found in less than 10% of the overall neuronal population.[35] Thus, cortical neurons expressing AMPA (or kainate) receptor-gated Ca^{2+}-permeable channels may often express a highly distinctive set of AMPA receptor subunits as well as a GABAergic inhibitory identity. This idea fits with recent observations of Monyer, Seeburg, and colleagues who found that stellate neurons in cortex layer 4 have reduced levels of GluR2 mRNA compared to layer 5 pyramidal neurons (personal communication). Special vulnerability of cortical GABAergic neurons to AMPA receptor-mediated excitotoxic death could underlie selective loss of circuit inhibition in certain disease states.

In addition to limitations of channel Ca^{2+} permeability, another characteristic of AMPA receptors that may limit their contribution to excitotoxic injury may be desensitization. If AMPA receptor desensitization is blocked with cyclothiazides,[37,38] an enhanced contribution results.[39] The identification of drugs capable of increasing AMPA receptor desensitization may be a useful pathway in the future for the development of new neuroprotective agents.

The effect of metabotropic receptor stimulation upon excitotoxic injury has not been well defined due to limitations of current pharmacology. However, this will clearly be an important area of future research. Rather paradoxically, the broad spectrum metabotropic receptor agonist trans-1-aminocyclopentyl-1,3-dicarboxylate (trans-ACPD) reduces rapidly triggered excitotoxicity.[40] The basis for this neuroprotective action remains to be determined, but recent experiments by Bruno and colleagues[41] are consistent with the possibility that it may be mediated by activation of inhibitory subtypes of metabotropic receptors, perhaps leading to reduction of presynaptic glutamate release as well as reduced postsynaptic Ca^{2+} influx through voltage-activated Ca^{2+} channels.

Amplification

Following the induction of glutamate toxicity, many events may act to further amplify resultant internal derangements, especially local elevations in intracellular free Ca^{2+} concentration, $[Ca^{2+}]_i$.[5] The impact of extracellular Ca^{2+} influx may be augmented by the release of Ca^{2+} from intracellular stores, as mediated by elevations in IP_3. Activation of key enzyme families, including C kinases and calmodulin-regulated enzymes, and genes, as well as consequent formation of the intercellular messengers arachidonic acid and nitric oxide, may act to increase the gain of the glutamate signaling system.[42–44] Under normal conditions, this increase serves the purpose of synaptic plasticity and long-term potentiation, but under excitotoxic conditions, this increase may worsen injury. Relevant enhancements may occur in Ca^{2+}-dependent glutamate release from nerve terminals, postsynaptic receptor responsiveness to glutamate, and Ca^{2+} influx through voltage-gated channels. Perhaps most importantly, a positive-feedback loop may be formed by glutamate efflux from injured neurons which may act to induce further excitotoxic injury on synaptic partners.[7,45]

The net contribution of these amplification steps to neuronal death may depend on the intensity of initial induction events. Following intense activation of NMDA receptors, enough Ca^{2+} may enter neurons directly through NMDA receptor-gated channels to produce lethal effects without subsequent amplification. Amplification may therefore be more important to slowly triggered excitotoxicity than to rapidly triggered excitotoxicity.

Expression

Ultimately, and perhaps largely due to sustained elevations in $[Ca^{2+}]_i$ at critical subcellular locations, events occur which bear direct responsibility for neuronal degeneration. It is parsimonious to hypothesize that many destructive events are common to both rapidly triggered excitotoxicity and slowly triggered excitotoxicity.

One set of important expression cascades may involve Ca^{2+}-activated catabolic enzymes. The Ca^{+}-activated neutral protease calpain is unleashed by glutamate receptor stimulation[46] and can degrade major neuronal structural proteins. Calpain inhibitors were recently found to be neuroprotective in gerbils subjected to global ischemia.[47] Elevated cytosolic Ca^{2+} also activates phospholipase A_2, capable of degrading membranes and liberating arachidonic acid, and endonucleases capable of degrading genomic DNA. Kure and colleagues[48] suggested that inhibition of endonucleases by aurintricarboxylic acid can attenuate glutamate-induced death of cultured cortical neurons; we confirmed this drug effect in our system, but have not observed evidence that it is explained by endonuclease inhibition[49] (see below).

Another set of expression cascades may involve oxygen-free radicals. These reactive molecules can initiate many destructive processes including lipid peroxidation[50]; once formed, they may promote further excitotoxic injury by promoting glutamate release.[51] Antioxidant approaches can attenuate excitotoxic injury,[52,53] and NMDA receptor activation has been specifically linked to superoxide radical formation in cerebellar granule cell cultures.[54] Kainate-induced radical formation is suggested by the studies of Puttfarcken and colleagues[55] as well as Bondy and Lee.[56]

Free radical production might be linked to loss of cellular Ca^{2+} homeostasis in several ways including: (1) Ca^{2+} activation of phospholipase A_2, leading to the liberation of arachidonic acid which upon further metabolism leads to free radical production[57]; (2) Ca^{2+} triggering the conversion of xanthine dehydrogenase to xanthine oxidase, a rich enzymatic source of free radicals[52]; and (3) stimulation of NMDA receptors leading to the activation of nitric oxide synthase and the release of nitric oxide,[44,58] which in turn can react with superoxide to form peroxynitrite and promote the production of hydroxyl radicals.[59] Pharmacological inhibition of nitric oxide synthase or removal of the nitric oxide precursor arginine from the culture medium was observed to reduce rapidly triggered excitotoxicity in rat cortical cultures,[60] leading those investigators to suggest that this third mechanism is of primary importance. Interestingly, ganglioside compounds that have been known for several years to inhibit excitotoxicity[61] were recently found to inhibit nitric oxide synthase.[62] Ca^{2+} activation of phospholipase A_2, via fatty acid liberation, might also lead to uncoupling of mitochondrial electron transport[63] and subsequent mitochondrial free radical production.[64]

In our unstimulated mixed neuronal-glial cortical cultures, nitric oxide generation does not appear to be a prominent factor in excitotoxic expression. However, if astrocytic inducible nitric oxide synthase (iNOS) is induced by prior cytokine exposure, then rapidly triggered NMDA receptor-mediated neurotoxicity is potentiated in a fashion dependent on nitric oxide production.[65]

EXCITOTOXIC DEATH: A NECROSIS?

Much attention has been drawn recently to the possible contributions of programmed cell death to neuronal loss in disease states. Although one study[48] suggested that glutamate neurotoxicity may occur via apoptosis, this study used very young (3 days *in vitro*) neuronal cultures. My colleagues and I found that cortical cultures of this age do not display the receptor-mediated (i.e., glutamate antagonist-blocked) excitotoxicity just outlined. We speculate that the glutamate-induced death studied by Kure and colleagues reflects oxidative stress secondary to inhibition of cystine uptake and consequent glutathione deprivation.[66,67] Both the morphology (prominent early cell swelling) and the disease linkages (e.g., to hypoxia-ischemia) of the glutamate receptor-mediated excitotoxicity just described favor its classification as a necrosis.[68] Rapidly triggered glutamate excitotoxicity in our cortical cell cultures is not associated with genomic DNA fragmentation and is not blocked by inhibitors of protein or RNA synthesis.[69]

CONCLUDING STATEMENT: HOW GENERALIZABLE IS THE IDEA OF EXCITOTOXIC CALCIUM OVERLOAD?

Excitotoxic neuronal death provides a useful example of death mediated by lethal cellular Ca^{2+} overload. Although the mechanistic underpinnings of excitotoxicity will likely turn out not to be unique, generalizations should proceed with caution. In particular, it is clear that not all deaths involve Ca^{2+} overload. Elevated intracellular free Ca^{2+} does accompany glucocorticoid-mediated apoptosis death of thymocytes,[70] but it is not generally a feature of apoptosis deaths.[71] Indeed, neuronal vulnerability to trophic factor-dependent apoptosis appears to be inversely related to intracellular free Ca^{2+} levels.[72] Thus, it is possible that interventions aimed at reducing cellular Ca^{2+} overload would be helpful against excitotoxic death, but harmful in settings in which certain forms of apoptosis predominate. In fact, prolonged exposure to voltage-gated Ca^{2+} channel blockers alone can kill cultured cortical neurons in a manner dependent on new protein synthesis.[73]

REFERENCES

1. OLNEY, J. W. 1969. Brain lesion, obesity and other disturbances in mice treated with monosodium glutamate. Science **164:** 719-721.
2. MELDRUM, B. 1985. Possible therapeutic applications of antagonists of excitatory amino acid neurotransmitters. Clin. Sci. **68:** 113-122.
3. OLNEY, J. W. 1986. Inciting excitotoxic cytocide among central neurons. Adv. Exp. Med. Biol. **203:** 631-645.
4. CHOI, D. W. 1988. Glutamate neurotoxicity and diseases of the nervous system. Neuron **1:** 623-634.
5. CHOI, D. W. 1990. Methods for antagonizing glutamate neurotoxicity. Cerebrovasc. Brain Metab. Rev. **2:** 105-147.
6. CHOI, D. W. 1992. Excitotoxic cell death. J. Neurobiol. **23:** 1261-1276.
7. HARTLEY, D. M. & D. W. CHOI. 1989. Delayed rescue of *N*-methyl-D-aspartate receptor-mediated neuronal injury in cortical culture. J. Pharmacol. Exp. Ther. **250:** 752-758.
8. BEAL, M. F., B. HYMAN & W. KOROSHETZ. 1993. Do defects in mitochondrial energy metabolism underlie the pathology of neurodegenerative diseases? Trends Neurosci. **16:** 125-131.
9. WATKINS, J. C., P. KROGSGAARD-LARSEN & T. HONORE. 1990. Structure-activity relationships in the development of excitatory amino acid receptor agonists and competitive antagonists. Trends Pharm. Sci. **11:** 25-33.
10. HOLLMANN, M., A. O'SHEA-GREENFIELD, S. W. ROGERS & S. HEINEMANN. 1989. Cloning by functional expression of a member of the glutamate receptor family. Nature **342:** 643-648.
11. MORIYOSHI, K., M. MASU, T. ISHII, R. SHIGEMOTO, N. MIZUNO & S. NAKANISHI. 1991. Molecular cloning and characterization of the rat NMDA receptor (see comments). Nature **354:** 31-37.

12. SOMMER, B. & P. H. SEEBURG. 1992. Glutamate receptor channels: Novel properties and new clones. Trends Pharmacol. Sci. **13:** 291-296.
13. MACDERMOTT, A. B., M. L. MAYER, G. L. WESTBROOK, S. J. SMITH & J. L. BARKER. 1986. NMDA-receptor activation increases cytoplasmic calcium concentration in cultured spinal cord neurones. Nature **321:** 519-522.
14. MASU, M., Y. TANABE, K. TSUCHIDA, R. SHIGEMOTO & S. NAKANISHI. 1991. Sequence and expression of a metabotropic glutamate receptor. Nature **349:** 760-765.
15. TANABE, Y., M. MASU, T. ISHII, R. SHIGEMOTO & S. NAKANISHI. 1992. A family of metabotropic glutamate receptors. Neuron **8:** 169-179.
16. HARTLEY, D. M., M. KURTH, L. BJERKNESS, J. H. WEISS & D. W. CHOI. 1993. Glutamate receptor-induced $^{45}Ca^{2+}$ accumulation in cortical cell culture correlates with subsequent neuronal degeneration. J. Neurosci. **13:** 1993-2000.
17. CHOI, D. W. 1987. Ionic dependence of glutamate neurotoxicity in cortical cell culture. J. Neurosci. **7:** 369-379.
18. ROTHMAN, S. M. 1985. The neurotoxicity of excitatory amino acids is produced by passive chloride influx. J. Neurosci. **5:** 1483-1489.
19. OLNEY, J. W., M. T. PRICE, L. SAMSON & J. LABRUYERE. 1986. The role of specific ions in glutamate neurotoxicity. Neurosci. Lett. **65:** 65-71.
20. CHOI, D. W., J. KOH & S. PETERS. 1988. Pharmacology of glutamate neurotoxicity in cortical cell culture: Attenuation by NMDA antagonists. J. Neurosci. **8:** 185-196.
21. KOH, J. & D. W. CHOI. 1991. Selective blockade of non-NMDA receptors does not block rapidly triggered glutamate induced neuronal death. Brain Res. **548:** 318-321.
22. FRANDSEN, A., J. DREJER & A. SCHOUSBOE. 1989. Direct evidence that excitotoxicity in cultured neurons is mediated via *N*-methyl-D-aspartate (NMDA) as well as non-NMDA receptors. J. Neurochem. **53:** 297-299.
23. KOH, J., M. P. GOLDBERG, D. M. HARTLEY & D. W. CHOI. 1990. Non-NMDA receptor-mediated neurotoxicity in cortical culture. J. Neurosci. **10:** 693-705.
24. CHOI, D. W. 1988. Calcium-mediated neurotoxicity: Relationship to specific channel types and role in ischemic damage. Trends Neurosci. **11:** 465-469.
25. MURPHY, S. N., S. A. THAYER & R. J. MILLER. 1987. The effects of excitatory amino acids on intracellular calcium in single mouse striatal neurons in vitro. J. Neurosci. **12:** 4145-4158.
26. TURETSKY, D. M., L. M. T. CANZONIERO, S. L. SENSI, J.H. WEISS, M. P. GOLDBERG & D. W. CHOI. 1994. Cortical neurons exhibiting kainate-activated Co^{2+} uptake are selectively vulnerable to AMPA/kainate receptor-mediated toxicity. Neurobiol. Dis., in press.
27. MARCOUX, F. W., J. E. GOODRICH, A. W. PROBERT & M. A. DOMINICK. 1988. Ketamine prevents glutamate-induced calcium influx and ischemia nerve cell injury. *In* Sigma and Phencyclidine-Like Compounds as Molecular Probes in Biology. E. F. Domino & J. Kamenka, Eds.: 735-746.
28. TYMIANSKI, M., M. P. CHARLTON, P. L. CARLEN & C. H. TATOR. 1993. Secondary Ca^{2+} overload indicates early neuronal injury which precedes staining with viability indicators. Brain Res. **607:** 319-323.
29. IINO, M., S. OZAWA & K. TSUZUKI. 1990. Permeation of calcium through excitatory amino acid receptor channels in cultured rat hippocampal neurones. J. Physiol. **424:** 151-165.
30. HUME, R. I., R. DINGLEDINE & S. F. HEINEMANN. 1991. Identification of a site in glutamate receptor subunits that controls calcium permeability. Science **253:** 1028-1031.
31. VERDOORN, T. A., N. BURNASHEV, H. MONYER, P. H. SEEBURG & B. SAKMANN. 1991. Structural determinants of ion flow through recombinant glutamate receptor channels. Science **252:** 1715-1718.
32. PRUSS, R. M., R. L. AKESON, M. M. RACKE & J. L. WILBURN. 1991. Agonist-activated cobalt uptake identifies divalent cation-permeable kainate receptors on neurons and glia. Neuron **7:** 509-518.

33. Turetsky, D. M., M. P. Goldberg & D. W. Choi. 1992. Kainate-activated cobalt uptake identifies a subpopulation of cultured cortical cells that are preferentially vulnerable to kainate-induced damage. Soc. Neurosci. Abstr. **18:** 81.
34. Brorson, J. R., P. A. Manzolillo & R. J. Miller. 1994. Ca^{2+} entry via AMPA/KA receptors and excitotoxicity in cultured cerebellar purkinje cells. J. Neurosci. **14:** 187-197.
35. Yin, H. A., D. Turetsky, D. W. Choi & J. H. Weiss. 1994. Cortical neurons with Ca^{2+} permeable AMPA/kainate channels display distinct receptor immunoreactivity and are GABAergic. Neurobiol. Dis., in press.
36. Bochet, P., E. Audinat, B. Lambolez, F. Crepel, J. Rossier, M. Iino, K. Tsuzuki & S. Ozawa. 1994. Subunit composition at the single-cell level explains functional properties of a glutamate-gated channel. Neuron **12:** 383-388.
37. Yamada, K. A. & S. M. Rothman. 1992. Diazoxide blocks glutamate desensitization and prolongs excitatory postsynaptic currents in rat hippocampal neurons. J. Physiol. **458:** 409-423.
38. Patneau, D. K., L. Vyklicky & M. L. Mayer. 1993. Hippocampal neurons exhibit cyclothiazide-sensitive rapidly desensitizing responses to kainate. J. Neurosci. **13:** 3496-3509.
39. Bateman, M. C., M. R. Bagwe, K. A. Yamada & M. P. Goldberg. 1993. Cyclothiazide potentiates AMPA neurotoxicity and oxygen-glucose deprivation injury in cortical culture. Soc. Neurosci. Abstr. **19:** 1643.
40. Koh, J. Y., E. Palmer & C. W. Cotman. 1991. Activation of the metabotropic glutamate receptor attenuates *N*-methyl-D-aspartate neurotoxicity in cortical cultures. Proc. Natl. Acad. Sci. USA **88:** 9431-9435.
41. Bruno, V. M. G., A. Copani, G. Battaglia, M. Marinozzi, B. Natalini, R. Pelliciari, A. P. Kozikowski, R. Giffard, D. W. Choi & F. Nicoletti. 1993. Effect of mGluR activation on different degenerative processes in cultured cells. Functional Neurol. Suppl. **4:** 12-13.
42. Barbour, B., M. Szatkowski, N. Ingledew & D. Attwell. 1989. Arachidonic acid induces a prolonged inhibition of glutamate uptake into glial cells. Nature **342:** 918-920.
43. Williams, J. H., M. L. Errington, M. A. Lynch & T. V. Bliss. 1989. Arachidonic acid induces a long-term activity-dependent enhancement of synaptic transmission in the hippocampus. Nature **341:** 739-742.
44. Bredt, D. S. & S. H. Snyder. 1992. Nitric oxide, a novel neuronal messenger. Neuron **8:** 3-11.
45. Rothman, S. M., J. H. Thurston & R. E. Hauhart. 1987. Delayed neurotoxicity of excitatory amino acids *in vitro*. Neuroscience **22:** 471-480.
46. Siman, R., J. C. Noszek & C. Kegerise. 1989. Calpain I activation is specifically related to excitatory amino acid induction of hippocampal damage. J. Neurosci. **9:** 1579-1590.
47. Lee, K. S., S. Frank, P. Vanderklish, A. Arai & G. Lynch. 1991. Inhibition of proteolysis protects hippocampal neurons from ischemia. Proc. Natl. Acad. Sci. USA **88:** 7233-7237.
48. Kure, S., T. Tominaga, T. Yoshimoto, K. Tada & K. Narisawa. 1991. Glutamate triggers internucleosomal DNA cleavage in neuronal cells. Biochem. Biophys. Res. Comm. **179:** 39-45.
49. Csernansky, C. A., L. M. T. Canzoniero & D. W. Choi. 1993. Delayed application of aurintricarboxylic acid reduces glutamate neurotoxicity. Soc. Neurosci. Abstr. **19:** 25.
50. Siesjo, B. K. 1989. Free radicals and brain damage. Cerebrovasc. Brain Metab. Rev. **1:** 165-211.
51. Pellegrini-Giampietro, D. E., G. Cherici, M. Alesiani, V. Carla & F. Moroni. 1988. Excitatory amino acid release from rat hippocampal slices as a consequence of free-radical formation. J. Neurochem. **51:** 1960-1963.

52. DYKENS, J. A., A. STERN & E. TRENKNER. 1987. Mechanism of kainate toxicity to cerebellar neurons *in vitro* is analogous to reperfusion tissue injury. J. Neurochem. **49:** 1222-1228.
53. MONYER, H., D. M. HARTLEY & D. W. CHOI. 1990. 21-Aminosteroids attenuate excitotoxic neuronal injury in cortical cell cultures. Neuron **5:** 121-126.
54. LAFON-CAZAL, M., S. PIETRI, M. CULCASI & J. BOCKAERT. 1993. NMDA-dependent superoxide production and neurotoxicity. Nature **364:** 535-537.
55. PUTTFARCKEN, P. S., R. L. GETZ & J. T. COYLE. 1993. Kainic acid-induced lipid peroxidation: Protection with butylated hydroxytoluene and U78517F in primary cultures of cerebellar granule cells. Brain Res. **624:** 223-232.
56. BONDY, S. C. & D. K. LEE. 1993. Oxidative stress induced by glutamate receptor agonists. Brain Res. **610:** 229-233.
57. CHAN, P. H., R. A. FISHMAN, S. LONGAR, S. CHEN & A. YU. 1985. Cellular and molecular effects of polyunsaturated fatty acids in brain ischemia and injury. Prog. Brain Res. **63:** 227-235.
58. GARTHWAITE, J., S. L. CHARLES & R. CHESS-WILLIAMS. 1988. Endothelium-derived relaxing factor release on activation of NMDA receptors suggests role as intercellular messenger in the brain. Nature **336:** 385-388.
59. BECKMAN, J. S., T. W. BECKMAN, J. CHEN, P. A. MARSHALL & B. A. FREEMAN. 1990. Apparent hydroxyl radical production by peroxynitrite: Implications for endothelial injury from nitric oxide and superoxide. Proc. Natl. Acad. Sci. USA **87:** 1620-1624.
60. DAWSON, V. L., T. M. DAWSON, E. D. LONDON, D. S. BREDT & S. H. SNYDER. 1991. Nitric oxide mediates glutamate neurotoxicity in primary cortical cultures. Proc. Natl. Acad. Sci. USA **88:** 6368-6371.
61. FAVARON, M., H. MANEV, H. ALHO, M. BERTOLINO, B. FERRET, A. GUIDOTTI & E. COSTA. 1988. Gangliosides prevent glutamate and kainate neurotoxicity in primary neuronal cultures of neonatal rat cerebellum and cortex. Proc. Natl. Acad. Sci. USA **85:** 7351-7355.
62. DAWSON, V. L., T. M. DAWSON, K. HUNG, J. P. STEINER & S. H. SNYDER. 1993. Gangliosides attenuate NMDA neurotoxicity by inhibiting nitric oxide synthase. Soc. Neurosci. Abstr. **19:** 25.
63. WOTJTCZAKC, L. 1976. Effect of long-chain fatty acids and acyl-CoA on mitochondrial permeability, transport and energy-coupling processes. J. Bioenerg. Biomem. **8:** 293-322.
64. DUGAN, L. L., S. L. SENSI, L. M. T. CANZONIERO, M. P. GOLDBERG, S. D. HANDRAN, S. M. ROTHMAN & D. W. CHOI. 1994. Imaging of mitochondrial oxygen radical production in cortical neurons exposed to NMDA. Soc. Neurosci. Abstr. **20:** 1532.
65. HEWETT, S. J., C. A. CSERNANSKY & D. W. CHOI. 1994. Selective potentiation of NMDA-induced neuronal injury following induction of astrocytic iNOS. Neuron **13:** 487-494.
66. MURPHY, T. H., M. MIYAMOTO, A. SASTRE, R. L. SCHNAAR & J. T. COYLE. 1989. Glutamate toxicity in a neuronal cell line involves inhibition of cystine transport leading to oxidative stress. Neuron **2:** 1547-1558.
67. RATAN, R. R., T. H. MURPHY & J. M. BARABAN. 1994. Oxidative stress induces apoptosis in embryonic cortical neurons. J. Neurochem. **62:** 376-379.
68. WYLLIE, A. H., J. F. R. KEER & A. R. CURRIE. 1990. Cell death: The significance of apoptosis. Int. Rev. Cytol. **68:** 251-306.
69. CSERNANSKY, C. A., L. M. T. CANZONIERO, S. L. SENSI, S. P. YU & D. W. CHOI. 1994. Delayed application of aurintricarboxylic acid reduces glutamate-induced cortical neuronal injury. J. Neurosci. Res. **38:** 101-108.
70. MCCONKEY, D. J., P. HARTZELL, P. NICOTERA & S. ORRENIUS. 1989. Calcium-activated DNA fragmentation kills immature thymocytes. FASEB J. **3:** 1843-1849.

71. UCKER, D. S. 1991. Death by suicide: One way to go in mammalian cellular development? The New Biologist **3:** 103-109.
72. JOHNSON, E. M. & T. L. DECKWERTH. 1993. Molecular mechanisms of developmental neuronal death. Annu. Rev. Neurosci. **16:** 31-46.
73. KOH, J. Y. & C. W. COTMAN. 1992. Programmed cell death: Its possible contribution to neurotoxicity mediated by calcium channel antagonists. Brain Res. **587:** 233-240.

Transcription Factors as Molecular Mediators in Cell Death

HOLLY D. SOARES, TOM CURRAN, AND
JAMES I. MORGAN

Roche Institute of Molecular Biology
Roche Research Center
Nutley, New Jersey 07110

Less than a decade ago, cell death was a subject of interest only to pathologists and embryologists. Today, however, the topic garners avid attention among a wide range of biological disciplines largely as new insights are gained into this subject through advances in cellular and molecular biology. Cell death used to be perceived as a relatively simple process; hence, only a few terms existed to describe it. Unfortunately, present terminology has failed to keep pace with recent rapid advances. As a result, the terms "programmed cell death," "apoptosis," and "necrosis" are applied with much abandon to very different forms of cellular death. Definitions for apoptosis and necrosis have been extensively discussed and will not be reviewed here.[1] The problems arise when programmed cell death and apoptosis are equated. Such a practice presumes that only one form of programmed cell death exists even though recent studies have suggested that this is not the case. For example, during nervous system development, programmed cell death exhibits a number of morphologic phenotypes that appear to fall outside of a strict definition for apoptosis.[2] In addition, not all apoptotic cell deaths result in DNA laddering which is often an index of apoptosis.[3,4] Furthermore, condensation of chromatin at the nuclear membrane is not always associated with endonuclease activation.[5,6] Although the terminology requires some modifications, it is becoming clear that multiple pathways for programmed cell death exist. Recent studies identifying cellular proteins responsible for carrying out a cell death sentence currently abound and have served to galvanize the field. The following treatise focuses upon specific molecular mediators thought to participate in cell death programs.

THE ANTITHESIS OF DEATH: BCL-2

A popular approach in the study of cell cycle abnormalities involves characterizing chromosomal rearrangements prevalent within hematopoietic malignancies.[7] Chromosomal translocations occur in 90% of chronic myelogenous leukemias,[8,9] in 80% of Burkitt's lymphomas,[10] and in 80% of follicular and 20% of diffuse B-cell lymphomas.[11–13] Typically, translocations are screened with probes to known oncogenes. However, no known oncogenes were found to be associated with either the t(11;14) translocation of chronic lymphocytic leukemia[14] or the t(14;18) translocation of most follicular lymphoma's.[11–13] The fact that these translocations contained band 14q32 proved fortuitous. Residing within 14q32 was the well characterized immunoglobulin

(Ig) heavy chain gene. Utilizing Ig as a probe, two novel genes were isolated within the t(11;14) and t(14;18) translocations. They were named *bcl*-1[14] and *bcl*-2[15–18] after B-cell leukemia.

For a brief time, the function of the *bcl*-2 gene product remained a mystery. It was discovered that t(14;18) translocations occurred during pre-B-cell development, resulting in the overproduction of a Bcl-2 fusion protein.[19–21] *In vitro* experiments reported overexpression of Bcl-2 and produced an arrest in G_0.[22] This observation raised the specter that B-cell malignancy could be attributed to cell survival rather than proliferation. Further studies went on to show that Bcl-2 prevented cell death following growth factor deprivation in hematopoietic[22–25] and neuronal cell lines.[26–30] Bcl-2 overexpression also prolonged the survival of B and T cells from transgenic animals[31–35] and attenuated death in transfected cells after irradiation, H_2O_2, or glucocorticoid exposure.[36–38] The latency membrane protein of Epstein-Barr virus has been shown to upregulate *bcl*-2, thereby blocking apoptosis.[39] Interestingly, the Epstein-Barr virus and the African swine fever virus both produce their own *bcl*-2 homologues.[40,41] Although, Bcl-2 appears to be versatile as a mediator of survival, Bcl-2 cannot prevent all forms of death. For example, Bcl-2 fails to prevent complement and cytotoxic T-cell killing.[42] In addition, Bcl-2 does not seem to impede the process of T-cell clonal deletion.[33]

Bcl-2 appears to be localized throughout the immune, endocrine, and nervous systems[43] and has been characterized as a 24-kD membrane-associated protein.[19,44] Upon ultrastructural examination, Bcl-2 can be found localized within membranes of the nuclear envelope, endoplasmic reticulum, and mitochondria.[44–47] Although its precise biochemical function is unknown, *in vitro* and knock-out experiments suggest a role for *bcl*-2 in antioxidant pathways.[38,48]

Two related *bcl*-2 family members have now been isolated. The *bcl-x* gene can be alternatively spliced, resulting in a protein that inhibits Bcl-2 function.[49] Another related member, BAX, directly binds Bcl-2 and can also inhibit Bcl-2's ability to enhance cell survival.[50] Surprisingly, overexpression of BAX accelerates cell death.[50] It has been hypothesized that BAX and Bcl-2 may interact with each other, perhaps balancing survival and death decisions. Although these genes have shown remarkable effects in the immune system, most of them are not highly expressed in the central nervous system. Therefore, their role in neuronal survival is currently a matter of some interest and debate.

THE GENETICS OF DEATH: CED GENES

While lymphoid translocations have proved especially fruitful in studying molecular mediators of cell survival, an equally fertile system has been *Caenorhabditis elegans*. During *C. elegans* development, 1,090 somatic cells form in the hermaphrodite, 131 of which undergo programmed cell death.[51,52] A total of 14 genes actively participate in nematode cell death programs.[53] The study of these genes in mutants reveals interesting characteristics of nematode programmed cell death. First, phagocytic engulfment, mediated by *ced*-1 and *ced*-2 gene products, is not a prerequisite for death.[54,55] Second, the lack of endodeoxyribonuclease activity found in *nuc*-1 mutants does not inhibit death[54,56] and DNA appears to persist in the remaining cellular bodies. Although the *nuc*-1 gene product has not been demonstrated to be related to the endonuclease involved in DNA laddering, cell death still transpires in *nuc*-1 mutants. Additional cell death genes include *ced*-3 and *ced*-4 which are required for developmentally regulated death.[57] Finally, programmed cell death can be suppressed by the *ced*-9 gene product.[58] The *ced*-9 gene shares some homology with Bcl-2, and expression of human Bcl-2 in *C. elegans* mutants can reduce programmed cell death.[59] These results suggest that inhibition of cell death may be a relatively well conserved mechanism across species.

The gene products of *ced*-3 and *ced*-4 have recently been identified. The *ced*-3 gene encodes a protein containing a serine-rich middle region and a highly conserved COOH-terminus.[60] The *ced*-3 product bears striking similarities to the human interleukin-1β-converting enzyme (ICE) and to the murine Nedd-2 protein.[61–63] Overexpression of Ced-3 in Rat-1 cells elicits death which can be inhibited by Bcl-2 and CrmA (an inhibitor of ICE).[64] The *ced*-4 gene product is a hydrophilic protein containing no obvious transmembrane domain. Although Ced-4 appears to be a novel protein, it does possess two regions exhibiting known calcium-binding functions.[65] As a result, Ced-4 has been hypothesized to participate in calcium-binding processes which may occur during programmed cell death.[65]

THE DOORWAYS TO DEATH: APO-1/FAS AND TNF-R1

The *bcl*-2 and *ced* experiments painted a scenario where intracellular mediators carry out life or death sentences. However, most cells communicate with their environment via surface receptors. Are there cell surface receptors for cell death?

Thus far, two receptors have been cloned that appear to induce cell death. These include the Fas/APO-1 antigen and tumor necrosis factor receptor type 1 (TNF-R1). Fas and APO-1 were discovered in separate labs[66,67] and later found to be identical.[68] In the mouse, the Fas/APO-1 antigen occurs in thymus, heart, liver, and ovary, but not in brain or spleen.[69] The Fas/APO-1 shows some similarities with nerve growth factor receptor, TNF-R1, as well as the presumptive receptors CD27, CD30, CD40, and OX40.[70,71] In the lymphoproliferation (*lpr*) mutant mouse, a strain that exhibits autoimmune-like illnesses, the Fas/APO-1 gene product is mutated.[72] Fas/APO-1 abnormalities in *lpr* mutants suggest that this antigen may participate in the negative selection of autoreactive thymic T cells.

Although there are two receptors for TNF,[73–75] only the 55-kD TNF-R1 mediates cytotoxicity.[76] The cytotoxic signal resides within an 80 amino acid stretch of the intracellular COOH-terminus. This so-called "death domain" is highly conserved with the COOH-terminus of Fas.[70,77] Occurring as a multiple, noncontiguous stretch of amino acids, the death domain may be a folded structure. Not surprisingly, overexpression of Bcl-2 can partially inhibit Fas/APO-1-induced extermination.[78] These data support the contention that some cell death programs use cell surface mediators. This raises the question of how these receptors signal the intracellular processes that lead to death.

TRANSCRIPTION FACTORS: THE SERVANTS OF DEATH?

A few studies have shown that cell death following trophic factor withdrawal can be inhibited when RNA or protein synthesis is suppressed.[79–81] Paradoxically, serum withdrawal from PC12 cells still induced cell death even when RNA and protein synthesis was prevented.[82] These results are not necessarily mutually exclusive. It is probable that some forms of cell death require transcription and translation whereas others do not. Those death pathways that require gene expression are presumed to recruit a transcriptional response. Indeed, a number of transcriptional mediators have been implicated either directly, or indirectly, in cell death.

p53

The tumor suppressor gene p53 was discovered as a protein closely associated with the simian virus 40 large tumor antigen.[83,84] p53 mutations commonly arise in human lung, breast, liver, and colon carcinomas.[85,86] p53 binds DNA in a sequence-specific manner and possesses transcriptional activity.[87–91] Under normal conditions, p53 intracellular levels are extremely low, due in part to its short half-life.[92] Overexpression of wild-type p53 in either human colon tumor cells or murine myeloid leukemic cells elicits cellular extermination.[93,94] Further evidence for the tumor suppressor's role in cell death comes from p53 knockout experiments. Cells containing p53 null mutations show resistance to chemotherapeutic agents and radiation.[95–97] Perhaps even more tantalizing is p53's participation during UV and ionizing radiation injury. Although thymocytes from p53 null mutants remained susceptible to glucocorticoids and T-cell-mediated killing, they proved resistant to injuries involving DNA strand breakage (ie, radiation).[96,97] These data imply that p53, either through transcriptional events[98] or through more direct means, plays a role in terminating cells whose DNA has been damaged.

c-*myc*

Historically, c-*myc* was cast more as a mediator of proliferation and differentiation than as a mediator of death.[99] Nevertheless, experiments have implicated c-*myc* in cell death processes. For example, Chinese hamster ovary cells will die after being exposed to heat shock only if they are first transfected with a construct containing c-*myc* under the control of a heat shock promoter.[100,101] Overexpression of the c-*myc* gene product also accelerates cell death in transformed fibroblasts.[102] During androgen-induced extermination of rat prostate cells, c-*myc* is consistently elevated.[103] Death in growth factor-deprived Rat-1 cells and interleukin-3 (IL-3)-deprived myeloid cells both require c-*myc* expression.[104,105] Finally, antisense c-*myc* oligonucleotides block expiration in T-cell hybridomas.[106] Myc contains both leucine zipper and basic helix-loop domains, and it has been proposed that specific DNA binding occurs via the latter.[99,107] A number of related family members exist that are capable of dimerizing with Myc, thereby influencing its transcriptional activity.[99,108,109] Myc's role in cell death may well depend upon which proteins dimerize with it.

c-*fos* and c-*jun*

c-*fos* and c-*jun* were isolated as the cellular homologues of transforming viral genes (v-*fos* and v-*jun*, respectively). Fos derives its name from the Finkel-Biskis-

Jinkins osteosarcoma virus,[110] whereas the term *jun* is a condensed form of "ju-nana," the Japanese word for 17. Jun was originally isolated from the avian sarcoma virus 17, hence its name.[111] Like Myc, Fos and Jun contain leucine zipper and DNA binding domains.[112–115] There are many related Fos and Jun proteins.[116] Jun family members can form homodimers and heterodimers, whereas Fos can only form heterodimers with Jun (and other) proteins.[116,117] Unable to bind DNA on its own, Fos requires a binding partner for transcriptional activity.[118–120]

Fos has been implicated in a myriad of physiologic roles.[116,117] Because of its versatility, it is perhaps not surprising to find studies linking Fos with cell death. Early work suggested that prostate cell death resulting from castration involved extensive c-*myc* and c-*fos* expression.[103] c-Fos protein was also observed during developmentally regulated programmed cell death in rats.[121] In the rat study, cytoplasmic Fos immunoreactivity appeared throughout developing cortical populations.[121] This cytoplasmic staining was not of phagocytic origin and was attributed to autophagic neurons that contained nuclear products within autolysosomal structures.[2]

Utilizing a c-*fos*-β-galactosidase (Fos-LacZ) construct containing all known c-*fos* regulatory sequences, our laboratory developed a transgenic mouse line capable of recapitulating normal c-*fos* induction.[122,123] In this model, continuous Fos expression occurred in developing populations of cells undergoing programmed cell death.[124] For example, Fos-LacZ was apparent within interdigital web cells, heart valve cells, and peridermal cells undergoing programmed cell death. In the adult, expression was associated with skin, hair follicles, and dying cells in the ovary.

To follow Fos induction in defined dying populations, our lab also examined Fos-LacZ expression utilizing an excitotoxic paradigm. Kainic acid is known to elicit profound neuronal degeneration within specific well-defined hippocampal populations.[125] Kainic acid administered to transgenic mice induced transient Fos-LacZ expression throughout the brain which disappeared 24 hours later. However, between 4 and 7 days postadministration, Fos-LacZ reappeared in degenerating CA1, CA3, and CA4 hippocampal pyramidal neurons.[124] Adjacent resistant CA2 neurons exhibited no Fos-LacZ induction. Interestingly, Fos cytoplasmic staining was also observed in degenerating neurons.

A number of mutant mice exist that exhibit neurodegenerative disorders. In *weaver* mutants, virtually all midline cerebellar granule cells undergo cell death.[126] To determine if *fos* induction was associated with stereotypical cerebellar degeneration of *weaver* mutants, Fos-LacZ mice were crossed with *weaver* mice. Analysis of the resultant *weaver* heterozygotes showed extensive spontaneous Fos-LacZ expression within degenerating cerebellar granule cells.[124]

Although previous studies implicate c-*fos* in cell death, they do not demonstrate a direct role. In an attempt to establish whether c-*fos* directly participated in cell death pathways, fibroblasts were transfected with an inducible c-*fos* construct.[124] Serum deprivation caused cell death in fibroblasts that were expressing Fos, whereas those that did not express Fos survived.[124] A more direct role for Fos in cell death was also shown using antisense oligonucleotides. Antisense oligonucleotides against either c-*fos* or c-*jun* inhibited cell death in both IL-6 and IL-2-deprived cell lines.[127]

The role of c-*jun* in cell death is currently more nebulous. Jun does appear in dying neurons after hypoxic-ischemic insult and after status epilepticus.[128] Interestingly, Jun also appears to be induced in regenerating neuronal populations,[129,130] suggesting that c-*jun* may also participate in "cell survival" programs.

CONCLUSIONS

Eukaryotic organisms have evolved complex mechanisms to ensure the correct balance of cell birth to cell death. In general terms, cell death can be viewed as having three components, namely, mechanisms that commit a cell to die, mechanisms that actually kill the cell, and processes that eliminate the cellular corpse. In some organisms, the genetic basis of these mechanisms is at least partially understood. From the perspective of our work, involvement of transcription factors is of particular importance. Immediate-early genes, such as c-*fos,* c-*jun,* and c-*myc,* are attractive candidates for cell death commitment genes. First, they are inducible transcription factors that respond to many types of stimuli including agents, such as TNF, that cause death. Thus, they could serve to control expression of target genes that might contribute in a more direct manner to death. Second, when overexpressed, several of these genes (c-*fos* and c-*myc*) can cause cell death following growth factor withdrawal. Third, blockade of function or expression of some of these genes can prevent cell death. Fourth, expression of Fos and Jun is temporally and spatially associated with cell death in mice. However, we must still question what precise role these immediate-early genes play in cell death. It is tempting to construct a hypothesis in which they are commitment genes, and culture experiments would sustain such a hypothesis. However, overexpression experiments in cell culture or antisense experiments are open to a range of criticism. Furthermore, the knockout of c-*fos* does not result in any overt changes in cell death (except perhaps in bone), and Fos and Jun (and Myc) are expressed in many cells that do not go on to die. Therefore, it would be prudent to maintain an open mind on this issue. Thus, Fos and Jun may play no direct role in cell death but rather be reacting to the molecules and signaling pathways that trigger death. Indeed, it is conceivable, and some experiments would support the notion, that Fos and Jun may have a protective role in cell death. Certainly there are intriguing associations between the induction of Fos and Jun and DNA damage that might suggest involvement in DNA repair. Such a notion would be consistent with the finding that p53, a gene intimately associated with the cell cycle and DNA repair, can modulate cell death.

The immediate-early gene response is a complex process involving many genes that are induced in a dynamic manner. This is particularly true for those members of the immediate-early gene class that belong to the basic-zipper superfamily of transcription factors. We have shown that neurotoxins, such as kainic acid, cause the staggered appearance and disappearance of proteins that can contribute to the transcription factor complex known as AP-1 (activator protein 1).[131,132] As just noted, Fos and Jun (and all known members of the basic-zipper superfamily) form dimers, and we therefore have to ask the question of whether the AP-1 complexes generated during the processes leading to cell death represent unique patterns of hetero- and homodimers. Preliminary evidence suggests that Fos and Jun are not always colocalized *in vivo* during normal development or in adult mice although they are both coinduced in some cell death models *in vivo* and in culture. Little to nothing is known about the other inducible basic-zipper proteins in these models. Therefore, further work is necessary to define the precise molecular composition of these complexes in standardized models of neuronal death.

The process of cell death is of importance not only in maintaining the size of a particular organ or cell population or allowing for cell selection in the immune system in the adult, but also in shaping the developing organism by eliminating transient structures and supernumerary cells. Controlled cell death is also necessary for the generation of a number of structures in the adult such as skin, hair, nail, and bone. In the nervous system, programmed cell death is essential to match neuronal input population size to its target, a process that appears to be mediated in some instances by competition for neurotrophic factors. Thus, programmed cell death contributes to the correct formation of the spinal cord as well as sympathetic and cranial ganglia. Cell death is also believed to be important in forming the corpus callosum and retina. For the neurobiologist, a critical issue is whether the processes that control normal, programmed cell death in the nervous system also contribute to neuropathologic conditions such as amyotrophic lateral sclerosis, Alzheimer's disease, or neuronal death following cerebral ischemia. We and others have established correlations between the expression of genes such as c-*fos* and c-*jun* in several instances of pathologic death in rodents that include hypoxia/ischemia, neurotoxin administration, sciatic nerve transection, and genetically dependent death in cerebellum and substantia nigra. However, to date, evidence in man is generally lacking in this regard. With the availability of specific reagents for many immediate-early gene products it would seem a reasonable goal to determine if their expression is associated with any of the human neurodegenerative disorders.

REFERENCES

1. Wyllie, A. H., J. F. R. Kerr & A. R. Currie. 1980. Int. Rev. Cytol. **68:** 251-306.
2. Clark, P. G. H. 1990. Anat. Embryol. **181:** 195-213.
3. Schnellman, R. G., A. R. Swagler & M. M. Compton. 1993. Am. J. Physiol. **256:** C485-490.
4. Schwartz, L. M., S. W. Smith, M. E. E. Jones & B. A. Osborne. 1993. Proc. Natl. Acad. Sci. USA **90:** 980-984.
5. Cohen, G. M., X. Sun, R. T. Snowden, D. Dinsdale & D. N. Skilleter. 1992. Biochem. J. **286:** 331-334.
6. Oberhammer, F., G. Fritsch, M. Schneid, M. Pauelka, D. Printz, T. Purchio, H. Lassman & R. Schult-Hermann. 1993. J. Cell. Sci. **104:** 317-326.
7. Yunis, J. J. 1983. Science **221:** 227-236.
8. Nowell, P. C. & D. A. Hungerford. 1960. Science **132:** 1497-1499.
9. Rowley, J. D. 1973. Nature **243:** 290-291.
10. Manolov, G. & Y. Manolova. 1972. Nature **237:** 33-34.
11. Fukuhara, S., J. D. Rowley, D. Varrakojis & H. M. Golomb. 1979. Cancer Res. **39:** 3119-3131.
12. Levine, E. G., D. G. Arthur, G. Frizzera, B. A. Peterson, D. D. Hurd & C. D. Bloomfield. 1985. Blood **66:** 1414-1422.
13. Yunis, J. J., G. Frizzera, M. M. Oken, J. McKenna, A. Theologides & M. Arnesen. 1987. N. Engl. J. Med. **316:** 79-84.
14. Tsujimoto, Y., J. Yunis, L. Onorato-Showe, J. Erikson, P. C. Nowell & C. M. Croce. 1984. Science **224:** 1403-1406.

15. BAKHSHI, A., J. P. JENSEN, P. GOLDMAN, J. J. WRIGHT, O. W. MCBRIDE, A. L. EPSTEIN & S. J. KORSMEYER. 1985. Cell **41:** 899-906.
16. CLEARY, M. L. & J. SKLAR. 1985. Proc. Natl. Acad. Sci. USA **82:** 7439-7443.
17. TSUJIMOTO, Y., L. R. FINGER, J. YUNIS, P. C. NOWELL & C. M. CROCE. 1984. Science **226:** 1097-1099.
18. TSUJIMOTO, Y., J. COSSMAN, E. JAFFE & C. M. CROCE. 1985. Science **228:** 1440-1443.
19. CHEN-LEVY, Z., J. NOURSE & M. L. CLEARY. 1989. Mol. Cel. Biol. **9:** 701-710.
20. GRANINGER, W. B., M. SETO, B. BOUTAIN, P. GOLDMAN & S. J. KORSMEYER. 1987. J. Clin. Invest. **80:** 1512-1515.
21. SETO, M., U. JAEGER, R. D. HOCKETT, W. GRANINGER, S. BENNETT, P. GOLDMAN & S. J. KORSMEYER. 1988. EMBO J. **7:** 123-131.
22. VAUX, D. L., S. CORY & J. M. ADAMS. 1988. Nature **335:** 440-442.
23. BORZILLO, G. V., K. ENDO & Y. TSUJIMOTO. 1992. Oncogene **7:** 869-876.
24. HOCKENBERY, D., G. NUÑEZ, C. MILLIMAN, R. D. SCHREIBER & S. J. KORSMEYER. 1990. Nature **348:** 334-336.
25. NUÑEZ, G., L. LONDON, D. HOCKENBERY, M. ALEXANDER, J. P. MCKEARN & S. J. KORSMEYER. 1990. J. Immunol. **144:** 3602-3610.
26. ALLSOP, T. E., S. WYATT, H. F. PATERSON & A. M. DAVIES. 1993. Cell **73:** 295-307.
27. BATISTATOU, A., D. E. MERRY, S. J. KORSMEYER & L. A. GREENE. 1993. J. Neurosci. **13:** 4422-4428.
28. GARCIA, I., I. MARTINOU, Y. TSUJIMOTO & J. MARTINOU. 1992. Science **258:** 302-304.
29. MAH, S. P., L. T. ZHONG, Y. LIU, A. ROGHANI, D. E. BREDESEN & R. H. EDWARDS. 1993. J. Neurochem. **60:** 1183-1186.
30. ZHONG, L. T., T. SARAFIAN, D. J. KANE, A. C. CHARLES, S. P. MAH, R. H. EDWARDS & D. E. BREDESEN. Proc. Natl. Acad. Sci. USA **90:** 4533-4537.
31. MCDONNELL, T. J., N. DEANE, F. M. PLATT, G. NUÑEZ, U. JAEGER, J. P. MCKEARN & S. J. KORSMEYER. 1989. Cell **57:** 79-88.
32. NEGRINI, M., E. SILINI, C. KOZAK, Y. TSUJIMOTO & C. M. CROCE. 1987. Cell **49:** 455-463.
33. SENTMAN, C. L., J. R. SUTTER, D. HOCKENBERY, O. KANAGAWA & S. J. KORSMEYER. 1991. Cell **67:** 879-888.
34. SIEGEL, R. M., M. KATSOMATA, T. MIYASHITA, D. C. LOUIE, M. I. GREENE & J. C. REED. 1992. Proc. Natl. Acad. Sci. USA **89:** 7003-7007.
35. STRASSER, A., A. W. HARRIS, M. L. BATH & S. CORY. 1990. Nature **348:** 331-333.
36. ALNEMRI, E. S., T. F. FERNANDES, S. HALDAR, C. C. CROCE & G. LITWAK. 1992. Cancer Res. **52:** 491-495.
37. MIYASHITA, I. & J. C. REED. 1992. Cancer Res. **52:** 5407-5411.
38. HOCKENBERY, D. M., Z. N. OITVAI, X. M. YIN, C. L. MILLIMAN & S. J. KORSMEYER. 1993. Cell **75:** 241-251.
39. HENDERSON, S., M. ROWE, C. GREGORY, D. CROOM-CARTER, F. WANG, R. LONGNECKER, E. KIEFF & A. RICKINSON. 1991. Cell **65:** 1107-1115.
40. HENDERSON, S., D. HUEN, M. ROWE, C. DAWSON, G. JOHNSON & A. RICKINSON. 1993. Proc. Natl. Acad. Sci. USA **90:** 8479-8483.
41. NEILAN, J. G., Z. LU, C. L. AFONSON, G. F. KUTISH, M. D. SUSSMAN & D. L. ROCK. 1993. J. Virol. **67:** 4391-4394.
42. STRASSER, A., A. W. HARRIS & S. CORY. 1992. Cell **67:** 889-899.
43. HOCKENBERY, D. M., M. ZUTTER, W. HICKEY, M. NAHM & S. J. KORSMEYER. 1991. Proc. Natl. Acad. Sci. USA **88:** 6961-6965.
44. CHEN-LEVY, Z. & M. L. CLEARY. 1990. J. Biol. Chem. **265:** 4929-4933.
45. HOCKENBERY, D., G. NUÑEZ, C. MILLIMAN, R. D. SCHREIBER & S. J. KORSMEYER. 1991. Nature **348:** 334-336.
46. JACOBSON, M. D., J. F. BURNE, M. P. KING, T. MIYASHITA, J. C. REED & M. C. RAFF. 1993. Nature **361:** 365-369.

47. Monaghan, P., D. Robertson, T. A. S. Amos, M. J. S. Dyer, D. Y. Mason & M. F. Greaves. 1992. J. Histochem. Cytochem. **40:** 1810-1825.
48. Veis, D. J., C. M. Sorenson, J. R. Shutter & S. J. Korsmeyer. 1993. Cell **75:** 229-240.
49. Boise, L. H., M. Gonzalez-Garcia, C. E. Postema, L. Ding, T. Lindsten, L. A. Turka, X. Mao, G. Nuñez & C. B. Thompson. 1993. Cell **74:** 597-608.
50. Oltvai, Z., C. L. Milliman & S. J. Korsmeyer. 1993. Cell **74:** 609-619.
51. Sulston, J. E. & H. R. Horvitz. 1977. Dev. Biol. **82:** 110-156.
52. Sulston, J. E., E. Schierenberg, J. G. White & N. Thompson. 1983. Dev. Biol. **100:** 64-119.
53. Ellis, R., J. Yuan & H. R. Horvitz. 1991. Ann. Rev. Cell. Biol. **7:** 663-698.
54. Hedgehock, E., J. E. Sulston & N. Thompson. 1983. Science **220:** 1277-1280.
55. Ellis, R. E., D. Jacobson & H. R. Horvitz. 1991. Genetics **112:** 591-603.
56. Sulston, J. E. 1976. Philos. Trans. R. Soc. London. Ser. B. **275:** 287-297.
57. Yuan, J. & H. R. Horvitz. 1990. Dev. Biol. **138:** 33-41.
58. Hengartner, M. O., R. E. Ellis & H. R. Horvitz. 1992. Nature **356:** 494-499.
59. Vaux, D. L., I. L. Weissman & S. K. Kim. 1992. Science **258:** 1955-1957.
60. Yuan, J., S. Shamam, S. Ledoux, H. M. Ellis & H. R. Horvitz. 1993. Cell **75:** 641-652.
61. Cerretti, D. P., C. J. Kozlosky, B. Mosley, N. Nelson, K. V. Ness, T. A. Greenstreet, C. J. March, S. R. Kronheim, T. Druck, L. A. Cannizzaro, K. Huebner & R. A. Black. 1992. Science **256:** 97-100.
62. Thornberry, N. A., H. G. Bull, J. R. Calaycay, K. T. Chapman, A. D. Howard, M. J. Kostura *et al.* 1992. Nature **356:** 768-774.
63. Kumar, S., Y. Tomooka & M. Noda. 1992. Biochem. Biophys. Res. Commun. **185:** 1155-1161.
64. Miura, M., H. Zhu, R. Rotello, E. A. Hartweig & J. Yuan. 1993. Cell **75:** 653-660.
65. Yuan, J. & H. R. Horvitz. 1992. Development **116:** 309-320.
66. Trauth, B. C., C. Klas, A. M. J. Peters, S. Matzku, P. Moller, W. Falk, K. M. Debatin & P. H. Krammer. 1989. Science **245:** 301-305.
67. Yonehara, S., A. Ishii & M. Yonehara. 1989. J. Exp. Med. **169:** 1747-1756.
68. Oehm, A., I. Behrann, W. Falk, M. Pawlita, G. Maier, C. Klas, M. Li-Weber, S. Richards, J. Dhein, B. C. Trauth, H. Posingl & P. H. Krammer. 1992. J. Biol. Chem. **267:** 10709-10715.
69. Watanabe-Fukunaga, R., C. I. Brannan, N. Itoh, S. Yonehara, N. G. Copeland, N. A. Jenkins & S. Nagata. 1992. J. Immunol. **148:** 1274-1279.
70. Itoh, N. & S. Nagata. 1993. J. Biol. Chem. **268:** 10832-10937.
71. Itoh, N., S. Yonehara, A. Ishii, M. Yonehara, S. Mizushima, M. Sameshima, A. Hase, Y. Seto & S. Nagata. 1991. Cell **66:** 233-243.
72. Watanabe-Fukunaga, R., C. I. Brannan, N. G. Copeland, N. A. Jenkins & S. Nagata. 1992. Nature **356:** 314-317.
73. Tartaglia, L. A. & D. V. Goeddel. 1992. Immunol. Today **13:** 151-153.
74. Brockhaus, M., H. J. Schoenfeld, E. J. Schlaeger, W. Hunzicker, W. Lesslauer & H. Loetscher. 1990. Proc. Natl. Acad. Sci. USA **87:** 3127-3131.
75. Hohmann, H. P., M. Brockhaus, P. A. Baeuerle, R. Remy, R. Kolbeck & A. P. G. M. Vanloon. 1990. J. Biol. Chem. **265:** 22409-22417.
76. Tartaglia, L. A., M. Rothe, Y. F. Hu & D. V. Goeddel. 1993. Cell **73:** 213-216.
77. Tartaglia, L. A., T. M. Ayres, G. H. W. Wong & D. V. Goeddel. 1993. Cell **74:** 845-853.
78. Itoh, N., Y. Tsujimoto & S. Nagata. 1993. J. Immunol. **151:** 621-627.
79. Martin, D. P., R. E. Schmidt, P. S. DiStefano, O. H. Lowry, J. G. Carter & E. M. Johnson, Jr. 1988. J. Cell Biol. **106:** 829-844.
80. Scott, S. A. & A. M. Davies. 1990. J. Neurobiol. **21:** 630-638.

81. OPPENHEIM, R. W., D. PREVETTE, M. TYTELL & S. HOMMA. 1990. Dev. Biol. **138:** 104-113.
82. RUKENSTEIN, A., R. E. RYDEL & L. A. GREENE. 1991. J. Neurosci. **11:** 2552-2563.
83. LINZER, D. & A. LEVINE. 1979. Cell **17:** 43-52.
84. LANE, D. & L. CRAWFORD. 1979. Nature **278:** 261-263.
85. HOLLSTEIN, M., D. SIDRANSKY, B. VOGELSTEIN & C. HARRIS. 1991. Science **253:** 49-53.
86. LEVINE, A. J., J. MAMAND & FINLAY. 1991. Nature **351:** 453-456.
87. FUNK, W. D., D. T. PAK, R. H. KARAS, W. E. WRIGHT & J. W. SHAY. 1992. Mol. Cell. Biol. **12:** 2866-2871.
88. KERN, S. E., K. W. KINZLER, A. BRUSKIN, D. JAROSZ, P. FRIEDMAN, C. PRIVES & B. VOGELSTEIN. Science **252:** 1708-1711.
89. BARGONETTI, J., P. N. FRIEDMAN, S. E. KERN, B. VOGELSTEIN & C. PRIVES. 1992. Cell **65:** 1083-1091.
90. FARMER, G., J. BARGONETTI, H. ZHU, P. FRIEDMAN, R. PRYWES & C. PRIVES. 1992. Nature **358:** 83-86.
91. SCHARER, E. & R. IGGO. 1992. Nucl. Acids. Res. **20:** 1539-1545.
92. OREN, M., W. MALTZMAN & A. J. LEVINE. 1981. Mol. Cel. Biol. **1:** 101-110.
93. SHAW, P., R. BOVEY, S. TARDY, R. SAHLI, B. SORDAT & J. COSTA. 1992. Proc. Natl. Acad. Sci. USA **89:** 4495-4499.
94. YONISH-ROUACH, E., D. RESNITSKY, J. LOTEM, L. SACHS, A. KIMCHI & M. OREN. 1991. Nature **352:** 345-347.
95. LOWE, S. W., H. E. RULEY, T. JACKS & D. E. HOUSMAN. 1993. Cell **74:** 957-967.
96. LOWE, S. W., E. M. SCHMITT, S. W. SMITH, B. A. OSBORNE & T. JACKS. 1993. Nature **362:** 847-849.
97. CLARKE, A. R., C. A. PURDIE, D. J. HARRISON, R. G. MORRIS, C. C. BIRD, M. L. HOOPER & A. H. WYLLIE. 1993. Nature **362:** 849-852.
98. LU, X. & D. P. LANE. 1993. Cell **75:** 765-778.
99. LUSCHER, B. & R. N. EISENMAN. 1990. Genes Dev. **4:** 2025-2035.
100. WURM, F. M., K. A. GWINN & R. E. KINGSTON. 1986. Proc. Natl. Acad. Sci. USA **83:** 5414-5419.
101. PALLAVINI, M. G., C. ROSETTE, M. REITSMA, P. S. DETERESA & J. W. GRAY. 1990. J. Cell. Physiol. **143:** 372-380.
102. WYLLIE, A. H., K. A. ROSE, R. G. MORRIS, C. M. STEEL, E. FOSTER & D. A. SPANDIDOS. 1987. Br. J. Cancer **56:** 251-259.
103. BUTTYAN, R., Z. ZAKERI, R. LOCKSHIN & D. WOLGMUTH. 1988. Mol. Endocrinol. **2:** 650-657.
104. EVAN, G. I., A. H. WYLLIE, C. S. GILBERT, T. D. LITTLEWOOD, H. LAND, M. BROOKS, C. M. WATERS, L. Z. PENN & D. C. HANCOCK. 1992. Cell **69:** 119-128.
105. ASKEW, D. S., R. A. ASHMUN, B. C. SIMMONS & J. L. CLEVELAND. 1991. Oncogene **6:** 1915-1922.
106. SHI, Y., J. M. GLYNN, L. J. GUILBERT, T. G. CUTTER, R. P. BISSONNETTE & D. R. GREEN. Science **257:** 212-214.
107. BLACKWELL, T., K. KRETZNER, E. M. BLACKWOOD, R. N. EISENMANN & H. WEINTRAUB. 1990. Science **250:** 1149-1151.
108. BLACKWOOD, E. M. & R. N. EISENMAN. 1991. Science **251:** 1211-1217.
109. AYER, D. E. & R. N. EISENMAN. 1993. Genes Dev. **7:** 2110-2119.
110. CURRAN, T. & N. M. TEICH. 1982. J. Virol. **42:** 114-122.
111. MAKI, Y., T. J. BOS, C. DAVIS, M. STARBUCK & P. K. VOGT. 1987. Proc. Natl. Acad. Sci. USA **84:** 2848-2852.
112. TURNER, R. & R. TJIAN. 1989. Science. **243:** 1689-1591.
113. SCHUERMANN, M., M. NEUBERG, J. B. HUNTER, T. JENUWEIN, R. R. RYSECK, R. BRAVO & R. MULLER. 1989. Cell **56:** 507-516.
114. GENTZ, R., F. J. RAUSCHER, C. ABATE & T. CURRAN. 1989. Science **243:** 1695-1699.

115. O'Shea, E. K., R. Rutkowski & P. S. Kim. 1992. Cell **68:** 699-708.
116. Angel, P. & M. Karin. 1991. Biochem. Biophys. Acta **1072:** 129-157.
117. Morgan, J. I. & T. Curran. 1991. Ann. Rev. Neurosci. **14:** 421-451.
118. Bohman, D., T. J. Bos, A. Admon, T. Nishimura, P. K. Vogt & R. Tjian. 1987. Science **238:** 1386-1392.
119. Rauscher, F. J., P. J. Voulalas, B. R. Franza & T. Curran. 1988. Genes Dev. **2:** 1687-1699.
120. Cohen, D. R., P. C. P. Ferreira, B. R. Franza & T. Curran. 1989. Genes Dev. **3:** 173-184.
121. Gonzàlez-Martin, C., I. DeDiego, D. Crespo & A. Fairen. 1992. Dev. Brain Res. **68:** 83-95.
122. Schilling, K., D. Luk, J. I. Morgan & T. Curran. 1991. Proc. Natl. Acad. Sci. USA **88:** 5665-5669.
123. Smeyne, R. J., K. Schilling, L. Robertson, D. Luk, J. Oberdick, T. Curran & J. I. Morgan. 1992. Neuron **8:** 13-23.
124. Smeyne, R. J., M. Vendrell, M. Hayward, S. Baker, G. G. Miao, K. Schilling, L. M. Robertson, T. Curran & J. I. Morgan. 1993. Nature **363:** 166-169.
125. Schwob, J. E., T. Fuller, J. L. Price & L. W. Olney. 1980. Neuroscience **5:** 991-1014.
126. Smeyne, R. J. & D. J. Goldowitz. 1989. J. Neurosci. **9:** 1608-1620.
127. Colotta, F., N. Polentarutti, M. Sironi & A. Mantovani. 1992. J. Biol. Chem. **267:** 18278-18283.
128. Dragunow, M., D. Young, P. Hughes, G. MacGibbon, P. Lawlor, L. Singleton, E. Sirimanne, E. Beilharz & P. Gluckman. 1993. Mol. Brain Res. **18:** 347-352.
129. Leah, J. D., T. Herdegen & R. Bravo. 1992. Brain Res. **566:** 198-207.
130. Jenkins, R., S. B. McMahon, A. B. Bond & S. P. Hunt. 1993. Eur. J. Neurosci. **5:** 751-759.
131. Sonnenberg, J. L., P. F. MacGregor-Leon, T. Curran & J. I. Morgan. 1989. Neuron **3:** 359-365.
132. Sonnenberg, J. L., C. Mitchelmore, P. F. MacGregor-Leon, J. Hempstead, J. I. Morgan & T. Curran. 1989. J. Neurosci. Res. **24:** 72-80.

Neurodegenerative Disease: Autoimmunity Involving Calcium Channels[a]

STANLEY H. APPEL,[b] R. GLENN SMITH, MARIA ALEXIANU, JOSEPH ENGELHARDT, DENNIS MOSIER, LUIS COLOM, AND ENRICO STEFANI

Departments of Neurology, Molecular Physiology and Biophysics
Baylor College of Medicine
Houston, Texas 77030

The neurodegenerative diseases—amyotrophic lateral sclerosis (ALS), Parkinson's disease, and Alzheimer's disease—are devastating clinical disorders that exhibit progressive and relatively selective neuronal injury and cell death. Although recent advances in molecular biology promise improved understanding of disease pathogenesis in familial cases, the etiologies of these disorders remain largely undefined. For example, in ALS, approximately 90% of cases are considered sporadic, whereas only 10% are inherited. Of the latter cases of familial ALS, approximately 25% (thus 2.5% of total cases) are due to mutations in Cu^{2+}/Zn^{2+} superoxide dismutase (SOD_1) that result in decreased enzyme activity.[1,2] Such alterations in SOD_1 in inherited cases presumably lead to an increase in free radical production (especially in levels of superoxide anion) and ultimately to motoneuron death. However, why motoneurons are selectively vulnerable to an enzyme deficiency present in many other neuronal cells as well as in other organ systems is far from clear. Furthermore, considering the similar clinical and pathologic phenotypes for familial and sporadic cases, it is puzzling how reductions in SOD_1 activity can provide insight into sporadic ALS pathogenesis, when levels of Cu^{2+}/Zn^{2+} SOD activity are normal.[3]

Similarly, in Alzheimer's disease, significant insights have developed from the demonstration that genetic defects in amyloid precursor protein can underlie the clinical expression of disease in cases of early onset,[4] whereas inheritance of the apolipoprotein E_4 allele may represent an important risk factor in Alzheimer's disease of late onset.[5] Thus, despite compelling evidence for cell injury initiated by mutations in amyloid precursor protein in these genetic cases and the large number of elegant studies documenting β-amyloid toxicity *in vitro*,[6-9] it is still not clear if β-amyloid is the key pathogenetic constituent in presumed sporadic cases of human Alzheimer's

[a] This work was supported by an Alzheimer's Disease Research Center Grant from the National Institute of Aging and grants from the Muscular Dystrophy Association and Cephalon, Inc.

[b] Address for correspondence: Stanley H. Appel, M.D., Professor and Chairman, Department of Neurology, Baylor College of Medicine, 6501 Fannin, NB302, Houston, Texas 77030.

disease or if it is only one of many secondary factors contributing to the cascade leading to cell injury and death. Furthermore, if increased β-amyloid protein is a primary factor in the pathogenesis of Alzheimer's disease, the relationship of β-amyloid toxicity to synaptic loss[10] and to modification by apolipoprotein E_4 must be better defined.[11]

Although our understanding of the etiology of sporadic neurodegenerative disease is limited, there is no shortage of models of cell injury relevant to pathogenesis and selective vulnerability in each of these disorders. Considerable circumstantial evidence from *in vitro* studies suggests a role for glutamate excitotoxicity or β-amyloid toxicity in neuron death, resulting from increased intracellular calcium, increased free radical and/or nitric oxide production, and activation of proteases, kinases, and endonucleases.[12–17] However, *in vivo* studies of sporadic neurodegenerative disease provide less compelling evidence as to what initiates disease, which pathways are actually operative in producing cell injury, and why certain neurons are selectively vulnerable in ALS, Parkinson's disease, or Alzheimer's disease.

EXCITOTOXICITY IN AMYOTROPHIC LATERAL SCLEROSIS

Early studies in sporadic ALS focused on glutamate excitotoxicity and ligand-gated calcium channels.[12] Glutamate, aspartate, and *N*-acetyl aspartyl glutamate concentrations are reduced in the spinal cord,[18,19] and glutamate transport is decreased in synaptosomes from ALS spinal cord and motor cortex.[20] However, no evidence exists for a primary defect in glutamate metabolism in ALS, and these alterations in excitotoxic amino acids could be simply the consequence of a loss of motoneurons. The latter interpretation is in accord with the decreased glutamate transport noted following motoneuron loss in MND mice.[21] Nevertheless, glutamate-mediated excitotoxicity might still contribute to motoneuron injury, even as a secondary factor, because neuronal AMPA (α-amino-3-hydroxy-5-methyl-4-isoxasole-proprionic acid)/kainate receptors can be activated to produce injury of motoneurons[22] as well as cortical neurons.[23]

IMMUNE-MEDIATED ANIMAL MODELS OF MOTONEURON DESTRUCTION

Our own studies of sporadic neurodegenerative disease have focused on ALS and on the specific etiologic and pathogenic role of autoimmune mechanisms in this disease. These studies were initiated with the development of two animal models of immune-mediated motoneuron disease. Experimental autoimmune motoneuron disease (EAMND) is a model of lower motoneuron destruction induced by inoculation of purified bovine spinal cord motoneurons.[24] Experimental autoimmune gray matter disease (EAGMD) is a model of both upper and lower motoneuron degeneration induced by inoculation of bovine spinal cord ventral horn homogenates.[25] In both models, high titers of antibodies to motoneurons are present in the serum, and IgG can be demonstrated at the neuromuscular junction and in motoneurons. Electrophysiologic studies demonstrate an increased resting release of acetylcholine from the motor nerve terminals with no alteration of postjunctional membrane properties.[26]

Human ALS resembles guinea pig EAGMD with respect to the loss of upper and lower motoneurons, the presence of inflammatory cells within the spinal cord,[27] and the presence of IgG within motoneurons.[28]

Most critical in establishing the importance of immune mechanisms was the ability of EAGMD IgG as well as IgG purified from humans with ALS (ALS IgG) to passively transfer physiologic changes to the mouse neuromuscular junction.[29] Mice injected with animal model and human ALS IgG demonstrated increased resting release of acetylcholine from motor nerve terminals with no effects on amplitude or time course of the miniature end-plate potential or muscle membrane potentials. Chronic application of ALS IgG altered both the rate of spontaneous neurotransmitter release and the quantal content of evoked release, and it produced axonal degeneration.[30]

AMYOTROPHIC LATERAL SCLEROSIS IgG INHIBITS SKELETAL L-TYPE VOLTAGE-GATED CALCIUM CHANNELS

Our early studies used single mammalian skeletal muscle fibers and demonstrated that ALS IgG reduced macroscopic dihydropyridine-sensitive L-type voltage-gated calcium current and charge movement without demonstrable effects on the Na^+-dependent action potential.[31,32] These actions of ALS IgG on L-type voltage-gated calcium channels (VGCCs) were qualitatively similar to the action of nifedipine on these channels. IgG controls from normal individuals as well as from patients with familial ALS, myasthenia gravis, multiple sclerosis, and chronic relapsing inflammatory polyneuropathy showed no alteration in calcium current kinetics; however, IgG from Lambert Eaton myasthenic patients (a disorder previously associated with antibodies to the VGCC) also inhibited calcium current. The effects of ALS IgG were lost with heat inactivation, proteolysis, and T-tubule preadsorption. F_{ab} fragments from ALS IgG produced even more rapid reduction in skeletal muscle calcium current than that observed using whole ALS IgG.[33] In phospholipid bilayers containing skeletal muscle T-tubule-derived L-type VGCCs, ALS IgG was shown to reduce calcium current by decreasing the mean channel open time and by diminishing the amplitude of voltage-dependent calcium current (FIG. 1).[34] Amyotrophic lateral sclerosis IgG addition also resulted in a reduction in the probability of channel opening. Again, normal control IgG and myasthenia gravis IgG had no effect on calcium channel activity. Finally, ALS IgG reduced calcium activity only when applied to the extracellular side of the calcium channel, as documented in both single mammalian skeletal muscle fiber studies and experiments using VGCCs in phospholipid bilayers.

ELISA ASSAY OF ANTIBODIES TO CALCIUM CHANNELS IN AMYOTROPHIC LATERAL SCLEROSIS

Using an ELISA assay with purified calcium channels, 75% of patients with sporadic ALS possessed antibodies to the L-type VGCC, and these antibody titers correlated with the rate of progression of the disease (FIG. 2).[35] Of the many IgG tested from patients with other diseases (including familial ALS), only 66% of LEMS

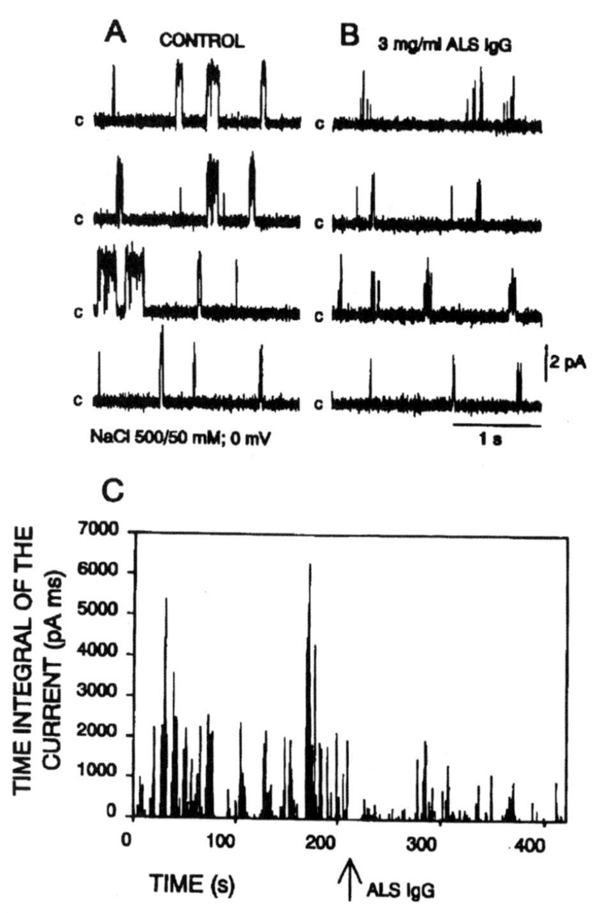

FIGURE 1. The addition of ALS IgG decreases activity of DHP-sensitive Ca^{2+} channels reconstituted into lipid bilayers. (**A**) Traces are recorded at 0 mV steady-state potential in a 500-50 mM *cis-trans* NaCl gradient. (**B**) Traces are recorded from the same channel after the addition of 3 mg/ml ALS IgG in the *trans* (extracellular-facing) chamber. (**C**) A diary plot of the experiment, documenting the time integral of the currents before and after the addition of ALS IgG.

IgG and 15% of Guillain-Barré IgG also possessed significant antibody titers to L-type voltage-gated calcium channels. Together with electrophysiologic data, these findings strongly suggested involvement of VGCC antibodies in ALS. However, they do not explain how ALS IgG-mediated inhibition of L-type voltage-gated calcium current noted in skeletal muscle and in artificial lipid bilayers could correlate with enhanced acetylcholine release observed from motoneuron terminals in passive transfer experiments, where calcium channel opening and increased calcium current would have been anticipated. A possible answer was that L-type VGCC from skeletal muscle may behave differently from neuronal VGCC.

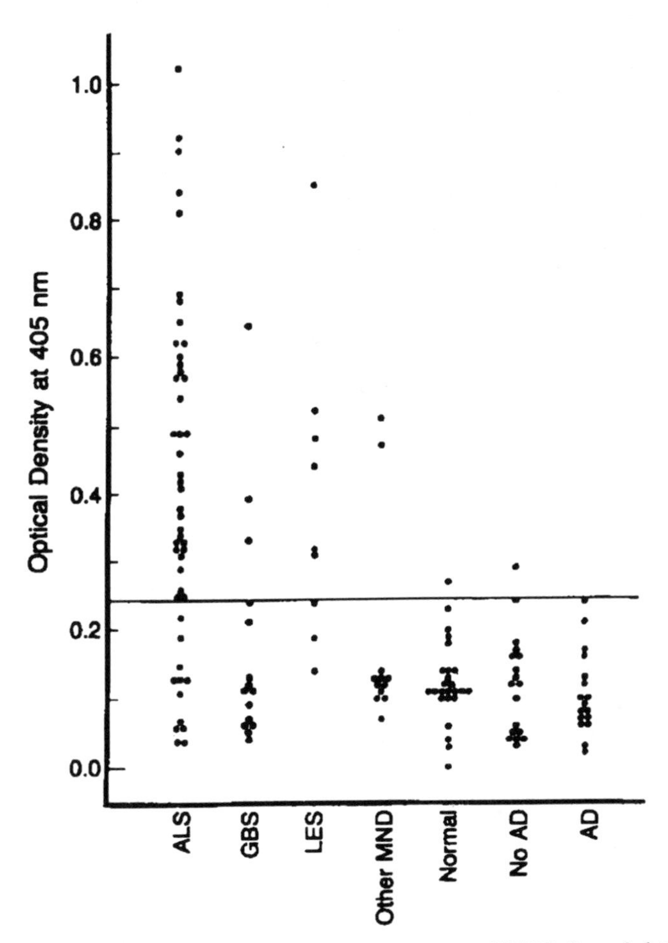

FIGURE 2. Quantitative comparison of serum binding to L-type VGCCs from skeletal muscle. Results of ELISA for individual citrate-treated serum samples (dilution, 1 : 1250) assayed against purified L-type VGCC antigen are shown. The *horizontal line* represents the value 2 SD above the pooled mean control values. GBS denotes Guillain-Barré syndrome, LES Lambert-Eaton syndrome. Other MND-familial or atypical motor neuron disease; Normal = normal controls; No AD = non-autoimmune disease; and AD = autoimmune disease.

EFFECTS OF AMYOTROPHIC LATERAL SCLEROSIS IgG ON NEURONAL CALCIUM CHANNELS

Because skeletal muscle is not a usual target in ALS, we recently expanded our studies to assay the effects of ALS IgG on neuronal calcium channels. A motoneuron cell line was developed by fusion of murine N18TG2 neuroblastoma cells with dissociated embryonic rat ventral spinal cord, employing techniques that had previously yielded a substantia nigra cell line in our laboratory.[36] Further selection for inducible choline acetyl transferase produced a cell hybrid, VSC 4.1, which after differentiation in the presence of dibutyryl cAMP and 0.1 μg/ml aphidocolin (a DNA polymerase inhibitor) produced cells with long-branched processes and a 3- to 10-fold increased somatic size. This line is morphologically similar to a motoneuron line previously produced by Cashman *et al.*[37] The differentiated VSC 4.1 cells possess high levels of immunohistochemically identified neuron-specific enolase, 200-kD neurofilament protein, synaptophysin, and choline acetyl transferase. In addition, the VSC 4.1 cells possess L-type VGCC (defined by binding of ^{3}H PN-200/110) and N-type VGCC (defined by binding of ^{125}I ω-conotoxin.

AMYOTROPHIC LATERAL SCLEROSIS IgG ARE CYTOTOXIC FOR DIFFERENTIATED VSC 4.1 CELLS

When ALS IgG was added to undifferentiated VSC 4.1 cells, a small (10-15 percent) reduction in cell number was noted, which was largely obscured by continued cell division. When ALS IgG was added to differentiated VSC 4.1 cells, significant cell loss began after 6-12 hours and reached maximal effect after 48-60 hours.[38] This effect was not observed with normal or disease control immunoglobulins and was not mediated by complement. Boiling or protease treatment of IgG completely inactivated the cytotoxic effect. When cell loss was quantitated by direct daily cell counts, a loss of 40-60% of cells was observed after 3 days of treatment. A close correspondence was observed between cell loss measured with whole cell counts and cell viability assayed with vital dyes.

After the addition of EGTA sufficient to buffer calcium concentrations to 10 μM-10 nM, cells could be maintained for several days without appreciable loss. After EGTA treatment for 6 hours, the addition of ALS IgG no longer produced cell loss, with complete protection from the effects of ALS IgG at 1 μM extracellular calcium. Amyotrophic lateral sclerosis IgG-specific toxicity could also be prevented by preincubation of immunoglobulins with purified L-type VGCC or with purified α_1 subunits of the L-type VGCC. However, neither α_2 nor β-subunit removes cytotoxicity from the ALS IgG. Nifedipine, an inhibitor of dihydropyridine-sensitive L-type VGCCs, did not affect ALS IgG-induced cytotoxicity, whereas ω-conotoxin (an N-type VGCC antagonist) completely blocked ALS IgG-induced VSC 4.1 cell loss. The ALS IgG blocking effect of ω-conotoxin was concentration dependent, saturating between 200 pM and 2 nM. The ALS IgG cytotoxicity was also greatly reduced by Aga-IVa toxin (a P-type VGCC antagonist) in a concentration-dependent fashion that saturated at 10-30 nM.

AMYOTROPHIC LATERAL SCLEROSIS IgG INCREASE CALCIUM CURRENT IN MOTONEURONS

In physiologic experiments, calcium currents in the VSC 4.1 motoneuron cell line were studied using whole cell patch-clamp. Initial studies using 10 nM barium documented the presence of high threshold calcium currents, which could be completely blocked with either 100 μM cadmium chloride or synthetic funnel web spider toxin (FTX). Ten minutes after adding ALS IgG (final concentration 1 mg/ml), a prominent increase in peak calcium amplitude was recorded. IgG from 5 of 6 patients with ALS documented a significant increase in calcium current in 50% of the cells tested, whereas no disease control IgG demonstrated any increase in calcium current.[39] Furthermore, the same ALS IgG significantly increased calcium current in P-type calcium channels in Purkinje cells and lipid bilayers.[40]

AMYOTROPHIC LATERAL SCLEROSIS IgG INCREASE CALCIUM ENTRY INTO MOTONEURONS

The third approach to examining selective vulnerability in VSC 4.1 cells employed an imaging workstation system and calcium-sensitive dyes. In initial studies, permeable fura-2 AM (380 and 353 nm excitation wavelengths, 510 nm emission wavelength) was loaded into VSC 4.1 cells by incubation for 30 minutes at room temperature. Brief exposure of the preparation to light minimized photobleaching. Images were recorded at 510 nm; differences in dye concentration were corrected by making ratios between excitations at 380 and 353 nm. Exposure to ALS IgG induced a progressive increase in intracellular calcium concentration, with different rates of response in different VSC 4.1 cell populations. This variation in cellular response to ALS IgG appeared to reflect both cell heterotopy in the VSC 4.1 hybrid and different stages of cell differentiation in responsive cells, with the most differentiated cells showing the greatest increase in calcium. Subsequent experiments have employed fluo-3 to detect small changes in calcium concentration. With this calcium-sensitive dye, ALS IgG specifically increased intracellular calcium in differentiated VSC 4.1 cells, whereas disease control IgG had no such effects. In the majority of cells, these changes were predictive of subsequent cell death.

FACTORS INFLUENCING SELECTIVE VULNERABILITY

Our studies documented the presence of antibodies to VGCCs in sporadic but not familial ALS. In a differentiated motoneuron cell line, the interaction of antibodies with neuronal calcium channels *in vitro* can enhance calcium current, increase intracellular calcium, and lead to cell death. Such data support the importance of abnormal calcium entry triggered by IgG and entering via neuronal VGCCs in motoneuron injury in ALS. However, because ALS IgG can also enhance calcium current through P-type channels in Purkinje cells,[40] and yet Purkinje cells are not compromised in ALS, it is unlikely that a specific neuronal channel per se is the critical factor in mediating selective vulnerability. Furthermore, in other *in vitro* studies, absolute levels of intracellular calcium appear less critical to the process of cell injury than does the route by which calcium enters and the cellular reactions triggered by such entry.[23,41,42] Thus, some factor other than the presence of specific calcium channels or the levels of intracellular calcium may likely explain the selective vulnerability of motoneurons in ALS.

CALCIUM BINDING PROTEINS IN AMYOTROPHIC LATERAL SCLEROSIS

In *in vitro* studies of cytotoxicity, the morphologically undifferentiated VSC 4.1 cells were far more resistant to cytotoxic effects of ALS IgG than were larger cells with branched processes produced by differentiation with dbcAMP and aphidocolin. Of greatest interest was that the immunoreactive levels of the intracellular calcium binding proteins calbindin D_{28k} and parvalbumin decreased significantly during differentiation of VSC 4.1 cells, similar to changes observed during motoneuron differentiation. Furthermore, cells containing significant immunohistochemical reactivity for these two calcium binding proteins (ie, undifferentiated VSC 4.1 cells and undifferentiated and differentiated mesencephalon-neuroblastoma MES 23.5 hybrid cells) were relatively resistant to the cytotoxic effects of ALS IgG, whereas differentiated VSC 4.1 cells with very little calcium binding protein immunoreactivity were extremely sensitive to the cytotoxic effects of ALS IgG. These findings are directly applicable to selective vulnerability in human ALS, because motoneurons destined to be destroyed early in ALS have decreased to absent calbindin D_{28k} and parvalbumin immunoreactivity (lower cranial nerve nuclei, spinal motoneurons and cortical motoneurons), while motoneurons that are relatively spared have increased levels of calcium binding proteins (cranial nerve nuclei III, IV and VI and Onuf's nucleus).[43]

CALCIUM-BINDING PROTEINS IN OTHER NEURODEGENERATIVE DISEASES

The two calcium binding proteins, calbindin D_{28k} and parvalbumin, may also play an important role in the selective vulnerability of neurons in Parkinson's disease and Alzheimer's disease. This hypothesis would suggest that neurons which contain these calcium-binding proteins should be relatively preserved in the neurodegenerative diseases and neurons which have minimal levels of such proteins might be more affected. However, there are many studies that would refute such a simplistic interpretation. One major problem is that most studies have been carried out using immunohistochemical techniques alone, and the actual levels or even functions of calbindin D_{28k} and parvalbumin may not be reflected in immunoreactivity levels. How calcium may regulate the synthesis, stability, and degradation of calcium-binding proteins in neuronal cells also remains to be adequately studied.

In Alzheimer's disease, an early study documented the loss of calbindin D_{28k}-positive neurons from the nucleus basalis of Meynert.[44] Subsequently, the loss of calbindin D_{28k}-positive reactivity on Western blot and mRNA in hippocampus, dorsal raphé nucleus, and nucleus basalis was reported by Iacopino and Christakos,[45] but a follow-up study suggested decreased calbindin D_{28k} gene transcription rather than neuronal loss.[46] The loss of parvalbumin-immunoreactive neurons from cortex was reported by Arai *et al.*,[47] but CNS tissues were examined at time intervals greater than 30 hours postmortem and the antibody reagents employed were poorly defined. In a carefully performed study, Hof *et al.*[48] demonstrated that parvalbumin-immunoreactive neurons in the neocortex were resistent to degeneration in Alzheimer's disease. They also documented that greater than 90% of the large pyramidal cells lost in

Alzheimer's disease did not contain immunoreactive parvalbumin, whereas the calbindin D_{28k}-immunoreactive interneurons in supragranular layers of the prefrontal cortex were unaffected in Alzheimer's disease. However, the calbindin D_{28k}-positive neurons in layer III were severely affected in Alzheimer's disease,[49] suggesting that calbindin D_{28k} may not be sufficient to convey resistance to cell death, even though it may possibly be downregulated before cell injury in the cells. Thus, at the present time, there is suggestive but not compelling evidence for the role of calbindin D_{28k} and parvalbumin in selective neuronal vulnerability in Alzheimer's disease. However, immunohistochemical techniques alone cannot define whether calbindin D_{28k}- or parvalbumin-immunoreactive neurons have been lost or even whether neurons have altered levels of these calcium-binding proteins. Such proof will require more definitive studies employing careful quantitation of calbindin D_{28k} and parvalbumin protein and mRNA levels, and assessment of how aging, chronic injury, or increased calcium might influence synthesis and degradation of these calcium-binding proteins.

Our studies in neurodegenerative disease, especially ALS, suggest that emphasis should be focused not only on the increase in intracellular calcium, but also on the specific triggers of calcium entry, the route of calcium entry, and the role of calcium-binding proteins and/or other intracellular second messengers in modulating the threshold for cell death. In sporadic ALS, antibodies to neuronal calcium channels trigger the entry of calcium and activate the cell death cascade. The calcium-binding proteins calbindin D_{28k} and parvalbumin appear to play a significant role in the selective vulnerability of motoneurons, possibly by influencing the extent and speed of calcium buffering or by regulating other aspects of calcium homeostasis.

REFERENCES

1. ROSEN, D. R., T. SIDDIQUE, D. PATTERSON, D. A. FIGLEVICZ, P. SAPP, A. HENTATI, D. DONALDSON, J. GOTO, J. P. O'REGAN, H.-X. DENG, Z. RAHMANI, A. KKRIZUS, D. MCKENNA-YASEK, A. CAYABYAB, S. M. GASTON, R. BERGER, R. E. TANZI, J. J. HALPERIN, B. HERZFELDT, R. VAN DEN BERGH, W.-Y. HUNG, T. BIRD, G. DENG, D. W. MULDER, C. SMYTH, N. G. LAING, E. SORIANO, M. A. PERICAK-VANCE, J. HAINES, G. A. ROULEAU, J. S. GUSELLA, H. R. HORVITZ & R. H. BROWN, JR. 1993. Mutations in Cu/Zn superoxide dismutase gene are associated with familial amyotrophic lateral sclerosis. Nature **362:** 59-62.
2. DENG, H., A. HENTATI, J. A. TAINER, Z. IGBAL, A. CAYABYAB, W. HUNG, E. D. GETZOFF, P. HU, B. HERZFELDT, R. P. ROOS, C. WARNER, G. DENG, E. SORIANO, C. SMYTH, H. E. PARGE, A. AHMED, A. D. ROSES, R. A. HALLEWELL, M. A. PERICAK-VANCE & T. SIDDIQUE. 1993. Amyotrophic lateral sclerosis and structural defects in Cu/Zn superoxide dismutase. Science **261:** 1047-1051.
3. BOWLING, A. C., J. B. SCHULZ, R. H. BROWN & M. F. BEAL. 1993. Superoxide dismutase activity, oxidative damage, and mitochondrial energy metabolism in familial and sporadic amyotrophic lateral sclerosis. J. Neurochem. **61:** 2322.
4. GOATE, A., M.-C. CHARTIER-HARLIN, M. MULLAN, J. BROWN, F. CRAWFORD, L. FIDANI, L. GIUFFRA, A. HAYNES, N. IRVING, L. JAMES, R. MANT, P. NEWTON, K. ROOKE, P. ROQUES, C. TALBOT, M. PERICAK-VANCE, A. ROSES, R. WILLIAMSON, M. ROSSOR, M. OWEN & J. HARDY. 1991. Segregation of a missense mutation in the amyloid precursor protein gene with familial Alzheimer's disease. Nature **349:** 704-706.
5. STRITTMATTER, W. J., A. M. SAUNDERS, D. SCHMECHEL, M. PERICAK-VANCE, J. ENGHILD, G. S. SALVESEN & A. D. ROSES. 1993. Apolipoprotein E: high avidity binding to

β-amyloid and increased frequency of type 4 allele in late-onset familial Alzheimer's disease. Proc. Natl. Acad. Sci. USA **90:** 1977-1981.
6. Selkoe, D. J. 1990. Deciphering Alzheimer's disease: The amyloid precursor protein yields new clues. Science **248:** 1058-1060.
7. Yankner, B. A. & M.-M. Mesulam. 1991. β-amyloid and the pathogenesis of Alzheimer's disease. N. Engl. J. Med. **325:** 1849-1857.
8. Neve, R. L., A. Kammescheidt & C. F. Hohmann. 1992. Brain transplants of cells expressing the carboxyl terminal fragment of the Alzheimer amyloid protein precursor cause specific neuropathology *in vivo*. Proc. Natl. Acad. Sci. USA **89:** 3448-3452.
9. Golde, T. E., S. Estus, L. H. Younkin, D. J. Selkoe & S. G. Younkin. 1992. Processing of the amyloid protein precursor to potentially amyloidogenic derivatives. Science **255:** 728-730.
10. Masliatt, E., M. Mallory, L. Hansen, R. Deteresa & R.-D. Terry. 1993. Quantitative synaptic alterations in the human neocortex during normal aging. Neurology **43:** 192-197.
11. Roses, A. D., M. A. Pericak-Vance, M. A. Saunders, D. Schmechel, D. Goldgaber & W. Strittmatter. 1994. Complex genetic disease: Can genetic strategies in Alzheimer's disease and neurogenetic mechanisms be applied to epilepsy? Epilepsia **35:** S20-S28.
12. Choi, D. W. 1988. Glutamate neurotoxicity and diseases of the nervous system. Neuron **1:** 623-634.
13. Siesjo, B. K., F. Bengtsson, W. Grampp & S. S. Theander. 1989. Calcium excitation and neuronal death in the brain. Ann. N.Y. Acad. Sci. **568:** 252-261.
14. Lipton, S. A., Y.-B. Choi, Z.-H. Pan, S. Z. Lei H. Chen, N. J. Sucher, J. Loscalzo, D. J. Singel & J. S. Stamler. 1993. A redox-based mechanism for the neuroprotective and neurodestructive effects of nitric oxide and related nitrosocompounds. Nature **364:** 626-632.
15. Coyle, J. T. & P. Puttfarken. 1993. Oxidative stress, glutamate and neurodegenerative disorders. Science **262:** 689-695.
16. Siman, R. & J. C. Noszek. 1988. Excitatory amino acids activate calpain I and induce structural protein breakdown *in vivo*. Neuron **1:** 279-287.
17. Loo, D. T., A. Copani, C. J. Pike, E. R. Whittemore, A. J. Walencewicz & C. W. Cotman. 1993. Apoptosis is induced by β-amyloid in cultured central nervous system neurons. Proc. Natl. Acad. Sci. USA **90:** 7951-7955.
18. Plaitakis, A. 1990. Glutamate dysfunction and selective motor neuron degeneration in amyotrophic lateral sclerosis: A hypothesis. Ann. Neurol. **28:** 3-8.
19. Rothstein, J. D., G. Tsai, R. W. Kuncl, L. Clawson, D. R. Cornblath, D. B. Drachman, A. Pestronk, B. L. Stauch & J. T. Coyle. 1990. Abnormal excitatory amino acid metabolism in amyotrophic lateral sclerosis. Ann. Neurol. **28:** 18-25.
20. Rothstein, J. D., L. J. Martin & R. W. Kuncl. 1992. Decreased glutamate transport by the brain and spinal cord in amyotrophic lateral sclerosis. N. Engl. J. Med. **326:** 1464-1468.
21. Battaglioli, G., D. L. Martin, J. Plummer & A. Messer. 1993. Synaptosomal glutamate uptake declines progressively in the spinal cord of a mutant mouse with motor neuron disease. J. Neurochem. **60:** 1567-1569.
22. Rothstein, J. D., L. Jin, H. Dykes-Hoberg & R. W. Kuncl. 1993. Chronic inhibition of glutamate uptake produces a model of slow neurotoxicity. Proc. Natl. Acad. Sci. USA **90:** 6591-6595.
23. Frandsen, A. & A. Schousboe. 1993. Excitatory amino acid-mediated cytotoxicity and calcium homeostasis in cultured neurons. J. Neurochem. **60:** 1202-1211.
24. Engelhardt, J., S. H. Appel & J. M. Killian. 1989. Experimental autoimmune motor neuron disease. Ann. Neurol. **26:** 368-376.
25. Engelhardt, J. I., S. H. Appel & J. M. Killian. 1990. Motor neuron destruction in guinea pigs immunized with bovine spinal cord ventral horn homogenate: Experimental autoimmune gray matter disease. J. Neuroimmunol. **27:** 21-31.

26. GARCÍA, J., J. ENGELHARDT, S. H. APPEL & E. STEFANI. 1990. Increased mepp frequency as an early sign of experimental immune mediated motor neuron disease. Ann. Neurol. **28:** 329-334.
27. ENGELHARDT, J. I., J. TAJTI & S. H. APPEL. 1993. Lymphocytic infiltrates in the spinal cord in amyotrophic lateral sclerosis. Arch. Neurol. **50:** 30-36.
28. ENGELHARDT, J. & S. H. APPEL. 1990. IgG reactivity in the spinal cord and motor cortex in amyotrophic lateral sclerosis. Arch. Neurol. **47:** 1210-1216.
29. APPEL, S. H., J. ENGELHARDT, J. GARCÍA & E. STEFANI. 1991. Immunoglobulins from animal models of motor neuron disease and human ASL passively transfer physiological abnormalities of the neuromuscular junction. Proc. Natl. Acad. Sci. USA **88:** 647-651.
30. UCHITEL, O. D., F. SCORNIK, D. A. PROTTI, C. G. FUMBERG, V. ALVAREZ & S. H. APPEL. 1992. Long-term neuromuscular dysfunction produced by passive transfer of amyotrophic lateral sclerosis immunoglobulins. Neurology **42:** 2175-2180.
31. DELBONO, O., J. GARCÍA, S. H. APPEL & E. STEFANI. 1991. Calcium current and charge movement of mammalian muscle: Action of lateral sclerosis immunoglobulins. J. Physiol. **444:** 723-742.
32. DELBONO, O., J. GARCÍA, S. H. APPEL & E. STEFANI. 1991. IgG from amyotrophic lateral sclerosis affects tubular calcium channels of skeletal muscle. Am. J. Physiol. **260:** C1347-C1351.
33. DELBONO, O., V. MAGNELLI, T. SAWADA, R. G. SMITH, S. H. APPEL & E. STEFANI. 1993. F_{ab} fragments from amyotrophic lateral sclerosis IgG affect calcium channels of skeletal muscle. Am. J. Physiol. **264** (Cell Physiol **33**): C537-C543.
34. MAGNELLI, V., T. SAWADA, O. DELBONO, R. G. SMITH, S. H. APPEL & E. STEFANI. 1993. Amyotrophic lateral sclerosis immunoglobulins action on single skeletal muscle Ca^{2+} channels. J. Physiol. **461:** 103-118.
35. SMITH, R. G., S. HAMILTON, F. HOFMANN, T. SCHNEIDER, W. NASTAINCZYK, L. BIRNBAUMER, E. STEFANI & S. H. APPEL. 1992. Serum antibodies to skeletal muscle-derived L-type calcium channels in patients with amyotrophic lateral sclerosis. N. Engl. J. Med. **327:** 1721-1728.
36. CRAWFORD, G. D., W.-D. LE, R. G. SMITH, W.-J. XIE, E. STEFANI & S. H. APPEL. 1992. A novel N18TG2 x mesencephalon hybrid expresses properties that suggest a dopaminergic cell line of substantia nigra origin. J. Neurosci. **12:** 3392-3398.
37. CASHMAN, N. R., H. D. DURHAM, J. K. BLUSZTAJN, K. ODA, T. TABIRA, I. T. SHAW, S. DAHROUGE & J. P. ANTEL. 1992. Neuroblastoma x spinal cord (NSC) hybrid cell lines resemble developing motor neurons. Dev. Dynamics **194:** 209-221.
38. SMITH, R. G., M. ALEXIANU, G. CRAWFORD, O. NYORMOI, E. STEFANI & S. H. APPEL. 1994. Cytotoxicity of immunoglobulins from amyotrophic lateral sclerosis patients on a hybrid motoneuron cell line. Proc. Natl. Acad. Sci. USA **91:** 3393-3397.
39. MOSIER, D. R., R. G. SMITH, O. DELBONO, S. H. APPEL & E. STEFANI. 1993. Effects of ALS immunoglobulins on calcium currents in hybrid neuroblastoma cells. Soc. Neurosci. Abstr. **19:** 196.
40. LLINAS, R., M. SUGIMORI, B. D. CHERKSEY, R. G. SMITH, O. DELBONO, E. STEFANI & S. H. APPEL. 1993. IgG from ALS patients increases current through P-type calcium channels in mammalian cerebellar Purkinje cells and in isolated channel protein in lipid bilayer. Proc. Natl. Acad. Sci. USA **90:** 11743-11747.
41. TYMIANSKI, M., M. P. CHARLTON, P. L. CARLEN & C. H. TATOR. 1993. Source specificity of early calcium neurotoxicity in cultured embryonic spinal neurons. J. Neurosci. **13:** 2085-2104.
42. BADING, H., D. D. GINTY & M. E. GREENBERG. 1993. Regulation of gene expression in hippocampal neurons by distinct calcium signaling pathways. Science **260:** 181-186.
43. ALEXIANU, M. E., R. G. SMITH & S. H. APPEL. The role of calcium binding proteins in selective motoneuron vulnerability in ALS. Ann. Neurol., in press.

44. Ichimiya, Y., P. C. Emson, C. Q. Mountjoy, D. E. M. Lawson & R. Iizuka. 1989. Calbindin-immunoreactive cholinergic neurons in the nucleus basalis of Meynert in Alzheimer-type dementia. Brain Res. **499:** 402-406.
45. Iacopino, A. M. & S. Christakos. 1990. Specific reduction of calcium binding protein (28-kd calbindin-D) gene expression in aging and neurodegenerative diseases. Proc. Natl. Acad. Sci. USA **87:** 4078-4082.
46. Iacopino, A., S. Christakos, D. German, P. K. Sonsalla & C. A. Altar. 1992. Calbindin D_{28k}-containing neurons in animal models of neurodegeneration: Possible protection from excitotoxicity. Mol. Brain Res. **13:** 251-261.
47. Arai, H., P. C. Emson, C. Q. Mountjoy, L. H. Carassco & C. W. Heizmann. 1987. Loss of parvalbumin-immunoreactive neurons from cortex in Alzheimer-type dementia. Brain Res. **418:** 164-169.
48. Hof, P. R., K. Cox, W. G. Young *et al.* 1991. Parvalbumin-immunoreactive neurons in the neocortex are resistant to degeneration in Alzheimer's disease. J. Neuropathol. Exp. Neurol. **50:** 451-462.
49. Hof, P. R. & J. H. Morrison. 1991. Neocortical neuronal subpopulations labeled by a monoclonal antibody to calbindin exhibit differential vulnerability in Alzheimer's disease. Exp. Neurol. **111:** 293-301.

Elevated Intracellular Calcium Blocks Programmed Neuronal Death

JAMES L. FRANKLIN AND
EUGENE M. JOHNSON, JR.

Department of Molecular Biology and Pharmacology
Washington University School of Medicine
St. Louis, Missouri 63110

A large portion of neurons produced during neurogenesis of the vertebrate nervous system die at about the time functional connections to target tissues are being made.[1] The primary purpose of this death apparently is to sculpt the developing nervous system by matching appropriately the number of neurons innervating a target tissue with the target size. This death, a normal component of development, is known as naturally occurring or programmed cell death (PCD) to distinguish it from pathological forms of death. Among the factors controlling neuronal survival during and subsequent to the period of PCD are target-derived neurotrophic factors. These factors are secreted in minute amounts by target tissues and bind to receptors on innervating nerve terminals where they are internalized and retrogradely transported with their receptors to the neuronal soma. A retrogradely transported signal, which may be carried by the ligand/receptor complex, activates mechanisms that maintain cellular survival and triggers a variety of trophic responses. Cells failing to obtain sufficient trophic factor are thought to undergo PCD, whereas those that acquire an adequate amount survive. This hypothesis is supported by experiments showing that mechanical insults separating neurons from their targets (axotomy and target removal) produce neuronal death that can be prevented by administering an appropriate trophic factor systemically[2,3] or, in the case of axotomy, at the site of the lesion.[4] Similarly, neuronal death can be induced by administering neutralizing antibodies to the trophic factor. Hence, naturally occurring, antibody-induced (e.g., immunosympathectomy), axotomy-induced, or target removal-induced neuronal death are all examples of death caused by trophic factor deprivation.

The most studied neuronal death caused by insufficient trophic factor is based on the physiological role of the prototypical neurotrophic factor, nerve growth factor (NGF). Depriving immature sympathetic[5] or certain sensory neurons[6] of NGF causes massive cell death *in vivo* and *in vitro*. In the case of sympathetic neurons, NGF deprivation results in extensive neuronal death even in adult animals.[7] In addition, NGF blocks the death caused by axotomy, target removal, and even certain chemical and virological insults to these neurons.[8,9] Therefore, death of sympathetic and sensory neurons caused by NGF deprivation can serve as a model for studying PCD. When sympathetic neurons are dissociated from embryonic rats and maintained in cell culture for 5-7 days in the presence of NGF, they hypertrophy and develop extensive neurites. Removal of NGF from the medium at this time causes condensation of chromatin, blebbing of neurites, neuritic fragmentation, and somatic atrophy prior to death. These morphological changes suggest that death occurs by apoptosis, a form of death common to many types of cells.[10] The same is true of other types of neurons undergoing PCD.[11] Electron micrographs of sympathetic neurons treated with neutralizing antibodies to NGF appear "apoptotic" at the ultrastructural level.[12] More compelling evidence that the PCD of these neurons occurs by apoptosis is that NGF deprivation causes the neuronal DNA to become fragmented into oligonucleosomes,[13,14] a hallmark of apoptosis. Additionally, inhibitors of RNA and protein synthesis completely prevent this death, another characteristic of apoptosis in some types of cells.[15] The observed requirement for ongoing or *de novo* macromolecular synthesis has been confirmed for several other types of neurons maintained *in vitro* by other neurotrophic factors.[16] Oppenheim and colleagues[17] showed that naturally occurring or axotomy-induced death of sensory and motor neurons in the chicken embryo is blocked by transcriptional and translational inhibitors, thus validating the *in vitro* paradigm.

ELEVATED POTASSIUM BLOCKS NEURONAL PCD *IN VITRO*

Neurotrophic factors are not the only agents capable of influencing survival of developing neurons. In 1970, Scott and Fisher[18] demonstrated that embryonic chicken dorsal root ganglion neurons can be maintained alive in cell culture if the K^+ concentration of the culture medium ($[K^+]_o$) is markedly elevated. Since then, a similar effect of elevated $[K^+]_o$ on survival has been noted for many other types of neurons from both the peripheral and the central nervous systems.[19] Indeed, the use of high $[K^+]_o$ has become a standard technique for maintaining *in vitro* viability of neurons when appropriate neurotrophic factors are unknown. An example of the ability of elevated $[K^+]_o$ to maintain rat sympathetic neurons in the absence of NGF is shown in FIGURE 1. Survival promotion by high $[K^+]_o$ is a long-term phenomenon; these cells will live and remain healthy in the presence of elevated $[K^+]_o$ for weeks or months without NGF.

The resting membrane potential of neurons is primarily determined by intracellular and extracellular K^+ concentrations. In cell culture, increasing $[K^+]_o$ causes depolarization to a new resting potential at which cells remain for as long as they are exposed to the elevated $[K^+]_o$.[20,21] This chronic depolarization is most likely responsible for the survival-promoting action of high $[K^+]_o$. A possible physiological correlate of this *in vitro* phenomenon is the role of afferent input in determining survival of neurons during the period of naturally occurring neuronal death (ie, depolarization may promote survival by mimicking survival-promoting effects of naturally occurring electrical activity[19]). The first work investigating possible mechanisms by which depolarization promotes survival demonstrated that enhancement of the survival of chicken ciliary ganglion neurons by high $[K^+]_o$ is blocked by the relatively nonselective Ca^{2+} channel antagonists D-600 (a verapamil derivative) and Mg^{2+}.[22] These findings suggest that influx of Ca^{2+} through voltage-gated channels is responsible for mediating the survival-promoting effects of depolarization. In the late 1980s, three groups examining different types of neurons, rat cerebellar granule cells,[23] chicken embryo

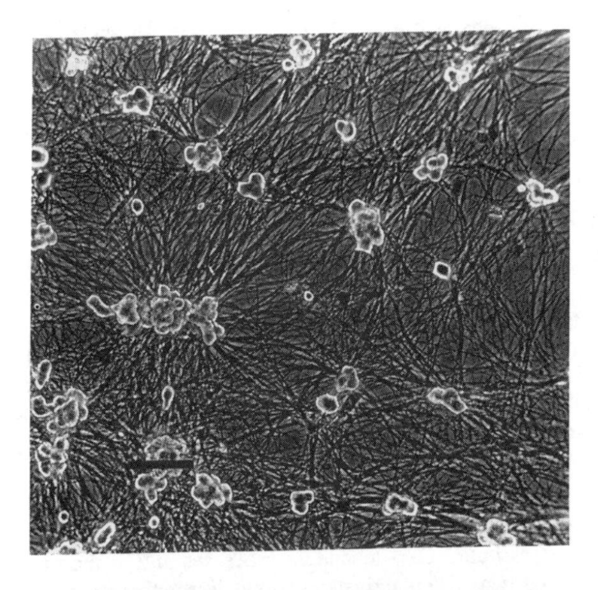

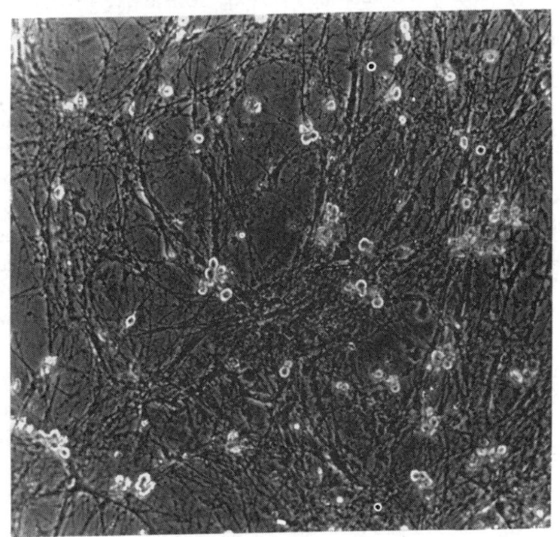

FIGURE 1. Chronic depolarization with elevated $[K^+]_o$ maintains survival of rat sympathetic neurons in culture after nerve growth factor deprivation. (**Top**) Superior cervical ganglion neurons dissociated from fetal rats on embryonic-day 21 and maintained for the first 7 days *in vitro* in medium containing NGF and 5 mM K^+. The cells were then deprived of NGF by incubating in medium containing no NGF, a NGF-neutralizing antibody, and 50 mM K^+ (isotonic). Neurons are alive and healthy. Photographs were taken 5 days after beginning of treatment. (**Bottom**) Neurons maintained for the first 7 days in culture in medium containing NGF and 5 mM K^+. The cells were then deprived of NGF as above in medium containing normal $[K^+]_o$ (5 mM). This treatment resulted in massive neurite fragmentation, somatic atrophy, and death. Photographs were taken 5 days after NGF deprivation. Scale bar is 215 μm.

neurons (from the ciliary, sympathetic, and dorsal root ganglia[24]), and rat sympathetic neurons,[25] used improved Ca^{2+} channel antagonists to investigate the role of voltage-sensitive Ca^{2+} channels in depolarization-enhanced survival. These groups showed that survival promotion by high $[K^+]_o$ could be prevented by specific dihydropyridine (DHP) Ca^{2+} channel antagonists such as nifedipine or isradipine (PN200-110). Conversely, cells were made more responsive to increased $[K^+]_o$ by DHP Ca^{2+} channel agonists such as Bay K 8644. These data suggest that in these types of neurons, chronic depolarization causes activation of DHP-sensitive or L-type Ca^{2+} channels[26] and that influx of Ca^{2+} through these channels is somehow responsible for the enhanced survival.

On the basis of the pharmacologic experiments, we suggested[19,25,27] that chronic depolarization (days or weeks) causes sustained elevation of intracellular free Ca^{2+} concentration ($[Ca^{2+}]_i$) that effectively makes neurons trophic-factor-independent. As a working model, we proposed a "Ca^{2+} set-point" hypothesis of neuronal survival and trophic factor dependence (FIG. 2). This hypothesis is a parallel construction to a similar idea developed by Kater and colleagues[28] concerning the role of varying levels of $[Ca^{2+}]_i$ on neurite outgrowth. The hypothesis posits four steady-state levels, or set-points, of $[Ca^{2+}]_i$ that influence neuronal survival. At one extreme is $[Ca^{2+}]_i$ too low to support essential Ca^{2+}-dependent processes and at the opposite extreme is markedly elevated $[Ca^{2+}]_i$ that activates various destructive processes within cells. Calcium levels at these extremes are incompatible with survival. Indeed, evidence supports the notion that large elevations of $[Ca^{2+}]_i$ (μM levels) mediate the death of neurons in response to excitotoxic or other insults[29] and that very low levels of $[Ca^{2+}]_i$ (near 0 $[Ca^{2+}]_i$, unpublished observation) are detrimental to neuronal growth and survival.[30] Between these extremes lie relatively low levels of $[Ca^{2+}]_i$ (100 nM, for example) with which neurons can survive but only when supplied with an appropriate neurotrophic factor. Elevating $[Ca^{2+}]_i$ further to levels that are sustained in cells with optimal survival-promoting chronic depolarizations (250 nM Ca^{2+} in rat sympathetic neurons, for example) allows survival in the absence of neurotrophic factor. Therefore, two $[Ca^{2+}]_i$ set-points (one too low, one too high) lead to neuronal death, a third set-point allows survival if trophic factor is present, and a fourth supports survival in the absence of trophic factor. In other words, within a certain range of concentrations, $[Ca^{2+}]_i$ determines trophic factor dependence. *In vivo,* afferent input may promote survival of developing neurons by causing frequent cellular depolarization and transient increases of $[Ca^{2+}]_i$, thus regulating the response of the developing neuron to target-derived trophic factors. This idea is supported by recent evidence showing that elevated $[K^+]_o$ and basic fibroblast growth factor have synergistic effects on the survival of chicken ciliary ganglion neurons in culture.[31]

TESTING THE VALIDITY OF THE Ca^{2+} SET-POINT HYPOTHESIS

The Ca^{2+} set-point hypothesis predicts that steady-state $[Ca^{2+}]_i$ in neurons chronically depolarized with high $[K^+]_o$ should be elevated above basal levels. Analysis of $[Ca^{2+}]_i$ in rat sympathetic,[32-34] chicken ciliary ganglion,[35] and rat myenteric neurons[21] with the Ca^{2+} sensitive dyes fura-2 or indo-1[36] has shown this to be the case. In all

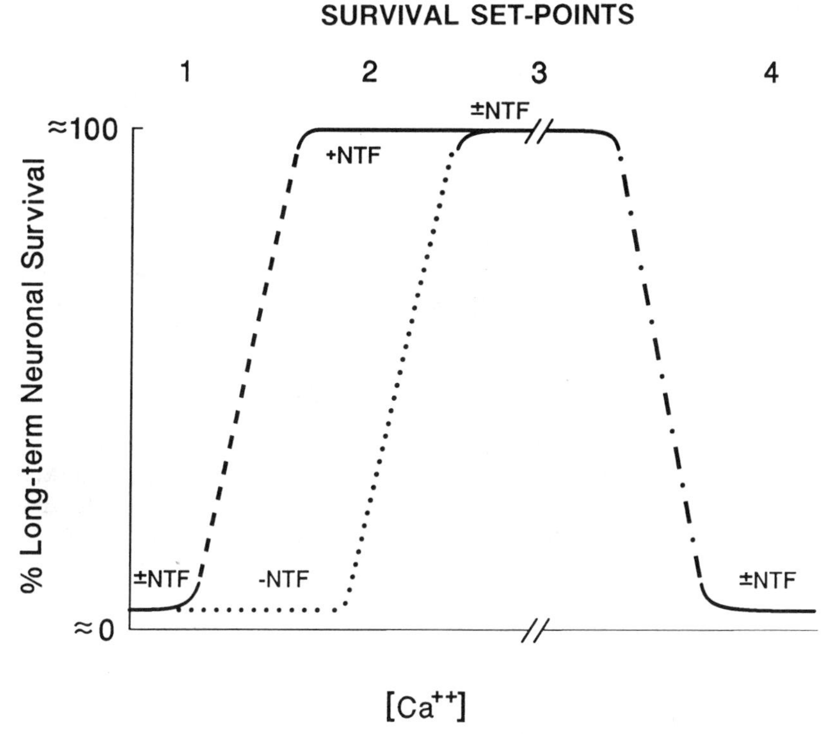

FIGURE 2. Ca^{2+} set-point hypothesis. Four levels of $[Ca^{2+}]_i$ that influence neuronal survival: (1) At very low $[Ca^{2+}]_i$ (essentially 0 mM Ca^{2+}), neurons cannot survive even in the presence of an appropriate neurotrophic factor (NTF). (2) At normal resting $[Ca^{2+}]_i$ (100 nM, for example) developing neurons can survive only if an adequate amount of neurotrophic factor, such as NGF, is present. (3) Steady-state $[Ca^{2+}]_i$ that is slightly elevated above baseline levels (one to several hundred nM) promotes survival (i.e., prevents PCD) in the complete absence of a neurotrophic factor. (4) Toxic $[Ca^{2+}]_i$ (sustained μM or higher levels) causes death of neurons under any condition. (Modified from ref. 19.)

three types of neurons, exposure to high $[K^+]_o$ causes a rapid increase of $[Ca^{2+}]_i$. Following the initial elevation of $[Ca^{2+}]_i$, continued exposure to increased $[K^+]_o$ causes a sustained rise in steady-state $[Ca^{2+}]_i$ that lasts for long periods (hours or days). In each of these three types of neurons DHP Ca^{2+} channel antagonists, which block survival promotion by chronic depolarization, reduce the sustained rise of $[Ca^{2+}]_i$ to basal or near basal levels. These results strongly suggest that the sustained increase of $[Ca^{2+}]_i$ caused by Ca^{2+} influx through voltage-gated channels is responsible for promoting survival rather than some other effect of depolarization or high $[K^+]_o$. This conclusion is further supported by experiments from our lab showing that the sesquiterpene lactone thapsigargin can promote survival of NGF-deprived rat sympathetic neurons.[37,38] Thapsigargin is a potent and selective inhibitor of a Ca^{2+} pump that is responsible for Ca^{2+} sequestration in a subset of intracellular Ca^{2+} stores.[39]

Inhibition of this pump by thapsigargin causes the rapid release of Ca^{2+} from these stores and results in a rapid rise in cytoplasmic $[Ca^{2+}]_i$. This increase is followed by a decline in $[Ca^{2+}]_i$ caused by Ca^{2+} being pumped out of the cell by plasma membrane Ca^{2+} pumps. In many types of cells $[Ca^{2+}]_i$ does not return to baseline levels, however, but remains elevated at a new steady-state level. This sustained rise of $[Ca^{2+}]_i$ is thought to be caused by Ca^{2+} influx through plasma membrane Ca^{2+} conductance activated by a messenger molecule released by depletion of Ca^{2+} stores.[40,41] We found that thapsigargin causes a DHP-insensitive sustained increase of $[Ca^{2+}]_i$ in rat sympathetic neurons.[37,38] By increasing extracellular Ca^{2+} in the presence of thapsigargin, we can increase $[Ca^{2+}]_i$ to levels similar to those associated with optimal survival-promoting $[K^+]_o$ without depolarizing cells to the potentials associated with those levels. This treatment saves most neurons from undergoing PCD during NGF deprivation.[37,38] Therefore, by using a method independent of depolarization and activation of L-type Ca^{2+} channels, we are able to cause a sustained increase in $[Ca^{2+}]_i$ that prevents PCD.

The sustained elevation of $[Ca^{2+}]_i$ that is reported to promote survival of neurons is small (from 100 nM above baseline in rat sympathetic neurons to several 100 nM in chicken ciliary ganglion neurons, differences possibly due to variations in calibration techniques) and seems to indicate that the Ca^{2+}-binding proteins responsible for promoting survival have a high affinity for Ca^{2+}. It is possible, of course, that the increases in steady-state $[Ca^{2+}]_i$ that have been measured in various depolarized neurons are not the actual $[Ca^{2+}]_i$ necessary for promoting survival. For example, neurotransmitter release appears to require $[Ca^{2+}]_i$ in the hundreds of μM range, concentrations reached only in a restricted region near the cytoplasmic side of the Ca^{2+} channel pore. Beyond this region, buffering rapidly lowers free $[Ca^{2+}]_i$ to low μM or nM levels,[42,43] concentrations measured by standard techniques. Conceivably, the small increases in $[Ca^{2+}]_i$ reported to promote survival are dim reflections of much larger changes in $[Ca^{2+}]_i$ in highly localized regions. In other words, perhaps the relevant Ca^{2+}-binding proteins are located near Ca^{2+} channel pores in a microdomain where $[Ca^{2+}]_i$ reaches high concentrations. Since the techniques used to measure Ca^{2+} in chronically depolarized neurons do not allow high-resolution determination of localized $[Ca^{2+}]_i$, the question of possible localization of the $[Ca^{2+}]_i$ signal has not been addressed. However, we consider localized high $[Ca^{2+}]_i$ to be an unlikely mechanism for promoting survival. Both depolarization and thapsigargin increase survival of NGF-deprived cells via similar levels of measured global steady-state $[Ca^{2+}]_i$.[33,34,37,38] Since the sustained elevation of $[Ca^{2+}]_i$ caused by thapsigargin is not DHP-sensitive, it is not caused by influx of Ca^{2+} through L-type channels as it is in depolarized cells. If localized high $[Ca^{2+}]_i$ near channel pores is the relevant factor determining survival, it is necessary to postulate that similar effectors are localized to both entry pathways. Therefore, although we cannot rule out a role for high $[Ca^{2+}]_i$ in restricted domains, we favor the idea that it is not required for the effect of $[Ca^{2+}]_i$ on survival.

It is interesting to speculate on the implications of the Ca^{2+} set-point hypothesis for maintenance of neuronal survival *in vivo* and particularly on its possible relevance to disease states. It has been suggested that a decrease in neurotrophic factor availability or the number of neurotrophic factor receptors might underlie the loss of neurons in some neurodegenerative diseases.[44] In light of the set-point hypothesis, it seems possible that effective deficits of neurotrophic factors might also occur in the presence of normal levels of factors and receptors. For example, afferent input, which is necessary for survival of at least some populations of developing neurons (see ref. 1 for review), might determine trophic-factor dependence *in vivo*. Synaptic input probably influences survival by causing transient increases of $[Ca^{2+}]_i$ in postsynaptic cells, a phenomenon mimicked by elevating steady-state $[Ca^{2+}]_i$ by chronic depolarization or thapsigargin. If input is decreased, or lost, a normally sufficient amount of neurotrophic factor may no longer be adequate for survival, and neurons might die.

MECHANISMS BY WHICH INCREASED $[Ca^{2+}]_i$ MAY PROMOTE SURVIVAL

Since intracellular free Ca^{2+} is a major second-messenger molecule, elevated $[Ca^{2+}]_i$ probably promotes survival by activating a signal transduction pathway, most likely the same one by which trophic factors affect survival. A great many Ca^{2+}-binding proteins have been discovered. Of these, the most ubiquitous and abundant is calmodulin, the principal intracellular Ca^{2+} receptor of all nonmuscle eukaryotic cells. Calmodulin has no enzymatic activity by itself, but when bound to Ca^{2+} it affects the activity of a large number of Ca^{2+}-dependent molecules that mediate a myriad of cellular events. Therefore, calmodulin, due to its abundance, ubiquity, and multiple effects, is a likely candidate molecule for the initial signal transduction event that leads to Ca^{2+}-promoted survival. The calmodulin antagonists calmidazolium and W7 block depolarization-promoted survival of rat sympathetic neurons at concentrations that do not affect survival in the presence of NGF and normal $[K^+]_o$. However, these antagonists appear to inhibit survival by blocking the sustained increase of Ca^{2+} caused by chronic depolarization rather than through an effect on calmodulin.[33,34] Therefore, a potential role for calmodulin in mediating survival of these cells cannot be investigated with available inhibitors because of their nonspecificity. A recent report[45] shows that calmodulin antagonists can block depolarization-enhanced survival of embryonic rat cerebellar ganglion neurons at concentrations that do not significantly affect Ca^{2+} influx. Moreover, this report shows that an antagonist of Ca^{2+}-calmodulin-dependent protein kinase (CAM-kinase) blocks depolarization-enhanced survival in these cells. These results suggest that a calmodulin-CAM-kinase pathway is responsible for the effect of Ca^{2+} on survival in these neurons. However, until the effect of the antagonists on the truly relevant parameter $[Ca^{2+}]_i$ in granule cells has been measured, it remains possible that the antagonists block survival through an effect on $[Ca^{2+}]_i$ rather than on calmodulin. Another protein that can be affected by $Ca,^{2+}$ protein kinase C (PKC), has been implicated in depolarization-enhanced survival of embryonic chicken sympathetic neurons in culture.[46] In these cells, depolarization increases PKC activity, and phorbol esters, which mimic PKC activation by diacyl glycerol, can substitute for chronic depolarization in promoting survival. However, phorbol esters have no effect on survival of NGF-deprived rat sympathetic neurons, suggesting that PKC is not involved in depolarization-enhanced survival of these cells.[47] The NGF receptor Trk also appears not to be involved in survival enhancement by depolarization of sympathetic neurons. While NGF causes constitutive phosphorylation of Trk on tyrosine residues, an indication of the activation of Trk tyrosine kinase activity, depolarization has no effect. However, a known component of the

NGF signaling pathway, MAP kinase (whose role in survival, if any, is not known) is tyrosine phosphorylated by both NGF and elevated $[K^+]_o$ in PC12[48] cells and probably rat sympathetic neurons.[34] Hence, although there are hints about possible signaling pathways by which Ca^{2+} promotes survival, very little is currently known about what these pathways actually are.

We recently found that NGF promotes both growth and survival of rat sympathetic neurons in culture, whereas chronic depolarization supports only survival.[34] Exposure of these cells to NGF results in survival, somatic hypertrophy, a linear rate of neurite outgrowth, and a linear rate of increase in total protein content. Removal of NGF in the presence of optimal survival-promoting $[K^+]_o$ (50 mM) maintains cells in a viable state, but growth almost completely ceases. There is little, if any, somatic hypertrophy, neurite outgrowth, or increase of protein content in depolarized cells lacking NGF. This finding strongly suggests that the signaling pathways for survival and growth are separable. There is currently much interest in using neurotrophic factors for treatment of neurodegenerative diseases.[49] One of the problems associated with clinical use of these factors to prevent neuronal death is that they may also promote unwanted growth of neurites. Agents that affect only the survival pathway, therefore, could be of considerable clinical importance. Since chronic depolarization appears to stimulate some, but not all, of the NGF signaling pathway, it may be a useful means of teasing out the survival signal transduction pathway from the signaling pathway that induces growth and may lead to the development of agents that could be used clinically to enhance neuronal survival without causing ectopic growth.

REFERENCES

1. Oppenheim, R. W. 1991. Cell death during development of the nervous system. Annu. Rev. Neurosci. **14:** 453-501.
2. Hendry, I. A. & J. Campbell. 1976. Morphometric analysis of rat superior cervical ganglion after axotomy and nerve growth factor treatment. J. Neurocytol. **5:** 351-360.
3. Hamburger, V., J. K. Brunso-Bechthold & J. W. Yip. 1981. Neuronal death in the spinal ganglia of the chick embryo and its reduction by nerve growth factor. J. Neurosci. **1:** 60-71.
4. Rich, K. M., J. R. Luszczynski, P. A. Osborne & E. M. Johnson, Jr. 1987. Nerve growth factor protects adult sensory neurons from cell death and atrophy caused by nerve injury. J. Neurocytol. **16:** 261-268.
5. Levi-Montalcini, R. & B. Booker. 1960. Destruction of the sympathetic ganglia in mammals by an antiserum to nerve-growth protein. Proc. Natl. Acad. Sci. USA **46:** 384-391.
6. Johnson, E. M., Jr., P. D. Gorin, L. D. Brandeis & J. Pearson. 1980. Dorsal root ganglion neurons are destroyed by exposure in utero to maternal antibody to nerve growth factor. Science **210:** 916-918.
7. Gorin, P. D. & E. M. Johnson, Jr. 1980. Effects of long-term nerve growth factor deprivation on the nervous system of the adult rat: An experimental autoimmune approach. Brain Res. **198:** 27-42.
8. Johnson, E. M., Jr. & L. Aloe. 1974. Suppression of the *in vitro* and *in vivo* cytotoxic effects of guanethidine in sympathetic neurons by nerve growth factor. Brain Res. **81:** 519-532.
9. Wilcox, C. L. & E. M. Johnson, Jr. 1987. Nerve growth factor deprivation results in the reactivation of latent herpes simplex virus *in vitro*. J. Virol. **61:** 2311-2315.
10. Duvall, E. & A. H. Wyllie. 1986. Death and the cell. Immunol. Today **7:** 115-119.

11. PILAR, G. & L. LANDMESSER. 1976. Ultrastructural differences during embryonic cell death in normal and peripherally deprived ciliary ganglia. J. Cell Biol. **68:** 339-356.
12. LEVI-MONTALCINI, R., F. CARAMIA & P. U. ANGELETTI. 1969. Alterations in the fine structure of nucleoli in sympathetic neurons following NGF-antiserum treatment. Brain Res. **12:** 54-73.
13. EDWARDS, S. N., A. E. BUCKMASTER & A. M. TOLKOVSKY. 1991. The death programme in cultured sympathetic neurones can be suppressed at the posttranslational level by nerve growth factor, cyclic AMP, and depolarization. J. Neurochem. **57:** 2140-2143.
14. DECKWERTH, T. L. & E. M. JOHNSON, JR. 1993. Temporal analysis of events associated with programmed cell death (apoptosis) of sympathetic neurons deprived of nerve growth factor. J. Cell Biol. **123:** 1207-1222.
15. MARTIN, D. P., R. E. SCHMIDT, P. S. DISTEFANO, O. H. LOWRY, J. G. CARTER & E. M. JOHNSON, JR. 1988. Inhibitors of protein synthesis and RNA synthesis prevent neuronal death caused by nerve growth factor deprivation. J. Cell Biol. **106:** 829-844.
16. SCOTT, S. A. & A. M. DAVIES. 1990. Inhibition of protein synthesis prevents cell death in sensory and parasympathetic neurons deprived of neurotrophic factor *in vitro.* J. Neurobiol. **21:** 630-638.
17. OPPENHEIM, R. W., D. PREVETTE, M. TYTELL & S. HOMMA. 1990. Naturally occurring and induced neuronal death in the chick embryo *in vivo* requires protein and RNA synthesis: Evidence for the role of cell death genes. Dev. Biol. **138:** 104-113.
18. SCOTT, B. S. & K. C. FISHER. 1970. Potassium concentration and number of neurons in cultures of dissociated ganglia. Exp. Neurol. **27:** 16-22.
19. FRANKLIN, J. L. & E. M. JOHNSON, JR. 1992. Suppression of programmed neuronal death by sustained elevation of cytoplasmic calcium. Trends Neurosci. **15:** 501-508.
20. CHALAZONITIS, A. & G. D. FISCHBACH. 1980. Elevated potassium induces morphological differentiation of dorsal root ganglionic neurons in dissociated cell culture. Dev. Biol. **78:** 172-183.
21. FRANKLIN, J. L., D. J. FICKBOHM & A. L. WILLARD. 1992. Long-term regulation of neuronal calcium currents by prolonged changes of membrane potential. J. Neurosci. **12:** 1726-1735.
22. NISHI, R. & D. K. BERG. 1981. Effects of high K^+ concentrations on the growth and development of ciliary ganglion neurons in cell culture. Dev. Biol. **87:** 301-307.
23. GALLO, V., A. KINGSBURY, R. BALÁZS & O. S. JØRGENSEN. 1987. The role of depolarization in the survival and differentiation of cerebellar granule cells in culture. J. Neurosci. **7:** 2203-2213.
24. COLLINS, F. & J. D. LILE. 1989. The role of dihydropyridine-sensitive voltage-gated calcium channels in potassium-mediated neuronal survival. Brain Res. **502:** 99-108.
25. KOIKE, T., D. P. MARTIN & E. M. JOHNSON, JR. 1989. Role of Ca^{2+} channels in the ability of membrane depolarization to prevent neuronal death induced by trophic-factor deprivation: Evidence that levels of internal Ca^{2+} determine nerve growth factor dependence of sympathetic ganglion cells. Proc. Natl. Acad. Sci. USA **86:** 6421-6425.
26. FOX, A. P., M. C. NOWYCKY & R. W. TSIEN. 1987. Kinetic and pharmacological properties distinguishing three types of calcium currents in chick sensory neurones. J. Physiol. (Lond) **394:** 149-172.
27. JOHNSON, E. M., JR., T. KOIKE & J. FRANKLIN. 1992. A "calcium set-point hypothesis" of neuronal dependence on neurotrophic factor. Exp. Neurol. **115:** 163-166.
28. KATER, S. B., M. P. MATTSON, C. COHAN & J. CONNOR. 1988. Calcium regulation of the neuronal growth cone. Trends Neurosci. **11:** 315-321.
29. CHOI, D. W. 1988. Calcium-mediated neurotoxicity: Relationship to specific channel types and role in ischemic damage. Trends Neurosci. **11:** 465-469.
30. TOLKOVSKY, A. M., A. E. WALKER, R. D. MURRELL & H. S. SUIDAN. 1990. Ca^{2+} transients are not required as signals for long-term neurite outgrowth from cultured sympathetic neurons. J. Cell Biol. **110:** 1295-1306.

31. Schmidt, M. F. & S. B. Kater. 1993. Fibroblast growth factors, depolarization, and substratum interact in a combinatorial way to promote neuronal survival. Dev. Biol. **158:** 228-237.
32. Koike, T. & S. Tanaka. 1991. Evidence that nerve growth factor dependence of sympathetic neurons for survival *in vitro* may be determined by levels of cytoplasmic free Ca^{2+}. Proc. Natl. Acad. Sci. USA **88:** 3892-3896.
33. Franklin, J. L., A. Juhasz, E. B. Cornbrooks, P. A. Lampe & E. M. Johnson, Jr. 1992. Promotion of sympathetic neuronal survival by chronic depolarization and increased $[Ca^{2+}]_i$ is a threshold phenomenon. Soc. Neurosci. Abstr. **18:** 304.
34. Franklin, J. L., C. Sanz-Rodriguez, A. Juhasz, T. L. Deckwerth & E. M. Johnson, Jr. 1994. Chronic depolarization prevents programmed death of sympathetic neurons *in vitro* but does not support growth: Requirement for Ca^{2+} influx but not Trk activation. J. Neurosci., in press.
35. Collins, F., M. F. Schmidt, P. B. Guthrie & S. B. Kater. 1991. Sustained increase in intracellular calcium promotes neuronal survival. J. Neurosci. **11:** 2582-2587.
36. Grynkiewicz, G., M. Poenie & R. T. Tsien. 1985. A new generation of Ca^{2+} indicators with greatly improved fluorescence properties. J. Biol. Chem. **260:** 3440-3450.
37. Lampe, P. A., E. B. Cornbrooks, A. Juhasz, J. L. Franklin & E. M. Johnson, Jr. 1992. Thapsigargin enhances survival of sympathetic neurons by elevating intracellular calcium concentration ($[Ca^{2+}]_i$). Soc. Neurosci. Abstr. **18:** 305.
38. Lampe, P. A., E. B. Cornbrooks, A. Juhasz, E. M. Johnson, Jr. & J. L. Franklin. 1994. Suppression of programmed neuronal death by a thapsigargin-induced Ca^{2+} influx. J. Neurobiol., in press.
39. Thastrup, O., P. J. Cullen, B. K. Drøbak, M. R. Hanley & A. P. Dawson. 1990. Thapsigargin, a tumor promoter, discharges intracellular Ca^{2+} stores by specific inhibition of the endoplasmic reticulum Ca^{2+}-ATPase. Proc. Natl. Acad. Sci. USA **87:** 2466-2470.
40. Hoth, M. & R. Penner. 1992. Depletion of intracellular calcium stores activates a calcium current in mast cells. Nature **355:** 353-355.
41. Putney, J. W., Jr. 1993. The signal for capacitative calcium entry. Cell **75:** 199-201.
42. Llinás, R., M. Sugimori & R. B. Silver. 1991. Imaging preterminal calcium concentration microdomains in the squid giant synapse. Biol. Bull. **181:** 316-317.
43. Augustine, G. J. & E. Neher. 1992. Neuronal Ca^{2+} signalling takes the local route. Current Opinion Neurobiol. **2:** 302-307.
44. Hefti, F. & P. A. Lapchak. 1993. Pharmacology of nerve growth factor in the brain. *In* Advances in Pharmacology. **24:** 239-273. Academic Press. New York.
45. Hack, N., H. Hidaka, M. J. Wakefield & R. Balázs. 1993. Promotion of granule cell survival by high K^+ or excitatory amino acid treatment and Ca^{2+} calmodulin-dependent protein kinase activity. Neuroscience **57:** 9-20.
46. Wakade, A. R., T. D. Wakade, R. K. Malhotra & S. V. Bhave. 1988. Excess K^+ and phorbol ester activate protein kinase C and support the survival of chick sympathetic neurons in culture. J. Neurochem. **51:** 975-983.
47. Martin, D. P., A. Ito, K. Horigome, P. A. Lampe & E. M. Johnson, Jr. 1992. Biochemical characterization of programmed cell death in NGF-deprived sympathetic neurons. J. Neurobiol. **23:** 1205-1220.
48. Tsao, H., J. M. Aletta & L. A. Greene. 1990. Nerve growth factor and fibroblast growth factor selectively activate a protein kinase that phosphorylates high molecular weight microtubule-associated proteins. J. Biol. Chem. **265:** 15471-15480.
49. Olson, L. 1993. NGF and the treatment of Alzheimer's disease. Exp. Neurol. **124:** 5-15.

AIDS-Related Dementia and Calcium Homeostasis[a]

STUART A. LIPTON[b]

Laboratory of Cellular & Molecular Neuroscience
Departments of Neurology
Children's Hospital
Beth Israel Hospital
Brigham & Women's Hospital
Massachusetts General Hospital
and
Program in Neuroscience
Harvard Medical School
Boston, Massachusetts 02115

Our laboratory has a long-standing interest in the relationship of neuronal viability and outgrowth to intracellular Ca^{2+} levels (reviewed in ref. 1). Glutamate, or a related excitatory amino acid (EAA), is the major excitatory neurotransmitter that controls the level of intracellular neuronal Ca^{2+} ($[Ca^{2+}]_i$). Escalating concentrations of glutamate have been measured *in vivo* following focal stroke and head injury (reviewed in refs. 2 and 3). As a result, there is a rapid rise in $[Ca^{2+}]_i$ which precedes neurotoxicity by ~24 hours. Although the elevation in $[Ca^{2+}]_i$ may not account by itself for the ensuing neuronal injury, several laboratories have now reported that prevention of the increase in $[Ca^{2+}]_i$ leads to amelioration of anticipated neuronal cell death (reviewed in refs. 2 and 3). Excessive intracellular Ca^{2+} is thought to contribute to the triggering of a series of potentially neurotoxic events leading to cellular necrosis or perhaps apoptosis.

Many mechanisms are involved in intracellular calcium homeostasis, and this subject is beyond the scope of this article (but see the review in ref. 4). Here we will consider only two modes of Ca^{2+} entry into neurons during these pathological processes. These two routes of entry of Ca^{2+} occur via ion channels that are permeable to Ca^{2+} and can be summarized as follows: (1) Glutamate or related EAAs trigger voltage-dependent calcium channels by depolarizing the cell membrane; a major voltage-dependent calcium channel subtype that is chronically activated by prolonged depolarizations is the L-type calcium channel (reviewed in ref. 5). (2) Glutamate or related EAAs activate ligand-gated ion channels directly; a predominant glutamate receptor-operated channel that is permeable to Ca^{2+} under these conditions is the *N*-

[a] This work was supported by National Institutes of Health grants HD29587, EY05477, EY09024, and NS07264, the American Foundation for AIDS Research, and an Established Investigator Award from the American Heart Association.

[b] Address for correspondence: 300 Longwood Avenue-Enders Building, Suite 361, Boston, MA 02115.

methyl-D-aspartate (NMDA) subtype, but other non-NMDA types may also contribute (reviewed in refs. 2 and 6).

We have shown that activation of these channel types can control neuronal plasticity during normal development, but our laboratory and many others have shown that in excessive amounts this stimulation can lead to neuronal death, for example, after a stroke (reviewed in ref. 1). Similar mechanisms may obtain in various neurodegenerative conditions. In fact, although not involved in the primary pathophysiology of a neurologic disorder, this mechanism may represent a final common pathway of neuronal injury. Most importantly, this pathway makes the disease process amenable to pharmacotherapy. This line of reasoning led us to think that this mechanism might be involved in acquired immunodeficiency syndrome (AIDS)-related neuronal injury.

NEURONAL LOSS IN THE CENTRAL NERVOUS SYSTEM OF AIDS PATIENTS

A substantial number of adults and children with AIDS eventually develop neurological manifestations of the disease, including dementia, myelopathy, and peripheral neuropathy; perhaps 50% of infected children have neurological deficits presenting as delayed developmental milestones. These deficits often occur even in the absence of superinfection with opportunistic organisms or associated malignancies and have been collectively grouped under the rubric human immunodeficiency virus type 1 (HIV-1)-associated cognitive-motor complex; more severe forms were originally labeled AIDS dementia complex.[7]

An excellent review of neurological syndromes in AIDS recently considered the epidemiology of these cognitive and motor deficits as well as the parameters of neuroinvasion (entry of the virus into the CNS), neurotropism (predilection for macrophages of particular strains of HIV-1 that invade the CNS), and neurovirulence (invasion of the CNS by only a subset of macrophage-tropic virus).[8] These topics will not be covered here.

Any theory of the pathogenesis of HIV-1 in the nervous system, however, must take into account the selective and predominant location of the virus in monocytic rather than in other cell types despite the neuropathological findings of widespread myelin pallor, reactive astrocytosis, and substantial injury to neurons. Several, but not all, groups agree that there is an 18–50% loss of cortical and retinal ganglion cell neurons in the CNS of patients with AIDS.[9–13] In addition, in the neocortex there is loss of complexity in dendritic arborization and presynaptic areas.[14]

gp120-INDUCED NEURONAL INJURY IS ATTENUATED BY CALCIUM CHANNEL ANTAGONISTS

As just mentioned, the major cell type infected with HIV-1 in the CNS is found in the macrophage/microglial lineage. These cells act as a reservoir for the virus and possibly release virus or viral proteins or protein fragments. Possibly accounting at least in part for the injury to neurons is the observation first made *in vitro* by Brenneman and colleagues[15] that picomolar concentrations of the envelope protein of HIV-1, gp120, can induce neuronal loss in rodent hippocampal cultures. Subsequently, our group demonstrated that in mixed cultures of neurons and glia, picomolar gp120 could increase $[Ca^{2+}]_i$ in a subpopulation of rodent hippocampal and retinal ganglion cell neurons within a few minutes of application.[16] Recently, similar findings were reported by Thayer's group[17] who were also able to resolve the increase in $[Ca^{2+}]_i$ into discrete oscillations by monitoring the calcium signal on a faster time scale. Within the next 24 hours, neuronal injury ensues.[16] Several groups[18] subsequently reported that picomolar concentrations of gp120 can cause injury in a variety of neuronal preparations *in vitro,* including rat cortical neurons[19,20] and cerebellar granule cells.[21] Both the early rise in $[Ca^{2+}]_i$ and the delayed neuronal injury can largely be prevented by antagonists of the L-type voltage-dependent calcium channel including nimodipine (100 nM in 5% serum or approximately 4 nM free drug).[16,20,21] Other antagonists of the L-type calcium channel are also effective to some degree.[22] Not only are rat retinal ganglion cells and cortical neurons *in vitro* partially protected by nimodipine and other voltage-dependent Ca^{2+} channel antagonists, but also in a rat pup animal model, stereotactic injection of gp120 into the cortex produces a lesion consisting of cellular infiltrates of foamy macrophages and putative neuronal injury that is ameliorated by concomitant intraperitoneal administration of nimodipine.[23] Additional *in vivo* evidence that low concentrations of gp120 are associated with neuronal injury has come from experiments of Brenneman, Hill, Ruff, Pert, and coworkers[24–26] who found that intraventricular injections of gp120 into rats result in dystrophic neurites in hippocampal pyramidal cells as well as behavioral deficits; moreover, cerebrospinal fluid of HIV-infected patients has gp120-like neurotoxic activity. Recently, gp120-transgenic mice were reported to have similar neuropathological findings to those found in the brains of AIDS patients.[27] This evidence points to a potential role of gp120 in a neurodegenerative process. Because of these developments, the AIDS Clinical Trials Groups (ACTG) of the National Institutes of Health Division of AIDS asked us to begin a multicenter, randomized double-blind, placebo-controlled clinical trial to test the effects of nimodipine in a small number of adult patients with HIV-1-associated cognitive/motor complex (a subset of which have the more debilitating AIDS dementia complex). The results of this study were sufficiently encouraging to suggest that a larger clinical trial for efficacy is indicated.

As in a variety of acute and chronic neurodegenerative diseases,[6] the excessive rise in neuronal Ca^{2+} triggered by gp120 is thought to contribute to a final common pathway of neurotoxic events leading to free radical formation, cellular necrosis, and possibly apoptosis; these events include overactivation of the neuronal enzymes protein kinase C, Ca^{2+}/calmodulin-dependent protein kinase II, phospholipases, proteases, protein phosphatases, xanthine oxidase, nitric oxide synthase, and endonucleases.

Nevertheless, these developments do not tell us the mechanism of action of gp120 on neurons, which more recent evidence has led us to believe is an indirect pathway via macrophages/microglia (see below). For example, we noted that only neurons clustered in groups, presumably with synaptic contacts, were vulnerable to gp120, and this fact suggested that cellular interactions were necessary to produce injury. The HIV envelope protein does not appear to act directly on calcium channels; in whole-cell and single-channel patch clamp recordings, picomolar gp120 does not increase calcium current per se (H.S.-V. Chen, M. Plummer, P. Hess, and S. A. Lipton, unpublished findings). Rather, it is possible that calcium channel antagonists ameliorate gp120-induced neuronal injury by reducing the overall intracellular Ca^{2+}

burden of the neurons. After all, Ca^{2+} can accumulate in neurons during normal activity with each action potential fired, and nimodipine may be only indirectly beneficial by helping offset an increased calcium load due to another mechanism.

INVOLVEMENT OF THE NMDA RECEPTOR IN gp120-INDUCED NEURONAL INJURY

As just outlined, another prominent mode of Ca^{2+} entry is via channels directly coupled to EAA/glutamate receptors. The glutamate receptor subtype that is primarily (but not exclusively) involved in this regard is the NMDA subtype. We reasoned that inasmuch as gp120 causes an early rise in $[Ca^{2+}]_i$ and delayed neurotoxicity, similar to glutamate acting at the NMDA receptor, perhaps glutamate or a closely related molecule was involved in HIV-related neuronal injury. Furthermore, it was well known that voltage-dependent calcium channel antagonists such as nimodipine could block some forms of glutamate neurotoxicity.[28–30] Therefore, it was certainly possible that glutamate or a related NMDA agonist was somehow involved in gp120-induced neuronal damage. Along these lines, Heyes and colleagues[31,32] had found that cerebrospinal fluid levels of quinolinate, a naturally occurring (albeit weak) NMDA agonist, was correlated with the degree of dementia in AIDS patients. To test the possibility that EAAs were involved, the following experiments were undertaken. NMDA antagonists were assessed for their ability to prevent gp120-induced neuronal injury. We found that MK-801 (dizocilpine), an open-channel blocker of NMDA receptor-coupled ion channels, prevented gp120-induced neuronal injury.[33,34] D-2-Amino-5-phosphonovalerate, a competitive antagonist at the glutamate binding site of the NMDA receptor, was partially effective in ameliorating this form of neuronal injury. Recently, other groups obtained similar results using NMDA antagonists.[19,21] Nitric oxide (NO·), in conjunction with superoxide anion ($O_2\cdot^-$), is believed to be involved in one of the toxic pathways activated by NMDA receptor stimulation,[35,36] and inhibitors of nitric oxide synthase were recently found to ameliorate gp120-induced neuronal damage *in vitro.*[20] In contrast, CNQX, a non-NMDA antagonist, did not protect from gp120-induced neuronal damage, at least to retinal ganglion cells.

Another possible link between the effects of gp120 and NMDA-receptor activation arises from the observation that one form of neuronal injury in both the brains of AIDS patients and the brains of rats injected with gp120 involves dystrophic neurites.[24] These neurites are excessively tortuous and display a paucity of branches. Some of these neurites may be retracting, giving them a "bald" appearance. We and others have found a similar pattern of dystrophic neurites, including retraction of growth cones, in response to sublethal concentrations of NMDA or glutamate in cultured rat retinal ganglion cells and hippocampal neurons.[1,37,38] Furthermore, these effects are dependent on influx of Ca^{2+} into the neurons. These findings indicate that the endpoints for neuronal injury related to gp120 or excitotoxicity should include more subtle changes than death, and these alterations in neuronal cytoarchitecture could have important consequences for neuronal function and plasticity.[1]

The simplest potential explanation for all of these findings is that gp120 might simulate an NMDA-evoked current or somehow augment such currents. To examine this idea, we used the patch-clamp technique to determine if gp120 affected membrane currents. However, in whole-cell recordings, using both conventional and perforated-patch techniques, no effect of picomolar gp120 was observed, even in recordings lasting tens of minutes. Similarly, enhancement of glutamate- or of NMDA-evoked currents was not encountered.[34] Interestingly, nanomolar concentrations of gp120 (a 1,000-fold excess over levels used in the aforementioned experiments) were reported to block NMDA receptor-operated ion channels, preventing NMDA-evoked increases in Ca^{2+} influx.[39] This finding may account, at least in part, for the dose-response curve of gp120-induced neuronal injury, which has an inverted "U" shape[15]; that is, at high nanomolar concentrations in contrast to picomolar concentrations, gp120 no longer induces neuronal cell injury. Nevertheless, during HIV-1 infection in the brain it is unlikely that nanomolar concentrations of gp120 actually occur because conventional ELISA analysis and Western blots are sensitive to these concentrations but have failed to detect their presence.

The next possible explanation that we considered is that endogenous levels of glutamate might become toxic in the presence of picomolar gp120. To test this hypothesis, the enzyme glutamate-pyruvate transaminase was used to degrade the endogenous glutamate in our retinal cultures. High performance liquid chromatography (HPLC) analysis of amino acids was used to verify glutamate degradation (~25 μM decreased to less than 5 μM). Under these conditions, the catabolism of endogenous glutamate *in vitro* protected neurons from gp120-induced injury to rat retinal ganglion cells.[33,34] Recently, Dawson *et al.*[20] also found that 25 μM glutamate was necessary for them to observe neurotoxicity in rat cortical cultures in the presence of 100 pM gp120.[20] Taken together, these data argue that concurrent activation of NMDA receptors is needed for neuronal injury by gp120 in AIDS. However, these experiments do not tell us if the action of gp120 is mediated directly on neurons or indirectly via an intervening cell type, such as astrocytes or macrophages/microglia, and this possibility is considered in the next section.

INDIRECT NEURONAL INJURY MEDIATED BY HIV-INFECTED OR gp120-STIMULATED MONOCYTIC CELLS

It is still not known definitively if the adverse effects of gp120 are mediated directly on neurons, via glial cells such as microglia and astrocytes, or by a combination of these mechanisms. To determine at least some of the cell types involved in neurotoxicity, we performed the following experiment. L-Leucine methyl ester was used to deplete monocytoid cells (verified by antibody labeling) from cultures of mixed glia and neurons. Under these conditions, gp120 no longer injured neurons, suggesting that at least under our culture conditions, macrophages/microglial were necessary to mediate the neurotoxic effects of gp120.[40] Along these lines, it is also interesting to note that an increase in microglial cell density in the brains of AIDS patients is correlated with the degree of cerebral atrophy, as reflected by ventricular dilatation.[41]

It is well known that gp120 binds to CD4 on human monocytic as well as lymphocytic cells, and, in fact, this appears to be the major (but probably not exclusive) route of entry of the virus into these cells. It has been argued that human macrophages, monocytes, and microglia, but apparently not rodent cells, possess the proper CD4 molecule to bind gp120; however, lack of known receptors does not, of course, rule out alternative mechanisms of binding or toxicity. In several laboratories, for instance, it has not been necessary to have human macrophages present to observe gp120-induced neuronal injury in cultures. Along these same lines, in our laboratory's cultures of rat retinal cells, anti-rat CD4 antibodies did not block the neuronal injury to retinal ganglion cells engendered by gp120 which follows an EAA pathway, whereas anti-gp120 antiserum completely blocked this toxic effect.[42] On the other hand, gp120 incubation with human blood monocytes or the monocytic cell line THP-1 also produced release of neurotoxins that follow an EAA pathway to cell injury; antibodies directed against the CD4-binding region of gp120, but not against the V3 loop of gp120, blocked this toxic effect.[43] Thus, both CD4- and possibly non-CD4-mediated mechanisms of gp120-induced neuronal injury exist, in a sense paralleling a similar situation concerning CD4- and non-CD4-mediated mechanisms for viral entry into T cells.

In conjunction with data from other laboratories, the aforementioned results suggest the following model of HIV-related neuronal injury (FIG. 1). HIV-infected macrophages[44,45] or gp120-stimulated macrophages[40] release potentially neurotoxic products. These neurotoxins include relatively small, possibly heat-stable compounds, which Gendelman and colleagues[46] have recently begun to characterize. They found that the products released by HIV-infected monocytes include products of phospholipase A_2 activity, such as platelet-activating factor (PAF) as well as arachidonic acid and its metabolites. Under their conditions, the release of these substances is enhanced in the presence of astrocytes, implying the existence of a positive feedback loop between astrocytic and monocytoid cells. HIV-infected macrophages also release the cytokines tumor necrosis factor-alpha (TNF-α) and interleukin-1β (IL-1β), which were shown to stimulate astrocyte proliferation,[47,48] another feature of HIV encephalitis.

Moreover, multiple, complex interactions and feedback loops affect cytokine and arachidonic acid metabolite production by macrophages and astrocytes. For example, TNF-α enhances IL-1 production in macrophages.[49] Arachidonic acid metabolites can influence the production of TNF-α and IL-1β in macrophages, and TNF in turn can amplify arachidonic acid metabolism. Platelet-activating factor (PAF) can enhance TNF and IL-1 production, and PAF synthesis in turn can be stimulated with TNF, IL-1β, or gamma interferon (IFN-γ) in human monocytes.[50–55] Finally, similar arachidonic acid metabolites and cytokines released by HIV-infected macrophages appear to be produced by gp120-stimulated monocytic cells. For example, this HIV glycoprotein induces the release of leukotrienes, TNF-α, and IL-1β as well as arachidonic acid itself from human monocytes.[56–58] It remains to be shown, however, if gp120-stimulated monocytic cells release PAF in the presence of astrocytes, similar to their HIV-infected monocyte counterparts.

Cytokines TNF-α and IL-1β in the amounts produced by HIV-infected or gp120-stimulated macrophages do not appear to be neurotoxic in and of themselves.[46] In conjunction with IFN-γ, however, IL-1β was recently shown to increase the activity of the inducible form of nitric oxide synthase in astrocytes,[59] thereby enhancing NO· production and lowering neuronal susceptibility to NMDA receptor-mediated toxicity.[60] It is possible, therefore, that this represents one pathway contributing to

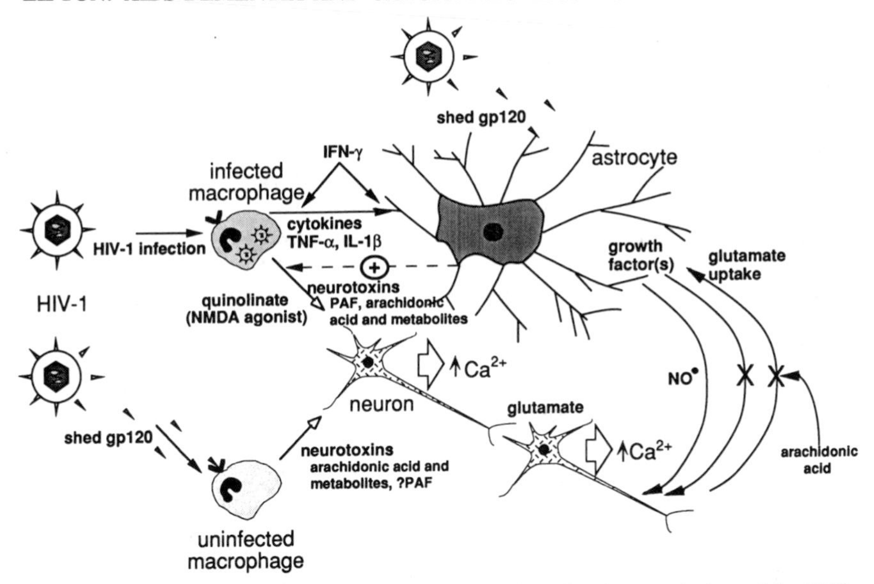

FIGURE 1. Current models of HIV-related neuronal injury. Previous work showed that HIV-infected macrophages/microglia release factors that lead to neurotoxicity.[44,45] These factors include platelet-activating factor (PAF), arachidonic acid and its metabolites, as well as cytokines and other as yet unidentified substances.[46,71] Macrophages and astrocytes have mutual feedback loops (signified by *reciprocal arrows*). The excitatory action of the macrophage factors may lead to an increase in neuronal Ca^{2+} and the consequent release of glutamate. In turn, glutamate overexcites neighboring neurons, leading to an increase in intracellular Ca^{2+}, neuronal injury and subsequent further release of glutamate. This final common pathway of neurotoxic action can be blocked by NMDA antagonists. For certain neurons, this form of damage can also be ameliorated to some degree by calcium channel antagonists or non-NMDA receptor antagonists.

The major pathway of entry of HIV-1 into monocytoid cells is via gp120 binding, and therefore it is not surprising that gp120 (or a fragment thereof) is capable of activating uninfected macrophages to release similar factors to those secreted in response to frank HIV infection.[56] Cytokines participate in this cellular network in several ways. For example, HIV infection or gp120 stimulation of macrophages enhances their production of TNF-α and IL-1β (*solid arrow*). The TNF-α and IL-1β produced by macrophages stimulate astrocytosis. Astrocytes appear to feed back (*dashed arrow*) onto monocytic cells by an as yet unknown mechanism to increase the macrophage production of these cytokines. TNF-α may also increase voltage-dependent calcium currents in neurons.[68] Gamma interferon (IFN-γ), elevated in the CNS of patients with AIDS,[76] can induce macrophage/microgliosis and macrophage production of quinolinate (an NMDA-like agonist)[77] and PAF[55]; in conjunction with IL-1β, IFN-γ can induce nitric oxide synthase (NOS) expression with consequent NO· production in cultured astrocytes,[59] and in this manner it potentiates NMDA receptor-mediated neurotoxicity in mixed neuronal-glial cultures.[60] Such cytokine stimulation of the inducible form of NOS in macrophages or astrocytes may thereby conceivably contribute to HIV-related neurotoxicity. In addition, the constitutive form of NOS (cNOS) has been implicated in gp120 neurotoxicity; the neuronal form of the enzyme (cNOS) is activated by a rise in intracellular Ca^{2+} after stimulation of the NMDA receptor, and inhibitors of this enzyme are reported to prevent gp120 neurotoxicity.[20] The coat protein gp120 may have additional direct or indirect effects on astrocytes, for example, to decrease growth factor production[80] or inhibit glutamate reuptake, for example, via arachidonic acid.[73] Arachidonate was also recently reported to enhance NMDA-evoked currents[126] and therefore could contribute to neurotoxicity not only by enhancing net glutamate efflux but also by increasing its effectiveness at the NMDA receptor.

HIV-related neuronal injury. However, a direct contribution to gp120-induced neuronal injury by NO· production of astrocytes has not yet been demonstrated.

There are other potential toxins that HIV-infected or gp120-stimulated macrophages might release. Activated macrophages can produce NO· and $O_2·^-$, which can react to form neurotoxic peroxynitrite ($ONOO^-$).[36] However, it is yet to be shown that this is an important pathway in neuronal injury under these conditions. Under some conditions, activated macrophages/microglia may also secrete glutamate or cysteine directly, which can act as NMDA agonists.[36,61–64] Although HIV-infected or gp120-stimulated monocytic cells have not been shown to release glutamate, we have recently demonstrated enhanced levels of cysteine.[65]

Could the arachidonic acid metabolites emanating from HIV-infected or gp120-stimulated macrophages be involved in neurotoxicity? In particular, PAF was recently shown to increase intracellular neuronal Ca^{2+} and, presumably via this mechanism, to increase glutamate release and hence excitatory neurotransmission.[66–68] TNF-α may also contribute to this process by increasing voltage-dependent Ca^{2+} currents.[69] In collaboration with Gendelman's group, our laboratory recently showed that the elevated levels of PAF measured in cultures of HIV-infected monocytic cells as well as in the cerebrospinal fluid of patients with the AIDS dementia complex are toxic to neurons *in vitro.*[70] Moreover, in these experiments, PAF-related neuronal injury is largely ameliorated by NMDA antagonists, similar to the pharmacology of neuroprotection observed in the face of HIV-infected or gp120-stimulated macrophages.[71] These recent developments suggest that *local* release of excessive glutamate with resultant overstimulation of NMDA receptors contributes to neuronal damage in AIDS patients with high concentrations of PAF. On the other hand, at more physiological levels, PAF can increase glutamate release perhaps more modestly, and this may contribute to the observation that PAF can induce long-term potentiation (LTP, a cellular correlate of learning and memory) in the hippocampus.[72] Thus, the spectrum of actions of PAF in exciting neurons may represent a bell curve with moderate levels of excitation enhancing normal physiological functions and excessive stimulation leading to neurotoxicity. The implications for such a mechanism are particularly important in the AIDS dementia complex because extreme elevations in PAF may thus lead to disruption of memory and cognitive function before actual neuronal cell death.

In addition, arachidonic acid is released from HIV-infected and gp120-stimulated monocytic cells.[46,56] Recently, arachidonic acid was shown to inhibit high-affinity uptake of glutamate into synaptosomes and astrocytes[73] and to potentiate NMDA receptor-activated currents by increasing open channel probability.[74] In conjunction with PAF, arachidonic acid may therefore contribute to excessive NMDA receptor stimulation by increasing the release of glutamate, inhibiting its reuptake, and enhancing its action at the NMDA receptor. Additional glutamate receptor activation may occur as a further consequence of these events as neurons are excited or injured and release their stores of glutamate onto neighboring neurons.[17,33,34,75] One line of evidence for this supposition lies in the finding, as just detailed, that NMDA antagonists or enzymatic degradation of glutamate ameliorates gp120-induced neuronal injury in mixed neuronal-glial cultures.[33,34] Intensive investigation in several laboratories is currently underway to study the potential pathways for neuronal injury that is triggered by HIV-infected or gp120-stimulated macrophages involving PAF, arachidonic acid, cysteine, nitric oxide, and potentially other substances.

Also, as already alluded to, another possible link between HIV-1 infection and EAA-induced neurotoxicity involves quinolinate, an endogenous NMDA agonist that is increased in the cerebrospinal fluid of patients with the AIDS dementia complex.[32] Quinolinate levels are known to be influenced by cytokines that are increased after HIV-1 infection. For example, it is known that IFN-γ, presumably released from cytotoxic T lymphocytes, is present in the brain of patients with AIDS,[76] and human macrophages activated by IFN-γ release substantial amounts of quinolinate.[77] In addition, under some conditions, such as after neuronal loss, quinolinate can also be produced by astrocytes.[78,79] Quinolinate, therefore, may also contribute to neuronal injury by activating NMDA receptors during HIV infection; however, this scenario apparently is also true for a variety of other CNS infections.

POSSIBLE INVOLVEMENT OF ASTROCYTES, OLIGODENDROCYTES, AND OTHER HIV-1 PROTEINS IN NEURONAL INJURY

Astrocytes and HIV-Related Neuronal Damage

In at least some model systems, the presence of astrocytes is necessary for HIV-infected macrophages to release substantial amounts of their neurotoxic factors.[46] In addition, astrocytes may be important in mediating HIV-related neuronal injury in other ways. For example, in murine hippocampal cultures Brenneman *et al.*[15] found that gp120-induced neurotoxicity can be prevented by the presence of vasoactive intestinal polypeptide (VIP) or by a five amino acid substance with sequence homology to VIP, peptide T. These workers also found that VIP acts on astrocytes to increase oscillations in intracellular Ca^{2+} and to release factors necessary for normal neuronal outgrowth and survival.[80] Thus, these results raise the possibility that gp120 may compete with endogenous VIP for a receptor, most likely on astrocytes, that is critical to normal neuronal function. The receptor may bear some resemblance to mouse CD4 because mouse anti-CD4 antibodies blocked the toxic effects of gp120 in this system.[15] This effect of gp120 is hypothesized to prevent the release of such astrocyte factors that are necessary to prevent neuronal injury and suggests that one pathway for neuronal damage is an indirect one that is mediated by astrocytes. Therefore, gp120 might interact with a receptor on astrocytes (FIG. 1); neurotoxicity may in part be realized by interfering with the normal function of astrocytes and their release of neuronal growth factor(s).[81] Our laboratory has gathered preliminary data suggesting that gp120 might also affect astrocyte function, either directly or indirectly, in another manner: gp120 inhibits the ability of cultured astrocytes (and possibly neurons) to take up glutamate, thus possibly contributing to EAA-induced neurotoxicity (E. B. Dreyer and S. A. Lipton, in preparation). Such an effect might account for the apparent increase in sensitivity of neurons to glutamate toxicity in the presence of gp120 and would also help explain the requirement for some glutamate (~25 μM) to be present in the culture medium in order to observe gp120-induced neurotoxicity.[20,33,34]

Neurotoxicity of Other HIV-1 Proteins

Besides gp120, two other HIV-1 proteins are reported to affect neurons or neuronal-like cells, raising the possibility of their involvement in HIV-related neuronal injury. The nuclear protein tat was shown to be toxic to glioma and neuroblastoma cell lines *in vitro* and to mice *in vivo*.[82,83] The basic region of the peptide (amino acid residues 49-57) appears to act nonspecifically to increase the leakage conductance of the membrane, thus altering cell permeability. Moreover, neurotoxicity of the related Maedi-Visna virus peptide was ameliorated by NMDA antagonists or by inhibitors of nitric oxide synthase,[84] reminiscent of the pharmacology of antagonism of the neurotoxic effects of gp120 and HIV-infected macrophages. Further work will be necessary to relate these findings with the tat peptide to the neuropathology encountered in the brains of patients with HIV-1-associated cognitive/motor complex.

Another HIV-1 protein, Nef, was also shown to affect neuronal cell function. Nef shares sequence and structural features with scorpion toxin peptides; both recombinant Nef protein and a synthetic portion of scorpion peptide increase total K^+ current in chick dorsal root ganglion cells.[85]

Direct Effects of HIV-Infected Macrophages on Neurons

Finally, HIV-infected monocytoid cells may have a cytopathic effect on neurons by direct contact.[86] However, this mechanism does not preclude an additional mechanism of neuronal injury mediated by soluble factors leading to excessive stimulation of NMDA receptors.[87]

EXCESSIVE STIMULATION OF NMDA RECEPTORS: A FINAL COMMON PATHWAY

From the foregoing, there apparently are at least two sites of potential interaction of HIV-related neurotoxins with NMDA receptors (FIG. 1). First, quinolinate emanating from macrophages may directly stimulate neurons. Second, after excitation contributed to by quinolinate, PAF, and possibly arachidonic acid, its metabolites, or other toxins, neurons would release glutamate onto second-order neurons. Additionally, astrocytes might fail to take up the glutamate. This hypothesis, in which one neuron acts as a "bad neighbor" by releasing excessive glutamate, is in some ways similar to that in which damage is thought to occur in the penumbra of a stroke–glutamate released by injured neurons contributes to further injury to neighboring neurons.

Moreover, NMDA antagonists ameliorate HIV-related neuronal injury induced by either HIV-infected macrophages[44] or, as mentioned earlier, gp120-activated macrophages.[19–21,33,34,88] Furthermore, in some cases calcium channel antagonists can attenuate this form of damage.[16,22] The pharmacology of neuroprotection from noxious agents generally depends on the repertoire and diversity of ion channel types in a particular class of neurons.[5] For example, neurons lacking NMDA receptors will obviously not be protected by NMDA antagonists. Conversely, if NMDA receptor-associated channels are the predominant channel in a specific neuronal cell type whereby Ca^{2+} enters the cell, then the lethal effects of excessive stimulation by glutamate may be ameliorated with NMDA antagonists. Some non-NMDA receptor-associated channels are directly permeable to Ca^{2+}, but most appear not to be (eg, those containing the GluR2 receptor subunit). However, depolarization of neurons by stimulation of non-NMDA receptors will trigger voltage-dependent Ca^{2+} channels. If sufficient L-type calcium channels exist on a particular neuronal cell type, then excessive influx of Ca^{2+} via these channels could contribute to toxic consequences. In fact, in some cell types such as hippocampal pyramidal cells, cortical neurons, and retinal ganglion cells, evidence indicates that calcium channel antagonists may attenuate damage due to activation of either NMDA or non-NMDA receptors.[28–30]

As just outlined, glutamate may be involved in the final common pathway of neuronal injury by HIV-infected macrophages or by gp120-stimulated macrophages. Thus, either NMDA or non-NMDA receptor activation may play a role in this form of toxicity depending on the exact repertoire of ion channels in a particular cell type. In fact, it was suggested that non-NMDA receptors could also be important in contributing to the neurotoxic events triggered by gp120.[20,89] Nevertheless, the majority of findings to date suggest that NMDA receptor-mediated neuronal injury plays a predominant role in the pathogenesis of the neurological manifestations of AIDS in the CNS.[23,75]

DEVELOPMENT OF CLINICALLY TOLERATED NMDA ANTAGONISTS FOR HIV-RELATED NEURONAL INJURY

NMDA receptors may be involved in HIV-related neurotoxicity at two separate sites, located (1) on the primary neuron injured by factors released from glial cells, and (2) on neurons that are secondarily affected (see above and FIG. 1). This fact has provided an impetus for our laboratory to being a drug development program for clinically tolerated NMDA antagonists. To understand our approach, we must first entertain a brief review of the possible mechanisms and sites of action of NMDA receptor antagonists.

Sites of Action of Potential, Clinically Tolerated NMDA Antagonists

Given the growing number of genes (at least 20) that have been cloned for various glutamate receptors, the complexity of EAA receptor pharmacology is great. Despite these concerns, we can consider currently available agents that appear to work on broad classes of these receptors. For this review, we will concentrate on NMDA antagonists that appear to be clinically tolerated and therefore can be considered for human trials.[90]

Several modulatory sites on the NMDA receptor-channel complex could potentially be used to modify the activity of receptor-operated ion channels and thus to prevent excessive influx of Ca^{2+} (FIG. 2). The first site is the glutamate or NMDA binding site. An antagonist acting here would be competitive in nature, that is, competing for the site with an EAA. For both theoretical and practical reasons, a competitive inhibitor might not be as desirable an antagonist as one that is not competitive for the glutamate binding site. A competitive antagonist would perforce eliminate the normal, physiological activity of the NMDA receptor even before it

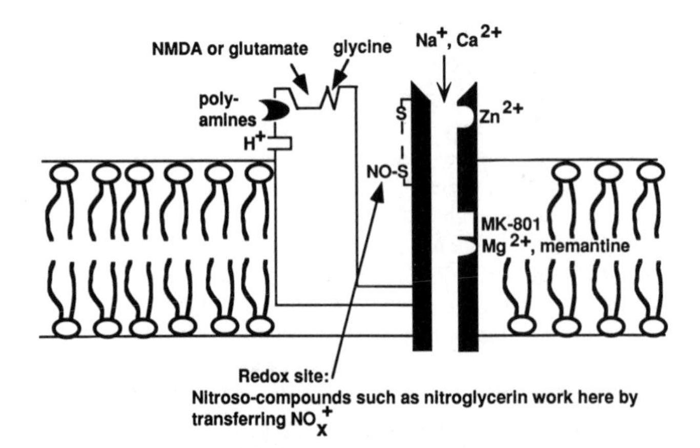

FIGURE 2. Sites of potential antagonist action on the NMDA receptor-channel complex. Competitive antagonists can compete with NMDA or glutamate for binding to the agonist site. Several antagonists to the glycine co-agonist site have been described that are chlorinated and sulfated derivatives of kynurenic acid. H^+ effects are transmitted through a modulatory site; decreasing pH acts to downregulate channel activity. Other sites for polyamines and Zn^{2+} can also be used to affect receptor-channel function. Sites that inhibit channel activity by binding Mg^{2+} or drugs such as MK-801, phencyclidine, and memantine are within the electric field of the channel, and these sites are only exposed when the channel is previously opened by agonist (termed uncompetitive antagonism). Finally, redox modulatory sites [probably disulfide bond(s), or at least a long-lasting covalent modification of a thiol group, that can be converted to free sulfhydryl groups (S-S $\rightleftarrows$ 2-SH)] is affected by chemical reducing and oxidizing agents. Oxidation can favor the disulfide conformation (S-S) over free thiol (–SH) groups and thus downregulate channel activity. In addition, several nitroso-compounds can downregulate channel activity by transferring an NO group to the thiol(s) of the NMDA receptor's redox site(s), producing RS-NO, a nitrosonium ion (NO^+) equivalent, which may lead to disulfide bond formation.

would affect potentially excessive levels of glutamate. Thus, cognition and memory, thought to be related to long-term potentiation, as well as other important functions mediated by excitatory transmission in the brain might be compromised. Even putting aside these concerns, as part of the disease process escalating levels of glutamate might be able to overcome or "out-compete" such an antagonist.

NMDA OPEN-CHANNEL BLOCKERS

In contrast, *modulatory* sites of the NMDA receptor-channel complex should be able to inhibit the effects of high levels of glutamate in compromised areas of the brain while leaving relatively spared the effects of normal neurotransmission in other regions of the brain.[91–93] For example, one site that appears to have this advantageous effect is located in the ion channel itself. There are drugs that only block the channel when it is open, that is, the antagonist can only gain access to the channel in the open state. On average, escalating levels of glutamate allow the channels to remain open for a greater fraction of time. Under these conditions, there is a better chance for an open-channel blocking drug to enter the channel and block it. The result of such a mechanism of action is that the untoward effects of greater (pathological) concentrations of glutamate are inhibited to a relatively greater extent than are lower (physiological) concentrations.[93] Unfortunately, some of these open-channel blockers, which include phencyclidine (angel dust) and MK-801 (dizocilpine), have neuropsychiatric side effects and probably cannot be administered safely.[94] Another concern with NMDA antagonists, such as phencyclidine and MK-801, is the development of neuronal vacuolization, albeit reversible.[95] One reason for the toxicity of a drug such as MK-801 is that once it enters an open channel, it leaves the channel only very slowly (half-time $\geq$ 1.5 hours). In practical terms this means that the degree of blockade builds up after the administration of MK-801 because each molecule of the antagonist entering a channel effectively does not leave.

Several members of this open-channel blocking class of agents, however, such as ketamine and dextromethorphan or the related molecule dextrorphan, are tolerated.[96–100] Unfortunately, it is not clear if these particular drugs are sufficiently potent NMDA antagonists at clinically tolerated doses. Nevertheless, the finding that certain members of this open-channel blocker family are clinically tolerated apparently is associated with their rapid kinetics of interaction with the channel (the kinetic parameters are composed of the on-rate and off-rate for channel blockade).[93,101,102] Most importantly, the safe drugs, such as memantine[103] (see discussion to follow), leave the channel promptly, with an off-rate of ~5 seconds.[93] Mg^{2+} also blocks open NMDA channels, and this may be the basis for its antiepileptic and neuroprotective effects.[104–106] These beneficial effects, however, may not be robust, probably because Mg^{2+} leaves the channel so quickly that it may not act effectively to offset toxic levels of glutamate. In addition, these charged channel-blocking drugs act to a lesser degree when neurons are depolarized, as under conditions of energy compromise.[107]

In summary, an agent that remains in the channel for at least some time is necessary to block the effects of glutamate overstimulation. Of the known NMDA open-channel blockers, memantine is one candidate for clinical trials to combat neurological disorders, such as HIV-1-associated cognitive/motor complex, that have a component of NMDA receptor-mediated neurotoxicity. Memantine has been used clinically with considerable safety in Germany for over 12 years in the treatment of Parkinson's disease and spasticity. Memantine is a congener of amantadine, the well-known antiviral and antiparkinsonian drug used in the United States. Amantadine, however, is considerably less potent on NMDA receptor-operated ion channels at clinically tolerated doses,[93] probably precluding its use for these other neurological diseases. It may be no accident that memantine both inhibits NMDA receptor responses and alleviates parkinsonian symptoms; one theory of Parkinson's disease is that neurons die, at least in part, because of a form of NMDA receptor-mediated toxicity.

NMDA RECEPTOR REDOX MODULATORY SITE(S)

Another modulatory site(s) on the NMDA receptor-channel complex of possible clinical utility in the near future was discovered several years ago in our laboratory and has been termed the redox modulatory site(s).[108] This site(s) consists of one or more sulfhydryl groups; these sulfhydryl groups may possibly be in close approximation and form a disulfide bond under oxidizing conditions. Under chemical reducing conditions that favor the formation of free thiol (–SH) groups over a disulfide, the opening frequency of NMDA receptor-associated channels increases,[108,109] and thus there is a net increase in Ca^{2+} influx through the channels[110,111] and an increase in the extent of NMDA receptor-mediated neurotoxicity.[112,113] Conversely, redox reagents that mildly oxidize the NMDA receptor, for example, to re-form disulfide bonds or form ligands on the free thiol groups, downregulate receptor-mediated neurotoxicity. Thus, these agents might prove useful in combating the myriad of neurological maladies resulting, at least in part, from a final common pathway of NMDA receptor-mediated neuronal damage.[114]

Indeed, several such redox reagents were recently reported, including surprisingly the common nitroso-compound, nitroglycerin.[115] One mechanism of nitroglycerin's action in this regard is mediated by a substance related to nitric oxide (NO·), but in a different redox state, for example, in the form of RS-NO_x (where x = 1 or 2, and the NO group represents nitrosonium ion equivalents, NO^+, having one less electron than NO·).[116] Nitric oxide (NO·) itself can participate in reactions with superoxide anion ($O_2·^-$) to form products that are toxic to nerve cells, such as peroxynitrite ($ONOO^-$) and its breakdown products including hydroxyl radical (HO·)-like compounds.[35,36,115,117–119] In other redox states, however, monoxides of nitrogen can interact with thiol groups, such as those comprising the redox modulatory site(s) of the NMDA receptor, by an *S*-nitrosylation reaction, resulting in transfer of the NO group to a thiol.[36,116] This action results in downregulation of NMDA receptor activity and protects neurons from excessive stimulation of the receptor.[115] Patients can be made tolerant of the cardiovascular effects of nitroglycerin within hours of continuous therapy. Under these conditions, our laboratory has shown in animal models that the extent of NMDA receptor-mediated neurotoxicity can be markedly attenuated in the absence of behavioral or systemic side effects of the drug.[120] Nevertheless, the exact dosing regimen for the chronic administration of nitroglycerin must be carefully worked out before attempts are made to apply this technique to humans. Alternatively, nitroso-compounds, such as nitroglycerin, can be administered acutely to affect NMDA receptor activity, but only if blood pressure is maintained with a pressor agent such as phenylephrine.

In addition, the NMDA receptor has other important modulatory sites, several of which are illustrated in FIGURE 2. Antagonists of each of these sites could possibly be useful in the treatment or prevention of NMDA receptor-mediated neurotoxicity. For this relatively brief review, I have highlighted only two of these, the ion channel and redox modulatory sites. The other modulatory sites may become therapeutically relevant, however, if clinically tolerated antagonists can be developed to interact with them. Intensive research efforts along these lines are now underway in both academic institutions and the pharmaceutical industry.

NMDA ANTAGONISTS FOR AIDS DEMENTIA?

A case can be made that the proven NMDA open-channel blocker memantine has already been in clinical use for years and is known to ameliorate some of the symptoms of Parkinson's disease. Furthermore, it is now known that the level of memantine (2-12 μM) achieved in the human brain during this form of treatment[121] can afford protection from NMDA receptor-mediated neurotoxicity both *in vitro* and *in vivo*.[93,122-125] Recently, our laboratory as well as another group reported that low micromolar levels of memantine can also protect neurons from damage induced by gp120 *in vitro*[19,88] and *in vivo* in an animal model.[23] These preliminary findings raise the possibility that a clinically-tolerated NMDA antagonist, memantine, might be useful in the treatment or prevention of the AIDS dementia complex.[90] Therefore, it has been proposed that memantine be studied as adjunctive therapy with antiretroviral drugs such as zidovudine or didanosine, and the AIDS Clinical Trials Group of the NIH is currently considering this option.

CONCLUSIONS

Although a complex web of cell interactions likely leads to neuronal loss in AIDS, HIV-infected macrophages or gp120-stimulated macrophages release toxins whose action apparently are mediated by a final common pathway involving over-stimulation of neurons by excitatory amino acids such as glutamate and quinolinate. In the final analysis, gp120 should be viewed as just one piece of a complicated network of factors that appear to contribute to neuronal damage in AIDS patients. The major pathway of entry of HIV-1 into monocytoid cells is via gp120 binding, and therefore it is not surprising that gp120 (or a fragment thereof) is capable of activating uninfected macrophages to release toxic factors similar to those secreted in response to frank HIV infection. Arachidonic acid, platelet-activating factor, free radicals (NO·, $O_2 \cdot^-$), glutamate, quinolinic acid, cysteine, cytokines (TNF-α, IL-1β), and as yet unidentified factors emanating from stimulated macrophages and/or reactive astrocytes may all contribute to the observed neuropathology (FIG. 1). A strong body of scientific evidence supports the premise that the mechanism for this form of HIV-related neuronal injury is similar to that currently thought responsible for a wide variety of acute and chronic neurological diseases.[2,3,75] Excitatory amino acids apparently exert this excitotoxic effect by engendering an excessive influx of Ca^{2+} into neurons. Currently, intensive investigation is being conducted to discover clinically tolerated drugs to combat the neurotoxic effects associated with excessive stimulation of glutamate receptors. One therapeutic approach has been to use glutamate receptor antagonists, and although several promising drugs are already in hand, additional agents are needed. With the possibility of a final common pathophysiology involving EAA receptors for many disorders of the central nervous system, including the AIDS dementia complex, future development and testing of safe and effective EAA antagonists should become a high priority.

SUMMARY

Approximately a third of adults and half of children with acquired immunodeficiency syndrome (AIDS) eventually suffer from neurological manifestations, including dysfunction of cognition, movement, and sensation. Among the various pathologies reported in the brain of patients with AIDS is neuronal injury and loss. A paradox arises, however, because neurons themselves are for all intents and purposes not infected by human immunodeficiency virus type 1 (HIV-1). This paper reviews evidence suggesting that at least part of the neuronal injury observed in the brain of AIDS patients is related to excessive influx of Ca^{2+}.

There is growing support for the existence of HIV- or immune-related toxins that lead indirectly to the injury or death of neurons via a potentially complex web of interactions between macrophages (or microglia), astrocytes, and neurons. Human immunodeficiency virus-infected monocytoid cells (macrophages, microglia, or monocytes), especially after interacting with astrocytes, secrete substances that potentially contribute to neurotoxicity. Not all of these substances are yet known, but they may include eicosanoids, that is, arachidonic acid and its metabolites, as well as platelet-activating factor. Macrophages activated by HIV-1 envelope protein gp120 also appear to release arachidonic acid and its metabolites. These factors can lead to increased glutamate release or decreased glutamate reuptake. In addition, gamma interferon (IFN-γ) stimulation of macrophages induces release of the glutamate-like agonist quinolinate. Human immunodeficiency virus-infected or gp120-stimulated macrophages also produce cytokines, including tumor necrosis factor-alpha and interleukin-1β, which contribute to astrogliosis. A final common pathway for neuronal susceptibility appears to be operative, similar to that observed in stroke, trauma, epilepsy, neuropathic pain, and several neurodegenerative diseases, possibly including Huntington's disease, Parkinson's disease, and amyotrophic lateral sclerosis. This mechanism involves the activation of voltage-dependent Ca^{2+} channels and *N*-methyl-D-aspartate (NMDA) receptor-operated channels, and therefore offers hope for future pharmacological intervention. This review focuses on clinically tolerated calcium channel antagonists and NMDA antagonists with the potential for trials in humans with AIDS dementia in the near future.

ACKNOWLEDGMENTS

I would like to thank my coworkers, Drs. E. B. Dreyer, N. J. Sucher, H.-S. V. Chen, P. K. Kaiser, M. Oyola, S. Z. Lei, J. Pellegrini, D. Zhang, and Y.-B. Choi, for insightful discussions, and Drs. D. Leifer and J. S. Stamler for comments on an earlier version of the manuscript.

REFERENCES

1. LIPTON, S. A. & S. B. KATER. 1989. Trends Neurosci. **12:** 265–270.
2. CHOI, D. W. 1988. Neuron **1:** 623–634.
3. MELDRUM, B. & J. GARTHWAITE. 1990. Trends. Pharmacol. Sci. **11:** 379–387.
4. MILLER, R. J. 1991. Prog. Neurobiol. **37:** 255–285.
5. LIPTON, S. A. 1991. Adv. Pharmacol. **22:** 271–291.
6. LIPTON, S. A. & P. A. ROSENBERG. 1994. N. Engl. J. Med. **330:** 613–622.
7. PRICE, R. W., B. BREW, J. SIDTIS, M. ROSENBLUM, A. C. SCHECK & P. CLEARLY. 1988. Science **239:** 586–592.
8. EPSTEIN, L. G. & H. E. GENDELMAN. 1993. Ann. Neurol. **33:** 429–436.
9. KETZLER, S., S. WEIS, H. HAUG & H. BUDKA. 1990. Acta Neuropathol. (Berl.) **80:** 90–92.

10. Wiley, C. A., E. Masliah, M. Morey, C. Lemere, R. M. DeTeresa, M. R. Grafe, L. A. Hansen & R. D. Terry. 1991. Ann. Neurol. **29:** 651-657.
11. Everall, I. P., P. J. Luthbert & P. L. Lantos. 1991. Lancet **337:** 1119-1121.
12. Tenhula, W. N., S. Z. Xu, M. C. Madigan, K. Heller, W. R. Freeman & A. A. Sadun. 1992. Am. J. Ophthalmol. **113:** 14-20.
13. Seilhean, D., C. Duyckaerts, R. Vazuex, F. Bolgert, P. Brunet, C. Katlama, M. Gentilini & J.-J. Hauw. 1993. Neurology **43:** 1492-1499.
14. Masliah, E., C. L. Achim, N. Ge, R. DeTeresa, R. D. Terry & C. A. Wiley. 1992. Ann. Neurol. **32:** 321-329.
15. Brenneman, D. E., G. L. Westbrook, S. P. Fitzgerald, D. L. Ennist, K. L. Elkins, M. Ruff & C. B. Pert. 1988. Nature **335:** 639-642.
16. Dreyer, E. B., P. K. Kaiser, J. T. Offermann & S. A. Lipton. 1990. Science **248:** 364-367.
17. Lo, T.-M., C. J. Fallert, T. M. Piser & S. A. Thayer. 1992. Brain Res. **594:** 189-196.
18. Pulliam, L., D. West, N. Haigwood & R. A. Swanson. 1993. AIDS Res. Hum. Retroviruses **9:** 439-444.
19. Müller, W. E. G., H. C. Schröder, H. Ushijima, J. Dapper & J. Bormann. 1992. Eur. J. Pharmacol.-Molec. Pharm. Sect. **226:** 209-214.
20. Dawson, V. L., T. M. Dawson, G. R. Uhl & S. H. Snyder. 1993. Proc. Natl. Acad. Sci. USA **90:** 3256-3259.
21. Savio, T. & G. Levi. 1993. J. Neurosci. Res. **34:** 265-272.
22. Lipton, S. A. 1991. Ann. Neurol. **30:** 110-114.
23. Lipton, S. A. & F. E. Jensen. 1992. Soc. Neurosci. Abstr. **18:** 757.
24. Mervis, R. F., J. M. Hill & D. E. Brenneman. 1990. Int. Conf. AIDS **6:** 184.
25. Glowa, J. R., L. V. Panlilio, D. E. Brenneman, I. Gozes, M. Fridkin & J. M. Hill. 1992. Brain Res. **570:** 49-53.
26. Buzy, J., D. E. Brenneman, C. B. Pert, A. Martin, A. Salazar & M. R. Ruff. 1992. Brian Res. **598:** 10-18.
27. Toggas, S. M., E. Masliah, E. M. Rockenstein, G. F. Rall, C. R. Abraham & L. Mucke. 1994. Nature **367:** 188-193.
28. Abele, A. E., K. P. Scholz, W. K. Scholz & R. J. Miller. 1990. Neuron **4:** 413-419.
29. Weiss, J. H., D. M. Hartley, J. Koh & D. W. Choi. 1990. Science **247:** 1474-1477.
30. Sucher, N. J., S. Z. Lei & S. A. Lipton. 1991. Brain Res. **551:** 297-302.
31. Heyes, M. P., D. Rubinow, C. Lane & S. P. Markey. 1989. Ann. Neurol. **26:** 275-277.
32. Heyes, M. P., B. J. Brew, A. Martin, R. W. Price, A. M. Salazqr, J. J. Sidtis, J. A. Yergey, M. M. Mouradian, A. Sadler, J. Keilp, D. Rubinow & S. P. Markey. 1991. Ann. Neurol. **29:** 202-209.
33. Lipton, S. A., P. K. Kaiser, N. J. Sucher, E. B. Dreyer & J. T. Offermann. 1990. Soc. Neurosci. Abstr. **16:** 289.
34. Lipton, S. A., N. J. Sucher, P. K. Kaiser & E. B. Dreyer. 1991. Neuron **7:** 111-118.
35. Dawson, V. L., T. M. Dawson, E. D. London, D. S. Bredt & S. H. Snyder. 1991. Proc. Natl. Acad. Sci. USA **88:** 6368-6371.
36. Lipton, S. A., Y.-B. Choi, Z.-H. Pan, S. Z. Lei, V. H.-S. Chen, N. J. Sucher, J. Loscalzo, D. J. Singel & J. S. Stamler. 1993. Nature **364:** 626-632.
37. Mattson, M. P., R. E. Lee, M. E. Adams, P. B. Guthrie & S. B. Kater. 1988. Neuron **1:** 865-876.
38. Offermann, J., K. Uchida & S. A. Lipton. 1991. Soc. Neurosci. Abstr. **17:** 927.
39. Sweetnam, P. M., O. H. Saab, J. T. Wroblewski, C. H. Price, W. Larbon & J. W. Ferkany. 1993. Eur. J. Neurosci. **5:** 276-283.
40. Lipton, S. A. 1992. NeuroReport **3:** 913-915.
41. Gelman, B. G. 1993. Ann. Neurol. **34:** 65-70.
42. Kaiser, P. K., J. T. Offermann & S. A. Lipton. 1990. Neurology **40:** 1757-1761.

43. GIULIAN, D., E. WENDT, K. VACA & C. A. NOONAN. 1993. Proc. Natl. Acad. Sci. USA **90:** 2769-2773.
44. GIULIAN, D., K. VACA & C. A. NOONAN. 1990. Science **250:** 1593-1596.
45. PULLIAM, L., B. G. HERNDLER, N. M. TANG & M. S. MCGRATH. 1991. J. Clin. Invest. **87:** 503-512.
46. GENIS, P., M. JETT, E. W. BERNTON, T. BOYLE, H. A. GELBARD, K. DZENKO, R. W. KEANE RESNICK, L. T. MIZRACHI, D. J. VOLSKY, L. G. EPSTEIN & H. E. GENDELMAN. 1992. J. Exp. Med. **176:** 1703-1718.
47. SELMAJ, K. N., M. FAROOQ, T. NORTON, C. S. RAINE & C. F. BROSMAN. 1990. J. Immunol. **144:** 129-135.
48. CHUNG, I. Y. & E. N. BENVENISTE. 1990. J. Immunol. **144:** 2999-3007.
49. MORGANATI-KOSSMANN, M. C., T. KOSSMANN & S. M. WAHL. 1992. Trends Pharmacol. Sci. **13:** 286-290.
50. CONTI, P., M. REALE, R. C. BARBACANE, M. BONGRAZIA, M. R. PANARA & S. FIORE. 1989. *In* Prostaglandins in Clinical Research: Cardiovascular System. Alan R. Liss, Inc. New York.
51. DUBOIS, C., E. BISSONNETTE & M. ROLA-PLESZCZYNSKI. 1989. J. Immunol. **143:** 964-970.
52. POUBELLE, P. E., D. GINGRAS, C. DEMERS, C. DUBOIS, D. HARBOUR, J. GRASSI & M. ROLA-PLESZCZYNSKI. 1991. Immunology **72:** 181-187.
53. PIGNOL, P., H. SYLVIE, J.-M. MENCIA-HUERTA & M. ROLA-PLESZCZYNSKI. 1987. Prostaglandins **33:** 931-939.
54. VALONE, F. H., R. PHILIP & R. J. DEBS. 1988. Immunology **64:** 715-718.
55. VALONE, F. H. & L. B. EPSTEIN. 1988. J. Immunol. **141:** 3945-3950.
56. WAHL, L. M., M. L. CORCORAN, S. W. PYLE, L. O. ARTHUR, A. HAREL-BELLAN & W. L. FARRAR. 1989. Proc. Natl. Acad. Sci. USA **86:** 621-625.
57. MERRILL, J. E., Y. KOYANAGI & I. S. Y. CHEN. 1989. J. Virol. **63:** 4404-4408.
58. MERRILL, J. E. & I. S. Y. CHEN. 1991. FASEB J. **5:** 2391-2397.
59. SIMMONS, M. L. & S. MURPHY. 1993. Eur. J. Neurosci. **5:** 825-831.
60. HEWETT, S. J. & D. W. CHOI. 1993. Soc. Neurosci. Abstr. **19:** 25.
61. PIANI, D., K. FREI, K. Q. DO, M. CUENOD & A. FONTANA. 1991. Neurosci. Lett. **133:**159-162.
62. PIANI, D., M. SPRANGER, K. FREI, A. SCHAFFNER & A. FONTANA. 1992. Eur. J. Immunol. **22:** 2429-2436.
63. GMÜNDER, H., H. P. ECK, B. BENNINGHOFF, S. ROTH & W. DRÖGE. 1990. Cell. Immunol. **129:** 32-46.
64. OLNEY, J. W., C. ZORUMSKI, M. T. PRICE & J. LABRYUERE. 1990. Science **248:** 596-599.
65. YEH, M. W., H. L. M. NOTTET, H. E. GENDELMAN & S. A. LIPTON. 1994. Soc. Neurosci. Abstr. **20:** 451.
66. BITO, H., M. NAKAMURA, Z. HONDA, T. ISUMI, T. IWATSUBO, T. SEYAMA, A. SGURA, Y. KIDO & T. SHIMIZU. 1992. Neuron **9:** 285-294.
67. MARCHESELLI, V. L. & N. G. BAZAN. 1993. Invest. Ophthalmol. Vis. Sci. **34:** 1048.
68. CLARK, G. D., L. T. HAPPEL, C. F. ZORUMSKI & N. G. BAZAN. 1992. Neuron **9:** 1211-1216.
69. SOLIVEN, B. & J. ALBERT. 1992. J. Neurosci. **12:** 2665-2671.
70. GELBARD, H. A., H. S. L. M. NOTTET, S. SWINDELLS, M. JETT, K. A. DZENKO, P. GENIS, R. WHITE, L. WANG, Y.-B. CHOI, D. ZHANG, S. A. LIPTON, W. W. TOURTELLOTTE, L. G. EPSTEIN & H. E. GENDELMAN. 1994. J. Virol. **68:** 4628-4635.
71. ZHANG, D., Y.-B. CHOI, J. OFFERMANN, H. E. GENDELMAN & S. A. LIPTON. 1993. Soc. Neurosci. Abstr. **19:** 1502.
72. WIERASZKO, A., G. LI, E. KORNECKI, M. V. HOGAN & Y. H. EHRLICH. 1993. Neuron **10:** 553-557.

73. Volterra, A., D. Trotti, P. Cassutti, C. Tromba, A. Salvaggio, R. C. Melcangi & G. Racagni. 1992. J. Neurochem. **59:** 600-606.
74. Miller, B., M. Sarantis, S. F. Traynelis & D. Attwell. 1992. Nature **355:** 722-725.
75. Lipton, S. A. 1992. Trends Neurosci. **15:** 75-79.
76. Tyor, W. R., J. D. Glass, J. W. Griffin, S. Becker, J. C. McArthur, L. Bezman & D. E. Griffin. 1992. Ann. Neurol. **31:** 349-360.
77. Heyes, M. P., K. Saito & S. P. Markey. 1992. Biochem. J. **283:** 633-635.
78. Speciale, C., E. Okuno & R. Schwarz. 1987. Brain Res. **436:** 18-24.
79. Kohler, C., L. G. Eriksson, E. Okuno & R. Schwarz. 1988. Neuroscience **27:** 49-76.
80. Brenneman, D. E., T. Nicol, D. Warren & L. M. Bowers. 1990. J. Neurosci. Res. **25:** 386-394.
81. Giulian, D., K. Vaca & M. Corpuz. 1993. J. Neurosci. **13:** 29-37.
82. Gourdou, I., K. Mabrouk, G. Harkiss, P. Marchot, N. Watt, F. Hery & R. Vigne. 1990. Cr. hebd. Séanc. Acad. Sci. Paris. **311** (Series III): 149-155.
83. Sabatier, J.-M., E. Vives, K. Mabrouk, A. Benjouad, H. Rochat, A. Duval, B. Jue & E. Bahraoui. 1991. J. Virol. **65:** 961-967.
84. Hayman, M., G. Arbuthnott, G. Harkiss, H. Brace, P. Filippi, V. Philippon, D. Thompson, R. Vigne & A. Wright. 1993. Neuroscience **53:** 1-6.
85. Werner, T., S. Ferroni, T. Saermark, R. Brack-Werner, R. B. Banati, R. Mayer, L. Steinaa, G. W. Kreutzberg & V. Erfle. 1991. AIDS. **5:** 1301-1308.
86. Tardieu, M., C. Héry, S. Peudenier, O. Boespflug & L. Montagnier. 1992. Ann. Neurol. **32:** 11-17.
87. Lipton, S. A. 1993. Ann. Neurol. **33:** 227-228.
88. Lipton, S. A. 1992. Neurology **42:** 1403-1405.
89. Dawson, V. L., T. M. Dawson, G. R. Uhl & S. H. Snyder. 1992. Soc. Neurosci. Abstr. **18:** 756.
90. Lipton, S. A. 1993. Trends Neurosci. **16:** 527-532.
91. Karschin, A., E. Aizenman & S. A. Lipton. 1988. J. Neurosci. **8:** 2895-2906.
92. Levy, D. I. & S. A. Lipton. 1990. Neurology **40:** 852-855.
93. Chen, H.-S. V., J. W. Pellegrini, S. K. Aggarwal, S. Z. Lei, S. Warach, F. E. Jensen & S. A. Lipton. 1992. J. Neurosci. **12:** 4427-4436.
94. Koek, W., J. H. Woods & G. D. Winger. 1988. J. Pharmacol. Exp. Ther. **245:** 969-974.
95. Olney, J. W., J. Labruyere & M. T. Price. 1989. Science **244:** 1360-1362.
96. MacDonald, J. F., Z. Miljkovic & P. Pennefather. 1987. J. Neurophysiol. **58:** 251-266.
97. Davies, S. N., S. T. Alford, E. J. Coan, R. A. Lester & G. L. Collingridge. 1988. Neurosci. Lett. **92:** 213-217.
98. O'Shaughnessy, C. T. & D. Lodge. 1988. Eur. J. Pharmacol. **153:** 201-209.
99. Choi, D. W. 1987. Brain Res. **403:** 333-336.
100. Choi, D. W., S. Peters & V. Viseskul. 1987. J. Pharmacol. Exp. Ther. **242:** 713-720.
101. Rogawski, M. A. & R. J. Porter. 1990. Pharmacol. Rev. **42:** 223-286.
102. Rogawski, M. A. 1993. Trends Pharm. Sci. **14:** 325-331.
103. Bormann, J. 1989. Eur. J. Pharmacol. **166:** 591-592.
104. Goldman, R. S. & S. M. Finkbeiner. 1988. N. Engl. J. Med. **319:** 1224-1225.
105. Wolf, G., G. Keilhoff, S. Fischer & P. Hass. 1990. Neurosci. Lett. **117:** 207-211.
106. Wolf, G., S. Fischer, P. Hass, K. Abicht & G. Keilhoff. 1991. Neuroscience **43:** 31-43.
107. Zeevalk, G. D. & W. J. Nicklas. 1992. J. Neurochem. **59:** 1211-1220.
108. Aizenman, E., S. A. Lipton & R. H. Loring. 1989. Neuron **2:** 1257-1263.
109. Tang, L. H. & E. Aizenman. 1993. J. Physiol. **465:** 303-323.
110. Sucher, N. J., L. A. Wong & S. A. Lipton. 1990. NeuroReport **1:** 29-32.

111. Reynolds, I. J., E. A. Rush & E. Aizenman. 1990. Br. J. Pharmacol. **101:** 178-182.
112. Levy, D. I., N. J. Sucher & S. A. Lipton. 1990. Neurosci. Lett. **110:** 291-296.
113. Aizenman, E. & K. A. Hartnett. 1992. Brain Res. **585:** 28-34.
114. Aizenman, E., K. A. Hartnett & I. J. Reynolds. 1990. Neuron **5:** 841-846.
115. Lei, S. Z., Z. H. Pan, S. K. Aggarwal, H. S. Chen, J. Hartman, N. J. Sucher & S. A. Lipton. 1992. Neuron **8:** 1087-1099.
116. Stamler, J. S., D. J. Singel & J. Loscalzo. 1992. Science **258:** 1898-1902.
117. Beckman, J. S., T. W. Beckman, J. Chen, P. A. Marshall & B. A. Freeman. 1990. Proc. Natl. Acad. Sci. USA **87:** 1620-1624.
118. Radi, R., J. S. Beckman, K. M. Bush & B. A. Freeman. 1991. J. Biol. Chem. **266:** 4244-4250.
119. Dawson, T. M., V. L. Dawson & S. H. Snyder. 1992. Ann. Neurol. **32:** 297-311.
120. Manchester, K. S., F. E. Jensen, S. Warach & S. A. Lipton. 1993. Neurology **43:** A365.
121. Wesemann, W., G. Sturn & E. W. Fünfgeld. 1980. J. Neural Transm. (Suppl.) **16:** 143-148.
122. Seif el Nasr, M., B. Perucher, C. Rossberg, H.-D. Mennel & J. Krieglstein. 1990. Eur. J. Pharmacol. **185:** 19-24.
123. Erdö, S. L. & M. Schäfer. 1991. Eur. J. Pharmacol. **198:** 215-217.
124. Keilhoff, G. & G. Wolf. 1992. Eur. J. Pharmacol. **219:** 451-454.
125. Osborne, N. N. & G. Quack. 1992. Neurochem. Int. **21:** 329-336.
126. Miller, B., M. Sarantis, S. F. Traynelis & D. Attwell. 1992. Nature **355:** 722-725.

Use of Cultured Fibroblasts in Elucidating the Pathophysiology and Diagnosis of Alzheimer's Disease[a]

HSUEH-MEEI HUANG,[b] RALPH MARTINS,[c]
SAM GANDY,[c] RENE ETCHEBERRIGARAY,[d]
ETSURO ITO,[d] DANIEL L. ALKON,[d] JOHN BLASS,[b]
AND GARY GIBSON[b,e]

[b]*Cornell University Medical College*
Department of Neurology and Neuroscience
Burke Medical Research Institute
785 Mamaroneck Avenue
White Plains, New York 10605

[c]*Cornell University Medical College*
Department of Neurology and Neuroscience
New York, New York 20021

[d]*Laboratory of Adaptive Systems*
National Institute of Neurological Disorders and Stroke
National Institutes of Health Building
Bethesda, Maryland 20892

Altered signal transduction systems in Alzheimer's disease (AD) are likely to be of pathophysiological importance, because these processes are critical to cellular and brain functions. In addition to their classic role of converting "environmental" stimuli into intracellular signals, these transduction systems modulate multiple aspects of cellular function, including posttranslational modification and processing of molecules such as amyloid precursor protein[1,2] and tau.[3,4] Alterations in those two proteins in particular are generally considered to play a critical role in the pathophysiology of aging and AD. Impaired oxidative metabolism is also known to accompany AD, and oxidative deficits can alter signal transduction systems.[5,6] One important difficulty in defining the relation of changes in signal transduction or oxidation to the causation of AD lies in determining if the abnormality is of pathophysiological importance or if it is an epiphenomenon occurring secondary to neurodegeneration. Thus, whether primary defects in oxidative metabolism or in a particular signal transduction system might cause the characteristic structural changes remains unclear. An understanding

[a]This work was supported by grants NIMH48325 (G.G.), AG08702 (G.G.), AG03853 (J.B.), AG11508 (S.G.), and the Altschul Foundation.

[e]To whom correspondence should be addressed.

of all of these alterations at cellular and molecular levels is required to answer these questions.

MEASUREMENTS OF AUTOPSIED BRAIN IN ALZHEIMER'S DISEASE AND AGING

Measurements of autopsied brain are a common, important strategy for studies of AD and aging, but this approach has serious limitations, especially for measures of dynamic processes such as signal transduction and oxidative processes. Two major sources of artifacts are postmortem delays and the agonal state. Although the former can partially be assessed with animal studies, the latter is almost impossible to evaluate. Even if these were not considerations, determining what comes first (ie, the classic chicken/egg dilemma) is impossible in autopsy brain. Also, any changes in pathologically affected areas may be secondary to neurodegenerative changes (eg, different numbers of dead neuronal cells or varying degrees of reactive gliosis). On the other hand, if abnormalities occur in affected and also pathoanatomically unaffected areas, the relationship of the changes to the disease (ie, the structural pathology) may become questionable, because structural pathology is considered critical. Furthermore, measurements such as receptor-mediated induction of signal transduction systems in autopsy brain may only be present at a fraction of their *in vivo* activity, if they are present at all. Finally, although a large number of AD brains are available, the availability of control brains from adequately characterized subjects is much more limited. Thus, there are many reasons to pursue alternatives to the use of autopsied brain tissue for elucidating the pathophysiology of aging and age-related disorders such as Alzheimer's disease.

One approach is to use peripheral tissues that are readily accessible. Red blood cells,[7] platelets,[8] and lymphocytes[9–12] or direct measurements on peripheral tissues such as skin[13] have been used productively. Many age-related changes in calcium homeostasis,[10–12,14] membrane fluidity,[8] and choline regulation[15,16] have been documented in animals and man with this approach. Changes in calcium homeostasis were also observed in lymphocytes from patients with AD.[9–12] A major difficulty with these studies is that the results may be influenced by the patients' diets and drugs. Furthermore, the amount of materials may be small and restricted to patients who are available at a particular research center, so that detailed studies of kindreds is difficult.

STUDY OF CULTURED CELLS FROM PATIENTS WITH ALZHEIMER'S DISEASE

The study of cultured cells from AD patients provides one possible solution to many of these problems, assuming that appropriate cell types (i.e., accessible cells that faithfully and reproducibly reflect some features of brain abnormalities) can be identified. Cultured cells offer many advantages for the study of changes in signal transduction systems with aging and AD. Although studies of primary neuronal cultures[17] or PC12 cells[2] have proven valuable in evaluating the role of calcium and various kinases in neuronal injury and cell death, these approaches do not test cells

from patients with the disease of interest and matched, disease-free controls. The advantage of cell lines over peripheral patient tissues not in culture is that the culture allows relatively complete control over multiple variables and avoids any effects of individuals' diets or drugs. Although fibroblasts[18] and lymphoblasts[19] have been utilized for studies of aging and AD, cultured fibroblasts are attractive for several reasons. For studies of signal transduction, fibroblasts have an advantage over lymphoblasts in that transformation of lymphocytes to lymphoblasts to allow serial culture may alter the cells' properties. For instance, calcium channels[20] are altered by transformation, and calcium has been implicated in the pathophysiology of AD.[21] Fibroblasts are relatively easy to obtain and to maintain under rigidly controlled conditions in tissue culture. Many cell lines are now available, so that results can readily be compared in laboratories around the world. This large number of available lines makes it possible to study particular kindreds. For example, cells from families that have a chromosome 14 abnormality linked to familial AD are available.[22] Another point favoring the study of fibroblasts is that interactions of numerous signal transduction systems are well documented in these cells (see below). Furthermore, fibroblasts have been used extensively to study basic properties of aging. For example, fibroblasts divide more and more slowly as they age,[23] and this "phase out" may be related to alterations in calcium binding proteins with aging.[24,25] A final advantage is that cells from Alzheimer and age-matched control subjects have shown similar growth patterns, so that Alzheimer/control differences may be easily assessed and are potentially distinguishable from any phenomenon of accelerated cellular aging in culture.[26–28]

Considerable precedent exists for the use of cultured fibroblasts in the study of neurological disorders. In Lesch-Nyhan disease, erythrocytes and cultured skin fibroblasts were used to determine that the primary deficiency is in an enzyme of purine metabolism.[29] In lysosomal storage diseases, such as Tay-Sachs disease, hexosaminidase A deficiency was characterized in cultured fibroblasts.[30] In disorders of energy metabolism, such as pyruvate dehydrogenase deficiencies, fibroblasts were used to elucidate the underlying biochemistry.[31] In lipidoses such as Refsum's disease, cultured fibroblasts were used to localize the defect to the α oxidation of phytanic acid.[32] Stanbury *et al.*[33] present numerous other neurological diseases in which fibroblasts have proven critical in the study of disease pathophysiology. Thus, considerable precedent exists for using cultured cells of patients with diseases which, like AD, may be systemic[34] and, at least in some forms, genetic.[35] The many abnormalities that have been observed in fibroblasts from AD patients suggest that the cultured fibroblast system can also be a productive approach to this disease.[36]

The advantages for utilization of fibroblasts outweigh the disadvantages. Many neuronal proteins are not normally present in these cells, so that direct examination of how abnormal signal transduction may alter processing of those neuron-specific molecules is not possible. However, some potentially relevant molecules are present in fibroblasts including MAP-4 (the primary MAP in fibroblasts) which has many homologies to tau. Amyloid precursor protein, a key component of plaques, exists and is processed in fibroblasts just as in other cells (see below). One of the most problematic disadvantages of fibroblast studies is that some observations have proven difficult to replicate between laboratories, and robust differences between AD and non-AD fibroblasts have disappeared upon extended study even within the same laboratory.[37–39] The lack of replicability may be related to unknown differences in

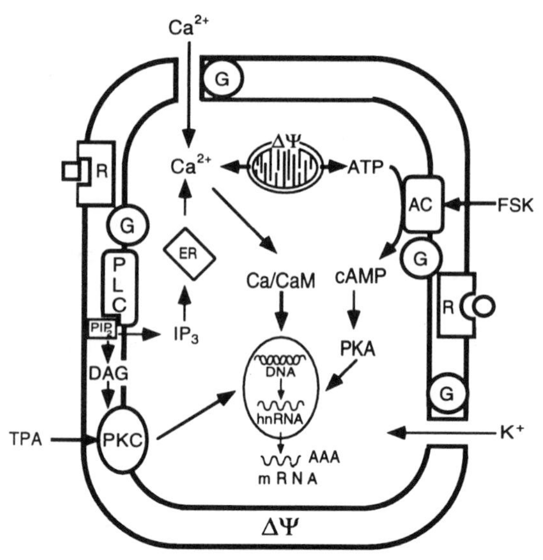

FIGURE 1. Overview of signal transduction systems altered in Alzheimer's disease. Each system shown is reported to be altered in Alzheimer's disease fibroblasts.

culture conditions or to selection of varied subpopulations over several generations. When culture conditions are carefully standardized, this problem diminishes. Hopefully, as experiments progress toward elucidating the primary molecular mechanism, subtle changes in culture conditions will no longer pose a problem.

ALTERED SIGNAL TRANSDUCTION SYSTEMS IN FIBROBLASTS FROM SUBJECTS WITH ALZHEIMER'S DISEASE

An overview of the signal transduction systems that are altered in AD and their interactions are summarized in FIGURE 1. Calcium homeostasis (calcium uptake, $[Ca^{2+}]_i$, calcium content, and subcellular pools), receptor activation of the phosphatidylinositide (PI) cascade and cyclic AMP (cAMP) system, and oxidative metabolism are altered in AD fibroblasts. The underlying basis of all of these dynamic changes and their possible interrelationship, if any, remains unknown. Elucidating these interactions and how they are altered with AD should be possible in living cells in culture and specifically in cultured fibroblasts. The changes in these systems and how they relate to each other and to abnormalities in brain are summarized.

ABNORMAL CALCIUM HOMEOSTASIS IN FIBROBLASTS FROM SUBJECTS WITH ALZHEIMER'S DISEASE

The hypothesis that altered calcium regulation accompanies AD has been tested directly in lymphocytes,[9–12] lymphoblasts,[19] and fibroblasts.[18,38–44] Calcium uptake declines with AD in fibroblasts[18] and lymphocytes.[9] Furthermore, fibroblasts from AD patients have increased calcium content.[44] Subsequent studies[38,39] demonstrated dramatic reductions in $[Ca^{2+}]_i$ in AD fibroblasts that had been serum deprived for 24 hours before the experiment, but these results have proven difficult to replicate despite considerable effort by several groups.[37,41,43] Other studies have focused on the $[Ca^{2+}]_i$ changes after application to cells of various agonists. For example, bradykinin stimulates $[Ca^{2+}]_i$ in a time- and dose-dependent manner. It produces an immediate $[Ca^{2+}]_i$ peak, which then declines to a new equilibrium value that is above the basal concentration. If calcium is omitted from the medium at the time bradykinin is added, the peak is unaffected, but the subsequent sustained elevation of $[Ca^{2+}]_i$ is abolished. Although initial examination revealed large AD-associated differences in bradykinin and other agonist-stimulated increases in $[Ca^{2+}]_i$,[38,39] these differences have proven difficult to replicate,[37,41] even though all experiments were performed with fura-2 as the calcium indicator. A recent report in which aequorin, a previous generation calcium dye with a lower affinity for calcium, was used indicated a diminished bradykinin and serum response in Alzheimer fibroblasts compared to controls.[42,43] One possible explanation for these differences is that various groups may have loaded internal pools of calcium to varying extents, because these can be altered by loading conditions. These discrepant findings suggest that the abnormality, if any, in $[Ca^{2+}]_i$ in AD fibroblasts has not been robustly demonstrated under the conditions used so far.[41] This inconsistency suggests that a change in the regulation of $[Ca^{2+}]_i$ is probably not the primary defect in these cells.

In assessing these observations, it is important to realize that measures of $[Ca^{2+}]_i$ do not necessarily reflect the internal pools of calcium that may be important in cell function and the processing of proteins.[45,46] For example, isoproterenol induces a large calcium exchange, but has no effect on overall $[Ca^{2+}]_i$.[47] Altered calcium regulation in the endoplasmic reticulum or Golgi apparatus may alter the processing of critical molecules[48–50] perhaps including tau and amyloid precursor protein. Mitochondrial calcium regulates the activity of key oxidative enzymes[51] and too much calcium can irreversibly inactivate them.[52] Characterization of the various internal calcium pools in AD and non-AD cells will help us to understand better whether these pools are involved in the pathophysiology of AD and to clarify the controversial reports on $[Ca^{2+}]_i$ in AD fibroblasts.

In our own laboratory, measures of $[Ca^{2+}]_i$ in combination with various treatments in calcium-free media have revealed changes in internal calcium stores in AD cells in which $[Ca^{2+}]_i$ or its response to bradykinin was comparable to that in non-AD controls. Stimulation with bradykinin abolished the response of $[Ca^{2+}]_i$ to the subsequent addition of bradykinin, but the subsequent addition of A23187 (+48%) or thapsigargin (+33%) to bradykinin-treated cells still increased $[Ca^{2+}]_i$ (percentage above basal values). Prior addition of thapsigargin reduced the $[Ca^{2+}]_i$ response to A23187 or bradykinin. Although the changes are small compared to bradykinin-induced elevations in $[Ca^{2+}]_i$, they likely represent large changes in individual pools of calcium. The A23187-induced change in $[Ca^{2+}]_i$ after bradykinin was nearly 100% larger in AD fibroblasts.[40] Other studies suggest that Alzheimer cells are more sensitive than control cells to low concentrations of bradykinin and bombesin.[53] Although the precise cellular localization of these calcium pools has not been established, $[Ca^{2+}]_i$ can be used as an endpoint to assess various discrete internal pools of calcium. Some

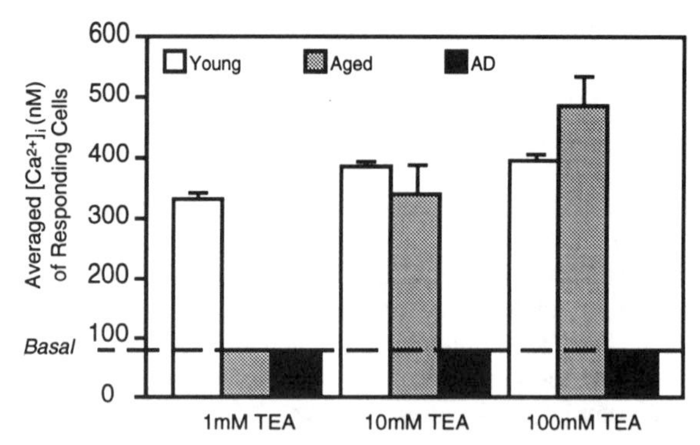

FIGURE 2. Alterations in K^+ channels with Alzheimer's disease as revealed by measurements of $[Ca^{2+}]_i$.[61]

of these appear to be promising in providing a consistent and robust alteration in AD.

The relation of these changes in calcium regulation in fibroblasts to those in brain is limited by the impossibility of making any dynamic measurements of calcium metabolism in autopsy brain. A variety of indirect evidence from brain (eg, increased calcium-dependent kinases[54]) and from neuronal tissue culture (eg, elevated calcium increases Alz-50 production[17]) supports an important role for calcium in the pathophysiology of AD. In aged animals, nerve terminals have reduced flux of calcium into the terminal and elevated $[Ca^{2+}]_i$,[55,57–59] whereas postsynaptic calcium cycling increases.[60] Such comparisons are more difficult to make with Alzheimer's disease, because the processes do not persist in autopsy tissue and the appropriate controls for studying biopsied brain are typically lacking. Thus, any inference of a role for calcium in AD brain is necessarily indirect and, as yet, tentative.

ALTERED K^+ CHANNELS IN FIBROBLASTS FROM SUBJECTS WITH ALZHEIMER'S DISEASE

Measurements of $[Ca^{2+}]_i$ in the presence or absence of the K^+ channel blocker tetraethylammonium (TEA) and patch clamp studies have been used to examine K^+ channels.[61] Patch clamp studies reveal two classes of K^+ channels (slope conductances of 113 and 166 pS) in cultured fibroblasts. The 166 pS channel is present in all groups and is observed with a slightly higher frequency in the AD group. A 113 pS K^+ channel is present in 56% of the cells from young donors (4 cell lines) and in 86% of the cells from aged donors (4 cell lines), but it is not present in cells from AD patients (4 cell lines). Patch clamp studies showed that those channels were equivalent to those that were blocked by 100 mM TEA. The addition of TEA increased $[Ca^{2+}]_i$ in 5 of 6 cell lines from young donors and in 10 of 10 cell lines from aged donors, but not in any of 13 cell lines from AD subjects (FIG. 2). The 113 pS K^+ channels are present in fibroblasts from patients with a variety of other disorders including schizophrenia, Parkinson's disease, Huntington's disease, and Wernicke-

Korsakoff syndrome. Thus, patch clamp studies and $[Ca^{2+}]_i$ in the presence of a K^+ channel blocker both indicate K^+ channel dysfunction that is apparently specific for AD.

Human brain K^+ channels cannot be assessed by patch clamp or $[Ca^{2+}]_i$. Measurements in human brain are largely limited to binding studies or measurements of mRNA or protein. Until it is determined at the molecular level which K^+ channels are selectively absent in AD fibroblasts, evaluating similar changes in brain will be difficult. Molecular biologists have identified numerous channels in brain, but it has not yet been possible to implicate one of these as being responsible for the deficit in AD fibroblasts. Along this line, however, one group reported reductions in apamin binding (a measure of calcium-dependent K^+ channels) in the subiculum and CA1 of the hippocampus.[62] On the other hand, other ATP-sensitive K^+ channels are associated with elements in the hippocampus that are preserved in AD.[63]

ABNORMAL PHOSPHATIDYLINOSITOL CASCADE IN FIBROBLASTS FROM SUBJECTS WITH ALZHEIMER'S DISEASE

Measurements of key enzymes and metabolites have been used to assess the phosphatidylinositol (PI) cascade in cultured fibroblasts. As shown independently by four groups, protein kinase C activity (PKC) decreases in AD fibroblast lines.[64–68] To date, this is the single, most consistently reproducible change that has been documented in these cells. Dynamic aspects of the PI cascade are also reported to be altered with AD. Our recent results show that the bradykinin-stimulated formation of IP_3 is enhanced by 78% in AD fibroblasts compared to that in age-matched controls.[69] The agonist specificity of these changes, their relation to the activity of phospholipase C and to the coupling of G-proteins, and the receptor number and affinity are under study. The methods used to examine these changes are important. For instance, in our early studies in which cells were in serum-free media for 4 days before measurement, the highly replicable Alzheimer-control differences (which are now robustly documented) were not observed.[41]

Changes in IP_3 and PKC may be related to each other. Our studies demonstrate that activation of PKC impairs bradykinin (BK)-induced IP_3 formation in fibroblasts (Huang and Gibson, unpublished results). In other cell types, downregulation of PKC accelerates IP_3 formation.[70] Thus, the reduced PKC activity in Alzheimer fibroblasts may underlie the accelerated BK response that we observe in fibroblasts from Alzheimer patients.

Evaluation of abnormalities of the PI cascade in the brain of AD patients further supports a role for this pathway in the pathophysiology of AD. Although no direct comparison with the cellular studies just described has been made in brain, a preliminary report suggests abnormal coupling of G-proteins to IP_3 formation in brain.[71,72] Protein kinase C, particularly the β isoform, is diminished in AD brains.[73,74] Overall brain PKC activity decreases, while activity in areas of degeneration actually increases. Protein kinase C βII immunostaining is increased in the neuronal cell body and neuropil of AD samples, particularly in association with diffuse plaques.[73] Although phospholipase C activity is not altered in AD brain,[75] phospholipase C δ accumulates abnormally in neurofibrillary tangles in AD brain.[76] Phosphatidylinositol[77] and IP_3 binding[78] are decreased in the brains of AD patients. In addition, an altered PI cascade can alter the processing or expression of amyloid precursor protein[1,2] or tau proteins.[17]

REDUCED β-ADRENERGIC STIMULATED FORMATION OF cAMP IN FIBROBLASTS FROM SUBJECTS WITH ALZHEIMER'S DISEASE

The isoproterenol-stimulated formation of cAMP is reduced approximately 80% in fibroblasts from AD subjects over a wide range of concentrations and numerous times points.[79] The decline in agonist-stimulated cAMP production paralleled the β-adrenergic potency of the agonists. The observation that changes occur over a wide range of β-agonist concentrations suggests that the deficit is probably not due to altered receptor affinity or number, but this has not been tested directly. Furthermore, the induction of cAMP by other receptor agonists (eg, a wide range of prostaglandin E_1 concentrations) was similar in non-AD and AD fibroblasts. The robustness of the result is also illustrated by the observation that the change occurs whether the measurement is made in phosphate-buffered saline or in complex media. Alterations in adenylate cyclase do not underlie the reduction, because stimulation of adenylate cyclase by several forskolin concentrations elevated cAMP production equally in AD and non-AD fibroblasts. The discrepancy with previous studies, which showed an increase[80] or no change[81] in agonist-stimulated cAMP formation by AD fibroblasts, may be due to differences in tissue culture conditions or other technical factors as well as the limited nature of the earlier studies. The basis of these discrepancies will likely become clearer as the underlying molecular mechanism is elucidated.

Alterations in β-adrenergic receptors also occur in brain and other tissues from patients with AD. *In vivo* experiments suggest that the abnormality in β-adrenergic stimulation is not just a tissue culture epiphenomenon. Skin vessel reactivity to isoproterenol is decreased significantly in AD subjects, whereas the response to the cholinergic agonist methacholine is entirely normal.[82] Region and receptor type (ie, β_1 and β_2) specific changes occur in the brain in AD.[83–89] For example, β_2 receptors increase in the hippocampus, frontal cortex,[83] and microvessels.[84,90] In olfactory neuroblasts, cAMP agonists are reported to diminish the excessive production of APP COOH-terminal fragments from AD subjects.[91] In AD brain, some reports suggest that basal and stimulated cAMP levels may diminish because of reduced adenylate cyclase,[92,93] whereas others suggest the reduction is related to altered G-protein coupling (see below). The relationship of altered β-adrenergic systems to the pathophysiology of AD has not been established. In the CNS, adrenergic receptors are important for control of cerebral blood flow and a wide variety of other functions including memory.[94]

ALTERED COUPLING OF G-PROTEINS TO RECEPTORS AND DEFICITS IN CYCLIC AMP PRODUCTION IN FIBROBLASTS FROM SUBJECTS WITH ALZHEIMER'S DISEASE

The coupling of G-proteins to receptors is often examined by the use of toxins. As shown in FIGURE 3, cholera toxin activates G_s to stimulate the effector enzyme

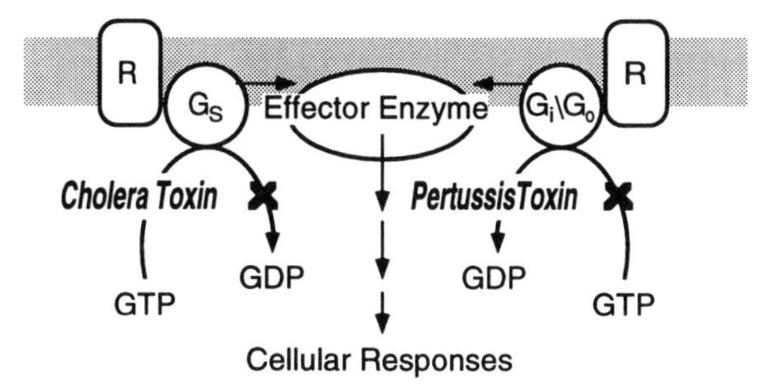

FIGURE 3. Use of toxins to reveal interactions of G-proteins with effector enzymes.

(adenylate cyclase), whereas pertussis toxin blocks G_i activation. Cholera toxin stimulates cAMP formation similarly in AD and non-AD fibroblasts. However, cholera toxin diminished the percent reduction in isoproterenol-stimulated formation of cAMP in AD cells from 80% to 40%. Pertussis toxin, on the other hand, did not alter the reduction of isoproterenol-induced formation of cAMP.[79] The results suggest several possible sites of an abnormality in AD fibroblasts: the receptor number and affinity, receptor biosynthesis, the coupling of G proteins to the receptor, or interactions with other signal transduction systems.[95] An example of the latter is that elevated production of $G_{i\alpha}$ reduces the β-adrenergic response.[96] Mutants of fibroblasts that overexpress $G_{i\alpha}$ have a reduced cAMP response.[96] On the other hand, expression of $G_{i\alpha}$ will activate phospholipase C[97] (ie, a result that we observed in AD fibroblasts). Alterations in G-proteins could conceivably explain the abnormalities in K^+ channels that occur in AD fibroblasts,[61] because G-proteins can directly gate ion channels.[98]

Altered coupling of G-proteins to adenylate cyclase also occurs in the brain of AD subjects. Research on the role of G-proteins in AD brain has up to now been limited. The results just cited suggest that this may be a fruitful area of research. *In situ* hybridization revealed marked regional increases in $G_{s\alpha}$ in hippocampal CA1, CA3, and CA4 regions and in the temporal cortex.[99] However, Western blotting did not reveal any differences in G_{i1}, G_{i2}, G_{sH}, G_{sL}, or G_o between AD and control brain.[100] Cowburn *et al.*[101] showed that G_s stimulation of adenylate cyclase was reduced in AD brain, whereas G_i was unaffected, a result that closely parallels the changes that we observed in fibroblasts. Warpman *et al.*[102] reported deficits in the coupling of muscarinic receptors to the G-proteins in AD brain. Neither somatostatin stimulation of adenylate cyclase nor somatostatin receptor G-protein coupling is altered in AD.[103]

PERSISTENT OXIDATIVE DEFICITS IN AD TISSUES IN CULTURE SUGGEST A PRIMARY DEFICIT

The occurrence of oxidative deficits in AD is unequivocal,[104–106] but their pathophysiological importance continues to be in question, as it has been difficult to establish whether diminished oxidation is a consequence or a cause of neurodegeneration. Numerous studies show that oxidative deficits persist in tissue culture. An oxidative deficit in peripheral tissues such as lymphocytes[107] and fibroblasts[44,108] sug-

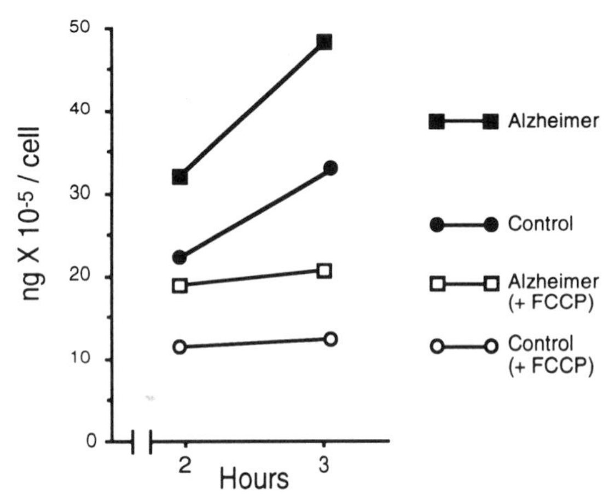

FIGURE 4. Estimation of mitochondrial function with rhodamine-123 in Alzheimer's disease and control fibroblasts.

gests that the decline does not just reflect deteriorating brain. Two reports have shown that glutamine oxidation is reduced in AD fibroblasts and that other abnormalities in carbohydrate oxidation persist in culture.[44,108] Furthermore, deficits in a key mitochondrial enzyme (α-ketoglutarate dehydrogenase) persist in tissue culture.[109]

The mitochondrial membrane potential is sensitive to a variety of metabolic insults[110,111] and is as an important integrator of cellular bioenergetics with cell function. Changes in brain function related to abnormal oxidative processes may be mediated by alterations in this potential. In brain slices, decreases in the mitochondrial membrane potential due to a variety of metabolic insults parallelled a decline in the synthesis of acetylcholine, protein, lipid, and RNA.[112] In part, the mitochondrial membrane potential regulates cytosolic events by modulating the buffering of $[Ca^{2+}]_i$.[5,113,114] In fibroblasts, the mitochondrial membrane potential determines the adenine nucleotide transport system[115] and is necessary for translocation of proteins into the mitochondria.[116] Abnormal oxidation can alter signal transduction, including acceleration of the PI cascade,[6,117] as well as expression of immediate early genes.[118]

Indirect measurements of mitochondrial membrane potentials suggest that they are altered with AD. Cellular membrane potentials were estimated by equilibration with rhodamine-123[119–121] over several hours, which represents a combination of plasma and mitochondrial membrane potentials. If equilibration was performed in high K^+ buffer to collapse the plasma membrane potential, the accumulation primarily reflects mitochondrial potential. The amount of cellular rhodamine-123 after a 3-hour loading period is much greater in AD than in non-AD lines (FIG. 4). The results suggest an exaggerated mitochondrial membrane potential with AD. Furthermore, the amount of dye released by FCCP, which collapses the mitochondrial membrane potential, is about 25% greater in the AD lines than in the controls. Treatment of cells with FCCP does not abolish the AD/non-AD difference, which suggests that a non-mitochondrial component also contributes to the AD/non-AD difference in rhodamine-123 accumulation. Although an exaggerated membrane potential was unexpected, this was observed in transformed cell lines[121] and with the drug oligomycin.[111]

Extensive studies show reduced oxidation in AD. Key oxidative enzymes are reduced in the brain in AD.[36,122] In addition, selective changes occur in various components of the electron transport chain[123] and in cytochrome oxidase.[124] At least some of these deficits persist in tissue culture, which implies that they may not be secondary to agonal or postmortem artifacts or to neurodegeneration.

USE OF CULTURED FIBROBLASTS IN UNDERSTANDING SIGNAL TRANSDUCTION SYSTEM INTERACTIONS IN ALZHEIMER'S DISEASE

The simplicity and isolation of cultured fibroblasts make them potentially amenable to our understanding of how AD alters the interactions of these various signal transduction systems. Now that multiple changes have been carefully established in both brain and fibroblasts, the challenge is to determine the common links and relationships. A criticism that was raised about the use of fibroblasts is that so many changes are occurring that determining the common link may be impossible. That point of view clearly applies to AD brain as well. Furthermore, in brain one has numerous cell types, alterations in blood flow, neurodegeneration, gliosis, dietary and drug effects, and major artifacts due to agonal and postmortem changes. All of these further complicate interpretation obtained with AD brain. In cells, the underlying basis of the various molecular changes can be determined by manipulations that are impossible to study in living brain or autopsy tissue. For example, activation of PKC alters the responsiveness of receptor-mediated processes, and this occurs in a receptor-specific manner. In tissue culture, PKC can be manipulated to determine the effects of activation or long-term downregulation of PKC. Mitochondria regulate calcium homeostasis in a variety of cells and nerve terminals, and calcium can activate PKC. Impairing oxidative metabolism with hypoxia alters many aspects of these signaling systems. For example, hypoxia can alter IP_3 and calcium levels and activate PKC. Thus, one interpretation of these direct experimental findings is that altered oxidative metabolism leads to signaling changes. Experimental systems have demonstrated that impaired oxidation can alter calcium regulation, the PI cascade, PKC, and the coupling of G-proteins. Multiple mitochondrial insults (eg, KGDH, complex I, complex IV, or free radical changes) could lead to these changes. In addition, impaired oxidation can impair the cholinergic system, which is an integral component of AD deficits, and as will be described, impaired oxidation can also lead to structural pathological changes.

SIGNAL TRANSDUCTION SYSTEMS CRITICAL TO ALTERED PROCESSING OF PROTEINS IN ALZHEIMER'S DISEASE

Two pathological hallmarks of AD brain are tangles and plaques. These are complex structures, but the primary proteins that are known to be abnormally pro-

TABLE 1. Altered Tangle Immunoreactivity with Alzheimer's Disease and a Mitochondrial Uncoupler[125]

Antibody	Alzheimer's Disease	Control	Control +CCCP
pAB-ICN	81	9	64
mAB 5-25	60	8	63
mAB Alz-50	68	0	34

cessed are tau and amyloid precursor protein, respectively. Both of these have been examined in fibroblasts from control and Alzheimer subjects.

Reactivity to tangle antibodies has been used in fibroblasts to ask questions about pathophysiology.[126] When grown in media that included multiple growth factors, fibroblasts from patients with Alzheimer's disease have much greater immunoreactivity to tangle antibodies than do fibroblasts from controls (TABLE 1). Furthermore, fibroblasts from controls grown under these conditions that are then treated with a mitochondrial uncoupler for days show enhanced immunoreactivity to tangle antibodies. These results suggest that oxidative changes in Alzheimer fibroblasts may underlie enhanced immunoreactivity.

Regulation of the processing of the amyloid precursor protein (APP) was also examined in these cells (FIG. 5). In standard pulse chase experiments, the 100-kD immature APP in cell lysate is converted to the 110-kD mature species. The latter

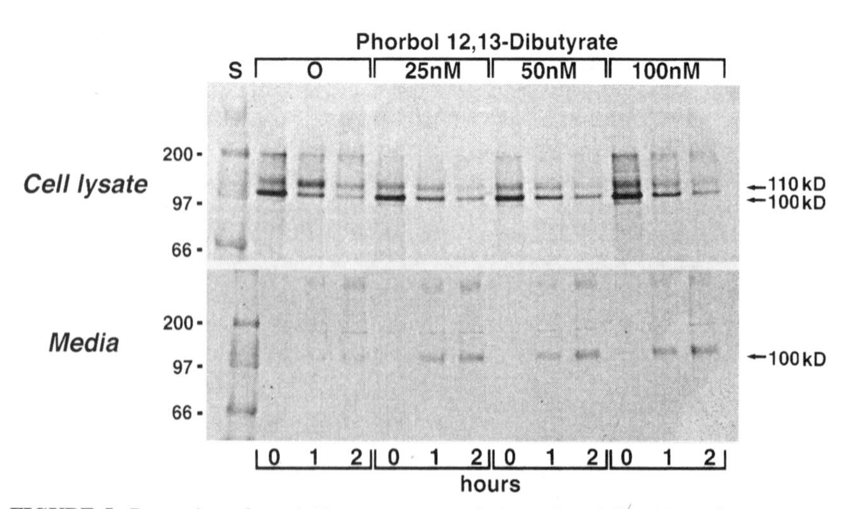

FIGURE 5. Processing of amyloid precursor protein by cultured fibroblasts from a patient with Alzheimer's disease. Cells were incubated with [^{35}S]methionine for 20 minutes. The radioactive medium was then removed and replaced with complete medium that contained the indicated concentration of phorbol ester (0 time).

migrates to the plasma membrane where it is clipped at its COOH-terminus, and the NH_2-terminal fragment is released as a 100-kD soluble protein. Treatment with phorbol ester stimulates this non-amyloidogenic processing, currently known as the α-secretase pathway, as evidenced by a decreased mature form in the cell lysate and an increased cleaved fragment in the media. The effects of different concentrations of phorbol ester were approximately equivalent. Furthermore, fold-stimulation of α-secretase processing of the protein was equivalent in all three Alzheimer and three control lines. Thus, in cells that have abnormalities in receptor-mediated signal transduction as well as altered oxidative metabolism, resting and phorbol ester-stimulated amyloid precursor protein processing via activation of the α-secretase pathway appears to be normal at the concentrations of drug used in these preliminary studies. Whether stimulation of pathways that are known to be abnormal in Alzheimer fibroblasts will lead to abnormal processing is unknown. Also under assessment is the effect of these signals on the production of soluble β-amyloid peptide by these cells.

USEFULNESS OF FIBROBLASTS IN DIAGNOSTIC TESTS

Many of the measurements that have been described show no overlap between control and AD patients. Thus, changes in K^+ channels, cAMP and IP_3, and the activity of the Krebs cycle enzyme α-ketoglutarate dehydrogenase can all potentially be used to identify AD patients. Only the K^+ channel has been tested for specificity as compared to other neurodegenerative disorders such as Parkinson's disease, Huntington's disease, Wernicke-Korsakoff syndrome, or schizophrenia. Culturing fibroblasts is generally more labor intensive than are other routine marker assay systems, but current evidence suggests that these cell biological markers may have diagnostic utility as well as pathophysiological importance. In addition, gene markers may not always be available or reliable. Mutations in the amyloid precursor gene appear to be exceedingly rare, and the chromosome 19 marker (APO E4) occurs in a high enough proportion of normal individuals to seriously limit its utility as a diagnostic marker. However, before fibroblast assays can be used as diagnostic indicators, extended prospective data are required, and it must be determined if these represent state- or trait-dependent markers (ie, markers for predisposition to the disease or markers for the presence of disease). It is likely that a panel of markers or risk factors may ultimately be used to define an individual's predisposition to AD.

CONCLUSION

Fibroblasts have multiple uses in the study of Alzheimer's disease. They may be useful in identifying pathophysiological mechanisms that reflect cerebral metabolism and pathobiology. They may provide a test for potential therapeutic approaches. Fibroblasts may also provide a system for assessing potential cell biological markers that have potential diagnostic value. These studies do not exclude brain and genetic approaches but complement them. Finally, in addition to explaining the standard structural neuropathological features of AD, hypotheses of the etiology and pathophysiology of this disease should also explain the existence of abnormalities in oxidative metabolism and signal transduction which are observed in cultured cells.

REFERENCES

1. Gandy, S., A. J. Czernik & P. Greengard. 1988. Phosphorylation of Alzheimer disease amyloid precursor peptide by protein kinase C and Ca^{2+}/calmodulin-dependent protein kinase II. Proc. Natl. Acad. Sci. USA **85:** 6218-6221.
2. Buxbaum, J. D., S. E. Gandy, P. Cicchetti, M. E. Ehrlich, A. J. Czernik, R. P. Fracasso, T. V. Ramabhadran, A. J. Unterbeck & P. Greengard. 1990. Processing of Alzheimer βA4 amyloid precursor protein; modulation by agents that regulate protein phosphorylation. Proc. Natl. Acad. Sci. USA **87:** 6003-6006.
3. Goedert, M., M. G. Spillantui, N. J. Cairns & R. A. Crowther. 1992. Tau proteins of Alzheimer paired helical filaments: Abnormal phosphorylation of all six brain isoforms. Neuron **8:** 159-168.
4. Papasozomenos, S. C., Y. Su. 1991. Altered phosphorylation of tau protein in heat-shocked rats and patients with Alzheimer disease. Proc. Natl. Acad. Sci. USA **88:** 4543-4547.
5. Gibson, G. E., L. Toral-Barza & H.-M. Huang. 1991. Cytosolic free calcium in synaptosomes during histotoxic hypoxia. Neurochem. Res. **16:** 461-467.
6. Huang, H. M. & G. E. Gibson. 1989. Phosphatidylinositol metabolism during *in vitro* hypoxia. J. Neurochem. **52:** 830-835.
7. Markesbery, W. R., P. K. Leung & D. A. Butterfield. 1988. Spin label and biochemical studies of erythrocyte membranes in Alzheimer's disease. J. Neurol. Sci. **45:** 223-230.
8. Zubenko, G. S. & R. E. Ferrer. 1988. Monozygotic twins concordant for probable Alzheimer's disease and increased membrane fluidity. Am. J. Med. Genet. **29:** 431-436.
9. Gibson, G. E., P. Nielsen, K. A. Sherman & J. P. Blass. 1987. Diminished mitogen-induced calcium uptake by lymphocytes from Alzheimer patients. Biol. Psychol. **22:** 1079-1086.
10. Grossmann, A., T. Bird, G. Martin & P. Rabinovitch. 1988. Intracellular calcium response is reduced in CD^{4+} lymphocytes in familial Alzheimer's disease. Alzheimer Dis. Assoc. Disorders **3:** 24.
11. Grossmann, A., J. A. Ledbetter & P. S. Rabinovitch. 1990. Aging-related deficiency in intracellular calcium response to anti-CD3 or concanavalin-A in murine T-Cell subsets. J. Gerontol. **45:** B81-B86.
12. Grossmann, A., L. Maggioprice, J. C. Jinneman & P. S. Rabinovitch. 1991. Influence of aging on intracellular free calcium and proliferation of mouse T-cell subsets from various lymphoid organs. Cell. Immunol. **135:** 118-131.
13. Joachim, C. L., H. Mori & D. J. Selkoe. 1989. Amyloid β-protein deposition in tissues other than brain in Alzheimer's disease. Nature **341:** 226-230.
14. Philosophe, B. & R. Miller. 1990. Diminished calcium signal generation in subsets of T lymphocytes that predominate in old mice. J. Gerontol. **45:** B87-B93.
15. Blass, J. P., I. Hanin, L. Barclay, U. Kopp & M. J. Reding. 1985. Red blood cell abnormalities in Alzheimer disease. J. Am. Geriatr. Soc. **33:** 401-405.
16. Sherman, K. A., G. E. Gibson & J. P. Blass. 1986. Human red blood cell choline uptake with age and Alzheimer's disease. Neurobiol. Aging **7:** 205-209.
17. Mattson, M. P., M. G. Engle & B. Rychlik. 1991. Effects of elevated intracellular calcium levels on the cytoskeleton and tau in cultured human cortical neurons. Mol. Chem. Neuropathol. **15:** 117-142.
18. Peterson, C., G. E. Gibson & J. P. Blass. 1985. Altered calcium uptake in cultured skin fibroblasts from patients with Alzheimer's disease. N. Engl. J. Med. **312:** 1063-1064.

19. GIBSON, G. E. & L. TORAL-BARZA. 1992. Cytosolic free calcium in lymphoblasts from young, aged and Alzheimer subjects. Mech. Ageing Dev. **63:** 1-9.
20. CHEN, C., M. J. CORBLEY, T. M. ROBERTS & P. HESS. 1988. Voltage sensitive calcium channels in normal and transformed 3T3 fibroblasts. Science **39:** 1024-1026.
21. GIBSON, G. E. & C. P. PETERSON. 1987. Calcium and the aging nervous system. Neurobiol. Aging **8:** 329-343.
22. SCHELLENBERG, G. D., T. D. BIRD, E. M. WIJSMAN, H. T. ORR, L. ANDERSON, E. NEMENS, J. A. WHITE, L. BONNYCASTLE, J. L. WEBER, M. E. ALONSO, H. POTTER, L. L. HESTON & G. M. MARTIN. 1992. Genetic linkage evidence for a familial Alzheimer's disease locus on chromosome 14. Science **258:** 668-671.
23. GOLDSTEIN, S. 1990. Replicative senescence: The human fibroblast comes of age. Science **249:** 1129-1133.
24. BROOK-FREDERICK, K. M., F. L. CIANCIARULO, S. R. RITTLING & V. J. CRISTOFALO. 1993. Cell cycle dependent regulation of Ca^{2+} in young and senescent WI-38 cells. Exp. Cell Res. **205:** 412-415.
25. CRISTOFALO, V. G. & R. J. PIGNOLO. 1993. Replicative senescence of human fibroblast-like cells in culture. Physiol. Rev. **73:** 617-638.
26. BALIN, A. K., A. C. BAKER, I. C. LEONG & J. P. BLASS. 1988. Normal replicative lifespan of Alzheimer skin fibroblasts. Neurobiol. Aging **9:** 195-198.
27. CARMELIET, G., R. HAUMAN, R. DOM, G. DAVID, J. P. FRYNS, H. VANDENBERGHE & J. J. CASSIMAN. 1990. Growth properties and *in vitro* life span of Alzheimer disease and Down syndrome fibroblasts—a blind study. Mech. Aging Dev. **53:** 17-33.
28. TESCO, G., M. VERGELLI, L. AMADUCCI & S. SORBI. 1993. Growth properties of familial Alzheimer skin fibroblasts during in vitro aging. Exp. Gerontol. **28:** 51-58.
29. SEEGMILLER, J. E., F. M. ROSENBLOOM & W. N. KELLEY. 1967. An enzyme defect associated with a sex-linked human neurological disorder and excessive purine synthesis. Science **155:** 1682-1686.
30. OKADA, S. & J. S. O'BRIEN. 1969. Tay-Sachs disease: Generalized absence of a beta—d-*N*-actylhexosaminidase component. Science **165:** 698-701.
31. BLASS, J. P. 1983. Inborn errors of pyruvate metabolism. *In* Metabolic Basis of Inherited Disease. 5th Ed. J. B. Stanbury, J. B. Wyngaarden, D. S. Fredrickson, J. L. Goldstein & M. S. Brown. Eds.: 193-200. McGraw Hill. New York.
32. HERNDON, J. H., D. STEINBERG, B. W. UHLENDORF & H. M. FALES. 1969. Refsum's disease: Characterization of the enzyme defect in cell culture. J. Clin. Invest. **48:** 1017-1022.
33. STANBURY, J. B., J. B. WYNGAARDEN, D. S. FREDRICKSON, J. L. GOLDSTEIN & M. S. BROWN. 1983. The Metabolic Basis of Inherited Disease, Fifth Ed. McGraw Hill. New York.
34. BLASS, J. P. & A. ZEMCOV. 1984. Alzheimer disease: A metabolic systems degeneration. Neurochem. Pathol. **2:** 103-114.
35. HARDY, J. 1990. Molecular genetics of Alzheimer's disease. Acta. Neurol. Scand. (Suppl.) **129:** 29-31.
36. BLASS, J. P. & G. E. GIBSON. 1993. Non-neural markers in Alzheimer's disease. Alzheimer's Dis. Assoc. Disorders **6:** 205-224.
37. BORDEN, L. A., F. R. MAXFIELD, J. E. GOLDMAN & M. L. SHELANSKI. 1991. Resting $[Ca^{2+}]_i$ and $[Ca^{2+}]_i$ transients are similar in fibroblasts from normal and Alzheimer's donors. Neurobiol. Aging **13:** 33-38.
38. PETERSON, C., R. R. RATAN, M. L. SHELANSKI & J. E. GOLDMAN. 1986. Cytosolic free calcium and cell spreading decrease in fibroblasts from aged and Alzheimer donors. Proc. Natl. Acad. Sci. USA **83:** 7999-8001.
39. PETERSON, C., R. R. RATAN, M. SHELANSKI & J. E. GOLDMAN. 1988. Altered response of fibroblasts from aged and Alzheimer donors to drugs that elevate cytosolic free calcium. Neurobiol. Aging **9:** 261-266.

40. GIBSON, G. E. & L. TORAL-BARZA. 1992. Characterization of internal calcium stores in calcium stores in cultured skin fibroblasts from Alzheimer and control subjects. Mol. Biol. Cell 3S: 841.
41. HUANG, H.-M., L. TORAL-BARZA, H. THALER, B. TOFEL-GREHL & G. E. GIBSON. 1991. Inositol phosphates and intracellular calcium after bradykinin stimulation in fibroblasts from young, normal aged and Alzheimer donors. Neurobiol. Aging **12:** 469-473.
42. MCCOY, K. R., R. D. MULLINS, T. G. NEWCOMB, G. M. NG, G. PAVLINKOVA, R. J. POLINSKY, L. E. NEE & J. E. SISKEN. 1993. Serum-induced and bradykinin-induced calcium transients in familial Alzheimer's fibroblasts. Neurobiol. Aging **14:** 447-455.
43. MCCOY, K. R., G. M. NG, T. G. NEWCOMB, R. D. MULLINS, L. NEE, R. POLINSKY & J. E. SISKEN. 1991. Calcium transients in fibroblasts from individuals with familial Alzheimer's disease. J. Cell Biol. **115:** 8a.
44. PETERSON, C. & J. E. GOLDMAN. 1986. Alterations in calcium content and biochemical processes in cultured skin fibroblasts from aged and Alzheimer donors. Proc. Natl. Acad. Sci. USA **83:** 2758-2762.
45. PETERSON, C., P. VANDERKLISH, P. SEUBERT, C. COTMAN & G. LYNCH. 1991. Increased spectrin proteolysis in fibroblasts from aged and Alzheimer donors. Neurosci. Lett. **121:** 239-243.
46. BYRON, K. L., G. BABINGG & M. L. VILLEREAL. 1992. Bradykinin-induced Ca^{2+} entry, release, and refilling of intracellular Ca^{2+} stores. J. Biol. Chem. **267:** 108-118.
47. MOORE, E. D. W., W. A. CARRINGTON, K. E. FOGARTY & F. S. FAY. 1991. Privileged access of the sodium-calcium exchanger to the ST in smooth muscle cells. J. Cell Biol. **115:** 380a.
48. BOOTH, C. & G. L. KOCH. 1989. Perturbation of cellular calcium induces secretion of luminal ER proteins. Cell **59:** 729-737.
49. SAMBROOK, J. F. 1990. The involvement of calcium in transport of secretory proteins from the endoplasmic reticulum. Cell **6:** 197-199.
50. WALZ, B. & O. BAUMANN. 1989. Calcium sequestering cell organelles: In situ localization, morphological and functional characterization. Prog. Histochem. Cytochem. **20:** 1-47.
51. HANSFORD, R. G. & F. CASTRO. 1985. Role of Ca^{2+} in pyruvate dehydrogenase interconversion in brain mitochondria and synaptosomes. Biochem. J. **227:** 129-136.
52. LAI, J. C. K., J. C. DILORENZO & K-FR. SHEU. 1988. Pyruvate dehydrogenase complex is inhibited in calcium-loaded cerebrocortical mitochondria. Neurochem. Res. **13:** 1043-1048.
53. ETCHEBERRIGARAY, R., E. ITO, K. OKA, B. TOFEL-GREHL, G. E. GIBSON & D. L. ALKON. 1993. A laboratory diagnostic test for Alzheimer's disease. Soc. Neurosci. Abstr. **19:** 221.
54. MCKEE, A. C., K. S. KOSIK, M. B. KENNEDY & N. W. KOWALL. 1990. Hippocampal neurons predisposed to neurofibrillary tangle formation are enriched in type II calcium/calmodulin dependent protein kinase. J. Neuropathol. Exp. Neurol. **49:** 40-63.
55. LESLIE, S. W., L. J. CHANDLER, E. M. BARR & R. P. FARRAR. 1985. Reduced calcium uptake by rat brain mitochondria and synaptosomes in response to aging. Brain Res. **329:** 177-183.
56. HUIDOBRO, A., P. BLANCO, M. VILLALBA, P. GOMEZPUERTAS, A. VILLA, R. PEREIRA, E. BOGONEZ, A. MARTINEZSERRANO, J. J. APARICIO & J. SATRUSTEGUI. 1993. Age-related changes in calcium homeostatic mechanisms in synaptosomes in relation with working memory deficiency. Neurobiol. Aging **14:** 479-486.
57. MICHAELIS, M., C. FOSTER & C. JAYAWICKREME. 1992. Regulation of calcium levels in brain tissue from adult and aged rats. Mech. Ageing Dev. **62:** 291-306.
58. PETERSON, C. & G. E. GIBSON. 1983. Aging and 3,4-diaminopyridine alter synaptosomal calcium uptake. J. Biol. Chem. **258:** 11482-11486.

59. Peterson, C., D. G. Nicholls & G. E. Gibson. 1985. Subsynaptosomal distribution of calcium during aging and 3,4-diaminopyridine treatment. Neurobiol. Aging **6:** 297-304.
60. Landfield, P. W. 1989. Calcium homeostasis in brain aging and Alzheimer's disease: 276-287. Springer-Verlag. Berlin, Heidelberg.
61. Etcheberrigaray, R., E. Ito, K. Oka, B. Tofel-Grehl, G. E. Gibson & D. L. Alkon. 1993. Potassium channel dysfunction in fibroblasts identifies patients with Alzheimer disease. Proc. Natl. Acad. Sci. USA **90:** 8209-8213.
62. Ikeda, M., D. Dewar & J. McCulloch. 1991. Selective reduction of [^{125}I]apamin binding sites in Alzheimer hippocampus: A quantitative autoradiographic study. Brain Res. Abstr. **567:** 51-56.
63. Ikeda, M., D. Dewar & J. McCulloch. 1993. High affinity hippocampal [^{3}H] glibenclamide binding sites are preserved in Alzheimer's disease. J. Neural Transmi. **5:** 177-184.
64. Bruel, A., G. Cherqui, S. Columelli, D. Margelin, M. Roudier, P. Sinet, M. Prieur, J. Perignon & J. Delabar. 1991. Reduced protein kinase C activity in sporadic Alzheimer's disease fibroblasts. Neurosci. Lett. **133:** 89-92.
65. Van Huynh, T., G. Cole, R. Katzman, K. P. Huang & T. Saitoh. 1989. Reduced protein kinase C immunoreactivity altered protein phosphorylation in Alzheimer's disease fibroblasts. Arch. Neurol. **46:** 1195-1199.
66. Govoni, S., S. Bergamaschi, M. Racchi, F. Battaini, G. Binetti, A. Bianchetti & M. Trabucchi. 1993. Cytosol protein kinase C down regulation in fibroblasts from Alzheimer's disease patients. Neurology. **43:** 2581-2586.
67. Liu, D., E. Trenkner & H. Wisniewski. 1993. Abnormalities of mitochondrial electrochemical gradient and protein kinase C activity in human skin fibroblasts from Alzheimer's disease patients. Soc. Neurosci. Abstr. **19:** 191.
68. Racchi, M., W. Wetsel, S. Govoni, M. Trabucchi, F. Battaini, G. Binetti, A. Bianchetti & S. Bergamaschi. 1990. Protein kinase C (PKC) immunoreactivity is decreased in fibroblasts from Alzheimer's disease (AD) patients. Soc. Neurosci. Abstr. **16:** 122.
69. Huang, H.-M., T. A. Lin, G. Y. Sun & G. E. Gibson. 1992. Increased bradykinin-induced inositol (1,4,5)trisphosphate formation in Alzheimer's disease fibroblasts. Mol. Biol. Cell **3**S: 836.
70. Xu, J., P. Zhang, S. Moore, C. Hsu & E. Hogan. 1993. Bradykinin-stimulated phosphoinositide turnover and prostacyclin production regulated by phorbol ester and calcium in cerebral endothelial cells. Soc. Neurosci. Abstr. **19:** 1389.
71. Ferraridileo, G. & D. D. Flynn. 1993. Diminished muscarinic receptor-stimulated [H^3]-PIP_2 hydrolysis in Alzheimer's disease. Life Sci. **53:** PL439-PL444.
72. Xiaohua, L., L. Song, M. Richard, S. Jope & R. Powers. 1993. Impaired cholinergic muscarinic receptor-stimulated phosphoinositide hydrolysis in Alzheimer's disease. Soc. Neurosci. Abstr. **19:** 1039.
73. Masliah, E., G. Cole, S. Shimohama, L. Hansen, R. Deteresa, R. D. Terry & T. Saitoh. 1990. Differential involvement of protein kinase-C isozymes in Alzheimer's disease. J. Neurosci. **10:** 2113-2124.
74. Shimohama, S., M. Narita, H. Matsushima, J. Kumura, M. O. Kameyama, M. Hagiwara, H. Hidaka & T. Taniguchi. 1993. Assessment of protein kinase-C isozymes by 2-site enzyme immunoassay in human brains and changes in Alzheimer's disease. Neurology **43:** 1407-1413.
75. Shimohama, S., S. Fujimoto, T. Taniguchi & J. Kimura. 1992. Phosphatidylinositol-specific phospholipase C activity in the postmortem human brain: No alteration in Alzheimer's disease. Brain Res. **579:** 347-349.
76. Shimohama, S., Y. Homma, T. Fujimoto, T. Suenaga, T. Taniguchi, W. Araki, Y. Yamaoka, H. Matsushima, T. Takenawa, T. Takenawa & J. Kimura. 1991.

Aberrant accumulation of phospholipase C δ in Alzheimer brains. Am. J. Pathol. **139:** 737-742.

77. STOKES, C. E. & J. H. HAWTHORNE. 1987. Reduced phosphoinositide concentrations in anterior temporal cortex of Alzheimer-diseased brains. J. Neurochem. **48:** 1018-1021.
78. YOUNG, L. T., S. J. KISH, P. P. LI & J. J. WARSH. 1988. Decreased brain [^{3}H] inositol 1,4,5-triphosphate binding in Alzheimer's disease. Neurosci. Lett. **94:** 198-202.
79. HUANG, H-M. & G. E. GIBSON. 1993. Altered β-adrenergic receptor stimulated cAMP formation in cultured skin fibroblasts from Alzheimer donors. J. Biol. Chem. **268:** 14616-14621.
80. MALOW, B. A., A. C. BAKER & J. P. BLASS. 1989. Cultured cells as a screen for novel treatments of Alzheimer's disease. Arch. Neurol. **46:** 1201-1203.
81. VOLICER, L., L. GREENE & M. SINEX. 1985. Epinephrine-induced cyclic AMP production in skin fibroblasts from patients with dementia of Alzheimer type and controls. Neurobiol. Aging **6:** 35-38.
82. HORNQVIST, R., R. HENRIKSSON, O. BACK, G. BUCHT & B. WINBLAD. 1987. Iontophoretic study of adrenergic and cholinergic skin vessel reactivity in normal ageing and Alzheimer's disease. Gerontology **33:** 374-379.
83. KALARIA, R. N., A. C. ANDORN, M. TABATON, P. J. WHITEHOUSE, S. I. HARK & J. R. UNNERSTALL. 1989. Adrenergic receptors in aging and Alzheimer's disease: Increased beta 2-receptors in prefrontal cortex and hippocampus. J. Neurochem. **53:** 1772-1781.
84. KALARIA, R. N., A. C. ANDORN & S. I. HARIK. 1989. Alterations in adrenergic receptors of frontal cortex and cerebral microvessels in Alzheimer's disease and aging. Prog. Clin. Biol. Res. **317:** 367-374.
85. JANSEN, K. L., R. L. FAULL, M. DRAGUNOW & B. L. SYNEK. 1990. Alzheimer's disease: Changes in hippocampal *N*-methyl-D-aspartate, quisqualate, neurotensin, adenosine, benzodiazepine, serotonin and opioid receptors—an autoradiographic study. Neuroscience **39:** 613-627.
86. BOWEN, D. M., S. J. ALLEN, J. S. BENTON, M. J. GOODHARDT, E. A. HAAN, A. M. PALMER, N. R. SIMS, C. T. SMITH, J. A. SPILLANE, M. M. ESIRI, D. NEARY, J. S. SNOWDON, G. K. WILCOCK & A. N. DAVISON. 1983. Biochemical assessment of serotonergic and cholinergic dysfunction and cerebral atrophy in Alzheimer's disease. J. Neurochem. **41:** 266-272.
87. CROSS, A. J., T. J. CROW, J. A. JOHNSON, E. K. PERRY, R. H. PERRY, G. BLESSED & B. E. TOMLINSON. 1984. Studies on neurotransmitter receptor systems in neocortex and hippocampus in senile dementia of the Alzheimer-type. J. Neurol. Sci. **64:** 109-117.
88. LEMMER, B., L. LANGER, T. OHM & J. BOHL. 1993. β-adrenoceptor density and subtype distribution in cerebellum and hippocampus from patients with Alzheimer's disease. Naunyn-Schmiedebergs Arch. Pharmacol. **347:** 214-219.
89. SHIMOHAMA, S., T. TANIGUCHI, M. FUJIWARA & M. KAMEYAMA. 1987. Changes in β-adrenergic receptor subtypes in Alzheimer-type dementia. J. Neurochem. **48:** 1215-1221.
90. KARLARIA, R. N. & S. K. HARIK. 1989. Increased alpha 2- and beta 2-adrenergic receptors in cerebral microvessels in Alzheimer disease. Neurosci. Lett. **106:** 233-238.
91. LITTLE, J., I. BASARIC-KEYS, R. LEBOVICS, M. CANTILLON, T. SUNDERLAND & B. WOLOZIN. 1993. Analysis of the cAMP signal transduction pathway in olfactory neuroblasts from Alzheimer's disease and control donors. Soc. Neurosci. Abstr. **19:** 1474.
92. OHM, T., J. BOHL & B. LEMMER. 1991. Reduced basal and stimulated (isoprenaline, Gpp (NH)p, forskolin) adenylate cyclase activity in Alzheimer's disease correlated with histopathological changes. Brain Res. **540:** 229-236.
93. OHM, T., J. BOHL & B. LEMMER. 1989. Reduced cAMP-signal transduction in postmortem hippocampus of demented old people. Prog. Clin. Biol. Res. **317:** 501-509.

94. Decker, M. W., T. M. Gill & J. L. McGaugh. 1990. Concurrent muscarinic and β-adrenergic blockade in rats impairs place-learning in a water maze and retention of inhibitory avoidance. Brain Res. **513:** 81-85.
95. Hadcock, J. R., J. D. Port & C. C. Malbon. 1991. Cross-regulation between G-protein-mediated pathways. J. Biol. Chem. **266:** 11915-11922.
96. Wong, Y. H., A. Federman, A. M. Pace, I. Zachary, T. Evans, J. Pouyssegur & H. R. Bourne. 1991. Mutant alpha subunits of Gi2 inhibit cyclic AMP accumulations. Nature **351:** 63-65.
97. Crouch, M. F. & E. G. Lapetina. 1988. A role for Gi in control of thrombin receptor-phospholipase C coupling in human platelets. J. Biol. Chem. **263:** 3363-3371.
98. Brown, A. M. & L. Birnbaumer. 1988. Direct G protein gating of ion channels. Am. J. Physiol. **254:** H401-H410.
99. Harrison, P. J., A. J. L. Barton, B. McDonald & R. C. A. Pearson. 1991. Alzheimer's disease-specific increases in a G-protein subunit (Gs-alpha) messenger RNA in hippocampal and cortical neurons. Mol. Brain Res. **10:** 71-81.
100. McLaughlin, M., B. M. Ross, G. Milligan, J. McCulloch & J. T. Knowler. 1991. Robustness of G proteins in Alzheimer's— An immunoblot study. J. Neurochem. **57:** 9-14.
101. Cowburn, R. F., C. O'Neill, C. J. Fowler, R. Ravid & B. Winblad. 1992. Selective impairment of Gs-protein stimulated adenylyl cyclase activity in Alzheimer's Disease brain. Neurobiol. Aging **13** (Suppl. 1): S60.
102. Warpman, U., I. Alafuzoff & A. Nordberg. 1992. Defect coupling of muscarinic receptors to GTP-proteins in Alzheimer Brains. Neurobiol. Aging **13** (Suppl. 1): S61.
103. Bergstrom, L., A. Garlind, L. Nilsson, I. Alafuzoff, C. J. Fowler, B. Winblad & R. F. Cowburn. 1991. Regional distribution of somatostatin receptor binding and modulation of adenylyl cyclase activity in Alzheimer's disease brain. J. Neuro. Sci. **105:** 225-233.
104. Blass, J. P. & G. E. Gibson. 1991. The role of oxidative abnormalities in the pathophysiology of Alzheimer's Disease. Rev. Neurol. **147:** 513-525.
105. Lin, F. H., R. Lin, H. M. Wisniewski, Y. W. Hwang, I. Grundke-Iqbal, G. Healy-Louie & K. Iqbal. 1992. Detection of point mutations in codon 331 of mitochondrial and dehydrogenase subunit 2 in Alzheimer's brains. Biochem. Biophys. Res. Comm. **182:** 238-246.
106. Mastrogiacomo, F., C. Bergeron, S. Dozic, L. Distefano & S. Kish. 1992. Brain α-ketoglutarate dehydrogenase activity: Influence of Alzheimer's disease and thiamine. Neurobiol. Aging **13** (Suppl. 1): S63.
107. Khansari, N., H. D. Whitten, Y. K. Chou & H. H. Fudenberg. 1985. Immunological dysfunction in Alzheimer's disease. J. Neuroimmunol. **7:** 279-285.
108. Sims, N. R., J. M. Finegan & J. P. Blass. 1987. Altered metabolic properties of cultured skin fibroblasts in Alzheimer's disease. Ann. Neurol. **21:** 451-457.
109. Sheu, K., A. Cooper, K. Koike, M. Koike, G. Lindsay & J. Blass. 1994. Abnormality of α-ketoglutarate dehydrogenase in fibroblasts from familial Alzheimer's disease. Ann. Neurology **35:** 312-318.
110. Farkus, D. L., M. Wei, P. Febbroriello, J. H. Caron & L. M. Loew. 1989. Simultaneous imaging of cell and mitochondrial membrane potentials. Biophys. J. **56:** 1053-1069.
111. Loew, L. M., W. A. Carrington, F. S. Fay, R. A. Tuft & M. D. Wei. 1991. Quantitative determination of membrane potential of individual mitochondria within a neurite. J. Cell Biol. **115:** 300a.
112. Gibson, G. E. & J. P. Blass. 1976. A relation between [NAD+]/[NADH] potentials and glucose utilization in rat brain slices. J. Biol. Chem. **251:** 4127-4130.
113. Akerman, K. E. O. & D. G. Nicholls. 1981. Intrasynaptosomal compartmentation calcium during depolarization-induced calcium uptake across the plasma membrane. Biochim. Biophys. Acta **645:** 41-48.

114. Scott, I. D., K. E. O. Akerman & D. G. Nicholls. 1980. Calcium-ion transport by intact synaptosomes. Intrasynaptosomal compartmentation and the role of the mitochondrial membrane potential. Biochem. J. **192:** 873-880.
115. Schild, L., P. V. Blair & E. J. Davis. 1991. Interdependence among [Ca^{2+}], [Mg^{2+}], membrane potential, respiration, and the accumulation and release of adenine nucleotide by mitochondria. J. Cell Biol. **115:** 300a.
116. Li, J. M., H. M. McBride, T. B. Shin & G. C. Shore. 1991. Reversing the orientation of an integral protein of the mitochondrial outer membrane. J. Cell. Biol. **115:** 254a.
117. Huang, H. M. & G. E. Gibson. 1989. Effects of in vitro hypoxia on depolarization stimulated accumulation of inositol phosphates in synaptosomes. Life Sci. **45:** 1443-1449.
118. Carroll, J. M., L. Toral-Barza & G. E. Gibson. 1992. Cytosolic free calcium and gene expression during chemical hypoxia. J. Neurochem. **59:** 1836-1843.
119. Chen, L. B. 1988. Mitochondrial membrane potential in living cells. Ann. Rev. Cell. Biol. **4:** 155-181.
120. Johnson, L. V., M. L. Walsh, B. J. Bockus & L. B. Chen. 1981. Monitoring of relative mitochondria membrane potential in living cells by fluorescence microscopy. J. Cell Biol. **88:** 526-535.
121. Nadakvukaren, K. K., J. J. Nadakvukaren & L. B. Chen. 1985. Increased rhodamine 123 uptake by carcinoma cells. Cancer Res. **45:** 6093-6099.
122. Gibson, G. E., K.-Fr. Sheu, J. P. Blass, A. Baker, K. C. Carlson, B. Harding & P. Perrino. 1988. Reduced activities of thiamine-dependent enzymes in brains and peripheral tissues of Alzheimer patients. Arch. Neurol. **45:** 836-840.
123. Mutisya, E., A. Bowling, L. Walker, D. Price, L. Cork, L. Vecsei & M. Beal. 1993. Impaired energy metabolism in aging and Alzheimer's Disease. Soc. Neurosci. Abstr. **19:** 1474.
124. Chandrasakaran, K., T. Giordano, D. Brady, J. Stoll, L. Martin & S. Rapaport. 1993. Impairments in mitochondrial gene expression in vulnerable brain regions in Alzheimer's Disease. Soc. Neurosci. Abstr. **19:** 222.
125. Blass, J. P., A. C. Baker, L. W. Ko & R. S. Black. 1990. Induction of Alzheimer antigens by an uncoupler of oxidative phosphorylation. Arch. Neurol. **47:** 864-869.

Molecular Mechanisms of Memory and the Pathophysiology of Alzheimer's Disease

RENÉ ETCHEBERRIGARAY,[a] GARY E. GIBSON,[b]
AND DANIEL L. ALKON[a,c]

[a]*Laboratory of Adaptive Systems*
National Institute of Neurological Disorders and Stroke
National Institutes of Health
Bethesda, Maryland 20892

[b]*Cornell Medical College*
Burke Medical Research Institute
White Plains, New York 10605

Research in our laboratory over two decades has implicated a sequence of biophysical and biochemical steps in the acquisition and storage of associative memory.[1-3] These steps, identified in both invertebrate and mammalian species, begin with temporally specific training stimuli in such paradigms as pavlovian conditioning, spacial water maze learning, and olfactory discrimination learning.[4-12] Learning-induced cellular changes occur when transduced sensory stimuli elicit electrical signals that converge on neuronal targets within brain networks. Neurotransmitter-gated (eg, GABA and NMDA) and voltage-dependent elevation of intracellular calcium, temporally related to diacylglycerol and arachidonic acid, causes activation of protein kinase C (PKC).[13] Membrane-associated PKC then phosphorylates specific substrates such as the membrane-traffic regulating G-protein Cp20. Cp20 phosphorylation, in turn, has subcellular consequences including: (1) Reduction of voltage-dependent K^+ channels and thus increased membrane excitability and enhanced electrical signaling[14]; (2) regulation of intraneuronal particle transport[15]; (3) rearrangements of synaptic branches[16]; and (4) activation of mRNA turnover.[17]

Because this sequence of events underlying memory storage is apparently conserved in highly diverse species, it may also have relevance for human memory which is characteristically associative.[3] We therefore considered the possibility that such a cellular cascade of memory-storing events might include targets of dysfunction due to the pathophysiology of a human disease, such as Alzheimer's disease, that is reasonably specific for memory loss early in its clinical course.

Hypothesizing that Alzheimer's disease has systemic expression,[18,19] we proposed that cellular changes in peripheral cells (such as skin fibroblasts, olfactory neuroblasts, and lymphocytes) might be analyzed with less interference by secondary pathological

[c] To whom correspondence should be addressed: Laboratory of Adaptive Systems, NINDS, Building 36, Room 4A-21, National Institutes of Health, Bethesda, MD 20892.

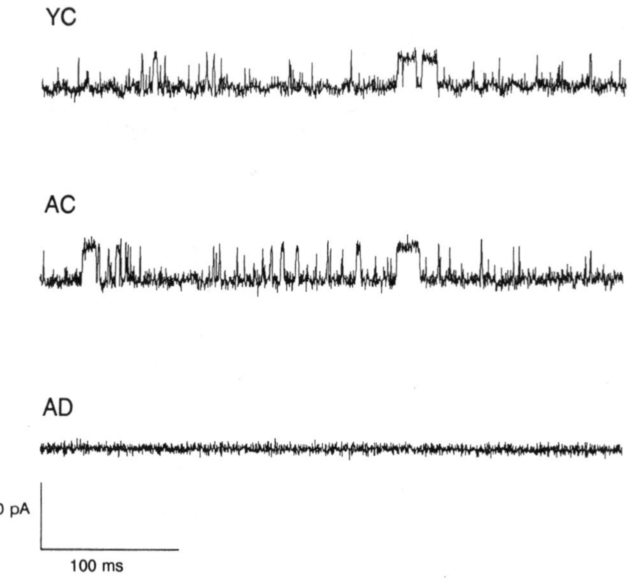

FIGURE 1. Patch clamp recordings from human fibroblasts. Sample traces (cell attached) of the 113-pS K^+ channel present in young (YC) and age-matched (AC) controls, but completely missing in Alzheimer's disease (AD) fibroblasts (*bottom trace*). The channel remains largely in the open state, exhibiting brief closures (*upward deflections*). The unitary current size at mV 0 pipette potential was ≈4.5 pA.

consequences of the disease such as those that occur with widespread plaque deposits in the brain. We undertook then to measure the same or similar biophysical and/or molecular events in peripheral cells of patients with Alzheimer's disease and control subjects as those we had implicated in memory storage. With surprising specificity and sensitivity, K^+ channel function, PKC-related calcium mobilization, and other memory-related cellular phenomena indeed appear involved in Alzheimer's disease, as the following observations indicate.

K^+ CHANNEL DYSFUNCTION IN ALZHEIMER'S DISEASE FIBROBLASTS

Using conventional patch clamp techniques,[20,21] we identified and studied two K^+ channels in human fibroblasts. A 113-pS conductance, tetraethylammonium (TEA)-sensitive K^+ channel, with virtually identical kinetics, was commonly observed in fibroblasts from young (YC, $n = 6$) and age-matched (AC, $n = 5$) controls. By contrast, this channel was found to be functionally absent in fibroblasts from patients with AD ($n = 7$)[22] (FIG. 1). Another channel of 166-pS conductance was observed with the same frequency in about 28% of the patches, in cells from control subjects as well as in AD fibroblasts. In addition, low percentage open time (≈10%) and lack of TEA sensitivity were common features for this 166-pS channel in control and AD

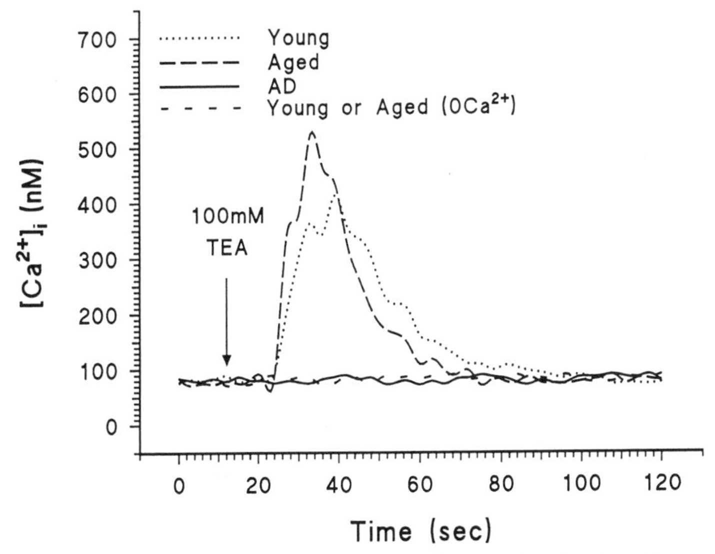

FIGURE 2. Tetraethylammonium (TEA)-induced $[Ca^{2+}]_i$ elevation. Bath application of 100 mM TEA (also 10 mM) caused a marked $[Ca^{2+}]_i$ increase, reaching peak elevation within 30 seconds. Note that responses of control fibroblasts (young [YC] and age-matched [AC] controls) are almost identical, whereas no responses (*flat solid trace*) were observed in Alzheimer's disease (AD) fibroblasts. In addition, external Ca^{2+} removal and/or the use of Ca^{2+} channel blockers eliminated the TEA-induced response (*flat broken trace*), indicating that Ca^{2+} influx (voltage dependent, also see text) accounts for the responses in control cells.

cells. Absence of the 113-pS channel in AD cells constituted the first clear evidence of K^+ channel alteration in AD. Nevertheless, the technique used in this study (patch clamp) only samples a minute fraction of the plasma membrane, and a relatively small number of cells can be examined in a short time. Therefore, we employed fluorescence imaging techniques[23] to measure intracellular Ca^{2+} ($[Ca^{2+}]_i$) elevation in response to K^+ channel blockade-induced depolarization as a method for assessing K^+ channel function that also explores the entire membrane surface. Because the 113 K^+ channel present in controls but not in AD cells was TEA sensitive, it was reasonable to assume that TEA will cause depolarization only in those cells with a significant population of functional TEA-sensitive 113-pS channels. Indeed, TEA induced a significant $[Ca^{2+}]_i$ elevation only in control cells (YC, n = 6; AC, n = 10), which included an additional group of cells from non-AD neurological and psychiatric conditions (NAD, n = 14). None of the AD (n = 13) cell lines examined had Ca^{2+} responses to TEA[22] (FIG. 2). High K^+-induced depolarization caused $[Ca^{2+}]_i$ elevation in control and AD cells, indicating preserved voltage-sensitive Ca^{2+} function in these cells. Removal of external Ca^{2+} and/or the use of Ca^{2+} channel blockers eliminated TEA responses in all cells, confirming that the TEA response arises from depolarization-induced Ca^{2+} influx. Two independent experimental methodologies, therefore, clearly indicated that K^+ channel dysfunction is present in AD fibroblasts, perhaps as a sign of early molecular changes in the disease process. Preliminary results of

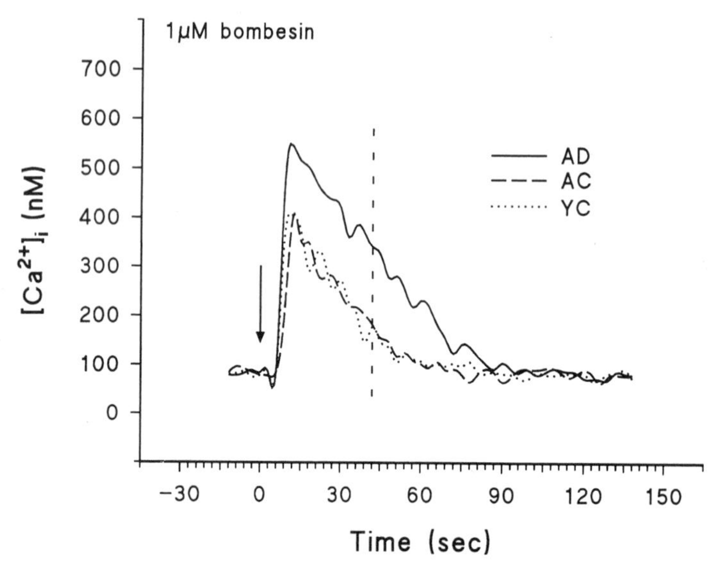

FIGURE 3. Intracellular Ca^{2+} release. Bombesin (1 μM) induced an enhanced IP_3-mediated Ca^{2+} release from intracellular stores (*solid line*) that was more evident in the absence of external Ca^{2+}. Maximum separation between Alzheimer's disease (AD) and controls was observed after peak responses. Young (YC) and age-matched (AC) control fibroblasts had significantly lower Ca^{2+} responses (*dotted and broken lines*), and virtually no differences were observed between the two controls.

an ongoing study in AD olfactory neuroblasts, which are closely related to central nervous system cells, also indicate that a related K^+ malfunction is present in these cells.

ALTERED INTRACELLULAR CA^{2+} RELEASE IN ALZHEIMER'S DISEASE FIBROBLASTS

Homeostasis of $[Ca^{2+}]_i$ has been the focus of significant attention in AD[24-32] research. These studies, although often contradictory, together with previous implications of $[Ca^{2+}]_i$ mobilization in memory and learning,[1] motivated us to also explore receptor-mediated $[Ca^{2+}]_i$ in AD fibroblasts. The same fibroblasts and cell culture conditions employed in K^+ channel function study were used to study IP_3-mediated $[Ca^{2+}]_i$ release in human fibroblasts. Bombesin, a peptide with multiple actions in various organs[33] and known to induce G-protein/phospholipase C-mediated IP_3 generation,[34-39] was used to stimulate $[Ca^{2+}]_i$ mobilization from intracellular stores in AD and control fibroblasts. Bath application of 1 μM bombesin in the absence of external Ca^{2+} caused a markedly enhanced transient Ca^{2+} response in AD (n = 10) fibroblasts (FIG. 3).[40] These enhanced responses were significantly different from those of all control groups that included YC (n = 6), AC (n = 8), and NAD (n = 10). When Ca^{2+} was added to the bath (2.5 mM), a small but significant late (≈90-100 seconds after

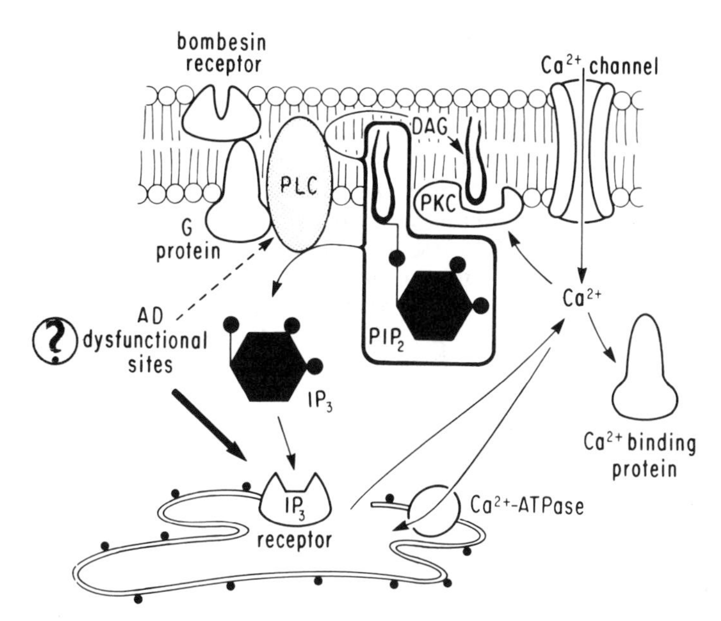

FIGURE 4. Model for bombesin-induced responses. Schematic representation, based on the work of Berridge and collaborators[38,39] as well as our own results, of the various steps involved in the bombesin-induced $[Ca^{2+}]_i$ release. The site(s) likely to be responsible for the observed AD-specific enhancement of the Ca^{2+} response in AD fibroblasts is indicated.

stimulation) Ca^{2+} entry was only observed in control cells, providing a second indicator of altered bombesin-induced Ca^{2+} response in AD fibroblasts. Biochemical analyses demonstrated that bombesin receptor numbers and/or affinity[40] cannot account for the observed differences in bombesin responses. Thapsigargin-induced Ca^{2+} release (IP_3-independent[41–43]) was similar in AD and control cells, indicating that $[Ca^{2+}]_i$ pools and buffering systems are also preserved in AD fibroblasts. Therefore, increased IP_3 generation and/or enhanced IP_3 receptor sensitivity are likely candidates to explain the observed differences in bombesin responses in AD fibroblasts. Recent evidence showing enhanced bradykinin-induced IP_3 generation in AD fibroblasts,[44] which, like bombesin, uses receptor/G-protein mechanisms, provides additional support for the aforementioned proposed altered site in AD fibroblasts (FIG. 4).

β-AMYLOID INDUCES K^+ CHANNEL DYSFUNCTION

Beta-amyloid deposits are neuropathological findings that help confirm the diagnosis of Alzheimer's disease.[45] It is also widely believed that this peptide fragment plays a major role in the pathophysiology of AD.[46–48] Nevertheless, no conclusive evidence supports a direct causal or mechanistic relationship between amyloid metabolism and AD genesis. Because our results clearly identified molecular AD-specific defects, we decided to study the potential relationships between these molecular changes and β-amyloid effects. Fibroblasts from control groups (YC and AC) with previously identified normal K^+ channels, normal TEA responses, and also normal bombesin responses were treated with low (10 nM) concentrations of β-amyloid. Cells were examined after 48 hours of incubation. Patch clamp experiments revealed that the 113-pS TEA-sensitive K^+, normally present in 60% of control cells, was missing in β-amyloid-treated cells mimicking the features of AD fibroblast K^+ channels.[49,50] The 166-pS channel, unaltered in AD, was also unaffected by β-amyloid treatment. Furthermore, the TEA response was completely absent or dramatically reduced in all YC and AC cells treated with 10 nM β-amyloid[49,50] (FIG. 5), whereas the non-AD-specific high K^+-induced response was conserved. Thus, in otherwise normal cells, β-amyloid changes normal K^+ channel function to an AD-like pattern. The bombesin response was not affected by treatment, perhaps indicating that the enhanced bombesin response observed in AD is a defect independent of β-amyloid actions. The enhanced bombesin response could also be the sign of a more advanced process in which additional cellular events became defective.

DISCUSSION

Our results clearly indicate that molecular substrates of associative memory, previously identified in animal models, are altered in AD. These changes might also represent molecular changes that are an integral part of pathophysiological events leading to cell death and clinical symptoms. The fact that soluble β-amyloid induced AD-specific features in K^+ channel function suggests that β-amyloid can affect normal memory mechanisms early in the disease before aggregation occurs. Plaque formation and cell death perhaps correspond to more advanced stages when not only is memory affected, but also a host of cognitive and neurological deficits are present.

Because changes in K^+ channel function (measured as the absence of the TEA response) and the enhanced bombesin response were so consistently AD specific, their use as laboratory tests for AD is suggested (TABLE 1). Finally, these results additionally support the hypothesis that AD may have systemic expression, and therefore the use of peripheral cells such as fibroblasts constitutes a valuable tool to study the pathophysiology of AD.

SUMMARY

Research on molecular and biophysical mechanisms of associative learning and memory storage identified a number of key elements that are phylogenetically conserved. In both vertebrates and invertebrates, K^+ channels, PKC, Cp20, and intracellular Ca^{2+} regulation play a fundamental role in memory mechanisms. Because memory loss is the hallmark and perhaps the earliest sign of Alzheimer's disease, we hypothesized that these normal memory mechanisms might be altered in AD. With the use of a variety of experimental methodologies, our results revealed that one of the critical elements in memory storage, K^+ channels, are dysfunctional in AD fibroblasts. Moreover, β-amyloid induced the same K^+ dysfunction in normal cells. Intracellular Ca^{2+} release, also associated with molecular memory mechanisms, was found altered

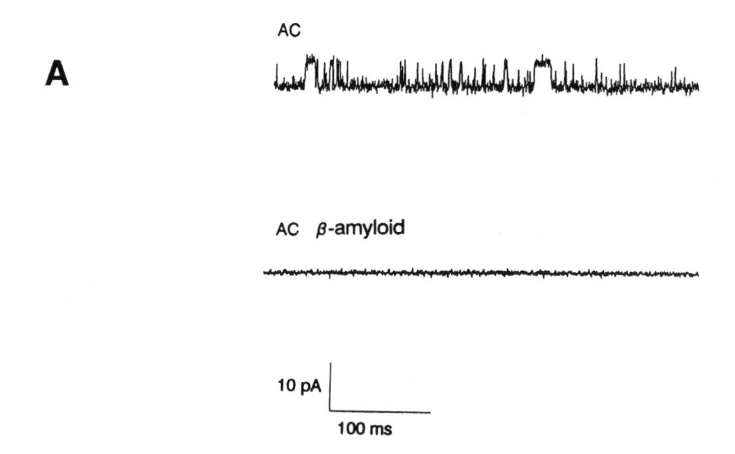

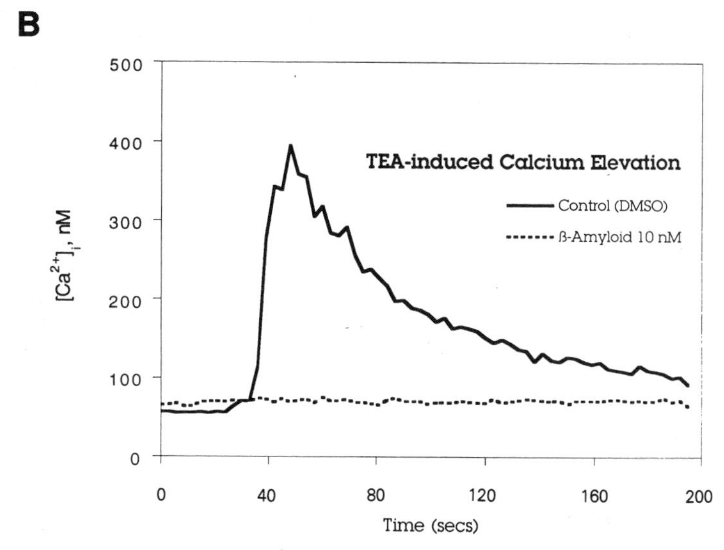

FIGURE 5. Beta-amyloid-induced Alzheimer's disease (AD)-like K^+ channel dysfunction. Incubation with 10 nM β-amyloid (1-40) of control cells induced the functional obliteration of the 113-pS channel (**A**), similar to the situation observed in AD fibroblasts (FIG. 1). The tetraethylammonium-induced response was also completely eliminated in otherwise normal, β-amyloid-treated cells (**B**).

in fibroblasts from patients with AD. The results therefore strongly suggest that biophysical and molecular mechanisms of associative learning could be altered in AD and that they may contribute to the memory loss observed early in the disease.

TABLE 1. Tetraethylammonium (TEA)- and Bombesin-Induced $[Ca^{2+}]_i$ Response in Alzheimer's and Control Fibroblasts[a]

Line No.	Age (yr) & Gender	Race	Condition	TEA	Bombesin
AG06840[b1]	56,M	W	Alzheimer's	–	+
AG06848[b2]	55,F	W	Alzheimer's*	–	+
AG07637[b]	55,F	W	Alzheimer's	–	+
AG08170[b]	56,M	W	Alzheimer's	–	+
AG06844[b]	59,M	W	Alzheimer's*	–	NT
AG04400[c]	61,F	W	Alzheimer's	–	NT
AG04401[c]	53,F	W	Alzheimer's*	–	+
AG05809	63,F	W	Alzheimer's	–	+
AG08243	72,M	W	Alzheimer's	–	+
AG07375	71,M	W	Alzheimer's	–	+
AG07376	59,M	W	Alzheimer's	–	+
AG06263	67,F	W	Alzheimer's	–	+
AG07377	59,M	W	Alzheimer's	–	NT
GM03524	67,F	B	Age-matched	+	–
AG06010	62,F	W	Age-matched	+	–
AG07603[b3]	61,F	W	Age-matched	+	–
AG09878	61,F	W	Age-matched	+	–
AG08044	58,F	B	Age-matched	+	–
AG06241	61,M	W	Age-matched	+	–
AG04560	59,M	W	Age-matched	+	–
GM04260	60,M	W	Age-matched	+	–
AG07141	66,F	W	Age-matched	+	NT
AG11363	74,F	W	Age-matched	+	NT
GM03652	24,M	W	Young	+	–
GM03651	25,F	W	Young	+	–
GM02987	19,M	W	Young	–	–
GM04390	23,F	W	Young	+	–
GM03377	19,M	W	Young	+	–

TABLE 1. Tetraethylammonium (TEA)- and Bombesin-Induced $[Ca^{2+}]_i$ Response in Alzheimer's and Control Fibroblasts[a] (*Continued*)

Line No.	Age (yr) & Gender	Race	Condition	TEA	Bombesin
GM08399	19,F	?	Young	+	–
AG08395	85,F	W	Parkinson's*	+	–
GM01835	27,F	W	Schizophrenia	+	–
GM02038	22,M	W	Schizophrenia	+	–
GM06274	56,F	W	Huntington's	+	–
GM02165	55,M	W	Huntington's	+	–
GM00305	56,F	W	Huntington's	–	–
GM01085	44,M	W	Huntington's	+	–
GM01061	51,M	W	Huntington's	+	–
GM05030	56,M	W	Huntington's	–	–
GM04777	53,M	W	Huntington's	+	–
7504	50,M	W	Wernicke-Korsakoff	+	–
7505	52,F	W	Wernicke-Korsakoff	+	–
7507	63,M	W	Wernicke-Korsakoff	+	–
7508	64,M	W	Wernicke-Korsakoff	+	–

[a] Alzheimer's fibroblasts were from familial ($n = 8$) and nonfamilial cases ($n = 5$). Five[b] are members of Canadian family 964; only 1 and 2 are immediate relatives (sibs). [c] Members (sibs) of family 747. Autopsy confirmed Alzheimer's disease in three cases (*). [b3] Unaffected member of Canadian family 964. All young control lines ($n = 6$) are from normal individuals and those without an AD family history. Criterion $[Ca^{2+}]_i$ responses (to 100 mM TEA) indicated as + were observed in all age-matched control (AC) lines used and in all but one of the young control (YC) lines. None of the AD lines exhibited a "positive" TEA response, $\chi^2 = 231.44$, $p < 0.001$. Enhanced bombesin responses (+) were observed in all 10 Alzheimer's disease (AD) cell lines tested, whereas none of the normal controls, AC ($n = 8$) and YC ($n = 6$), had enhanced bombesin responses. Fibroblasts from a Parkinson's disease donor had normal TEA (+) and did not significantly differ from responses observed in the age-matched control group. Fibroblasts from two schizophrenic patients also had normal TEA. Normal TEA responses were observed in five of seven patients with Huntington's disease. Fibroblasts from patients with Wernicke-Korsakoff disease ($n = 4$) had also normal TEA responses. The TEA responses are significantly different from those of AD fibroblasts to the level of $p < 0.0001$ (Fisher's exact test). Bombesin responses in these non-AD neuropsychiatric ($n = 14$) conditions were identical to those of normal controls. The bombesin response "correctly" identified the three control cases (one YC and two with Huntington's disease) that had an AD-like TEA response. NT indicates cell line/condition that were not tested.

REFERENCES

1. ALKON, D. L. 1987. Memory Traces in the Brain. Cambridge University Press. New York.
2. ALKON, D. L. 1989. Sci. Am. **260:** 42-50.
3. ALKON, D. L. 1992. Memory's Voice. Harper Collins. New York.
4. ALKON, D. L. 1984. Science **226:** 1037-1045.
5. ALKON, D. L., S. NAITO, M. KUBOTA, C. CHEN, B. BANK, J. SMALLWOOD, P. GALLANT & H. RASMUSSEN. 1988. J. Neurochem. **51:** 903-917.
6. COLLIN, C., H. IKENO, J. F. HARRIGAN, I. LEDERHENDLER & D. L. ALKON. 1988. Biophys. J. **55:** 955-960.
7. ETCHEBERRIGARAY, R., D. L. MATZEL, I. I. LEDERHENDLER & D. L. ALKON. 1992. Proc. Natl. Acad. Sci. USA **89:** 7184-7188.
8. BANK, B., A. DEWEER, A. M. KUZIRIAN, H. RASMUSSEN & D. L. ALKON. 1988. Proc. Natl. Acad. Sci. USA **85:** 1988-1992.
9. OLDS, J. L., M. L. ANDERSON, D. L. MCPHIE, L. D. STANTEN & D. L. ALKON. 1989. Science **245:** 866-869.
10. SÁNCHEZ-ANDRÉS, J. V. & D. L. ALKON. 1991. J. Neurophysiol. **65:** 796-807.
11. OLDS, J. L., S. GOLSKI, D. L. MCPHIE, D. OLTON, M. MISHKIN & D. L. ALKON. 1990. J. Neurosci. **10:** 3707-3713.
12. OLDS, J. L., U. S. BAHALLA, D. L. MCPHIE, D. S. LESTER, J. BOWER & D. L. ALKON. 1994. Behav. Brain Res. **61:** 37-46.
13. LESTER, D. S. & D. L. ALKON. 1991. Prog. Brain Res. **89:** 235-261.
14. NELSON, T. J., C. COLLIN & D. L. ALKON. 1990. Science **247:** 1479-1483.
15. MOSHIACH, S., T. NELSON, J. V. SANCHEZ-ANDRES, M. SAKAKIBARA & D. ALKON. 1993. Brain Res. **605:** 298-304.
16. ALKON, D. L., H. IKENO, J. DWORKIN, D. L. MCPHIE, J. OLDS, I. LEDERHENDLER, L. MATZEL, B. SCHREURS, A. KUZIRIAN, C. COLLIN & E. YAMOAH. 1990. Proc. Natl. Acad. Sci. USA **87:** 1611-1614.
17. NELSON, T. J., J. L. OLDS, J. KIM & D. L. ALKON. Submitted.
18. BAKER, A. C., L-W. KO & J. P. BLASS. 1988. Age **11:** 60-65.
19. SCOTT, R. B. 1993. J. Am. Geriatr. Soc. **41:** 268-276.
20. SAKMANN, B. & E. NEHER. 1983. Single-Channel Recordings. Plenum. New York.
21. FRENCH, A. S. & L. L. STOCKBRIDGE. 1988. Proc. R. Soc. Lond. **232:** 395-412.
22. ETCHEBERRIGARAY, R., E. ITO, K. OKA, B. TOFEL-GREHL, G. E. GIBSON & D. L. ALKON. 1993. Proc. Natl. Acad. Sci. USA **90:** 8209-8213.
23. CONNOR, J. A. 1993. Cell Calcium **14:** 185-200.
24. PETERSON, C., R. R. RATAN, M. L. SHELANSKI & J. E. GOLDMAN. 1986. Proc. Natl. Acad. Sci. USA **83:** 7999-8001.
25. GIBSON, G. E., P. NIELSEN, K. A. SHERMAN & J. P. BLASS. 1987. Biol. Psychiatry **22:** 1079-1086.
26. PETERSON, C., R. R. RATAN, M. L. SHELANSKI & J. E. GOLDMAN. 1988. Neurobiol. Aging **9:** 261-266.
27. BORDEN, L. A., F. R. MAXFIELD, J. E. GOLDMAN & M. L. SHELANSKI. 1991. Neurobiol. Aging **13:** 33-38.
28. HUANG, H.-M., L. TORAL-BARZA & G. E. GIBSON. 1991. Biochim. Biophys. Acta **1091:** 409-416.
29. HUANG, H.-M., L. TORAL-BARZA, H. THALER, B. TOFEL-GREHL & G. E. GIBSON. 1991. Neurobiol. Aging **12:** 469-473.
30. MATTSON, M. P., B. CHENG, D. DAVIS, K. BRYANT, I. LIEBERBURG & R. E. RYDEL. 1992. J. Neurosci. **12:** 376-389.
31. MCCOY, K. R., R. D. MULLINS, T. G. NEWCOMB, G. M. NG, G. PAVLINKOVA, R. J. POLINSKY, L. E. NEE & J. E. SISKEN. 1993. Neurobiol. Aging **14:** 447-455.
32. MATTSON, M. P., S. W. BARGER, B. CHENG, I. LIEBERBURG, V. L. SMITH-SWINTOSKY & R. E. RYDEL. 1993. TINS **16:** 406-414.

33. SPINDEL, E. R., E. GILADI, T. P. SEGERSON & S. NAGALLA. 1993. Recent Prog. Hormone Res. **48:** 365-391.
34. LLOYD, A. C., S. A. DAVIES, I. CROSSLEY, M. WHITAKER, M. D. HOUSLAY, A. HALL, C. J. MARSHALL & M. J. O. WAKELAM. 1989. Biochem. J. **260:** 813-819.
35. MOTOZAKI, T., W-Y. ZHU, Y. TSUNODA, B. GÖKE & J. A. WILLIAMS. 1991. Am. J. Physiol. **260:** G858-G864.
36. MURPHY, A. C. & E. ROZENGURT. 1992. J. Biol. Chem. **267:** 25296-25303.
37. STREB, H., R. F. IRVINE, M. J. BERRIDGE & I. SCHULTZ. 1983. Nature **306:** 67-69.
38. BERRIDGE, M. J. & R. F. IRVINE. 1989. Nature **341:** 197-205.
39. BERRIDGE, M. J. 1993. Nature **361:** 315-325.
40. ITO, E., K. OKA, R. ETCHEBERRIGARAY, B. TOFEL-GREHL, G. E. GIBSON & D. L. ALKON. 1994. Proc. Natl. Acad. Sci. USA **91:** 534-538.
41. THASTRUP, O., H. LINNEBJERG, P. J. BJERRUM, J. B. KNUDSEN & S. B. CHRISTENSEN. 1987. Biochim. Biophys. Acta **927:** 65-73.
42. TAKEMURA, H., A. R. HUGHES, O. THASTRUP & J. W. PUTNEY, JR. J. Biol. Chem. **264:** 12266-12271.
43. THASTRUP, O., P. J. CULLEN, B. J. DRØBAK, M. R. HANLEY & A. P. DAWSON. 1990. Proc. Natl. Acad. Sci. USA **87:** 2466-2470.
44. HUANG, H.-M., T-A. LIN, G. Y. SUN & G. E. GIBSON. 1992. Trans. Am. Soc. Neurochem. **23:** 30a.
45. KATZMAN, R. 1986. N. Engl. J. Med. **314:** 964-973.
46. JOACHIM, C. L. & D. J. SELKOE. 1992. Alzheimer Dis. Assoc. Disord. **6:** 7-34.
47. SELKOE, D. J. 1993. TINS **16:** 403-409.
48. ARISPE, N., E. ROJAS & H. B. POLLARD. 1993. Proc. Natl. Acad. Sci. USA **90:** 567-571.
49. ETCHEBERRIGARAY, R., E. ITO, C. S. KIM & D. L. ALKON. 1994. Biophys. J. (Abstr.) A256.
50. ETCHEBERRIGARAY, R., E. ITO, C. S. KIM & D. L. ALKON. Science, in press.
51. UEDA, K., G. COLE, M. SUNDSMO, R. KATZMAN & T. SAITO. 1989. Ann. Neurol. **25:** 246-251.
52. GOVONI, S., S. BERGAMASCHI, M. RACCHI, F. BATTAINI, G. BINETTI, A. BIANCHETTI & M. TRABUCCHI. 1993. Neurology **43:** 2581-2586.

The Ability of Amyloid β-Protein [AβP (1-40)] to Form Ca^{2+} Channels Provides a Mechanism for Neuronal Death in Alzheimer's Disease

NELSON ARISPE,[a] HARVEY B. POLLARD, AND EDUARDO ROJAS

Laboratory of Cell Biology and Genetics
NIDDK
National Institutes of Health
Bethesda, Maryland 20892–0840

Alzheimer's disease (AD) is a dementia pathologically characterized by the presence of plaques in the brain, intraneuronal neurofibrillary tangles, and neuronal death.[1-8] Although the ultimate cause of neuronal death in AD remains to be elucidated, the discovery that a major component of brain plaques is a 38- to 42-residue peptide termed amyloid β-protein[2,3,9] suggested the possibility that the amyloid β-proteins *per se* could be the causal factor in neuronal death.[10] Several reports of neuronal death induced by injecting exogenous amyloid β-protein directly into the brain[4] or into cultured brain tissue[11-17] support this hypothesis. In addition, other studies of neuronal cultures showed that β-amyloid peptides induce neurodegeneration.[18,19] Taken together these studies led to the conclusion that the direct interaction between β-amyloid molecules and neurons is probably the cause of cell death in AD. Although some groups failed to show a clear deleterious effect of synthetic amyloid β-protein in monkey cerebral cortex[20] and in rat brain[21] or of AβP(25-35) peptide in rat hippocampal neurons,[22] the ensuing controversy[17,23] stimulated further work to test whether β-amyloid molecules could be neurotoxic. A crucial experiment that shed light on the molecular mechanism for AβP (1-40) neurotoxicity showed that when amyloid β-protein molecules are allowed to aggregate in solution, they become neurotoxic.[18,19] Thus, it is the physical state of β-amyloid molecules in solution that determines the neurotoxicity. Monomeric forms, lacking β-pleated sheet conformations, appear relatively inert, whereas aggregation leads to enhanced toxicity. Because cell survival in cultures treated with β-amyloid is significantly reduced,[24,25,30] it is now clear that under suitable conditions β-amyloid molecules are indeed neurotoxic. Furthermore, our results showing that AβP (1-40) forms Ca^{2+} channels in bilayer membranes provided a unifying mechanism for amyloid β-protein toxicity.[26,27] We also learned recently that AβP (1-40) molecules form multi-conductance channels which can undergo transitions between different large conductance states.[27] Our work with AβP (1-40) allowed us to speculate that the formation of such channels in brain

[a] Corresponding author.

cells would disrupt the cationic gradients across the plasma membrane of target cells. Indeed, the ensuing Ca^{2+} influx would cause a dramatic intracellular Ca^{2+} concentration $[Ca^{2+}]_i$ rise, as shown by others.[28–31] It should not be forgotten that this mechanism of AβP (1-40) toxicity requires the presence of external Ca^{2+}.[29] Presumably the maintained Ca^{2+} influx exceeds the intracellular buffering capacity,[32] and the elevated $[Ca^{2+}]_i$ is the ultimate cause of cell death.[33]

Analysis of amyloid β-protein molecules by physical chemistry provided several important bits of information regarding assembly and aggregation.[34] These include analysis of conformational states of the protein using circular dichroism spectra,[35] and demonstration by low angle X-ray diffraction and other methods of a hydrophobic conformation with high affinity for lipid-like environments.[35–38] These new data also provided the basis for our mechanistic hypothesis that the AβP molecules themselves might be an integral part of an unregulated Ca^{2+} entry pathway which kills brain cells in AD, possibly including endothelial, glial, and neuronal cells. Indeed, Ca^{2+} entry into other non-neuronal cells could explain the large variety of cellular disorders reported in AD.

THE AβP (1-40) PROTEIN FORMS CATION-SELECTIVE CHANNELS

From these and other data, a new and fascinating clue has emerged that not only changed our understanding of the pathology in AD but has also provided a mechanism for β-amyloid neurotoxicity. We found and review here that the peptide AβP(1-40) inserts itself into lipid bilayer membranes and forms ionic channels.[26,27] AβP(1-40) channels are selective for Ca^{2+}, but other monovalent cations are also permeable. AβP (1-40) channel activity can be acquired by direct incorporation of soluble β-amyloid molecules from the solution into a lipid bilayer formed at the tip of a patch pipette (FIG. 1). After a synthetic 1-palmitoyl 2-oleoyl phosphatidylethanolamine (POPE) monomolecular film was spread onto the surface of the solution filling a glass trough, bilayer membranes were readily formed by dipping the pipette twice into the solution (25 mM $CaCl_2$ on both sides of the bilayer). FIGURE 1A depicts eight superimposed records of the AβP (1-40)-channel activity after the addition of β-amyloid (50 mM NaHepes at pH 7.4 was the carrier solution) to the chamber to a final concentration of 0.4 μM. Before and after the application of rectangular pulses (C), channel events are observed as upward deflections of the current trace (A) at a holding potential of −20 mV. Current records in response to the application of positive pulses (60 ms) of increasing magnitude (10-mV increments) are observed as downward deflections of the current trace for positive membrane potentials during the pulses. As expected for a system with symmetrical solutions on either side of the bilayer, no events are apparent during the application of the pulse that took the transmembrane potential from the V_{hold} to 0 mV (second current trace).

AβP (1-40) CHANNELS ARE BLOCKED BY TROMETHAMINE

We previously showed that tromethamine blocks AβP (1-40) channels which have been incorporated into artificial membranes by allowing liposomes containing

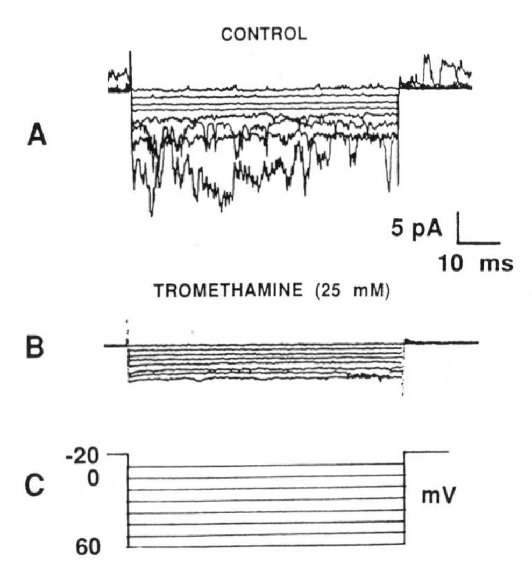

FIGURE 1. Channel activity induced by direct incorporation of Aβ (1-40) is also blocked by tromethamine. Symmetrical $CaCl_2$ (in mM: 25 $CaCl_2$, 1 NaHepes at pH 7.4) solutions were used both in the glass chamber and inside the pipette. (**A**) Control: family of AβP (1-40)-channel currents in response to eight rectangular (60-ms) voltage pulses. (**B**) Channel currents in the presence of tromethamine (25 mM). (**C**) Transmembrane potential during pulses is given in mV. The potential was held at −20 mV between pulses.

AβP (1-40) to fuse with the planar bilayer.[26] FIGURE 1B shows that AβP (1-40) channels directly incorporated from the solution are also sensitive to this drug.[27] It may be seen that AβP (1-40)-channel activity is almost absent from each of the eight superimposed records (B) which were made 2-5 minutes after the addition of tromethamine (20 mM), and the current-voltage relationship calculated from records made in the presence of the drug is linear (not shown), corresponding to a leakage resistance of about 10 GΩ.

AβP (1-40) CHANNELS ARE CATION SELECTIVE

Ion substitution studies revealed that AβP (1-40) forms cation-selective channels.[26] From these studies we obtained the following permeability sequence:

$$P_{Cs} > P_{Li} > P_{Ca} = P_K > P_{Na}.$$

FIGURE 2 illustrates AβP (1-40)-channel activity recorded using a planar lipid bilayer made of synthetic phosphatidylserine (PS) and POPE. The bilayer was exposed to a suspension of PS liposomes (made up in the presence of AβP [1-40]) exposed to asymmetrical solutions, that is, 37.5 mM CsCl on the *cis* side and 25 mM $CaCl_2$ on the *trans* side (FIG. 2A, inset). When liposomes are allowed to fuse with the bilayer and there is incorporation of AβP (1-40) molecules, recorded current jumps indicate the formation of a cationic channel across the membrane. FIGURE 2A depicts

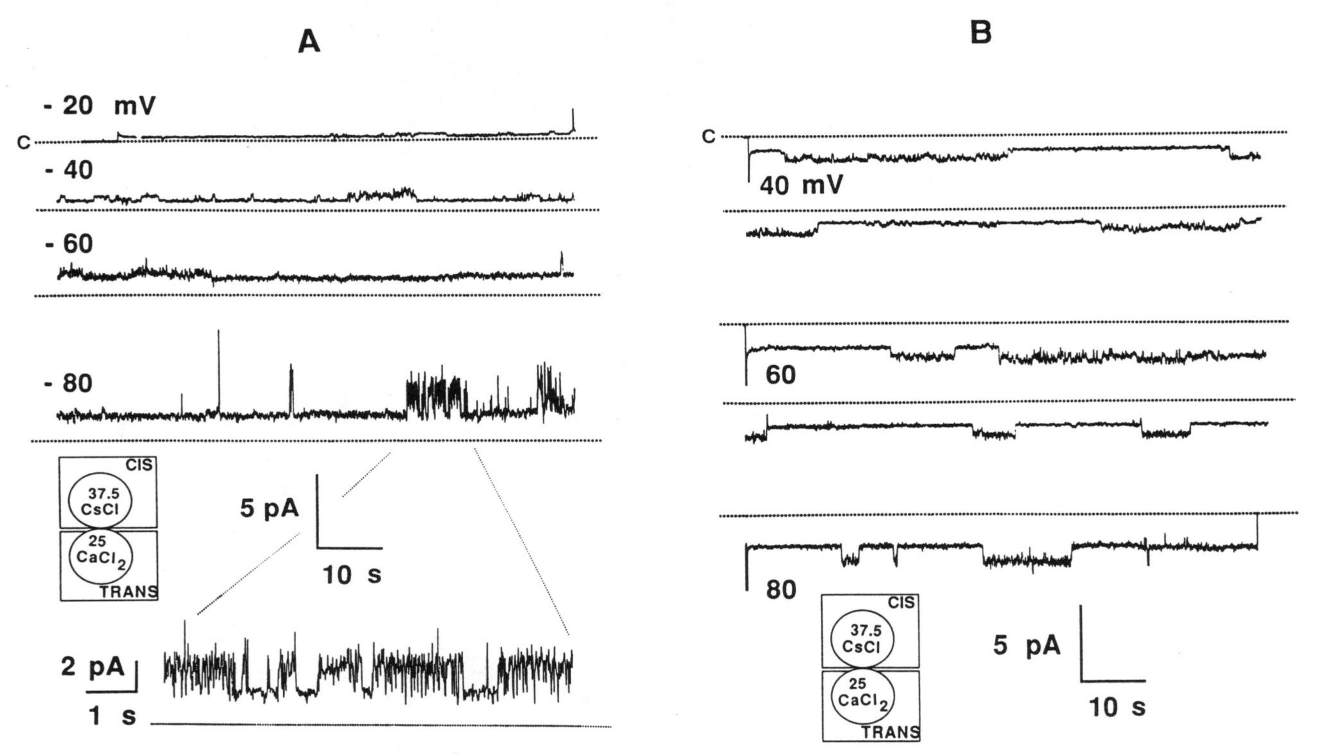

FIGURE 2. AβP (1-40) channels are permeable to Ca^{2+} and Cs^{+}. Asymmetrical solutions (for composition see insets). **(A)** Ca^{2+} as charge carrier moving from the *trans* to the *cis* side. **(B)** Cs^{+} as charge carrier moving in the opposite direction.

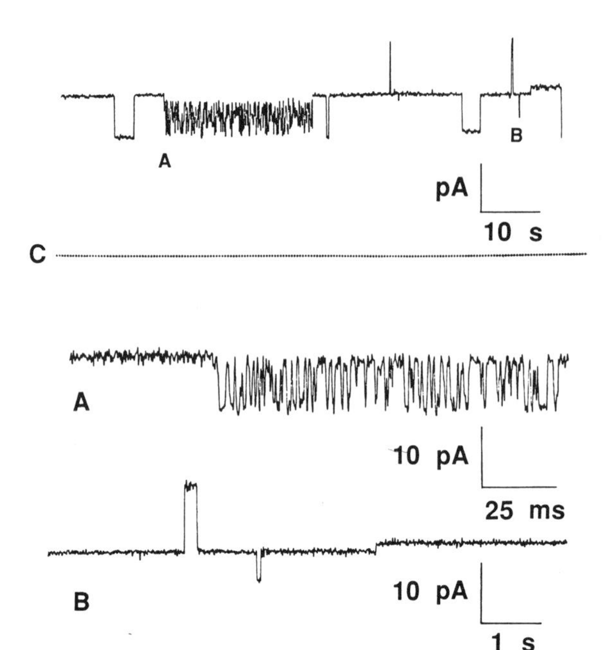

FIGURE 3. The AβP (1-40) channel undergoes high frequency transitions between different conductance levels. Symmetrical K^+ solutions (in mM: 140 KCl, 10 KHepes at pH 7.4). The channel was incorporated from a liposome complex into a planar bilayer. **A** and **B** represent segments of the upper record on an expanded time base. Transmembrane potential was 25 mV throughout.

channel events at negative transmembrane potentials (–20, –40, –60, and –80 mV). Upward deflections of current traces represent a flow of Ca^{2+} from the *trans* to the *cis* side. At positive transmembrane potentials (40, 60, and 80 mV) downward deflections of current traces represent a flow of Cs^+ from the *cis* to the *trans* side (FIG. 2B).

Although transmembrane potential is maintained at a fixed level, discrete jumps of current to different levels are observed throughout the records. This is better illustrated on an expanded time scale in FIGURE 2A (bottom). Increasing the driving force for Ca^{2+} by setting the potential of the *cis* side at –80 mV induced bursts of high frequency. Incomplete closures suggest that the AβP (1-40) channel operates between multiple conductance levels.

THE AβP (1-40) CHANNEL EXHIBITS MULTIPLE LEVELS OF CONDUCTANCE

Current jumps between different conductance levels depicted in FIGURE 2A (bottom record) can also be due to the presence of more than one active channel in the bilayer. However, as illustrated in FIGURE 3, high frequency flickering occurs between two well-defined conductance levels, suggesting rapid conformational transitions of a single channel. Segments A and B taken from the upper trace at the times indicated in FIGURE 3 are shown on an expanded time base. We also observed rapid transitions between small (picosiemens range) and large conductance levels (nanosiemens range). These transitions apparently occur in a random manner, with no particular preference for a specific level.

THE AβP (1-40) ACQUIRES CONFORMATIONAL STATES OF LARGE CONDUCTANCE

Occasionally we found the AβP (1-40) channel in the open state exhibiting long-lasting stable conductance, in the nanosiemens range. FIGURE 4 depicts a family of records of a single AβP (1-40)-channel activity in symmetrical K^+ (150 mM) at different transmembrane potentials (–2, –3, –4, and –10 mV). Frequent interconversions between conductance levels occur in this modality of channel activity, each level of conductance being stable and long lasting. At each transmembrane voltage the AβP (1-40) channel exhibited brief but complete closures (FIG. 4). Furthermore, the maximum value of the current at each transmembrane potential was a linear function of voltage with a slope of 4.2 nS (not shown). Linear functions of voltage were also obtained by drawing straight lines (2.5-, 2.3-, and 1.8-nS slopes) to fit the remaining current levels at different potentials. We also noted that in the nanosiemen conductance modality, the AβP (1-40) channel remained open for prolonged periods, suggesting that AβP-amyloid molecules have the ability to form extremely stable AβP complexes in the bilayer and thereby allow enormous fluxes across the membrane.

CONCLUSIONS

Taken together these data strongly suggest the participation of AβP-amyloid molecules in the etiology of AD. Furthermore, our results[26,27] and other recently published work clearly indicate that AβP-amyloid molecules are cell killers. We therefore concluded that AβP-amyloid peptide molecules, possibly in an intermediate state of aggregation in which only a few molecules acquire the appropriate conformation to effectively interact with the target bilayers and not in their state of insoluble plaques, are the causal effect of cell death in the brain in AD.

These ideas are summarized in the scheme illustrated in FIGURE 5. AβP-peptides are produced by enzymatic cleavage of the amyloid precursor protein APP, and they are released into the extracellular space.[39–41] AβP-amyloid molecules in free solution form aggregates. The rate of aggregation appears to depend on conditions such as AβP-amyloid concentration, pH, and temperature, and the process has been modeled *in vitro* as "AβP-amyloid aging."[18] Indeed, physicochemical analysis of AβP-peptide solutions showed that molecules aggregate, forming dimers and larger oligomers.[35,36] It is interesting that theoretical models for AβP-amyloid aggregates, based on conformational energy levels and dimensional configuration of the constitutive amino acids, have confirmed that stable configurations are those found experimentally, in particular the hexamer form.[42–44] Furthermore, the molecular configuration in the aggregate resembles a channel with a pore at the center of the aggregate. It is thus possible

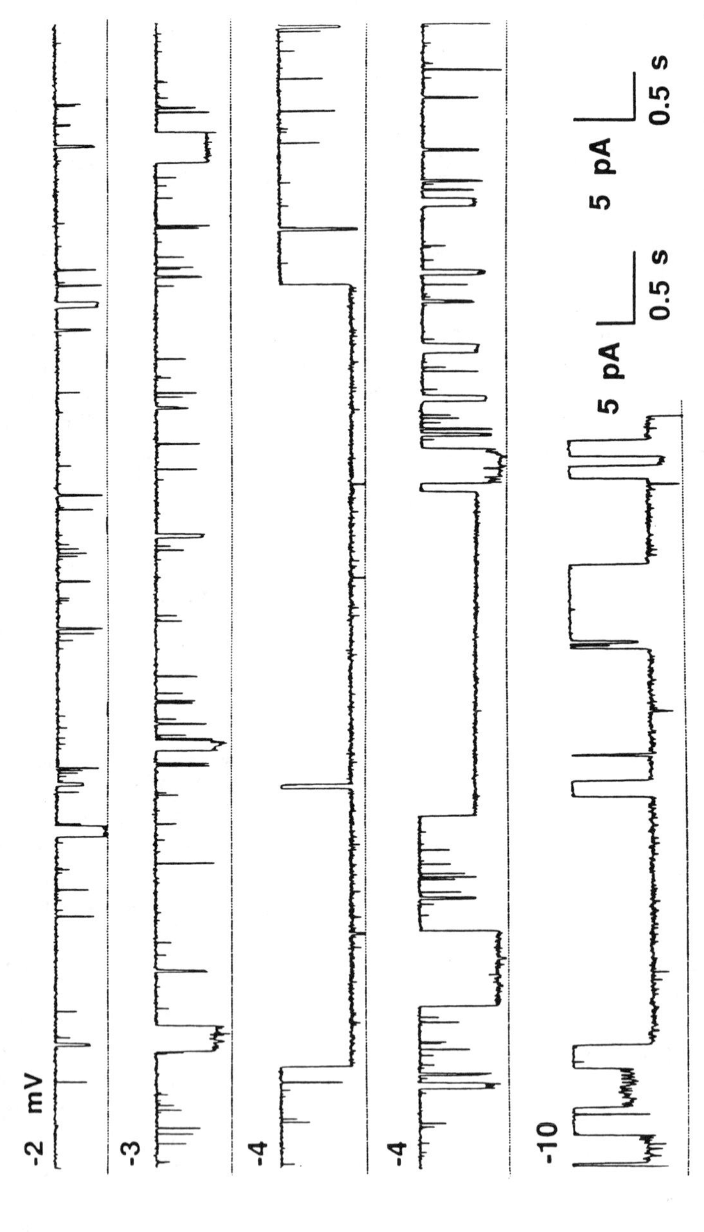

FIGURE 4. Multiple conductance levels of the AβP (1–40) channel in the nanosiemens range. Conditions and solutions as for FIGURE 3.

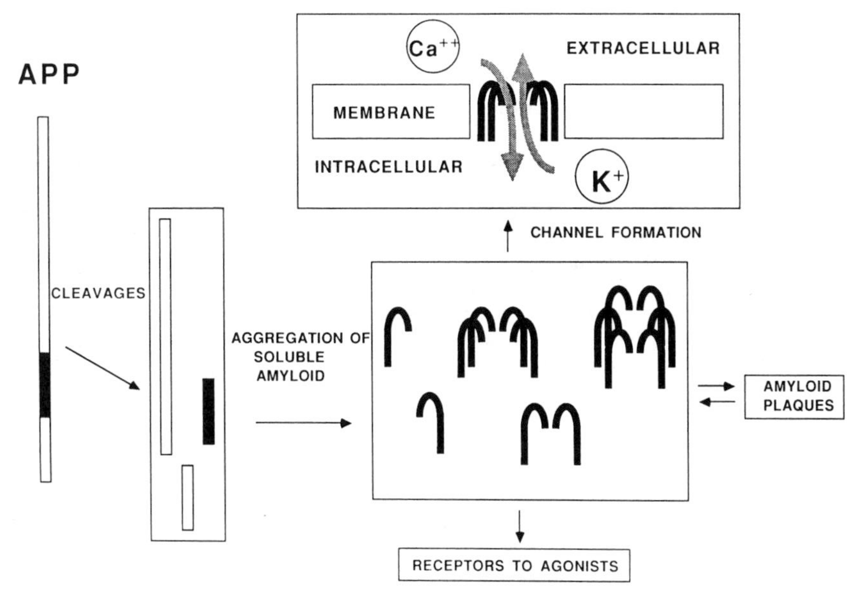

FIGURE 5. Amyloid-channel hypothesis for neurotoxicity.[45]

that AβP molecules may interact with the target membrane in two ways. Direct interaction with the bilayer can lead to the formation of amyloid Ca^{2+} channels. Alternatively, interaction with the membrane may also cause changes in the function of other types of channels. These channels allow ion fluxes down the corresponding electrochemical gradient to occur. Because of the large Ca^{2+} electrochemical gradient, Ca^{2+} influx will prevail. For a nerve cell with a single AβP (1-40) channel (FIG. 4) of large conductance (in the nanosiemens range), the corresponding Ca^{2+} influx will induce a $[Ca^{2+}]_i$ rise at a rate of 5 μM/s. This Ca^{2+} influx will rapidly saturate the Ca^{2+}-buffering capacity of any nerve cell.[32]

This process of aggregation of AβP-amyloid molecules explains not only the formation of Ca^{2+} channels but also the deposition of insoluble plaques. It is also possible that both β-amyloid deposition and AβP-channel formation proceed independently, and in many instances it is possible to have plaque formation without neuronal death caused by the neurotoxic effects of AβP-amyloid aggregates.

REFERENCES

1. REISBERG, B., ED. 1983. Alzheimer's Disease. The Free Press. New York.
2. GLENNER, G. G. & C. W. WONG. 1984. Alzheimer's disease: Initial report of the purification and characterization of a novel cerebrovascular amyloid protein. Biochem. Biophys. Res. Commun. **120:** 885-890.
3. MASTERS, C. L., G. SIMMS, N. A. WEINMAN, G. MULTHAUP, B. L. MCDONALD & K. BEYREUTHER. 1985. Amyloid plaque core protein in Alzheimer disease and Down syndrome. Proc. Natl. Acad. Sci. USA **82:** 4245-4249.

4. FRAUTSCHY, S. A., A. BAIRD & G. M. COLE. 1991. Effects of injected Alzheimer beta-amyloid cores in rat brain. Proc. Natl. Acad. Sci. USA **88:** 8362-8366.
5. NEVE, R. L., L. R. DAWES, B. A. YANKNER, L. L. BENOWITZ, W. RODRIGUEZ & G. A. HIGGINS. 1990. Genetics and biology of the Alzheimer amyloid precursor. Prog. Brain Res. **86:** 257-267.
6. SELKOE, D. J. 1991. The molecular pathology of Alzheimer's disease. Neuron **6:** 487-498.
7. HARDY, J. A. & G. A. HIGGINS. 1992. Alzheimer's disease: The amyloid cascade hypothesis. Science **256:** 780-783.
8. SELKOE, D. J. 1993. Physiological production of the β-amyloid protein and the mechanism of Alzheimer's disease. Trends Neurosci. **16**(10): 403-409.
9. KANG, J., H.-G. LEMAIRE, A. UNTERBECK, J. M. SALBAUM, C. L. MASTERS, K.-H. GRZESCHIK, G. MULTHAUP, K. BEYREUTHER & B. MULLER-HILL. 1987. The precursor of Alzheimer's disease amyloid A4 protein resembles a cell-surface receptor. Nature **325:** 733-736.
10. YANKNER, B. A., L. R. DAWES, S. FISHER, L. VILLA-KOMAROFF, M. L. OSTER-GRANITE & R. L. NEVE. 1989. Neurotoxicity of a fragment of the amyloid precursor associated with Alzheimer's disease. Science **245:** 417-420.
11. YANKNER, B. A., A. CACERES & L. K. DUFFY. 1990. Nerve growth factor potentiates the neurotoxicity of β amyloid. Proc. Natl. Acad. Sci. USA **87:** 9020-9023.
12. YANKNER, B. A., L. K. DUFFY & D. A. KIRSCHNER. 1990. Neurotrophic and neurotoxic effects of β amyloid protein: Reversal by tachykinin neuropeptides. Science **250:** 279-282.
13. ROHER, A. E., M. J. BALL, S. V. BHAVE & A. R. WAKADE. 1991. β-amyloid protein from Alzheimer's disease brain inhibits sprouting and survival of sympathetic neurons. Biochem. Biophys. Res. Commun. **174:** 572-579.
14. KOWALL, N. W., A. C. MCKEE, B. A. YANKNER & M. F. BEAL. 1992. *In vivo* neurotoxicity of beta amyloid [β (1-40)] and the β(25-35) fragment. Neurobiol. Aging **13**(5): 537-542.
15. MALOUF, A. T. 1992. Effect of beta amyloid peptides on neurons in hippocampal slice cultures. Neurobiol. Aging **13**(5): 543-551.
16. EMRE, M., C. GEULA, B. J. RANSIL & M. M. MESULAM. 1992. The acute neurotoxicity and effects upon cholinergic axons of cerebrally injected β-amyloid in the rat brain. Neurobiol. Aging **13**(5): 553-559.
17. YANKNER, B. A. 1992. Commentary and perspective on studies of beta amyloid neurotoxicity. Neurobiol. Aging **13**(5): 615-616.
18. PIKE, C., A. J. WALENCEWICZ, C. G. GLABE & C. W. COTMAN. 1991. *In vitro* aging of β-amyloid protein causes peptide aggregation and neurotoxicity. Brain Res. **563:** 311-314.
19. PIKE, C., D. BURDICK, A. J. WALENCEWICZ, C. G. GLABE & C. W. COTMAN. 1993. Neurodegeneration induced by β-amyloid peptides *in vitro:* The role of peptide assembly state. J. Neurosci. **13**(4): 1676-1687.
20. PODLISNY, M. B., D. T. STEPHENSON, M. P. FROSCH, I. LIEBERBURG, J. A. CLEMENS & D. J. SELKOE. 1992. Synthetic amyloid β-protein fails to produce specific neurotoxicity in monkey cerebral cortex. Neurobiol. Aging **13**(5): 561-567.
21. GAMES, D., K. M. KHAN, F. G. SORIANO, P. S. KEIM, D. L. DAVIS, K. BRYANT & I. LIEBERBURG. 1992. Lack of Alzheimer's pathology after β-amyloid protein injections in rat brain. Neurobiol. Aging **13**(5): 569-576.
22. STEIN-BEHRENS, B., K. ADAMS, M. YEH & R. SAPOLSKY. 1992. Failure of beta-amyloid protein fragment 25-35 to cause hippocampal damage in the rat. Neurobiol. Aging **13**(5): 577-579.
23. COTMAN, C. W., C. J. PIKE & A. COPANI. 1992. β-amyloid neurotoxicity: A discussion of *in vitro* findings. Neurobiol. Aging **13**(5): 587-590.

24. SIMMONS, M. A. & C. R. SCHNEIDER. 1993. Amyloid β-peptides act directly on single neurons. Neurosci. Lett. **150:** 133-136.
25. BUSCIGLIO, J., J. YEH & B. A. YANKNER. 1993. β-amyloid neurotoxicity in human cortical culture is not mediated by excitotoxins. J. Neurochem. **61**(4): 1565-1568.
26. ARISPE, N., E. ROJAS & H. B. POLLARD. 1993. Alzheimer disease amyloid β protein forms calcium channels in bilayer membranes: Blockade by tromethamine and aluminum. Proc. Natl. Acad. Sci. USA **90:** 567-571.
27. ARISPE, N., H. B. POLLARD & E. ROJAS. 1993. Giant multilevel cation channels formed by Alzheimer disease amyloid β-protein [AβP-(1-40)] in bilayer membranes. Proc. Natl. Acad. Sci. USA **90:** 10573-10577.
28. KHACHATURIAN, Z. S. 1989. The role of calcium regulation in brain aging: Reexamination of a hypothesis. Aging (Milano) **1**(1): 17-34.
29. MATTSON, M. P., B. CHENG, D. DAVIS, K. BRYANT, I. LIEBERBURG & R. RYDEL. 1992. β-amyloid peptides destabilize calcium homeostasis and render human cortical neurons vulnerable to excitotoxicity. J. Neurosci. **12**(2): 376-386.
30. MATTSON, M. P., K. J. TOMASELLI & R. E. RYDEL. 1993. Calcium-destabilizing and neurodegenerative effects of aggregated β-amyloid peptide are attenuated by basic FGF. Brain Res. **621:** 35-49.
31. MATTSON, M. P., S. W. BARGER, B. CHENG, I. LIEBERBURG, V. L. SMITH-SWINTOSKY & R. RYDER. 1993. β-amyloid precursor protein metabolites and loss of neuronal Ca^{2+} homeostasis in Alzheimer's disease. Trends Neurosci. **16:** 409-414.
32. CHOI, D. W. 1988. Calcium-mediated neurotoxicity: Relationship to specific channel types and role in ischemic damage. Trends Neurosci. **11**(10): 465-469.
33. MILLER, R. J. 1991. The control of neuronal Ca^{++} homeostasis. Prog. Neurobiol. **37:** 255-285.
34. BURDICK, D., B. SOREGHAN, M. KWON, J. KOSMOSKI, M. KNAUER, A. HENSCHEN, J. YATES, C. COTMAN & C. GABLE. 1992. Assembly and aggregation properties of synthetic Alzheimer's A4/β amyloid peptide analogs. J. Biol. Chem. **267:** 546-554.
35. BARROW, C. J., A. YASUDA, P. T. M. KENNY & G. ZAGORSKI. 1992. Solution conformations and aggregational properties of synthetic amyloid β-peptides of Alzheimer's disease. Analysis of circular dichroism spectra. J. Mol. Biol. **225:** 1075-1093.
36. HILBICH, C., B. KISTERS-WOIKE, J. REED, C. MASTER & K. BEYREUTHER. 1991. Aggregation and secondary structure of synthetic amyloid βA4 peptides of Alzheimer's disease. J. Mol. Biol. **218:** 149-163.
37. MASON, R. P., W. J. SHOEMAKER, L. SHAJENKO & L. G. HERBETTE. 1993. X-ray diffraction analysis of brain lipid membrane structure in Alzheimer's disease and β-amyloid peptide interactions. Ann. N.Y. Acad. Sci. **695:** 231-234.
38. JARRET, J. T. & T. LANSBURY, JR. 1993. Seeding "one dimensional crystalization" of amyloid: A pathogenic mechanism in Alzheimer's disease and scrapie. Cell **73:** 1055-1058.
39. SHOJI, M., T. E. GOLDE, J. GHISO, T. T. CHEUNG, S. ESTUS, L. M. SHAFFER, X-D. CAI, D. M. MCKAY, R. TINTNER, B. FRANGIONE & S. G. YOUNKIN. 1992. Production of the Alzheimer's amyloid β protein by normal processing. Science **258:** 126-129.
40. SEUBERT, P., T. OLTERSDORF, M. G. LEE, R. BARBOUR, C. BLOMQUIST, D. L. DAVIS, K. BRYANT, L. C. FRITZ, D. GALASKO, L. J. THAL, I. LIEBERBURG & D. B. SCHENK. 1993. Secretion of β-amyloid precursor protein cleaved at the amino terminus of the β-amyloid peptide. Nature **361:** 260-263.
41. HAASS, C. & D. J. SELKOE. 1993. Cellular processing of β-amyloid precursor protein and the genesis of amyloid β-peptide. Cell **75:** 1039-1042.
42. FRASER, P. E., J. T. NGUYEN, W. K. SUREWICZ & D. A. KIRSCHNER. 1992. pH-dependent structural transitions of Alzheimer amyloid peptides. Biophys. J. **60:** 1190-1201.

43. INOUYE, H., P. E. FRASER & D. A. KIRSCHNER. 1993. Structure of β-crystallite assemblies formed by Alzheimer β-amyloid protein analogues: Analysis by x-ray diffraction. Biophys. J. **64:** 502-519.
44. DURELL, R. S., R. H. GUY, N. ARISPE, E. ROJAS & H. B. POLLARD. 1994. Theoretical models of the ion channel structure of amyloid β-protein. Biophys. J., in press.
45. POLLARD, H. B., E. ROJAS & N. ARISPE. 1994. β-amyloid in Alzheimer disease. Therapeutic implications CNS Drugs **2:** 1-6.

Molecular Pharmacology of Voltage-Gated Calcium Channels

DAVID J. TRIGGLE

State University of New York
School of Pharmacy
Buffalo, New York 14260

Calcium is a cation of critical cellular significance. Its movements and storage are regulated at a number of sites (FIG. 1). Defects in these control processes are believed to be associated with a variety of acute and chronic losses of cellular function and with or causal to a number of experimental and clinical disease states including those of the central nervous system.[1-3]

The control of Ca^{2+} movements is a fundamental preoccupation of the cell dictated by both the physiology and the pathology of Ca^{2+}.[4] The control of Ca^{2+} movements by specific drugs will therefore have important therapeutic implications. In principle, such drugs should be available for all of the control sites depicted in FIGURE 1. In practice, however, therapeutically available agents are available only for one site, the voltage-gated Ca^{2+} channel.[5-7]

VOLTAGE-GATED CALCIUM CHANNELS

Voltage-gated Ca^{2+} channels, activated and inactivated by changes in membrane potential, exist as members of a superfamily of transmembrane proteins, homologous to Na^{+} and K^{+} channels.[8] At least four major classes of voltage-gated Ca^{2+} channel exist that are distinguished by electrophysiological, permselectivity, and pharmacological criteria and by their distribution and postulated cellular functions (TABLE 1).[9-12] The therapeutically available Ca^{2+} channel antagonists, including the prototypical verapamil, nifedipine, and diltiazem (FIG. 2) and the second-generation agents amlodipine, felodipine, isradipine, nicardipine, and nimodipine, interact at the L-type channels whose properties dominate the cardiovascular system where their large, long-lasting conductances are appropriate for excitation-contraction coupling processes. The T channel has transient characteristics that underlie its likely pacemaking functions; no truly selective drugs are yet known. The N and P channels and other pharmacologically less well characterized channels are of major importance in neural tissues, and they likely represent major strategic targets for subsequent therapeutic intervention in diseases of the central nervous system. Potent and specific toxins, including the conotoxins and agatoxins, are known for N- and P-type channels.[13,14] Ca^{2+} channels may thus be considered a class of pharmacological receptors with specific drug-binding sites.[15]

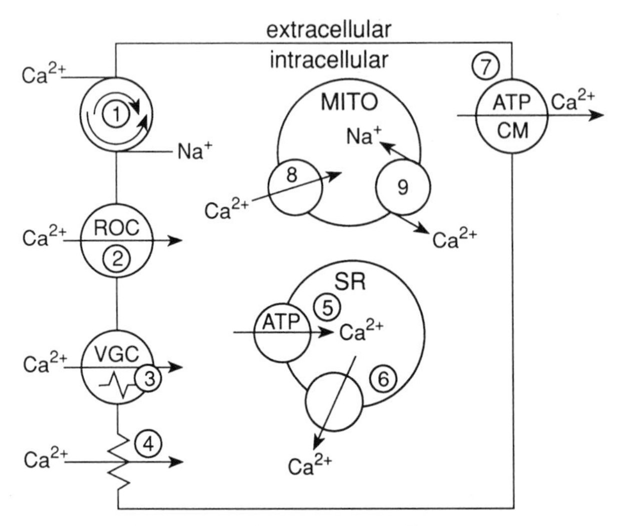

FIGURE 1. Regulation of cellular calcium: (1) $Na^+ : Ca^{2+}$ exchanger; (2) receptor-operated channels; (3) voltage-gated channels; (4) leak pathways; (5) Ca^{2+} uptake into sarcoplasmic reticulum; (6) Ca^{2+} release channel; (7) plasma membrane Ca^{2+}-ATPase; (8,9) mitochondrial uptake and release processes.

PHARMACOLOGY OF THE CALCIUM CHANNEL

Because of the therapeutic availability of drugs active at the L-type channel, our understanding of the molecular pharmacology of this channel is particularly detailed (reviewed in refs. 5-7). The L-type channel is associated with at least three discrete categories of binding sites that serve as receptors for the phenylalkylamine (verapamil), 1,4-dihydropyridine (nifedipine), and benzothiazepine (diltiazem) classes of drug (FIG. 2). However, probably at least six or more drug-binding sites are associated with this channel class including those for the diphenylbutylpiperidines, lactamimides, benzoylpyrroles, indolizines, and other synthetic structures.[15,16] The pharmacology of these additional sites has not been well explored.

The available first and second generation Ca^{2+} antagonists are employed in cardiovascular disorders including angina, some cardiac arrhythmias, and hypertension. Although these agents have a common target and a common endpoint–the inhibition of Ca^{2+} current through L-type channels–they clearly exhibit quantitative and qualitative differences in their pharmacological behavior and clinical efficacy.[17,18] Under most conditions of clinical use these agents do not produce significant neuronal effects, although L-type channels are widely distributed in the peripheral and central nervous systems.[19,20] However, under some experimental and clinicopathological conditions neuronal effects are observed that may be of clinical benefit (TABLE 2).[21-25] Thus, factors that determine the selectivity of action of the Ca^{2+} antagonists are of particular interest in determining their clinical profile.

TABLE 1. Classification of Voltage-Gated Calcium Channels

Property	L	T	N	P
Conductance (pS)	25	8	12–20	10–12
Activation threshold	High	Low	High	Moderate
Inactivation rate	Slow	Fast	Moderate	Rapid
Permeation	$Ba^{2+} > Ca^{2+}$	$Ba^{2+} = Ca^{2+}$	$Ba^{2+} > Ca^{2+}$	$Ba^{2+} > Ca^{2+}$
Function	E-C coupling cardiovascular system, smooth muscle, endocrine cells, and some neurons	Cardiac sinoatrial node: neuronal spiking repetitive spike activity in neurons and endocrine cells	Neuronal only; neurotransmitter release	Neuronal only?; neurotransmitter release
Pharmacologic sensitivity: 1,4-Dihydropyridines (activators/antagonists) Phenylalkylamines Benzothiazepines	Sensitive	Insensitive	Insensitive	Insensitive
ω-Conotoxin	Sensitive? (some)	Insensitive	Sensitive	Insensitive
Octanol, amiloride	Insensitive?	Sensitive	Insensitive	?
Funnel web spider toxin	Insensitive	Insensitive	Insensitive	Insensitive

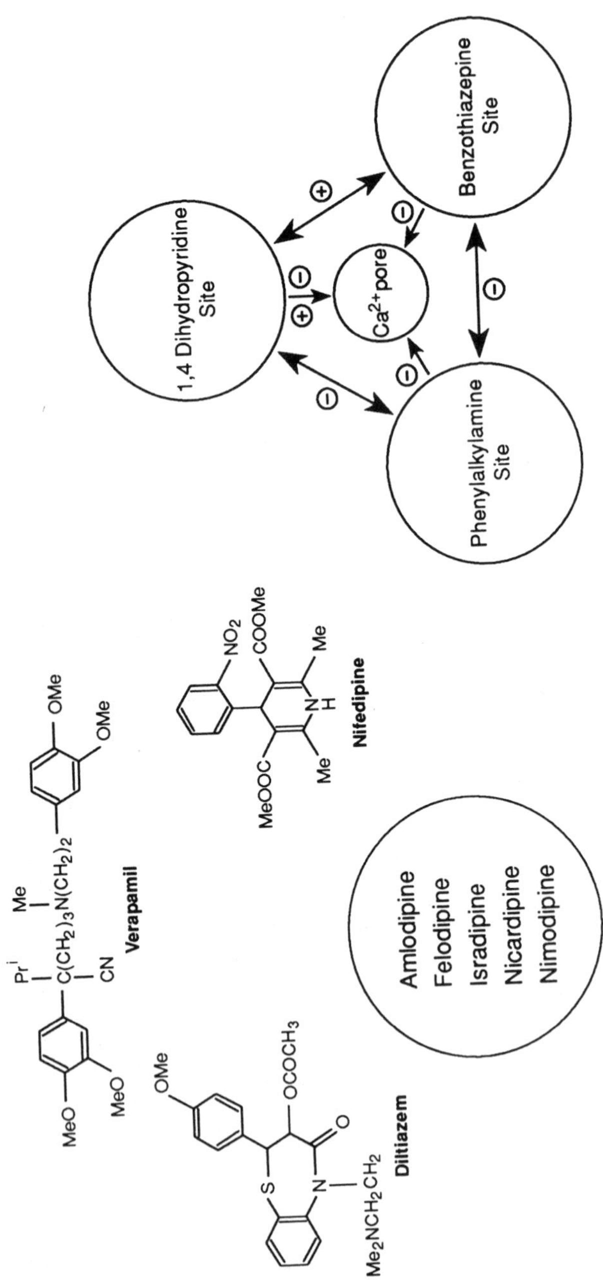

FIGURE 2. Organization of drug-binding sites at the L-type voltage-gated Ca^{2+} channel.

TABLE 2. Experimental and Clinical Roles of Calcium Antagonists in the Central Nervous System

Clinical Disorder	References
Dementias	Ban *et al.*[26]; Tollefson[27]; Scriabine *et al.*[25]
Depression	Eccleston and Cole[28]; Lewis and Bacher[29]
Epilepsy	Schmidt and Reid[30]; deFalco *et al.*[31]
Ischemia	Gelmers *et al.*[32]; Wong and Haley[33]; Grotta[34]; Levene *et al.*[35]
Mania	Dubovsky *et al.*[36]; Giannini *et al.*[37,38]; Brunet *et al.*[39]
Migraine	Greenberg[40]; Solomon[41]; Anderson and Vinge[42]
Panic	Goldstein[43]
Stuttering	Brady *et al.*[44]
Tardive dyskinesia	Adler *et al.*[45]; Buck and Havey[46]; Reiter *et al.*[47]; Kushnie and Ratner[48]
Tourette's syndrome	Micheli *et al.*[49]

SELECTIVITY OF ACTION OF CALCIUM ANTAGONISTS

Several factors contribute to the observed selectivity of actions of the Ca^{2+} antagonists including the following:[5,15,18] (1) Pharmacokinetics: distribution, absorption, metabolism, or elimination; (2) Mode of Ca^{2+} mobilization: voltage-gated and receptor-operated channels and intracellular release; (3) Class and subclass of voltage-gated channel; (4) Channel localization; (5) State-dependent processes: frequency- and voltage-dependent interactions; and (6) Disease state: influence of pathology on channel number and function. All of these factors contribute to the observed selectivity pattern, but the influence of membrane potential and pathological state on the regulation of drug interactions with Ca^{2+} channels are of particular interest.[15,50–52]

STATE-DEPENDENT INTERACTIONS

Voltage-gated ion channels exist in three families of states, resting, open, and inactivated (FIG. 3). According to the modulated receptor hypothesis, the affinities and/or access of drugs to their specific sites may vary according to channel state.[53–55] The equilibrium between these states is determined by several factors including membrane potential as established by the frequency and intensity of depolarizing stimuli.

At the L-type channel, verapamil and diltiazem exhibit prominent frequency-dependent interactions and nifedipine (and other 1,4-dihydropyridines) prominent voltage-dependent interactions whereby activity increases with increasing frequency of stimulation or increasing maintained depolarization, respectively.[15,50,56–60] These observations underlie the general cardiovascular profile of these Ca^{2+} antagonists.[17,18]

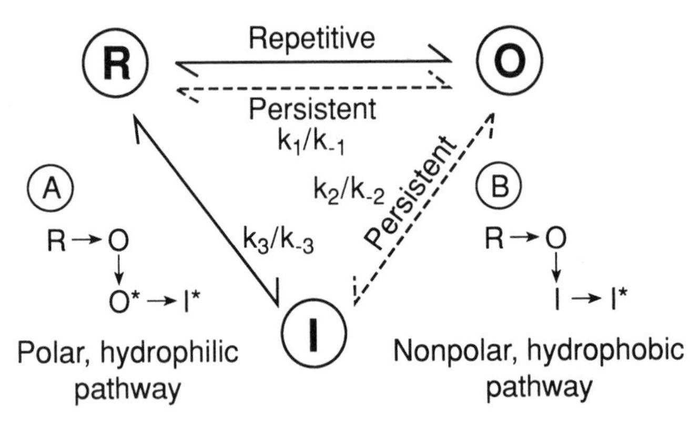

FIGURE 3. Voltage-gated calcium channel and interconversion between resting, open, and inactivated states. **A** and **B** represent hydrophilic and hydrophobic pathways of drug access to the inactivated channel state.

Verapamil and diltiazem are antiarrhythmic and less vascular-selective than are nifedipine and other 1,4-dihydropyridines; these differences in cardiovascular profile likely represent selective interactions through the open and inactivated channel states, respectively. Verapamil and diltiazem are charged species at physiological pH, and the 1,4-dihydropyridines are dominantly hydrophobic species with high membrane : water partition coefficients.[61] The binding sites for verapamil and the 1,4-dihydropyridines are located at the internal and external domains of the $alpha_1$ subunit of the Ca^{2+} channel; accordingly, access pathways for the drug and cell membrane composition that may influence drug partitioning likely contribute to the observed selectivity profile.[62]

Direct evidence for state-dependent interactions of Ca^{2+} channel antagonists derives from both electrophysiological and radioligand binding studies. Electrophysiological studies indicate a higher affinity of the drugs for the inactivated or open states of the channel (TABLE 3). More limited radioligand binding data indicate also that the affinities of the 1,4-dihydropyridine antagonists are increased by maintained depolarization (TABLE 4).

In principle, drug binding to resting, open, or inactivated states may mediate channel inhibition. Drugs that do not exhibit state-dependent interactions, including ω-conotoxin at N channels,[73] are not selective agents and will produce channel blockade independent of channel state as regulated by membrane potential. In contrast, antagonists at the L channel with marked state-dependence of interactions will block preferentially channels in the open or inactivated states. When channel opening is brief, the kinetics of drug binding will not permit drug-channel equilibrium to be achieved at therapeutically significant concentrations of the drug. Thus, at plasma concentrations that mediate cardiovascular actions, the Ca^{2+} antagonists may have little significant neuronal impact. However, pathological or experimental conditions that generate abnormal neuronal firing patterns or prolonged depolarization will increase Ca^{2+} antagonist efficacy and demonstrate neuronal antagonism (TABLE 2). Consistent with this interpretation the 1,4-dihydropyridine Ca^{2+} channel activator Bay

TABLE 3. State-Dependent Interactions of Calcium Channel Antagonists

Tissue	Antagonist	K_D, M			Ref.
		Resting	Open	Inactivated	
Cardiac purkinje	Nisoldipine	1.3×10^{-6}	—	1.0×10^{-9}	59
Ventricle	Nitrendipine	7.3×10^{-7}	—	2.5×10^{-9}	60
Mesenteric artery	Nisoldipine	1.2×10^{-8}	—	7.0×10^{-11}	63
	Nitrendipine	2.2×10^{-7}	—	4.6×10^{-10}	64
Dorsal root ganglion	Nimodipine	4.0×10^{-8}	—	1.3×10^{-9}	65
		2.0×10^{-6}	—	3.0×10^{-9}	66
GH_3 cells	(+)RBay k8644	$>1.0 \times 10^{-8}$	—	4.0×10^{-9}	67
GH_4C_1 cells	Nimodipine	7.0×10^{-6}	5.0×10^{-10}	—	68

K 8644 (and related agents) produces a variety of readily observed neuronal effects from neurotransmitter release to behavioral syndromes.[74–76]

CHANNEL REGULATION

That voltage-gated Ca^{2+} channels are regulated by homologous and heterologous influences and altered in expression and function in a number of experimental and clinical disease states in both cardiac and neuronal preparations is now well known (reviewed in refs. 51 and 52). Thus, neuronal and cardiac L-type channels are up- and downregulated during chronic channel antagonist and activator administration,[77–79] during hyper- and hypothyroidism,[80] and neuronal channels are upregulated after chronic morphine or alcohol administration.[81–83] Neuronal lesions including kainic acid[84,85] and 6-hydroxydopamine[86,87] induce down- and upregulation, respectively.

TABLE 4. Calcium Channel Antagonist Binding in Polarized and Depolarized Preparations

Tissue	Antagonist	K_D, M		Ref.
		Polarized	Depolarized	
Rat cardiomyocytes	(+) [^{3}H]Isradipine	3.5×10^{-9}	6.3×10^{-11}	69
Rat aorta	(+) [^{3}H]Isradipine	2.5×10^{-10}	6.8×10^{-11}	70
Rat mesenteric	(+) [^{3}H]Isradipine	2.5×10^{-10}	4.1×10^{-11}	71
PC12 cells	(+) [^{3}H]Isradipine	2.7×10^{-10}	5.5×10^{-11}	72
Rat cerebellar granule cells	(+) [^{3}H]Isradipine	3.8×10^{-9}	2.5×10^{-10}	[a]
GH_4C_1	(+) [^{3}H]Isradipine	2.2×10^{-9}	1.1×10^{-10}	[a]

[a] Data from J. Liu, R. Bangalore, A. Rutledge, and D. J. Triggle.

Channels are regulated in both experimental and clinical disease states. Cardiac channels are downregulated in rodent[88] and human congestive heart failure.[89] In Lambert-Eaton syndrome, a myasthenic-like disorder often associated with small cell lung carcinoma, autoantibodies are expressed against nerve terminal Ca^{2+} channels with resultant loss of channel function.[90,91] Antibodies against L-type channels are found in serum from patients with amyotrophic lateral sclerosis, and antibody titer correlates with the severity of the disease.[92,93]

Among the regulatory factors depolarization is of particular importance because it is a physiological signal, but one that can also assume a pathological role during conditions of ischemia when extracellular K^+ accumulates. These roles are of particular importance in neuronal systems because membrane potential and Ca^{2+} movements may be critically important to neuron survival, neurite development, and neuronal death. Substantial evidence exists that bioelectric activity has a marked influence on neuronal development and survival,[94,95] and neuronal survival in culture depends on elevated K^+ concentrations that are presumably mimicking *in vivo* activity.[96,97] This depolarization-mediated neuronal survival is Ca^{2+}-dependent and presumably reflects the requirement for Ca^{2+} in neuronal growth and differentiation.[98] It may be assumed that a well-regulated Ca^{2+} mobilization process, including voltage-gated Ca^{2+} channels, occurs during neuronal growth and development. Disruptions in this process are likely to be detrimental to the survival of the cell.[99,100] Alterations in L-type channel density have been reported subsequent to neuronal ischemia. Transient forebrain ischemia in rats is associated with a major reduction in 1,4-dihydropyridine sites, and muscarinic and adenosine receptors, in CA_1 neurons.[101,102] Similarly, ischemic gerbil brain shows reductions in 1,4-dihydropyridine sites, and adenosine and glutamate receptors, in the vulnerable CA_1 and CA_3 regions.[103,104] However, in global ischemia subsequent to cardiac arrest a significant increase of some 25% of 1,4-dihydropyridine sites occurs in beagle brain.[105]

Chronic depolarization downregulates L-type channel number and function in PC 12 cells,[106,107] chick retinal neurons,[108] rat myenteric neurons,[109] and molluscan neurons.[110] In contrast, chronic depolarization of rat cardiac myocytes did not induce channel downregulation.[108] It is probable that several Ca^{2+} channel types are all regulated by membrane potential, but that underlying mechanisms may be different. In rat myenteric neurons the sustained and 1,4-dihydropyridine-sensitive Ca^{2+} current was reduced more slowly and recovered more rapidly from chronic depolarization than did the decaying component of current.[109] Downregulation of L-type current in chick retinal neurons is well characterized; it is time-, K^+-, and Ca^{2+}-dependent and can be mimicked by Ca^{2+} ionophores and prevented by Ca^{2+} antagonists.[108] Changes in both membrane potential and intracellular Ca^{2+} are likely involved in this downregulation process.

The long-term adaptive responses of Ca^{2+} channels to depolarization are likely a part of the general neuronal strategy for growth and development. It is also tempting to associate the downregulation of Ca^{2+} channels in neurons subsequent to long-term depolarization with protection against cellular Ca^{2+} overload. This may, however, be an oversimplification. The process may not be fast enough to protect against cellular injury caused by Ca^{2+}, cardiac cell channels are not apparently similarly affected, and, furthermore, neuronal survival and growth may be similarly dependent on cell

TABLE 5. Short-Term Depolarization-Mediated Regulation of Neuronal L-Type Channels[a]

Cell	K^+_{ext} (mM)	[^{3}H]1,4-Dihydropyridine Binding K_D ($\times 10^{-9}$ M)	B_{max} (fmol/mg)
Cerebellar granule	5	2.15	214.0
	50	0.11	24.0
GH^4C^1	5	3.75	364.2
	50	0.25	70.6

[a] Data from J. Liu, R. Bangalore, A. Rutledge, and D. J. Triggle. Cells were incubated for 90 minutes under polarizing or depolarizing conditions during the radioligand binding assay.

depolarization. However, it is now established that Ca^{2+} channels in neurons and neurosecretory cells are also subject to short-term regulation on depolarization.

Chick retinal neurons[108] and cerebellar granule, GH_3, and GH_4C_1 cells in culture all exhibit L-type 1,4-dihydropyridine-sensitive Ca^{2+} channels. Short-term depolarization (<120 minutes) with elevated K^+ produces a time- and concentration-dependent loss of binding sites (TABLE 5) and Ca^{2+} uptake which are rapidly restored on repolarization.[111] Whole cell and total cell homogenate binding studies demonstrate that 1,4-dihydropyridine binding sites are transiently internalized from the cell surface.

These demonstrations of short- and long-term regulation of voltage-gated Ca^{2+} channels resemble similar events in conventional receptor systems including the beta-adrenoceptor.[112,113] In the beta-adrenoceptor system there are two short-term desensitization processes, both reversible and independent of protein synthesis, mediated through receptor phosphorylation and internalization in a light membrane fraction and a long-term downregulation process that involves receptor internalization and lysosomal processing.

CONCLUSIONS

Voltage-gated Ca^{2+} channels are pharmacological receptors. They can be classified according to a variety of criteria including their sensitivity to synthetic and natural drugs and toxins. The L-type 1,4-dihydropyridine-sensitive channel is the best characterized, primarily because of the clinical availability of these selective drugs in cardiovascular therapies. However, these channels are abundant in the central nervous system, and under some conditions L-type antagonists do exhibit neuronal effects that may translate into clinical efficacy. There is active interest in synthetic nonpeptide molecules active at the N- and P-type channels that appear to be specific to neurons.

The L-type channels in neurons and neurosecretory cells are regulated by a variety of homologous and heterologous influences including depolarization. Regulation by depolarization shows analogies to other receptor systems because both rapid and slow processes occur. These processes are likely to be important to mechanisms of neuronal plasticity and development and to neuronal protection.

REFERENCES

1. CAMPBELL, A. K. 1983. Intracellular Calcium. Its Universal Role as Regulator. J. Wiley & Sons Inc. New York.
2. FERRANTE, J. & D. J. TRIGGLE. 1990. Drug- and disease-induced regulation of voltage-dependent calcium channels. Pharmacol. Rev. **42:** 29-42.
3. TRIGGLE, D. J. 1992. Calcio, canali del calcio e calcio antagonisti nelle pathologie del sistema nervoso centrale. Antagonisti del calcio, Fisiopatol. Farmacol. Terapia **3:** 35-48.
4. KRETSINGER, R. 1976. Calcium in biological systems. Coord. Chem. Rev. **18:** 29-124.
5. JANIS, R. A., P. J. SILVER & D. J. TRIGGLE. 1987. Drug action and cellular calcium regulation. Adv. Drug Res. **16:** 309-591.
6. GODFRAIND, T. R., R. MILLER & M. WIBO. 1986. Calcium antagonism and calcium entry blockade. Pharmacol. Rev. **38:** 321-416.
7. EPSTEIN, M., ED. 1992. Calcium Antagonists in Clinical Medicine. Hanley and Belfus. Philadelphia.
8. CATERALL, W.A. This volume.
9. TSIEN, R. W., P. HESS, E. W. MCCLESKEY & R. L. ROSENBERG. 1987. Calcium channels: Mechanisms of selectivity, permeation and block. Ann. Rev. Biophys. Biophys. Chem. **16:** 265-290.
10. BEAN, B. P. 1989. Classes of calcium channels in vertebrate cells. Ann. Rev. Physiol. **51:** 367-384.
11. HESS, P. 1990. Calcium channels in vertebrate cells. Ann. Rev. Neurosci. **13:** 337-356.
12. MILLER, R. J. 1992. Voltage-sensitive Ca^{2+} channels. J. Biol. Chem. **267:** 1403-1406.
13. OLIVERA, B. M., J. RIVIER, J. K. SCOTT, D. R. HILLYARD & L. J. CRUZ. 1991. Conotoxins. J. Biol. Chem. **266:** 22067-22070.
14. JACKSON, H. & T. N. PARKS. 1989. Spider toxins: Recent applications in neurobiology. Ann. Rev. Neurosci. **12:** 405-414.
15. JANIS, R. A. & D. J. TRIGGLE. 1991. Drugs acting on calcium channels. *In* Calcium Channels: Their Properties, Functions, Regulation and Clinical Relevance. L. Hurwitz, L. D. Partridge, J. F. Leach, Eds.: 195-249. CRC Press. Boca Raton, FL.
16. RAMPE, D. & D. J. TRIGGLE. 1993. New synthetic ligands for L-type voltage-gated calcium channels. Prog. Drug Res. **40:** 191-238.
17. TRIGGLE, D. J. 1992. Calcium channel antagonists: Mechanisms of action, vascular selectivities and clinical relevance. Cleveland Clin. J. Med. **59:** 617-627.
18. TRIGGLE, D. J. 1992. Biochemical and pharmacologic differences among calcium channel antagonists: Clinical implications. *In* Calcium Antagonists In Clinical Medicine. M. Epstein, Ed.: 1-27. Hanley and Belfus. Philadelphia.
19. WESTENBROEK, R. E., M. K. ANIJANIAN & W. A. CATTERALL. 1990. Clustering of L-type Ca^{2+} channels at the base of major dendrites in hippocampal pyramidal neurons. Nature **347:** 281-283.
20. HULLIN, R., M. BIEL, V. FLOCKERZI & F. HOFMANN. 1993. Tissue-specific expression of calcium channels. Trends Cardiovasc. Med. **3:** 48-53.
21. DUBOVSKY, S. L. & R. D. FRANKS. 1983. Intracellular calcium ions in affective disorders: A review and hypothesis. Biol. Psychiatry **18:** 781-797.
22. TIETZE, K. J., M. L. SCHWARTZ & P. H. VLASSES. 1987. Calcium antagonists in cerebral/peripheral disorders. Current status. Drugs **32:** 531-538.
23. WAUQUIER, A. 1988. On the possible central effects of calcium antagonists. Acta Otolaryngol. (Suppl.) **460:** 80-86.
24. SCRIABINE, A., T. SCHUURMAN & J. TRABER. 1989. Pharmacological base for the use of nimodipine in central nervous system disorders. FASEB J. **3:** 1799-1806.
25. HOSCHL, C. 1991. Do calcium antagonists have a place in the treatment of mood disorders? Drugs **42:** 721-729.
26. BAN, T. A., L. MOREY & E. AGUGLIA. 1991. Nimodipine in the treatment of old age dementias. Prog. Neuropsychopharmacol. Biol. Psychiatry **14:** 525-551.

27. TOLLEFSON, G. D. 1990. Short-term effects of the calcium channel blocker nimodipine (Bay e 9736) on the management of primary degenerative dementia. Biol. Psychiatry **27:** 1133-1142.
28. ECCLESTON, D. & A. J. COLE. 1990. Calcium channel blockade and depressive illness. Brit. J. Psychiatry **156:** 889-891.
29. LEWIS, H. A. & N. M. BACHER. 1984. Verapamil and depression. Am. J. Psychiatry **141:** 613.
30. SCHMIDT, D. & S. REID. 1989. Klinische Relvanz von Calcium-Antagonisten in der Behandlung von Epilepsine. Arzneim. Forsch. **39:** 156-158.
31. DEFALCO, F. A., V. BARTIMORO, L. MAJELLO, G. DIGERONIMO & P. MUNDO. 1992. Calcium antagonist nimodipine in intractable epilepsy. Epilepsia **33:** 343-345.
32. GELMERS, H. J., K. GORTER, C. J. DEWEERDT & H.-J. A. WIEZER. 1988. A controlled trial of nimodipine in acute ischemic stroke. N. Engl. J. Med. **318:** 203-207.
33. WONG, M. C. W. & E. C. HALEY. 1989. Calcium antagonists: Stroke therapy coming of age. Curr. Conc. Cerebrovasc. Dis.-Stroke **24:** 31-36.
34. GROTTA, J. C. 1991. Clinical aspects of the use of calcium antagonists in cerebrovascular disease. Clin. Neuropharmacol. **14:** 373-390.
35. LEVENE, M. I., N. A. GIBSON, A. C. FENTON, E. PAPATHOMA & D. BARNETT. 1990. The use of a calcium channel blocker, nicardipine, for severely asphyxiated newborn infants. Dev. Med. Child. Neurol. **32:** 567-574.
36. DUBOVSKY, S. L., R. D. FRANKS, M. LIFSCHITZ & R. COEN. 1982. Effectiveness of verapamil in the treatment of a manic patient. Am. J. Psychiatry **139:** 502-504.
37. GIANNINI, A. J., W. L. HOUSER, R. H. LOISELLE, M. G. GIANNINI & W. A. PRICE. 1984. Antimanic effects of verapamil. Am. J. Psychiatry **141:** 1602-1603.
38. GIANNINI, A. J., R. H. LOISELLE, W. A. PRICE & M. G. GIANNINI. 1985. Comparison of antimanic efficacy of clonidine and verapamil. J. Clin. Pharmacol. **25:** 307-308.
39. BRUNET, G., B. CERLICH, P. ROBERT, S. DUMAS, E. SOVETRE & G. DARCOUNT. 1990. Open trial of a calcium antagonist, nimodipine, in acute mania. Clin. Neuropharmacol. **13:** 224-228.
40. GREENBERG, D. A. 1986. Calcium channel antagonists and the treatment of migraine. Clin. Neuropharmacol. **9:** 311-328.
41. SOLOMON, G. D. 1989. Verapamil in migraine prophylaxis in a five year review. Headache **29:** 425-427.
42. ANDERSON, K. E. & E. VINGE. 1990. B-Adrenoceptor blockers and calcium antagonists in the prophylaxis and treatment of migraine. Drugs **39:** 355-373.
43. GOLDSTEIN, J. A. 1985. Calcium channel blockers in the treatment of panic disorder. J. Clin. Psychiatry **46:** 12.
44. BRADY, J. P., T. R. PRICE, T. W. MCCALLISTER & K. DETRICH. 1989. A trial of verapamil in the treatment of stuttering in adults. Biol. Psychiatry **25:** 626-630.
45. ADLER, L., E. DUNCAN, S. REITER, B. ANGRIST, E. PESELOW & J. ROTROSEN. 1988. The effects of calcium channel antagonists on tardive dyskinesia and psychoses. Psychopharmacol. Bull. **24:** 421-425.
46. DUCK, O. D. & R. HAVEN. 1988. Treatment of tardive dyskinesia with verapamil. J. Clin. Psychopharmacol. **8:** 303-304.
47. REITER, S., L. ADLER, B. ANGRIST, E. PESELORO & J. ROTROSEN. 1989. Effects of verapamil on tardive dyskinesia and psychoses in schizophrenic patients. J. Clin. Psychiatry **50:** 26-27.
48. KUSHNIE, S. L. & J. T. RATNER. 1989. Calcium channel blockers for tardive dyskinesia in geriatric psychiatric patients. Am. J. Psychiatry **146:** 1218-1219.
49. MICHELI, F., M. GATBO, E. LEKHUNIEO, C. MANGONE, M. FERNANDEZ-PARDAL, R. PIKIELNY & I. C. PARERA. 1990. Treatment of Tourette's syndrome with calcium antagonists. Clin. Neuropharmacol. **12:** 77-83.
50. TRIGGLE, D. J. 1989. Structure-function correlations of 1,4-dihydropyridine calcium channel antagonists and activators. *In* Molecular and Cellular Mechanisms of Antiar-

rhythmic Agents. L. M. Hondeghem, Ed.: 269-291. Futura Publishing. Mt. Kisco, NY.

51. FERRANTE, J. & D. J. TRIGGLE. 1990. Homologous and heterologous regulation of voltage-dependent calcium channels. Biochem. Pharmacol. **39:** 1267-1270.
52. GOPALAKRISHNAN, M. & D. J. TRIGGLE. 1990. The regulation of receptors, ion channels and G proteins in congestive heart failure. Cardiovasc. Drug Rev. **8:** 255-302.
53. HILLE, B. 1977. The pH-dependent rate of action of local anesthetics on the node of Ranvier. J. Gen. Physiol. **69:** 475-496.
54. HONDEGHEM, L. M. & B. G. KATZUNG. 1977. Time- and voltage-dependent interaction of antiarrhythmic drugs with cardiac sodium channels. Biochim. Biophys. Acta **472:** 373-398.
55. STARMER, C. F., A. O. GRANT & H. C. STRAUSS. 1984. Mechanisms of use-dependent block of sodium channels in excitable membranes by local anesthetics. Biophys. J. **46:** 15-27.
56. HONDEGHEM, L. M. & B. G. KATZUNG. 1985. Antiarrhythmic drugs: The modulated receptor mechanism of action of sodium and calcium channel blocking drugs. Ann. Rev. Pharmacol. **24:** 387-423.
57. BAYER, R., R. HENNCKES, R. KAUFMANN & R. MANNHOLD. 1975. Inotropic and electrophysiological actions of verapamil and D 600 in mammalian myocardium. Naunyn-Schmied. Arch. Pharmacol. **290:** 49-68.
58. MCDONALD, T. F., D. PELZER & W. TRAUTWEIN. 1984. Cat ventricle muscle treated with D600: Characteristics of calcium channel block and unblock. J. Physiol. **352:** 217-244.
59. SANGUINETTI, M. C. & R. S. KASS. 1984. Voltage-dependent block of calcium channel current in calf cardiac Purkinje fibers by dihydropyridine calcium channel antagonists. Circ. Res. **55:** 336-348.
60. BEAN, B. P. 1984. Nitrendipine block of cardiac calcium channels: High affinity binding to the inactivated state. Proc. Natl. Acad. Sci. USA **81:** 6388-6392.
61. MASON, R. P., D. G. RHODES & L. G. HERBETTE. 1991. Reevaluating equilibrium and kinetic binding parameters for lipophilic drugs based on a structural model for drug interaction with biological membranes. J. Med. Chem. **34:** 869-877.
62. CATTERALL, W. A. & J. STRIESSNIG. 1992. Receptor sites for calcium antagonists. Trends Pharmacol. Sci. **13:** 256-262.
63. NELSON, M. T. & J. F. WORLEY. 1989. Dihydropyridine inhibition of single calcium channels and contraction in rabbit mesenteric artery depends on voltage. J. Physiol. **412:** 65-91.
64. BEAN, B. P., M. STUREK, A. PUGA & K. HERMSMEYER. 1986. Calcium channels in muscle cells isolated from rat mesenteric arteries. Modulation by dihydropyridine drugs. Circ. Res. **59:** 229-235.
65. MCCARTHY, R. T. 1989. Nimodipine block of L-type calcium channels in dorsal root ganglion cells. *In* Nimodipine and Central Nervous System Function: New Vistas. J. Traber and W. Gispen, Eds.: 35-49. Schattauer. Stuttgart.
66. BOLAND, L. M. & R. W. DINGLEDINE. 1990. Multiple components of both transient and sustained barium currents in a rat dorsal root ganglion cell line. J. Physiol. **422:** 223-236.
67. MCCARTHY, R. T. & C. J. COHEN. 1986. The enantiomers of Bay K 8644 have differential effects on Ca channel gating in rat anterior pituitary cells. Biophys. J. **49:** 432a.
68. COHEN, C. J. & R. T. MCCARTHY. 1987. Nimodipine block of calcium channels in rat anterior pituitary cells. J. Physiol. **387:** 195-225.
69. WEI, X.-Y., A. RUTLEDGE & D. J. TRIGGLE. 1989. Voltage-dependent binding of 1,4-dihydropyridine Ca antagonists and activators in cultured neonatal rat ventricular myocytes. Mol. Pharmacol. **35:** 541-552.

70. MOREL, N. & T. GODFRAIND. 1991. Characterization in rat aorta of the binding sites responsible for blockade of noradrenaline-evoked calcium entry by nisoldipine. Brit. J. Pharmacol. **102:** 467-477.
71. MOREL, N. & T. GODFRAIND. 1988. Selective modulation by membrane potential of the interaction of some calcium entry blockers with calcium channels in rat mesenteric artery. Brit. J. Pharmacol. **95:** 252-258.
72. GREENBERG, D., C. A. CARPENTER & R. O. MESSING. 1986. Depolarization-dependent binding of the calcium channel antagonist (+) [^{3}H] PN 200 110 to intact cultured PC 12 cells. J. Pharmacol. Exp. Therap. **238:** 1021-1027.
73. WANG, X., S. N. TREISTMAN & J. R. LEMOS. 1992. Two types of high threshold calcium currents inhibited by ω-conotoxin in nerve terminals of rat neurohypophysis. J. Physiol. **445:** 181-199.
74. MIDDLEMISS, D. N. 1985. The calcium channel activator, Bay K 8644, enhances K^+-evoked efflux of acetylcholine and noradrenaline from rat brain slices. Naunyn-Schmied. Arch. Pharmacol. **331:** 114-116.
75. BOLGER, G. T., B. A. WEISSMAN & P. SKOLNICK. 1985. The behavioral effects of the calcium agonist Bay K 8644 in the mouse: Antagonism by the calcium antagonist nifedipine. Naunyn-Schmied. Arch. Pharmacol. **328:** 373-377.
76. MOGILNICKA, E., A. CZYRAK & J. MAJ. 1988. Bay K 8644 enhances immobility in the mouse behavioral despair test, an effect blocked by nifedipine. Eur. J. Pharmacol. **151:** 307-311.
77. PANZA, G., J. A. GREBB, E. SANNA, A. G. WRIGHT & I. HANBAUER. 1985. Evidence for down-regulation of [^{3}H]nitrendipine recognition sites in mouse brain after long-term treatment with nifedipine or verapamil. Neuropharmacology **34:** 1113-1117.
78. GENGO, P., A. SKATTEBOL, J. F. MORAN, S. GALLANT, M. HAWTHORN & D. J. TRIGGLE. 1989. Regulation by chronic drug administration of neuronal and cardiac calcium channel, beta-adrenoceptor and muscarinic receptor levels. Biochem. Pharmacol. **37:** 627-633.
79. SKATTEBOL, A., D. J. TRIGGLE & A. M. BROWN. 1989. Homologous regulation of voltage-dependent Ca^{2+} channels by 1,4-dihydropyridines. Biochem. Pharmacol. **160:** 929-936.
80. HAWTHORN, M., P. GENGO, X.-Y. WEI, A. RUTLEDGE, J. F. MORAN, S. GALLANT & D. J. TRIGGLE. 1988. Effect of thyroid status on beta-adrenoceptors and calcium channels in rat cardiac and vascular tissue. Naunyn-Schmied. Arch. Pharmacol. **337:** 539-544.
81. RAMKUMAR, V., E. E. EL-FAKAHANY. 1988. Prolonged morphine treatment increases rat brain dihydropyridine binding sites: Possible involvement in the development of morphine dependence. Eur. J. Pharmacol. **146:** 73-83.
82. DOLIN, S., H. LITTLE, M. HUDSPITH, C. PAGONIS & J. LITTLETON. 1987. Increased dihydropyridine-sensitive calcium channels in rat brain may underlie ethanol physical dependence. Neuropharmacology **26:** 275-279.
83. MESSING, R. O., C. L. CARPENTER, C. L. DIAMOND & D. A. GREENBERG. 1985. Ethanol regulates calcium channels in clonal neural cells. Proc. Natl. Acad. Sci. USA **83:** 613-615.
84. SANNA, E., G. A. HEAD & I. HANBAUER. 1986. Evidence for a selective localization of voltage-sensitive Ca^{2+} channels in nerve cell bodies of corpus striatum. J. Neurochem. **47:** 1552-1557.
85. SKATTEBOL, A., R. E. HRUSKA, M. HAWTHORN & D. J. TRIGGLE. 1988. Kainic acid lesions decrease striatal dopamine receptors and 1,4-dihydropyridine sites. Neurosci. Lett. **89:** 85-89.
86. BOLGER, G. T., A. S. BASILE, A. J. JANOWSKY, S. M. PAUL & P. SKOLNICK. 1987. Regulation of dihydropyridine calcium antagonist binding sites in the rat hippocampus following neurochemical lesions. J. Neurol. Res. **17:** 285-290.

87. SKATTEBOL, A. & D. J. TRIGGLE. 1986. 6-Hydroxydopamine treatment increases beta-adrenoceptors and Ca^{2+} channels in rat heart. Eur. J. Pharmacol. **127:** 287-289.
88. GOPALAKRISHNAN, M., D. J. TRIGGLE, A. RUTLEDGE, Y. W. KWON, J. A. BAUER & H. L. FUNG. 1991. Regulation of K^+ and Ca^{2+} channels in experimental cardiac failure. Am. J. Physiol. **261:** H1979-1987.
89. TAKAHASHI, K., P. D. ALLEN, R. V. LACRO, A. R. MARKS, A. R. DENNIS, F. J. SCHOEN, W. GROSSMAN, J. D. MARSH & S. IZUMO. 1992. Expression of dihydropyridine receptor (Ca^{2+} channel) and calsequestrin genes in the myocardium of patients with end-stage heart failure. J. Clin. Invest. **90:** 927-935.
90. PEERS, C., B. LANG, J. NEWSOM-DAVIS & D. W. WRAY. 1990. Selective action of myasthenic syndrome antibodies on calcium channels in a rodent neuroblastoma cell line. J. Physiol. **421:** 293-308.
91. HEWETT, S. J. & W. D. ATCHISON. 1992. Specificity of Lambert-Eaton Myasthenic Syndrome immunoglobulin for nerve terminal calcium channel. Brain Res. **599:** 324-332.
92. SMITH, R. G., S. HAMILTON, F. HOFMANN, T. SCHNEIDER, W. NASTAINCZYK, L. BIRNBAUMER, E. STEFANI & S. H. APPEL. 1992. Serum antibodies to L-type calcium channels in patients with amyotrophic lateral sclerosis. N. Engl. J. Med. **327:** 1721-1728.
93. MAGNELLI, V., T. SAWADA, O. DELBONO, R. G. SMITH, S. H. APPEL & E. STEFANI. 1993. The action of amyotrophic lateral sclerosis immunoglobulins on mammalian single skeletal muscle channels. J. Physiol. **461:** 103-118.
94. WEISEL, T. N. & D. H. HUBEL. 1963. Effects of visual deprivation on morphology and physiology of cells in the cat's lateral geniculate body. J. Neurophysiol. **26:** 978-993.
95. BLACK, I. B. 1978. Regulation of autonomic development. Ann. Rev. Neurosci. **1:** 183-224.
96. SCOTT, S. & K. C. FISHER. 1970. Potassium concentration and the number of neurons in cultures of denervated ganglia. Exp. Neurol. **27:** 16-22.
97. GALLO, V., A. KINGSBURY, R. BALZS & O. S. JORGENSEN. 1987. The role of depolarization in the survival and differentiation of cerebellar granule cells in culture. J. Neurosci. **7:** 2203-2213.
98. KATER, S. B. & L. R. MILLS. 1991. Regulation of growth cone behavior by calcium. J. Neurosci. **11:** 891-899.
99. MEYER, F. B. 1989. Calcium, neuronal hyperexcitability and ischemic injury. Brain Res. Rev. **14:** 227-243.
100. SIESJO, B. K. 1990. Calcium in the brain under physiological and pathological conditions. Eur. Neurol. **30** (Suppl. 2): 3-9.
101. ONODERA, H. & K. KOGURE. 1990. Calcium antagonist, adenosine A1, and muscarinic binding in rat hippocampus after transient ischemia. Stroke **21:** 771-776.
102. HOGAN, M., A. GJERDE & A. HAKIM. 1991. Activity of dihydropyridine calcium channels following cerebral ischemia. Arzneim. Forsch. **41:** 332-333.
103. ARAKI, T., H. KATO, K. KOGURE & T. SAITO. 1991. Postischemic alteration of muscarinic acetylcholine, adenosine A1 and calcium antagonist binding sites in selectively vulnerable areas: An autoradiographic study of gerbil brain. J. Neurol. Sci. **106:** 206-212.
104. ARAKI, T., H. KATO, K. KOGURE & Y. KANAI. 1992. Long-term changes in gerbil brain neurotransmitter receptors following transient cerebral ischemia. Brit. J. Pharmacol. **107:** 437-442.
105. HOEHNER, P. J., T. J. J. BLANCK, R. ROY, R. E. ROSENTHAL & G. FISKUM. 1992. Alterations in voltage-dependent calcium channels in canine brain during global ischemia and perfusion. J. Cereb. Blood Flow Metab. **12:** 418-424.
106. DELORME, E. M. & R. MCGEE. 1986. Regulation of voltage-dependent Ca^{2+} channels by chronic changes in membrane potential. Brain Res. **397:** 189-192.

107. DeLorme, E. M., C. S. Rabe & R. McGee. 1988. Regulation of the number of functional voltage-sensitive Ca^{2+} channels on PC12 cells by chronic changes in membrane potential. J. Pharmacol. Exp. Therap. **244:** 838-847.
108. Ferrante, J., D. J. Triggle & A. Rutledge. 1991. The effects of chronic depolarization on L-type 1,4-dihydropyridine-sensitive, voltage-dependent Ca^{2+} channels in chick neural retina and rat cardiac cells. Can. J. Physiol. Pharmacol. **69:** 914-920.
109. Franklin, J. L., D. J. Fickbohm & A. L. Willard. 1992. Long-term regulation of neuronal calcium currents by prolonged changes of membrane potential. J. Neurosci. **12:** 1726-1735.
110. Berdan, R. C., J. C. Easaw & R. Wang. 1993. Alterations in membrane potential after axotomy at different distances from the soma of an identified neuron and the effect of depolarization on neurite outgrowth and calcium channel expression. J. Neurophysiol. **69:** 151-164.
111. Liu, J., R. Bangalore, A. Rutledge & D. J. Triggle. 1994. Modulation of L-type Ca^{2+} channels in clonal rat pituitary cells by membrane depolarization. Mol. Pharmacol. **45:** 1198-1206.
112. Perkins, J. P., W. P. Hausdorff & R. J. Lefkowitz. 1991. Mechanisms of ligand-induced desensitization of beta-adrenergic receptors. *In* Beta-Adrenergic Receptors. J. P. Perkins, Ed. Humana Press. Clifton, NJ.
113. Lefkowitz, R. J., W. P. Hausdorff & M. C. Caron. 1990. Role of phosphorylation in desensitization of the beta-adrenoceptor. Trends Pharmacol. Sci. **11:** 190-194.

Differential Phosphorylation, Localization, and Function of Distinct α_1 Subunits of Neuronal Calcium Channels

Two Size Forms for Class B, C, and D α_1 Subunits with Different COOH-Termini [a]

JOHANNES W. HELL, RUTH E. WESTENBROEK, ELICIA M. ELLIOTT, AND WILLIAM A. CATTERALL

Department of Pharmacology
University of Washington
Seattle, Washington 98195

Calcium influx through voltage-gated calcium channels controls a variety of neuronal functions such as membrane excitability and neurotransmitter release. Four types of voltage-gated calcium channels, designated T, L, N, and P, can be distinguished physiologically and pharmacologically.[1-4] Activation of the T-type channel requires only a small depolarization, whereas a stronger depolarization is necessary to activate the high-threshold L-, N-, and P-type channels. L-type channels are inhibited by dihydropyridines, phenylalkylamines, and benzothiazepines (see ref. 5 for review), N-type channels by ω-conotoxin GVIA,[6,7] and P-type channels by ω-agatoxin IV A.[8,9]

L-type channels from skeletal muscle are a complex of five subunits: α_1, α_2, β, γ, and δ.[10-12] cDNAs for each of these subunits have been cloned and sequenced.[13-16] The α_2 and δ subunits are encoded by the same gene and created by posttranslational proteolytic processing.[17,18] The two size forms of the α_1 subunit differ at their COOH-terminal ends, and this is probably also the result of a specific proteolytic cleavage step *in vivo*.[19,20] Expression studies in *Xenopus* oocytes or mammalian cells demonstrate that the α_1 subunit is sufficient to form a functional calcium channel,[21,22] but coexpression of other subunits affects its functional properties.[23-25]

The subunit composition of neuronal L- and N-type calcium channels is similar to that of the skeletal muscle L-type channel and includes α_1, $\alpha_2\delta$, and β subunits.[26-31] Utilizing skeletal muscle α_1 subunit cDNA for homologous screening allowed the identification of five classes (A-E) of neuronal calcium channel α_1 subunits.[32-44] Two different L-type channels are encoded by class C and D cDNAs. They

[a] This work was supported by National Institutes of Health Research Grant R01-NS22625 (to W. A. C.), by a postdoctoral fellowship from the Deutsche Forschungs Gemeinschaft (to J. W. H.), by a postdoctoral fellowship from the National Institutes of Health (to E. M. E.), and by the W. M. Keck Foundation.

are more closely related to the skeletal muscle L-type channel and to each other than to classes A, B, and E. Accordingly, the latter ones form the non-L-type subfamily of calcium channels. The class B α_1 subunit encodes the ω-conotoxin GVIA-sensitive N-type channel,[39,40] the class A α_1 subunit a high threshold channel with incompletely defined functional properties,[33] and the class E α_1 subunit a channel with an intermediate voltage activation threshold when expressed in *Xenopus* oocytes.[44]

Polyclonal antibodies were raised against peptides derived from sequences specific to class B, C, and D α_1 subunits. Binding sites of antibodies CNB1, CNC1, and CND1 are located in the cytoplasmic loop between domain II and III, and recognition sequences of antibodies CNB3, CNC2, and CND2 are derived from the COOH-terminal ends of full-length cDNAs of class B, C, and D α_1 subunits, respectively.[40,45–48] Antibody CND2 was produced against the peptide KYSHRQDYELQDF-GPGYSD, which represents amino acids 2123-2139 in the COOH-terminal region of the class D L-type α_1 subunit as cloned and sequenced from cDNA.[38] NH_3-terminal amino acids lysine and tyrosine are not found in the channel sequence, but were added for cross-linking and labeling purposes. (See ref. 45 for details of the antibody production.) Anti-CP(1382-1400) was raised against the highly conserved amino acid sequence following transmembrane region IVS6 of the skeletal muscle α_1 subunit.[49] This antibody recognizes not only L-type but also N-type channel α_1 subunits.[28] The monoclonal antibody MANC1 was obtained after mice were immunized with microsomal membranes prepared from rabbit skeletal muscle.[27] MANC1 recognizes the α_2 subunit of skeletal muscle and neuronal L-type channels and, to a lesser degree, that of the N-type channel from the rabbit.[27,28]

THE TWO SIZE FORMS OF CLASS B, C, AND D α_1 SUBUNITS DIFFER AT THEIR COOH-TERMINI

Immunoblotting with CNB1, CNC1, and CND1 revealed an additional level of multiplicity of neuronal calcium channels. Two size forms of class B, class C, and class D α_1 subunits were detected by corresponding antibodies (FIG. 1, lanes 1, 3, and 5). It seems unlikely that *in vitro* proteolysis of larger forms gives rise to shorter forms, because extensive precautions to prevent *postmortem* proteolysis did not inhibit the appearance of the shorter forms. (See ref. 45 for a detailed discussion.) Since the two size forms of the skeletal muscle L-type α_1 subunit differ at their COOH-terminal ends,[19,20] we used antibodies CNB3, CNC2, and CND2 for immunoblotting, which bind specifically to COOH-terminal sequences of class B, C, and D α_1 subunits. As demonstrated earlier,[46,48] CNB3 and CNC2 recognized only the longer form, not the shorter one (FIG. 1, lanes 2 and 4). As shown in FIGURE 1 (lane 6), the same is true for CND2. The specificity of the interaction of CND2 with the longer form of the class D α_1 subunit was demonstrated by preblocking the antibody with 50 μM CND2 peptide, which completely prevented binding of the antibody to its antigen (FIG. 1, lane 7). This block was not due to a nonspecific effect of the peptide, because the CNB3 peptide, which possesses a size and charge similar to that of the CND2 peptide, did not inhibit binding of the CND2 antibody to the long form of the class D α_1 subunit at a concentration of 50 μM (FIG. 1, lane 8). These results indicate that COOH-terminal sequences recognized by CNB3, CNC2, or CND2 are not present in the shorter forms of the corresponding α_1 subunits.

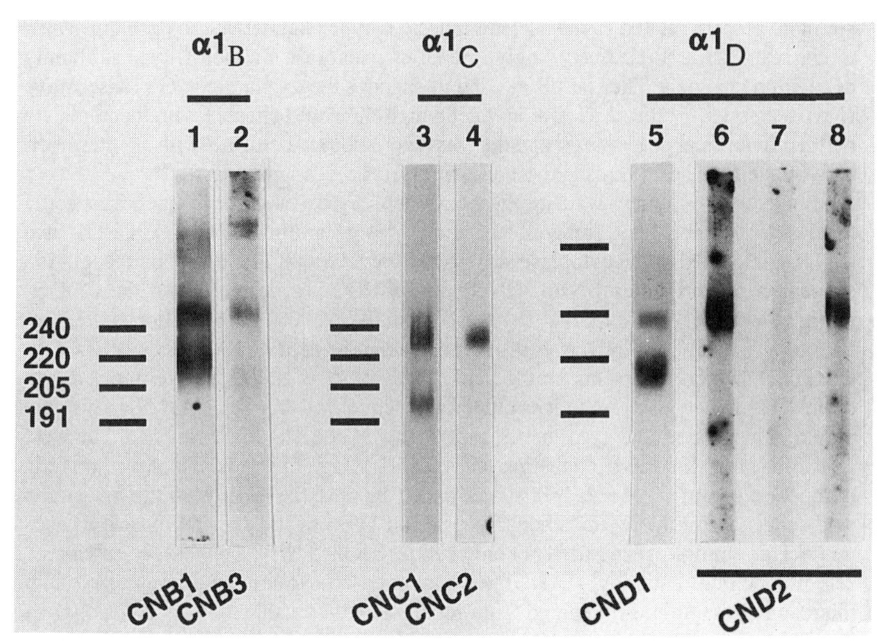

FIGURE 1. Immunoblotting reveals two size forms with different COOH-termini for class B, C, and D calcium channel α_1 subunits. Neuronal calcium channels were enriched from solubilized rat brain membranes by wheat germ agglutinin chromatography and sucrose gradient centrifugation. Sucrose gradient fractions were extracted by precipitation with heparin-agarose followed by SDS-polyacrylamide gel electrophoresis (PAGE) and immunoblotting with either CNB1 (*lane 1*), CNB3 (*lane 2*), CNC1 (*lane 3*), CNC2 (*lane 4*), CND1 (*lane 5*), or CND2 (*lanes 6-8*). For blocking experiments, CND2 was preincubated overnight with 50 μM CND2 peptide (*lane 7*) or 50 μM of the unspecific CNB3 peptide (*lane 8*). All antibodies used for probing the blots were affinity-purified on the corresponding peptide. (See ref. 45 for experimental details.) Alpha spectrin (240 kD), β spectrin (220 kD), myosin (205 kD), and α_2 macroglobulin (covalently linked to Coomassie brilliant blue; 191 kD) were used as molecular mass markers; their migration positions are indicated at the left side of each blot.

The existence of two size forms of the human class B α_1 subunit of 251 kD and 262 kD, differing only in their COOH-terminal regions, was suggested by cloning and sequencing of human cDNA.[39] The polypeptide with an apparent molecular mass of about 250 kD as detected by immunoblotting with CNB1 and CNB3 likely corresponds to the longer form of human class B α_1 subunit cDNA[39] as well as to the only cloned form of rat and rabbit class B α_1 subunit cDNA.[40,43] No sequence which could encode a peptide recognized by CNB3 is found in the shorter form of human class B α_1 subunit cDNA, which may therefore encode the polypeptide with 220 kD. Complementary DNAs encoding two forms of the class D α_1 subunit as well as class A and class E α_1 subunits have been identified. In all these cases, the two forms possess COOH-terminal ends of different length, which might be due to differential splicing of mRNA.[33,34,38,42] Our findings suggest that the longer and shorter forms of class B, C, and D α_1 subunits may be created by differential splicing of mRNAs in the 3′ region of their coding sequence or by specific posttranslational proteolytic processing analogous to the skeletal muscle L-type α_1 subunit.

DIFFERENTIAL PHOSPHORYLATION OF THE TWO SIZE FORMS OF CLASS B AND CLASS C CALCIUM CHANNEL α_1 SUBUNITS BY SECOND MESSENGER-ACTIVATED PROTEIN KINASES

To examine the phosphorylation of the two size forms of the class B and C α_1 subunits by second messenger-activated protein kinases, solubilized calcium channels were affinity purified by immunoprecipitation with CNB2 or CNC1 from an enriched fraction and incubated with different kinases in the presence of [γ^{32}P]ATP. To ensure the specificity of the precipitation and to identify the resulting phosphorylated polypeptides unequivocally as α_1 subunits, the protein A sepharose (PAS)/antibody/channel complexes were dissociated with sodium dodecylsulfate (SDS) and dithiothreitol. After SDS was neutralized with an excess of Triton X 100, anti-CP(1382-1400), which precipitates L- and N-type channels (as just described), was then used for a second round of immunoprecipitation. Following immunoprecipitation by CNB1 and phosphorylation with either cAMP-dependent protein kinase (cA-PK) or protein kinase C (PKC), two labeled polypeptides with relative molecular masses corresponding to the two size forms of the class B α_1 subunit as detected by immunoblotting appeared (FIG. 2, lanes 1 and 2). The polypeptides were not detectable when control antibodies were used for immunoprecipitation (data not shown). In contrast, calcium- and calmodulin-dependent protein kinase II (CaM kinase II) phosphorylated only the higher form of the α_1 subunit (FIG. 2, lane 3). These experiments show that cA-PK and PKC phosphorylate both size forms of the class B α_1 subunit, whereas CaM kinase II preferentially phosphorylates the full-length form.

Similarly, both forms of the class C α_1 subunit were phosphorylated by PKC (FIG. 2, lane 5). In contrast to class B, however, PKA phosphorylated only the longer form (FIG. 2, lane 4), and CaM kinase II both forms of the class C α_1 subunit (FIG. 2, lane 6).

The most likely interpretation for the differential phosphorylation of the class B α_1 subunit by CaM kinase II is that this kinase phosphorylates one or more sites in the COOH-terminal sequence of the long form of this α_1 subunit, which are missing in the short form. Similarly, the phosphorylation site or sites of the class C α_1 subunit for cA-PK may be in the COOH-terminal region of the longer α_1 subunit. Our results, however, do not rule out that differences in the amino acid sequence in the COOH-terminus of the short forms cause conformational changes that prevent phosphorylation of the class B α_1 subunit by CaM kinase II or of class C α_1 subunit by cA-PK.

The identification of two size forms of several types of calcium channels multiplies the diversity of these channels. Remarkably, all neuronal calcium channels cloned to date possess unique COOH-terminal ends.[51] Our results suggest the possibility of differential regulation of the different size forms of members of both the L-type subfamily and the non-L-type subfamily of calcium channels by phosphorylation by second messenger-activated serine/threonine kinases.

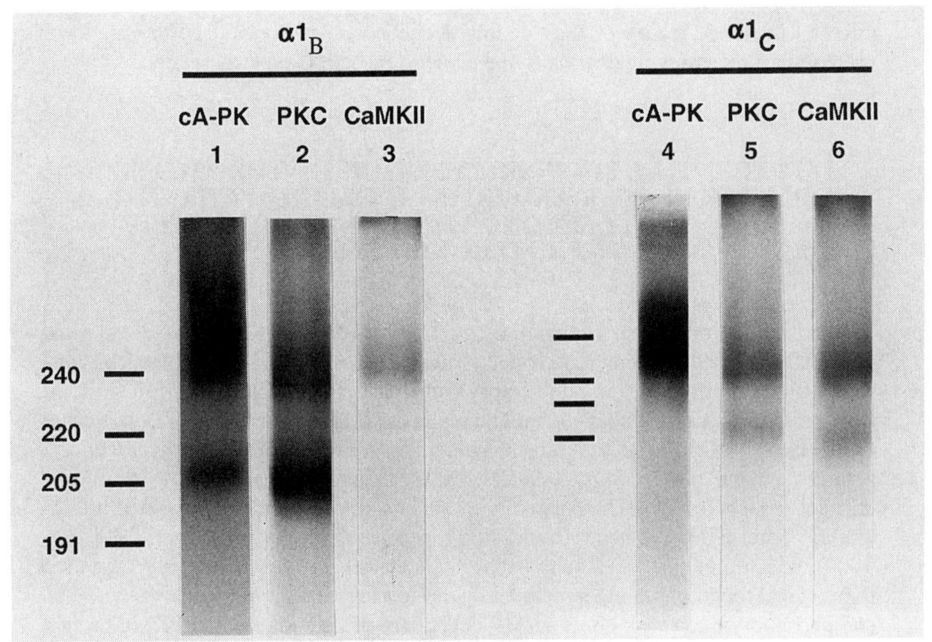

FIGURE 2. Differential phosphorylation of full-length and truncated forms of class B and C calcium channel α_1 subunits. Calcium channels were solubilized from rat brain membranes, enriched by wheat germ agglutinin chromatography and sucrose gradient centrifugation, and immunoprecipitated with affinity-purified CNB1 (*lanes 1-3*) or CNC1 (*lanes 4-6*). The immunocomplexes were incubated with cA-PK (*lanes 1 and 4*), PKC (*lanes 2 and 5*), or CaM kinase II (*lanes 3 and 6*) in the presence of 0.2 μM [γ^{32}P]ATP, washed, and dissociated by treatment with 1% SDS and 5 mM dithiothreitol at 50-60 °C for 30 minutes. SDS was neutralized by adding eight volumes of 1% Triton X-100 together with phosphatase and protease inhibitors, and dissociated calcium channel α_1 subunits were immunoprecipitated with 20 μg anti-CP(1382-1400). (See ref. 46 for experimental details.) Molecular mass markers are given as in FIGURE 1.

AN α_2 SUBUNIT IMMUNOLOGICALLY RELATED TO THE SKELETAL MUSCLE α_2 SUBUNIT IS ASSOCIATED WITH THE NEURONAL CLASS C AND THE CLASS D α_1 SUBUNIT

We previously reported the distribution of an α_2 subunit associated mainly with L-type calcium channels using monoclonal antibody MANC1.[27,50] Similar to CNC1 and CND1, MANC1 labels predominantly cell somata and proximal dendrites of neurons throughout the rostral-caudal extent of the rat brain. We therefore tested if the α_2 subunit recognized by MANC1 is associated with the class C or class D α_1 subunit. Calcium channels were solubilized from rabbit brain, enriched by sucrose gradient centrifugation, precipitated, and used for immunoblotting with CNC1. If precipitation was performed with heparin agarose, which binds to all calcium channels so far identified by molecular cloning and sequencing (see refs. 46 and 47 and unpublished results), CNC1 bound to two polypeptides on the immunoblot with

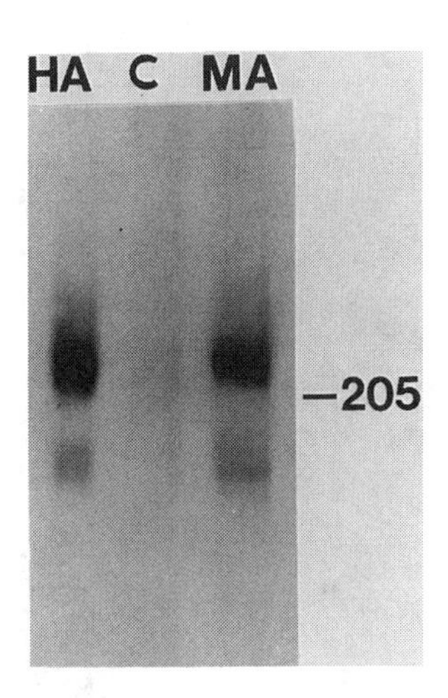

FIGURE 3. MANC1, an antibody recognizing an α_2 subunit of rabbit brain, immunoprecipitates both size forms of the class C α_1 subunit as detected with CNC1 by immunoblotting. Calcium channels were enriched from solubilized rabbit cerebral cortex membranes by sucrose gradient centrifugation and precipitated with heparin agarose (HA; *left lane*), control antibody (C; *middle lane*), or MANC1 (MA; *right lane*) as described for rat brain calcium channels (FIG. 1). After SDS-PAGE and transfer onto nitrocellulose, the immunoblot was probed with affinity-purified CNC1. The migration position for myosin heavy chain (205 kD) is indicated at the right side.

relative molecular masses similar to those of the long and short forms of rat brain class C α_1 subunits (FIG. 3, left lane). The same two bands were detected by CNC1 when MANC1 was used for precipitation (FIG. 3, right lane). No band was detected if a control antibody was used instead of MANC1, proving the specificity of immunoprecipitation (FIG. 3, middle lane). Evidently, the neuronal α_2 subunit as recognized by MANC1 is associated with some class C α_1 subunits.

The association of the class C α_1 subunit with the neuronal α_2 subunit detected with MANC1 was further corroborated; membranes from rabbit dorsal cortex and hippocampus were labeled with the L-type ligand [^{3}H]PN200-110 and solubilized as described for rat brain,[45] cleared by centrifugation, and incubated with either CNC1, MANC1 or control antibody and PAS. After 5-6 hours, the PAS/antibody/antigen complex was sedimented and the supernatant saved. The original pellet was washed as described and counted to determine how much [^{3}H]PN200-110 receptor was removed from the sample by this first round of immunoprecipitation. Both CNC1 and MANC1 precipitated about 40% of the [^{3}H]PN200-110 receptor at the antibody concentration applied. Use of control antibodies always resulted in less than 4% of the receptor being bound. The supernatants of the first spin were again incubated with either CNC1 or MANC1 and treated as described before. During the second round of immunoprecipitation MANC1 precipitated significantly less [^{3}H]PN200-110 receptor from samples pre-precipitated with either CNC1 or MANC1 itself than from samples pretreated with control antibodies. Pre-precipitation with CNC1 or MANC1 caused a reduction by 60% of the [^{3}H]PN200-110 receptor as recognized by MANC1 at the second precipitation as compared to the samples pretreated with control antibodies (FIG. 4). Analogous results were obtained when CNC1 was applied during the second round of immunoprecipitation; preincubation with MANC1 or CNC1 removed about 50% of CNC1-precipitable [^{3}H]PN200-110 in comparison with control-pretreatment (FIG. 4). These experiments confirm that the α_2 subunit recognized by MANC1 is partially precipitated by CNC1 (and vice versa) and therefore forms a complex with the class C α_1 subunit.

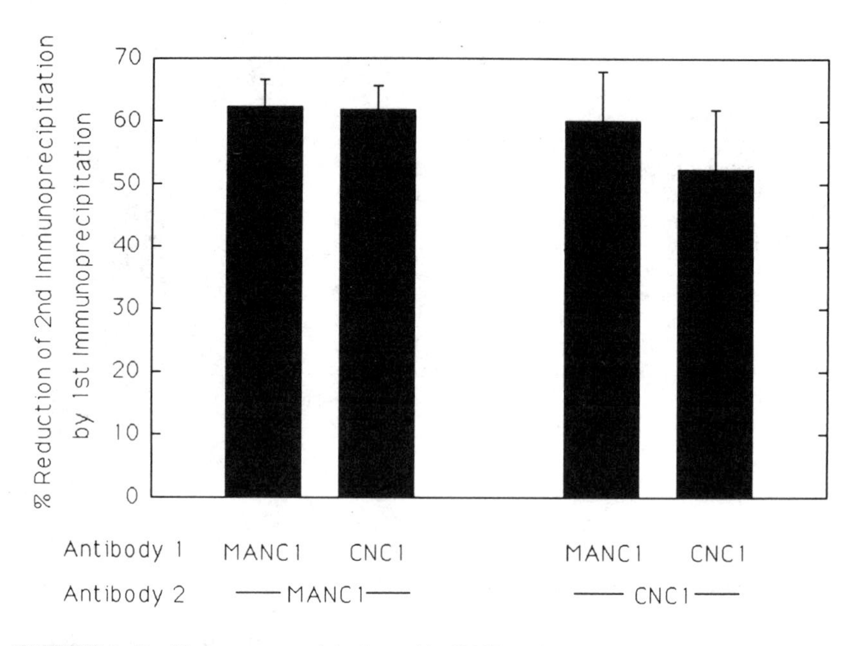

FIGURE 4. Double-immunoprecipitation with CNC1 and MANC1. 150 μl aliquots of the [^{3}H]PN200-110-labeled, solubilized membrane fraction from rabbit cerebral cortex were incubated with affinity-purified CNC1 (60 μg) or MANC1 (2 μg) or an equal amount of control antibodies (chromatographically purified rabbit or mouse IgG, respectively) and precipitated with a saturating amount of PAS which had been preincubated with 25 μg rabbit anti-mouse IgG for MANC1 precipitations. After 5 hours the samples were spun and the supernatants re-precipitated with either MANC1 or CNC1. The values are given in terms of percentage reduction of immunoprecipitation of [^{3}H]PN200-110 during the second round of precipitation by pre-precipitation with CNC1 or MANC1 in relation to pre-precipitation with control antibodies, the latter one corresponding to 0% reduction.

Unfortunately, MANC1 does not immunoprecipitate rat brain calcium channels. Furthermore, in contrast to CNC1, CND1 did not detect calcium channels from rabbit brain by either immunoblotting or immunoprecipitation (Hell *et al.*, unpublished results). Therefore, analogous experiments could not be performed with CND1. We instead performed double-labeling experiments with MANC1 and CND1. As shown for the dorsal cortex, MANC1 and CND1 immunoreactivity is very well colocalized; neurons with relatively strong MANC1 reactivity also show strong labeling for CND1, and neurons with a low level of CND1 labeling are only faintly stained with MANC1 (FIG. 5). Furthermore, both antibodies label predominantly cell somata and proximal dendrites. Staining of the medial and distal parts of dendrites is hardly detectable for MANC1 and only weakly visible for CND1. The precise colocalization of CND1 and MANC1 immunoreactivity indicates that the class D α_1 subunit is also associated with the α_2 subunit identified by MANC1.

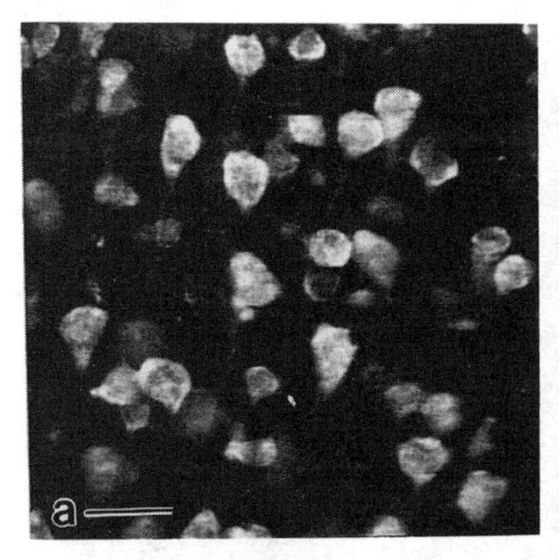

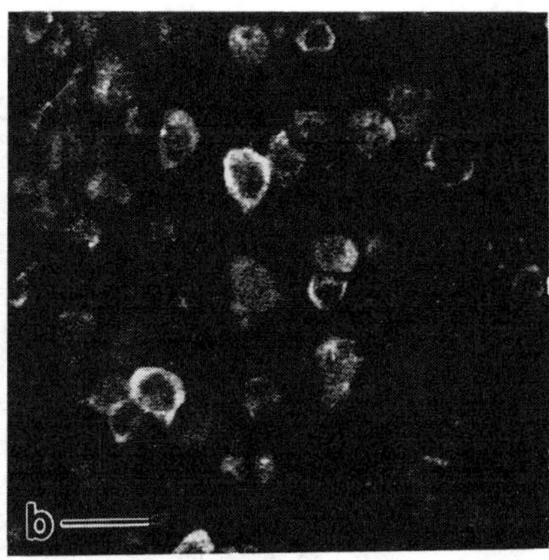

FIGURE 5. Double labeling of α_1 and α_2 subunits of L-type calcium channels with CND1 and MANC1. Sagittal sections of adult rat brain were stained with CND1 and MANC1 antibodies using the immunofluorescence technique described previously[47] except that both CND1 and MANC1 were applied to the tissue simultaneously as were the secondary antibodies (goat anti-rabbit IgG-FITC and goat anti-mouse IgG-rhodamine which were used to recognize CND1 and MANC1, respectively). This figure demonstrates the overlap in distribution of CND1 and MANC1 immunoreactivity in the cell body and proximal dendrites of cortical neurons.

DIFFERENTIAL SUBCELLULAR LOCALIZATION OF CLASS B, C, AND D CALCIUM CHANNELS

Throughout the rostral-caudal extent of the rat brain, N-type calcium channels recognized by CNB1 antibodies are localized in dendrites and presynaptic terminals forming synapses on dendrites and, to a lesser extent, on cell bodies. Smooth as well as punctate staining is observed along the surface of dendrites and cell bodies. This pattern of immunoreactivity is illustrated for the dorsal cerebral cortex in FIGURE 6A. Shown are cells located in layer V of the dorsal cortex, which exhibit relatively weak cell body staining and relatively strong dendritic immunoreactivity. In addition to smooth labeling of the plasma membrane reflecting an even distribution of a low density of the class B N-type channel, more intense staining of small, punctate structures along the length of dendrites having the appearance of synapses can also be seen.

By contrast, immunocytochemical studies using CNC1 and CND1 reveal that α_1 subunits of both L-type calcium channels are localized mainly in neuronal cell bodies and proximal dendrites (FIG. 6B and C). In addition, relatively dense labeling is observed at the base of major dendrites in many neurons. Class D calcium channels are distributed evenly over most of the surface membrane of cell bodies with accumulations at the base of major dendrites of some neurons. Class C calcium channels are concentrated in clusters, mainly on cell bodies and proximal dendrites. These results define complementary distributions of N- and L-type calcium channels in dendrites, cell bodies, and nerve terminals of most central neurons and support distinct functional roles in intracellular calcium regulation, neurotransmitter release, and calcium-dependent electrical activity for these two channel subtypes.

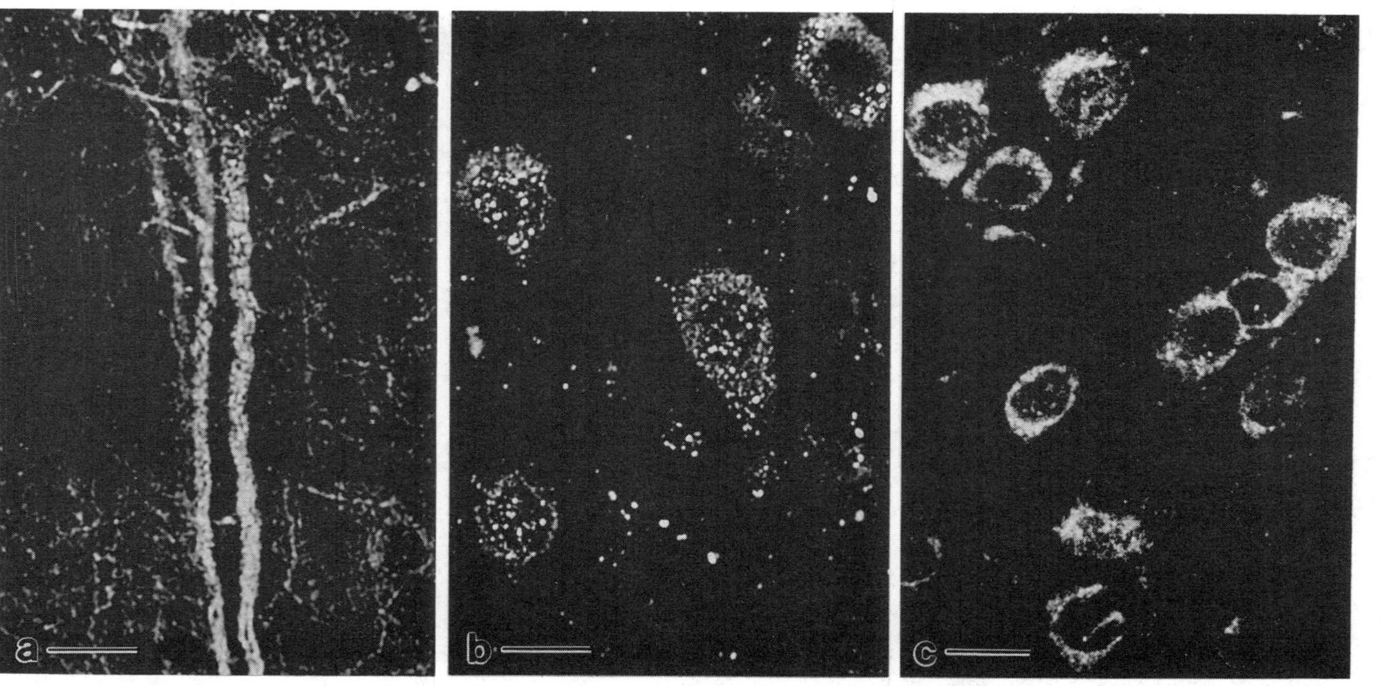

FIGURE 6. Class B, C, and D α_1 subunits show different subcellular distributions in the dorsal cortex. Sagittal sections of adult rat brain were stained with CNB1, CNC1, or CND1 as described previously.[45] Class B N-type calcium channels (**A**) are detected mainly in the dendrites and punctate structures along the length of dendrites, with relatively less staining in the cell body. In comparison, class C L-type calcium channels (**B**) are found in clusters mainly in cell bodies and proximal dendrites of neurons. This pattern is similar to that of class D L-type calcium channels (**C**) which are also localized mainly in the cell body and proximal dendrites of most neurons. Class D, however, is distributed in a more even pattern along the cell surface and does not result in a punctate appearance of immunoreactivity.

DISTINCT ROLES OF N- AND L-TYPE CALCIUM CHANNEL TRANSIENTS

As a first step towards defining the distinct roles of differentially localized calcium channel subtypes in neuronal calcium entry, we measured calcium transients in the cell bodies of single CA3 hippocampal pyramidal cells in organotypic cultures of hippocampal slices from neonatal rats.[52] Individual neurons were injected with fura-2 for recording cytosolic calcium concentrations. Cells were stimulated with trains of electrical pulses (20 Hz for 10 seconds) from a bipolar extracellular electrode placed near the cell body or in the distal dendritic field approximately 200-300 μm from the cell body.

The peak of the calcium transients in the cell body elicited by somal stimulation was reduced by the dihydropyridines nimodipine and isradipine to less than half of its control value. In contrast, ω-conotoxin GVIA had no effect. Thus, as expected from their distinct localization, L-type calcium channels are important determinants of calcium transients in the cell body, whereas N-type calcium channels are not.

When cell body calcium transients are elicited by stimulation of the apical dendrites, a different picture emerges. Dihydropyridine antagonists of L-type calcium channels still reduce peak calcium transients in the cell body by more than half, but ω-conotoxin GVIA also reduces peak calcium transients by 30%. The effects of ω-conotoxin GVIA are completely occluded by inhibition of excitatory synaptic transmission by the combined application of the glutamate receptor antagonists 2-amino-5-phosphonopentanoic acid and 6-cyano-7-nitroquinoxaline-2,3-dione. Thus, the effects of N-type calcium channels on calcium transients are due to their role in the release of glutamate from presynaptic terminals. N-type calcium channels in the dendritic membrane are obviously not required to transmit the synaptic signal from the distal dendrites to the cell body. These results show that the differential localization of N-type and L-type calcium channels leads to distinct roles in neuronal signal transduction.

ACKNOWLEDGMENTS

We wish to thank Drs. Eric I. Rotman and Brian J. Murphy for generously providing cA-PK and PKC, Drs. D. A. Brickey and T. Soderling, Vollum Institute, for the generous supply of CaM kinase II, and Anita A. Colvin and Dr. Karen S. De Jongh, Molecular Pharmacology Facility, Department of Pharmacology, University of Washington, for synthesizing peptides.

REFERENCES

1. Bean, B. P. 1989. Annu. Rev. Physiol. **51:** 367-384.
2. Llinas, R., M. Sugimori, J.-W. Lin & B. Cherksey. 1989. Proc. Natl. Acad. Sci. USA **86:** 1689-1693.
3. Hess, P. 1990. Annu. Rev. Neurosci. **13:** 337-356.
4. Tsien, R. W., P. T. Ellinor & W. A. Horne. 1991. Trends Pharmacol. Sci. **12:** 349-354.
5. Glossman, H. & J. Striessnig. 1990. Rev. Physiol. Biochem. Pharmacol. **114:** 1-105.

6. PLUMMER, M. R., D. E. LOGOTHETIS & P. HESS. 1989. Neuron **2:** 1453-1463.
7. AOSAKI, T. & H. KASAI. 1989. Pfluegers Arch. **414:** 150-156.
8. MINTZ, I. M., M. E. ADAMS & B. P. BEAN. 1992. Neuron **9:** 85-95.
9. MINTZ, I. M., V. J. VENEMA, K. M. SWIDEREK, T. D. LEE, B. P. BEAN & M. E. ADAMS. 1992. Nature **355:** 827-829.
10. TAKAHASHI, M., M. J. SEAGAR, J. F. JONES, B. F. REBER & W. A. CATTERALL. 1987. Proc. Natl. Acad. Sci. USA **84:** 5478-5482.
11. CAMPBELL, K. P., A. T. LEUNG & A. H. SHARP. 1988. Trends Neurosci. **11:** 425-430.
12. CATTERALL, W. A. 1988. Science **242:** 50-61.
13. TANABE, T., H. TAKESHIMA, A. MIKAMI, V. FLOCKERZI, H. TAKAHASHI, K. KANGAWA, M. KOJIMA, H. MATSUO, T. HIROSE & S. NUMA. 1987. Nature **328:** 313-318.
14. ELLIS, S. B., M. E. WILLIAMS, N. R. WAYS, R. BRENNER, A. H. SHARP, A. T. LEUNG, K. P. CAMPBELL, E. MCKENNA, W. J. KOCH, A. HUI, A. SCHWARTZ & M. M. HARPOLD. 1988. Science **241:** 1661-1664.
15. RUTH, P., A. RÖHRKASTEN, M. BIEL, E. BOSSE, S. REGULLA, H. E. MEYER, V. FLOCKERZI & F. HOFMANN. 1989. Science **245:** 1115-1118.
16. JAY, S. D., S. B. ELLIS, A. F. MCCUE, M. E. WILLIAMS, T. S. VEDVICK, M. M. HARPOLD & K. P. CAMPBELL. 1990. Science **248:** 490-492.
17. DE JONGH, K. S., C. WARNER & W. A. CATTERALL. 1990. J. Biol. Chem. **265:** 14738-14741.
18. JAY, S. D., A. H. SHARP, S. D. KAHL, T. S. VEDVICK, M. M. HARPOLD & K. P. CAMPBELL. 1991. J. Biol. Chem. **266:** 3287-3293.
19. DE JONGH, K. S., D. K. MERRICK & W. A. CATTERALL. 1989. Proc. Natl. Acad. Sci. USA **86:** 8585-8589.
20. DE JONGH, K. S., C. WARNER, A. A. COLVIN & W. A. CATTERALL. 1991. Proc. Natl. Acad. Sci. USA **88:** 10778-10782.
21. PEREZ-REYES, E., H. S. KIM, A. E. LACERDA, W. HORNE, X. Y. WEI, D. RAMPE, K. P. CAMPBELL, A. M. BROWN & L. BIRNBAUMER. 1989. Nature **340:** 233-236.
22. MIKAMI, A., K. IMOTO, T. TANABE, T. NIIDOME, Y. MORI, H. TAKESHIMA, S. NARUMIYA & S. NUMA. 1989. Nature **340:** 230-233.
23. WEI, X., E. PEREZ-REYES, A. E. LACERDA, G. SCHUSTER, A. M. BROWN & L. BIRNBAUMER. 1991. J. Biol. Chem. **266:** 21943-21947.
24. VARADI, G., P. LORY, D. SCHULTZ, M. VARADI & A. SCHWARTZ. 1991. Nature **352:** 159-162.
25. SINGER, D., M. BIEL, I. LOTAN, V. FLOCKERZI, F. HOFMANN & N. DASCAL. 1991. Science **253:** 1553-1557.
26. TAKAHASHI, M. & W. A. CATTERALL. 1987. Science **236:** 88-91.
27. AHLIJANIAN, M. K., R. E. WESTENBROEK & W. A. CATTERALL. 1990. Neuron **4:** 819-832.
28. AHLIJANIAN, M. K., J. STRIESSNIG & W. A. CATTERALL. 1991. J. Biol. Chem. **266:** 20192-20197.
29. MCENERY, M. W., A. M. SNOWMAN, A. H. SHARP, M. E. ADAMS & S. H. SNYDER. 1991. Proc. Natl. Acad. Sci. USA **88:** 11095-11099.
30. SAKAMOTO, J. & K. P. CAMPBELL. 1991. J. Biol. Chem. **266:** 18914-18919.
31. WITCHER, D. R., M. DE WAARD, J. SAKAMOTO, C. FRANZINI-ARMSTRONG, M. PRAGNELL, S. D. KAHL & K. P. CAMPBELL. 1993. Science **261:** 486-489.
32. SNUTCH, T. P., J. P. LEONARD, M. M. GILBERT, H. A. LESTER & N. DAVIDSON. 1990. Proc. Natl. Acad. Sci. USA **87:** 3391-3395.
33. MORI, Y., T. FRIEDRICH, M.-S. KIM, A. MIKAMI, J. NAKAI, P. RUTH, E. BOSSE, F. HOFMANN, V. FLOCKERZI, T. FURUICHI, K. MIKOSHIBA, K. IMOTO, T. TANABE & S. NUMA. 1991. Nature **350:** 398-402.
34. HUI, A., P. T. ELLINOR, O. KRIZANOVA, J.-J. WANG, R. J. DIEBOLD & A. SCHWARZ. 1991. Neuron **7:** 35-44.

35. Starr, T. V. B., W. Prystay & T. P. Snutch. 1991. Proc. Natl. Acad. Sci. USA **88:** 5621-5625.
36. Snutch, T. P., W. J. Tomlinsom, J. P. Leonhard & M. M. Gilbert. 1991. Neuron **7:** 45-57.
37. Chin, H., C. A. Kozak, H.-L. Kim, B. Mock & O. W. McBride. 1991. Genomics **11:** 914-919.
38. Williams, M. E., D. H. Feldman, A. F. McCue, R. Brenner, G. Velicelebi, S. B. Ellis & M. M. Harpold. 1992. Neuron **8:** 71-84.
39. Williams, M. E., P. F. Brust, D. H. Feldman, S. Patthi, S. Simerson, A. Maroufi, A. F. McCue, G. Velicelebi, S. B. Ellis & M. M. Harpold. 1992. Science **257:** 389-395.
40. Dubel, S. J., T. V. B. Starr, J. Hell, M. K. Ahlijanian, J. J. Enyeart, W. A. Catterall & T. P. Snutch. 1992. Proc. Natl. Acad. Sci. USA **89:** 5058-5062.
41. Seino, S., L. Chen, M. Seino, O. Blondel, J. Takeda, J. H. Johnson & G. Bell. 1992. Proc. Natl. Acad. Sci. USA **89:** 584-588.
42. Niidome, T., M.-S. Kim, T. Friedrich & Y. Mori. 1992. FEBS Lett. **308:** 7-13.
43. Fujita, Y., M. Mynlieff, R. T. Dirksen, M.-S. Kim, T. Niidome, J. Nakai, T. Friedrich, N. Iwabe, T. Miyata, T. Furuichi, D. Furutama, K. Mikoshiba, Y. Mori & K. G. Beam. 1993. Neuron **10:** 585-598.
44. Soong, T. W., A. Stea, C. D. Hodson, S. J. Dubel, S. R. Vincent & T. P. Snutch. 1993. Science **260:** 1133-1136.
45. Westenbroek, R. E., J. W. Hell, C. Warner, S. J. Dubel, T. P. Snutch & W. A. Catterall. 1992. Neuron **9:** 1099-1115.
46. Hell, J. W., C. T. Yokoyama, S. T. Wong, C. Warner, T. P. Snutch & W. A. Catterall. 1993. J. Biol. Chem. **268:** 19451-19457.
47. Hell, J. W., R. E. Westenbroek, C. Warner, M. K. Ahlijanian, W. Prystay, M. M. Gilbert, T. P. Snutch & W. A. Catterall. 1993. J. Cell Biol. **123:** 949-962.
48. Hell, J. W., S. M. Appleyard, C. T. Yokoyama, C. Warner & W. A. Catterall. 1994. J. Biol. Chem. **269:** 7390-7396.
49. Striessnig, J., H. Glossmann & W. A. Catterall. 1990. Proc. Natl. Acad. Sci. USA **87:** 9108-9112.
50. Westenbroek, R. E., M. K. Ahlijanian & W. A. Catterall. 1990. Nature **347:** 281-284.
51. Snutch, T. P. & P. R. Reiner. 1992. Curr. Opin. Neurobiol. **2:** 247-253.
52. Elliott, E. M., A. T. Malouf & W. A. Catterall. 1993. Soc. Neurosci. Abstr. **19:** 1752.

Biophysical and Pharmacological Characterization of a Class A Calcium Channel

WILLIAM A. SATHER,[a,b] TSUTOMU TANABE,[a,c] JI-FANG ZHANG,[a] AND RICHARD W. TSIEN [a]

[a]*Department of Molecular & Cellular Physiology*
Stanford University
Stanford, California 94305

[c]*Howard Hughes Medical Institute*
and
Department of Cellular & Molecular Physiology
Yale University
New Haven, Connecticut 06536

Voltage-gated calcium channels function primarily as signaling elements in a variety of neuronal processes, most prominently in the release of neurotransmitter. Four major subtypes of calcium channels (L-, T-, N-, and P-types) have been distinguished on the basis of functional criteria,[1-6] and these subtypes apparently are specialized to subserve different physiological tasks in neurons. Until recently, it was generally difficult to study particular calcium channel subtypes in isolation. This has impeded progress in both characterizing the detailed biophysical and pharmacological properties of previously described channels and identifying new calcium channel subtypes. The advent of molecular cloning of calcium channel genes has overcome these problems by making it possible to study pure populations of calcium channels in heterologous expression systems.

Voltage-gated calcium channels are composed of a single α_1 subunit, which houses the voltage-sensor, gating machinery, and channel pore, and several ancillary subunits that modify the behavior of the α_1 subunit.[7-14] Genes encoding five different calcium channel α_1 subunits have been identified in mammalian brain[11,15-22] and, in one system of nomenclature, have been labeled classes A, B, C, D, and E.[6,15] Based on sequence homology, these α_1 subunit genes are organized into two major subfamilies as shown in FIGURE 1. One subfamily comprises the dihydropyridine-sensitive L-type channels, including those encoded by class C and class D clones[9,19] as well as the L-type channel of skeletal muscle encoded by the class S clone.[8] The other major subfamily includes classes A, B, and E. The class B clone clearly encodes an ω-conotoxin-GVIA-sensitive N-type calcium channel.[18,20,21] Information about the functional properties of class A and class E clones is more limited, but neither encodes

[b] Address for correspondence: William A. Sather, Department of Molecular & Cellular Physiology, B101 Beckman Center, Stanford University Medical Center, Stanford, CA 94305-5426.

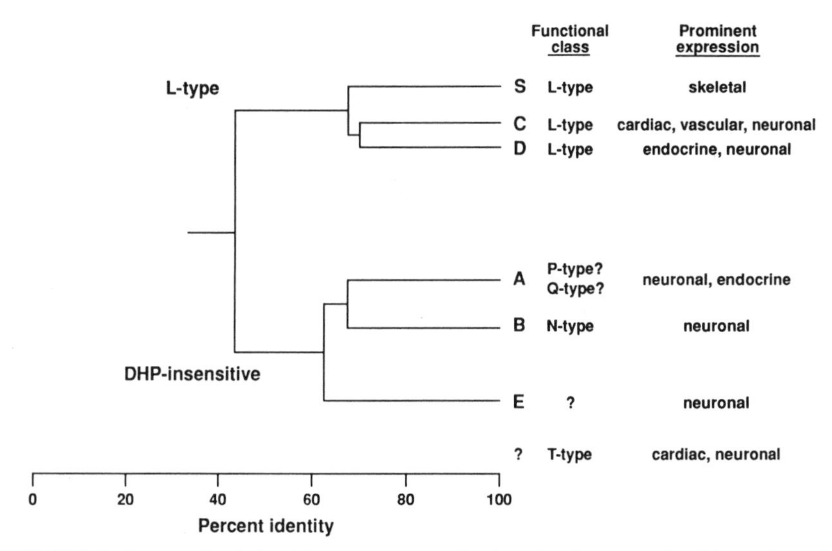

FIGURE 1. Structural relationships among α_1 subunits of voltage-gated calcium channels based on cDNA sequence identity. The correspondence between α_1 subunit classes and functional calcium channel types is shown on the *right side of the figure* along with an indication of those tissues expressing high levels of particular classes of α_1 subunits. The nomenclature for the cloned calcium channel α_1 subunits follows that of Snutch *et al.*[15] Adapted from Zhang *et al.*[34]

classical T-, N-, or L-type channels.[11,16,17,22] Indeed, no functionally defined calcium channel type has yet been firmly established as the class E gene product. Based on highly suggestive observations, the class A clone has been linked with the P-type channel. That class A might encode the P channel α_1 subunit was initially suggested both by the abundance of class A transcripts in the cerebellum (which expresses a high level of P-type current) and by the sensitivity of class A channels heterologously expressed in *Xenopus* oocytes to the crude venom of the funnel web spider.[11] Additionally, the level of class A transcripts in the cerebellum was greatly reduced in Purkinje cell-deficient mice,[11] and because nearly all calcium current in cerebellar Purkinje cells is P-type,[23,24] this finding lent further support to the view that the class A clone corresponds to the α_1 subunit of the P-type channel.

Closer examination, however, reveals some clear differences between the properties of the class A channel in a heterologous expression system and those of the P-type channel in cerebellar Purkinje cells.[4,25] For example, class A channels, when expressed in *Xenopus* oocytes together with the ancillary α_2/δ and β_1 subunits, exhibit distinct inactivation of macroscopic currents. FIGURE 2A shows a family of class A currents elicited by step depolarizations, and inactivation is profound and relatively rapid.

We investigated the effects of auxiliary subunits on class A channel inactivation because subunit composition was demonstrated to influence the kinetics of cloned L-type calcium channels.[12–14,26,27] Because only a single α_2/δ species has been identified so far, the behavior of this subunit is not a potential source of complexity in the

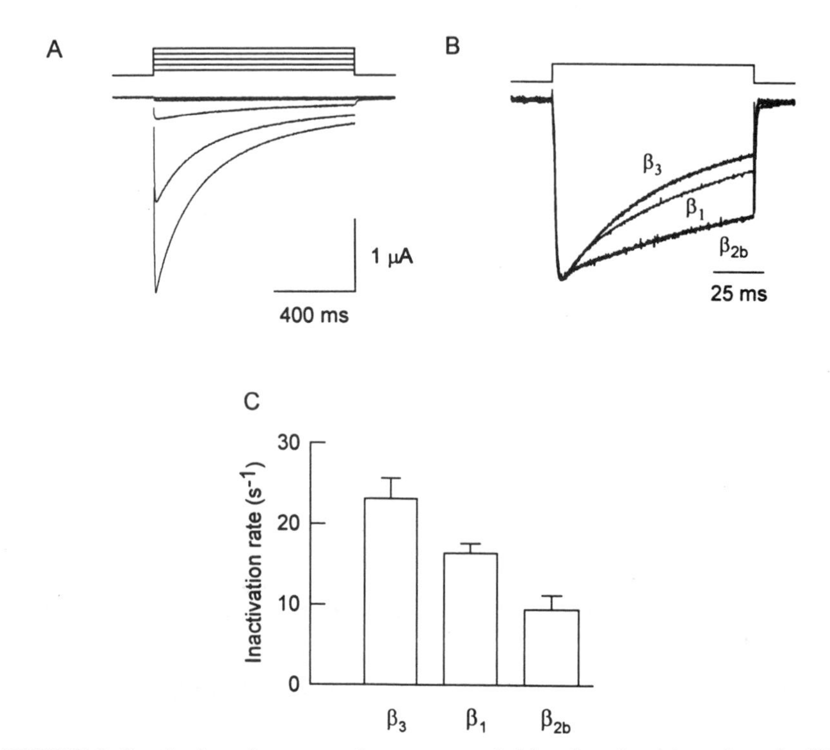

FIGURE 2. Inactivation of macroscopic currents carried by class A calcium channels. (**A**) Family of voltage-clamp records obtained from *Xenopus* oocytes coinjected with synthetic RNA encoding α_{1A}, α_2/δ and β_1 subunits. The holding potential was −80 mV, and test potentials ranged from −50 mV to −10 mV in 10-mV increments. (**B**) Effect of various β subunits on inactivation kinetics of the class A channel. Oocytes were coinjected with synthetic RNAs encoding α_{1A}, α_2/δ, and β_1 subunits. The currents, which were evoked by steps from −80 mV to −10 mV, are superimposed and normalized to their peak amplitudes to facilitate comparison of their waveforms. (**C**) Mean inactivation rates (±SEM) for combinations of α_{1A}, α_2/δ, and various β subunits. Inactivation rates (τ^{-1}) were determined from single exponential fits to the decay phases of the currents, and non-zero steady-state currents were allowed. In **A, B,** and **C,** data were obtained using a two-electrode voltage clamp and oocytes bathed in 2 mM Ba^{2+}, 0 Cl^- solutions. Records were leak-subtracted, and remaining capacitance artifacts were blanked. Reproduced with permission from Sather *et al.*[25]

experiments described here. However, several genes encoding β subunits have been cloned.[26-28] Examples of normalized current waveforms for combinations of α_{1A} with various β subunits are illustrated in FIGURE 2B; FIGURE 2C shows mean inactivation rates for the various subunit combinations. Inactivation is most rapid for the combination containing the β_3 subunit and slowest for the combination containing the β_{2b} subunit. Although the species of β subunit clearly affects inactivation kinetics, a salient point is that class A channels inactivate regardless of the species of β subunit present.

Inactivation was examined at the single channel level to be certain that the apparent inactivation observed at the whole-cell level did not arise from inadequate

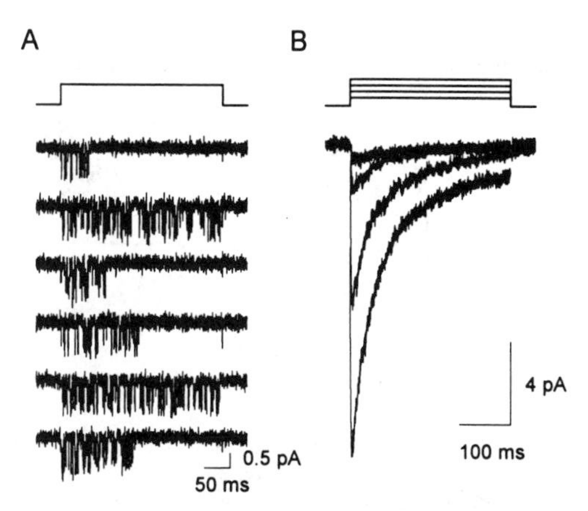

FIGURE 3. Inactivation of class A channels at the single channel level. (**A**) Single channel records obtained from a cell-attached patch on an oocyte coinjected with RNA encoding α_{1A}, α_2/δ, and β_1 subunits. Channel activity was elicited by voltage steps from a holding potential of −80 mV to a test potential of 0 mV. (**B**) Ensemble-averaged currents from a cell-attached patch showing multiple channel activity. The oocyte was coinjected with RNA encoding α_{1A}, α_2/δ, and β_1 subunits. The holding potential was −80 mV, and test potentials ranged from −10 mV to +20 mV in 10-mV increments, with Ba^{2+} current increasing with larger depolarizations. Ensemble averages were constructed from 23–43 records at each potential. In **A** and **B,** currents were carried by 110 mM Ba^{2+}. Records were filtered at 2 kHz, sampled at 10 kHz, and leak subtracted using a P/4 protocol. Reproduced with permission from Sather *et al.*[25]

voltage clamp in whole oocytes or from superimposition of other currents on class A currents. FIGURE 3A shows a series of cell-attached patch records of single class A channel activity. Channel activity, initiated at the onset of step depolarization, often ceases during the voltage pulse, indicating entry of the channel into an inactivated state. The time course of inactivation at the single channel level can be visualized more easily by ensemble averaging single channel records, particularly ones that show multiple channel activity. FIGURE 3B illustrates such ensemble averages, and like whole-cell currents, pronounced and relatively rapid inactivation is evident. Thus, inactivation appears to be a consistent property of the class A gene product in this expression system. P-type current, in sharp contrast, shows essentially no inactivation (FIG. 4B), even over a period of seconds.[5,23,24,29]

The second distinct contrast between class A channels and P-type current is a pharmacological one; class A channels are less sensitive to ω-Aga-IVA than are P-type currents. FIGURE 4 illustrates the effect of this peptide toxin on class A channels expressed in oocytes and on P-type currents recorded from cerebellar Purkinje neurons. Almost all of the P-type current can be blocked by 20 nM ω-Aga-IVA (FIG. 4B),[23,24] whereas only about half the current carried by the class A channels can be blocked by a 10-fold higher concentration of ω-Aga-IVA (FIG. 4A).[25]

Do these biophysical and pharmacological differences between class A channels and P-type currents arise as artifacts of the oocyte expression system or do they in fact reflect the existence of two molecularly distinct channel types? Comparison of

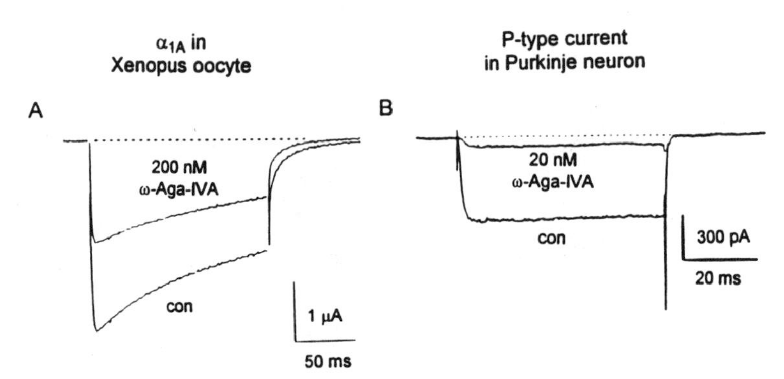

FIGURE 4. Comparison of ω-Aga-IVA sensitivity and inactivation kinetics for the class A channel expressed in *Xenopus* oocytes and the P-type channel of cerebellar Purkinje neurons. (**A**) An example of the block of class A channels by 200 nM ω-Aga-IVA.[25] The oocyte was coinjected with RNA encoding α_{1A}, α_2/δ, and β_1 subunits and was perfused with a 2-mM Ba^{2+}, 0 Cl^- solution supplemented with 0.1 mg/ml cytochrome C to saturate nonspecific peptide binding sites. ω-Aga-IVA was applied in a solution that was identical to the control perfusion solution except for the presence of the ω-Aga-IVA itself. The holding potential was –90 mV and the test potential was 0 mV. (**B**) An example of ω-Aga-IVA block of P-type current in an acutely dissociated cerebellar Purkinje neuron.[23] Whole-cell patch clamp recording was carried out in 5 mM Ba^{2+} solution containing 1 mg/ml cytochrome C, and currents were evoked by steps from –90 mV to –20 mV. Parts **A** and **B** are adapted from refs. 23, 25, and 34.

the available information (TABLE 1) shows that the two channels are alike in many ways, including similar sensitivity to the recently described peptide toxin, ω-CTx-MVIIC.[25,30,31] On the other hand, they clearly differ in the two ways already described (inactivation, ω-Aga-IVA sensitivity) and also apparently in single-channel current-voltage properties.[25,29] If class A channels are truly distinguishable from P-type, then a component of Ca^{2+} current with the class A phenotype would be expected to be present in the brain, particularly in the cerebellum because it is rich in class A transcripts.[11] Indeed, pharmacological dissection of currents in cerebellar granule neurons has revealed a current exhibiting key characteristics of the class A channel when expressed in oocytes.[32] Like both P-type current and class A channels heterologously expressed in oocytes, this current is high-voltage activated, insensitive to dihydropyridines, insensitive to ω-CTx-GVIA, and blocked by ω-CTx-MVIIC. This granule cell current, which has been tentatively designated "Q-type",[32] exhibits an inactivating waveform and can be half-blocked by ~200 nM ω-Aga-IVA, very much like the class A channel expressed in oocytes and unlike the P-type current. It is worth noting that classical P-type current (non-inactivating, half-blocked by ~1 nM ω-Aga-IVA) is also present in cerebellar granule neurons. Q-type current was also found very recently at the CA3-CA1 synapse in the hippocampus,[33] where it is apparently the dominant calcium channel type supporting neurotransmission. It will be interesting to determine if this channel is generally localized to presynaptic membranes.

It appears then that the class A channel may indeed be distinct from the P-type channel. However, the pursuit by many laboratories of brain calcium channel genes

TABLE 1. Comparison of α_{1A} in Oocytes and Native P-Type Channels

Property	α_{1A} (in Oocytes)	P Type (Native)[5,23,29]
Activation		
Voltage range	HVA	HVA
Time to peak	8 ms	~4-6 ms
Inactivation		
Rate of onset	Faster	Slower
Single channel properties		
Conductance[a]	16 pS	9, 14, 19 pS
Unitary current (0 mV)	0.8 pA	0.8, 1.1, 1.4 pA
DHP sensitivity		
Nifedipine	Insensitive	Insensitive
Nimodipine	~Insensitive	Insensitive
ω-Aga-IIIA	~40% block (at 50 nM)	Max. block: 40% (at 200 nM)
ω-Aga-IVA (IC_{50})[b]	200 nM	0.7 nM
ω-CTx-MVIIC (IC_{50})[c]	<150 nM	~10 nM[31]
ω-CTx-GVIA	Insensitive[11]	Insensitive

ABBREVIATION: HVA = high voltage activated.
[a] Value of dominant conductance state for α_{1A}.
[b] 2 mM Ba^{2+} for α_{1A}, 5 mM Ba^{2+} for P type.
[c] 2 mM Ba^{2+}.

has been ardent, and it would seem surprising, given the amount of P-type current in the mammalian brain, that this channel's gene has yet to be found. One possible explanation is that perhaps the P-type channel has little homology with the other calcium channels cloned so far, despite sharing with them the functional attribute of high-voltage activation. An alternative, and maybe more attractive, hypothesis in these circumstances is that the class A gene encodes both the Q-type current described here as well as the classical P-type channel. The differences in functional properties between these two members of a hypothesized "P-type subfamily," among many possibilities, might stem from alternative RNA splicing, divergent posttranslational processing, or preferential association with different, as yet uncharacterized, auxiliary subunits.

As putative targets of environmental factors or disease processes, calcium channels represent potentially complex elements in the aging process; control of expression of the diverse α_1 and ancillary subunit species, regulation of appropriate subunit co-assembly, maintenance of spatial cues specifying correct subcellular localization of particular channel types, and activity of second messenger systems that influence delivery of Ca^{2+} to the cytoplasm by calcium channels are all among the many processes that may malfunction and thereby be involved in the etiology of aging and dementia. Further information on calcium channel subtypes, their distinctive biophysical and pharmacological properties, and their cellular and subcellular localization will very likely be helpful in understanding the precise role of calcium channels in the destabilization of Ca^{2+} homeostasis that is thought to occur during aging and dementia.

REFERENCES

1. TSIEN, R. W., D. LIPSCOMBE, D. V. MADISON, K. R. BLEY & A. P. FOX. 1988. Multiple types of neuronal calcium channels and their selective modulation. Trends Neurosci. **11:** 431-438.
2. BEAN, B. P. 1989. Classes of calcium channels in vertebrate cells. Annu. Rev. Physiol. **51:** 367-384.
3. HESS, P. 1990. Calcium channels in vertebrate cells. Annu. Rev. Neurosci. **13:** 1337-1356.
4. TSIEN, R. W., P. T. ELLINOR & W. A. HORNE. 1991. Molecular diversity of voltage-dependent Ca^{2+} channels. Trends Pharmacol. Sci. **12:** 349-354.
5. LLINÁS, R., M. SUGIMORI, D. E. HILLMAN & B. CHERKSEY. 1992. Distribution and functional significance of the P-type, voltage-dependent Ca^{2+} channels in the mammalian central nervous system. Trends Neurosci. **15:** 351-355.
6. SNUTCH, T. P. & P. B. REINER. 1992. Ca^{2+} channels: Diversity of form and function. Curr. Opinion Neurobiol. **2:** 247-253.
7. CAMPBELL, K. P., A. T. LEUNG & A. H. SHARP. 1988. The biochemistry and molecular biology of the dihydropyridine-sensitive calcium channel. Trends Neurosci. **11:** 425-430.
8. TANABE, T., H. TAKESHIMA, A. MIKAMI, V. FLOCKERZI, H. TAKAHASHI, K. KANGAWA, M. KOJIMA, H. MATSUO, T. HIROSE & S. NUMA. 1987. Primary structure of the receptor for calcium channel blockers from skeletal muscle. Nature **328:** 313-318.
9. MIKAMI, A., K. IMOTO, T. TANABE, T. NIIDOME, Y. MORI, H. TAKASHIMA, S. NARUMIYA & S. NUMA. 1989. Primary structure and functional expression of the cardiac dihydropyridine-sensitive calcium channel. Nature **340:** 230-233.
10. PEREZ-REYES, E., H. S. KIM, A. E. LACERDA, W. HORNE, X. WEI, D. RAMPE, K. P. CAMPBELL, A. M. BROWN & L. BIRNBAUMER. 1989. Induction of calcium currents by the expression of the α_1-subunit of the dihydropyridine receptor from skeletal muscle. Nature **340:** 233-236.
11. MORI, Y., T. FRIEDRICH, M.-S. KIM, A. MIKAMI, J. NAKAI, P. RUTH, E. BOSSE, F. HOFMANN, V. FLOCKERZI, T. FURUICHI, K. MIKOSHIBA, K. IMOTO, T. TANABE & S. NUMA. 1991. Primary structure and functional expression from complementary DNA of a brain calcium channel. Nature **350:** 398-402.
12. SINGER, D., M. BIEL, I. LOTAN, V. FLOCKERZI, F. HOFMANN & N. DASCAL. 1991. The roles of the subunits in the function of the calcium channel. Science **253:** 1553-1557.
13. LACERDA, A. E., H. S. KIM, P. RUTH, E. PEREZ-REYES, V. FLOCKERZI, F. HOFMANN, L. BIRNBAUMER & A. M. BROWN. 1991. Normalization of current kinetics by interaction between the α_1 and β subunits of the skeletal muscle dihydropyridine-sensitive Ca^{2+} channel. Nature **352:** 527-530.
14. VARADI, G., P. LORY, D. SCHULTZ, V. VARADI & A. SCHWARTZ. 1991. Acceleration of activation and inactivation by the beta-subunit of the skeletal muscle calcium channel. Nature **352:** 159-162.
15. SNUTCH, T. P., J. P. LEONARD, M. M. GILBERT, H. A. LESTER & N. DAVIDSON. 1990. Rat brain expresses a heterogeneous family of calcium channels. Proc. Natl. Acad. Sci. USA **87:** 3391-3395.
16. STARR, T. V. P., W. PRYSTAY & T. P. SNUTCH. 1991. Primary structure of a calcium channel that is highly expressed in the rat cerebellum. Proc. Natl. Acad. Sci. USA **88:** 5621-5625.

17. NIIDOME, T., M.-S. KIM, T. FRIEDRICH & Y. MORI. 1992. Molecular cloning and characterization of a novel calcium channel from rabbit brain. FEBS Lett. **308:** 7-13.
18. DUBEL, S. J., T. V. B. STARR, J. HELL, M. K. AHLIJANIAN, J. J. ENYEART, W. A. CATTERALL & T. P. SNUTCH. 1992. Molecular cloning of the α-1 subunit of an ω-conotoxin-sensitive calcium channel. Proc. Natl. Acad. Sci. USA **89:** 5058-5062.
19. WILLIAMS, M. E., D. H. FELDMAN, A. F. MCCUE, R. BRENNER, G. VELIÇELEBI, S. B. ELLIS & M. M. HARPOLD. 1992. Structure and functional expression of α_1, α_2, and β subunits of a novel human neuronal calcium channel subtype. Neuron **8:** 71-84.
20. WILLIAMS, M. E., P. F. BRUST, D. H. FELDMAN, P. SARASWATHI, S. SIMERSON, A. MAROUFI, A. F. MCCUE, G. VELIÇELEBI, S. B. ELLIS & M. M. HARPOLD. 1992. Structure and functional expression of an ω-conotoxin-sensitive human N-type calcium channel. Science **257:** 389-395.
21. FUJITA, Y., M. MYNLIEFF, R. T. DIRKSEN, M.-S. KIM, T. NIIDOME, J. NAKAI, T. FRIEDRICH, N. IWABE, T. MIYATA, T. FURUICHI, D. FURUTAMA, K. MIKOSHIBA, Y. MORI & K. G. BEAM. 1993. Primary structure and functional expression of the ω-conotoxin-sensitive N-type channel from rabbit brain. Neuron **10:** 585-598.
22. SOONG, T. W., A. STEA, C. D. HODSON, S. J. DUBEL, S. R. VINCENT & T. P. SNUTCH. 1993. Structure and functional expression of a member of the low voltage-activated calcium channel family. Science **260:** 1133-1136.
23. MINTZ, I. M., V. J. VENEMA, K. SWIDEREK, T. LEE, B. P. BEAN & M. E. ADAMS. 1992. P-type calcium channels blocked by the spider toxin ω-Aga-IVA. Nature **355:** 827-829.
24. MINTZ, I. M., M. E. ADAMS & B. P. BEAN. 1992. P-type calcium channels in rat central and peripheral neurons. Neuron **9:** 85-95.
25. SATHER, W. A., T. TANABE, J.-F. ZHANG, Y. MORI, M. E. ADAMS & R. W. TSIEN. 1993. Distinctive biophysical and pharmacological properties of class A (BI) calcium channel α_1 subunits. Neuron **11:** 291-303.
26. HULLIN, R., D. SINGER-LAHAT, M. FREICHEL, M. BIEL, N. DASCAL, F. HOFMANN & V. FLOCKERZI. 1992. Calcium channel β subunit heterogeneity: Functional expression of cloned cDNA from heart, aorta and brain. EMBO J. **11:** 885-890.
27. PEREZ-REYES, E., A. CASTELLANO, H. S. KIM, P. BERTRAND, E. BAGGSTROM, A. E. LACERDA, X. WEI & L. BIRNBAUMER. 1992. Cloning and expression of a cardiac/brain β subunit of the L-type calcium channel. J. Biol. Chem. **267:** 1792-1797.
28. RUTH, P., A. RÖHRKASTEN, M. BIEL, E. BOSSE, S. REGULLA, H. E. MEYER, V. FLOCKERZI & F. HOFMANN. 1989. Primary structure of the β subunit of the DHP-sensitive calcium channel from skeletal muscle. Science **245:** 1115-1118.
29. USOWICZ, M., M. SUGIMORI, B. CHERKSEY & R. LLINÁS. 1992. P-type calcium channels in the somata and dendrites of adult cerebellar Purkinje cells. Neuron **9:** 1185-1199.
30. HILLYARD, D. R., V. D. MONJE, I. M. MINTZ, B. P. BEAN, L. NADASDI, J. RAMACHANDRAN, G. MILJANICH, A. AZIMI-ZOONOOZ, J. M. MCINTOSH, L. J. CRUZ, J. S. IMPERIAL & B. M. OLIVERA. 1992. A new Conus peptide ligand for mammalian presynaptic Ca^{2+} channels. Neuron **9:** 69-77.
31. SWARTZ, K. B., I. M. MINTZ & B. P. BEAN, personal communication.
32. RANDALL, A. D., B. WENDLAND, F. SCHWEIZER, G. MILJANICH, M. E. ADAMS & R. W. TSIEN. 1993. Five pharmacologically distinct high voltage-activated Ca^{2+} channels in cerebellar granule cells. Soc. Neurosci. Abstr. **19:** 1478.
33. WHEELER, D. B., R. W. TSIEN & A. D. RANDALL. 1993. A calcium channel with novel pharmacology supports synaptic transmission and plasticity in the hippocampus. Soc. Neurosci. Abstr. **19:** 909.
34. ZHANG, J.-F., A. D. RANDALL, P. T. ELLINOR, W. A. HORNE, W. A. SATHER, T. TANABE, T. L. SCHWARZ & R. W. TSIEN. 1993. Distinctive pharmacology and kinetics of cloned neuronal Ca^{2+} channels and their possible counterparts in mammalian CNS neurons. Neuropharmacology **32:** 1075-1088.

Derangements in Calcium-Dependent Membrane Currents in Senescent Human Fibroblasts Are Associated with Overexpression of a Novel Gene Sequence[a]

SAMUEL GOLDSTEIN,[b] SHI LIU,
CHARLES K. LUMPKIN, JR., MING HUANG,
DAVID LIPSCHITZ,[c] AND RAY THWEATT

University of Arkansas for Medical Sciences
John L. McClellan Memorial Veterans' Hospital
4300 West 7th Street–182LR
Little Rock, Arkansas 72205

Biological aging in metazoans is accompanied by functional decline and a rising incidence of degenerative and neoplastic diseases.[1] Although the mechanism(s) remains unknown, the progressive loss of cellular replicative capacity in several tissues is intimately involved.[2–5] Human diploid fibroblasts (HDF) possess a limited replicative lifespan which can be monitored by the cumulative number of mean population doublings (MPD) before senescence.[3,6] The maximum MPD attained by a variety of HDF cultures show an inverse proportion to donor age, and HDF derived from individuals with inherited disorders that display premature aging, such as Werner syndrome (WS), show premature replicative senescence.[2,7] Hybrid fusions between senescent and early-passage (young) HDF demonstrate a dominant inhibitory effect of the senescent cell nucleus on DNA synthesis in the nucleus of the young cell, which can be abolished by chemical agents that block protein and RNA synthesis.[5,8,9] Moreover, microinjection of polyA$^+$ RNA derived from senescent HDF into young

[a]This work was supported in part by grants AG08708 (to S.G.) and AG09458 (to D.A.L.) from the National Institutes of Health, institutional funds from the University of Arkansas for Medical Sciences (to S.L.), grants from the National Institute on Alcohol Abuse and Alcoholism (07550) (to C.L.), the Arkansas Experimental Program to Stimulate Competitive Research (funded by the National Science Foundation, the Arkansas Science and Technology Authority, and the University of Arkansas for Medical Sciences) and the Department of Veterans Affairs (to S.G.), and a National Research Service Award from the NIA (AG05554) and an American Federation for Aging Research (to R.T.).

[b]The staff of the Geriatric Research, Education and Clinical Center at the John L. McClellan Memorial Veterans' Hospital and the Division on Aging at the University of Arkansas for Medical Sciences would like to dedicate this manuscript to the memory of Dr. Goldstein who died suddenly on July 4, 1994.

[c]Corresponding author.

HDF inhibits DNA synthesis.[10] Taken together, the results strongly suggest that HDF senescence is mediated by the overexpression or new expression of genes that lead to the inhibition of DNA synthesis.[3,5]

OVEREXPRESSION OF WS3-10 mRNA IN SENESCENT HUMAN FIBROBLASTS

Based on the rationale that senescent arrest of cellular replication is dominant and therefore likely to be accompanied by overexpression of one or more genes that are causally connected to senescence, we constructed and screened a cDNA library from WS fibroblasts.[11,12] We identified 18 independent gene sequences, nine of them known and nine unknown. Among the known genes were several such as $\alpha 1(I)$ and $\alpha 2(I)$ procollagens, fibronectin, IGF-binding protein-3, osteonectin, plasminogen activator inhibitor type 1, and thrombospondin. The nine unknown clones included two novel gene sequences, WS3-10 and WS9-14, and seven other gene sequences containing both novel segments along with the highly repetitious family of interspersed Alu nuclear elements; five of these seven sequences also contained the LINE family of repetitive elements. Overexpression of RNA transcripts during the cell cycle corresponding to three of the known genes and two of the novel genes is shown in FIGURE 1. The remainder of this report focuses on the WS3-10 cDNA which encodes a novel human cytoplasmic protein and whose cognate gene is located near the tip of the q arm of chromosome 19 (R. Thweatt, D. Dao, and S. Goldstein, unpublished data). The following is a review of our present knowledge of the structure and function of WS3-10 cDNA and insights into the role of this gene sequence in HDF senescence.

COMPARISON BETWEEN WS3-10 AMINO ACID SEQUENCE AND HOMOLOGOUS PROTEINS

The WS3-10 cDNA sequence was determined[13] and found to encode a protein of 22.5 kD, which showed similarities to the sequences of four proteins present in GenBank and EMBL databases (FIG. 2). There was 34% identity to MP20, a flight-muscle protein in drosophila[14] 38% identity to the 22-kD amino terminal ends of turkey gizzard calponin[15] and chicken gizzard calponin[16]; and 84% identity to SM22, a smooth-muscle protein isolated from chicken gizzard but known to reside also in several smooth-muscle-containing tissues such as intestine, uterus, esophagus, and aorta.[17,18] MP20 is a cytoplasmic protein apparently restricted to the synchronous flight muscles of drosophila and believed to harbor two calcium binding sites of the E-F hand variety presumably involved in a sarcoplasmic reticulum Ca^{2+} pump.[14] The WS3-10 amino acid sequence appears degenerate in the first Ca^{2+} binding site at residues 27 (GLU) to 38 (CYS) but is identical at 8 of the 12 residues in the second Ca^{2+} binding site (amino acid residues encompassing positions 108-LYS to 119-GLU) and over 75% homologous when conserved sequences are counted.[13] Using the prototypical Ca^{2+}, our attempts to demonstrate Ca^{2+} binding to the WS3-10 fusion protein by Western ligand blotting with $^{45}Ca^{2+}$ have been unsuccessful so far (V. Grigoriev, R. Thweatt, and S. Goldstein, unpublished data). Failure to demonstrate Ca^{2+} binding by this method is not unusual for proteins that have low capacity and/

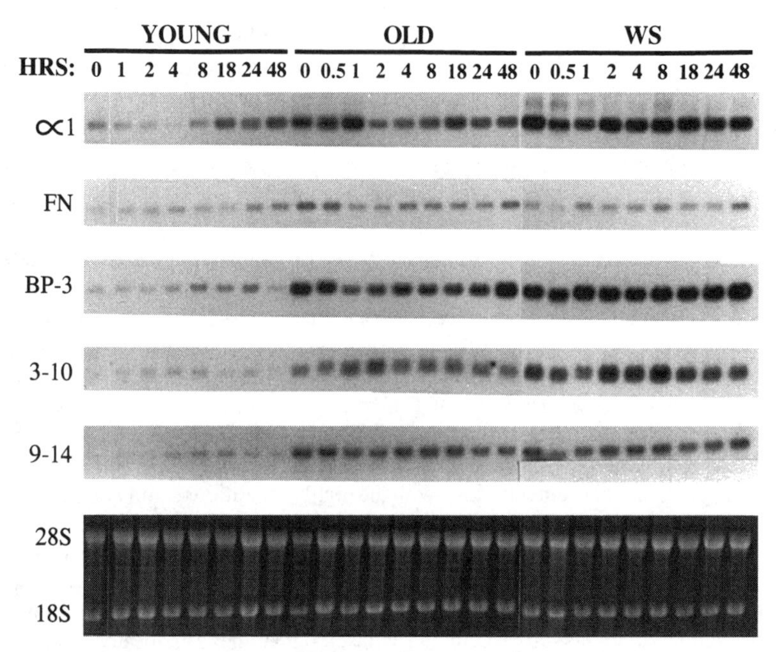

FIGURE 1. Northern analysis of five mRNA sequences in Werner syndrome and normal young and senescent fibroblasts. Cells exposed to serum-depleted medium for 5 days were repleted with regular growth medium. Total cellular RNA was isolated at the intervals shown, fractionated on agarose-formaldehyde gels, transferred to nylon membranes, and hybridized to each of the five ^{32}P-labeled cDNA inserts. Gels stained with ethidium bromide were photographed before transfer to demonstrate equivalent sample loading and RNA integrity as judged by 28S and 18S rRNA bands. $\alpha1 = \alpha1(1)$ procollagen; FN = fibronectin; BP-3 = insulin-like growth factor-binding protein-3. Reproduced from ref. 12 with permission of the American Society of Microbiology.

or low affinity Ca^{2+}-binding sites, and definitive proof of Ca^{2+} binding will require more sensitive techniques.

FUNCTIONAL STUDIES OF THE WS3–10 NOVEL GENE SEQUENCE

We first attempted to demonstrate inhibitory activity on DNA synthesis after transfection of WS3-10 cDNA in a eukaryotic expression vector into young proliferatively competent HDF. However, these studies gave negative results (Thweatt *et al.*, unpublished data). Similarly after microinjection of WS3-10 mRNA transcribed *in vitro* into young HDF, we were again unable to demonstrate inhibition of DNA synthesis.[13,19] Thus, it may be that forced expression of the WS3-10 gene alone cannot inhibit DNA synthesis directly or more likely that such inhibition requires the concerted action of one or more proteins encoded in the several overexpressed

```
DMP20   -------MSLERA*RA**AS*RNP*MDKEAQ***EAIIAEKFPA*QSYEDV*K  046
TCALP   -M*R**A**L*A**KN*LAQ***PQT*RQ*RV**EGAT*RRI--*DN------  043
CSM22   ******A*****D**********D***D******VA***SS*--******R*-  050
WS310   MANKGPSYGMSREVQSKIEKKYDEELEERLVEWIIVQCGPDV--GRPDRGPL-  050

DMP20   DG**LC*LINV**PNAVPK***SG---------------GQ**F**NINN*Q*  084
TCALP   -*XDG**D****ME*----I*K*Q*GSVQK*ND*V----QNWHKL*NIGN**R  087
CSM22   *********IV**Q*----*************I*DS**T********I*****  099
WS310   GFQVWLKNGVILSKL----VNSLYPDGSKPVKVPENPPSMVFKQMEQVAQFLK  099

DMP20   *LKE***PDI*V******Y*K**I*N*TN*IF***-R*TY*HADFKGPFLG--  134
TCALP   *IKH***KPH*I*EAN****NTNHTQ**S**I**A*Q*K**GNNVGL*V----  136
CSM22   *******V***********LA**********V***************H******  152
WS310   AAEDYGVIKTDMFQTVDLFEGKDMAAVQRTLMALGSLAVTKNDGHYRGDPNWF  152

DMP20   P*P*D*C**D***E**KA*QTIV***A***K**T***QNL-*AG*K*LLGK    184
TCALP   -*Y*EKQQ*R*QPEK*R**RNI******T*KF***Q***A*              176
CSM22   ***********S****K***NI******T*K*******S-*******        198
WS310   MKKAQEHKREFTESQLQEGKHVIGLQMGSNRGASQAGMTGYGRPRQIIS      201
```

FIGURE 2. Alignment of amino acid sequences encoding human WS3-10, chicken SM22, drosophila MP20, and the amino-terminal end of turkey calponin. Inasmuch as the turkey and chicken calponin sequences are 98% homologous in this region, only the turkey calponin sequence is shown. Identical residues are denoted by *stars*; *dashes* represent gaps introduced into the protein sequence to obtain the best fit. The two underlined regions in DMP20 represent proposed calcium-binding regions.[14] Reproduced from ref. 13 with permission of Academic Press.

gene sequences identified in the WS cDNA library[12] as well as other senescence-related proteins.[20,21]

ANALYSIS OF MEMBRANE CURRENTS BY WHOLE-CELL PATCH CLAMPING

Because of our high index of suspicion that the WS3-10 protein sequence possesses a Ca^{2+}-binding site and hence plays a role in Ca^{2+}-mediated cell processes, we elected to study membrane currents in senescent HDF.[19] It is known that cultured fibroblasts contain a major Ca^{2+}-mediated K^+ current.[22–25] Thus, we first explored such membrane currents as they exist spontaneously in young and senescent HDF. When young HDF were voltage clamped at –40 mV and subjected to 300-ms voltage pulses to potentials between –100 and +40 mV, we observed families of major time-independent whole-cell membrane currents (FIG. 3A). In striking contrast, membrane currents became markedly depressed in young HDF internally dialyzed with the same pipette solution but containing 5 mM EGTA (FIG. 3B). This confirmed earlier reports indicating that these whole-cell membrane currents in young HDF are Ca^{2+} dependent.[22–25] The findings in senescent HDF were remarkably similar to control young HDF dialyzed with 5 mM EGTA, that is, severe suppression of these Ca^{2+}-dependent membrane

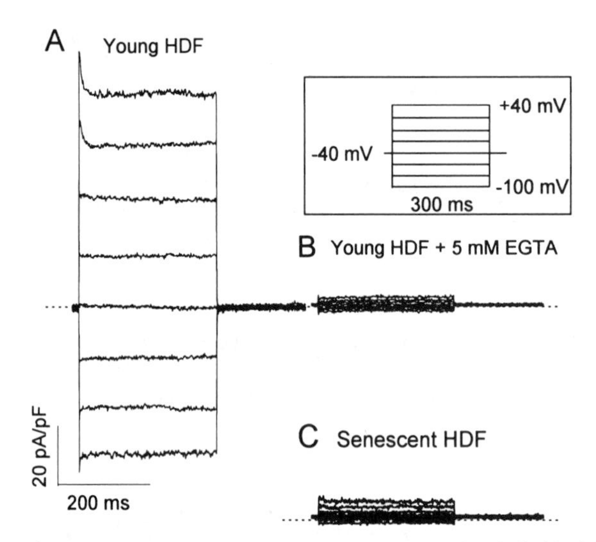

FIGURE 3. Ca^{2+}-dependent membrane currents in representative individual human diploid fibroblasts (HDF). The rectangular inset depicts the protocol used to elicit families of membrane currents by applying 300-ms voltage pulses between −100 and +40 mV from a holding potential of −40 mV in 20-mV increments normal HDF (strain A39, 51-year-old donor). (**A**) Superimposed current families recorded from a young cell. (**B**) Superimposed current families recorded from a young cell after equilibration with a pipette solution containing 5 mM EGTA. (**C**) Superimposed current families recorded from a senescent cell using identical voltage-pulse protocol and control pipette solutions as for young HDF (**A**). *Dotted lines* indicate zero current level. The same scale (pA/pF and ms) applies to panels **A, B,** and **C.** Reproduced from ref. 19.

currents (FIG. 3C). Similar spontaneous reductions of membrane currents were also observed in a second normal strain of senescent HDF and in WS fibroblasts about half-way through their severely curtailed replicative lifespan.[19] These results can also be depicted by plotting the magnitude of currents measured at the end of 300-ms pulses against the test potentials (FIG. 4). In short, these current-voltage relationships again indicate a spontaneous suppression of Ca^{2+}-dependent membrane currents in senescent HDF, whether they be derived from normal individuals or from WS cells, in comparison to young HDF.

SUPPRESSION OF CA^{2+}-DEPENDENT CURRENTS AFTER MICROINJECTION OF WS3-10 mRNA

We synthesized and capped WS3-10 mRNA *in vitro* using full-length WS3-10 cDNA as template, and microinjected this mRNA into young HDF followed by whole-cell patch clamping (FIG. 5). Membrane currents of HDF injected with WS3-10 mRNA were severely suppressed in comparison to uninjected young HDF. This suppression following mRNa injection apparently depended on the coding function of polymeric mRNA because microinjection of hydrolyzed WS3-10 mRNA produced

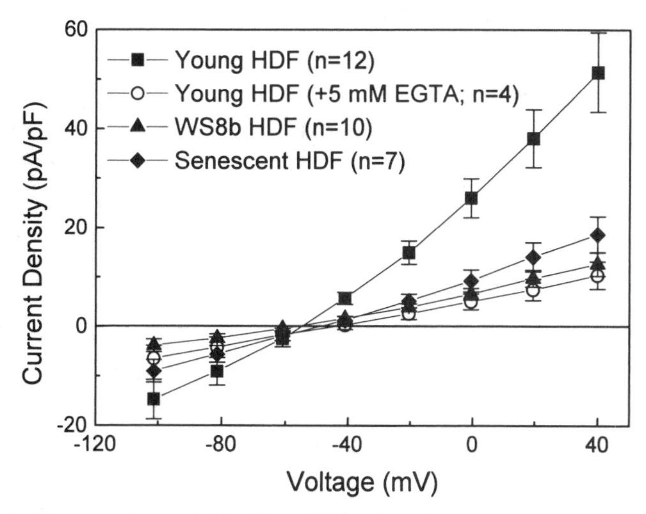

FIGURE 4. Current-voltage relationships of Ca^{2+}-dependent currents in human diploid fibroblasts (HDF). Data were obtained by plotting the magnitude of current measured near the end of 300-ms pulses versus the applied test potentials illustrated in the FIGURE 3 inset. Pooled data are expressed as means ± SE (n = number of cells examined). Reproduced from ref. 19.

no significant change in membrane currents compared to those in uninjected young HDF (FIG. 5). Similarly, microinjection of young HDF with antisense WS3-10 produced no significant change versus that in uninjected controls,[19] and microinjection of an irrelevant mRNA, the αB crystallin mRNA, showed no suppression of membrane current (data not shown). Translation of WS3-10 mRNA in a system of rabbit reticulocyte lysates yielded a protein of the expected M_r of 22.5 kD.[19] Thus, it is likely that microinjection of intact WS3-10 mRNA into young HDF is followed by forced expression of the cognate protein and in turn suppression of the Ca^{2+}-dependent current. Studies to examine the levels of endogenous WS3-10 protein in young and senescent HDF and in a variety of tissues *in vivo* are now underway using a monospecific polyclonal rabbit antibody we recently raised against a fusion protein of WS3-10 (V. Grigoriev, R. Thweatt, and S. Goldstein, work in progress).

STUDIES OF CA^{2+} METABOLISM IN SENESCENT FIBROBLASTS

In view of the apparent derangement of Ca^{2+}-dependent membrane currents, we initiated studies on Ca^{2+} metabolism in HDF (M. Huang, D. Lipschitz, and S. Goldstein, work in progress). Early results demonstrated only a mild (5-10%) reduction in basal levels of intracellular free Ca^{2+} ($[Ca^{2+}]_i$) in senescent HDF versus young controls. However, following stimulation of HDF with a variety of mitogens such as PDGF, EGF, thrombin, and other agents such as bradykinin and the Ca^{2+} ionophore ionomycin, there was a consistently reduced response in senescent HDF measured as the ratio of peak elevations of $[Ca^{2+}]_i$/basal $[Ca^{2+}]_i$. FIGURE 6 indicates that $[Ca^{2+}]_i$ failed to

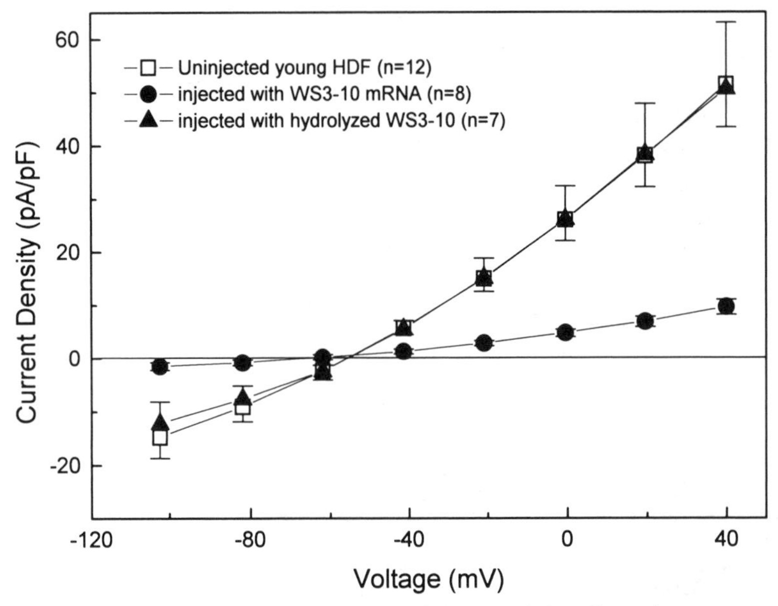

FIGURE 5. Effect of microinjecting WS3-10 mRNA on whole-cell membrane currents of young human diploid fibroblasts (HDF). Current-voltage relationships of Ca^{2+}-dependent currents in uninjected young HDF and young HDF injected with intact WS3-10 mRNA or with hydrolyzed WS3-10 mRNA. Data are expressed as the mean ± SE. *Error bars* were within the symbols for cells injected with WS3-10 mRNA. Note that symbols for cells injected with hydrolyzed mRNA are superimposed upon those for uninjected cells. Reproduced from ref 19.

increase in senescent HDF after thrombin treatment, whereas young cells showed a mean 52% peak increase above basal levels. Additional studies are now underway to define the nature of deranged Ca^{2+} metabolism including Ca^{2+} entry from the extracellular pool and Ca^{2+} release from intracellular pools.

DISCUSSION

We describe herein the spontaneous suppression of Ca^{2+}-dependent membrane currents in senescent normal HDF and prematurely senescent WS HDF. Indeed, the suppression of these membrane currents after injection of young HDF with intact WS3-10 mRNA strongly suggests that overexpression of the cognate endogenous gene during HDF senescence is causally connected to these membrane current reductions.

Several physiologic processes are governed by Ca^{2+} flux in nonexcitable cells including gene expression, initiation of DNA synthesis, entry into mitosis, and cellular contractility and motility.[26-29] The similarly suppressed membrane currents in senescent and WS HDF are strongly suggestive of a reduced level of $[Ca^{2+}]_i$ in both of these cell types, and this presumption is supported by our recent studies on $[Ca^{2+}]_i$ levels by image analysis (FIG. 6, and Huang *et al.*, manuscript in preparation). In

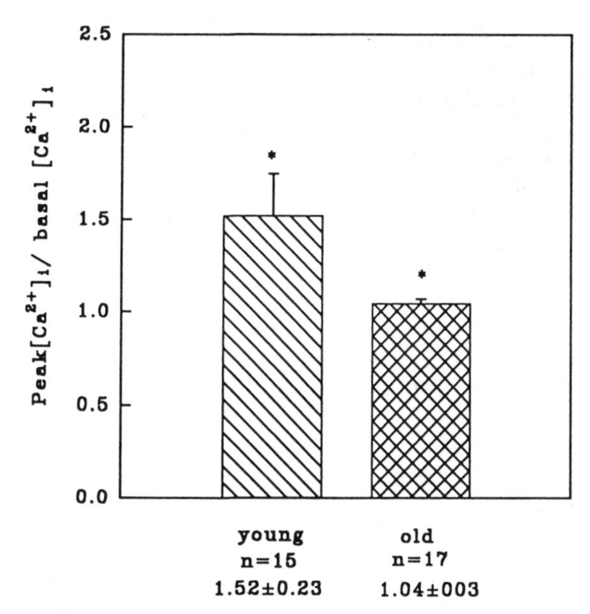

FIGURE 6. Response of $[Ca^{2+}]_i$ in young and senescent human fibroblasts after thrombin treatment. Normal skin fibroblasts (strain A25, 9-year-old donor) were loaded with 3 μM fluo-3/AM, then washed and maintained in 10 mM HEPES-buffered Hank's balanced salt solution. $[Ca^{2+}]_i$ was measured in individual cells using ACAS 570 interactive laser cytometry. Thrombin (5 units/ml) was added to young cells (n = 15) and senescent cells (n = 17) at 30 seconds, and $[Ca^{2+}]_i$ was monitored for a further 180 seconds. The peak increase in young cells occurred at 60 seconds, that is, 30 seconds after the addition of thrombin, whereas old cells failed to respond at any time (data not shown). Results are expressed as the mean ratio of the peak increase/basal $[Ca^{2+}]_i$ ± SEM (* significance of difference, $p < 0.05$).

two previous reports, basal $[Ca^{2+}]_i$ levels were unchanged during senescence of a fetal lung strain of HDF (WI-38), but peak levels after bradykinin stimulation were reduced in one study[30] but not another.[31] In contrast, basal $[Ca^{2+}]_i$ of several young HDF strains derived from skin were decreased significantly as a function of donor age.[32] In this context the relationship of the WS3-10 gene product to calmodulin,[33] the prototypic intracellular Ca^{2+}-binding protein, is unknown. However, calmodulin maintained constant levels in senescent WI-38 cells following stimulation by fetal bovine serum, whereas the characteristic calmodulin profile in young HDF displays an initial decrease followed by a sharp peak preceding entry into S phase after serum stimulation.[31,33,34]

CONCLUSIONS

The salient phenotypic feature of senescent HDF is their inability to initiate DNA synthesis due to an apparent block in the late G1 phase of the cell cycle (see ref. 3). Although the basis for the replicative block is unknown, there is clearly repression of several critical genes such as c-*fos,* p^{34cdc2} kinase, cyclins A and B, and the apparent inability to phosphorylate the Rb protein (ref. 3 and references therein). More recently, D and E cyclins were overexpressed, but the kinase activity supported by complexes of these cyclins with the CDK genes is reduced,[35] perhaps because of the newly discovered senescence-related and p53-inducible 21-kD protein variously known as sdi-1,[20] cip-1,[36] and waf-1.[37] Because many of these factors play a central role in initiating and completing DNA synthesis in many cells and require Ca^{2+} for activation,[29,38] chronically suppressed Ca^{2+}-mediated currents likely reflect a basic defect in the ability of senescent cells to regulate Ca^{2+} metabolism. However, because forced expression of WS3-10 cDNA after microinjection of its cognate mRNA into young HDF failed to inhibit DNA synthesis in short-term assays,[13,19] it would appear that if overexpression of WS3-10 is involved in the G1 senescence block, it would have to exert its effect over a protracted time and/or in concert with several of the other overexpressed genes.

Our observation that a similar constellation of genes is overexpressed in senescent normal HDF and prematurely senescent HDF from WS, along with the similar suppression of membrane currents in these senescent cell types, strongly suggests that a final common pathway of cellular senescence exists *in vitro.* Future studies will need to characterize further the repercussions of WS3-10 gene overexpression, including their relationships to suppression of Ca^{2+}-dependent membrane currents and, putatively, the attenuation of Ca^{2+} flux in senescent HDF. We will also need to explore the role of this novel gene in nonproliferative functions *in vitro* and *in vivo* which should provide powerful insights into the origins of the functional decline and age-dependent diseases that accompany biological aging.

ACKNOWLEDGMENTS

We thank Elena J. Moerman and Richard A. Jones for technical assistance and Patricia L. Spies for manuscript preparation.

REFERENCES

1. GOLDSTEIN, S. 1989. Cellular senescence. *In* Endocrinology, Vol 3. L. J. DeGroot, L. Martini, J. Potts, D. Nelson, A. Winegrad, W. Odell & G. Cahill. Eds.: 2525-2549. Grune & Stratton. New York.
2. GOLDSTEIN, S. 1978. Human genetic disorders which feature accelerated aging. *In* The Genetics of Aging. E. L. Schneider, Ed.: 171-224. Plenum Press. New York.
3. GOLDSTEIN, S. 1990. Replicative senescence: The human fibroblast comes of age. Science **249:** 1129-1133.
4. MARTIN, G. M., C. A. SPRAGUE & C. J. EPSTEIN. 1970. Replicative life-span of cultivated human cells. Effects of donor's age, tissue, and genotype. Lab. Invest. **23:** 86-92.
5. NORWOOD, T. H., J. R. SMITH & G. H. STEIN. 1990. Aging at the cellular level: The human fibroblastlike cell model. *In* Handbook of the Biology of Aging. E. L. Schneider & J. W. Rowe, Eds.: 131-154. Academic Press. San Diego, CA.
6. HAYFLICK, L. 1965. The limited in vitro lifetime of human diploid cell strains. Exp. Cell Res. **37:** 614-636.
7. THWEATT, R. & S. GOLDSTEIN. 1993. Werner Syndrome and biological aging: A molecular genetic hypothesis. BioEssays **15:** 421-426.

8. NORWOOD, T. H., W. A. PENDERGRASS, C. A. SPRAGUE & G. M. MARTIN. 1974. Dominance of the senescent phenotype in heterokaryons between replicative and post-replicative human fibroblast-like cells. Proc. Natl. Acad. Sci. USA **71:** 2231-2235.
9. BURMER, G. C., C. J. ZIEGLER & T. H. NORWOOD. 1982. Evidence for endogenous polypeptide-mediated inhibition of cell-cycle transit in human diploid cells. J. Cell Biol. **94:** 187-192.
10. LUMPKIN, C. K., JR., J. K. MCCLUNG, O. M. PEREIRA-SMITH & J. R. SMITH. 1986. Existence of high abundance and antiproliferative mRNA's in senescent human diploid fibroblasts. Science **232:** 393-395.
11. GOLDSTEIN, S., S. MURANO, H. BENES, E. J. MOERMAN, R. A. JONES, R. THWEATT, R. J. SHMOOKLER REIS & B. H. HOWARD. 1989. Studies on the molecular genetic basis of replicative senescence in Werner syndrome and normal fibroblasts. Exp. Gerontol. **24:** 461-468.
12. MURANO, S., R. THWEATT, R. J. SCHMOOKLER-REIS, R. A. JONES, E. J. MOERMAN & S. GOLDSTEIN. 1991. Diverse gene sequences are overexpressed in Werner syndrome fibroblasts undergoing premature replicative senescence. Mol. Cell. Biol. **11:** 3905-3914.
13. THWEATT, R., C. K. LUMPKIN, JR. & S. GOLDSTEIN. 1992. A novel gene encoding a smooth muscle protein is overexpressed in senescent human fibroblasts. Biochem. Biophys. Res. Comm. **187:** 1-7.
14. AYME-SOUTHGATE, A., P. LASKO, C. FRENCH & M. PARDUE. 1989. Characterization of the gene for MP20: A drosophila muscle protein that is not found in asynchronous oscillatory flight muscle. J. Cell Biol. **108:** 521-531.
15. VANCOMPERNOLLE, K., M. GIMONA, M. HERZOG, J. VAN DOMME, J. VANDEKERCKHOVE & V. SMALL. 1990. Isolation and sequence of a tropomyosin-binding fragment of turkey gizzard calponin. FEBS Let. **274:** 146-150.
16. TAKAHASHI, K. & B. NADAL-GINARD. 1991. Molecular cloning and sequence analysis of smooth muscle calponin. J. Biol. Chem. **266:** 13284-13288.
17. PEARLSTONE, J. R., M. WEBER, J. P. LEES-MILLER, M. R. CARPENTER & L. B. SMILLIE. 1987. Amino acid sequence of chicken gizzard smooth muscle SM22. J. Biol. Chem. **262:** 5985-5991.
18. LEES-MILLER, J. P., D. H. HEELEY & L. B. SMILLIE. 1987. An abundant and novel protein of 22 kDa (SM22) is widely distributed in smooth muscles. Biochem. J. **244:** 705-709.
19. LIU, S., R. THWEATT, C. K. LUMPKIN, JR. & S. GOLDSTEIN. 1994. Suppression of calcium-dependent membrane currents in human fibroblasts by replicative senescence and forced expression of a novel gene sequence. Proc. Natl. Acad. Sci. USA, in press.
20. NODA, A., Y. NING, S. F. VENABLE, O. M. PEREIRA-SMITH & J. R. SMITH. 1994. Exp. Cell Res., in press.
21. HUNTER, T. 1993. Braking the cycle. Cell **75:** 839-841.
22. NELSON, P. G., J. PEACOCK & J. MINNA. 1972. An active electrical response in fibroblasts. J. Gen. Physiol. **60:** 58-71.
23. HOSOI, S. & C. L. SLAYMAN. 1985. Membrane voltage, resistance, and channel switching in isolated mouse fibroblasts (L cells): A patch-electrode analysis. J. Physiol. Lond. **367:** 267-290.
24. FRENCH, A. S. & L. L. STOCKBRIDGE. 1988. Potassium channels in human and avian fibroblasts. Proc. R. Soc. Lond. B **232:** 395-412.
25. STOCKBRIDGE, L. L. & A. S. FRENCH. 1989. Characterization of a calcium-activated potassium channel in human fibroblasts. Can. J. Physiol. Pharmacol. **67:** 1300-1307.
26. RASMUSSEN, H. & P. Q. BARRATT. 1984. Calcium messenger system: An integrated view. Physiol. Rev. **64:** 938-984.
27. TSIEN, R. W. & R. Y. TSIEN. 1990. Calcium channels, stores, and oscillations. Annu. Rev. Cell Biol. **6:** 715-760.
28. DAVIS, T. N. 1992. What's new with calcium? Cell **71:** 557-564.

29. BERRIDGE, M. J. 1993. Inositol triphosphate and calcium signalling. Nature **361:** 315-325.
30. TAKAHASHI, Y., T. YOSHIDA & S. TAKASHIMA. 1992. The regulation of intracellular calcium ion and pH in young and old fibroblast cells (WI-38). J. Gerontol. BS **47:** B65-B70.
31. BROOKS-FREDERICH, M., F. L. CIANCIARULO, S. R. RITTLING & V. J. CRISTOFALO. 1993. Cell cycle-dependent regulation of Ca^{2+} in young and senescent WI-38 cells. Exp. Cell Res. **205:** 412-415.
32. PETERSON, C., R. RATAN, M. SHELANSKI & J. GOLDMAN. 1989. Changes in calcium homeostasis during aging and Alzheimer's disease. Ann. N.Y. Acad. Sci. **568:** 262-270.
33. MEANS, A. R., M. F. A. VANBERKUM, I. BAGCHI, K. P. LU & C. D. RASMUSSEN. 1991. Regulatory functions of calmodulin. Pharmacol. Ther. **50:** 225-270.
34. CRISTOFALO, V. J., D. L. DOGGETT, K. M. BROOKS-FREDERICH & P. D. PHILLIPS. 1989. Growth factors as probes of cell aging. Exp. Gerontol. **24:** 367-374.
35. DULIC, V., L. F. DRULLINGER, E. LEES, S. I. REED & G. H. STEIN. 1992. Altered regulation of G1 cyclins in senescent human diploid fibroblasts: Accumulation of inactive cyclin E-Cdk2 and cyclin D1-Cdk2 complexes. Proc. Natl. Acad. Sci. USA **90:** 11034-11038.
36. HARPER, J. W., G. R. ADAMI, N. WEI, K. KEYOMARSI & S. J. ELLEDGE. 1993. The p21 Cdk-interacting protein cip1 is a potent inhibitor of G1 cyclin-dependent kinases. Cell **75:** 805-816.
37. EL-DEIRY, W. S., T. TOKINO, V. E. VELCULESCU, D. B. LEVY, R. PARSONS, J. M. TRENT, D. LIN, W. E. MERCER, K. W. KINZLER & B. VOGELSTEIN. 1993. WAF-1, a potential mediator of p53 tumor suppression. Cell **75:** 817-825.
38. PARDEE, A. B. 1989. G1 events and regulation of cell proliferation. Science **246:** 603-608.

Calcium Imaging in Hippocampal Neurons using Confocal Microscopy[a]

THOMAS H. BROWN[b] AND DAVID B. JAFFE

Departments of Psychology and
Cellular and Molecular Physiology and
Yale Center for Theoretical and Applied Neuroscience
Yale University
PO Box 208205
New Haven, Connecticut 06520-8205

Calcium ions are ubiquitous second messengers that mediate numerous processes in both excitable and nonexcitable cells. In neurons, changes in intracellular Ca^{2+} concentration ($[Ca^{2+}]_i$) trigger neurotransmitter release, regulate membrane excitability, affect gene expression, and govern short- and long-term forms of synaptic plasticity. In addition, growing evidence supports the hypothesis that disregulation of calcium homeostasis is a key factor in aging and dementia.[1-8]

We have therefore been interested in methods that can furnish insights into calcium dynamics in neurons under both normal and pathological conditions. Here we describe an approach that offers great promise for understanding calcium dynamics in identifiable neurons from acute brain slices. The approach, which combines electrophysiology with simultaneous confocal laser scanning microscopy (CLSM), can be applied to any brain region. Furthermore, high spatial resolution can be achieved without resorting to culture methods. The CLSM system is conveniently combined with video-enhanced, differential-interference contrast (DIC) microscopy in a way that enables visually guided recording from preselected cells.

This approach has already led to new discoveries in calcium dynamics in hippocampal neurons and has obvious relevance to the calcium hypothesis for aging and dementia. This hypothesis demands an explanation for the specificity of neuropathology, which in turn raises the need for a comparative analysis of calcium metabolism. In particular, the hypothesis requires an explanation of why some cells are at more risk than others.[1,4] Thus, we would like to be able to record simultaneously both electrical and calcium signals from visually preselected classes of neurons. The methods described here enable this to be done routinely with either human or animal tissue.

[a] This research was supported by Office of Naval Research, National Institute of Mental Health, and the National Institutes of Health.

[b] Address for correspondence: Thomas H. Brown, Department of Psychology, PO Box 208205, New Haven, CT 06520-8205.

METHODS AND RATIONALE

In our preliminary video microscopic studies,[9] we noted four obvious factors that improved the quality of the optical images of amygdala and hippocampal brain slices. First, thin slices (150-250 μm) were optically superior to thick ones (350-450 μm). However, an obvious trade-off exists between optics and cell viability. The thinnest slices tend to have the fewest living cells because the dendritic arbors of most of the cell types are large and are therefore badly damaged. Second, sectioning with a vibratome gave optically and physiologically superior slices than did sectioning with a conventional tissue chopper. Third, more translucent slices were obtained from younger rather than older animals. Finally, we replicated the observation of MacVicar[10] that infrared improves resolution and/or contrast of cells lying deeper in the tissue.

In what follows we describe the general methods and the rationale underlying several of the key decision points.

Preparation of Acute Hippocampal Brain Slices

In the following experiments, brain slices were vibratome sliced (300 μm) from 15-25-day-old Sprague-Dawley rats. The brain was oriented for transverse sections through the hippocampus. Dissection and slicing were performed in cold (4 °C), artificial cerebrospinal fluid (aCSF) containing (in mM): 124 NaCl, 2.5 KCl, 26 NaH_2PO_4, 2 $MgCl_2$, 1.25 NaH_2PO_4, 2 $CaCl_2$, and 10 dextrose. The slices were then placed in a holding chamber containing aCSF bubbled with 95% O_2/5% CO_2. Individual slices were transferred as needed to a recording chamber perfused at 1 ml/min with oxygenated aCSF.

Neurophysiological Stimulation and Recording Methods

General methods for stimulation of synaptic inputs and for recording electrical responses are described elsewhere.[11] Recorded cells were typically 50-100 μm from the surface of the slice. Recordings were performed using both sharp and whole-cell electrodes. Tips of the sharp microelectrodes were filled with 2-6 mM of Calcium green (CaG) or Fluo-3 and 200 mM KAcetate (see also TABLE 1). Shanks were then

TABLE 1. Long Wavelength Ca^{2+} Indicator Dyes[a]

Dye	Optimum Excitation (nm)	Maximum Emission (nm)	K_d (nM)
Ca^{2+} green	506	534	189
Fluo-3	506	525	316
Ca^{2+} orange	554	575	328
Ca^{2+} green 5N	505	531	3300

[a] Krypton-Argon laser lines: 488, 568, and 670 nm.

backfilled with 4 M KAcetate. The dye was iontophoresed into the neuron for 10-15 minutes using about 1 nA steady hyperpolarizing current.

Patch micropipette tips contained 150 mM K-Gluconate, 10 mM HEPES, 1 mM EGTA, 2 mM $MgCl_2$, and 2 mM Na_2ATP. In some experiments CsCl was substituted for K-Gluconate to block K^+ channels. The shank was then filled with the same solution plus 0.1-0.2 mM CaG or Fluo-3. Dialyzing the neuron with the pipette saline solution for 5-10 minutes was sufficient to fill the dendrites with dye, making iontophoresis unnecessary. Electrical signals were amplified using an Axoclamp 2a under current- or voltage-clamp modes. Recordings with sharp electrodes were performed "blind," whereas whole-cell recordings were done under direct guidance using video-enhanced contrast DIC microscopy. An example is shown in FIGURE 1A, where the somatic electrode is entering the field from the right (*arrow*).

Patch Clamping Visually Preidentified Hippocampal Neurons in Brain Slices

It was evident from our original video-DIC microscopy of living brain slices[9] that considerable cellular and subcellular detail could easily be seen, even in relatively thick tissue. Video microscopy is now regularly employed in our lab to record from identified neurons. It is a relatively simple matter to visualize both the soma and the dendrites (FIG. 1A) and to patch-clamp either or both. General methods used to patch clamp neurons using video microscopy are described in detail elsewhere.[12,13]

To combine video microscopy with confocal imaging, we use an upright microscope (Zeiss Axioskop) with a dual-port head. One port (with 2.5× additional magnification) contains a video camera (Hamamatsu C2400) that is sensitive into the near infrared (IR) range. Use of IR illumination has two advantages. First, IR enables better (higher contrast) images of deeper cells as well as dendrites.[9,10,13] Second, it makes it possible to image and record from preselected cells without exciting the fluorophores of interest (TABLE 1). In this way, phototoxicity and/or photobleaching can be avoided or reduced during phases of an experiment when video microscopy is used. The second port on the microscope, which has a glass-free path to the objective, is attached to our laser scanner (to be described).

Neurons were visualized beneath a water-immersion Ziess 40× objective (NA = 0.75) with a working distance of 1.5 mm. "Healthy" cells, selected for by their appearance, were patched under direct visual control.[13] Once a giga-ohm seal was obtained, the system was switched to CLSM. Whole-cell access to the neuron was then achieved by applying additional negative pressure to the pipette. Access resistances were between 7 and 20 MΩ.

Confocal Imaging of Neurons in Living Hippocampal Brain Slices

Equipment and Procedures. A Biorad MRC 600 equipped with a krypton-argon laser, which has peaks at 488, 568, and 670 nm (TABLE 1), was used in our experiments. The calcium-sensitive dyes were excited by the 488-nm laser line, and emissions greater than 510 nm were detected. A neutral density filter of 90 or 95% was always used to limit exposure of the neuron to laser light. We have used several of the

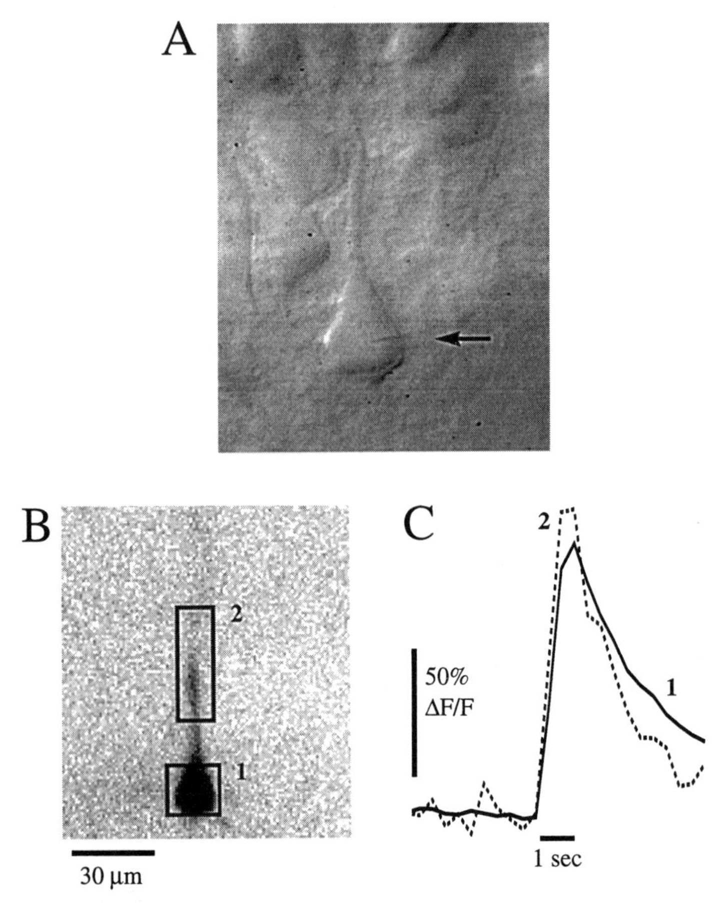

FIGURE 1. Imaging during simultaneous recordings. (**A**) Video microscopy of a rat hippocampal neuron from the CA1 region. A patch-clamp pipette, entering from the right (*arrow*), has formed a giga-Ω seal and is recording in the whole-cell mode. (**B**) CLSM image of a cell filled with fluo-3. The rectangles indicate regions over which measurements were made of the time course of calcium transients. (**C**) Time course of calcium transients produced by a depolarizing current step that elicited a train of action potentials. The signal is plotted as the change in fluorescence divided by resting fluorescence (corrected for background fluorescence), as described in Jaffe and Brown.[11] Plots 1 and 2 correspond, respectively, to the signals obtained from rectangles 1 and 2.

longer wavelength calcium indicator dyes, which differ mainly in their affinity (TABLE 1). For the experiments to be described, we used the high affinity dye CaG and the lower affinity dye Fluo-3, as mentioned previously. In the latter case, to image the cell using CLSM it is necessary to increase the resting calcium level. For example, the neuron shown in FIGURE 1B was difficult to image before it was depolarized beyond the firing threshold. Images were made as 128 × 128 pixel frames sampled at 500-ms intervals for a total of 24 frames. Scanning and electrical stimulation were all under the control of a Master-8 (A.M.P.I.) stimulus generator. High-time resolution line scans were made every 4 ms over the soma and primary apical dendrite. Images were analyzed using custom software written with PV-WAVE (Precision Visuals).

Advantages of Confocal Laser Scanning Microscopy Imaging. Confocal laser scanning microscopy offers several important advantages for the types of experiments that we perform on acute brain slices. First, it has higher spatial resolution than does a conventional widefield microscope. For the latter, the resolution d is given by

$$d = \frac{\lambda}{(NA_{obj} + NA_{con})},$$

where NA_{obj} and NA_{con} are the numerical apertures of the objective and condenser, respectively, and λ is the wavelength of light. In a confocal microscope, $NA_{obj} = NA_{con}$, because the lens serves as the condenser, and the resolution is given by

$$d = \frac{\lambda}{1.4(NA_{obj} + NA_{con})} = \frac{\lambda}{1.4(2 \cdot NA_{obj})}.$$

In practice, the improvement with CLSM is commonly greater than suggested by the factor of 1.4 in the denominator because in a typical nonconfocal microscope the numerical aperture of the condenser is usually less than that of the lens. For a suitable water-immersion lens, the maximum $NA_{obj} \approx 0.9$, which means that for physiological experiments using an upright CLSM system and visible light, the practical limit of resolution (TABLE 1) is about

$$d \approx \frac{500\text{ nm}}{1.4(2 \cdot 0.9)} = \frac{500\text{ nm}}{2.52} \approx 0.2\ \mu\text{m}.$$

Of course, objects smaller than this can be seen, but one cannot trust the diameter measurements. In practice, optical resolution in living brain slices decreases with distance from the surface of the slice. In slices from young animals, the resolution is normally best within 50 μm of the surface.

A second important advantage is that CLSM eliminates out-of-focus fluorescence by taking thin optical sections. The thickness of the optical section is a function of both the lens and the pinhole aperture. In practice, it is commonly the micron range so that the resolution in the Z axis is considerably less than in the X and Y axes. With our 40x objective (0.75 NA), the measured Z half-width resolution can be selected for ranges of approximately 1.5-6.0 μm (via an adjustable detector aperture). This resolution in Z is vastly superior to what is achieved with nonconfocal systems and is sufficient to limit observations to specific subcellular structures such as dendritic spines. Even greater Z-axis resolution can be achieved with a higher NA objective. Stacks of these thin optical sections can be combined electronically into a three-dimensional image that can be rotated and analyzed in various ways using volume rendering software (VoxelView, Vital Images).

One obvious concern in any non-ratiometric method of calcium imaging is that spatial differences in the magnitude of fluorescence changes could reflect differences in dye concentration or resting calcium levels. We think that this is not a problem for the types of information that we have been collecting for two reasons. First, our results[11,14] on the spatial distribution of fluorescence changes are qualitatively similar to those obtained using (non-CLSM) ratiometric methods.[15,16] Second, extremely nonuniform dye distribution (compartmentalization) should be obvious when using high affinity dyes, such as CaG, which produce a substantial signal throughout resting neurons. The potential problem of dye saturation always exists when using a high-affinity dye, which is why we routinely compare the effects of dyes with different affinities (TABLE 1).

Special Considerations in Measuring Calcium Transients in Brain Slices

In evaluating calcium dynamics in neurons from acute brain slices, we insist on knowing something about the "health" of the cells and the electrical events occurring during changes in calcium concentration. In an acutely sliced hippocampus we cannot assume that most of the cells are "healthy." To the contrary, it is obvious from our video imaging, neurophysiology, and analysis of calcium transients that many cells are dead or dying. As the cells begin to deteriorate (based on conventional neurophysiological criteria), calcium dynamics become very abnormal. For example, calcium signals tend to remain elevated for prolonged periods. This is almost always a sure sign that the cell is in serious trouble.[6,11]

We therefore require simultaneous electrophysiological measurements as an independent criterion of cell viability. There are other reasons as well. For example, neurophysiological data are critical for relating electrical activity to changes in calcium concentration. Knowing this relationship is often necessary for determining the cause of specific changes in calcium. Furthermore, electrical recording methods allow the investigator to manipulate directly electrical activity in the neuron in a more natural and controlled manner than can be done, for example, with bath-applied drugs or elevated potassium.[11,17–21]

A second important consideration in measuring calcium dynamics in hippocampal neurons is the spatial resolution of the system. It is becoming increasingly clear that various calcium microdomains[6] may be regulated very differently.[19] Obviously absolute $[Ca^{2+}]_i$ levels cannot be measured accurately if the spatial resolution of the system is low relative to variations in the microdomains of interest. These microdomains can be as small as the distal portion of a dendritic spine. The high spatial resolution of the CLSM system is very important in this respect. Furthermore, calcium indicators with a range of affinities are useful in delineating different microdomains as well.[6] For example, Llinas[6] used low affinity calcium dyes to great advantage in helping to localize the calcium channels responsible for exocytotic neurotransmitter release.

For those who insist on ratiometric measurements, we should point out that it is possible to do this using CLSM. The obvious way is to use an ultraviolet confocal laser in conjunction with a ratiometric dye such as Fura 2. An alternative method is to use two dyes with different emission properties, such as Fluo-3 and Fura-Red.[22] We are currently exploring this and other alternatives to the use of UV.

RESULTS

Here we summarize briefly three Ca^{2+} sources that we have begun to examine using CLSM and simultaneous neurophysiology, influx via voltage-gated calcium channels (VGCCs), influx through synaptically gated channels, and release from intracellular stores.

Voltage-Gated Calcium Influx

We found that action potentials increase intracellular calcium levels through voltage-gated calcium influx in both the soma and the dendrites of every type of mammalian CNS neuron that we have examined thus far. These include hippocampal pyramidal neurons, interneurons of the hippocampal formation, dentate hilar interneurons, dentate granule cells, pyramidal neurons of the neocortex, and amygdala neurons.

Examples of calcium transients using Fluo-3 (TABLE 1) are illustrated in FIGURE 1. The CLSM image of the fluorescence during repetitive spiking produced by an outward current step is shown in FIGURE 1B. Due to the relatively low affinity of the dye (TABLE 1), the cell cannot easily be imaged in the resting state (without increasing intracellular calcium by depolarizing the cell). Rectangle 1 is placed over the soma and rectangle 2 is aligned along the primary apical dendrite, proximal to the soma. FIGURE 1C shows the time course of the fluorescence change produced by a depolarizing current step that elicited a train of action potentials in the cell. The results are plotted as the change in (ΔF) fluorescence divided by the resting fluorescence (F) after subtracting background fluorescence (which was negligible). Curves labeled 1 and 2 represent, respectively, signals measured in the areas designated by rectangles 1 and 2.

In hippocampal CA1 pyramidal neurons, the magnitude of the calcium signal depended on the number of action potentials and the specific anatomical region of the neuron.[11] Single action potentials were sufficient to produce detectable increases in calcium, whereas calcium summated during trains of action potentials. The largest calcium signals produced by action potentials usually occurred in the proximal dendrites. Smaller increases were typically observed in the soma, presumably as a result of differences in surface to volume ratios. Somatically generated action potentials produce significantly less voltage-gated calcium entry into distal dendrites (>100 μm from the soma). The attenuation of action potentials with distance along the apical dendrite has been proposed to account for the failure of spikes to elevate distal dendritic calcium.[20]

Under normal conditions, elevated calcium levels decayed back to baseline. Calcium increases measured with high-time resolution (4-ms sample rate) were time-locked with action potentials. The decay of calcium began immediately after each individual spike. In our experiments, the decay of calcium in the dendrites usually had a time constant of 1-2 seconds. Because of the buffering effects of the calcium indicators and EGTA, it is difficult to determine what the decay time constant for calcium removal is in the absence of dye.[19,23,24]

When dendrites are sufficiently close to the surface of the slice, we have been able to measure changes in calcium within very high spatial resolution, at the level of the dendritic spine. Preliminary work suggests that voltage-gated calcium channel activation triggers calcium signals in dendritic spines.[19] We are currently following up this interesting possibility using higher resolution methods combined with ratiometric methods.

In contrast to trains of somatically generated action potentials, orthodromic synaptic stimulation (100 Hz stimulus trains delivered for 1 second) produced larger calcium increases in the more distal part of the dendritic arbor.[11] The largest orthodromically produced signals were clearly and directly associated with firing of the cell. In the absence of cell firing, orthodromic stimulation produced very small calcium signals, and it was suggested that much of this synaptically mediated calcium signal may be associated with conventional voltage-gated calcium channels.[19,20]

Ligand-Gated Calcium Influx

N-methyl-D-aspartate (NMDA) receptors have a large calcium permeability[25] and are activated by synaptic transmission.[26] Therefore, NMDA receptors may significantly contribute to the dendritic calcium signal produced by orthodromic high frequency stimulation. However, voltage-gated calcium entry appears to dominate synaptically mediated calcium signals, possibly contaminating highly localized increases in calcium.[21] We therefore attempted to isolate experimentally the synaptically mediated calcium signals due to voltage-gated calcium entry.

Using whole-cell recording methods, CA1 pyramidal neurons were voltage-clamped at potentials at or above the synaptic reversal potential (0 to +10 mV) using Cs-filled pipettes.[11] This procedure limited the activation of voltage-gated calcium entry during high frequency stimulation in two ways. First, holding cells depolarized should inactivate, in a voltage-dependent manner, a significant proportion of calcium channels. Second, the synaptic potential produced during high frequency stimulation (100 Hz stimulation for 1 second) should be close to zero or even hyperpolarizing. Under these conditions, high frequency stimulation should not trigger voltage-gated calcium entry. In addition, we chose to examine synapses onto the basal dendrites by placing stimulating electrodes in stratum oriens. Basal dendrites are electrotonically shorter than apical dendrites, affording better voltage control.[27]

Depolarizing CA1 neurons to the synaptic reversal potential resulted in a transient increase in calcium throughout the dendrites (both apical and basal) through voltage-gated calcium channel activation. Calcium signals did not remain elevated, consistent with voltage-dependent channel inactivation. Under these conditions (0 to +10 mV holding potential) high frequency stimulation produced highly localized increases in calcium. Assuming that the membrane potential at all synapses was at or above the reversal potential, these signals reflect changes in calcium independent of voltage-gated calcium entry. The ligand-stimulated calcium transients measured in these experiments always returned to pre-stimulation levels at a rate similar to that produced by voltage-gated calcium entry.

Intracellular Calcium Release

In addition to the ligand- and voltage-gated calcium entry just described, a third mechanism for raising intracellular calcium involves release of calcium from intracellular stores. There are two primary receptor mechanisms for releasing calcium, inositol trisphosphate (IP3) receptors and ryanodine receptors. Both of these receptor types are found in many types of mammalian CNS neurons and can serve as calcium-induced calcium release mechanisms.

Metabotropic glutamate (mGlu) receptors can trigger intracellular calcium release in CA1 hippocampal neurons. Brief application of the specific mGlu receptor agonist ACPD to the dendrites produced an initially local increase in dendritic calcium.[14] Increases in calcium quickly propagated from the site of ACPD application throughout the neuron, including the soma and much of the dendritic tree. The speed of these "waves" (approximately 40 μm/s) was consistent with calcium waves produced by calcium-induced calcium release mechanisms described for other "active" cell types. Small outward membrane currents were associated with mGlu-stimulated increases in calcium, presumably due to the activation of calcium-dependent potassium currents. Inward currents, indicative of ionotropic glutamate receptor activation, were not observed.

An interesting property of mGlu-stimulated calcium signals was that repeated APCD applications failed to trigger further release of calcium.[14] Our experiments suggest that this effect was due to depletion of intracellular stores rather than receptor inactivation. Simply allowing time for reactivation did not restore the effects of mGlu receptor activation. We found that calcium entry during trains of action potentials was needed to replenish or "prime" the neuron for subsequent ACPD application to trigger the release of calcium. Apparently a threshold number of action potential trains was needed to "prime" ACPD-stimulated calcium release, suggesting a nonlinear relationship between voltage-gated calcium entry, calcium sequestration, and mGlu-triggered calcium release.

DISCUSSION

We have described methods and results that appear relevant to the calcium hypothesis for aging and dementia. The general approach combines IR video microscopy, CLSM, and simultaneous recording and control of electrical events. The procedures seem appropriate for use on both human and animal tissue and have already been applied to numerous cell types in a variety of different kinds of brain slices.

Multiple Sources of Calcium

Using these methods, we have studied multiple sources of calcium changes in CNS neurons. For CA1 pyramidal neurons, voltage-gated calcium entry clearly produces the most widespread increases in calcium in both the soma and the proximal dendrites, where action potential amplitude is believed to be largest.[11,14,20] In contrast, orthodromic synaptic stimulation can produce larger distal dendritic calcium signals.[11,16,21] A proportion of this calcium signal appears to be independent of voltage-gated calcium activation, presumably entering through ligand-gated channels such as NMDA receptors.[11,18] Ligand-gated calcium entry is highly localized and may sometimes be restricted to dendritic spines.[19,28] This may compartmentalize high calcium concentrations (in ranges that could prove toxic) to very local domains.[29–31]

The third source of calcium we investigated is released from intracellular stores.[14] We found that focal stimulation of mGlu receptors can trigger an initially local increase in dendritic calcium that rapidly propagates throughout the neuron, possibly by a mechanism of calcium-induced calcium release. These calcium waves could be triggered repeatedly as long as the cell was "primed" by a threshold number of action potential trains. Our results suggest that voltage-gated calcium, presumably through a sequestering mechanism, was required to replenish intracellular calcium stores. This source of calcium may be tightly coupled to the regulation of intracellular calcium dynamics.

Calcium Regulation and Cell Health

Resting calcium concentration in neurons is maintained at very low levels, approximately 20-200 nM.[15] Changes in intracellular calcium concentration produced by any of the three sources just described always decay back to baseline levels (also see ref. 32). The decay of calcium results from the contribution of endogenous calcium buffers, calcium extrusion, and calcium sequestration.[7] The latter two mechanisms ultimately rely on the metabolic viability of the neuron.

There are several indicators of neuronal deterioration: the loss of resting membrane potential, the decline of input resistance and membrane time constant, the inability to fire overshooting action potentials at a high frequency, and the appearance of dendritic "blebbing." Video microscopy also reveals that the optical properties of neurons characteristically change with degeneration and death.[13] All of these characteristics of pathology are associated with large and persistent elevations in intracellular calcium.

Reduction or failure in calcium homeostatic mechanisms (calcium buffering, sequestration, and extrusion) has been proposed to account for neuronal degeneration and cell death.[4] Conditions that produce abnormally large amounts of calcium entry (glutamate toxicity, anoxia, and epileptiform activity) may precipitate this failure by depleting metabolic resources regulating intracellular calcium levels. This is in addition to the deleterious effects of calcium-dependent proteases. Our empirical observations are consistent with this hypothesis. When cells are considered healthy, based on electrophysiological criteria, calcium regulation is effective in returning calcium to baseline levels.

Calcium Hypothesis for Aging and Dementia

Disregulation of calcium homeostatic mechanisms has been proposed as a primary neuronal factor in aging and dementia.[1–5,7,8] The methods just described now allow us to explore the regulation of calcium dynamics within individual CNS neurons under controlled conditions.

Test of the Calcium Hypothesis. In this report, we described our experiments with CA1 pyramidal neurons, but the same methods can be used to examine the physiology and calcium dynamics of many other cell types throughout the CNS. We have been comparing calcium dynamics in hilar interneurons and neurons in the amygdala complex. Thus, we can now look for differences in intracellular calcium regulation among different types of neurons[14,33] and as a function of the age of the animal.[5] This type of work will be especially important for investigating the vulnerability of different types of neurons to various forms of insult, such as glutamate toxicity, anoxia, or intracellular poisons, at different stages of development. The calcium hypothesis requires an explanation for the differential susceptibility of different cell types as a function of aging.[1,4] As we learn more about which calcium regulatory mechanisms are vulnerable to insult and aging, we can begin to explore ways of correcting or preventing the failure of calcium regulatory mechanisms that occur with age or specific pathophysiological conditions.

Calcium Regulation and Cognition. Calcium regulation has been related to cognition, and alterations in calcium regulation offer a very promising therapeutic approach to cognitive dysfunction.[3,5] One possible explanation for this connection involves the synaptic modifications that are responsible for cognitive learning. The leading candidate synaptic substrate for this type of learning, called long-term potentiation, involves a calcium-dependent mechanism.[26,34]

Thus, regulation of intracellular calcium levels is likely to be very relevant to the endogenous induction of long-term potentiation.[3,5,14,19,26,29–31,34] The methods and techniques described here provide a significant step toward providing us with new information regarding long-term potentiation mechanisms that could be relevant to cognitive and emotional learning. Such information could be very important in understanding the causes and appropriate treatment of age-related dementia.

REFERENCES

1. APPEL, S. 1994. Neurodegenerative diseases: Autoimmune disease involving ion channels. This volume.
2. CHOI, D. 1994. Calcium and excitotoxicity. This volume.
3. DISTERHOFT, J. 1994. Neural calcium and learning in aging brain. This volume.
4. KHACHATURIAN, Z. 1994. Calcium hypothesis—review and update. This volume.
5. LANDFIELD, P. 1994. Stress hormones and calcium currents in brain aging and Alzheimer's disease. This volume.
6. LLINAS, R. 1994. Calcium currents in aging neurons. This volume.
7. MICHAELIS, M. 1994. Ca^{2+} transport systems in aging neurons. This volume.
8. SIESJO, B. 1994. Calcium-mediated processes in neuronal degeneration. This volume.
9. KEENAN, C. L., P. F. CHAPMAN, V. C. CHANG & T. H. BROWN. 1988. Videomicroscopy of acute brain slices from amygdala and hippocampus. Brain Res. Bull. **21:** 373-383.
10. MACVICAR, B. A. 1984. Infrared video microscopy to visualize neurons in the in vitro brain slice preparation. J. Neurosci. Methods **12:** 133-139.
11. JAFFE, D. B. & T. H. BROWN. 1994. Confocal imaging of dendritic Ca^{2+} transients in hippocampal brain slices during simultaneous current- and voltage-clamp recording. Microscopy Res. Tech. **29:** in press.
12. EDWARDS, F. A., A. KONNERTH, B. SAKMANN & T. TAKAHASHI. 1989. A thin slice preparation for patch clamp recordings from neurons of the mammalian central nervous system. Pflugers Arch. **414:** 600-612.
13. STUART, G. J., H. U. DODT & B. SAKMANN. 1993. Patch-clamp recordings from the soma and dendrites of neurons in brain slices using infrared video microscopy. Pflugers Arch.
14. JAFFE, D. B. & T. H. BROWN. 1994. Metabotropic glutamate receptor activation induces calcium waves within hippocampal dendrites. J. Neurophysiol. **72:** 471-474.
15. REGEHR, W. G., J. A. CONNOR & D. W. TANK. 1989. Optical imaging of calcium accumulation in hippocampal pyramidal cells during synaptic activation. Nature **341:** 533-536.
16. REGEHR, W. G. & D. W. TANK. 1992. Calcium concentration dynamics produced by synaptic activation of CA1 hippocampal pyramidal cells. J. Neurosci. **12:** 4202-4223.
17. KNOPFEL, T. & B. H. GAHWILER. 1992. Activity-induced elevations of intracellular calcium concentration in pyramidal and nonpyramidal cells of the CA3 region of rat hippocampal slice cultures. J. Neurophysiol. **68:** 961-963.

18. ALFORD, S., B. G. FRENGUELLI, J. G. SCHOFIELD & G. L. COLLINGRIDGE. 1993. Characterization of Ca^{2+} signals induced in hippocampal CA1 neurones by the synaptic activation of NMDA receptors. J. Physiol. (Lond.) **469:** 693-716.
19. JAFFE, D. B., S. A. FISHER & T. H. BROWN. 1994. Confocal laser scanning microscopy reveals voltage-gated calcium signals within hippocampal dendritic spines. J. Neurobiol. **25:** 220-233.
20. JAFFE, D. B., D. JOHNSTON, N. LASSER-ROSS, J. E. LISMAN, H. MIYAKAWA & W. N. ROSS. 1992. The spread of Na^{+} spikes determines the pattern of dendritic Ca^{2+} entry into hippocampal neurons. Nature **357:** 244-246.
21. MIYAKAWA, H., V. LEV-RAM, N. LASSER-ROSS & W. N. ROSS. 1992. Calcium transients evoked by climbing fiber and parallel fiber synaptic inputs in guinea pig cerebellar Purkinje neurons. J. Neurophysiol. **68:** 1178-1189.
22. LIPP, P. & E. NIGGLI. 1993. Ratiometric confocal Ca^{2+}-measurements with visible wavelength indicators in isolated cardiac myocytes. Cell Calcium **14:** 359-372.
23. NEHER, E. & G. J. AUGUSTINE. 1992. Calcium gradients and buffers in bovine chromaffin cells. J. Physiol. (Lond.) **450:** 273-301.
24. HERNANDEZ-CRUZ, A., F. SALA & P. R. ADAMS. 1990. Subcellular calcium transients visualized by confocal microscopy in a voltage-clamped vertebrate neuron. Science **247:** 858-862.
25. SCHNEGGENBURGER, R., Z. ZHOU, A. KONNERTH & E. NEHER. 1993. Fractional contribution of calcium to the cation current through glutamate receptor channels. Neuron **11:** 133-143.
26. BLISS, T. V. P. & G. L. COLLINGRIDGE. 1993. A synaptic model of memory: Long-term potentiation in the hippocampus. Nature **361:** 31-39.
27. TSAI, K., N. T. CARNEVALE & BROWN. 1994. Efficient mapping from neuroanatomical to electronic space. Network: Comp. Neural Syst. **5:** 21-46.
28. PERKEL, D. J., J. J. PETROZZINO, R. A. NICOLL & J. A. CONNOR. 1993. The role of Ca^{2+} entry via synaptically activated NMDA receptors in the induction of long-term potentiation. Neuron **11:** 817-823.
29. MULLER, W. & J. A. CONNOR. 1991. Dendritic spines as individual neuronal compartments for synaptic Ca^{2+} responses. Nature **354:** 73-76.
30. ZADOR, A., C. KOCH & T. H. BROWN. 1990. Biophysical model of a Hebbian synapse. Proc. Natl. Acad. Sci. USA **87:** 6718-6722.
31. HARRIS, K. M. & S. B. KATER. 1994. Dendritic spines: Cellular specializations imparting both stability and flexibility to synaptic function. Ann. Rev. Neurosci. **18:** 341-371.
32. CONNOR, J. A., W. J. WADMAN, P. E. HOCKBERGER & R. K. S. WONG. 1988. Sustained dendritic gradients of Ca^{2+} induced by excitatory amino acids in CA1 hippocampal neurons. Science **240:** 649-653.
33. SLOVITER, R. S., G. VALIQUETTE, G. M. ABRAMS, E. C. RONK, A. L. SOLLAS, L. A. PAUL & S. NEUBORT. 1989. Selective loss of hippocampal granule cells in the mature rat brain after adrenalectomy. Science **243:** 535-538.
34. BROWN, T. H., P. F. CHAPMAN, E. W. KAIRISS & C. L. KEENAN. 1988. Long-term synaptic potentiation. Science **242:** 724-728.

Modulation of Voltage-Dependent Calcium Channels in Cultured Neurons[a]

HUGH A. PEARSON,[b] VERONICA CAMPBELL, NICK BERROW, ANATOLE MENON-JOHANSSON,[c] AND ANNETTE C. DOLPHIN

Department of Pharmacology
Royal Free Hospital School of Medicine
Rowland Hill Street
London NW3 2PF, UK

In many types of excitable cells there are several classes of voltage-dependent calcium channels (VDCC) as determined by the characteristic properties of single channel activity. A clear distinction exists between low conductance channels, activated by moderate depolarizations (low voltage-activated, LVA), and high conductance channels activated by large depolarizations (high voltage-activated, HVA).[1,2] In cardiac tissue these were termed T and L channels, respectively, because they were found to be pharmacologically as well as biophysically distinct.[1] Because of this, they can also be clearly differentiated in whole-cell current recordings.[1,3] Evidence also exists for a third class of single channel conductance (N type) whose biophysical properties were originally described as being intermediate between T and L and which appears to be expressed only in cells of neuronal origin.[3-5] The sensitivity of N-type channels to irreversible block by ω-conotoxin GVIA (ω-CgTx)[5] and the large number of ω-CgTx binding sites in neuronal tissue indicate that they are likely to be important for neuronal function.[6] High threshold Ca^{2+} currents insensitive to both ω-CgTx and dihydropyridines have been reported,[5,7] and a selective blocker for at least part of this current is the peptide toxin from *Agelenopsis aperta,* ω-agatoxin IVA (ω-aga IVA). The current inhibited by this toxin has been termed P current, because Purkinje cells express calcium channel currents that are largely resistant to ω-CgTx and 1,4-dihydropyridines (DHPs),[7] and these currents are sensitive to ω-aga IVA.[8] It is difficult to distinguish these currents by biophysical means at the whole cell level,[5,8,9] and although estimates of their single channel conductances indicate differences, this is complicated by the existence of subconductance states.[10] However, prolongation of single channel open times by DHP agonists remains diagnostic of L channels.[11]

[a] The work in this laboratory was supported by the Wellcome Trust and MRC.

[b] Present address: Department of Pharmacology, Worsley Medical & Dental Building, University of Leeds, L52 9JT, UK.

[c] Present address: Department of Molecular & Cellular Physiology, Beckman Center, Stanford University Medical Center, Stanford, CA 94305.

SUBTYPES OF CALCIUM CHANNEL IN DORSAL ROOT GANGLION NEURONS

The calcium channel currents in dorsal root ganglion (DRG) neurons have been examined in a number of species, including chick and rat.[1] As well as LVA currents, DRGs express ω-CgTx-sensitive, DHP-sensitive, and ω-agaIVA-sensitive currents.[8,12–14] On the basis of these pharmacological experiments, in acutely dissociated DRGs, N-type current represents about 50%, P-type current about 25%, and L-type current about 20% of the whole cell current, with a proportion of current remaining unblocked by all these antagonists. In cultured rat DRGs the proportion of L-type current may be higher (about 40%),[12] but such determinations hinge on the uncertain specificity of μM concentrations of DHP antagonists for L-type channels.[9,14] Interestingly, in acutely axotomized cultured DRGs, L current appears to be absent, possibly suggesting localization on neurites.[15] Evidence suggests that at least for nodose sensory neurons, functional L-type calcium channels appear after Na^+ and K^+ channels in development.[16]

INHIBITION OF VDCCs BY PTX SUBSTRATE G-PROTEINS IN DRGS

The direct interaction of PTX substrate G-proteins with ion channels is now well-established particularly for K^+ channels.[17] Initial evidence concerning VDCCs came from studies of calcium currents in DRG neurons in which noradrenaline acting at α_2-adrenergic receptors and GABA acting at (–)-baclofen-sensitive $GABA_B$ receptors inhibit the current (FIG. 1), and this can be mimicked by GTPγS and inhibited by GDPβS and PTX.[18–20] It is now clear that numerous receptors on DRGs, all acting via PTX-sensitive G-proteins, are able to modulate VDCCs; these include adenosine A_1, μ, and κ opioid, $5\text{-}HT_{1A}$, and NPY receptors.[21–25] Similar modulation occurs in many other neurons and neuronal cell lines and also in neurosecretory cells, such as the pituitary cell line At-T-20, in which the calcium current is inhibited by somatostatin and by GTP analogs.[26] The most marked effect, particularly of GTP analogs, is to slow current activation (FIG. 2)[20] which has been interpreted as inhibition of an inactivating N current. (For a review, see Tsien *et al.*[4]) However, pituitary secretory cells have not generally been shown to possess N current, but a similar slowing of current activation is observed.[26,27] More recently, it was suggested that this slowed current activation is a result of voltage-dependent interaction between the activated G-protein and the calcium channel.[28,29] Evidence has also been put forward that in addition there are indirect pathways possibly involving a soluble second messenger.[30]

Our evidence in cultured DRGs that have been acutely replated is that N current is preferentially modulated by $GABA_B$ receptor stimulation,[15] and this is also the conclusion of Cox and Dunlap.[31] However, in DRGs grown in culture for 2 weeks (–)-baclofen is still able to inhibit calcium channel current following blockade of the ω-conotoxin GVIA-sensitive component, and this may be attributable to inhibitions of P and/or L channels (A. C. Dolphin, unpublished results).

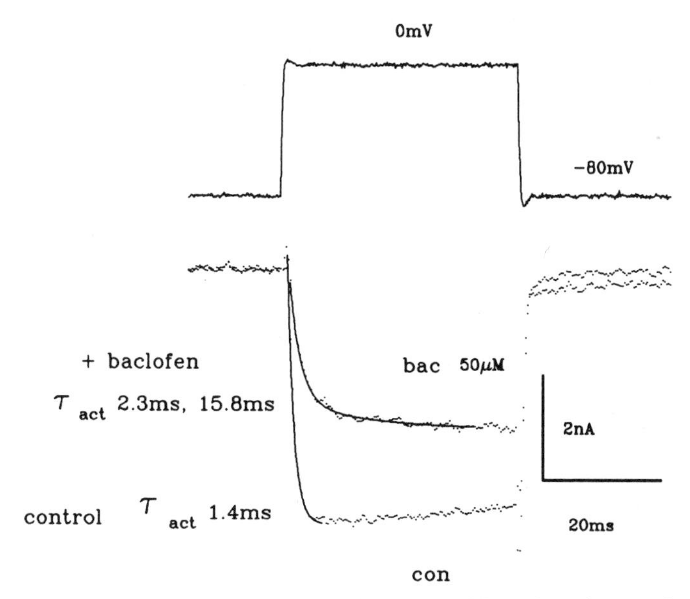

FIGURE 1. Effect of the $GABA_B$ agonist (–)-baclofen on calcium channel current in cultured rat DRGs. Calcium channel currents were recorded as described in Menon-Johansson *et al.*[15] using 1 mM Ba^{2+} as the charge carrier. (–)-Baclofen (50 μM) was applied by pressure application from a puffer pipette. The current is slowed and the amplitude reduced by (–)-baclofen. The activation phase of the control current can be fitted by a single exponential (**solid line**) with a time constant τ_{act} of 1.4 ms. The activation phase of the current in the presence of (–)-baclofen can be fitted by a double exponential (**solid line**) with τ_{act} 2.3 and 15.8 ms.

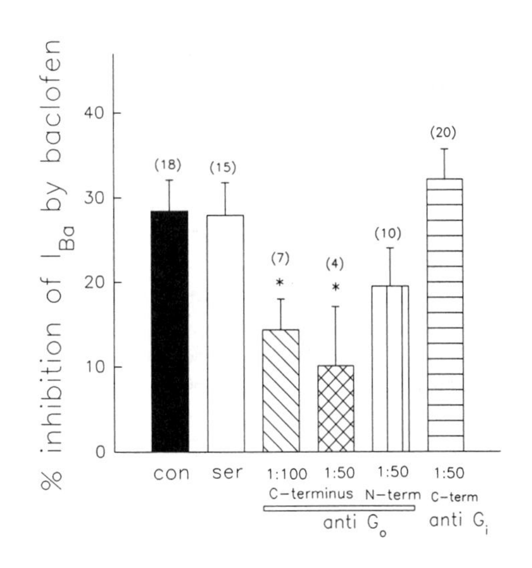

FIGURE 2. Effect of anti-G_o and anti-G_i on the inhibition of calcium channel currents by (–)-baclofen. Calcium channel currents (I_{Ba}) were recorded as described in Menon-Johansson *et al.*[15] The anti-G-protein antibodies were obtained from Dr. G. Milligan, Glasgow University, and were introduced into the DRGs at the time of their acute replating by a form of scrape-loading, as described in Menon-Johansson *et al.*[15] Only the antibody directed against the COOH-terminus of $G\alpha_o$ significantly inhibited the effect of (–)-baclofen. We have evidence that all antibodies used disrupt (–)-baclofen-stimulated GTPase.[53]

THE NATURE OF THE G-PROTEIN INVOLVED IN RECEPTOR-CALCIUM CHANNEL COUPLING IN DRGs

Identification of the G-protein involved in coupling receptors to Ca^{2+} current inhibition has been aided by experiments in which G-proteins are included in the patch pipette when recording Ca^{2+} currents in PTX-treated cells. Exogenously added G-proteins can restore the ability of neurotransmitters to inhibit Ca^{2+} currents. These experiments have generally suggested that G_o protein mediates inhibition of neuronal calcium currents by neurotransmitters.[32–34] We investigated the ability of anti-G-protein antibodies to prevent the effect of the $GABA_B$ agonist (–)-baclofen and GTPγS in cultured rat DRGs. The cells were replated before use by nonenzymatic removal from the culture dish. When this was done in the presence of IgG, it resulted in the entry of antibodies into the cells by a form of scrape-loading. This was confirmed immunocytochemically. We showed that an anti-G_o antipeptide antibody against the COOH-terminus of G_o (OC1, provided by G. Milligan, Glasgow, Scotland), when loaded into DRGs in this way, reduced the ability of (–)-baclofen to inhibit I_{Ba} (FIG. 2). In contrast, antibodies against the NH_2-terminus of G_o or antibodies against the COOH-terminus of G_i, recognizing all G_i species, were ineffective,[35] and the anti G_o antibody OC1 was unable to prevent GTPγS from modulating the calcium channel current.[15] Additional evidence for a role of G_o was that internal application of a COOH-terminal peptide sequence from G_o also reduced the ability of (–)-baclofen to inhibit I_{Ba}. Presumably this peptide mimics G_o and interferes with coupling between the $GABA_B$ receptor and native G_o.[15] In these replated DRGs, in contrast to nonreplated DRGs with neurites, L-type calcium channels were absent and ω-CgTx inhibited approximately 30% of the current. No inhibition of the current by (–)-baclofen was observed following irreversible inhibition of the ω-CgTx-sensitive portion of the current, suggesting that only N channels are modulated by (–)-baclofen in this preparation.[15] However, it is clear from other secretory cell types that receptors are able to modulate L channels, again via G_o.[36]

Further evidence that G_o and not G_i is involved in coupling $GABA_B$ receptors to VDCCs in DRGs comes from experiments in which antisense oligonucleotides complementary to mRNA sequences for G_o and G_i were microinjected into DRGs and selectively depleted either G-protein. Evidence was obtained from confocal microscopy of the extent and selectivity of the G-protein depletion. Only depletion of G_o reduced the effectiveness of (–)-baclofen (FIG. 3).[37]

Thus $GABA_B$ receptors are able to couple both to G_i and to G_o, but only coupling to G_o produces inhibition of calcium channels. This may indicate that activated α_o interacts directly with calcium channels or produces a second messenger that does so, or that α_o couples to specific βγ subunits that have the same effect (FIG. 4). Other evidence has been obtained in secretory cells that G_o is involved in mediating the effects of a number of neuromodulators on calcium channels, and further that different subtypes of G_o containing various combinations of α, β, and γ subunits specifically transduce signals from different receptors. (For a review see Hescheler and Schultz.[38])

INTERACTION OF Mg^{2+} WITH CALCIUM CHANNELS

Modulation and/or regulation of ion channels may be effected by means other than pathways involving G-protein activation in neurons. One factor that has attracted

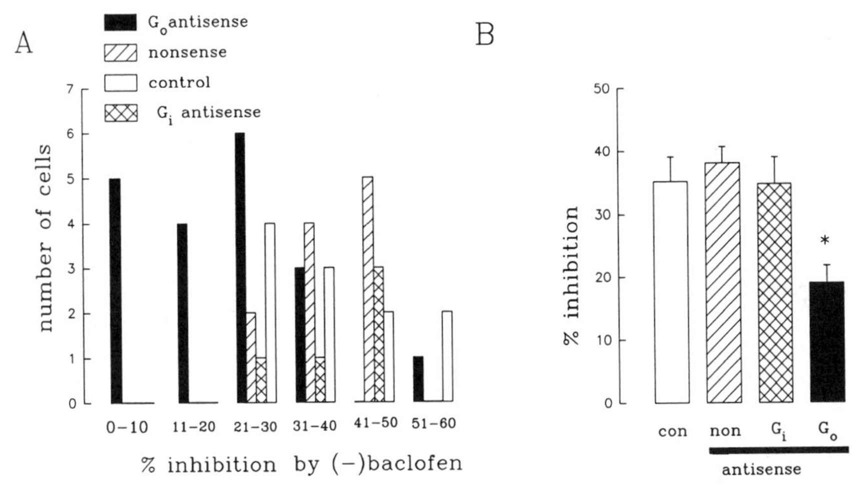

FIGURE 3. Effect of G-protein antisense on inhibition of calcium channel currents by (−)-baclofen. Cells were microinjected with phosphorothioate antisense DNA oligonucleotides corresponding to sequences unique to $G\alpha_o$ or $G\alpha_i$ as described by Campbell *et al.*[37] Calcium channel currents were recorded 24-32 hours later, following depletion of $G\alpha_o$ or $G\alpha_i$. The ability of (−)-baclofen to inhibit the current was examined. (**A**) The percentage response to (−)-baclofen has been grouped, and the number of cells responding in each 10 percentile is plotted. The effect of $G\alpha_o$ antisense is to reduce the ability of (−)-baclofen to inhibit I_{Ba}. This is also shown in **B,** where the mean percentage inhibition ± SEM is given for each treatment.

increasing interest of late has been intracellular magnesium ion concentration. The presence of magnesium ions in the cytoplasm (1 mM) has been shown to inhibit the outward movement of K^+ through certain K^+ channels, and physiological levels of internal Mg^{2+} are responsible for their inward rectification.[39] Mg^{2+} (1.5 mM) has also been shown to inhibit the release of Ca^{2+} from intracellular stores by blocking Ca^{2+} release channels in the sarcoplasmic reticulum of skeletal muscle.[40] In cardiac myocytes, voltage-dependent Ca^{2+} channels were found to be inhibited by an increase in the level of intracellular Mg^{2+} (1-3 mM).[41] Furthermore, voltage-independent Ca^{2+} channels in invertebrate neurons have similarly been shown to be sensitive to changes in internal Mg^{2+} levels.[42] The entry of Ca^{2+} ions through NMDA receptors is also dependent on intracellular Mg^{2+}.[43] All these actions suggest the possibility of an important role for internal Mg^{2+} in modifying and regulating the excitability and activity of a range of excitable cells.

Recent work in our laboratory suggests that intracellular free Mg^{2+} can selectively inhibit the ω-CgTX GVIA-sensitive component of the calcium channel current in cultured rat cerebellar granule neurons. The investigation was prompted by the observation that the use of two different intracellular solutions for whole-cell patch clamp recordings gave rise to Ca^{2+} channel currents that were remarkably different in terms of both the amplitude of the whole-cell current and the sensitivity of the currents to ω-CgTX.[44] Systematic manipulation of the concentrations of $MgCl_2$ and K_2ATP added to the pipette solution, such that the Mg.ATP concentration was kept constant whilst

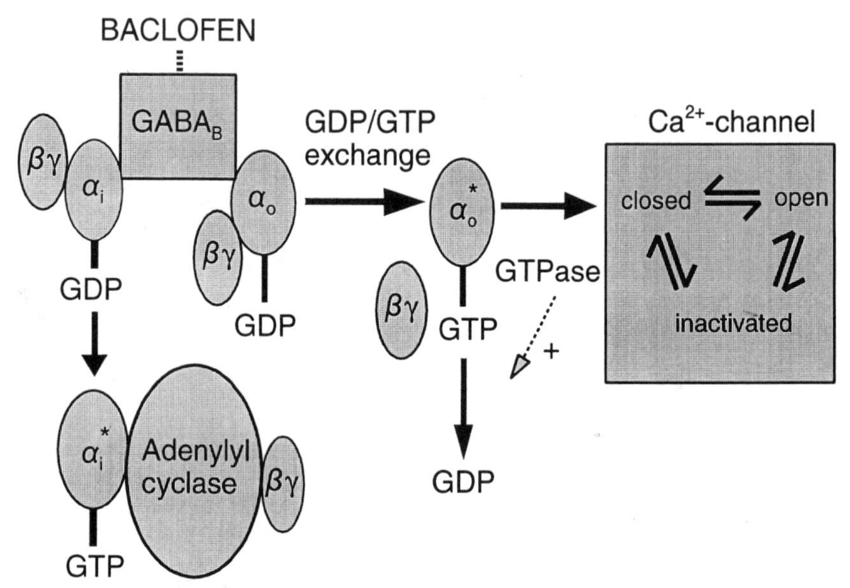

FIGURE 4. Signal transduction involving the $GABA_B$ receptor. Coupling via G_i leads to adenylyl cyclase inhibition, whereas coupling via $G\alpha_o$ leads to neuronal voltage-dependent calcium channel inhibition. Preliminary evidence also suggests that calcium channels may feedback on the activated G_o protein and enhance its GTPase activity.[53]

the Mg^{2+} concentration was varied, revealed an inhibitory effect of increasing intracellular Mg^{2+} concentration on the amplitude of Ca^{2+} channel currents in these cells (FIG. 5A and B). Accompanying these changes in amplitude was a negative shift in the voltage at which 50% of the calcium channels were activated (FIG. 5C), which was interpreted to be due to the surface charge effect of divalent cations on the internal cell plasma membranes.[45]

The sensitivity of Ca^{2+} channel currents to ω-CgTX was also tested over the same range of internal Mg^{2+} concentrations (FIG. 6). Approximately 40% of the calcium current was irreversibly inhibited by ω-CgTX in the presence of 0.13 mM Mg^{2+} (FIG. 6A). As intracellular Mg^{2+} was increased to 0.5 mM, the proportion of current inhibited by ω-CgTX declined, with a parallel reduction in calcium current amplitude (FIG. 6B). These data clearly indicate that almost all of the ω-CgTX-sensitive current was blocked by nominal Mg^{2+} concentrations of 0.5 mM or greater, whereas the ω-CgTX-resistant component was less affected, exhibiting only a slight decline at a Mg^{2+} concentration of 1 mM. Our results indicate an IC_{50} for Mg^{2+} inhibition of the ω-CgTx-sensitive component of current in the range of 150-200 μM. This inhibitory effect of intracellular Mg^{2+} on N-type calcium channel currents is not confined to cerebellar granule neurons, because a similar inhibitory effect on calcium channel amplitude can be seen in DRGs (A. C. Dolphin, unpublished observations). As with the granule neurons, this inhibition is accompanied by a selective (although partial) loss of ω-CgTX sensitivity.

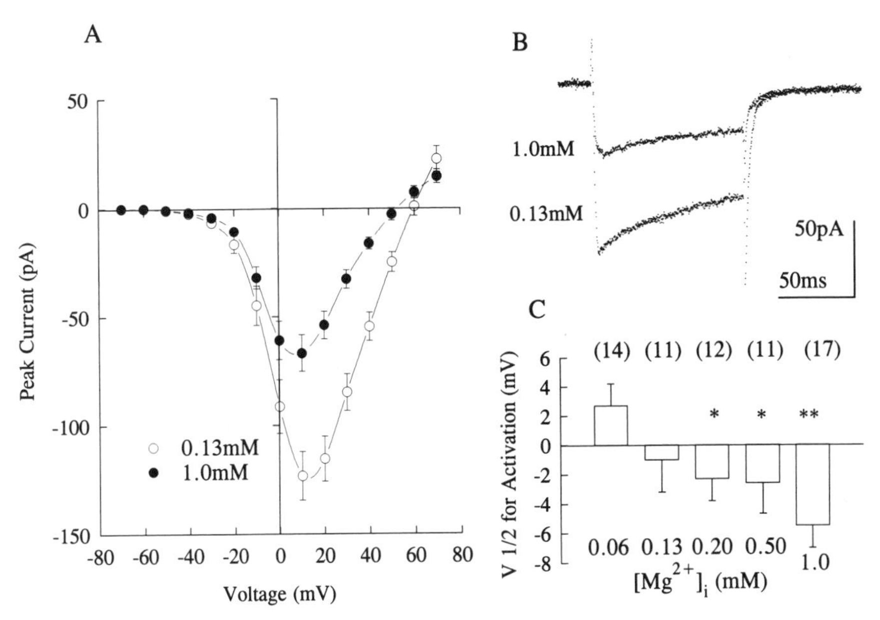

FIGURE 5. Raising $[Mg^{2+}]_i$ inhibits whole-cell Ca^{2+} channel currents in cerebellar granule neurons. (**A**) Mean I-V relationships for 11 cells with 0.13 mM Mg^{2+} internally (○) and 17 cells with 1.0 mM Mg^{2+} internally (●). $[Mg.ATP]_i$ was 0.6 mM in both cases. (**B**) Averaged currents from the 11 cells with 0.13 mM $[Mg^{2+}]_i$ and the 17 cells with 1.0 mM $[Mg^{2+}]_i$. *Solid lines* are least squares fits to the inactivation of the currents and have time constants of 60.8 ms (0.13 mM $[Mg^{2+}]_i$) and 50.0 ms (1.0 mM $[Mg^{2+}]_i$). (**C**) The half-voltage for activation of calcium currents is shifted in the negative direction by increasing intracellular Mg^{2+} concentration. Numbers in parentheses indicate number of cells tested at each concentration. * $p < 0.05$; ** $p < 0.01$, as compared to values obtained in the presence of 0.13 mM intracellular Mg^{2+}.

Estimates of $[Mg^{2+}]_i$ in neuronal preparations suggest values of around 3 mM when measured in squid giant axon[46] and 660 μM in neurons from *Lymnea stagnali.*[47] High, millimolar concentrations of Mg^{2+} may not be typical of levels seen in vertebrate central neurons. Intracellular Mg^{2+} levels measured in the neuroblastoma x glioma hybrid cell line NG108-15 using the Mg^{2+}-sensitive fluorescent dye Mag-Indo suggested levels of less than 140 μM.[48] However, a more recent study in cultured cortical neurons using magfura-2 gave mean resting values of 630 μM.[49] Our results indicate that at the lower of these concentrations of Mg^{2+}, ω-CgTx-sensitive channels would not be inhibited and might be expected to contribute to the Ca^{2+} channel currents in these cells. A relatively small increase in $[Mg^{2+}]_i$ from this resting level would lead to a significant inhibition of the ω-CgTx-sensitive Ca^{2+} channels. Little is known about the regulation of neuronal $[Mg^{2+}]_i$ under different conditions (for a review, see ref. 50), but it is likely that Mg^{2+} would rise significantly when ATP is depleted and pH falls under conditions of anoxia or glucose depletion.[51] Anoxia also leads to release of glutamate with a consequent rise in intracellular Ca^{2+} concentration which in turn leads to rises in intracellular Mg^{2+}.[49] Blockade of N-type channels by Mg^{2+}

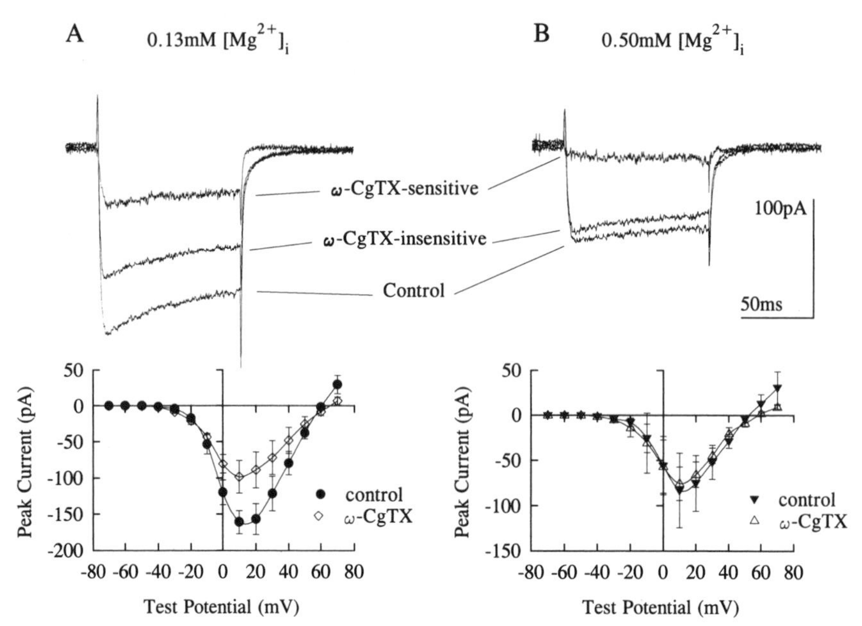

FIGURE 6. Inhibitory effect of ω-CgTX on I-V relationships. (**A**) **Top:** Averaged current records from eight cells showing control current, current sensitive to ω-CgTX (1 μM), and the component of current that was unaffected by ω-CgTX application. Test potential was +10 mV, holding potential was –80 mV. **Bottom:** Mean I-V relationships for five cells under control conditions and following application of 1 μM ω-CgTX. Free internal Mg^{2+} was 0.13 mM and Mg.ATP was 0.6 mM. (**B**) as for **A,** but in this case 0.5 mM Mg^{2+} was present internally. Currents are averages from five cells and mean I-V relationships were obtained from three cells.

might contribute to Ca^{2+} channel inhibition and blockade of synaptic responses under these conditions[52] and may thus be a means of reducing further Ca^{2+} influx, thereby limiting damage to ischemic tissues.

ACKNOWLEDGMENTS

Thanks are due to J. Hendy for his help in preparing the manuscript.

REFERENCES

1. Bean, B. P. 1985. Two kinds of calcium channels in canine atrial cells. J. Gen. Physiol. **86:** 1-30.
2. Nilius, B., P. Hess, J. B. Lansman & R. W. Tsien. 1985. A novel type of cardiac calcium channel in ventricular cells. Nature **316:** 443-446.
3. Fox, A. P., M. C. Nowycky & R. W. Tsien. 1987. Kinetic and pharmacological properties distinguishing three types of calcium currents in chick sensory neurones. J. Physiol. **394:** 149-172.

4. TSIEN, R. W., D. LIPSCOMBE, D. V. MADISON, K. R. BLEY & A. P. FOX. 1988. Multiple types of neuronal calcium channels and their selective modulation. TiNS **11:** 431-438.
5. PLUMMER, M. R., D. E. LOGOTHETIS & P. HESS. 1989. Elementary properties and pharmacological sensitivities of calcium channels in mammalian peripheral neurons. Neuron **2:** 1453-1463.
6. MARTIN-MOUTOT, N., M. SEAGAR & F. COURAUD. 1990. Subtypes of voltage-sensitive calcium channels in cultured rat neurons. Neurosci. Letts. **115:** 300-306.
7. LLINÁS, R., M. SUGIMORI, J.-W. LIN & B. CHERKSEY. 1989. Blocking and isolation of a calcium channel from neurons in mammals and cephalopods utilizing a toxin fraction (FTX) from funnel-web spider poison. Proc. Natl. Acad. Sci. USA **86:** 1689-1693.
8. MINTZ, I. M., M. E. ADAMS & B. P. BEAN. 1992. P-type calcium channels in rat central and peripheral neurons. Neuron **9:** 85-95.
9. JONES, S. W. & L. S. JACOBS. 1990. Dihydropyridine actions on calcium currents of frog sympathetic neurons. J. Neurosci. **10:** 2261-2267.
10. USOWICZ, M. M., M. SUGIMORI, B. CHERKSEY & R. LLINÁS. 1992. P-type calcium channels in the somata and dendrites of adult cerebellar Purkinje cells. Neuron **9:** 1185-1199.
11. HESS, P., J. B. LANSMAN & R. W. TSIEN. 1984. Different modes of Ca channel gating behaviour favoured by dihydropyridine Ca agonists and antagonists. Nature **311:** 538-544.
12. DOLPHIN, A. C. & R. H. SCOTT. 1989. Interaction between calcium channel ligands and guanine nucleotides in cultured rat sensory and sympathetic neurones. J. Physiol. **413:** 271-288.
13. SCOTT, R. H., A. C. DOLPHIN, V. P. BINDOKAS & M. E. ADAMS. 1990. Inhibition of neuronal Ca^{2+} channel currents by the Funnel Web spider toxin ω-Aga-1A. Mol. Pharmacol. **38:** 711-718.
14. REGAN, L. J., D. W. Y. SAH & B. P. BEAN. 1991. Ca^{2+} channels in rat central and peripheral neurons: High-threshold current resistant to dihydropyridine blockers and ω-conotoxin. Neuron **6:** 269-280.
15. MENON-JOHANSSON, A., N. S. BERROW & A. C. DOLPHIN. 1993. G_o transduces $GABA_B$ receptor modulation of N-type calcium channels in cultured dorsal root ganglion neurones. Pflugers Arch. **425:** 335-343.
16. LARMET, Y., A. C. DOLPHIN & A. M. DAVIES. 1992. Intracellular calcium regulates the survival of early sensory neurons before they become dependent on neurotrophic factors. Neuron **9:** 563-574.
17. BREITWIESER, G. E. & G. SZABO. 1985. Uncoupling of cardiac muscarinic and β-adrenergic receptors from ion channels by a guanine nucleotide analogue. Nature **317:** 538-540.
18. HOLZ, G. G., S. G. RANE & K. DUNLAP. 1986. GTP binding proteins mediate transmitter inhibition of voltage-dependent calcium channels. Nature **319:** 670-672.
19. SCOTT, R. H. & A. C. DOLPHIN. 1986. Regulation of calcium currents by a GTP analogue: potentiation of (–)-baclofen-mediated inhibition. Neurosci. Letts. **69:** 59-64.
20. DOLPHIN, A. C. & R. H. SCOTT. 1987. Calcium channel currents and their inhibition by (–)-baclofen in rat sensory neurones: modulation by guanine nucleotides. J. Physiol. **386:** 1-17.
21. DOLPHIN, A. C., S. R. FORDA & R. H. SCOTT. 1986. Calcium-dependent currents in cultured rat dorsal root ganglion neurones are inhibited by an adenosine analogue. J. Physiol. **373:** 47-61.
22. MARSZALEC, W., R. S. SCROGGS & E. G. ANDERSON. 1988. Serotonin-induced reduction of the calcium-dependent plateau in frog dorsal root ganglion cells is blocked by serotonergic agents acting at 5-hydroxytryptamine$_{1A}$ sites. J. Pharm. & Exp. Therap. **247:** 399.
23. AOSAKI, T. & H. KASAI. 1989. Characterization of two kinds of high-voltage-activated Ca-channel currents in chick sensory neurons. Pflugers Arch. **414:** 150-156.

24. Ewald, D. A., I.-H. Pang, P. C. Stenweis & R. J. Miller. 1989. Differential G protein-mediated coupling of neurotransmitter receptors to Ca^{2+} channels in rat dorsal root ganglion neurons *in vitro*. Neuron **2:** 1185-1193.
25. Schroeder, J. E., P. S. Fischbach, D. Zheng & E. W. McCleskey. 1991. Activation of μ opioid receptors inhibits transient high- and low-threshold Ca^{2+} currents, but spares a sustained current. Neuron **6:** 13-20.
26. Lewis, D. L., F. F. Weight & A. Luini. 1986. A guanine nucleotide binding protein mediates the inhibition of voltage-dependent calcium current by somatostatin in a pituitary cell line. Proc. Natl. Acad. Sci. USA **83:** 9035-9039.
27. Schmidt, A., J. Hescheler, S. Offermanns, K. Spicher, K.-D. Hinsch, F.-J. Klinz, J. Codina, L. Birnbaumer, H. Gausepohl, R. Frank, G. Schultz & W. Rosenthal. 1991. Involvement of pertussis toxin-sensitive G-proteins in the hormonal inhibition of dihydropyridine-sensitive Ca^{2+} currents in an insulin-secreting cell line (RINm5F). J. Biol. Chem. **266:** 18025-18033.
28. Grassi, F. & H. D. Lux. 1989. Voltage-dependent GABA-induced modulation of calcium currents in chick sensory neurons. Neurosci. Letts. **105:** 113-119.
29. Scott, R. H. & A. C. Dolphin. 1990. Voltage-dependent modulation of rat sensory neurone calcium channel currents by G protein activation: Effect of a dihydropyridine antagonist. Br. J. Pharmacol. **99:** 629-630.
30. Beech, D. J., L. Bernheim & B. Hille. 1992. Pertussis toxin and voltage dependence distinguish multiple pathways modulating calcium channels of rat sympathetic neurons. Neuron **8:** 97-106.
31. Cox, D. H. & K. Dunlap. 1992. Pharmacological discrimination of N-type from L-type calcium current and its selective modulation by transmitters. J. Neurosci. **12:** 906-914.
32. Harris-Warwick, R. M., C. Hammond, D. Paupardin-Tritsch, V. Homburger, B. Rouot, J. Bockaert & H. M. Gerschenfeld. 1988. An α40 subunit of a GTP binding protein immunologically related to G_o mediates a dopamine-induced decrease of a Ca^{2+} current in snail neurones. Neuron **1:** 17-32.
33. McFadzean, I., I. Mullaney & D. A. Brown. 1989. Antibodies to the GTP binding protein, G_o, antagonize noradrenaline-induced calcium current inhibition in NG108-15 hybrid cells. Neuron **3:** 177-182.
34. Lledo, P. M., V. Homburger, J. Bockaert & J.-D. Vincent. 1992. Differential G protein-mediated coupling of D_2 dopamine receptors to K^+ and Ca^{2+} currents in rat anterior pituitary cells. Neuron **8:** 455-463.
35. Kleuss, C., J. Hescheler, C. Ewel, W. Rosenthal, G. Schultz & B. Wittig. 1991. Assignment of G-protein subtypes to specific receptors inducing inhibition of calcium currents. Nature **353:** 43-48.
36. Campbell, V., N. S. Berrow & A. C. Dolphin. 1993. $GABA_B$ receptor modulation of Ca^{2+} currents in rat sensory neurones by the G protein G_o: Antisense oligonucleotide studies. J. Physiol. **470:** 1-11.
37. Hescheler, J. & G. Schultz. 1993. G-proteins involved in the calcium channel signalling system. Curr. Opin. Neurobiol. **3:** 360-367.
38. Matsuda, H., A. Saigusa & H. Irisawa. 1987. Ohmic conductance through the inwardly rectifying K channel and blocking by internal Mg^{2+}. Nature **325:** 156-159.
39. Smith, J. S., R. Coronado & G. Meissner. 1986. Single-channel measurements of the calcium release channel from skeletal muscle sarcoplasmic reticulum. Activation by calcium ATP and modulation by magnesium. J. Gen. Physiol. **88:** 573-588.
40. White, R. E. & H. C. Hartzell. 1988. Effects of intracellular free magnesium on calcium current in isolated cardiac myocytes. Science **239:** 778-780.
41. Strong, J. A. & S. A. Scott. 1992. Divalent-selective voltage-independent calcium channels in *Lymnaea* neurons: Permeation properties and inhibition by intracellular magnesium. J. Neurosci. **12:** 2993-3003.
42. Johnson, J. W. & P. Ascher. 1990. Voltage dependent block by intracellular Mg^{2+} of *N*-methyl-D-aspartate activated channels. Biophys. J. **57:** 1085-1090.

43. PEARSON, H. A. & A. C. DOLPHIN. 1993. Inhibition of ω-conotoxin-sensitive calcium channel currents by internal Mg^{2+} in cultured rat cerebellar granule neurones. Pflugers Arch. **425:** 518-527.
44. HILLE, B. 1992. Ionic Channels of Excitable Membranes: 390-422. Sinauer Associates Inc. Massachusetts.
45. BRINLEY, F. J. & A. SCARPA. 1975. Ionized magnesium concentration in axoplasm of dialyzed squid axons. FEBS Lett. **50:** 82-85.
46. ALVAREZ-LEEFMANS, F. J., S. M. GAMINO & T. J. RINK. 1984. Intracellular free magnesium in neurones of *Helix aspersa* measured with ion-selective electrodes. J. Physiol. **354:** 303-317.
47. ROBBINS, J., R. CLOUES & D. A. BROWN. 1992. Intracellular Mg^{2+} inhibits the IP_3-activated $I_{K(Ca)}$ in NG108-15 cells. [Why intracellular citrate can be useful for recording $I_{K(Ca)}$]. Pflugers Arch. **420:** 347-353.
48. BROCARD, J. B., S. RAJDEV & I. J. REYNOLDS. 1993. Glutamate-induced increases in intracellular free Mg^{2+} in cultured cortical neurons. Neuron **11:** 751-757.
49. ROMANI, A. & A. SCARPA. 1992. Regulation of cell magnesium. Arch. Biochem. Biophys. **298:** 1-12.
50. MURPHY, E., E. E. FREUDENRICH & M. LIEBERMAN. 1991. Cellular magnesium and Na/Mg exchange in heart cells. Ann. Rev. Physiol. **53:** 273-287.
51. KRNJEVIC, K. & W. WALZ. 1990. Acidosis and blockade of orthodromic responses caused by anoxia in rat hippocampal slices at different temperatures. J. Physiol. **422:** 127-144.
52. SWEENEY, M. I. & A. C. DOLPHIN. 1992. 1,4-Dihydropyridines modulate GTP hydrolysis by G_o in neuronal membranes. FEBS Lett. **310:** 66-70.

Neuropharmacology of Nimodipine: From Single Channels to Behavior

RICHARD J. FANELLI,[a] RICHARD T. McCARTHY, AND JANE CHISHOLM

Institute for Dementia Research
Miles Inc.
West Haven, Connecticut 06516

Nimodipine is a 1,4-dihydropyridine that is approved in the United States for the treatment of neuronal deficits associated with subarachnoid hemorrhage and is currently under investigation for the treatment of Alzheimer's disease (AD). This compound selectively blocks L-type calcium channel current with high affinity in sensory neurons.[1] There is abundant evidence that calcium channel antagonists have beneficial effects in various disorders of the central nervous system, including ischemia, brain damage, and neuronal degeneration.[2,3] An alteration in neuronal calcium homeostasis with age has been hypothesized to contribute to brain pathology and reduced function.[4,5] Therefore, an agent such as nimodipine, with calcium channel-blocking properties selective for the depolarized state, might be expected to provide amelioration of the physiological and behavioral consequences of brain aging. Indeed, nimodipine has been shown to prevent age-associated motor deficits in rats,[6] enhance acquisition of conditioned eyeblink in aging rabbits,[7] and improve recent memory function in aging primates.[8] The objectives of the present investigations were to characterize the mechanism of action of nimodipine in neurons as well as investigate the effects of nimodipine in *in vitro* and *in vivo* models of neurodegeneration and dementia. In addition, these studies were designed to investigate the role of neuronal calcium in aging and dementia.

METHODS

Primary Culture

Long-Evans rat pups within 48 hours of birth were decapitated, and the brain was placed into an oxygenated sterile dissection buffer containing (in mM): NaCl, 137; KCl, 5.3; $MgCl_2$, 1; sorbitol, 25; HEPES, 10; $CaCl_2$, 3; pH 7.2. The hippocampal region of the brain was removed, cleaned of microvessels and fascia, and digested in a papain solution (Earl's balanced salt solution containing [in mM]: $CaCl_2$, 1.5; EDTA, 0.5; L-cysteine, 0.2 mg/ml; papain 7.5 u/ml) for 30 minutes at 37°C. The tissue was then placed into two changes of triturating buffer, diluting out the papain

[a] Address for correspondence: Richard J. Fanelli, PhD, Institute for Dementia Research, Miles Inc., 400 Morgan Lane, West Haven, CT 06516.

solution, and triturated. The triturating buffer contained: 1.5 mg of trypsin inhibitor and 1.5 mg of bovine serum albumin per milliliter of dissection buffer. Cells were then counted and plated onto coated (poly-L-ornithine/laminin) coverslips at a density of 300,000 cells/well. Cells were maintained in a medium containing DMEM: Ham's F12 (1 : 1), fetal calf serum 10%, penicillin (100 U/ml), streptomycin (100 μg/ml), and Fungizone (0.5 μg/ml), in an environment ventilated with 5% O_2/95% CO_2 at 37°C. Cells were fed 24 hours after plating.

Patch Voltage-Clamp Experiments

Hippocampal cultured cells were rinsed in a solution containing (in mM): NaCl, 150; KCl, 4.0; $CaCl_2$, 9.0; $MgCl_2$, 0.5; dextrose, 5; HEPES, 10; pH 7.5. In cell-attached patch recordings, the pipette solution contained 110 mM $BaCl_2$, 10 mM HEPES, and 200 nM TTX (pH 7.4 w/$Ba[OH]_2 \cong 5$ mM) and a bath solution of (in mM): 140 K^+-aspartate; 10 K-EGTA; 1 $MgCl_2$; 10 HEPES (pH 7.4 w/KOH), was used to zero the membrane potential. Cultured cells were voltage-clamped at room temperature (21.0-23.3°C) using a List EPC-7 (Medical Systems, Greenvale, New York) in the cell-attached patch configuration. Patch electrodes were made from Corning glass (#7052) and had resistances between 0.6 and 2.6 MΩ. Patch pipettes were insulated within 100 μm of the tip with Sylgard 184 (Dow Corning) and fire-polished with a glass-coated, heating platinum wire. Voltage pulse generation, data acquisition, and analysis were performed using the ADAS-1 data acquisition system (Axon Instr., Foster City, California) on an IBM compatible 80386 PC (25 MHz) coupled to a TL-1-125 interface. Membrane current was sampled at 5 kHz and filtered with an 8-pole low-pass Bessel filter (Frequency Devices, Haverhill, Massachusetts) with a cutoff frequency (–3 dB) of 1 kHz. In cell-attached recordings, ensemble leak traces were accumulated either from null sweeps to the same potentials as the displayed records or from sweeps without openings with equivalent voltage steps. These sweeps were averaged and then digitally subtracted.

Measurement of $[Ca^{2+}]_i$ Using Calibrated Fura-2

On the day of the experiment, cells on coverslips were rinsed with HBS II and incubated with 5 μM fura-2 AM for 30 minutes at 37°C. Coverslips were rinsed free of unaccumulated dye and placed in a nonperfusion microscope slide chamber of 200 μl volume. Measurements of $[Ca^{2+}]_i$ were performed at room temperature (22°C). Individual pyramidal-type neurons were identified and imaged on an inverted epifluorescent microscope (Nikon Diaphot) through a 40X UV-passing objective (Cf-fluor oil immersion, NA 1.3). Monolayers were excited alternately with 340 or 380 nm filtered light (Omega Optical, Brattleboro, Vermont) from a xenon source and passed through a liquid light guide (Oriel Corp., Stratford, Connecticut) to the microscope. Fluorescence intensity at 480-520 nm, collected through both high pass and interference filters from single cells centered in a circular aperture 10 μm in diameter (approximate cell diameter), was quantified using a photomultiplier tube (Hamamatsu Photonics). This method integrates the output of the entire cell soma at a resolution of 1 second (300 ms/excitation λ). Individual neurons were sampled separately. Drug or peptide applications were made directly on the slide chamber, leaving a readily identifiable artifact at the time of addition on the recording. Drug or peptide persisted in the chamber until it was washed out. Calibration was performed in individual cells by permeabilization of the plasma membrane to Ca^{2+}. Application of 100 μM ionomycin in the presence of saturating $[Ca^{2+}]$ (10 mM) provided a measurement of maximum bound dye, and measurement of minimum bound dye was obtained during chelation with EGTA. Alternatively, fluorescent ratio was calibrated to absolute $[Ca^{2+}]_i$ levels using a cell-free calibration curve obtained with buffers of known ionized calcium concentration. In fura-2-loaded cells, residual fluorescence in the presence of 10 mM Mn^+ was not higher than background noise of unloaded cells, and therefore no mathematical correction for unhydrolyzed or compartmentalized dye was performed.

$[Ca^{2+}]_i$ Measurements Using Laser Scanning Confocal Microscopy

Coverslips of fluo-3-loaded hippocampal monolayers were maintained at room temperature and placed in the same buffer and chamber as used for the fura-2 measurements just described. The chamber was positioned on an epifluorescent microscope (Zeiss Axiovert) fitted with a Biorad MRC 600 laser scan head and a Neofluar 40X objective. A krypton/argon laser was filtered for excitation at the 488-nm line. Images were collected through a dichroic mirror at a 2-second resolution and integrated. Additions of vehicle, with or without βA4(25-35), were applied directly to the open slide chamber.

Source of Peptides and Dyes

No difference was found in response to βA4 peptides from different sources. The peptides βA4(1-40) and βA4(25-35) were obtained commercially from Bachem California (Torrence, California) or Sigma Chemical Co. (St. Louis, Missouri) or synthesized on site at Miles, Inc. (West Haven, Connecticut). Peptides from all three sources were found active. Both the "reverse" peptide βA4(35-25) and the D-amino acid analog [D-βA4(25-35)] were synthesized at Miles. Fluo-3 and fura-2 were obtained from Molecular Probes (Eugene, Oregon).

Rabbit Classically Conditioned Eyeblink Procedure

Female New Zealand albino rabbits were received as 30-month-old retired breeders and habituated to standard restraining boxes for 5 days before training. A delay conditioning paradigm was used, with an 8.5-dB, 6-kHz tone of 500 ms duration as the conditioned stimulus (CS) and a 2.5 psi corneal air puff as the unconditioned stimulus (UCS), presented during the last 150 ms of the CS. Eyeblink responses were measured noninvasively using a light-emitting diode and phototransistor device held, along with the tube used to deliver the air puff, at a distance ≅ 1 cm from the right cornea. Rabbits received 100 conditioning trials daily with an intertrial interval ≅ 30 seconds. Initial training continued for 7 days or until animals reached performance levels of at least 30% conditioned responses (CRs) in a training session. An unconditioned response (UCR) was defined as an eyeblink that occurred after the onset of the UCS, and a CR was any response occurring after CS onset but before UCS onset. After initial training, pairs of rabbits with similar acquisition performance were assigned to treatment with either nimodipine (5 mg/kg po bid) or vehicle (6% sucrose and 1% methylcellulose in a ratio of 7:3) during a 3-week retention interval during which there was no additional training. Following this interval, rabbits were again trained for 2 days on the delay conditioning procedure (reacquisition) while drug treatment continued (30 minutes before training).

Morris Water Maze Performance after Medial Septal Lesions

Male young adult Long-Evans rats were used in this study. The behavioral apparatus used was a circular galvanized steel tank that was divided into equivalent quadrants in each of which could be placed an escape platform.[9] The maze was filled with water (≅ 26°C), made opaque with powdered milk, to a depth 1 cm above the escape platform. Numerous stationary visual cues were present in the testing room. Data were recorded using the Multiple Zone Distance Travelled program of the Videomex-V analysis system (Columbus Instruments International Corp., Columbus, Ohio). Following 1 week of acclimation to the animal facility, rats were given a 30-second free swim during which no escape platform was present. Next, acquisition training began. Rats were assigned a goal quadrant in which the escape platform was located. A total of 10 acquisition trials were given 2 each day for 5 days. No drug treatments were given during this initial acquisition training. Rats were placed into the maze and allowed 120 seconds to find the goal platform. If the platform was not found before 120 seconds, the animal was placed on the platform by the experimenter for 25 seconds. The second trial of each day began approximately 25 seconds after completion of the first trial. Immediately after the second acquisition trial on day 5 of training, rats were given a 30-second probe trial, Probe 1. During the probe trial, no escape platform was present and the time spent in each quadrant was measured. Two days after completion of Probe 1, rats were randomly assigned to one of six groups: (1) sham-vehicle (n = 10); (2) sham-10 mg/kg nimodipine (n = 10); (3) lesion-vehicle (n = 8); (4) lesion-3 mg/kg nimodipine (n = 9); (5) lesion-10 mg/kg nimodipine (n = 8); or (6) lesion-30 mg/kg nimodipine (n = 7). Rats in the lesion groups received an electrolytic lesion of the medial septal area. Rats in sham groups received the same surgery with no current passing through the electrode. Coordinates, determined from the atlas of Paxinos and Watson,[10] were at midline and 0.7 mm anterior of bregma. Once a hole had been drilled in the skull, the electrode (Radionics TC 4112; David Kopf Instruments, Tujunga, California) was lowered to a depth of 5.0 mm ventral from the dura. For sham animals, the electrode was left in place for 60 seconds without current. For lesioned animals, the lesion generator (Radionics RFG-4A; David Kopf Instruments) was turned on for 60 seconds. Timing began once electrode temperature reached 70°C. Lesions produced in this experiment were generally small and did not result in complete destruction of the medial septal area. Following completion of sham or lesion surgery, rats were given a 7-day recovery period. Rats received a single dose of either vehicle or nimodipine after surgery. For

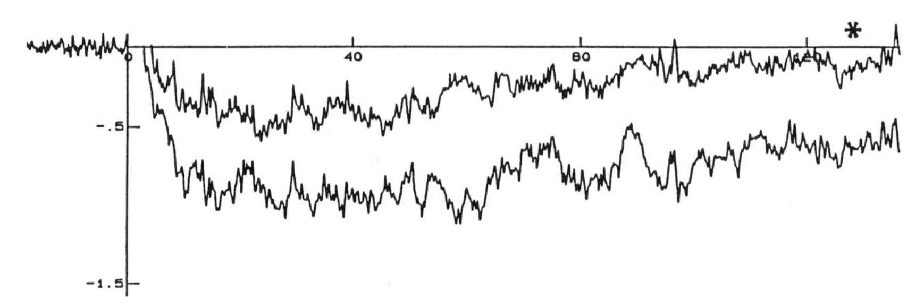

FIGURE 1. Macroscopic averages of 40 consecutive sweeps of single L-type calcium channel activity in the absence (*lower trace*) and presence (*upper trace marked with asterisk*) of 10 nM nimodipine.

the next 4 days of recovery, rats were given their respective treatments twice per day. Nimodipine was suspended in 1% methylcellulose, and rats not treated with nimodipine received a 1-ml/kg dose of the vehicle. After the 7-day recovery period, behavioral testing in the water maze resumed. Rats were given a single probe trial, Probe 2, followed by four additional training trials. The additional training trials were started 24 hours after Probe 2. During the additional trials, the goal quadrant remained the same as that used during initial training, but only 1 trial per day was conducted. Twenty-four hours after completion of the additional training trials, rats were given another probe trial, Probe 3. On days of behavioral testing conducted after surgery, rats were treated with a single dose of drug or vehicle 1 hour before behavioral testing.

RESULTS

The selectivity of nimodipine block for L-type calcium channels was investigated by comparing macroscopic averages of single channel recordings from cultured rat hippocampal neurons in the presence and absence of nimodipine. L-type calcium channel activity could largely be isolated from activity of other channel types by holding the cell at relatively depolarized potentials (Vh = −50 mV). Under our recording conditions, L-type calcium channels were specifically identified by unitary slope conductance (22.1 pS; see McCarthy and TanPiengco[1]), sensitivity to the calcium channel agonist BAY K 8644, and resistance to voltage-dependent inactivation by less negative holding potentials. Although the holding potentials chosen for these experiments are less well polarized than are normal healthy neurons, they may be consistent with potentials reached in compromised tissue during cerebral ischemia. Furthermore, and well suited to our purposes, depolarized potentials promote high affinity nimodipine block.

In FIGURE 1, macroscopic averages of consecutive sweeps of single L-type calcium channel activity in the presence and absence of 10 nM nimodipine were constructed and superimposed. FIGURE 1 confirms that L-type calcium channel activity was markedly suppressed by nimodipine. Control activity was more than half maximally reduced by just 10 nM nimodipine. The reduction in L-type calcium channel current is most prominent near the end of the 160-ms test depolarization.

FIGURE 2 shows a histogram of current amplitude obtained from a patch exposed to Bay K 8644 (an L-type calcium channel agonist) and nimodipine. Pretreatment of the cell with Bay K 8644 induced prolonged L-type calcium channel openings that were easily resolved and well suited to construction of amplitude histograms. In the upper panel at 0 mV, two current peaks can be seen corresponding to at least two L-type calcium channels opening singly (1.06 pA) and simultaneously (2.10 pA). Following the presence of nimodipine (10 nM) in the bath, L-type channel activity is strongly inhibited.

Nimodipine also inhibited elevated intracellular free calcium ($[Ca^{2+}]_i$) throughout the soma in depolarized cultured rat hippocampal neurons. FIGURE 3A shows the typical biphasic response of a neuronal soma to depolarization evoked by elevations of extracellular $[K^+]$ from a basal level of 5 mM to 55 mM. This elevated level is consistent with that reported during experimental ischemia. The peak $[Ca^{2+}]_i$ response, which was partially inhibitable by nimodipine pretreatment, varied greatly from cell to cell. This component decreased in size during repeated challenge and recovery periods in the same cell, consistent with contribution from a depletable source. The subsequent sustained plateau phase of the $[Ca^{2+}]_i$ response (350 nM ± 49 SEM, n = 20) averaged 250% of basal (99 nM ± 10). The plateau returned to the same elevated level during several independent challenges of the same cell, consistent with a nondepletable source. This plateau elevation of $[Ca^{2+}]_i$ was sustained for the full duration of depolarization (over 1 hour in some experiments). Nimodipine inhibited a majority component (70%) of this elevation whether the drug was applied during the plateau phase, as in FIGURE 3, or as a pretreatment. The nimodipine-inhibitable component was unaffected by pretreatment with the N-channel blocking agent, ω-conotoxin GVIA in these sustained somatic measurements and was considerably larger than the latter. Some cells exhibited a sustained component that was resistant to combined block by ω-conotoxin GVIA and nimodipine, but the plateau could be completely eliminated by chelation of extracellular Ca^{2+} with EGTA. The ED_{50} of 0.5 nM for inhibition by nimodipine (FIG. 3B) is consistent with both its K_d for the dihydropyridine binding site and its inhibition of L-channel activity, supporting a dominant role for this channel in somatic calcium homeostasis during sustained depolarization.

Synthetic β-amyloid (βA4) fragments, which are toxic to neurons in culture, also produced some nimodipine-inhibitable elevations in $[Ca^{2+}]_i$ in these neurons. The βA4 protein, a major component of the plaque formations of Alzheimer's disease, varies in length from 39 to 43 amino acids. Responses to βA4(1-42) and smaller fragments, compromising amino acids 25-35, and 1-40 were studied. Calcium responses, typical of those in FIGURE 4, were measured in 49 of 50 cultured rat hippocampal neurons tested. Responses to βA4(25-35) were concentration-dependent and reversible with removal of the peptide. Responses were mimicked by βA4(1-40) and βA4(1-42), but not the reverse peptide βA4(35-25), nor a peptide formed from exclusively D-amino acids [D-βA4(25-35)]. Responses varied quantitatively and qualitatively from neuron to neuron. Rechallenge of the same neuron produced responses that were occasionally decreased in magnitude, suggesting contribution from a depletable (intracellular) store. These repeated responses were qualitatively similar to those of the initial challenge however, suggesting that the qualitative aspects

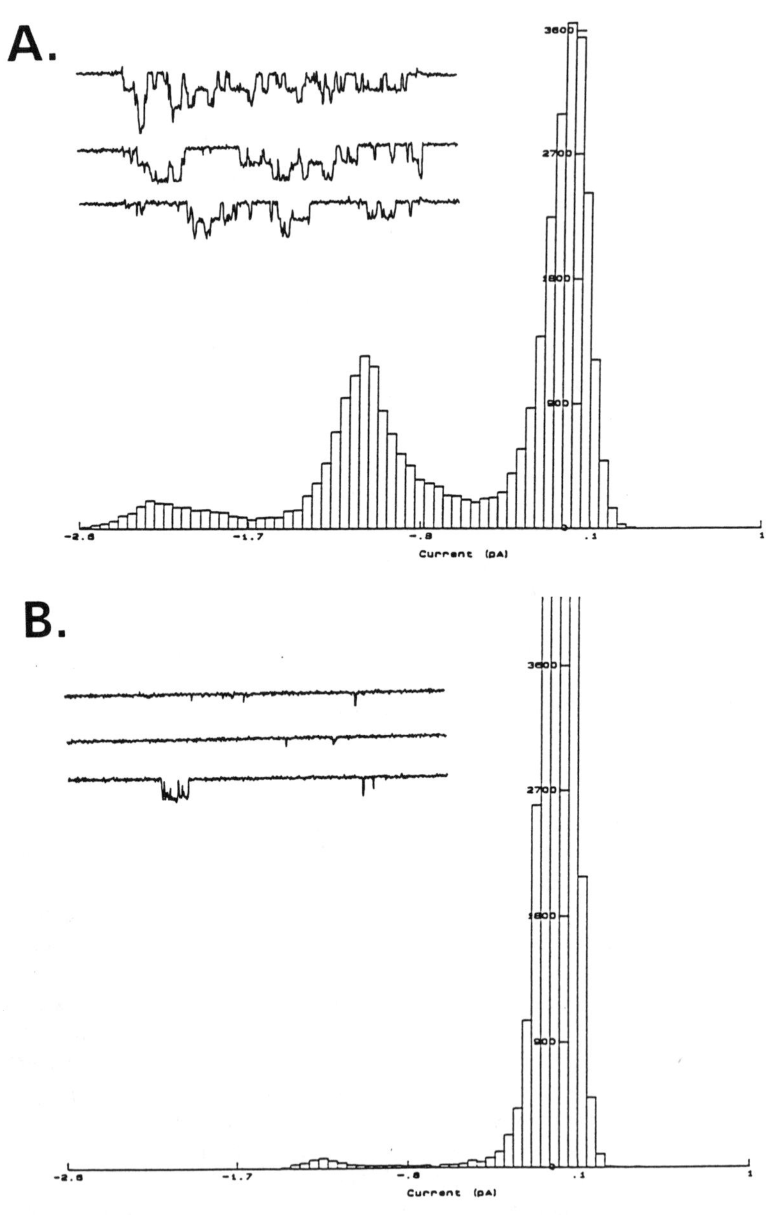

FIGURE 2. Histograms of current amplitude obtained from a patch exposed to: (**A**) Bay K 8644 or (**B**) nimodipine. *Inserts* demonstrate representative channel openings from three successive sweeps.

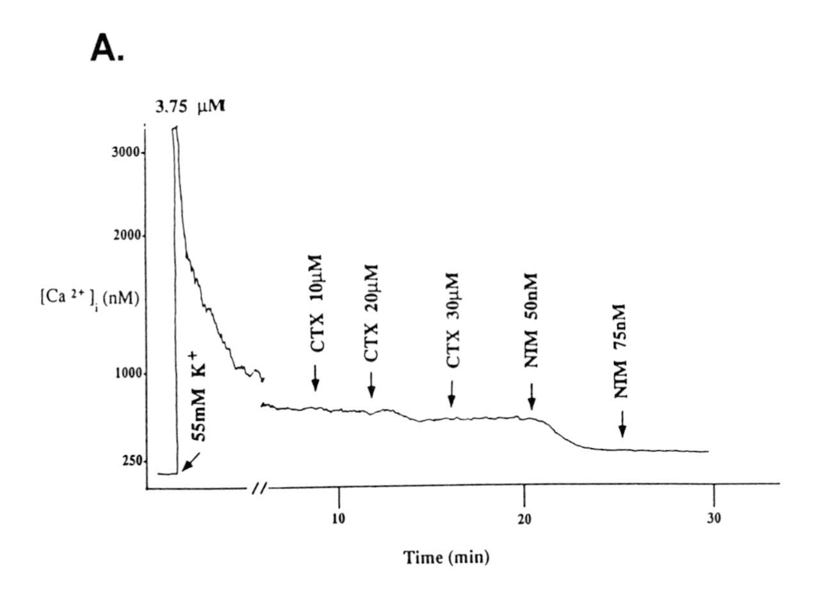

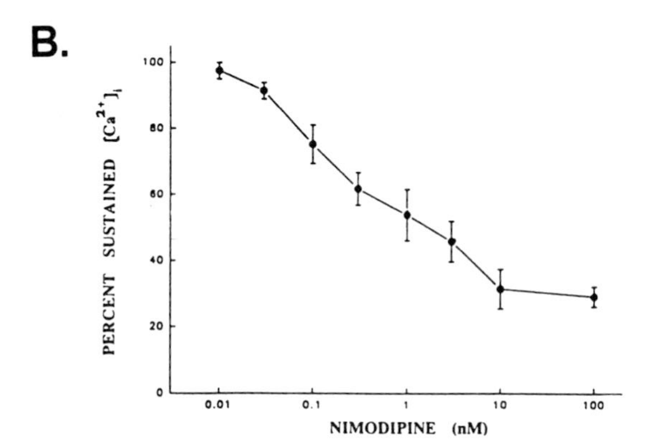

FIGURE 3. (**A**) Whole cell $[Ca^{2+}]_i$ integrated across the soma of a neuron. Quantification was derived from ratioed fura-2 measurements taken at 1-second intervals and transformed to absolute $[Ca^{2+}]_i$ by calibration of the same cell at the end of the experiment. Concentrations of ω-conotoxin and nimodipine are cumulative. Buffer composition (in mM): 118 NaCl, 4.6 KCl, 0.4 CaCl, 10 glucose, 20 HEPES, pH to 7.2 with NaOH. This trace is representative of 12 similar experiments. (**B**) Concentration-response relationship for nimodipine block of depolarization-induced elevations in $[Ca^{2+}]_i$. Nimodipine was applied before depolarization of single neurons whose sustained plateau elevation in $[Ca^{2+}]_i$ was previously determined in the absence of drug. Points represent mean ± SEM of 2-4 determinations. (Adapted from Chisholm *et al.*[14])

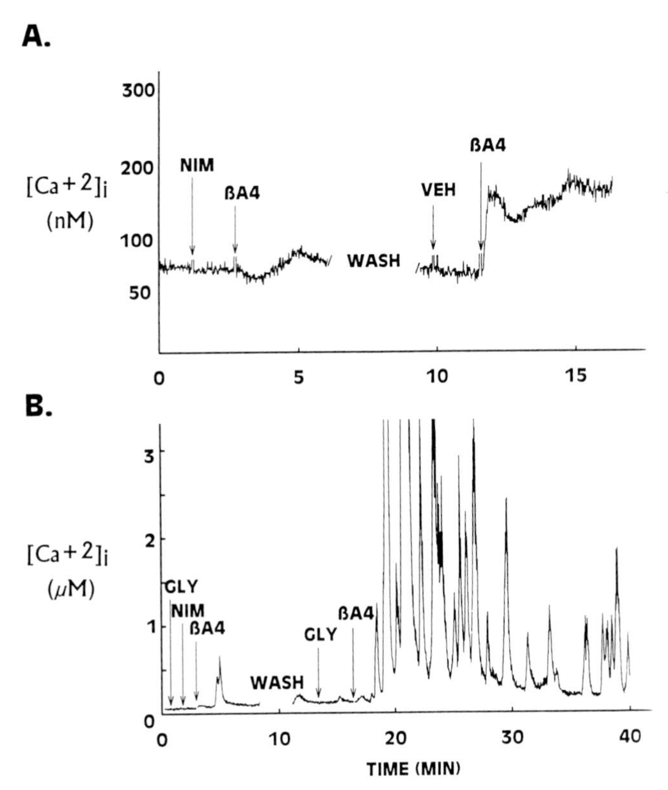

FIGURE 4. Nimodipine block of $[Ca^{2+}]_i$ (measured as in FIG. 3) responses to βA4(25-35) (abbreviated "βA4" in figure) in a rat hippocampal neuron. Nimodipine (NIM, 100 nM) inhibited response to βA4 in two different neurons in the absence (**A**) or presence (**B**) of glycine (GLY, 100 nM). Buffer composition (in mM): 118 NaCl, 5.4 KCl, 1.3 CaCl, 5.6 glucose, 20 HEPES, pH to 7.2 with NaOH. This experiment was repeated eight times with similar results.

of the response were relevant to some characteristic of the individual cell's calcium homeostasis. Response was often biphasic, with both transient(s) and sustained components. Unlike responses to elevated $[K+]_e$, however, there occasionally were multiple transients, occasionally no sustained component was detectable, and nimodipine was not as effective when applied after the peptide as it was in pretreatment paradigms. Although βA4(25-35) produced responses on its own, the incidence of repetitive transients increased during coadministration with the NMDA agonist glycine, suggesting possible interaction with endogenous glutamatergic activity. Nimodipine inhibited $[Ca^{2+}]_i$ responses to βA4(25-35) alone or in combination with glycine (FIG. 4). Inhibition by nimodipine was dose-dependent (EC_{50} between 10 and 20 nM) and mimicked by nifedipine (60 nM), and the calcium chelator EGTA applied extracellularly. Image analysis of the response to βA4(25-35) demonstrated that the elevation in $[Ca^{2+}]_i$ occurred throughout the entire neuronal soma, as shown in the scanning

nanomolar physiological concentrations that may produce trophic effects. The ability of nimodipine to block the $[Ca^{2+}]_i$ responses induced by βA4 peptides implicates a role for this drug in attenuating the effects of βA4 mediated by increases in $[Ca^{2+}]_i$.

In both of the behavioral paradigms, in both aging rabbits and brain-damaged rats, nimodipine was able to enhance retention at therapeutically relevant doses. The findings using the conditioned eyeblink in aging rabbits extends previously reported work[7] to include effects on retention. This result is significant not only for delineating the scope of cognitive benefit from nimodipine in this model, but also for reporting an effect on retention, a function generally affected in aging and dementing disorders. The results from the lesion study are consistent with previous reports that nimodipine can improve lesion-induced cognitive deficits.[12,13] The present work extends these findings to include effects following lesions of the septal nucleus and provides evidence on the specific aspect of cognitive function that is improved with nimodipine treatment after this brain damage.

In summary, in hippocampal neurons in culture, nimodipine was shown to block L-channel activity, and calcium influx in response to direct depolarization or to βA4 peptides. In behavioral studies, nimodipine produced enhanced retention in aging rabbits and protected against medial septal lesion-induced retention deficits in spatial learning. These findings support the therapeutic usefulness of nimodipine in the treatment of aging and dementia, and they are consistent with the view that calcium regulation is important in neurodegeneration.

SUMMARY

To supplement the existing pharmacological evidence describing the effects of nimodipine, a 1,4-dihydropyridine with calcium channel blocking properties, our group has used a multidisciplinary approach. This work attempts to characterize the mechanism of action of nimodipine in neurons as well as investigate the effects of nimodipine in models of neurodegeneration and dementia. Patch voltage clamp studies demonstrated high-affinity nimodipine block of voltage-dependent L-type calcium channel activity in central neurons from primary cultures of neonatal rat hippocampus. Nimodipine potently blocks depolarization-induced increases in free calcium throughout the soma of these hippocampal neurons. In addition, somatic free calcium elevations induced by acute βA4(25-35) exposure are also potently blocked by nimodipine. In behavioral studies, nimodipine produced enhanced retention in aging rabbits on eyeblink conditioning and also was shown to protect against medial septal lesion-induced retention deficits in a spatial learning task. These findings, from channel to behavioral effects, support the therapeutic usefulness of nimodipine in the treatment of aging and dementia and are consistent with the view that calcium regulation is important in disorders of neuronal degeneration.

REFERENCES

1. McCarthy, R. T. & P. E. TanPiengco. 1992. Multiple types of high-threshold calcium channels in rabbit sensory neurons: High-affinity block of neuronal L-type by nimodipine. J. Neurosci. **12:** 2225-2234.

2. Scriabine, A., T. Schuurman & J. Traber. 1989. Pharmacological basis for the use of nimodipine in central nervous system disorders. FASEB J. **3:** 1799-1806.
3. Traber, J. & W. H. Gispen, Eds. 1989. Nimodipine and Central Nervous System Function New Vistas. Schattauer Verlag. Stuttgart, Germany.
4. Khachaturian, Z. S. 1989. The role of calcium regulation in brain aging: Re-examination of a hypothesis. Aging **1:** 17-34.
5. Landfield, P. 1989. Calcium homeostasis in brain aging and Alzheimer's disease. *In* Diagnosis and Treatment of Senile Dementia. M. Bergener & B. Reisberg, Eds.: 276-287. Springer-Verlag. Berlin, Germany.
6. Schuurman, T., H. Klein, M. Beneke & J. Traber. 1987. Nimodipine and motor deficits in the aged rat. Neurosci. Res. Comm. **1:** 9-15.
7. Deyo, R. A., K. T. Straube & J. F. Disterhoft. 1989. Nimodipine facilitates trace conditioning of the eye-blink response in aging rabbits. Science **243:** 809-811.
8. Sandin, M., S. Jasmin & T. E. LeVere. 1990. Aging and cognition: Facilitation of recent memory in aged nonhuman primates by nimodipine. Neurobiol. Aging **11:** 567-571.
9. McMonagle-Strucko, K. & R. J. Fanelli. 1993. Enhanced acquisition of reversal training in a spatial learning task in rats treated with chronic nimodipine. Pharmacol. Biochem. Behav. **44:** 827-835.
10. Paxinos, G. & C. Watson. 1986. The Rat Brain in Stereotaxic Coordinates. Academic Press. Sydney, Australia.
11. Westenbroek, R. E., M. K. Ahlijanian & W. A. Catterall. 1990. Clustering of L-type Ca^{2+} channels at the base of major dendrites in hippocampal pyramidal neurons. Nature **347:** 281-284.
12. LeVere, T. E., T. Brugler, M. Sandin & S. Gray-Silva. 1989. Recovery of function after brain damage: Facilitation by the calcium entry blocker nimodipine. Behav. Neurosci. **103:** 561-565.
13. Nelson, C., J. Bawa & S. Finger. 1992. Radial maze performance after hippocampal lesions: Beneficial effects of nimodipine. Restor. Neurol. Neurosci. **4:** 33-40.
14. Chisholm, J. C., E. J. Hunnicutt, Jr. & J. N. Davis. 1993. Neuronal $[Ca^{2+}]_i$ responses to different classes of calcium-lowering agents. *In* Drugs in Development, Vol. 2: Ca^{2+} Antagonists in the CNS. A. Scriabine, R. A. Janis & D. J. Triggle, Eds.: 117-125. Neva Press. Branford, CT.
15. Bannon, A. W., K. McMonagle-Strucko & R. J. Fanelli. 1993. Nimodipine prevents medial septal lesion-induced performance deficits in the Morris water maze. Psychobiology **21:** 209-214.

Increased Hippocampal Ca^{2+} Channel Activity in Brain Aging and Dementia

Hormonal and Pharmacologic Modulation[a]

PHILIP W. LANDFIELD[b]

Department of Pharmacology
University of Kentucky
College of Medicine
Chandler Medical Center MS 305
Lexington, Kentucky 40536-0084

The role of calcium (Ca^{2+}) influx in neuropathologic and cardiotoxic conditions, particularly those resulting from ischemic/hypnoxic states, has been recognized for over 15 years. (See review in ref. 1.) However, it was not until the early 1980s that a role for altered Ca^{2+} regulation specifically in brain aging or Alzheimer's disease (AD) was formally proposed.[2] Over the last 10 years several variations and further modifications of this concept, involving different regulatory and pathogenetic aspects of Ca^{2+} homeostasis, also have been suggested.[3–7]

The initial impetus for this hypothesis arose from several separate lines of evidence that were beginning to point to alterations in Ca^{2+} homeostasis as consistent factors in the mammalian brain aging process. Among these several lines of evidence were (1) electrophysiological studies of rat hippocampal slices that pointed to excess voltage-dependent Ca^{2+} influx with aging,[8–10] (2) studies of synaptosomes showing aging-impaired extrusion and buffering of Ca^{2+}[11,12] and (3) studies of synaptosomes and/or slices indicating reduced Ca^{2+} uptake in aged animals.[4] Moreover, Ca^{2+} homeostasis in fibroblasts from AD subjects seemed to be altered, suggesting that disturbances in Ca^{2+} regulation also may be found outside of the brain.[13] In addition, evidence was growing that amino acid-induced excitotoxicity[14] depended in part on Ca^{2+} influx through glutaminergic receptor-operated channels.[15,16] Evidence consistent with the general Ca^{2+} and brain aging hypothesis was reviewed recently (cf ref. 3).

On the basis of findings linking excitotoxicity and Ca^{2+} influx,[14–16] many recent considerations of Ca^{2+} neurotoxicity have focused on Ca^{2+} entry via excitatory amino acid-activated channels (eg, the NMDA receptor-operated channel). However, Ca^{2+} influx through voltage-activated Ca^{2+} channels, such as the long-opening L-type channel, may actually represent a still larger pathway of Ca^{2+} entry. That is, during

[a] Much of the work summarized here was supported by grants from the National Institute on Aging and Miles, Inc.

[b] Address for correspondence: Philip W. Landfield, PhD, Department of Pharmacology, University of Kentucky, College of Medicine, Chandler Medical Center, 800 Rose Street, Lexington, KY 40536-0084.

depolarization, the widespread postsynaptic distribution and high density of L-type channels[17] and their large conductance and relative resistance to inactivation provide the capacity for a substantial Ca^{2+} influx throughout the postsynaptic cell body and dendrites (cf reviews in refs. 18-20). Conversely, ligand-gated channels, such as the NMDA receptor, are generally restricted to regions of synaptic contact. In fact, increasing evidence indicates that a significant amount of the Ca^{2+} influx that accompanies amino acid activation may in fact enter the neuron via voltage-activated Ca^{2+} channels that are depolarized during synaptic (amino acids) activation rather than through the receptor-operated channels themselves. For example, Ca^{2+} influx through voltage-activated L-type channels recently was found to be the main pathway for Ca^{2+} influx during NMDA-stimulated long-term potentiation in hippocampal field CA1.[21] (Also see Brown, this volume.)

These arguments do not rule out a potentially major role for ligand-gated Ca^{2+} influx in neurotoxicity, particularly in regions of high glutaminergic activation (eg, dendritic spines). Instead, they emphasize that any consideration of Ca^{2+} influx-mediated neurotoxicity, even that induced by amino acids, must also take into account the major Ca^{2+} entry pathway through voltage-activated channels.

These arguments also do not diminish the strong likelihood that Ca^{2+} buffering/extrusion processes[22,23] also play important roles in neuronal Ca^{2+} dyshomeostasis during brain aging and possibly AD. It seems highly likely that alterations in one major entry or extrusion mechanism of Ca^{2+} regulation would ripple throughout the cell and disrupt other major components of Ca^{2+} regulatory systems. At present, it is unclear which aspect of neuronal Ca^{2+} regulation may be altered first with aging.

However, the present article focuses primarily on evidence related to aging changes in the activity of *voltage-activated Ca^{2+} channels* in hippocampal neurons and summarizes our working hypothesis that such changes may play a major pathogenetic role in brain aging/AD.

IMPAIRED HIPPOCAMPAL SYNAPTIC PLASTICITY IN AGING

The first electrophysiological evidence that pointed to aging changes in voltage-activated Ca^{2+} channels with aging actually arose not from direct studies of postsynaptic Ca^{2+} potentials and currents, but rather from studies of synaptic function that only indirectly reflect Ca^{2+} regulation. In the mid-late 1970s, we began a series of electrophysiological studies that found consistent evidence of impaired synaptic plasticity (potentiation) in the hippocampus of aging rats. These effects of aging were observed using microelectrode recording in both the *in vitro* slice[24] and the intact, anesthetized rat[25] preparations. Moreover, several of our experiments showed that the impaired frequency potentiation was correlated closely with learning/memory impairments (cf review in ref. 26).

The electrophysiological deficits in synaptic plasticity were most prominent during repetitive synaptic stimulation, which showed that the amount of *frequency potentiation* (FP, the increase in synaptic responses during a train of 3-20-Hz repetitive activation) of hippocampal responses was markedly reduced in aged rats.[24-26] The deficit appears to involve both pre- and postsynaptic components; moreover, alterations in postsynaptic membrane hyperpolarization may well contribute to impaired transmission, particularly during periods of high frequency activity.[27]

In some experimental protocols, age-related deficits also have been found in hippocampal *long-term potentiation* (LTP, the sustained increase in synaptic responses *following* a train of repetitive activation).[25,28–30] However, in most studies aging differences in LTP were less pronounced than in FP and were found primarily in the rising or decay time courses of LTP[24,28,31] or were not seen at all.[26] Nonetheless, it has been proposed that under natural conditions, FP acts as a trigger for LTP by the mechanism of "amplifying" postsynaptic depolarization.[26] If this were the case, and if the FP mechanism were impaired with aging, then FP-dependent LTP under physiological conditions also would be reduced with aging.

ROLE OF INCREASED CALCIUM INFLUX IN AGING-IMPAIRED SYNAPTIC PLASTICITY

In a series of studies, the calcium channel blocker magnesium (Mg^{2+}) was used to study the role of Ca^{2+} in the synaptic deficits. In these studies, the extracellular magnesium/calcium ratio was varied in the bathing medium of hippocampal slices. Our results showed that a very modest elevation of the Mg^{2+}/Ca^{2+} ratio strongly improved FP in hippocampal slices, particularly in those from aged rats. Conversely, elevated Ca^{2+} reduced FP even in slices from young animals.[32] Other studies showed that in intact animals a high Mg^{2+} diet, which elevated plasma Mg^{2+}, both improved FP and counteracted a deficit in reversal maze learning in aged rats.[9]

Intracellular recordings showed that Mg^{2+} was not acting by blockade of NMDA potentials, but rather by a combination of other effects, including prevention of EPSP decay, and the reduction of postsynaptic hyperpolarization.[32] Because FP itself is Ca^{2+}-dependent, it might on the surface appear paradoxical that excess Ca^{2+} can reduce FP. However, there are several possible mechanisms through which elevated Ca^{2+} might impair FP of the EPSP, which include: (1) accelerated transmitter depletion; (2) increased Ca^{2+}-dependent hyperpolarization of the axon terminals or dendrites, resulting in action potential failure or a "shunt" of the dendritic EPSP; or (3) Ca^{2+}-dependent inactivation of subsequent Ca^{2+} influx.[32] Quantitative ultrastructural analyses of synaptic vesicle distribution and density during FP indicated that transmitter depletion was not the basis of the aging deficit.[33,34] Therefore, excess Ca^{2+} influx appears to act by activating K^+ currents or inducing Ca^{2+}-dependent inactivation in terminals, or both. Other mechanisms are also possible.[32–34]

Thus, evidence that moderate elevation of Mg^{2+}, a Ca^{2+} channel antagonist, could enhance FP, whereas elevating extracellular Ca^{2+} impaired FP suggested that the aging-dependent impairment of FP in aged neurons might be due to excess voltage-dependent Ca^{2+} influx.[8,32–34] Despite this evidence, however, synaptic processes are highly complex and membrane currents cannot yet be consistently recorded at synaptic contacts in brain. Therefore, it is difficult to draw definitive conclusions about aging effects on Ca^{2+} currents or potentials in brain synaptic elements. Consequently, we undertook a series of studies that directly examined well-defined Ca^{2+}-dependent potentials and Ca^{2+} currents in the soma, as described.

CALCIUM CURRENTS IN NONDISSOCIATED HIPPOCAMPAL SLICE NEURONS

Because our primary focus in this research was the study of Ca^{2+} currents in aged brain neurons, the types of preparations suitable for such studies were limited. Clearly, embryonic cell cultures and dissociated neuron preparations (generally obtained from very young animals), as are often used in patch clamp studies of Ca^{2+} currents, would not be useful for studies of aging brain tissues. Dissociated neurons from adult and aged animals have been used occasionally (cf Verkhratsky, this volume), but these approaches may result in differential yields or trauma to neurons from aged animals. A few studies have used patch-clamp techniques in adult animal tissue slices (eg, see refs. 35 and 36), and such techniques appear promising. Nevertheless, currently the most well-characterized method for statistical comparison studies of intracellular electrophysiological variables in relatively nontraumatized adult brain tissues is the sharp electrode voltage clamp technique in brain slices.[37]

In our studies of aging processes to date, we have used these relatively physiological (nondissociated) hippocampal slice preparations for studies of membrane potentials and currents in neurons from adult and aged rats. During recording, neurons are disrupted only by penetration with a single sub-micron-tip pipette and are bathed in artificial cerebrospinal fluid. Under these conditions, conventional intracellular voltage recordings allow measurements of the long Ca^{2+}-dependent K^+-mediated afterhyperpolarization (AHP) that follows sodium (Na^+) spikes elicited by depolarizing pulses injected through the intracellular electrode. Another electrophysiological indicator of voltage-activated Ca^{2+} influx in these neurons is the Ca^{2+} action potential (spike). Ca^{2+} action potentials are relatively "pure" reflections of Ca^{2+} influx, and they also can be elicited by intracellular depolarizing current injection, providing Na^{2+} action potentials have been blocked with tetrodotoxin (TTX).[38,39] In addition, blocking the K^+ currents underlying the AHP with cesium loading of the neuron through the pipette[37] reveals a large secondary plateau or "hump" on the Ca^{2+} spike.[40]

However, direct measurements of membrane currents under voltage clamp conditions yield more detailed information on voltage-activated Ca^{2+} currents. When tetraethylammonium application, which blocks a wide range of other K^+ currents, is combined with TTX and cesium applications, it is also feasible to study Ca^{2+} membrane currents in slices using the sharp electrode (switching) voltage clamp. Under these conditions, we have been able to identify three apparent functional types of whole cell Ca^{2+} currents in CA1 hippocampal neurons.[6,40–42]

In our conditions, pyramidal cells that are cesium-loaded and treated with tetrodotoxin and tetraethylammonium exhibit a low amplitude noninactivating Ca^{2+} current at the lowest activation voltages. It is relatively resistant both to blockade by several Ca^{2+} blockers and to Ca^{2+}-dependent inactivation, and we have initially termed it "R" (for resistant). This current may have some overlap with or be similar to P type current.[43] At somewhat higher depolarizing voltages, a rapidly inactivating (N- or T-like) current is activated, and at voltage levels only slightly more positive, a larger, slowly inactivating L-type current is activated (cf refs 18-20 for reviews of functional channel types).

VERY LONG CALCIUM TAIL CURRENTS

To our surprise, activation of this L-like current in slice neurons was consistently associated with a very long (several hundred milliseconds) tail-like Ca^{2+} current.[40–42] Although long tails are often thought to reflect inadequate space clamp, this repolarization current was highly sensitive to selective inactivation and therefore did not appear to be artifactual. Consequently, we have been working for several years to define the nature of this long postactivation current. Considerable evidence from studies in which we severed dendrites and altered chloride loading indicates that the repolarization current is not the result of inadequate space clamp or Ca^{2+} activated chloride current.[41]

The most convincing evidence of this, however, arises from recent studies of single Ca^{2+} channels in our laboratory.[44] These studies found that repolarization openings of (nimodipine-sensitive) L-type Ca^{2+} channels after a depolarizing step can occur with high frequency under relatively physiological conditions (eg, lower Ba^{2+} or Ca^{2+} in the cell-attached pipette). Although single Ca^{2+} channel openings during the repolarization phase had been seen previously,[45,46] they had not been found to occur with high frequency.

The major experimental manipulation that enabled us to find high frequency single Ca^{2+} channel correlates of the tail current came about when we lowered the pipette Ba^{2+} concentration used to study Ca^{2+} channels from the typical 90-110 mM Ba^{2+} to 20 mM Ba^{2+}. This resulted in a remarkable increase in the frequency of repolarization openings (FIG. 1). The high concentrations of extracellular Ba^{2+} typically employed to study single Ca^{2+} channels apparently suppress the repolarization openings. Thus, these repolarization openings likely account for the repolarization tail-like Ca^{2+} current as well as the AHP and late plateau or "hump" phase of the Ca^{2+} action potential. Furthermore, these openings appear to have major physiological implications because they may reflect a major Ca^{2+} entry pathway that is more active at physiological levels of extracellular divalent cations.[44]

EFFECTS OF AGING ON HIPPOCAMPAL CALCIUM POTENTIALS AND CURRENTS

To test the hypothesis of elevated voltage-activated Ca^{2+} influx more directly, extensive quantitative studies of the effects of aging on Ca^{2+}-dependent potentials and Ca currents were performed on hippocampal slice neurons that met strict criteria for stable recordings and good cell "health." Aged neurons in this preparation appear to be as viable as young neurons. As noted, hippocampal pyramidal cells are characterized by long (100-500 ms) Ca^{2+}-dependent, K^+-mediated AHPs following a Na^{2+} action potential. In our initial studies of Ca^{2+}-mediated potentials in aging rat neurons, we quantified the Ca^{2+}-dependent AHP duration and amplitude for a given number of current-elicited Na^+ spikes. These experiments showed a highly consistent aging-dependent increase in the duration and amplitude of the AHP (eg, FIG. 2), which we replicated in two independent studies.[10,47]

However, an increase in the AHP could be due to altered K^+ channels or to altered Ca^{2+} buffering/extrusion rather than to an increase in voltage-activated Ca^{2+} influx

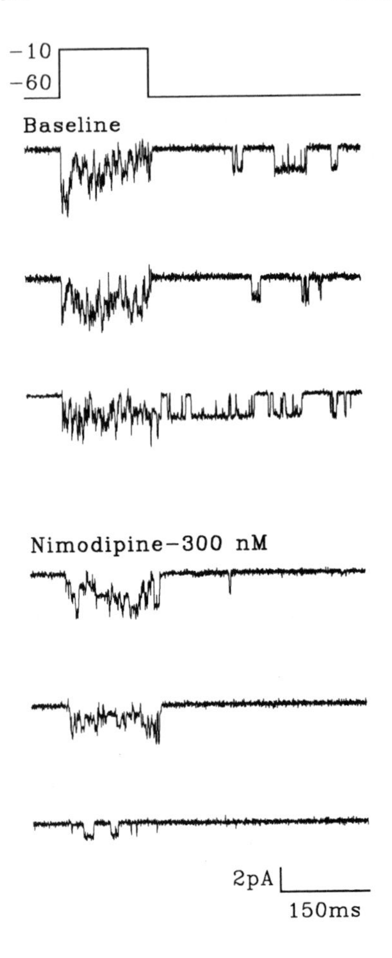

FIGURE 1. Repolarization openings of L-type channels. Cell-attached patch recordings of single Ca^{2+} channels from a multi-channel patch of a hippocampal neuron in culture. BAY K 8644 was added to the pipette solution to enrich L-type Ca^{2+} channel activity in the patch. The charge carrier in the pipette is 20 mM Ba^{2+}, which in comparison to 110 mM Ba^{2+}, facilitates the occurrence of repolarization openings after a depolarization pulse. Those repolarization openings may underlie long Ca^{2+}-dependent processes such as the AHP, as they are increased at increasingly physiological concentrations of divalent cations.[45] **Baseline:** Several depolarizing step episodes showing multiple Ca^{2+} channel activity during the pulse, followed by substantial Ca^{2+} channel activity during the repolarization phase. **Nimodipine:** During application of 300 nM of nimodipine, there is substantial inhibition of Ca^{2+} channel activity during the pulse. Repolarization openings are also suppressed, indicating their origin from L-type channels. (Unpublished data from O. Thibault, N. M. Porter and P. W. Landfield.)

per se. Consequently, we also studied Ca^{2+} action potentials (spikes) in current clamp studies of tetrodotoxin-treated, cesium-loaded hippocampal neurons and again found a highly consistent aging-dependent increase in the size and duration of the late "hump" phase of the Ca^{2+} spike (FIG. 3).[48]

Because the Ca^{2+} spike potential reflects relatively "pure" Ca^{2+} currents, these studies point directly to an increase in voltage-activated Ca^{2+} influx with aging. Nevertheless, they do not clarify which specific types of Ca^{2+} currents are involved or the voltage dependence of the aging changes.

Therefore, aging comparisons also have been performed on tetrodotoxin-treated and cesium-loaded CA1 neurons also treated with tetraethylammonium under sharp electrode voltage-clamp conditions. Aged neurons exhibited a larger Ca^{2+} current measured at the end of a 200-ms step than did young neurons, but the rate of Ca^{2+}-dependent inactivation during a 5-pulse train of depolarizing steps was approximately the same for both young and aged neurons.[6,49,50] In addition, the very long tail currents, which, as noted, probably underlie the AHP and the plateau phase of the Ca^{2+} spike,

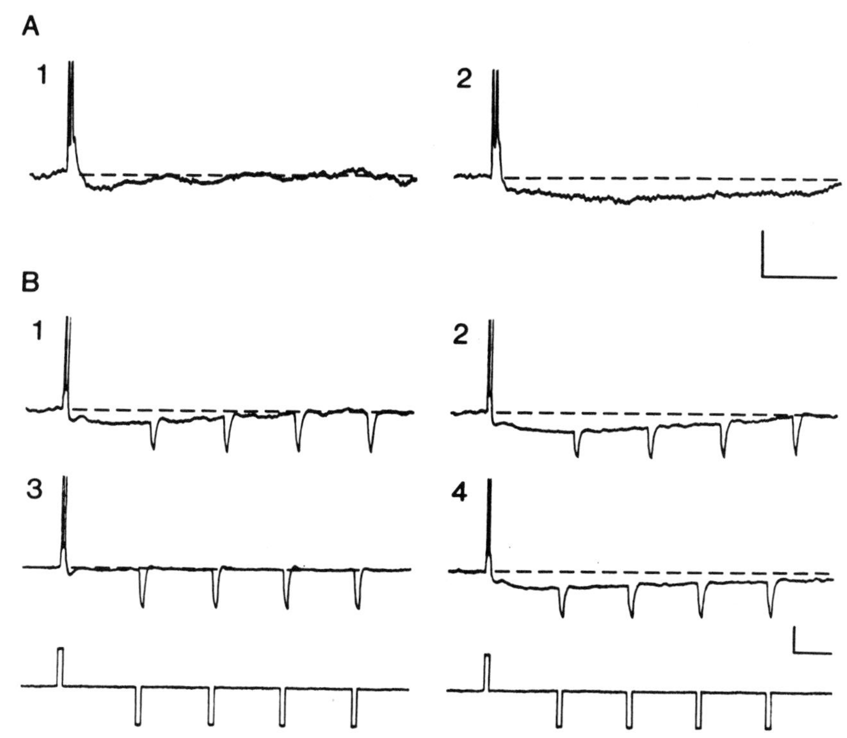

FIGURE 2. Intracellular current-induced bursts of Na^{2+} action potentials and subsequent AHPs in CA1 neurons of hippocampal slices from young and aged rats. **(A)** AHPs after current-induced burst of two Na^{+} spikes. **(A_1)** Cell from a young rat. **(A_2)** Cell from an aged rat. **(B)** AHPs and concomitant conductance increase after a 0.4 nA current-induced burst of three spikes. **(B_1)** Young rat cell in normal Ca^{2+}. **(B_2)** Aged rat cell in normal Ca^{2+}. **(B_3)** Young rat cell in low Ca^{2+} high Mg^{2+}. **(B_4)** Young rat cell in high Ca^{2+}, low Mg^{2+}. *Dashed lines* show resting potentials before the burst. *At the bottom* are shown the initial intracellular depolarizing current pulse used to induce a spike burst and the subsequent 2-Hz train of 0.4-nA hyperpolarizing pulses used to assess input conductance during the AHP, for cells shown in **B**. (Reprinted from ref. 10 with permission.)

were significantly increased in the aged neurons.[51] Thus, data from each of these approaches are consistent with the hypothesis[5–7] that, with aging, there is greater voltage-activated Ca^{2+} influx in hippocampal neurons. Moreover, this increased influx appears to occur primarily during a late, prolonged phase of activation of L-type channels.

Interestingly, similar results for both the AHP and the Ca^{2+} spike were recently described for a different mammalian family (rabbits) (cf, Disterhoft, this volume; also see ref. 52). This has highly important implications for the interpretation of these changes as potential pathogenetic factors in brain aging. That is, any basic underlying aging process presumably should be generalized widely across mammalian species.

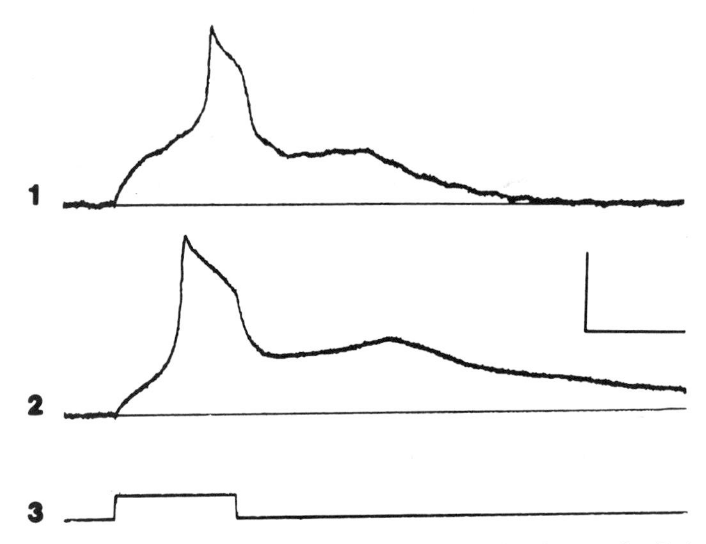

FIGURE 3. Ca^{2+} spikes recorded from cesium-loaded, tetrodotoxin-treated cells in normal medium. (1) Young rat cell. (2) Aged rat cell. (3) Current trace showing depolarizing pulse used to trigger the Ca^{2+} spike. Calibration: 25 mV, 50 ms. (Reprinted with permission from ref. 48.)

SENSITIVITY TO DIHYDROPYRIDINES OF CALCIUM CURRENTS/POTENTIALS AFFECTED BY AGING

Dihydropyridines (DHPs) are specific Ca^{2+} channel activator and blockers that are viewed as specific for the L-type of Ca^{2+} channel.[53] Nimodipine, which is a DHP L-channel blocker relatively specific for brain sites, reduced the current activated from cells from aged animals almost to the level seen in neurons from young animals.[49,50] However, nimodipine did not eliminate all high voltage-activated (L-like) current in these neurons. Instead, nimodipine blocked approximately 30% of the high voltage-activated Ca^{2+} current, much as it was shown to do in studies of isolated CNS neurons.[54]

Moreover, nimodipine blocks, and BAY K 8644, a DHP Ca^{2+} channel activator, increases specific aspects of Ca^{2+}-mediated voltage potentials that have been found to increase with aging. That is, nimodipine blocks the AHP in CA1 neurons[55] and blocks the slow plateau ("hump") phase of the Ca^{2+} spike.[50] This slow hump is the Ca^{2+} spike component most affected by aging.[48] FIGURE 1 shows also that nimodipine blocks L-type Ca^{2+} channel repolarization openings, which may underlie many of the long duration Ca^{2+}-dependent potentials that are influenced by aging. Again, very similar effects of nimodipine on the AHP and the Ca^{2+} spikes have been found in rabbits, and in those studies a differential aging sensitivity to Ca^{2+} channel antagonists was also apparent (ref. 52 and Disterhoft, this volume).

Taken together, these data imply that under physiological conditions, DHP Ca^{2+} antagonists may act to reduce long-lasting Ca^{2+} currents at the soma, thereby reducing the AHP, increasing excitability of hippocampal neurons, and making these cells more readily available for participation in cognitive functions. Substantial behavioral and neuronal excitability data consistent with this view were obtained by several investigators (cf refs. 52, 53, and 56). Thus, an important conclusion from the DHP studies is that L-type Ca^{2+} channels are importantly involved in age-related alterations in Ca^{2+} channel activity.

GLUCOCORTICOID EFFECTS ON CALCIUM CURRENTS IN HIPPOCAMPAL NEURONS

Despite the specificity of DHPs, little is known with certainty about endogenous modulators of Ca^{2+} channels. Although it is clear that a variety of drugs, neurotransmitters, and second messengers can influence Ca^{2+} channels,[18–20] the relative importance of these factors is not understood. Of interest, therefore, is recent evidence that peripheral glucocorticoid hormones which can bind to brain receptors can also modulate Ca^{2+} channel activity over extended periods.[42,47,57]

An extensive literature also implicates glucocorticoids as one putative causal factor in the aging of the brain, particularly of the hippocampus, which is extremely rich in glucocorticoid receptors (cf review in ref. 58). Consequently, we tested the hypothesis that glucocorticoids and Ca^{2+} current mechanisms in brain aging might be linked, and that glucocorticoid effects on brain aging might be mediated by altering Ca^{2+} currents. In studies with the hippocampal slice, we[47] and others[57] found that glucocorticoids could increase the Ca^{2+}-dependent AHP. Conversely, adrenalectomy, which is neuroprotective during aging,[58] substantially reduced the AHP and the Ca^{2+} spike. Moreover, in another study, the highly specific type II glucocorticoid receptor agonist RU 28362 dramatically increased Ca^{2+} action potentials recorded from CA1 neurons of young-mature rats adrenalectomized several days earlier. Cycloheximide blocked this effect of RU 28362 in adrenalectomized animals, suggesting that *de novo* protein synthesis is involved in the effect of glucocorticoids on the Ca^{2+} spike potential (FIG. 4 and ref. 42). Thus, it appears that type II glucocorticoid receptor activation can increase voltage-sensitive Ca^{2+} spikes/currents through a cycloheximide-sensitive mechanism. Moreover, the effect of glucocorticoids on Ca^{2+}-dependent AHPs apparently increases with aging.[47] Therefore, not only do steroids modulate voltage-sensitive Ca^{2+} channels, but in addition the glucocorticoids-Ca^{2+} channel linkage may represent two phases of a common aging-related neurodegenerative process.

Clearly, many other endogenous factors also modulate Ca^{2+} channel activity. However, one of the important implications of the glucocorticoid effect is that such modulation can occur over extended time frames and that the set-point of activity apparently may be partially modulated by genomic action.

CONCLUSIONS AND FUTURE CHALLENGES

These studies in aged rat neurons have indicated that Ca^{2+}-mediated potentials and Ca^{2+} currents appear to be enhanced in aging. Moreover, a particular component of L-type Ca^{2+} currents (a prolonged, or repolarization, phase) seems to be particularly

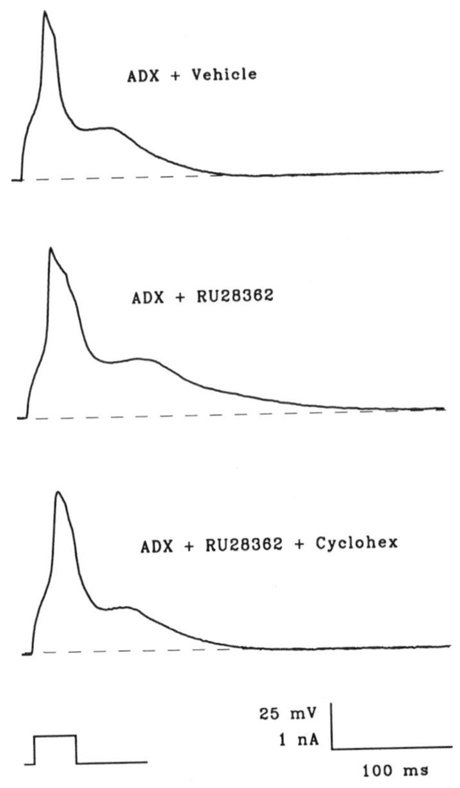

FIGURE 4. Representative Ca^{2+} action potentials for cesium chloride-loaded tetrodotoxin-treated neurons in hippocampal slices exposed to either vehicle, a specific glucocorticoid agonist, or the agonist and cycloheximide. Action potentials were triggered by depolarizing intracellular pulses and have an initial large amplitude, fast phase and a late slow phase. Neurons exposed to a saturating dose of the agonist exhibited wider initial phases and longer-duration and larger-amplitude slow phases than did control neurons. Cycloheximide blocked the effect. (Reprinted with permission from ref. 42.)

affected by the aging process. The fact that a series of studies in another mammalian family (rabbits) has resulted in generally similar findings for Ca^{2+} potentials (Disterhoft, this volume) indicates that this phenomenon may be generalized across aging mammals.

The potential relevance of studies on normal mammalian brain aging to Alzheimer's disease and other age-related neurodegenerative disorders is clearly indicated by the observations that aging is the major risk factor for AD and that up to 50% of those over age 85 years may have some form of the condition.[59] Thus, something about the brain aging process vastly increases the susceptibility to neurodegenerative conditions such as AD. Moreover, most of the pathological hallmarks of AD brain also are found, with less frequency, in the brain of normal aged subjects.[59]

The specific version of the general Ca^{2+}/brain aging hypothesis that we have proposed[5–7] is that age-related alterations in the activity of voltage-activated Ca^{2+} channels play a critical role in the rapidly increasing vulnerability with advancing age of the hippocampus and other brain structures to AD. This view is of course fully compatible with the possibility that other aspects of Ca^{2+} homeostasis are disrupted by aging.

One of the major problems for future research will be to attempt to define the causes of the age changes in Ca^{2+} channels and other membrane molecules. Analyzing the relationships of molecular-level phenomena to complex human brain disorders will be highly challenging. However, such approaches will likely prove essential for advancing our understanding of basic aging processes and for the development of new therapies for human neurodegenerative disorders.

ACKNOWLEDGMENTS

The author wishes to acknowledge the essential contributions of former and present coworkers, including Thomas Pitler, Michael Applegate, D. Steven Kerr, Su-Yang Hao, Lee Campbell, Olivier Thibault, Mary Mazzanti, Nada Porter, and Eric Blalock. The author also thanks Lisa Lowery for excellent preparation of the manuscript.

REFERENCES

1. SIESJÖ, B. K. & F. BENGTSSON. 1989. Calcium fluxes, calcium antagonists, and calcium-related pathology in brain ischemia, hypoglycemia, and spreading depression: A unifying hypothesis. J. Cereb. Blood Flow Metab. **9:** 127-140.
2. KHACHATURIAN, Z. S. 1984. Toward theories of brain aging. *In* Handbook of Studies on Psychiatry and Old Age. D. S. Kay & G. W. Burrow, Eds.: 7-30. Elsevier. Amsterdam.
3. KHACHATURIAN, Z. S. 1989. The role of calcium regulation in brain aging: Reexamination of a hypothesis. Aging **1:** 17-34.
4. GIBSON, G. E. & C. PETERSON. 1987. Calcium and the aging nervous system. Neurobiol. Aging **8:** 329-344.
5. LANDFIELD, P. W. 1987. "Increased calcium-current" hypothesis of brain aging. Neurobiol. Aging **8:** 346.
6. LANDFIELD, P. W., L. W. CAMPBELL, S.-Y. HAO & D. S. KERR. 1989. Aging-related increases in voltage-sensitive, inactivating calcium currents in rat hippocampus: Implications for mechanisms of brain aging and Alzheimer's disease. *In* Calcium, Membranes, Aging, and Alzheimer's Disease. Z. S. Khachaturian, C. W. Cotman & J. W. Pettegrew, Eds. Ann. N. Y. Acad. Sci. **568:** 95-105.
7. LANDFIELD, P. W., O. THIBAULT, M. L. MAZZANTI, N. M. PORTER & D. S. KERR. 1992. Mechanisms of neuronal death in brain aging and Alzheimer's disease: Role of endocrine-mediated calcium dyshomeostasis. J. Neurobiol. **9:** 1247-1260.
8. LANDFIELD, P. W. 1983. Mechanisms of altered neural function during aging. *In* Aging of the Brain. W. H. Gispen & J. Traber, Eds.: 51-71. Elsevier. Amsterdam.
9. LANDFIELD, P. W. & G. MORGAN. 1984. Chronically elevating plasma Mg^{2+} improves hippocampal frequency potentiation and reversal learning in aged and young rats. Brain Res. **322:** 167-171.
10. LANDFIELD, P. W. & T. A. PITLER. 1984. Prolonged Ca^{2+}-dependent afterhyperpolarizations in hippocampal neurons of aged rats. Science **226:** 1089-1092.

11. MICHAELIS, M. L., K. JOHE & T. E. KITOS. 1984. Age-dependent alterations in synaptic membrane systems for Ca regulation. Mech. Aging Dev. **25:** 215-225.
12. MARTINEZ-SERRANO, A., P. BLANCO & J. SATRÙSTEGUI. 1992. Calcium binding to the cytosol and calcium extrusion mechanisms in intact synaptosomes and their alterations with aging. J. Biol. Chem. **267:** 4673-4679.
13. PETERSON, C., G. GIBSON & J. P. BLASS. 1985. Altered calcium uptake in cultured skin fibroblasts from patients with Alzheimer's disease. N. Engl. J. Med. **312:** 1063-1065.
14. ROTHMAN, S. & J. W. OLNEY. 1987. Glutamate and the pathophysiology of hypoxic-ischemic brain damage. Ann. Neurol. **19:** 105-111.
15. CHOI, D. W. 1987. Ionic-dependence of glutamate neurotoxicity. J. Neurosci. **7:** 369-379.
16. MATTSON, M. P. 1990. Antigenic changes similar to those seen in neurofibrillary tangles are elicited by glutamate and Ca^{2+} influx in cultured hippocampal neurons. Neuron **2:** 105-117.
17. ABLIJANIAN, M. K., R. E. WESTENBROEK & W. A. CATTERALL. 1990. Subunit structure and localization of dihydropyridine-sensitive calcium channels in mammalian brain, spinal cord, and retina. Neuron **4:** 819-832.
18. TSIEN, R. W., D. LIPSCOMBE, D. V. MADISON, K. R. BLEY & A. P. FOX. 1988. Multiple types of neuronal Ca channels and their selective modulation. Trends Neurosci. **11:** 431-438.
19. BEAN, B. P. 1989. Classes of calcium channels in vertebrate cells. Annu. Rev. Physiol. **51:** 367-384.
20. MILLER, R. J. 1992. Voltage-sensitive Ca^{2+} channels. J. Biol. Chem. **267:** 1403-1406.
21. JOHNSTON, D., S. WILLIAMS, D. JAFFE & R. GRAY. 1992. NMDA-receptor-independent long-term potentiation. Annu. Rev. Physiol. **54:** 489-505.
22. MICHAELIS, M. L., C. T. FOSTER & C. JAYAWICKREME. 1992. Regulation of calcium levels in brain tissue from adult and aged rats. Mech. Ageing Dev. **62:** 291-306.
23. SMITH, D. O. 1988. Muscle-specific decrease in presynaptic calcium dependence and clearance during and neuromuscular transmission in aged rats. J. Neurophysiol. **59:** 1069-1082.
24. LANDFIELD, P. W. & G. S. LYNCH. 1977. Impaired monosynaptic potentiation in *in vitro* hippocampal slices from aged, memory-deficient rats. J. Gerontol. **32:** 523-533.
25. LANDFIELD, P. W., J. L. MCGAUGH & G. S. LYNCH. 1978. Impaired synaptic potentiation processes in the hippocampus of aged memory-deficient rats. Brain Res. **150:** 85-101.
26. LANDFIELD, P. W. 1988. Hippocampal neurobiological mechanisms of age-related memory dysfunction. Neurobiol. Aging **9:** 571-579.
27. PITLER, T. A. & P. W. LANDFIELD. 1987. Postsynaptic membrane shifts during frequency potentiation of the hippocampal EPSP. J. Neurophysiol. **58:** 866-882.
28. BARNES, C. A. 1979. Memory deficits associated with senescence: A neurophysiological and behavioral study in the rat. J. Comp. Physiol. Psychol. **93:** 74-104.
29. DEUPREE, D. L., J. BRADLEY & D. A. TURNER. 1991. Age-related alterations in potentiation in the CA1 region in F344 rats. Neurobiol. Aging **14:** 249-258.
30. TIELEN, A. M., W. J. MOLLEVANGER, F. H. LOPES DA SILVA & C. F. HOLLANDER. 1983. *In* Aging of the Brain. W. H. Gispen & J. Traber, Eds.: 73-84. Elsevier. Amsterdam.
31. BARNES, C. A. & B. L. MCNAUGHTON. 1985. An age comparison of the rates of acquisition and forgetting of spatial information in relation to long-term enhancement of hippocampal synapses. Behav. Neurosci. **99:** 1040-1048.
32. LANDFIELD, P. W., T. A. PITLER & M. D. APPLEGATE. 1986. The effects of high Mg^{2+}-to Ca^{2+} ratios on frequency potentiation in hippocampal slices of young and aged rats. J. Neurophysiol. **56:** 797-811.
33. APPLEGATE, M. D. & P. W. LANDFIELD. 1988. Synaptic vesicle redistribution during hippocampal frequency potentiation and depression in young and aged rats. J. Neurosci. **8:** 1096-1111.

34. LANDFIELD, P. W., M. D. APPLEGATE, T. A. PITLER & D. S. KERR. 1988. Presynaptic mechanisms in hippocampal short- and long-term potentiation: Relevance to brain aging. *In* Long-Term Potentiation: From Biophysics to Behavior. P. W. Landfield & S. A. Deadwyler, Eds.: 377-408. Liss. New York.
35. GRAY, R., R. FISHER, N. SPRUSTON & D. JOHNSTON. 1990. Acutely exposed hippocampal neurons: A preparation for patch clamping neurons from adult hippocampal slices. *In* Preparations of Vertebrate Central Nervous System in Vitro. H. Jahnsen, Ed.: 3-24. John Wiley & Sons Ltd. New York.
36. SAKMANN, B., E. EDWARDS, A. KONNERTH & T. TAKAHASHI. 1989. Patch clamp techniques used for studying synaptic transmission in slices of mammalian brain. Q. J. Exp. Physiol. **74:** 1107-1118.
37. JOHNSTON, D., J. J. HABITZ & W. A. WILSON. 1980. Voltage clamp discloses slow inward current in hippocampal burst-firing neurons. Nature **286:** 391-393.
38. LLINAS, R. & R. HESS. 1976. Tetrodotoxin-resistant dendritic spikes in avian Purkinje cells. Proc. Natl. Acad. Sci. USA **73:** 2520-2523.
39. SCHWARTZKROIN, P. A. & M. SLAWSKY. 1977. Probable calcium spikes in hippocampal neurons. Brain Res. **168:** 299-309.
40. PITLER, T. A. & P. W. LANDFIELD. 1987. Probable Ca^{2+}-mediated inactivation of Ca^{2+}-currents in mammalian brain neurons. Brain Res. **410:** 147-153.
41. CAMPBELL, L. W., O. THIBAULT, S.-Y. HAO & P. W. LANDFIELD. 1990. Properties of two long-lasting calcium currents in non-dissociated, adult hippocampal CA1 pyramidal cells. Soc. Neurosci. (Abstr. 2838) **16**(part 1): 676.
42. KERR, D. S., L. W. CAMPBELL, O. THIBAULT & P. W. LANDFIELD. 1992. Hippocampal glucocorticoid receptor activation enhances voltage-dependent calcium conductance: Relevance to brain aging. Proc. Natl. Acad. Sci. USA **89:** 8527-8531.
43. USOWICZ, M. M., M. SUGIMORI, B. CHERKSEY & R. LLINAS. 1992. P-type calcium channels in the somata and dendrites of adult cerebellar Purkinje cells. Neuron **9:** 1185-1199.
44. THIBAULT, O., N. M. PORTER & P. W. LANDFIELD. 1993. Low Ba^{2+} and Ca^{2+} induce a sustained high frequency of repolarization openings of L-type Ca^{2+} channels in hippocampal neurons: Physiological implications. Proc. Natl. Acad. Sci. USA **90:** 11792-11796.
45. FISHER, R. E., R. GRAY & D. JOHNSTON. 1990. Properties and distribution of single voltage-gated calcium channels in adult hippocampal neurons. J. Neurophysiol. **64:** 91-104.
46. SLESINGER, P. A. & J. B. LANSMAN. 1991. Reopening of Ca^{2+} channels in mouse cerebellar neurons at resting membrane potentials during recovery from inactivation. Neuron **7:** 755-762.
47. KERR, D. S., L. W. CAMPBELL, S.-Y. HAO & P. W. LANDFIELD. 1989. Corticosteroid modulation of hippocampal potentials: Increased effect with aging. Science **245:** 1505-1509.
48. PITLER, T. A. & P. W. LANDFIELD. 1990. Aging-related prolongation of calcium spike duration in rat hippocampal slice neurons. Brain Res. **508:** 1-6.
49. CAMPBELL, L. W., S.-Y. HAO & P. W. LANDFIELD. 1989. Aging-related increases in L-like calcium currents in rat hippocampal slices. Soc. Neurosci. (Abstr.) **15:** 106.1.
50. LANDFIELD, P. W. 1989. Nimodipine modulation of aging-related increases in hippocampal calcium currents. *In* Nimodipine and Central Nervous System Function: New Vistas. J. Traber & W. H. Gispen, Eds. Stuttgart-New York. Schattauer.
51. LANDFIELD, P. W., O. THIBAULT, L. W. CAMPBELL & E. M. BLALOCK. 1993. Aging-dependent increases in very long calcium tail currents in rat hippocampal slice CA1 neurons. Soc. Neurosci. (Abstr. #7175) **19**(part II): 1743.
52. DISTERHOFT, J. F., J. R. MOYER, JR., L. T. THOMPSON & M. KOWALSKA. 1993. Functional aspects of calcium-channel modulation. Clin. Neuropharmacol. **16**(1): S12-S24.
53. SCRIABINE, A., T. SCHUURMAN & J. TRABER. 1989. Pharmacological basis for the use of nimodipine in central nervous system disorders. FASEB J. **3:** 1799-1806.

54. REGAN, L. J., D. W. Y. SAH & B. P. BEAN. 1991. Ca^{2+} channels in rat central and peripheral neurons: High-threshold current resistant to dihydropyridine blockers and ω-contotoxin. Neuron **6:** 269-280.
55. MAZZANTI, M. L., O. THIBAULT & P. W. LANDFIELD. 1991. Dihydropyridine modulation of normal hippocampal physiology in young and aged rats. Neurosci. Res. Comm. **9:** 117-126.
56. MCMONAGLE-STRUCKO, K. & R. J. FANELLI. 1993. Enhanced acquisition of reversal training in a spatial learning task in rats treated with chronic nimodipine. Pharmacol. Biochem. Behav. **44:** 827-835.
57. JOËLS, M. & E. R. DE KLOET. 1989. Effects of glucocorticoids and norepinephrine on the excitability of the hippocampus. Science **245:** 1502-1505.
58. LANDFIELD, P. W. & J. C. ELDRIDGE. 1991. The glucocorticoid hypothesis of brain aging and neurodegeneration: Recent modifications. Acta Endocrinol. **125:** 54-64.
59. KATZMAN, R. & T. SAITOH. 1991. Advances in Alzheimer's disease. FASEB J. **5:** 278-286.

Age-Dependent Changes in Calcium Currents and Calcium Homeostasis in Mammalian Neurons[a]

ALEXEJ VERKHRATSKY,[b] ANATOLY SHMIGOL, SERGEJ KIRISCHUK, NINA PRONCHUK, AND PLATON KOSTYUK

Bogomoletz Institute of Physiology
Bogomoletz St. 4
Kiev-24, GSP 252601, Ukraine

Calcium ions serve as a ubiquitous second messenger in all eukaryotic cells. Under resting conditions the cytoplasmic calcium concentration ($[Ca^{2+}]_{in}$) is maintained at a level about four orders of magnitude lower than that in extracellular solution. Cellular activation induces transient fluctuations of $[Ca^{2+}]_{in}$ which regulates various intracellular events, including neurotransmitter release, excitability, synaptic plasticity, and gene expression.[1,2] The increase in $[Ca^{2+}]_{in}$ is determined by calcium entry through plasmalemmal membrane channels and calcium liberation from intracellular stores.[3,4] The Ca^{2+} signal is terminated by removal of calcium ions from the cytoplasm by a variety of mechanisms, namely, mitochondrial sequestration, accumulation into the nonmitochondrial internal calcium stores, buffering by Ca-binding proteins, and Ca^{2+} extrusion into the extracellular milieu via plasmalemmal ATP-dependent Ca^{2+} pumps and/or Na^+/Ca^{2+} exchangers. (For a review see refs. 5 and 6.)

Evidence is accumulating that brain aging is accompanied by substantial alteration in the mechanisms of calcium homeostasis.[7-11] In particular, the increase in cytosolic calcium concentration as well as the decrease in voltage-induced calcium influx associated with aging were reported from experiments on brain synaptosomes.[2,12,13] Unfortunately, direct measurement of $[Ca^{2+}]_{in}$ in aged mammalian neurons was not done yet. Similarly, measurements of calcium currents revealed controversial results; some authors described a decrease in calcium currents in senile hippocampal neurons,[14] whereas others observed an increase in this current in hippocampal[15,16] and molluscan neurons.[17]

In the present paper the results of experiments designed to investigate the cytoplamic calcium concentration as well as the activity of voltage-activated calcium channels in aged neurons are reported.

METHODS

Male Wistar rats of three age groups (neonatal, 2-7 days; adults, 7 months; and old, 30 months) were used throughout this study. For studying calcium homeostasis

[a] This research was partially supported by a Bayer AG research grant (to A.V. and P.K.).

[b] To whom all correspondence should be addressed

in aged brain two experimental models were used: (1) freshly isolated neurons from various portions of rat nervous system, and (2) primarily cultured neurons from the lumbar dorsal root ganglia (DRG) of rats.

Cultures of DRG neurons were prepared as described elsewhere.[18,19] Freshly isolated cells were obtained from different portions of the rat nervous system, namely, dorsal root ganglia, frontal neocortex, CA1 hippocampal region, and nucleus cuneatus. Cell isolation was achieved by enzymatic treatment of brain slices that contained regions of interest and subsequent mechanical isolation under microscopic control.

For $[Ca^{2+}]_{in}$ measurements neurons were loaded with calcium-sensitive probes indo-1/AM or fura-2/AM as described elsewhere.[20,21] Fluorescent signals were recorded with a pair of photomultipliers[21,22] or video imaging microscopy based on image dissector tube technology.[18]

Calcium currents were studied using the patch clamp technique in a whole cell or cell-attached configuration.[23] Current and voltage signals were amplified with conventional electronics (EPC-7 amplifier, List Electronics, Darmstadt, Germany), filtered at 1-3 kHz (8 pole Bessel, -3dB), and sampled at 2-10 kHz by a 12-bit, 125-kHz DMA Labmaster interface (Axon Instruments, USA) connected to an AT-compatible computer system. Recordings were performed at room temperature.

For controlling neuron viability at the beginning of the experiment immediately after establishing a whole cell recording configuration and before the cell was dialyzed, the resting membrane potential (RP) of the neuron under study was determined in a current clamp mode. Similarly, after the end of $[Ca^{2+}]_{in}$ measurements, RP was determined using a conventional microelectrode technique. If the resting potential was lower than -40 mV, the cell was discarded.

Basic Tyrode solution contained (in mM): NaCl, 140; KCl, 5.4; $CaCl_2$, 1.8; $MgCl_2$, 1.1; HEPES/NaOH, 10; pH 7.4. The sodium-free extracellular solution contained (in mM): $CaCl_2$, 1.8; $MgCl_2$, 2; *N*-methyl-D-glucamine chloride or choline chloride, 130; TEA-Cl, 10; HEPES/TrisOH, 10; pH 7.4. In calcium-free solution $CaCl_2$ was omitted, $MgCl_2$ was increased to 2 mM, and 0.2 mM EGTA was added, yielding an estimated Ca^{2+} concentration of about 30 nM. High potassium solution containing different K^+ concentrations was equivalent to the Tyrode solution just described except that part of Na^+ was replaced by K^+. The pipette solution dialyzing the cell interior contained (in mM): Cs-Aspartate, 60; CsCl, 60; $MgCl_2$, 4; HEPES/CsOH, 10; EGTA, 10; pH 7.2. For some recordings we used a pipette solution that had been optimized to prevent the run-down of Ca^{2+} currents; it contained, in addition, cyclic AMP (10^{-5} M), Na_2ATP (3 mM), and $MgCl_2$ increased to 5 mM. For monitoring resting and action potentials in cultured DRG neurons, patch pipettes were filled with potassium intracellular solution (in mM): KCl, 140; $MgCl_2$, 4; EGTA, 10; HEPES/KOH, 10; pH 7.2

For experiments in cell-attached patch clamp configuration the pipette solution was as follows (in mM): $CaCl_2$, 60; $MgCl_2$, 1.1; TEA-Cl, 50; HEPES/CsOH, 10; pH 7.2. Fura-2/AM and Indo-1/AM were obtained from Molecular probes, Eugene, Oregon, and all other chemicals were from Sigma Chemical Co. (USA).

RESULTS

Resting Free Calcium Level in Adult and Old Neurons

The level of intracellular free calcium was significantly higher in neurons isolated from old rats, whereas a minor difference in resting $[Ca^{2+}]_{in}$ value was noted between cells obtained from neonatal and adult animals. FIGURE 1 summarizes the age-dependent changes in resting $[Ca^{2+}]_{in}$ levels recorded from freshly isolated peripheral and central neurons. In cultured sensory neurons a similar increase in resting $[Ca^{2+}]_{in}$ was found; the mean value of $[Ca^{2+}]_{in}$ in the cytoplasm of cultured sensory neurons obtained from neonatal rats was 77 ± 12 nM (n = 57) and from adult animals, 96 ± 23 nM (n = 17), whereas the level of free intracellular Ca^{2+} estimated for DRG neurons isolated from old rats was 207 ± 37 nM (n = 21).[18] In both adult and old neurons the resting free calcium was distributed practically uniform through the cellular soma without any sign of "hot spots" of Ca^{2+}.

Depolarization-Induced $[Ca^{2+}]_{in}$ Transients

The influx of calcium ions through voltage-operated channels is one of the most important pathways underlying generation of the calcium signal. In our experiments, depolarization-induced calcium influx was produced by either changing the bath solution from normal Tyrode solution containing 5.6 mM potassium to high K^+ (50 mM of potassium) solution or extracellular application of 200-400 μM of excitatory amino acid glutamate.

Cell challenging by high K^+ solution brings the membrane potential to -20-15 mV, thereby opening voltage-operated calcium channels. In our experiments the opening of these channels in response to high K^+ depolarization caused a fast and prominent increase in $[Ca^{2+}]_{in}$ in both peripheral and central neurons. To specify the source of Ca^{2+} ions producing the observed $[Ca^{2+}]_{in}$ transients, several experiments were performed. Potassium depolarization in Ca^{2+}-free media failed to elicit any rise in $[Ca^{2+}]_{in}$ in all types of neurons. Furthermore, organic Ca^{2+} channel blockers nifedipin and verapamil effectively inhibited elevation of $[Ca^{2+}]_{in}$ produced by application of high K^+ solution. These findings clearly show that depolarization-induced calcium transients are produced by calcium influx through voltage-operated calcium channels.

The application of high K^+ extracellular solution increased $[Ca^{2+}]_{in}$ in neurons obtained from all three age groups, although the quantitative characteristics of depolarization-induced $[Ca^{2+}]_{in}$ transients markedly differed in adult and old nervous cells.

The rise in $[Ca^{2+}]_{in}$ in response to potassium depolarization in neonatal and adult neurons was considerably faster and reached significantly higher levels than did cells from old animals (FIGS. 2 and 3). The increase in intracellular calcium concentration following application of high K^+ solutions reached micromolar levels in neurons from young rats, whereas in neurons isolated from old animals the amplitudes of depolarization-induced $[Ca^{2+}]_{in}$ transients were two to three times lower. The initial phase of depolarization-triggered $[Ca^{2+}]_{in}$ transients was also significantly steeper in neurons obtained from neonatal and adult rats than in old ones.

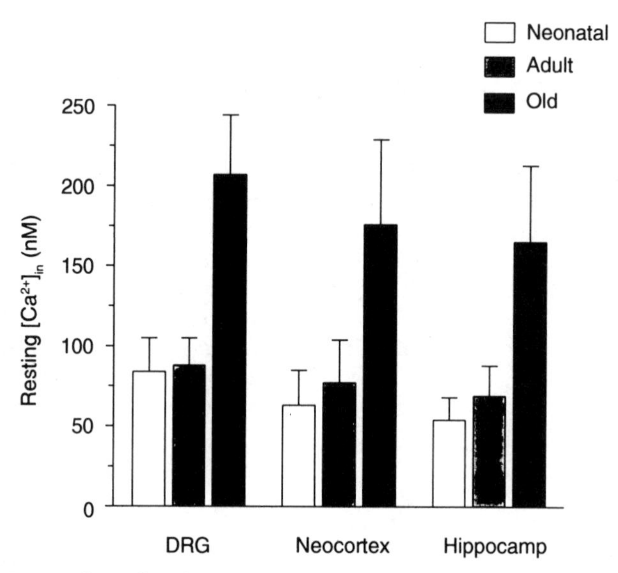

FIGURE 1. Mean values of basal cytoplasmic calcium concentration in freshly isolated neurons from rats of different age groups. Bars correspond to SD. Number of cells tested was: (1) for DRG neurons: neonatal $n = 47$; adult, $n = 56$; old, $n = 54$; (2) for neocortical neurons: neonatal, $n = 220$; adult, $n = 94$; old, $n = 103$; and (3) for hippocampal neurons: neonatal, $n = 207$; adult, $n = 87$; old, $n = 137$.

As an alternative way to induce depolarization of central neurons we used the extracellular application of 200-400 μM of glutamate. The latter is believed to open cationic channels[24] generating an inward current and subsequent neuronal depolarization. The glutamate-induced $[Ca^{2+}]_{in}$ transients in old neocortical and hippocampal neurons were significantly smaller and slower than those in neurons isolated from neonatal and adult animals (FIG. 2C).

On the basis of all these findings we suggest that the observed decrease in the depolarization-induced $[Ca^{2+}]_{in}$ elevation in old neurons indicates specific age-dependent changes in calcium currents.

Downregulation of Calcium Channel Expression in Late Ontogenesis

Lack of Low-Threshold Calcium Current in Old Neurons. The coexistence of multiple types of Ca^{2+}-selective channels was shown in various types of mammalian neurons.[25-27] We examined the types of calcium currents present in cultured dorsal root ganglion neurons from different age groups using conventional voltage-clamp protocol for separation of high-threshold (I_{HT}) and low-threshold (I_{LT}) calcium current components.

The main finding was an age-dependent disappearance in low-threshold calcium current: neurons from neonatal and adult animals possessed both I_{LT} and I_{HT}, whereas neurons from old rats lacked I_{LT}. FIGURE 4 shows calcium currents measured in neonatal DRG neurons in response to increasing step depolarizations from a holding potential of -80 mV together with corresponding current-voltage curves. They evi-

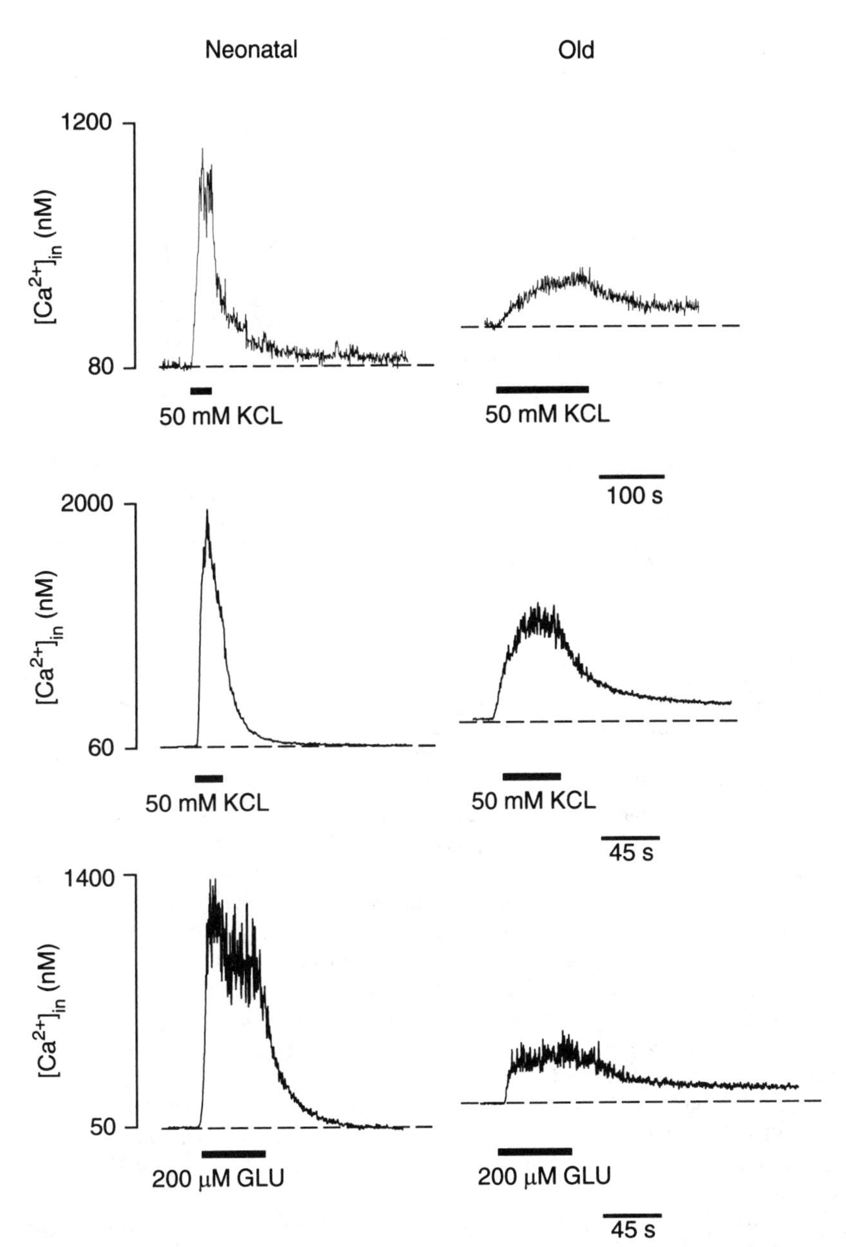

FIGURE 2. Examples of depolarization-induced $[Ca^{2+}]_{in}$ transients recorded from rat sensory **(A)** and neocortical **(B,C)** neurons from neonatal and old rats. Cell depolarization was achieved by extracellular application of high-potassium solutions **(A,B)** or external administration of 200 μM glutamate **(C)**.

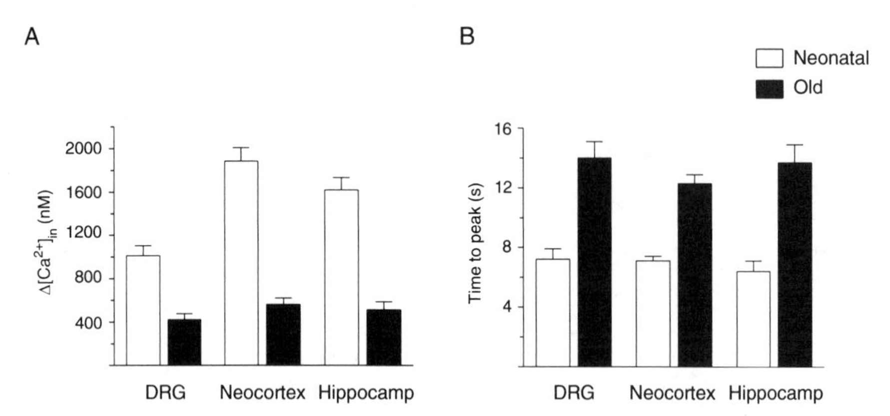

FIGURE 3. Mean values of amplitudes **(A)** and time of reaching the peak **(B)** of depolarization-induced calcium transients in cultured DRG and freshly isolated neocortical and hippocampal neurons from neonatal and old rats. *Bars* correspond to SEM. Number of cells tested was: (1) for DRG neurons: neonatal, $n = 21$; old, $n = 16$; (2) for neocortical neurons: neonatal, $n = 47$; old, $n = 43$; and (3) for hippocampal neurons: neonatal, $n = 59$; old, $n = 39$.

dently show the presence of two calcium current components with different kinetics and voltage dependence. The low-threshold component could be activated at membrane potentials more positive than -40 mV and reached a maximum at -20mV. This current had relatively fast inactivation kinetics and was strongly dependent on the holding potential (V_h); shifting V_h to values more positive than -50 mV led to its complete disappearance. Such an effect is clearly shown in FIGURE 3C which represents a family of whole-cell currents in response to ramp depolarizations from progressively decreasing V_h. The disappearance of the first hump on the ramp response indicates depression of I_{LT} because of steady-state inactivation. The high-threshold component (I_{HT}) of calcium current activated with depolarizations to potentials positive to -20 mV and reached maximal values between 0 and +10 mV. It displayed much slower inactivation and it remained when the cell was held at -50 or -40 mV. About 65% of all examined neurons from neonatal rats possessed both I_{LT} and I_{HT}; the amplitude of I_{LT} reached in average 40% of amplitude of high-threshold calcium current. Some neurons expressed only the I_{LT}.

In the second age group (7-month-old rats), expression of I_{LT} channels was lower; only 30% of neurons examined displayed this calcium current component. However, in this age group it was still possible to find a number of neurons that expressed only the low-threshold calcium current.

Finally, in the group of aged neurons we found not a single cell (from 66 tested) with measurable low-threshold calcium current. Peak current-voltage relations as well as ramp I-V curves (FIG. 5) revealed the presence of only a high-threshold calcium current.

From these observations we conclude that a progressive decrease in low-threshold calcium channel expression occurs during aging, which in the long run causes complete disappearance of I_{LT} in aged sensory neurons.

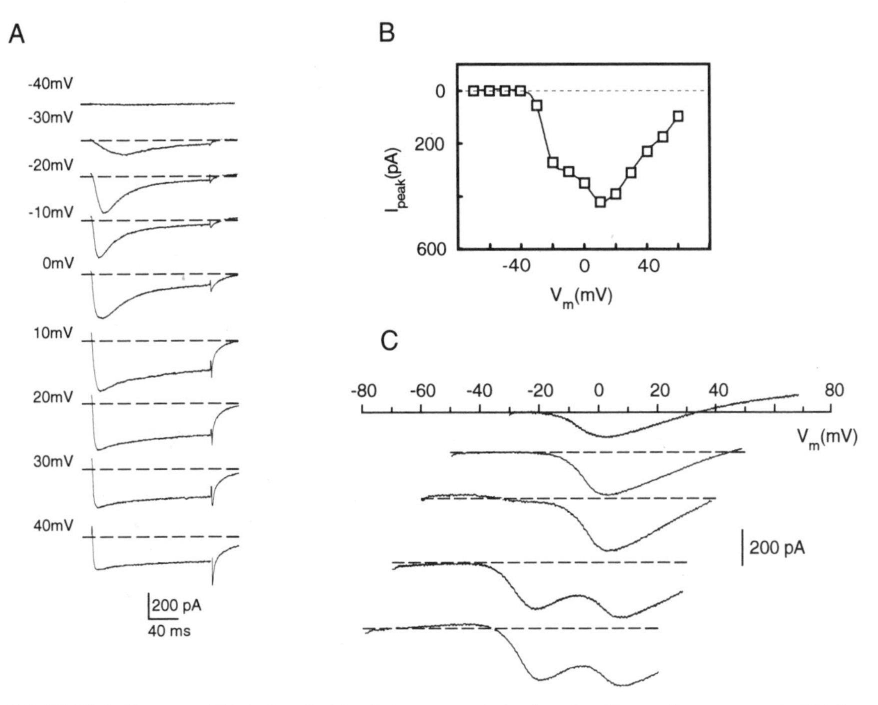

FIGURE 4. Low- and high-threshold calcium currents in dorsal root ganglion neuron isolated from neonatal rat (from ref. 19). **(A)** Representative family of calcium current records. Holding potential -80 mV, 200 ms voltage clamp pulses to potentials indicated near current traces. Note clearly distinguished transient and sustained current components. **(B)** Peak current (I_{peak}) to voltage (V_m) relationship for the cell illustrated in **A**. **(C)** Family of currents recorded in response to ramp stimulation (amplitude 100 mV, duration 200 ms) from holding potentials between -80 and -30 mV. Note the disappearance of the first hump in the ramp I-V curves at more positive holding potentials.

Age-Dependent Decrease in High-Threshold Calcium Current

To compare the main characteristics of the high-threshold calcium currents we chose neurons that expressed only the I_{HT}. The I_{HT} of dorsal root ganglion neurons were recorded at holding potentials between -80 and -60 mV. The activation threshold for I_{HT} in neurons from all age groups was around -20 mV, and I-V curves peaked at about +10 mV. No age-dependent differences in kinetics of high-threshold calcium currents were observed.

However, the density of I_{HT} in aged DRG neurons was significantly lower; the high-threshold Ca^{2+} current density was 28.4 ± 6.3 pA/pF (n = 54) in neonatal, 39.1 ± 7.2 pA/pF in adult (n = 62), and only 11.0 ± 4.6 pA/pF (n = 64) in aged DRG neurons. Changes in the density of low- and high-threshold calcium channels in ontogenesis are summarized in FIGURE 6.

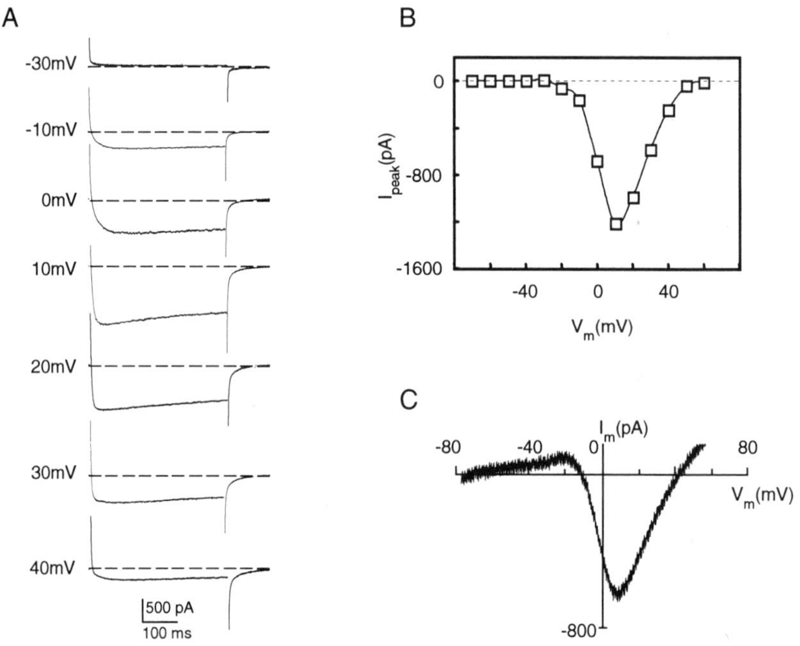

FIGURE 5. One component of calcium current in the membrane of dorsal root ganglion neuron, isolated from 30-month-old rat (from ref. 19). **(A)** Representative family of calcium current records. Holding potential -80 mV, 400 ms voltage clamp pulses to potentials indicated near current traces. Note the existence of only one current component. **(B)** Peak current (I_{peak}) to voltage (V_m) relationship for the cell illustrated in **A**. **(C)** Ramp current to voltage curve from the holding potential -80 mV.

Elementary Calcium Currents in Neurons from Different Age Groups

To check the possible alterations in the properties of calcium channels with aging, we also analyzed the basic properties of elementary calcium currents in neurons from different age groups. These experiments showed practically no changes in Ca^{2+} channel characteristics with aging (FIG. 7). Single channel conductance was (for 60 mM of Ca^{2+} in recording micropipette) 16.0 ± 2.7 pS (n = 9) in neonatal, 16.2 ± 1.7 pS (n = 11) in adult, and 16.4 ± 1.2 pS (n = 12) in old neurons. We also found no significant difference in the kinetic properties of unitary high-threshold calcium currents in various age groups.

Consistent with data from whole-cell observations, single-channel experiments also failed to reveal low-threshold activated elementary calcium currents in neurons isolated from old animals (in 21 tested patches), whereas it was possible to record regularly the activity of single low-threshold calcium channels in the membrane of dorsal root ganglion neurons obtained from neonatal and adult animals (FIG. 7A).

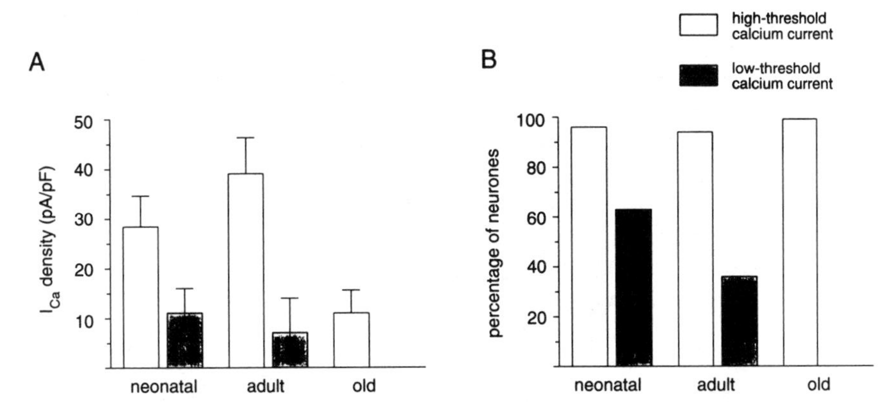

FIGURE 6. Density of low- and high-threshold calcium currents in rat DRG neurons isolated from animals of different age groups (**A**), and percentage of neurons that expressed these components of calcium current (**B**) (from ref. 19).

Calcium Channels in Old Neurons Are Not Regulated by Cytoplasmic Phosphorylation

The functional activity of calcium channels is controlled by cytoplasmic metabolically dependent processes. Exchange of soluble components between the cytoplasm and the recording pipette during intracellular dialysis leads to the disappearance of this metabolic support of calcium channels and a progressive fall (rundown) of I_{HT} amplitude.[26,27] In various cell types this decrease in calcium current amplitude during intracellular perfusion could be slowed down or even reversed by the addition into the intracellular solution of substances involved in channel-protein phosphorylation. We found that metabolic dependence of high-threshold calcium current which is very pronounced in dorsal root ganglion neurons isolated from neonatal rats became insignificant in neurons obtained from old animals. Moreover, the rundown of I_{HT} was clearly slower in neurons obtained from old animals. Results of such experiments are shown at FIGURE 8 where the rundown of calcium currents measured in the presence or absence of cyclic AMP and ATP in intracellular solution are compared. Mean rundown half-time was practically unaffected by cyclic AMP and ATP in aged DRG neurons, whereas it was clearly slower in neonatal neurons treated by intracellularly applied cyclic AMP and ATP. These findings are in line with the data of Fedulova *et al.*[28] who also described a decrease in the ability of cyclic AMP to restore calcium currents in most adult neurons in comparison with neonatal ones. Obviously, with aging such an effect becomes even more prominent.

The data obtained clearly show that neuronal aging is associated with a prominent decrease in voltage-induced calcium influx due to the reduction in the number of active calcium channels. This reduction in transmembrane calcium influx results in depression of the Ca^{2+} rising phase of the cytoplasmic calcium signal. Therefore, it became of interest to investigate also the age-dependent changes in the mechanisms of calcium signal termination.

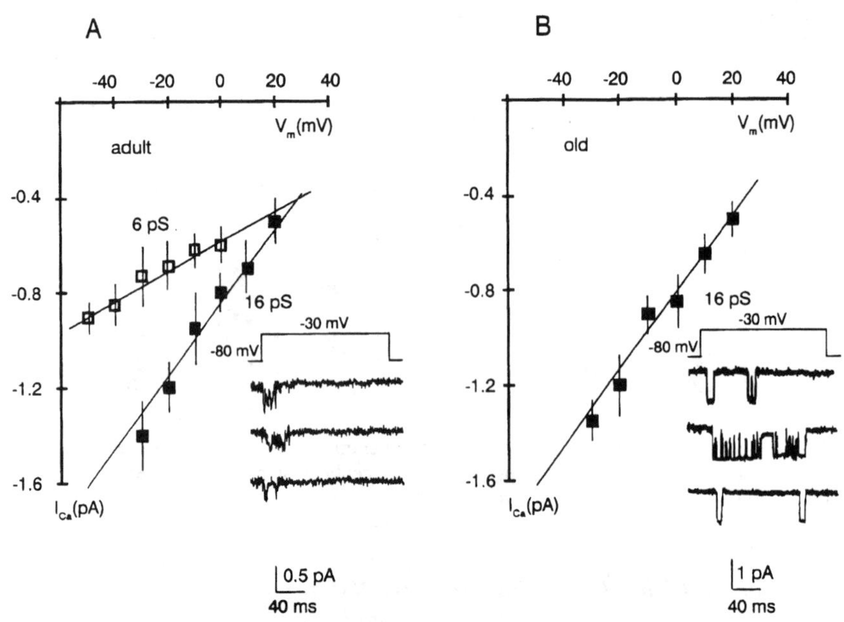

FIGURE 7. Conductance of elementary calcium currents in DRG neurons isolated from adult and old rats (from ref. 19). Single calcium currents were measured in a cell-attached patch clamp configuration at 60 mM of Ca^{2+} in recording micropipette. Resting potential of neurons was zeroed by cell incubation in hyperpotassium external solution. Unitary current amplitudes are plotted against membrane potential. **(A)** I-V curves for low- and high-threshold single channel currents measured in adult DRG neurons. *Solid lines* indicate the slope conductance; slope conductances were 6.1 ± 0.6 pA (n = 9) for I_{LT} channels and 16.2 ± 1.7 pS (n = 11) for I_{HT} channels. **Inset:** Examples of activity of low-threshold single calcium channels measured in response to step depolarization as indicated. **(B)** I-V curves for high-threshold single channel currents measured in old DRG neurons; slope conductance (16.4 ± 1.2 pS, n = 12) indicated by *solid line.* **Inset:** Examples of activity of the high-threshold single calcium channels measured in response to step depolarization as indicated.

Age-Dependent Changes in $[Ca^{2+}]_{in}$ Recovery after Depolarization

The recovery of $[Ca^{2+}]_{in}$ to the basal level after depolarization-triggered calcium elevation was much slower in neurons obtained from old rats (Figs. 2 and 9). Moreover, the restitution of $[Ca^{2+}]_{in}$ after repolarization in cells from neonatal and adult rats was complete; in contrast, in old neurons $[Ca^{2+}]_{in}$ remained above the resting level (~120%) during a long time. Usually we had to wait up to 10–15 minutes to achieve complete recovery of $[Ca^{2+}]_{in}$ after the end of depolarization.

The recovery of $[Ca^{2+}]_{in}$ after its elevation is achieved by a combination of its efflux through the cellular membrane (via Ca^{2+} pumps and/or Ca^{2+} exchangers) and sequestration by intracellular stores. The latter could be directly investigated using pharmacological tools known to modulate their functional state.

Caffeine-Induced $[Ca^{2+}]_{in}$ Transients in Adult and Old Neurons

To study the age-dependent changes in neuronal calcium stores, we used caffeine, the widely known Ca^{2+} liberator from endoplasmic reticulum. It is generally accepted that caffeine interacts with Ca^{2+}-activated Ca^{2+} channels of the endoplasmic reticulum, making them more sensitive to Ca^{2+}, so that the Ca^{2+} release mechanism can be activated even at resting levels of $[Ca^{2+}]_{in}$.[21,29,30] Caffeine in concentrations of 10-20 mM applied to both adult and old DRG neurons caused a fast and pronounced elevation of $[Ca^{2+}]_{in}$. The mean values of the caffeine-induced increase in intracellular Ca^{2+} concentration was 810 ± 46 nM (n = 14) in adult and 927 ± 81 nM (n = 17) in old DRG neurons. Although this difference was not significant, the amplitude of caffeine-induced $[Ca^{2+}]_{in}$ transients was somewhat larger in old neurons, suggesting probably a higher content of releasable calcium in caffeine-sensitive stores.

In contrast to DRG in central neurons we found prominent age-dependent differences in the behavior of caffeine-induced $[Ca^{2+}]_{in}$ transients. Previously we showed that in neurons isolated from the central nervous system of young rats caffeine-sensitive stores have only a minute amount of releasable calcium under resting conditions; nevertheless, they can be rapidly but transiently charged by depolarization-triggered calcium entry.[22] This is demonstrated in FIGURE 10A. The initial application of caffeine to adult neocortical neurons evoked only a low-amplitude $[Ca^{2+}]_{in}$ elevation, whereas after cell depolarization the amplitude of subsequent caffeine-induced $[Ca^{2+}]_{in}$ transients drastically increased presumably due to the charging of intracellular stores by releasable calcium. In old neurons the releasable calcium content of caffeine-sensitive stores initially was high enough to produce $[Ca^{2+}]_{in}$ transient in response to caffeine, and the charging depolarization did not substantially change the amplitudes of these transients (FIG. 10B and C).

The velocity of rise in caffeine-induced $[Ca^{2+}]_{in}$ transients was practically the same in both adult and old neurons, whereas the recovery of caffeine-evoked $[Ca^{2+}]_{in}$ elevation was usually slower in old neurons.

DISCUSSION

In the present study we performed a direct comparison of intracellular calcium concentration and voltage-activated calcium currents in neurons of neonatal, adult, and old rats. Our experiments demonstrated prominent changes in the mechanisms of calcium homeostasis associated with neuronal aging. Using the calcium-sensitive dyes in combination with microfluorometric recordings, we demonstrated a significant increase in basal calcium concentration in the cytoplasm of old peripheral and central neurons. Moreover, we found that mechanisms responsible for both raising phase and termination of calcium signal are markedly altered in aged neurons. Voltage-dependent calcium mobilization in old neurons is greatly reduced due to a decrease in the number of functional calcium channels. This finding is in close agreement with previously reported data from experiments on brain synaptosomes, which demonstrated an age-dependent increase in resting $[Ca^{2+}]_{in}$ as well as a decrease in voltage-induced calcium influx.[12,31,32] We can speculate that the persistently elevated intracellular calcium concentration in old neurons may serve as a signal that induces the

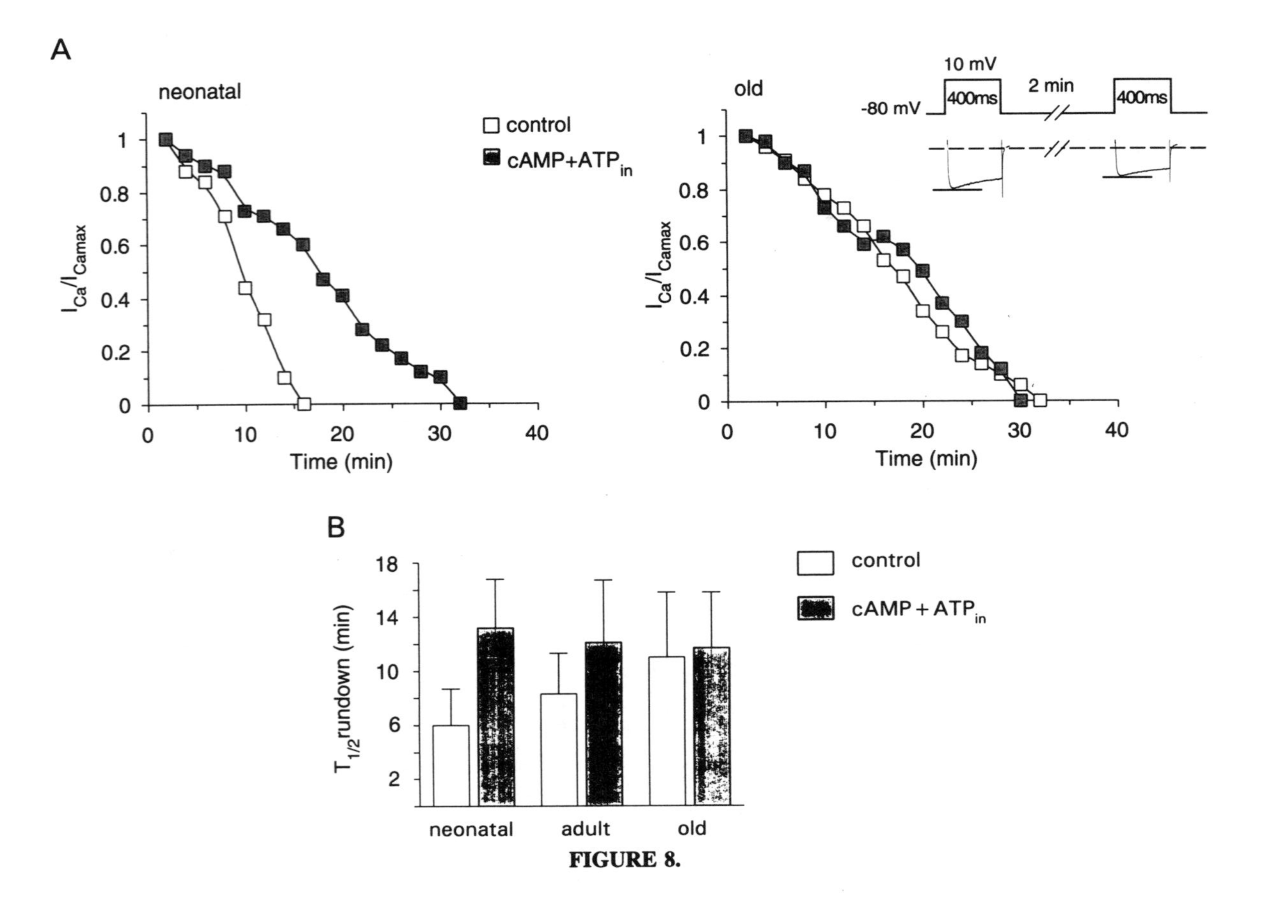
A
neonatal
old
control
cAMP+ATP$_{in}$
I_{Ca}/I_{Camax}
Time (min)
10 mV
-80 mV
400ms
2 min
B
$T_{1/2}$rundown (min)
neonatal
adult
old

FIGURE 8.

FIGURE 8. Rundown of high-threshold calcium currents and effect of cyclic AMP (modified from ref. 19). **(A)** Comparison of the rundown of the high-threshold calcium current in DRG neurons obtained from neonatal and old rat in the absence (*open squares*) and presence (*filled squares*) of cyclic AMP and ATP in intracellular solution. Currents were recorded in 2-minute intervals after establishment of the whole cell recording configuration. Currents were activated by voltage jumps to +10 mV (corresponding to the maximum of I_{HT} I-V-curve) from holding potential of -80 mV (stimulation protocol is shown in the **inset**). Peak currents were normalized to the current measured after establishment of the whole cell recording configuration and complete washout of potassium currents and plotted as a function of recording time. **(B)** Dependence of the rundown half-time (times at which a 50% decrease in I_{HT} occurred) from the presence of cyclic AMP and ATP in the intracellular solution in DRG neurons isolated from rats of different age groups.

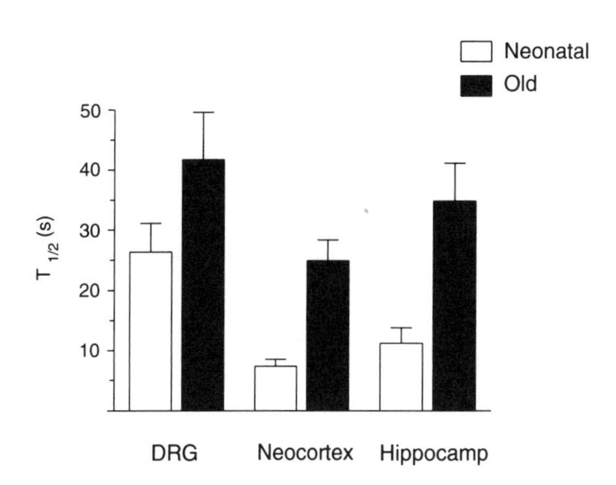

FIGURE 9. Mean values of depolarization-induced $[Ca^{2+}]_{in}$ transients recovery half-times in rat sensory and central neurons obtained from neonatal and old rats. Bars correspond to SD. Number of cells tested was: (1) for DRG neurons: neonatal, n = 21; old, n = 16; (2) for neocortical neurons: neonatal, n = 47; old, n = 43; (3) for hippocampal neurons: neonatal, n = 59; old, n = 39.

downregulation of Ca^{2+} channel expression, as shown in some model experiments.[33] We also found that in old neurons calcium channels lose their sensitivity to cytoplasmic phosphorylation. In neonatal and in adult DRG neurons the cyclic AMP-dependent phosphorylation seems to prevent the transition of calcium channels between active and ''silent'' forms, whereas in aged cells this mechanism does not work. We demonstrated also that the low threshold component of calcium current disappeared in aged sensory neurons which might significantly change kinetics of the calcium signal.

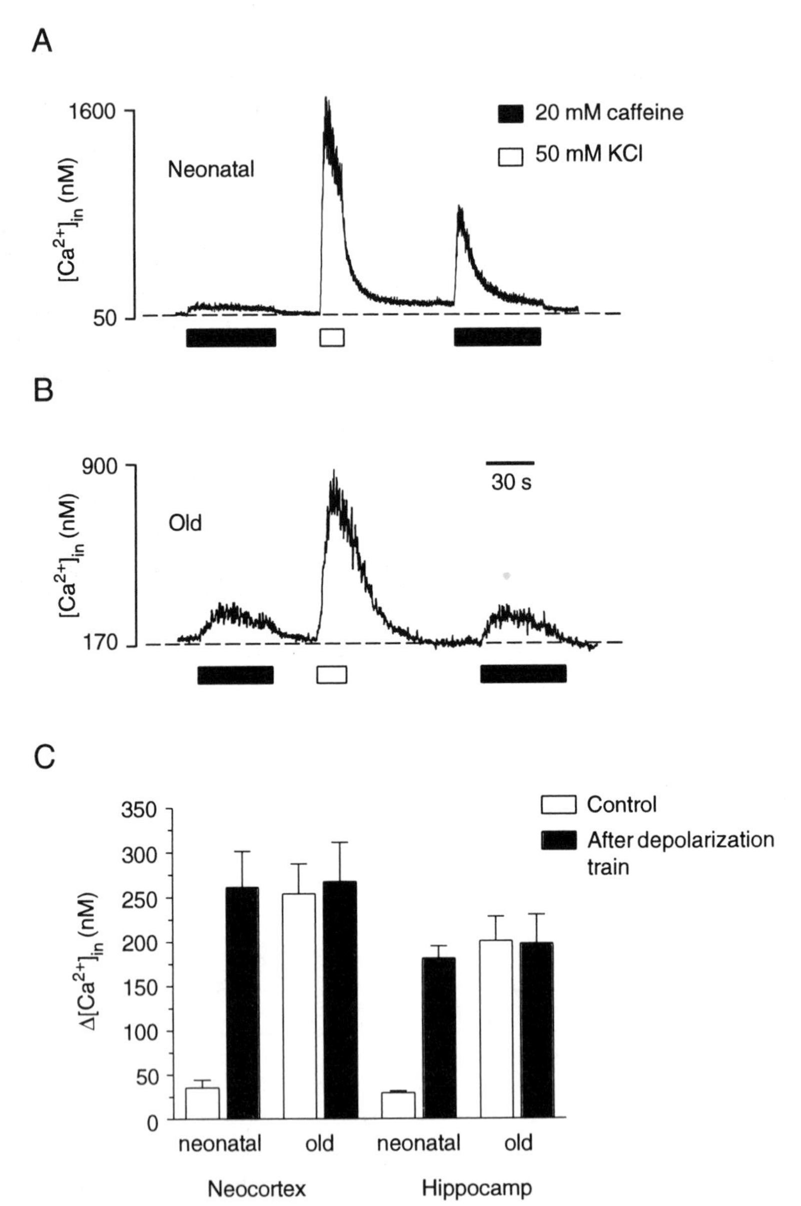

FIGURE 10. Caffeine-induced calcium transients in central neurons obtained from neonatal and old rats. Caffeine-induced $[Ca^{2+}]_{in}$ transients recorded before and after charging depolarization in neocortical neonatal (**A**) and old (**B**) neurons are shown. In **C** the mean amplitudes of initial caffeine-triggered $[Ca^{2+}]_{in}$ transients and the same amplitudes after cell depolarization are summarized for neocortical and hippocampal neurons obtained from neonatal and old rats. Bars correspond to SEM. Number of cells tested was: (1) for neocortical neurons: neonatal, n = 23; old, n = 19; (2) for hippocampal neurons: neonatal, n = 49; old, n = 27.

The recovery of cytoplasmic calcium to the resting level appeared to be much slower in aged neurons. This could be explained in terms of age-dependent changes in calcium accumulation by intracellular stores, inhibition of plasmalemmal calcium extrusion, and reduction of cytoplasmic calcium buffer capacity. Our data suggest that the amount of calcium stored in the caffeine-sensitive compartment of endoplasmic reticulum is significantly higher in old central neurons, presumably due to overload of these stores by releasable calcium in aged neurons. We may assume that overloaded stores could not effectively participate in calcium removal after periods of neuronal activity. Indeed, depolarization-induced calcium entry caused a significant loading of caffeine-sensitive stores in neurons from young animals, whereas it failed to increase the releasable calcium content in neurons obtained from old rats. Therefore, we suggest that the reduction in calcium sequestration by internal stores as well as the previously reported age-dependent decrease in cytoplasmic Ca^{2+} buffering, mitochondrial Ca^{2+} accumulation, and plasmalemmal calcium efflux by Ca^{2+} pumps and Ca^{2+} exchangers[11,34,35] are responsible for age-dependent alteration in $[Ca^{2+}]_{in}$ recovery in neuronal cells. However, a detailed estimation of the importance of these systems in alteration of calcium homeostasis in old neurons needs further investigation.

REFERENCES

1. SMITH, S. J. & G. J. AUGUSTUNE. 1988. Calcium ions, active zones, and synaptic transmitter release. TINS **11:** 458-464.
2. MALENKA, R. C., J. A. KAUER, D. J. PERKEL & R. A. NICOLL. 1989. The impact of postsynaptic calcium on synaptic transmission: Its role in long term potentiation. TINS **12:** 444-450.
3. HENZI, V. & A. B. MCDERMOTT. 1992. Characteristics and function of Ca^{2+} and inositol 1,4,5-trisphosphate-releasable stores of Ca^{2+} in neurones. Neuroscience **46:** 251-274.
4. TSIEN, R. W. & R. Y. TSIEN. 1990. Calcium channels, stores and oscillations. Annu. Rev. Cell. Biol. **6:** 715-760.
5. CARAFOLI, E. 1987. Intracellular calcium homeostasis. Annu. Rev. Biochem. **56:** 395-433.
6. PETERSEN, O. H., C. C. H. PETERSEN & H. KASAI. 1994. Calcium and hormone action. Annu. Rev. Physiol. **56**, in press.
7. MICHAELIS, M. L., K. JOHE & T. E. KITOS. 1984. Age-dependent alterations in synaptic membrane systems for Ca^{2+} regulation. Mech. Aging. Dev. **25:** 215-225.
8. LANDFIELD, P. 1989. Calcium homeostasis in brain ageing and Alzheimer's disease. *In* Diagnosis and Treatment of Senile Dementia. M. Bergener & B. Reisberg, Eds. :276-28. Springer. Berlin.
9. GIBSON, G. E. & C. PETERSON. 1987. Calcium and the ageing nervous system. Neurobiol. Ageing **8:** 329-343.
10. PETERSON, C., R. RATAN, M. SHELANSKI & J. GOLDMAN. 1989. Changes in calcium homeostasis during ageing and Alzheimer's disease. Ann. N. Y. Acad. Sci. **568:** 262-270.
11. MARTINEZ-SERRANO, A., P. BLANCO & J. SATRUSTEGUI. 1992. Calcium binding to the cytosol and calcium extrusion mechanisms in intact synaptosomes and their alteration with ageing. J. Biol. Chem. **267:** 4672-4679.
12. PETERSON, C. & G. E. GIBSON. 1983. Aging and 3,4-diaminopyridine alter synaptosomal calcium uptake. J. Biol. Chem. **258:** 11482-11486.
13. LESLIE, S. W., L. J. CHANDLER, E. M. BARR & R. P. FARRAR. 1985. Reduced calcium uptake by rat brain mitochondria and synaptosomes in response to ageing. Brain Res. **329:** 177-183.

14. REYNOLDS, J. N. & P. L. CARLEN. 1989. Diminished calcium currents in aged hippocampal dentate gyrus granule neurones. Brain Res. **508:** 384-390.
15. LANDFIELD, P. W., L. W. CAMPBELL, S.-Y. HAO & D. S. KERR. 1989. Aging-related increases in voltage-sensitive, inactivating calcium currents in rat hippocampus: Implications for mechanisms of brain aging and Alzheimer's disease. Ann. N. Y. Acad. Sci. **568:** 95-105.
16. PITLER, T. A. & P. W. LANDFIELD. 1990. Aging-related prolongation of calcium spike duration in rat hippocampal slice neurones. Brain Res. **508:** 1-6.
17. FROLKIS, V. V., O. A. MARTYNENKO & A. N. TIMCHENKO. 1991. Potential-dependent Ca channels of neurones in the mollusc *Limnea stagnalis* in ageing: Effect of norepinephrine. Mech. Aging. Dev. **58:** 75-83.
18. KIRISCHUK, S., N. PRONCHUK & A. VERKHRATSKY. 1992. Measurements of intracellular calcium in sensory neurons of adult and old rats. Neuroscience **50:** 947-951.
19. KOSTYUK, P., N. PRONCHUK, A. SAVCHENKO & A. VERKHRATSKY. 1993. Calcium currents in aged rat dorsal root ganglion neurones. J. Physiol. (Lond.) **461:** 467-483.
20. GRYNKIEWICZ, G., M. POENIE & R. Y. TSIEN. 1985. A new generation of Ca^{2+} indicators with greatly improved fluorescent properties. J. Biol. Chem. **260:** 3440-3450.
21. USACHEV, Y., A. SHMIGOL, N. PRONCHUK, P. KOSTYUK & A. VERKHRATSKY. 1993. Caffeine-induced calcium release from internal stores in cultured rat sensory neurones. Neuroscience **57:** 845-859.
22. SHMIGOL, A., S. KIRISCHUK, P. KOSTYUK & A. VERKHRATSKY. 1994. Different properties of caffeine-sensitive Ca^{2+} stores in peripheral and central mammalian neurones. Pflügers Arch. **426:** 174-176.
23. HAMILL, O. P., A. MARTY, E. NEHER, B. SAKMANN & F. J. SIGWORTH. 1981. Improved patch clamp techniques for high-resolution current recording from cell and cell-free membrane patches. Pflügers Arch. **391:** 85-100.
24. MAYER, M. L. & G. L. WESTBROOK. 1987. The physiology of excitatory amino acids in the vertebrate central nervous system. Prog. Neurobiol. **28:** 197-276.
25. BEAN, B. P. 1989. Classes of calcium channels in vertebrate cells. Annu. Rev. Physiol. **51:** 367-384.
26. KOSTYUK, P. G. 1989. Diversity of calcium ion channels in cellular membranes. Neuroscience **28:** 253-261.
27. HESS, P. 1990. Calcium channels in vertebrate cells. Annu. Rev. Neurosci. **13:** 337-356.
28. FEDULOVA, S. A., P. G. KOSTYUK & N. S. VESELOVSKY. 1991. Ionic mechanisms of electrical excitability in rat sensory neurones during postnatal ontogenesis. Neuroscience **41:** 303-309.
29. SITSAPESAN, R. & A. J. WILLIAMS. 1990. Mechanisms of caffeine activation of single calcium-release channels of sheep cardiac sarcoplasmic reticulum. J. Physiol. (Lond.) **423:** 425-439.
30. FRIEL, D. D. & R. W. TSIEN. 1992. A caffeine- and ryanodine-sensitive Ca^{2+} store in bullfrog sympathetic neurones modulates effects of Ca^{2+} entry on $[Ca^{2+}]_{in}$. J. Physiol. (Lond.) **450:** 217-246.
31. MARTINEZ, A., J. VITORICA & J. SATRUSTEGUI. 1988. Cystolic free calcium levels increase with age in rat brain synaptosomes. Neurosci. Lett. **88:** 336-342.
32. MARTINEZ-SERRANO, A., E. BOGONEZ, J. VITORICA & J. SATRUSTEGUI. 1989. Reduction of K^+-stimulated^{45} Ca^{2+} influx in synaptosomes with age involves inactivating and noninactivating calcium channels and is correlated with temporat modifications in protein dephosphorylation. J. Neurochem. **52:** 576-584.

33. DeLorme, E. M., C. S. Rabe & R. McGee, Jr. 1988. Regulation of the number of functional voltage-sensitive Ca^{++} channels on PC12 cells by chronic changes in membrane potential. J. Pharmacol. Exp. Ther. **244:** 838-843.
34. Vitorica, J. & J. Satrustegui. 1986. The influence of age on the calcium efflux pathway and matrix calcium buffering power in brain mitochondria. Biochim. Biophys. Acta **851:** 209-216.
35. Vitorica, J. & J. Satrustegui. 1986. Involvement of mytochondria in the age-dependent decrease in calcium uptake in rat brain synaptosomes. Brain Res. **378:** 36-48.

The Calcium Rationale in Aging and Alzheimer's Disease

Evidence from an Animal Model of Normal Aging[a]

JOHN F. DISTERHOFT,[b] JAMES R. MOYER, JR., AND LUCIEN T. THOMPSON

Department of Cell and Molecular Biology
and
the Institute for Neurosciences
Northwestern University Medical School
Chicago, Illinois 60611-3008

Calcium is required for the function of all cells in the body, including neurons. Considerable research has described the function of calcium in the regulation of numerous processes including neurotransmitter release, cytoarchitecture and growth, and activation of enzyme systems including kinases and phosphatases. Calcium is intimately involved in a variety of "plastic" changes in the brain. For example, during adaptive processes such as learning and development, changes in transmembrane calcium fluxes correlate with changes in neuronal excitability and structural connectivity. Calcium thus is likely to have key roles in the cellular processes underlying aging-related changes in the brain, including normal age-associated memory impairments as well as more severe dementias, including Alzheimer's disease.

The pivotal role of calcium in so many neuronal processes dictates the need for precise regulation of its intracellular levels. Any dysregulation, however subtle, could lead to dramatic changes in normal neuronal function. Recent studies from our laboratory and those of others have implicated altered calcium influx with aging-related changes at both the behavioral and the neurophysiological levels. These findings led to and continue to support the calcium hypothesis[1,2] which posits that in the aging brain, transient or sustained increases in the average concentration of intracellular free calcium contribute to impaired function, eventually leading to cell death. The hypothesis suggests that the final common pathway that may contribute to cognitive deterioration of aging vertebrates, including persons with Alzheimer's disease or other aging-related dementias, is increased free calcium within neurons. The functional impairment that characterizes a patient at a particular time in the aging-related disease process may be relieved by reducing excessive calcium influx. Additionally, because calcium dysregulation terminating in cell death is likely to be

[a] The work was supported by NIA R01 AG08796.

[b] Address for correspondence: Dr. John F. Disterhoft, Department of Cell & Molecular Biology, Northwestern University Medical School, 303 E. Chicago Avenue, Chicago, IL 60611-3008.

a continuous process, with many cells exhibiting varying degrees of impairment, the rate of additional loss of neurons resulting from altered calcium influx may be reduced pharmacologically by calcium antagonists.

This chapter focuses on data from our laboratory illustrating the involvement of calcium in the cellular mechanisms of learning and memory. In addition, we provide evidence linking altered calcium influx with learning deficits and changes in the electrophysiological properties of hippocampal neurons using an animal model of normal aging. We shall consider the potential for calcium channel antagonists to ameliorate these aging-related changes and the relevance of these issues both to normal aging and to Alzheimer's disease.

REGULATION OF INTRACELLULAR CALCIUM AND AGING

Numerous cellular processes depend upon the maintenance of appropriate calcium concentrations for their normal function. A variety of intracellular mechanisms exist for maintaining calcium homeostasis in neurons, some of which are summarized in FIGURE 1. There are at least three major transmembrane sources of calcium influx: (1) voltage-gated calcium channels, of which there are at least four classes; (2) the NMDA receptor channel complex; and (3) activation of the Na^+/Ca^{2+} exchanger. Additional sources of free intracellular calcium come from release of intracellular stores (including organelles such as calciosomes, the endoplasmic reticulum, and mitochondria) and release from Ca^{2+}-binding proteins such as calmodulin, calbindin, and parvalbumin. Efflux of intracellular calcium occurs via ATP-driven pumps and the Na^+/Ca^{2+} exchangers. The calcium hypothesis for Alzheimer's disease posits that dysregulation of free intracellular calcium is causally linked to some of the processes that underlie the neuronal deterioration occurring in Alzheimer's disease as well as in normal aging.[1,2] The root cause for Alzheimer's disease, whether genetic, environmental, or some interaction between the two, has not been definitively established. It has been postulated that excessive influx, raised levels, or poor buffering of intracellular calcium are results, not causes, of the molecular and cellular mechanisms leading to the development of Alzheimer's disease. If the calcium hypothesis is correct, the reason that some individuals have accelerated age-related deterioration should also be amenable to empirical determination and treatment. It is critically important to empirically test the hypothesis, both in animal models and in clinical settings, and to modify the hypothesis if necessary to fit the empirical data.

Considerable evidence shows that calcium is intimately involved in cell deterioration and death in a variety of disease processes (FIG. 2). One intensively studied example is glutamate neurotoxicity that accompanies ischemic episodes.[3,4] Calcium influx is enhanced through NMDA-receptor channels and through voltage-gated calcium channels, both of which require sustained depolarization for opening. Calcium is also released from intracellular stores as a result of activation of metabotropic glutamate receptors. Free cytosolic calcium levels would be raised as a result of these processes, resulting in cellular damage from various causes including reduced cellular energy metabolism[5] and activation of catabolic enzymes including a calcium-activated neutral protease (calpain) which can degrade neuronal structural proteins.[6,7] Sustained elevation of intracellular calcium would also lead to a number of destructive cascades:

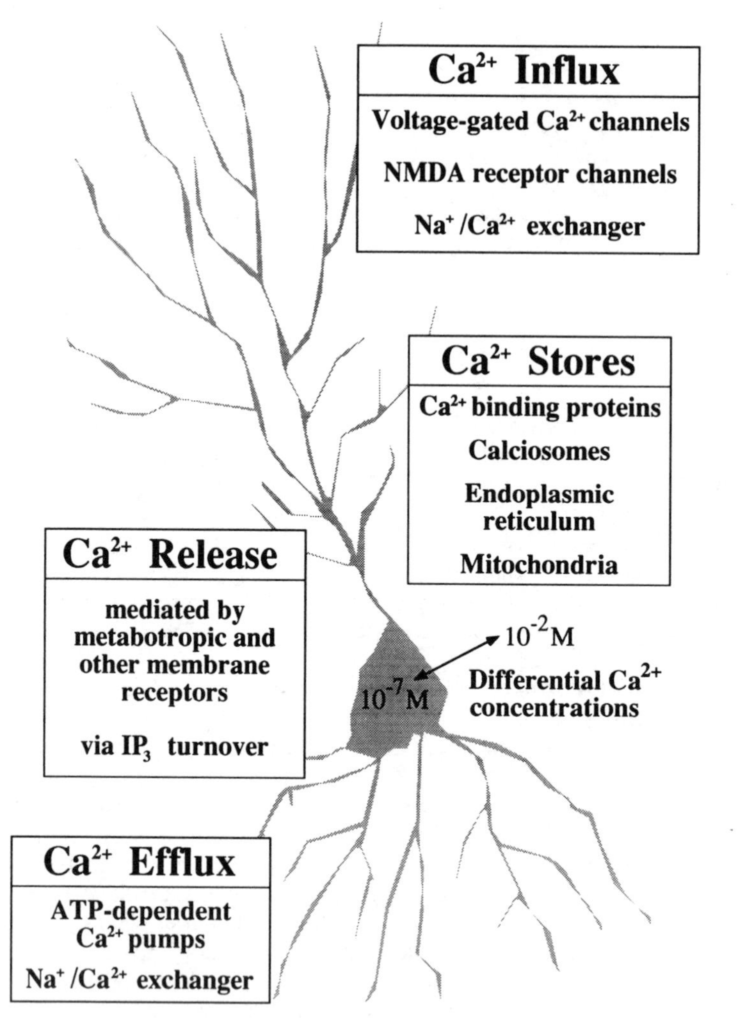

FIGURE 1. Calcium homeostasis in young neurons involves buffering of calcium from a number of sources (both intra- and extracellular) to maintain an intracellular free Ca^{2+} concentration in the nanomolar range. Calcium influx occurs via voltage-operated calcium channels (VOCCs, including the L-type channels blocked by nimodipine), via one type of excitatory amino acid receptor (*N*-methyl-D-aspartate receptors, NMDA-Rs), and via Na^+/Ca^{2+} exchange proteins. Calcium is stored intracellularly in a variety of ways, including binding with Ca^{2+}-binding proteins (CaBPs including calmodulin, calbindin, and parvalbumin), and sequestration in cellular organelles including calciosomes, the endoplasmic reticulum, and mitochondria. A number of membrane receptors activate second messenger cascades that liberate Ca^{2+} from intracellular stores, and Ca^{2+}-dependent intracellular Ca^{2+} release has been described. Calcium efflux both via energy-dependent Ca^{2+} pumps and via Na^+/Ca^{2+} exchange proteins also serve to regulate intracellular free Ca^{2+} concentrations. Disruptions of any of these processes can have profound consequences to both the neuron and the organism to which it belongs.

(1) the generation of phospholipase A_2 and then superoxide radicals, leading to cellular damage; (2) activation of phospholipase C which activates protein kinase C, leading to increased calcium channel activity, and to IP_3 generation, releasing additional calcium from internal stores; (3) generation of calcium-calmodulin-dependent protein kinase II, which phosphorylates presynaptic synapsin I, leading to further glutamate release; (4) activation of endonucleases that cause DNA fragmentation; and (5) production of nitric oxide synthase, which inhibits mitochondrial respiration, the citric acid cycle enzyme aconitase, and DNA synthesis.[3,4] Thus, it is likely that even a small disruption of normal calcium homeostasis could have devastating consequences for neurons. These kinds of cascades and the evidence for them are discussed at length in other chapters within this volume.

At the same time, numerous studies suggest a role for calcium channel blockers as treatments for various neuropathologies, including the functional sequelae of ischemic episodes.[5] For example, preclinical studies demonstrate that discrimination learning and memory deficits associated with prenatal hypoxia in developing rats is markedly reduced in pups whose mothers received oral nimodipine (a 1,4-dihydropyridine, L-type calcium channel antagonist) during induction of prenatal hypoxia.[8] In addition, animals whose mothers received nimodipine showed normal densities of serotonin and of acetylcholine staining fibers in their hippocampi, whereas hypoxic controls did not. In two different clinical trials, it was shown that nimodipine has a beneficial effect, especially on functions involving learning, in patients with chronic cerebrovascular disorder or vascular dementia.[9,10] These preclinical and clinical studies suggest that calcium channel blockers may be useful in countering the effects of the neurotoxic events that accompany ischemic episodes. As will be discussed, these findings also suggest that Ca^{2+} antagonists may be useful adjuncts to treatments for age-associated impairments including Alzheimer's disease.

CALCIUM-MEDIATED PROCESSES IN NEURONS

A variety of processes are governed by the concentration of available intracellular free Ca^{2+}. For example, during development, neurons extend growth cones for generating and extending new processes, and recent research has shown that calcium is vital to this process.[11] Studies in cultured neurons further suggest that calcium mobilization from caffeine-sensitive intracellular stores may be important in growth cone activity.[12] It is thought that similar processes may be involved in the formation of new connections and/or the strengthening of preexisting connections that occur during learning and memory in the adult brain.[13] Other examples in which neurons require intracellular calcium include the regulation of membrane currents and second messenger cascades as well as the activation of various kinases and phosphatases. Disturbances in calcium regulation have been implicated in a number of neuropathological syndromes.

When Ca^{2+} concentrations reach high levels, calcium-activated regulatory processes are stimulated to restore calcium to resting levels. A good example of such a calcium-activated process is the afterhyperpolarization (AHP), a hyperpolarizing voltage shift from the resting membrane potential that occurs after a burst of action potentials.[14,15] The burst depolarization activates voltage-dependent calcium channels. The calcium influx through these channels in turn activates outward calcium-dependent potassium currents. When these outward currents hyperpolarize the neuronal membrane, voltage-sensitive calcium currents are inhibited. This feedback sequence prevents uncontrolled firing activity in healthy neurons.

It should be recognized that calcium-dependent systems can show saturation. Under pathological conditions, intracellular calcium concentrations can in theory swamp homeostatic control mechanisms, rapidly leading to edema and cell death. As currently formulated, the calcium hypothesis does not presume that massive acute calcium overloads are the primary source of age-associated pathologies. Both aging and Alzheimer's disease may represent the cumulative effects of small disturbances in calcium homeostasis, with some disturbances resulting in further small but significant breakdowns of the regulatory process.

AGING-ASSOCIATED LEARNING DEFICITS

Aging mammals are clearly impaired in learning a variety of tasks. Learning in our behavioral model task, eyeblink conditioning, is impaired for both aging rabbits and aging humans.[16–19] Eyeblink conditioning is further impaired in Alzheimer's patients as compared to age-matched controls.[20,21] We used trace eyeblink conditioning to study age-associated learning deficits of aging rabbits and aging humans. In this paradigm, the subject must form a short-term memory trace of a brief tone (the conditioned stimulus) in order to appropriately time the occurrence of a conditioned eyeblink response. We chose this paradigm because trace eyeblink conditioning depends on an intact hippocampus for its successful acquisition[22,23] (FIG. 3A). Early work from our own laboratory, to be discussed in more detail, demonstrated that age-associated deficits in learning this task could be reversed by treatment with a centrally active Ca^{2+} channel antagonist (FIG. 3B). Consequently, we pursued a convergent series of studies examining both behavioral and physiological effects of Ca^{2+} channel blockade in aging rabbits.

It has been hypothesized that hippocampally dependent learning tasks are impaired in aging animals and humans owing to the early involvement of the hippocampus in aging-associated changes that affect learning and plasticity.[17,24] In our animal experiments, aging rabbits are impaired in acquisition of trace eyeblink conditioning. A recent detailed analysis of aging-related changes revealed that rabbits show a systematic age-related decline in their learning ability, beginning as early as 24 months of age with a plateau at around 30-36 months of age.[17] Additionally, with increasing age an increasingly larger percentage of the rabbit population was severely impaired in learning the trace conditioning task (ie, exhibiting much more severe learning deficits than did their age-matched cohorts). This inherent heterogeneity among the aging rabbit population is comparable to that seen in aging humans, where most individuals acquire the task, but where learning deficits become more obvious in older individuals.[18,21,25]

We have not restricted our model system studies solely to eyeblink conditioning, but have also evaluated the performance of aging rabbits in an "open field" environment.[26] Recall that rabbits are prey animals. Thus, it is not surprising that young rabbits, when placed in an open field, tended to stay on the border next to the walls, not explore much, and spent considerably more time sitting and observing their

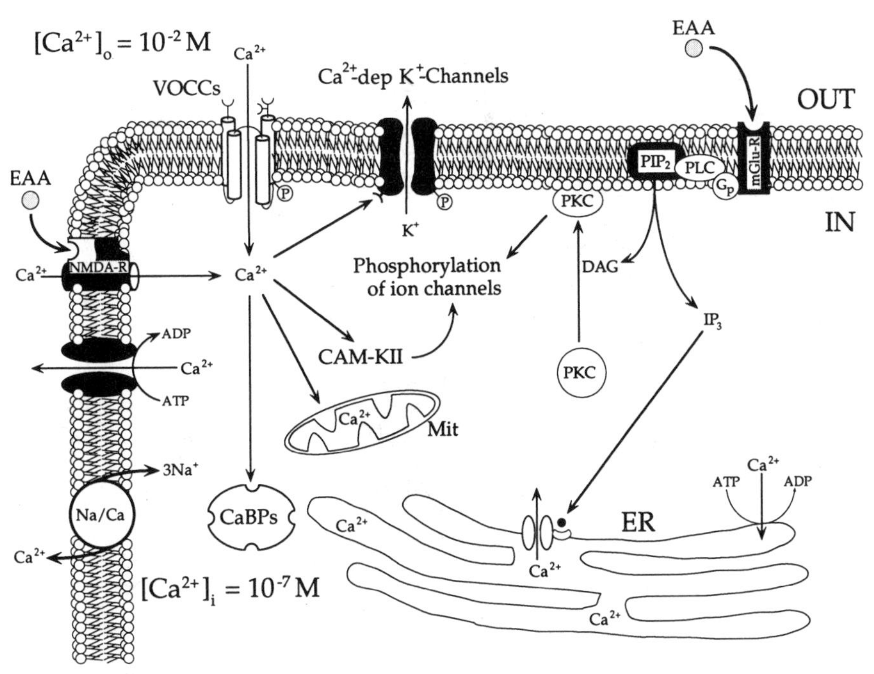

FIGURE 2. Altered intracellular free Ca^{2+} concentrations need not be directly excitotoxic in order to severely compromise neuronal functioning. As seen, excess Ca^{2+} influx (from sources such as voltage-operated calcium channels [VOCCs] and *N*-methyl-D-aspartate receptors [NMDA-Rs]) can rapidly and reversibly alter responses such as the postburst AHP generated by a Ca^{2+}-dependent K^+ conductance. Increased intracellular free Ca^{2+} would also alter numerous second-messenger cascades. For example, increased activity of the calcium-dependent enzyme protein kinase C (PKC) would increase protein phosphorylation. Excess phosphorylation of the microtubule-associated tau protein is one consequence observed in Alzheimer's brains. Excess phosphorylation of receptor and ion channel proteins could further amplify initial increases in free Ca^{2+} concentrations. Phosphorylation of calcium-binding proteins (CaBPs including calbindin, calmodulin, and parvalbumin) may reduce their capacity for Ca^{2+} buffering. Increased intracellular free Ca^{2+} would also increase the activity of calcium-dependent enzymes, including calcineurin, calpains I and II, phospholipase A2 (PLA2), Ca^{2+}-calmodulin-dependent protein kinase (CAM-KII), and phospholipase-C (PLC), altering metabolism of both proteins and phospholipids. Multiple ligand-gated membrane receptors, including metabotropic receptors (mGlu-R) could activate PLC and liberate inositol triphosphate (IP_3) and diacylglycerol (DAG) via G-protein (G_p)-mediated breakdown of phosphatidylinositol 4,5-bisphosphate (PIP_2). Increased IP_3 production would increase release of Ca^{2+} from intracellular stores such as endoplasmic reticulum (ER). Amplified depolarizing responses to excitatory amino acids (EAAs) would increase Na^+ influx, altering activity of the Na^+/Ca^{2+}-exchange antiporter. Altered metabolic demands and an overload of intracellular free Ca^{2+} could overwhelm the calcium pump (Ca^{2+}/Mg^{2+}-ATPase). As noted, the effects of both aging and Alzheimer's disease are cumulative, and the severity of the functional deficits observed is the product of multiple dysfunctions of varying severities.

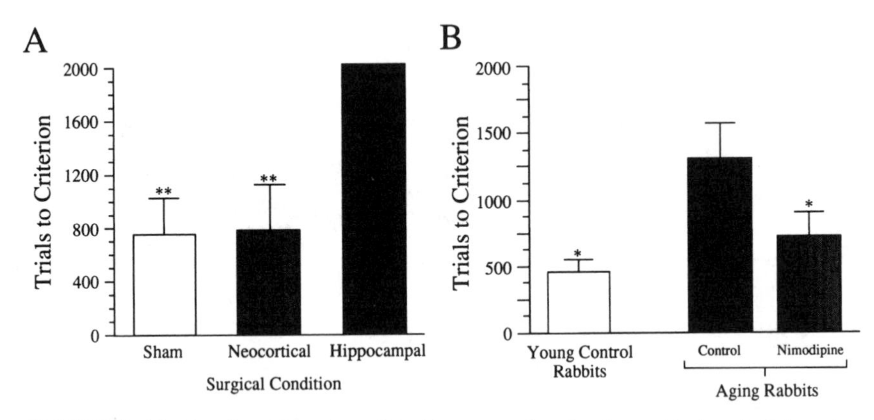

FIGURE 3. The functional integrity of the hippocampal region is notably impaired by aging and is severely altered in Alzheimer's disease. Hippocampal lesions severely impair or block acquisition of 500 msec trace eyeblink conditioning in young rabbits (**A**; Moyer *et al.*[22]). Aging rabbits are impaired in trace conditioning as compared to young controls, but daily ingestion of nimodipine-treated food pellets (860 ppm) largely reversed this deficit (**B**; Straube *et al.*[69]). Since Ca^{2+} channels are particularly abundant within the hippocampus and since Ca^{2+} serves important roles in a number of cellular functions, we have utilized a broad range of techniques to assess the functional consequences of altered Ca^{2+} influx in a primary cell type of the hippocampus, the CA1 pyramidal neuron, and to delineate links between Ca^{2+} and aging-associated learning or memory impairments.

environment. On the other hand, aging control rabbits wandered aimlessly around the open field. Their tendency to expose themselves in the center of the open space, and to move around a lot, would make them easy prey in the wild. The open field test has components of general cognitive functioning; it evaluates awareness of the rabbit to its surroundings and its level of alterness. As will be discussed later, calcium appears to be an important contributer to these and other age-related learning deficits.

CALCIUM AND CELLULAR MECHANISMS OF LEARNING

The calcium hypothesis argues that intracellular free Ca^{2+} plays a significant role in many forms of neural plasticity, including the models of learning studied in our laboratory. For example, single neuron recording *in vivo* demonstrated that the activity of hippocampal pyramidal neurons is increased during and after acquisition of the conditioned eyeblink response.[27] We used intracellular recording techniques and the *in vitro* rabbit hippocampal slice preparation to delineate the cellular mechanisms involved in this observed increase in excitability *in vivo* (FIG. 4). By using hippocampally dependent trace eyeblink conditioning,[22] we demonstrated that both the afterhyperpolarization (AHP; mediated by a Ca^{2+}-dependent K^+ current) which follows a burst of action potentials and spike frequency accommodation are reduced in CA1 and CA3 pyramidal neurons after acquisition.[28,29] These reductions increase neuronal excitability and are well correlated with behavioral acquisition as they are not observed

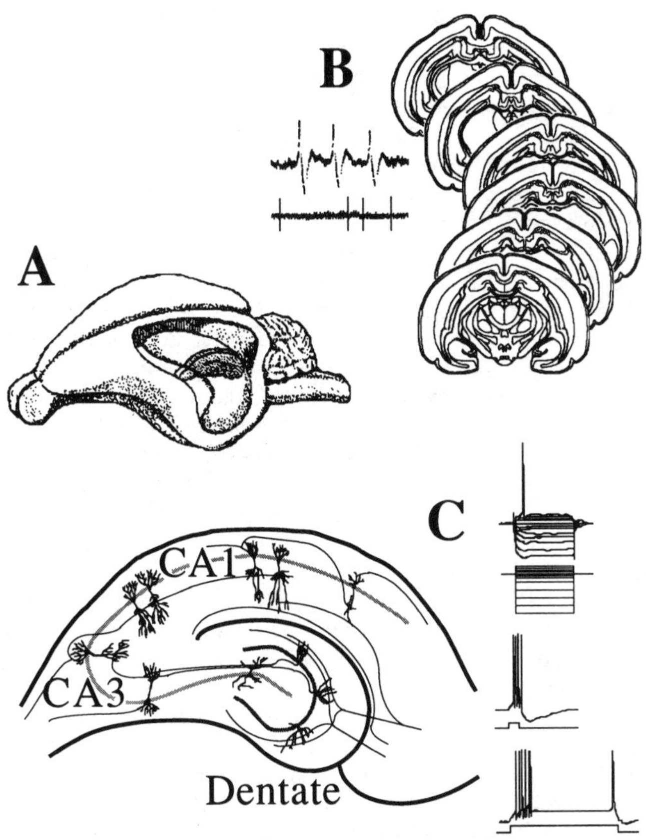

FIGURE 4. The structure and circuitry of the hippocampus lend themselves well to neuroscientific study. A large bilateral telencephalic structure (**A**), the hippocampus has a long and well-documented history investigating its functional role in behavioral plasticity (learning). Pyramidal neurons are arranged in curved sheets, with aligned basal and apical dendritic fields receiving afferents from defined sources. This arrangement facilitates replicable recordings of field potentials evoked by mass activation of afferent fiber bundles, a method that spawned the study of long-term potentiation of excitatory synaptic transmission within the hippocampus. It is also possible to place extracellular microelectrodes into specified regions and replicably isolate and record pyramidal neuron activity from a number of individual neurons across individual subjects (**B**). Pyramidal cell activity is distinguished extracellularly by occasional complex-spike (decremental bursting) activity, by waveform analysis, and by typical slow spontaneous rates of activity. The anatomical arrangement of hippocampal neurons yields a relatively uniform *in vitro* preparation, the transverse hippocampal slice (**C**). Using current-clamp protocols, a number of different biophysical measurements can be obtained from CA1 pyramidal cells, including current-voltage relationships, the afterhyperpolarization (AHP) following burst firing (typically, the AHP is measured following a burst of four action potentials elicited by injection of a short pulse of depolarizing current), and accommodation (or spike frequency adaptation, the decremental bursting observed with a longer depolarizing current injection).

in hippocampal slices taken from pseudoconditioned or poorly conditioned rabbits (FIG. 5). Furthermore, these alterations are localized to the hippocampus, as they are observed using *in vitro* slices separated from their normal afferent and efferent connections.[30,31] They are also postsynaptic, as they are evoked by intracellular current injection and persist even after block of sodium spike-dependent synaptic transmission.[32] Our evidence strongly supports involvement of Ca^{2+} in the learning-related changes observed in the hippocampus after trace eyeblink conditioning.

Calcium also appears to be critically involved in several other measures of hippocampal plasticity. For example, long-term potentiation (LTP)[33] has been extensively studied as an experimental model for synaptic changes that occur during learning and memory. The most commonly studied (and best understood) form of LTP depends upon the activation of NMDA (*N*-methyl-D-aspartate) receptors for its induction. Evidence has shown that NMDA channels are permeable to Ca^{2+}, whose influx is necessary for induction of LTP. Several other forms of plasticity, including those that depend on the hippocampus and temporal lobe for their acquisition, also require NMDA-dependent transmission.[34] When neurons are sufficiently depolarized by summated inputs, NMDA channels open, resulting in additional calcium influx. This calcium flows into very localized zones in dendritic spines, and calcium-dependent postsynaptic processes, such as stimulation of kinases and phosphatases, or mobilization of calcium from intracellular stores are activated. These result in synaptic changes hypothesized by many researchers to contribute to memory storage.[34] For example, in CA1 neurons, PKC activity associated with the membrane fraction was increased (while PKC activity associated with the cytosolic fraction was decreased) after eyeblink conditioning.[35] Immunocytochemical data from our laboratory further demonstrated that staining for the PKC-gamma subunit is significantly increased after hippocampally-dependent trace eyeblink conditioning.[36] Subsequent western blot or biochemical analyses should allow us to determine whether the PKC-gamma subunit staining observed after learning results from altered *de novo* synthesis of the protein, from translocation of the protein, or from changes in the conformation of the protein. Since activation of PKC by phorbol esters reduces the AHP,[37] these data suggest that the AHP reductions we observed after learning result from phosphorylation of calcium-dependent potassium channels, possibly by PKC or other kinases.

CALCIUM AND CELLULAR MECHANISMS OF NORMAL AGING

The calcium hypothesis posits that during normal aging, neurons become more vulnerable to Ca^{2+} dysregulation, with possible elevation of intracellular free calcium.[38] Our current understanding of neurotoxicity suggests that this Ca^{2+} dysregulation would make neurons vulnerable to other insults and disease processes, including the multiple factors contributing to Alzheimer's disease. In its most controversial form, the calcium hypothesis asserts that a breakdown in calcium homeostasis is the primary cause of aging-associated pathologies, including Alzheimer's disease. It should be noted that this extreme calcium hypothesis is probably incorrect and that elevated intracellular free Ca^{2+} is only one causative factor in the development of Alzheimer's disease. Even so, if neurons can be protected by appropriate calcium channel blockers against

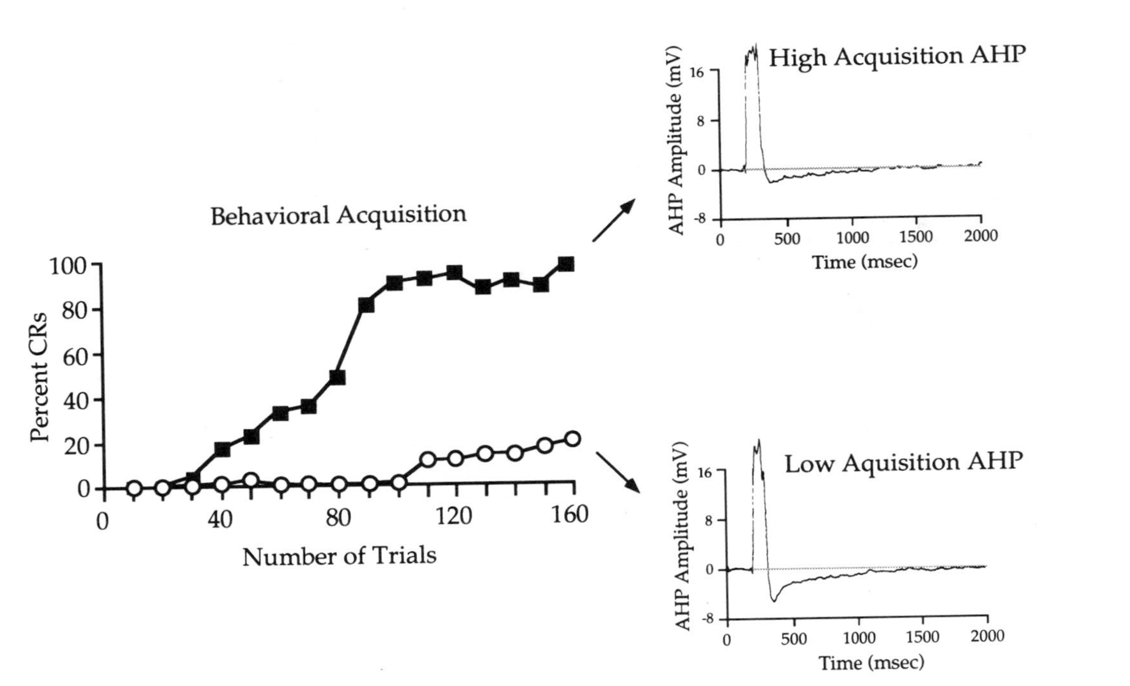

FIGURE 5. The postburst AHP reductions are learning-specific. (***Left panel***) Learning curves for rabbits receiving delay conditioning for two consecutive sessions. Note that some rabbits (high acquisition, ■) reached the criterion of 80% CRs during the second session while others (low acquisition, ○) showed relatively few CRs. CA1 pyramidal cells in slices taken from the high acquisition population (***top right panel***) exhibited a significantly reduced postburst AHP as compared to cells from the low acquisition population (***lower right panel***). Both the peak amplitude and the integrated area of the AHP were reduced after acquisition. (Adapted from Disterhoft *et al.*[31]).

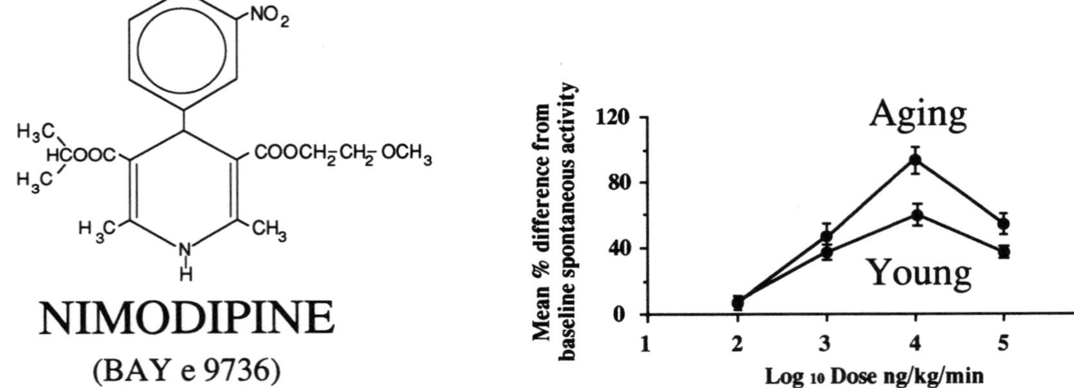

FIGURE 6. Calcium channel antagonists vary in their specificity for blocking different voltage-operated calcium channels (VOCCs) and in their tissue specificity (eg, cardiac, smooth muscle, or neuronal). Both factors are thought to be related to their physical structure and influenced by a number of factors, including their lipophilicity. Nimodipine (**A**), a 1,4-dihyrodyridine with high specific activity in cerebrovascular smooth muscle, also crosses the blood brain barrier more readily than do many other calcium antagonists, giving it ready access to the central nervous system after peripheral administration. Intravenous administration of varying doses of nimodipine (**B**) significantly enhanced the spontaneous firing activity of hippocampal neurons in awake rabbits. Other calcium antagonists that increase cerebral blood flow (as does nimodipine) but do not readily cross the blood brain barrier or that cross the blood brain barrier but block non-L-type Ca^{2+} channels had no effect on CA1 pyramidal cell firing activity. The effects of nimodipine were much greater in aging (mean 47.3-month-old) rabbits than in young (mean 4.1-month-old) rabbits. These results were suggestive of differential Ca^{2+} regulation in the brains of aging rabbits. (Adapted from Thompson *et al.*[41])

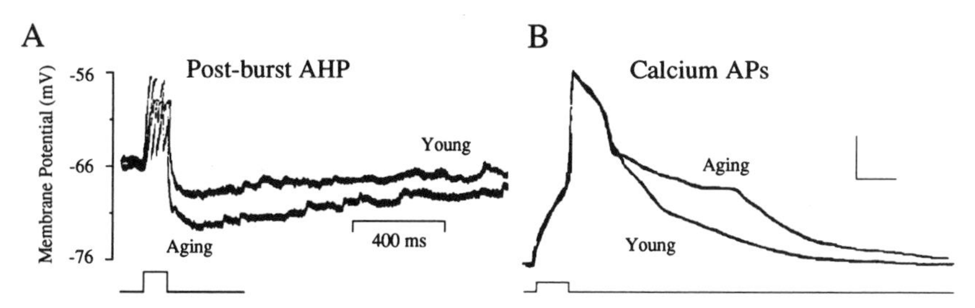

FIGURE 7. The slow postburst AHPs of aging CA1 pyramidal neurons were significantly enhanced in comparison with those of young neurons at the same resting potentials, being both longer in duration and greater in amplitude (**A**). Interestingly, the slow plateau phase of Ca^{2+} action potentials of hippocampal neurons was also greatly enhanced in aging (**B**), indicating a significant increase in calcium influx during depolarizing events. Sustained increases in Ca^{2+} influx are a likely basis for the enhanced calcium-dependent AHP observed in aging hippocampal neurons. (Adapted from Moyer *et al.*[44] and Moyer and Disterhoft.[45])

factors in aging that make them more susceptible to insult and disease, their function should be enhanced and they should be less vulnerable to other processes contributing to cell death.

Magnesium, a divalent calcium channel antagonist, has been shown to facilitate maze reversal learning and to improve hippocampal frequency potentiation in aging rats.[39] In addition, blocking calcium influx by altering Mg^{2+} to Ca^{2+} ratios improved intracellular and extracellular measures of frequency potentiation in hippocampal slices prepared from aging rats.[40] These experiments strongly suggest that reducing calcium influx in aging hippocampal neurons can reverse some of the functional and learning deficits observed and that behavioral and physiological effects of calcium channel blockade go hand in hand. As described, we studied the effects of Ca^{2+} channel blockers on eyeblink conditioning and on performance of other behaviors. Parallel work has examined the effects of calcium antagonists on a number of physiological measures, ranging from single-unit recording in the same intact awake rabbit preparation used for our eyeblink conditioning studies, to biophysical studies of neuronal excitability and of calcium influx in the isolated hippocampal slice or in acutely dissociated neurons. Convergent data, consistent with the calcium hypothesis of aging, was obtained and is summarized below.

We reasoned that the resting activity of hippocampal neurons in conscious aging rabbits should be influenced by calcium influx, if our hypothesis was correct. Reducing Ca^{2+} influx should increase the excitability of hippocampal neurons, especially in aging animals, by reducing feedback activation of the Ca^{2+}-dependent AHP. We tested this corollary, with the result that nimodipine dose- and age-dependently increased the baseline activity of single CA1 hippocampal neurons (Fig. 6).[41] Nimodipine increased neuronal baseline firing rates at the same doses that facilitated acquisition of trace eyeblink conditioning in aging rabbits.[42] The increase was much larger in aging rabbits than in young animals. So, a physiological effect of nimodipine was to enhance pyramidal neuron firing rates, an effect that might be expected to contribute to enhanced eyeblink conditioning in aging animals.

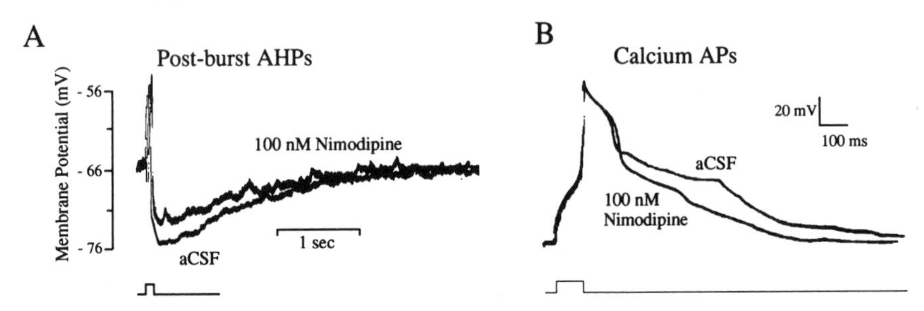

FIGURE 8. Although greatly enhanced in aging CA1 neurons, both postburst AHPs (**A**) and Ca^{2+} action potentials (**B**) were significantly reduced by low concentrations (100 nM) of nimodipine. Again, concentrations in the micromolar range were required for effect in young CA1 neurons. Treatment with nimodipine restored Ca^{2+} influx to levels typical of young neurons. The Ca^{2+}-dependent AHP was also restored to levels typical of young pyramidal cells. (Adapted from Moyer *et al.*[44] and Moyer and Disterhoft.[45])

The Ca^{2+}-dependent afterhyperpolarization is prolonged in the hippocampus of aged mammals due to excess calcium influx through voltage-activated, L-type calcium channels and/or to tonically elevated levels of free intracellular calcium in neurons[43,44] (FIG. 7A). The postburst afterhyperpolarization is mediated by an outward calcium-dependent potassium current, activated by the Ca^{2+} influx during the burst of action potentials. It seemed likely that the enhanced AHP in aging CA1 neurons results from an increase in the amount of calcium entering the neuron during the burst. Thus, we next determined that the calcium action potential was indeed enhanced in CA1 neurons from aging rabbits[45] (FIG. 7B). Note that the enhanced depolarizing plateau of the calcium action potential is ideally timed to underlie the enhanced slow AHP seen in aging neurons.

The prolonged AHP of hippocampal principal neurons of aged mammals causes this structure to be less excitable and would act to impede information transfer through the hippocampus. This impaired function could contribute to age-associated learning impairments.[16,17] The hippocampus, critical for learning and memory processes, is often the first neuronal structure to be affected by the pathologies associated with Alzheimer's dementia.[46] Thus, learning and memory deficits are often a cardinal feature of Alzheimer's disease even in its early stages. Prolonged elevation of free Ca^{2+} during the period following bursts of action potentials in aging neurons is linked to larger AHPs and to impaired learning. Tonically enhanced intracellular Ca^{2+} within hippocampal neurons over time, even if slight, could contribute to cell death in this pivotal region.[1,2]

We next explored whether nimodipine could reduce the postburst AHP, carried by a Ca^{2+}-activated outward potassium current, an effect that would increase aging neuronal excitability. Again, we were particularly interested in this type of functional change, as we have shown that the postburst AHP is reduced in pyramidal neurons after eyeblink conditioning in both young[28–32] and aging rabbits.[47] Our recent studies indicate that this AHP reduction in both CA1 and CA3 neurons is related to the acquisition and consolidation of the trace eyeblink conditioned response rather than its longer term storage during the learning process.[28,29] We also know that the slow AHP is markedly enhanced in CA1 pyramidal neurons from aging rats and rabbits, an enhancement that has been suggested would reduce hippocampal neuron functional excitability and thus slow learning rate.[43,44]

Nimodipine, in concentrations as low as 100 nM, reduced the slow afterhyperpolarization of aging CA1 neurons in rabbit hippocampal slices (FIG. 8A).[44] Concentrations this low had no effect on the AHPs of CA1 neurons from young rabbits. We similarly showed that nimodipine, in concentrations as low as 100 nM, significantly reduced the Ca^{2+} action potential in aging but not young neurons (FIG. 8B).[45] It is important to note that nimodipine affected the late depolarizing plateau of the calcium action potential, that same portion that was noticeably enhanced in aging neurons. This slow component, apparently mediated by L-type channels, is ideally suited temporally to underlie the enhanced slow calcium-activated postburst AHPs observed in aging neurons.

Concentrations of nimodipine as low as 10 nM were also effective in reducing spike frequency accommodation to a long depolarizing current injection in aging but not young CA1 neurons (FIG. 9).[44] This is another measure of excitability change in hippocampal neurons, one more sensitive to neuronal depolarizations due to the repeated action potentials that are evoked by sustained depolarization. Spike frequency accommodation is also reduced after trace eyeblink conditioning in CA1 and CA3 pyramidal neurons.[28,29] The final portion of our physiological analyses which has been completed indicates that nimodipine markedly reduces the high threshold, noninactivating component of the Ca^{2+} current which is presumably the L-type current, as evaluated with voltage-clamp recordings of neurons acutely dissociated from young hippocampus.[48]

Thus, a lipophilic dihydropyridine, nimodipine, which has ready access to the hippocampal neurons presumably involved in the learning process, may facilitate learning in aging rabbits by reducing age-associated increases in Ca^{2+} influx. Nimodipine restored these hippocampal neurons to a level of excitability comparable to that observed in young hippocampal neurons. An extension of these findings, to be discussed, is whether long-term treatment with a centrally active calcium antagonist would have long-term beneficial effects in aging.

The functional characteristics of L-type calcium channels (those blocked by dihydropyridines) change with age. As discussed, there is evidence that the conductance through these channels increases in aging hippocampal CA1 neurons and is selectively blocked by nimodipine.[45,49] L-type Ca^{2+} channels may be especially important in age-associated changes, because they are open a long time and pass a relatively large amount of current.[50] Thus, their alteration can contribute in a major fashion to tonic increases in free intracellular Ca^{2+} that may lead to neuronal degeneration. Note that L-channels are concentrated on the neuron soma and proximal dendrites and are not directly involved in synaptic transmission.[51] This is important to stress because it implies that an L-type calcium channel blocker will not slow synaptic transmission. Instead, L-type calcium channel blockers will influence cellular function through their effects on postsynaptic excitability and neuronal calcium concentrations at low doses that will not interfere with synaptic transmitter release. More specifically, if cellular excitability is reduced as a result of excess Ca^{2+} or larger AHPs, an L-type

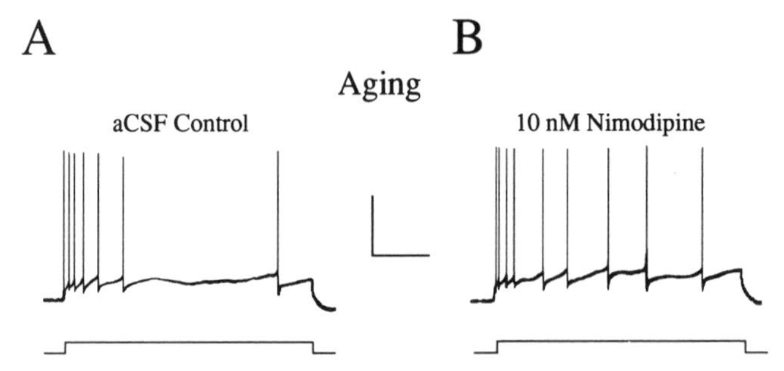

FIGURE 9. Bath application of nimodipine to CA1 pyramidal neurons in slices from aging (mean 39.3-month-old) rabbits reduced accommodation to prolonged depolarization at concentrations as low as 10 nM. Concentrations this low were ineffective in slices from young rabbits (effects were seen on young neurons only at concentrations in the micromolar range). This increase in cellular excitability *in vitro* models that observed *in vivo,* with differential sensitivity exhibited by aging neurons. Because hippocampal pyramidal cell excitability increases during conditioning, the findings summarized in this and the previous figure provide a hypothetical mechanism for the ability of centrally active calcium channel antagonists to restore learning capacity in aging animals.

calcium channel blocker could enhance neuronal function in an age-dependent fashion at levels that will have no deleterious side effects on neuronal function.[41,44,45]

Potential interactions of Ca^{2+} influx through NMDA-receptor channels with Ca^{2+} influx through L-type calcium channels or with release of intracellular calcium by other excitatory amino acids are important when considering the calcium hypothesis of aging and Alzheimer's disease. The calcium hypothesis posits that Ca^{2+} channel blockers can enhance postsynaptic excitability by reducing depolarization-induced Ca^{2+} influx and thus restore neuronal function to a level more like it was in young neurons. As mentioned above, NMDA-receptor channels are known to be critically important for some forms of learning and act to enhance Ca^{2+} entry into the cell. This may seem counterintuitive, unless it is considered that L-type calcium channels and NMDA glutamate receptors are located at two very different spatial sites on neurons, the cell body and proximal dendrites for L-type channels and out on the spines of distal dendrites for NMDA channels, respectively.[51] Thus, L-type Ca^{2+} antagonists can enhance cellular excitability by reducing calcium influx in the relatively large soma and not affect Ca^{2+} influx in synaptic regions where the NMDA receptor complex is located. Interactions between these two significant sources of cellular calcium influx obviously require further study.

LINK BETWEEN NORMAL AGING AND ALZHEIMER'S DISEASE

As just noted, aging animals show dramatic deficits in learning many hippocampally dependent tasks,[17] and calcium channel blockade effectively reverses these learning deficits. Recent findings from our laboratory confirm that aging humans are also impaired in acquisition of the hippocampally dependent eyeblink conditioning

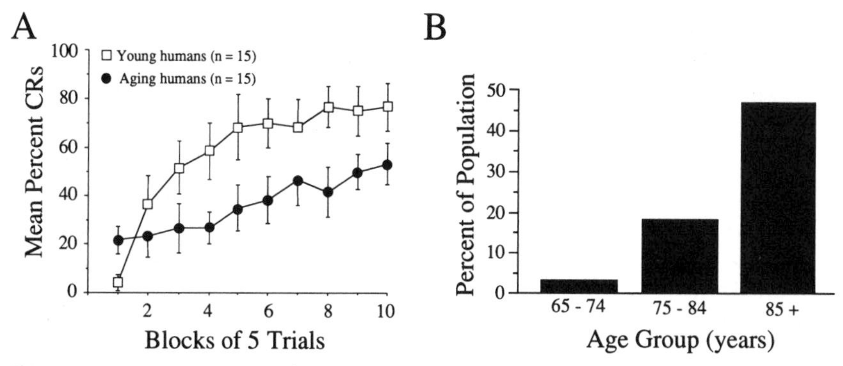

FIGURE 10. Treatment with Ca^{2+} channel antagonists may be useful in reversing the behavioral deficits associated with both normal aging and Alzheimer's disease. Even relatively simple measures of human learning ability, such as eyeblink conditioning, indicate impairments in aging (mean 65 years) as compared to young controls (mean 25 years)[25] (**A**). Alzheimer's incidence in the human population increases significantly with increasing age (**B**), with subsequent and eventually disabling impairment in learning and in cognitive functions required to perform even simple daily tasks.[52] Because Ca^{2+} antagonists can restore neuronal function in aging to mimic that seen in young cells, there is considerable hope that they can also be valuable in impeding the progression or even reversing some of the symptoms of Alzheimer's disease.

task used in our rabbit studies[18,25] (FIG. 10A). Impaired learning was observed early in training and throughout the duration of the experiment. The preponderance of evidence suggests a progressive decline in learning ability with increasing age, with considerable individual variability. The data can be extrapolated to hypothesize that all members of the human population would eventually develop Alzheimer's disease, if they survived long enough. Although this extreme view is untestable, several lines of evidence suggest that aging and Alzheimer's disease may be causally linked:

1. *The incidence of Alzheimer's disease increases exponentially in the 65+ age range.* For example, an important study[52] evaluated the prevalence of Alzheimer's disease in aging individuals living in a community setting. This study estimated that approximately 10% of individuals over 65 had cognitive deficits related to Alzheimer's disease. The estimated incidence of probable Alzheimer's disease increased from 3% in the 65-74-year age range to 47% for the group over 85 years old (FIG. 10B). Some feel that this study overestimated the incidence of Alzheimer's disease; however, even more conservative estimates are unsettling. For example, a 1992 Framingham study[53] estimated that the incidence of Alzheimer's disease increased from 0.5% in the 65-74-year age range to 13.1% in the 84+ age group. This is actually a larger proportional increase than the estimate of Evans *et al.*[52] Estimates of dementia from all causes in the aging population are similarly disturbing. One study[54] diagnosed 14% of a sample of 3,777 community residents in France over 65 years old as demented. The Framingham study[53] estimated that the cause for dementia in 55% of the aging population was Alzheimer's disease. If this estimate is generalizable to other parts of the world, somewhere between 7 and 8% of the aging population over 65 years of age have Alzheimer's disease.

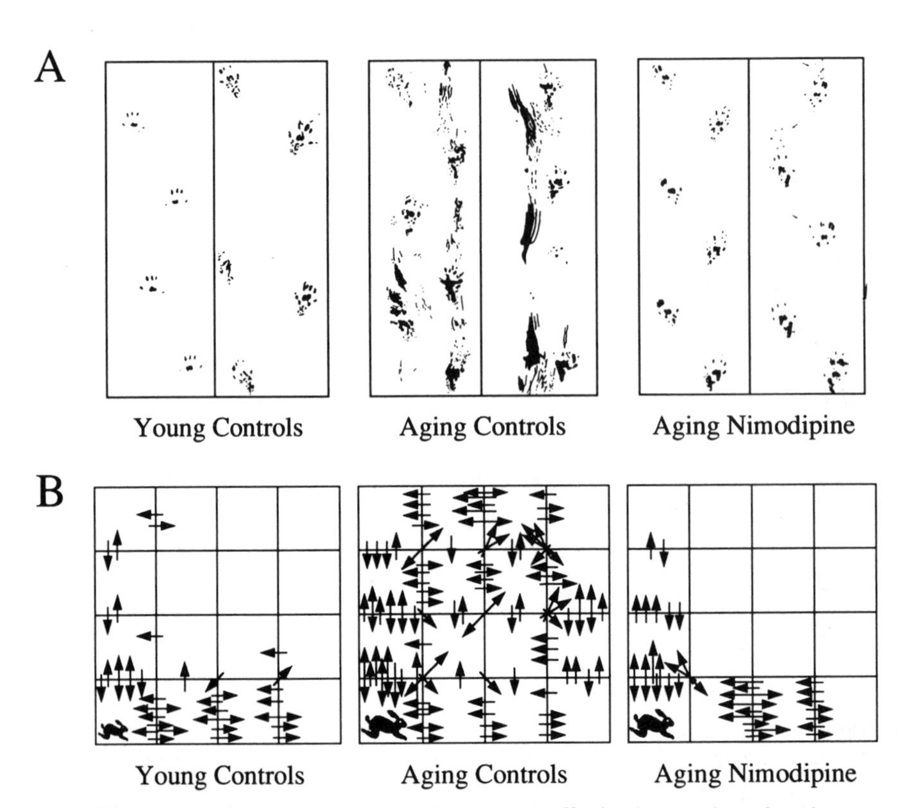

FIGURE 11. Calcium channel antagonists have proven effective in reversing a broad range of behavioral deficits associated with aging. For example, treatment with nimodipine, an L-type Ca^{2+} channel antagonist, restored a youthful gait to aging-impaired rats (**A;** Scriabine *et al.*[70]). Work in our laboratory has shown that daily oral treatment with nimodipine reversed the impairments in open-field behavior exhibited by aging (more than 36-month-old) rabbits (**B;** Deyo *et al.*[26]).

2. *Age-related changes in calcium influx may be an important risk factor for Alzheimer's disease.* As just discussed, calcium action potentials are enhanced in aging hippocampal CA1 pyramidal neurons.[45] This enhancement may be due to a reduction in the amount of calcium-mediated inactivation of calcium currents which occurs in aging neurons[55] or to an increase in the number of repolarization openings of L-type calcium channels.[56] The portion of the calcium action potential that is most increased in aging neurons is the late depolarizing plateau phase. In aging CA1 neurons, a likely consequence of this increase is the reported enhancement of the calcium-activated slow afterhyperpolarization following a burst of action potentials.[43,44] An increased AHP can markedly reduce neuronal excitability and thus, presumably, should impair hippocampal function. Poor buffering of intracellular Ca^{2+} or impaired inactivation of Ca^{2+} channels in aging neurons may contribute to these observations. Indirect evidence for impaired calcium buffering stems from observed

decreases in immunoreactivity for the calcium binding protein calbindin-28K in aging rat and rabbit hippocampus.[57,58] Calbindin-28K is also reduced in several cortical regions, including temporal cortex, in the brains of demented individuals diagnosed as having either Alzheimer's or non-Alzheimer's dementia.[59] Increased calcium influx has been demonstrated in soleus muscle motor nerve terminals in aging mice, resulting in increased quantal content at the terminals.[60] Finally, synaptosomal Ca^{2+} uptake has been demonstrated to decrease in aging rats.[61] A subsequent study[62] demonstrated that both the mitochondrial (digitonin-resistant) and non-mitochondrial (digitonin-labile) compartments of the synaptosomes showed reduced calcium uptake in aging mice, which could lead to increased intracellular free Ca^{2+} in axon terminals. Thus, multiple lines of evidence indicate that free intracellular calcium may not be well regulated in aging neurons. Further research is required to determine if this imbalance is qualitatively or quantitatively different in persons with Alzheimer's disease. Lack of a solid animal model for Alzheimer's disease has certainly not facilitated progress in this area.

3. *Pathological and functional characteristics of Alzheimer's disease have been observed in the normal aging brain.* It has been reported that beta amyloid-protein plaques and neurofibrillary tangles occur, although to a lesser degree, in normal aging and that the difference between the Alzheimer's brain and the normal aging brain may be quantitative, not qualitative.[63] Although this view of the development of Alzheimer's disease is speculative, in early stages of the disease process, neurofibrillary tangles and plaques are especially concentrated in the subiculum and the hippocampus.[64] These structures are known to be essential for normal learning and memory functions.[22,23,65,66] Because the earliest neuropathological changes in Alzheimer's disease tend to be concentrated in these areas, learning deficits are an early and sustained cardinal feature of the disease. Further evidence that processes mediated by the hippocampus and the temporal lobe, such as learning, are critically affected by the aging process comes from animal studies that illustrate that physiological properties such as frequency potentiation and LTP are altered by aging[24,35] and that such changes are especially pronounced in animals demonstrating the largest age-related learning deficits.[67]

In fact, it is difficult to discriminate Alzheimer's disease from advanced aging or from other disease processes. Careful psychological testing, including a demonstrable memory loss and gradual onset of the disease, must be combined with postmortem neuropathological studies of brain tissue to determine the prevalence and distribution pattern of currently accepted markers such as β-amyloid plaques and neurofibrillary tangles for a definitive diagnosis of the disease. Even with these precautions, there are a variety of cases in which the diagnosis is not clear.[46] Although numerous animal models of Alzheimer's disease have been proposed, it should be emphasized that none has full attributes of the disease. Extensive research efforts must continue to develop and explore different animal models of aging and Alzheimer's disease. This is critical for researchers to understand how the disease develops, to identify persons at risk, and to provide hope for reasonable treatment.

CLINICAL APPLICATIONS OF CALCIUM ANTAGONISTS

There are several families of calcium channel blockers, including the dihydropyridines nifedipine and nitrendipine in current clinical use for the management of hypertension. Are these drugs likely to reduce excess calcium levels in neurons and to counteract neuronal dysfunction and possibly cell death? For a majority of such compounds, the answer is probably no, because a necessary feature to subserve such function is the capacity to cross the blood brain barrier when given systemically. Peripherally active dihydropyridines such as nifedipine and nitrendipine cross the blood brain barrier with low efficiency. Centrally acting dihydropyridines like nimodipine, on the other hand, are more lipophilic and can readily cross the blood brain barrier.[68] Other compounds also exhibit this type of selectivity with reference to their compartments of distribution and may also exhibit clinical efficacy when tested.

Several studies have used Ca^{2+} channel blockers to examine the hypothesis that an inability to reduce the calcium-dependent AHP and thus modulate neuronal excitability contributes to learning deficits in aging animals. Elevation of plasma magnesium (a competitive inhibitor of calcium) improved reversal learning in both aged and young rats.[39] Because of its ability to cross the blood brain barrier,[68] the calcium antagonist nimodipine has been tested in a variety of learning and behavioral tasks in aging mammals. A series of studies from our own laboratory (see FIG. 3, for example) indicate that nimodipine markedly facilitates acquisition of the trace eyeblink conditioned response in aging-impaired rabbits,[16,41,69] with improvements in learning whether nimodipine was given orally or intravenously. Optimal behavioral effects were obtained at the same dose that maximally enhanced hippocampal neuronal activity *in vivo,*[41] with calculated circulating concentrations equivalent to those that enhanced hippocampal excitability *in vitro.*[44] Nimodipine (FIG. 11) also improved sensorimotor behaviors in aging rats[70]; reversed open field deficits in aging rabbits[26]; improved delayed matching-to-sample performance in aging primates[71]; and improved spatial learning in aging rats.[72]

Double-blind, placebo-controlled clinical trials indicate that nimodipine successfully reverses some of the cognitive deficits of dementia and Alzheimer's disease or slows the progression of these diseases in aging populations. With fairly low oral doses (30 mg tid) cognitive improvements are seen. A large double-blind placebo-controlled study of 228 patients separated into probable primary degenerative dementia (a large percentage of whom were likely to have Alzheimer's disease) and into multi-infarct dementia was summarized by Fischhof *et al.*[73] After 12 weeks of treatment, patients receiving nimodipine showed improved cognitive abilities. Schmage *et al.*[74] did a so-called "metanalysis," pooling data from 11 double-blind studies in which a total of 391 patients diagnosed with organic brain syndrome received nimodipine and 398 patients received placebo. In 7 of the studies, a low dose of 30 mg tid of nimodipine was used. The nimodipine-treated subjects improved in 9 of the 11 studies. Improvement was particularly marked on the cognitive subtests of the Sandoz clinical assessment geriatric scale.

In another double-blind multicenter trial of 178 aging patients with primary degenerative dementia, multi-infarct dementia, or a mixed syndrome, 30 mg tid of nimodipine was found to be superior to placebo during a 12-week trial.[75] Nimodipine-treated patients improved on the Wechsler memory scale, the Folstein mini-mental status examination, the Sandoz clinical assessment geriatric scale, the Plutchick geriatric rating scale, the severity of illness and global improvement scales of the clinical global impression test, and the Hamilton psychiatric rating scale for depression. In another example in which low doses of nimodipine were used, 30 patients characterized as having organic brain syndrome were compared to an age-matched control group of 20 patients.[76] A high percentage of the nimodipine-treated patients improved on a series of psychometric tests for cognitive function, memory capacity, concentration, orientation, and degree of autonomy, whereas similar improvement was not observed in control patients.

Tollefson[77] summarized a multicenter trial of 227 patients who received nimodipine (30 mg tid) or placebo for 12 weeks following a 2-week placebo baseline. Nimodipine slowed the cognitive decline associated with Alzheimer's disease progression. Assessments that were improved relative to the other two groups included the Buschke long-term storage and retrieval test, first and last names recall test, activities of daily living subtests on personal care, housing and communications, and the relative's assessment of global symptomatology subtests on anxiousness and speech. It should be noted, however, that one of the site investigators involved with this study was concerned to stress its limitations in terms of the magnitude of the behavioral effects.[78] In other words, although improvement was noted, it should not be construed that nimodipine treatment "cured" the patients or restored them to a state approximating normal function. A higher dose of nimodipine (60 mg tid) produced no significant improvements in cognition.

We recently completed a double-blind placebo-controlled study evaluating the effects of nimodipine on eyeblink conditioning in both normal aging and young human subjects.[79] The subjects in this 3-month trial were assessed for normal learning ability. We found that aging, but not young, people who received nimodipine (60 mg tid for 3 months) acquired conditioned eyeblink responses faster than did their age-matched placebo controls. This effect was a trend after 1 month and was statistically significant after 3 months of treatment. This study demonstrates facilitation of learning by a pharmacological agent using exactly the same learning task in both humans[25] and an animal model.[16,43,69] The results of these studies are consistent with the calcium hypothesis of aging, as discussed above and in other chapters throughout this volume.

It should be stressed that in all of the nimodipine clinical trials, the adverse side effects reported were relatively few and very mild. Similarly, essentially no drug-related changes were noted in patients' normal physiological functions. From a clinical perspective, nimodipine appears to be an extremely safe agent for use in humans and exhibits some degree of efficacy.

Because nimodipine is so safe and well tolerated, the possibility of using it in a prophylactic context is appealing. Unfortunately, it is possible that L-type calcium channel blockers would have only small therapeutic effects on patients with advanced stages of Alzheimer's disease owing to the likelihood that significant irreversible cell loss in the hippocampus and other brain regions may have already occurred. If therapy can begin early enough in the development of the disease, it is theoretically possible to slow progression of degenerative disease processes with a calcium channel blocker. This could either slow or reverse the progression of the disease and enhance the quality of life in these patients.

COMMENTS AND SUMMARY

We have attempted to summarize both hypothetical and experimental relationships between the calcium hypothesis, our current understanding of the neurobiological basis of aging, and proposed mechanisms leading to the etiology of Alzheimer's disease. Our own work has attempted to relate alterations in calcium-mediated events in neurons during the aging process to their potential impact on behavioral learning. As can be seen in the various chapters contained in this volume, a variety of approaches can be taken relating calcium, aging, and dementia. Our own approach uses an animal model of normal aging, with a behavioral assay designed to tap hippocampal/temporal lobe function. We know that the temporal lobe is importantly involved in learning and its impairments, including those that accompany aging, and therefore have concentrated our work on cellular changes in this region.

Our preclinical work has used trace eyeblink conditioning to study learning in aging rabbits. We have shown that acquisition of this task is slowed in aging animals. We have also found that calcium action potentials and calcium-mediated AHPs are augmented in hippocampal neurons in the same age ranges in which learning performance is impaired. These effects can be reduced using a lipophilic dihydropyridine calcium antagonist, nimodipine, at concentrations that we have also shown accelerate learning in aging rabbits. It is quite striking that this same Ca^{2+} channel blocker can reduce the behavioral effects of cerebral ischemia during development; can improve performance in a variety of learning tasks in aging animals; can improve learning in ischemic or demented adults; and, as we have shown, can enhance the ability of normal aged humans to learn our eyeblink conditioning task.

The fact that aging and Alzheimer's disease are so intimately related is probably not accidental. The calcium hypothesis asserts that increases in free intracellular Ca^{2+} during aging likely contribute to the cellular deterioration leading to age-associated dementias. The preclinical data that we and others have gathered *in vitro* and *in vivo* are tantalizing, because they suggest it may be possible to reverse or retard the impact of aging both behaviorally and physiologically by reducing calcium influx. More work is needed to explore this hypothesis. It is necessary to determine mechanisms that lead to excess Ca^{2+} influx and to explore potential dysfunctional changes in regulatory mechanisms that may contribute to a breakdown in calcium homeostasis in aging. Further work is needed to determine if calcium antagonists can be effective over very long-term periods (years) with prophylactic administration. Finally, more work is needed to meet the challenge of designing more effective pharmaceutical agents than are currently available for treatment of learning impairments in normal aging and in states of dementia such as Alzheimer's disease.

ACKNOWLEDGMENTS

The figures in this chapter were originally designed for a proposed supplement for Alzheimer's Disease and Associated Disorders.

REFERENCES

1. KHACHATURIAN, Z. S. 1989. The role of calcium regulation in brain aging: Reexamination of a hypothesis. Aging **1:** 17–34.

2. LANDFIELD, P. W. 1987. 'Increased calcium current' hypothesis of brain aging. Neurobiol. Aging **8:** 346-347.
3. CHOI, D. W. 1988. Glutamate neurotoxicity and diseases of the nervous system. Neuron **1:** 623-634.
4. MELDRUM, B. & J. GARTHWAITE. 1990. Excitatory amino acid neurotoxicity and neurodegenerative disease. TiPS **11:** 379-387.
5. SIESJO, B. K. 1988. Historical overview: Calcium, ischemia and death of brain cells. Ann. N.Y. Acad. Sci. S**22:** 638-661.
6. LYNCH, G., J. LARSON & M. BAUDRY. 1986. Proteases, neuronal stability, and brain aging: A hypothesis. *In* Treatment Development Strategies for Alzheimer's Disease. T. Crook, R. Bartus, S. Ferris & S. Gershon, Eds: 119-139. Mark Powley. Madison, CT.
7. SIMAN, R. & J. C. NOSZEK. 1988. Excitatory amino acids activate calpain I and induce structural protein breakdown *in vivo.* Neuron **1:** 279-287.
8. NYAKAS, C., E. MARKEL, R. KRAMERS, E. GASPAR, B. BOHUS & P. LUITEN. 1989. Effects of nimodipine on hypoxia-induced learning and memory deficits. *In* Nimodipine & Central Nervous Function: New Vistas. J. Traber & W. Gispen, Eds.: 175-208. Schattauer. Stuttgart.
9. BONO, G., E. SINFORIANI, M. TRUCCO, A. CAVALLINI, G. C. ACUTO & G. NAPPI. 1985. Nimodipine in cCVD patients: Clinical and neuropsychological results of a double-blind cross-over study. *In* Nimodipine—Pharmacological and Clinical Properties. E. Betz, K. Deck & F. Hoffmeister, Eds.: 275-285. Schattauer-Verlag. Stuttgart.
10. TOBARES, N., A. PEDROMINGO & J. BIGORRA. 1989. Nimodipine treatment improves cognitive functions in vascular dementia. *In* Diagnosis & Treatment of Senile Dementia. M. Bergener & B. Reisberg, Eds.: 360-365. Springer-Verlag. Berlin.
11. KATER, S. B., M. P. MATTSON, C. COHAN & J. CONNOR. 1988. Calcium regulation of the neuronal growth cone. TINS **11:** 315-321.
12. BANDTLOW, C. E., M. F. SCHMIDT, T.D. HASSINGER, M. E. SCHWAB & S. B. KATER. 1993. Role of intracellular calcium in NI-35-evoked collapse of neuronal growth cones. Science **259:** 80-83.
13. WALLACE, C. S., N. HAWRYLAK & W. T. GREENOUGH. 1991. Studies of synaptic structural modifications after long-term potentiation and kindling: Context for a molecular morphology. *In* Long-Term Potentiation—A Debate of Current Issues. M. Baudry & J. L. Davis, Eds.: 189-232. MIT Press. Cambridge, MA.
14. HOTSON, J. R. & D. A. PRINCE. 1980. A calcium-activated hyperpolarization follows repetitive firing in hippocampal neurons. J. Neurophysiol. **43:** 409-419.
15. LANCASTER, B. & P. R. ADAMS. 1986. Calcium-dependent current generating the afterhyperpolarization of hippocampal neurons. J. Neurophysiol **55:** 1268-1282.
16. DEYO, R. A., K. STRAUBE & J. F. DISTERHOFT. 1989. Nimodipine facilitates trace conditioning of the eye-blink response in aging rabbits. Science **243:** 809-811.
17. THOMPSON, L. T., J. R. MOYER & J. F. DISTERHOFT. 1994. Trace eyeblink conditioning demonstrates heterogeneity of learning ability both between and within age groups. Behav. Neurosci., submitted.
18. DISTERHOFT, J. F., S. W. CONROY, L. T. THOMPSON, B. J. NAUGHTON & J. D. E. GABRIELI. 1991. Age affects eyeblink conditioning and response discrimination in humans. Soc. Neurosci. Abstr. **17:** 476.
19. WOODRUFF-PAK, D. S. & R. F. THOMPSON. 1985. Classical conditioning of the eyelid response in rabbits as a model system for the study of brain mechanisms of learning and memory in aging. Exp. Aging Res. **11:** 109-122.
20. SOLOMON, P. R., E. LEVINE, T. BEIN & W. W. PENDLEBURY. 1991. Disruption of classical conditioning in patients with Alzheimer's disease. Neurobiol. Aging **12:** 283-287.
21. WOODRUFF-PAK, D. S., R. G. FINKBINER & D. K. SASSE. 1990. Eyeblink conditioning discriminates Alzheimer's patients from non-demented aged. NeuroReport **1:** 45-49.
22. MOYER, J. R., R. A. DEYO & J. F. DISTERHOFT. 1990. Hippocampectomy disrupts trace eye-blink conditioning in rabbits. Behav. Neurosci. **104:** 243-252.

23. Solomon, P. R., E. R. van der Schaaf, D. J. Weisz & R. F. Thompson. 1986. Hippocampus and trace conditioning of the rabbit's classically conditioned nictitating membrane response. Behav. Neurosci. **100:** 729-744.
24. Barnes, C. A. 1979. Memory deficits associated with senescence: A neurophysiological and behavioral study in the rat. J. Comp. Physiol. Psychol. **93:** 74-104.
25. Carillo, M. C., L. T. Thompson, B. J. Naughton, J. Gabrieli & J. F. Disterhoft. 1993. Aging impairs trace conditioning in humans independent of changes in the unconditioned response. Soc. Neurosci. Abstr. **19:** 386.
26. Deyo, R. A., K. T. Straube, J. R. Moyer & J. F. Disterhoft. 1990. Nimodipine ameliorates aging-related changes in open-field behaviors of the rabbit. Exp. Aging Res. **15:** 169-175.
27. Berger, T. W. & R. F. Thompson. 1978. Neuronal plasticity in the limbic system during classical conditioning of the rabbit nictitating membrane response. I. The hippocampus. Brain Res. **145:** 323-346.
28. Thompson, L. T., J. R. Moyer & J. F. Disterhoft. 1994. Learning (not performance or memory) increases *in vitro* excitability of hippocampal CA3 neurons. Soc. Neurosci. Abstr. **20:** 796.
29. Moyer, J. R., L. T. Thompson & J. F. Disterhoft. 1994. The hippocampus as an intermediate storage buffer after associative learning: *In vitro* evidence from rabbit CA1. Soc. Neurosci. Abstr. **20:** 796.
30. Disterhoft, J. F., D. A. Coulter & D. L. Alkon. 1986. Conditioning-specific membrane changes of rabbit hippocampal neurons measured *in vitro.* Proc. Natl. Acad. Sci. USA **83:** 2733-2737.
31. Disterhoft, J. F., D. T. Golden, H. R. Read, D. A. Coulter & D. L. Alkon. 1988. AHP reductions in rabbit hippocampal neurons during conditioning are correlated with acquisition of the learned response. Brain Res. **462:** 118-125.
32. Coulter, D. A., J. J. LoTurco, M. Kubota, J. F. Disterhoft, J. W. Moore & D. L. Alkon. 1989. Classical conditioning reduces the amplitude and duration of the calcium-dependent afterhyperpolarization in rabbit hippocampal pyramidal cells. J. Neurophysiol. **61:** 971-981.
33. Bliss, T. V. P & G. L. Collingridge. 1993. A synaptic model of memory: Long-term potentiation in the hippocampus. Nature **361:** 31-39.
34. Morris, R. G. M., S. Davis & S. P. Butcher. 1991. Hippocampal synaptic plasticity and *N*-methyl-D-aspartate receptors: A role in information storage? *In* Long-term potentiation: A debate of current issues. M. Baudry & J. L. Davis, Eds.: 267-300. MIT Press. Cambridge, MA.
35. Bank, B, A. DeWeer, A. M. Kuzirian, H. Rasmussen & D. L. Alkon. 1988. Classical conditioning induces long-term translocation of protein kinase C in rabbit hippocampal CA1 cells. Proc. Natl. Acad. Sci. USA **85:** 1988-1992.
36. van der Zee, E., M. Kronforst & J. F. Disterhoft. 1994. Hippocampally-dependent trace eyeblink conditioning changes the immunoreactivity for muscarinic acetylcholine receptors and PKCgamma in the rabbit hippocampus. Soc. Neurosci. Abstr. **20:** 1433.
37. Malenka, R. C., D. V. Madison, R. Andrade & R. A. Nicoll. 1986. Phorbol esters mimic some cholinergic actions in hippocampal pyramidal neurons. J. Neurosci. **6:** 475-480.
38. Mattson, M. P. 1992. Calcium as sculptor & destroyer of neural circuitry. Exp. Gerontol. **27:** 29-49.
39. Landfield, P. W. & G. Morgan. 1984. Chronically elevating plasma Mg^{2+} improves hippocampal frequency potentiation and reversal learning in aged and young rats. Brain Res. **322:** 167-171.
40. Landfield, P., T. Pitler & M. Applegate. 1986. The effects of high Mg^{2+}/Ca^{2+} ratios on *in vitro* hippocampal frequency potentiation in young & aged rats. J. Neurophysiol. **56:** 797-811.

41. THOMPSON, L. T., R. A. DEYO & J. F. DISTERHOFT. 1990. Nimodipine enhances spontaneous activity of hippocampal pyramidal neurons in aging rabbits at a dose that facilitates learning. Brain Res. **535:** 119-130.
42. KOWALSKA, M. & J. F. DISTERHOFT. 1994. Dose- and concentration-dependent effect of nimodipine on learning rate of aging rabbits. Exp. Neurol. **127:** 159-166.
43. LANDFIELD, P. W. & T. A. PITLER. 1984. Prolonged Ca^{2+}-dependent afterhyperpolarizations in hippocampal neurons of aged rats. Science **226:** 1089-1092.
44. MOYER, J. R., JR., L. T. THOMPSON, J. P. BLACK & J. F. DISTERHOFT. 1992. Nimodipine increases excitability of rabbit CA1 pyramidal neurons in an age- and concentration-dependent manner. J. Neurophysiol. **68:** 2100-2109.
45. MOYER, J. R., JR. & J. F. DISTERHOFT. 1994. Nimodipine decreases calcium action potentials in an age- and concentration-dependent manner. Hippocampus **4:** 11-18.
46. VAN HOESEN, G. W. & A. R. DAMASIO. 1987. Neural correlates of cognitive impairment in Alzheimer's disease. *In* Handbook of Physiology Section 1: The Nervous System; Volume V. Higher Functions of the Brain, Part 2. V. B. Mountcastle, F. Plum & S. R. Geiger, Eds.: 871-898. American Physiological Society. Bethesda, MD.
47. DISTERHOFT, J. F., J. R. MOYER, L. T. THOMPSON, F. B. CUTTING & J. M. POWER. 1994. *In vitro* analyses of aging-related learning deficits. Soc. Neurosci. Abstr. **20:** 476.
48. MOYER, J. R., J. F. DISTERHOFT, J. P. BLACK & J. Z. YEH. 1994. Dihydropyridine-sensitive calcium channels in acutely-dissociated hippocampal CA1 neurons. Neurosci. Res. Comm. **15:** 39-48.
49. MAZZANTI, M. L., O. THIBAULT & P. W. LANDFIELD. 1991. Dihydropyridine modulation of normal hippocampal physiology in young and aged rats. Neurosci. Res. Comm. **9:** 117-126.
50. TSIEN, R. W., D. LIPSCOMBE, D. V. MADISON, K. R. BLEY & A. P. FOX. 1988. Multiple types of neuronal calcium channels and their selective modulation. TINS **11:** 431-437.
51. WESTENBROEK, R. E., M. K. AHLIJANIAN & W. A. CATTERALL. 1990. Clustering of L-type Ca^{2+} channels at the base of major dendrites in hippocampal pyramidal neurons. Nature **347:** 281-284.
52. EVANS, D. A., H. H. FUNKENSTEIN, M. S. ALBERT, P. A. SCHERR, N. R. COOK, M. J. CHOWN, L. E. HEBERT, C. H. HENNEKENS & J. O. TAYLOR. 1989. JAMA **262:** 2551-2556.
53. SELKOE, D. J. 1992. Aging brain, aging mind. Sci. Amer. **267:** 135-142.
54. DARTIGUES, J. F., M. GAGNON, J. M. MAZAUX, P. BARBERGER-GATEAU, D. COMMENGES, L. LETENNEUR & J. M. ORGOGOZO. 1992. Occupation during life and memory performance in nondemented French elderly community residents. Neurology **42:** 1697-1701.
55. PITLER, T. A. & P. W. LANDFIELD. 1987. Probable Ca^{2+}-mediated inactivation of Ca^{2+} currents in mammalian brain neurons. Brain Res. **410:** 147-153.
56. THIBAULT, O., N. M. PORTER & P. W. LANDFIELD. 1993. Low Ba^{2+} and Ca^{2+} induce a sustained high probability of repolarization openings of L-type Ca^{2+} channels in hippocampal neurons: Physiological implications. Proc. Natl. Acad. Sci. USA **87:** 11792-11796.
57. DUTAR, P., B. POTIER, Y. LAMOUR, P. EMSON & M. SENUT. 1991. Loss of calbindin-28K immunoreactivity in hippocampal slices from aged rats: A role for calcium? Eur. J. Neurosci. **3:** 839-849.
58. NABER, P. A., L. T. THOMPSON, P. G. M. LUITEN & J. F. DISTERHOFT. 1993. Age-related expression of calbindin D-28K and parvalbumin immunoreactivity in the rabbit hippocampus. Soc. Neurosci. Abstr. **19:** 388.
59. MCLACHLAN, D., L. WONG, C. BERGERON & K. G. BAIMBRIDGE. 1987. Calmodulin and calbindin D28K in Alzheimer disease. Alz. Dis. & Assoc. Disorders **1:** 171-179.
60. ALSHUAIB, W. B. & M. A. FAHIM. 1990. Aging increases calcium influx at motor nerve terminal. Int. J. Dev. Neurosci. **8:** 655-666.
61. PETERSON, C. & G. E. GIBSON. 1983. Aging and 3,4-diaminopyridine alter synaptosomal calcium uptake. J. Biol. Chem. **258:** 11482-11486.

62. PETERSON, C., D. G. NICHOLLS & G. E. GIBSON. 1985. Subsynaptosomal distribution of calcium during aging and 3,4-diaminopyridine treatment. Neurobiol. Aging **6:** 297-304.
63. BALL, M. J. 1977. Neuronal loss, neurofibrillary tangles and granulovacuolar degeneration in the hippocampus with aging and dementia. Acta Neuropathol. **37:** 111-118.
64. HYMAN, B. T., G. W. VAN HOESEN, A. R. DAMASIO & C. L. BARNES. 1984. Alzheimer's disease: Cell-specific pathology isolates the hippocampal formation. Science **225:** 1168-1170.
65. DISTERHOFT, J. F., J. BLACK, J. R. MOYER & L. T. THOMPSON. 1991. Calcium-mediated changes in hippocampal neurons and learning. Brain Res. Rev. **16:** 196-198.
66. SQUIRE, L. R. 1987. Memory and Brain. Oxford. New York.
67. GEINISMAN, Y., L. DE TOLEDO-MORRELL & F. MORRELL. 1986. Loss of perforated synapses in the dentate gyrus: Morphological substrate of memory deficit in aged rats. Proc. Natl. Acad. Sci. USA **83:** 3027-3031.
68. VAN DEN KERCKHOFF, W. & L. R. DREWES. 1989. Transfer of nimodipine and another calcium antagonist across the blood-brain barrier and their regional distribution *in vivo*. *In* Diagnosis and Treatment of Senile Dementia. M. Bergener & B. Reisberg, Eds.: 308-321. Springer-Verlag. Berlin.
69. STRAUBE, K. T., R. A. DEYO, J. R. MOYER, JR. & J. F. DISTERHOFT. 1990. Dietary nimodipine improves associative learning in aging rabbits. Neurobiol. Aging **11:** 659-661.
70. SCRIABINE, A., T. SCHUURMAN & J. TRABER. 1989. Pharmacological basis for the use of nimodipine in central nervous system disorders. FASEB J. **3:** 1799-1806.
71. SANDIN, M., S. JASMIN & T. E. LEVERE. 1990. Aging and cognition: Facilitation of recent memory in aged nonhuman primates by nimodipine. Neurobiol. Aging **11:** 573-575.
72. LEVERE, T. E. & A. WALKER. 1991. Aging and cognition: Enhancement of recent memory in rats by the calcium channel blocker nimodipine. Neurobiol. Aging **13:** 63-66.
73. FISHHOF, P. K., G. WAGNER, L. LITTSCHAUER, E. RUTHER, M. APECECHEA, R. HIERSEMENZEL, J. ROHMEL, F. HOFFMEISTER & N. SCHMAGE. 1989. Therapeutic results with nimodipine in primary degenerative dementia and multi-infarct dementia. *In* Diagnosis & Treatment of Senile Dermentia. M. Bergener & B. Reisberg, Eds.: 350-359. Springer-Verlag. Berlin.
74. SCHMAGE, N., K. BOEHME, J. DYCKA & H. SCHMITZ. 1989. Nimodipine for psychogeriatric use: Methods, strategies, & considerations based on experience with clinical trials. *In* Diagnosis & Treatment of Senile Dementia. M. Bergener & B. Reisberg, Eds.: 374-381. Springer-Verlag. Berlin.
75. BAN, T. A., L. MOREY, E. AGUGLIA, O. ASSARELLI, F. BALSANO, V. MARIGLIANO, N. CAGLIERIS, M. STERLICCHIO, A. CAPURSO, N. A. TOMASI, G. CREPALDI, D. VOLPE, G. PALMIERI, G. AMBROSI, E. POLLI, M. CORTELLARO, C. ZANUSSI & M. FROLDI. 1990. Nimodipine in the treatment of old age dementias. Prog. Neuro-Psychopharmacol. Biol. Psychiatry **14:** 525-551.
76. PITTERA, A., G. M. BASILE & S. CIANCITTO. 1990. Effect of nimodipine on chronic organic brain syndrome: Analysis of performance tests during medium-term treatment with the calcium antagonist. Curr. Ther. Res. **48:** 707-715.
77. TOLLEFSON, G. D. 1990. Short-term effects of the calcium channel blocker nimodipine (Bay-e-9736) in the management of primary degenerative dementia. Biol. Psychiatry **27:** 1133-1142.
78. JARVIK, L. 1991. Calcium channel blocker nimodipine for primary degenerative dementia. Biol. Psychiatry **30:** 1171-1172.
79. CARRILLO, M. C., L. T. THOMPSON, J. D. E. GABRIELI, B. J. NAUGHTON, A. HELLER & J. F. DISTERHOFT. 1994. Nimodipine enhances trace eyeblink conditioning in aging humans. Soc. Neurosci. Abstr. **20:** 387.

Ion Transport Systems and Ca^{2+} Regulation in Aging Neurons

MARY L. MICHAELIS

Department of Pharmacology
University of Kansas
Lawrence, Kansas 66045

Our knowledge of the mechanisms by which all cells, but particulary nerve cells, use Ca^{2+} pulses for intracellular signaling has increased exponentially in the last decade, and this advance has opened up the possibility for determining if altered Ca^{2+} regulation indeed plays some role in the age-dependent decline in some cellular functions. Certainly peripheral regulation of Ca^{2+} homeostasis, which is controlled by at least three systemic hormones—vitamin D, calcitonin, and parathyroid hormone—is well documented to change with age.[1-3] However, the picture is not so clear with intracellular Ca^{2+} handling mechanisms. This is due in part to the fact that there are so many systems that participate in determining the localized free Ca^{2+} concentrations in all types of cells at any given point in time, and we are still in the process of developing detailed information about each system and integrating that into a coherent picture of "Ca^{2+} regulation."[4] Since Ca^{2+} signaling in neurons is so crucial for neurotransmitter release and intercellular communication, new insights into mechanisms underlying neuronal Ca^{2+} homeostasis are particularly important for assessing the effects of the aging process on overall brain function.

Recent advances in protein chemistry, molecular biology, and neurophysiology have now provided us with powerful analytical tools for studying the behavior of the voltage-gated Ca^{2+} channels in nerve cells and eventually determining with confidence whether the aging process leads to alterations in the characteristics of these entities in neurons.[5,6] Similarly, the Ca^{2+} channels that exist on intraneuronal vesicular structures, namely, the IP_3 receptor and the ryanodine receptor, have now been isolated and the genes cloned, and we are well on our way to obtaining a clearer picture of how these systems operate in Ca^{2+} signaling activities.[7,8] Although these plasma membrane and intracellular membrane Ca^{2+} channels certainly account for a significant portion of the Ca^{2+} used in intracellular signaling, a number of other proteins participate in this process because of their ability to function as Ca^{2+} switches, such as calbindin and calmodulin,[9] or because of their actions as Ca^{2+} effector or mediator proteins, such as the annexins, the synaptic terminal proteins involved in transmitter release.[10,11] Among the neuronal proteins that play crucial, if not yet fully characterized, roles in regulating Ca^{2+} signaling processes, are the multiple membrane ion-translocating systems. The principal ion transport systems that contribute either directly or indirectly to the regulation of free intracellular Ca^{2+} concentrations $[Ca^{2+}]_i$ in synaptic terminals are represented schematically in FIGURE 1. Essentially the same systems are operative in other regions of the neuron, but their activity in synaptic terminals will be the principal focus of the following discussion.

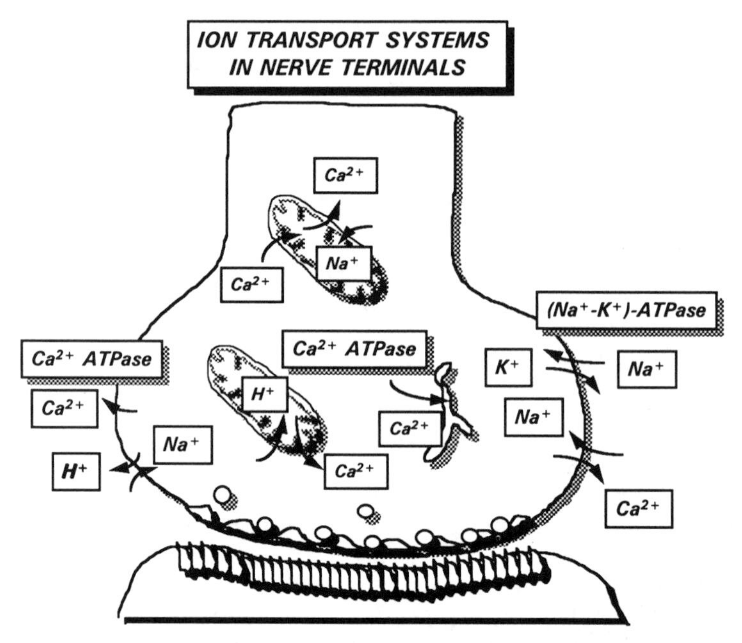

FIGURE 1. Synaptic terminal ion transport systems that influence regulation of intrasynaptsomal Ca^{2+}. The plasma membrane contains a Ca^{2+}-ATPase which pumps Ca^{2+} to the extracellular medium, plus Na^+/H^+ and Na^+/Ca^{2+} exchangers which can affect $[Ca^{2+}]_i$ by affecting the Na^+ gradient across the membrane. The Na^+ gradient ultimately depends on the Na^+/K^+-ATPase in the plasma membrane. Within the terminal Ca^{2+} can be released from the mitochondria via a Na^+/Ca^{2+} exchanger or reversal of the ATP synthase and taken up by the Ca^{2+} uniporter. A Ca^{2+}-ATPase in the endoplasmic reticulum can also remove Ca^{2+} from the cytosol.

INTRACELLULAR CA^{2+} TRANSPORT PROCESSES

As evident in FIGURE 1, several ion transporters are located on different organelles within the synaptic terminal, including the mitochondria, endoplasmic reticulum (ER), and plasma membrane. Depending on moment-to-moment intracellular conditions, these ion transporters participate in either decreasing the free $[Ca^{2+}]_i$ by removing it from the cytosol or bringing about an increase in $[Ca^{2+}]_i$ through several different mechanisms. Perhaps the most striking observation that becomes obvious in this figure is that Ca^{2+} regulation is very closely tied to the concentration of Na^+ at specific sites within the neuron and to the translocation of Na^+ across intracellular organelle and plasma membranes. Both the mitochondrial membrane and the plasma membrane have Na^+/Ca^{2+} exchangers that directly couple the translocation of Ca^{2+} to that of Na^+. The actual driving force for the plasma membrane exchanger is the electrochemical gradient created by the activity of the $(Na^+ + K^+)$-ATPase. Thus, neuronal distribution of Na^+ has a very significant impact on the activity of Ca^{2+}.

The actual contribution of each transport system just depicted is certain to vary under different conditions. Furthermore, even though the activity of each of these systems can be monitored in isolated preparations *in vitro,* the complex interactions among the ion regulators and the lack of potent, selective inhibitors for each have limited our understanding of the overall coordination of intracellular Ca^{2+} regulation in intact cells or intact nerve terminals. For example, despite several decades of work on the role of mitochondria in Ca^{2+} regulation, there is still argument about the significance of mitochondrial Ca^{2+} uptake in neuronal "Ca^{2+} buffering" under conditions of normal physiological activity.[4,12,13] However, the discovery by Denton, McCormack, and colleagues that Ca^{2+} transport in mitochondria may actually serve to link fluctuations in cytosolic Ca^{2+} directly to the activity of key matrix enzymes involved in oxidative metabolism has added an important new dimension to our conception of the relationship between mitochondria and Ca^{2+} signaling.[14,15] The possibility that Ca^{2+} may play a crucial role in mobilizing the mitochondria to respond to metabolic demands has potentially significant, unexplored implications for mechanisms underlying enhanced vulnerability to metabolic stress in aged organisms.

The membranous ER-like structures observed in the soma, axons, dendrites, and presynaptic terminals of neurons are believed to sequester cytosolic Ca^{2+} in a manner similar to that in the sarcoplasmic reticulum of muscle.[4,12,13,16] The uptake of Ca^{2+} occurs via a Ca^{2+}-ATPase belonging to the SERCA family, that is, sarcoplasmic-endoplasmic reticulum Ca^{2+} ATPases. The K_{act} for Ca^{2+} uptake is in the 0.2 μM range, suggesting this process may be an important homeostatic regulator in nerve cells.[12,16] More recent work on intracellular storage sites, particularly that with the SERCA inhibitor thapsigargin, suggests that this ATPase transports Ca^{2+} into the vesicular structures from which intracellular messengers such as inositol trisphosphate (IP_3) stimulate its release to amplify signaling events from the plasma membrane.[17] Actually, good evidence now exists for the presence of both IP_3 receptors and ryanodine receptors (or Ca^{2+}-induced Ca^{2+} release channels) on ER membranes.[18,19] These Ca^{2+} release sites are presumably localized on two different but related subtypes of organelles, with the one normally activated by IP_3 to release Ca^{2+} and the other possibly activated by Ca^{2+} and the newly identified messenger cyclic ADP ribose.[20] It has even been suggested that two different SERCAs may be present on the two subtypes of organelles.[21] The functional consequences of these differences are not yet known, but it is possible that specific conditions lead to the reuptake of Ca^{2+} into one pool rather than another. In fact, one of the most complex and interesting problems currently being investigated with regard to these ATPase-containing structures is the process by which depletion of Ca^{2+} from these vesicles signals a mechanism for extracellular Ca^{2+} influx which leads to refilling of the vesicles.[22] The data are clear that refilling from extracellular stores is required under conditions of repetitive activation of neurons, but neither the mechanism nor the mediators of this filling process have yet been identified. Moreover, the possibility that neuronal aging is accompanied by alterations in these processes has not yet been examined at all.

PLASMA MEMBRANE CA^{2+} TRANSPORT MECHANISMS

Ultimately, to preserve the signaling function of Ca^{2+}, a portion of the Ca^{2+} that enters the cytosol from the extracellular medium, particularly that entering via the voltage-gated channels, needs to be extruded across the plasma membrane. The

TABLE 1. Characteristics of Plasma Membrane Ca^{2+} ATPase Activity

Properties	References
Unidirectional Ca^{2+} translocation	16, 23
P-type ATPase with several isoforms and multiple genes	29-32
Very high affinity for Ca^{2+} K_{act} ~0.2-0.7 μM Ca^{2+}	16, 23
Low capacity transport system V_{max} ~200 pmol/mg protein/min	23
Calmodulin increases affinity for Ca^{2+} and maximal transport	28
Activated by acidic phospholipids	28

neuronal plasma membrane contains two major systems to carry out this extrusion function, a typical plasmalemmal ATP-dependent Ca^{2+} pump and an Na^+/Ca^{2+} exchanger or antiporter,[16,23,24] and these two extrusion mechanisms have been the focus of much work in this laboratory and the subject of our studies with synaptic membranes from the brains of aged rats.[26,27] The general characteristics of the CA^{2+}-translocating activity of these two transporters in both neurons and muscle cells have now been described, and this information has provided some leads into understanding the role that each of these systems plays in regulating intracellular Ca^{2+} concentrations. TABLE 1 summarizes some of the properties of the plasma membrane (Ca^{2+} + Mg^{2+})-ATPase, although not all characteristics have been specifically demonstrated in neuronal preparations. This enzyme is a member of the P-type ATPase family, all of which undergo phosphorylation of an aspartyl residue in their cytoplasmic domains as part of the ion transporting mechanism. Because this enzyme has a relatively high affinity for Ca^{2+}, it is believed to be involved in continuous maintenance of homeostatic levels of $[Ca^{2+}]_i$, rather than the extrusion of large quantities of Ca^{2+} which can enter with repetitive depolarizing stimuli. Nevertheless, the activity of this transporter can be enhanced by calmodulin, which associates strongly with the enzyme when Ca^{2+} levels are elevated.[28] The presence of the calmodulin binding site on this protein was used in calmodulin affinity chromatography procedures to purify the plasma membrane Ca^{2+} pump from a number of different tissues, and this has now led to the cloning of genes for the Ca^{2+} pumps.[29,30] Three isoforms of Ca^{2+}-ATPase are expressed in brain (PMCA1, 2, and 3), and they have >80% sequence homology. The PMCA 1 is widely expressed in many tissues, whereas PMCA 2 is expressed primarily in brain and heart, and PMCA 3 is expressed primarily in brain and skeletal muscle.[31] It is likely that some significant differences exist in the regulation of this transporter within specific brain regions. This likelihood of local differences in regulation is also supported by the existence of splice variants which have differences in the calmodulin binding domain and COOH-terminal regions.[32] However, we must await an understanding of the functional consequences of these variations in sequences.

The second plasma membrane Ca^{2+} transporter in neurons is the Na^+/Ca^{2+} exchanger, and the general properties of this system are summarized in TABLE 2. This

TABLE 2. Characteristics of Plasma Membrane Na^+/Ca^{2+} Exchanger Activity

Properties	References
Electrogenic transport (3 Na^+:1 Ca^{2+})	34
Bidirectional, depending on Na^+ and Ca^{2+} gradients and membrane potential	12, 24, 35
Selective for Na^+	24, 35
pH-sensitive—active from 7 to >9.5	48
Low affinity transport system K_{act} 10–40 μM Ca^{2+}	12, 24, 35
High capacity transport system V_{max} at 23°C = 15–24 nmol/mg protein/min with 40-fold Na^+ gradient	25, 48

transport system is believed to be the one that translocates large amounts of Ca^{2+} from inside the cell to the extracellular medium, particularly because its maximal transport capacity is about 100 times greater than that of the Ca^{2+}-ATPase. When this system was initially described in plasma membrane vesicles from heart and brain, the very low affinity for Ca^{2+} (10–40 μM) caused doubts about its physiological significance. However, the recent work of Llinas and colleagues has demonstrated clearly that in the environment of the presynaptic plasma membrane, where both voltage-gated Ca^{2+} channels and the Na^+/Ca^{2+} exchanger are located, the Ca^{2+} concentrations reached with repeated activation of neurons are orders of magnitude higher (200 μM) than the general cytosolic levels previously thought to represent the $[Ca^{2+}]_i$ (400–800 nM).[33] Observations such as these have made the reported exchanger affinities for Ca^{2+} well within the range of Ca^{2+} concentrations achieved just inside the plasma membrane of excitable cells where the high concentrations of Ca^{2+} channels also exist. Nevertheless, the fact that there are no known truly selective inhibitors of the exchanger has made it difficult to assess the contribution it makes to overall neuronal Ca^{2+} regulation under resting or activating conditions.

On the basis of studies with cardiac myocytes, it is clear that Na^+/Ca^{2+} exchange transport process is electrogenic, that is, 3 Na^+ ions are moved in exchange for 1 Ca^{2+} ion, and a measurable current is produced.[34] Transport activity depends on the gradients for Na^+ and Ca^{2+} across the membrane and on the membrane potential. When the potential is negative, operation of the exchanger is thought to produce outward movement of Ca^{2+}, but as the membrane depolarizes and the magnitude of the Na^+ gradient across the membrane is reduced, Ca^{2+} extrusion stops and the elevated intracellular Na^+ may lead to Ca^{2+} influx in exchange for intracellular Na^+. Potential models for the *in vivo* operation are discussed by Blaustein.[12,35]

Both cardiac sarcolemmal and retinal rod outer segment exchangers have now been cloned after exchanger proteins were isolated from each of these tissues.[36,37] Interestingly, little sequence homology is found between these two proteins, although their predicted structures within the membrane appear to be similar, that is, two sets

TABLE 3. Age-Related Changes in Ca^{2+}-Regulating Systems in Rat Brain Synaptic Terminals

Ca^{2+} Activity	Nature of Age-Related Change		References
Ca^{2+} ATPase activity	K_{act}	NC[a]	27, 39
	V_{max}	↓↓[b]	
Na^+/Ca^{2+} exchanger	K_{act}	↑	26, 39
	V_{max}	↓	
Depolarization-induced synaptosomal ^{45}Ca influx		↓↓↓	42, 43
Mitochondrial Ca^{2+} uptake		↓↓↓	43, 44
Intrasynaptosomal resting free $[Ca^{2+}]_i$		↑↑	40, 45
Rise in intrasynaptosomal $[Ca^{2+}]_i$ with depolarization		↑↑↑	40, 45
Rate of Ca^{2+} clearance following depolarization		↓↓	45, 46

[a] NC = no change.
[b] ↓ = <10% change; ↓↓ = 10–20% change; ↓↓↓ = >20% change.

of six transmembrane helices separated by a large cytoplasmic loop. A brain cDNA related to the cardiac exchanger was also recently discovered, but the actual brain protein has not been isolated.[38] As will be discussed, much of our recent work has been devoted to identifying brain protein(s) responsible for Na^+/Ca^{2+} exchange activity in synaptic terminals. The development of molecular information about this system, akin to that which is becoming available for the Ca^{2+} pump, is essential for us to determine whether the kinetic characteristics or expression of specific isoforms of the transporter change in the aging nervous system.

SYNAPTOSOMAL CA^{2+} REGULATION IN AGING NEURONS

Several years ago we suggested that perhaps subtle changes in the activity of plasma membrane Ca^{2+} extrusion systems could occur with aging of postmitotic cells such as neurons and myocytes.[26] Such alterations could lead to a decline in cellular function, because both types of cell are so dependent on their Ca^{2+} signaling and effector mechanisms and both cell types must last throughout the lifespan of an organism. Thus, even though all the essential molecular tools for characterizing age-dependent alterations in Ca^{2+}-regulating proteins are not yet available, a number of studies have been performed to see if *in vitro* assays of the activity of various proteins suggest age-dependent alterations.[26,27,39] In terms of the two plasma membrane Ca^{2+} extrusion systems just described, we used isolated synaptic plasma membrane vesicles from brains of rats at different ages to determine if any changes in the kinetic properties of these transporters could be detected.[26,27] The results of these studies are summarized along with related studies from other laboratories in TABLE 3. Our examination of Na^+/Ca^{2+} exchanger activity in aged synaptic membranes revealed rather subtle differences between adult and aged brain membranes, primarily in the form of a lower affinity for Ca^{2+} and a small (~10%) reduction in V_{max}, which was not statistically significant. Studies of the plasma membrane Ca^{2+} pump, including measurement of both Ca^{2+}-activated ATP hydrolysis and ATP-dependent $^{45}Ca^{2+}$ transport, have revealed consistent reductions in the V_{max} of this system in membranes from aged animals. Even more dramatic age-related decreases in Ca^{2+} efflux mediated by the Ca^{2+} pump and the Na^+/Ca^{2+} exchanger in intact synaptosomes were recently reported by Satrústegui and colleagues.[39]

In other efforts to study the effects of aging on brain function, several investigators used synaptosomes, an organelle that contains most of the critical Ca^{2+} signaling and regulating entities, to probe a number of dynamic aspects of Ca^{2+} regulation.[39–42] From these various studies it appears that voltage-activated ^{45}Ca uptake is decreased in synaptosomes from aged animals,[42,43] that the contribution of mitochondria to calcium homeostasis is also decreased,[43,44] that the cytosolic Ca^{2+} levels under resting conditions are elevated above those in young adult animals, and that the rise in cytosolic Ca^{2+} with depolarization is greater and lasts longer.[40,45] The elevated resting Ca^{2+} and the enhanced rise with depolarization could be due to alterations in mitochondrial Ca^{2+} regulation, to decreased Ca^{2+} extrusion across the plasma membrane, or to both. At this point we do not know the nature of the altered Ca^{2+} handling activity of the mitochondria nor do we know, at the molecular level, what if any changes have occurred in the plasma membrane transporting proteins. Nevertheless, the observations with intact synaptosomes are consistent with a number of observations in other systems in which Ca^{2+} regulation appears to be altered in the aging nervous system.[46,47]

All of these observations should be viewed with a couple of important caveats in mind. The first is that most of these *in vitro* assays with either synaptic vesicles or synaptosomes have been performed under conditions designed to optimize activity, that is, excess ATP, a large transmembrane Na^+ gradient, and so on. Such optimal conditions are not necessarily present at all times *in vivo,* and the fact that age-related differences could even be observed suggests that substantial changes in these transporters or their membrane environments do occur in aging brain. The second consideration to keep in mind is that any observed alterations in the activities of the transporters or other aspects of Ca^{2+} regulation may be an adaptive response to some other age-related processes taking place in the brain and therefore should not be assumed to reflect a necessarily deleterious loss of Ca^{2+} regulation. Our understanding of the aging process is too limited at this point to attach this interpretation to such observations.

STUDIES OF THE SYNAPTIC PLASMA MEMBRANE NA^+/Ca^{2+} EXCHANGER

As mentioned, the neuronal plasma membrane (Ca^{2+} + Mg^{2+})-ATPase pump was initially cloned in 1988 and additional isoforms were identified shortly thereafter. Since that time, a great deal of new information has been obtained on these transporters, and it should now be possible to use some of this information and these molecular tools to determine if aging leads to alterations in the expression of specific isoforms of the pump or if the observed decreases in activity are the result of some other process. The development of information about the Na^+/Ca^{2+} exchanger has not yet progressed this far. Much recent work in my laboratory has been devoted to isolating the protein(s) responsible for exchanger activity in synaptic terminals, and some of our preliminary results will be summarized along with information about the cloned exchanger from cardiac cells.

We recently reported on the solubilization, purification, and reconstitution of a protein fraction from synaptic membranes that contained two protein species (one ~50 kD and the other ~36 kD based on SDS-PAGE) and resulted in a substantial enrichment in specific exchanger activity.[48] Antibodies were raised against the ~36-kD protein, and the antibodies were shown to immuno-extract >90% of the exchanger activity present in a solubilized synaptic membrane preparation, suggesting that this protein is indeed involved in exchanger activity in synaptic terminals.[48] Although the antibodies were raised against the 36-kD protein, they also reacted with the ~50-kD protein in the purified fraction with activity, indicating that the two proteins were likely to be closely related or that the ~36-kD protein might be a proteolytic fragment of the larger protein. Recently completed immunohistochemical studies in rat brain confirmed that the antibodies most heavily labeled antigenic sites in synaptic terminal rich brain regions, such as the *stratum radiatum* and *stratum lucidum* of the CA1 and CA3 regions of the hippocampus, respectively, and the molecular layer of the cerebellum (manuscript in preparation). The cell bodies and the dendritic fields showed much less staining. Confocal microscopy studies with primary neurons in culture confirmed that most of the antigenic sites are located on the major neuritic processes believed to be the developing axons. In addition, a series of recently completed studies with primary neurons in culture and neuronal and non-neuronal cell lines revealed that the neurons had much higher exchanger activity than did other cells, and the intensity of labeling by the anti-36-kD antibodies on immunoblots was highly correlated with the actual level of exchanger activity present in various cells tested.[49] Thus, these observations all support the contention that the 36-kD synaptic membrane protein is associated with exchanger activity.

The antibodies raised against the 36-kD protein were used to screen a brain cDNA expression library, and three positive clones were found. One of these has now been partially characterized. At this point we determined that one of the positive clones (11-1) contained a 1.7-kb cDNA insert, and initial DNA sequencing indicated an open reading frame that codes for a 417 amino acid protein with an M_r of 48 kD. The insert has regions of homology with several Na^+ and Ca^{2+} channels and with the syntaxin family of proteins. *In vitro* transcription and translation experiments yielded a labeled product of approximately 50 kD. A restriction fragment (400 bp) of the cDNA was used to probe a brain poly A^+ RNA preparation and hybridized most strongly with a 1.9-kb transcript, with fainter bands at ~4.5 and 5.5 kb. Expression of cDNA in a bacterial host under the control of an inducible T7 polymerase promoter revealed enrichment of both the 50- and 36-kD proteins recognized by the antibodies on Western blots after partial purification of the protein from the bacteria. Attempts are underway to express the protein in a mammalian system that has low endogenous exchanger activity. Meanwhile, assessment of the exchanger activity in various preparations from the bacterial host cells indicated that cells transfected with the 1.7-kb insert had activity that was approximately 3-4 times higher than that in bacteria treated with the vector only. Clearly many issues are yet to be resolved, but our initial attempts to clone DNA for synaptic membrane protein suggest that it may indeed be an exchanger or at least intimately involved with synaptic exchanger activity.

The cloned Na^+/Ca^{2+} exchangers from heart and retinal rod outer segment share little homology with each other or with the protein we have been cloning.[36,37] The recently reported brain clone was identified by screening with probes based on the cardiac sequence and is highly homologous with the cardiac protein.[38] It is possible that there are multiple families of exchangers in various tissues and that the protein we have been studying is one that primarily exists in axonal and synaptic terminal membranes. Our current studies are focused on using antisense oligonucleotides to decrease expression of both our protein and the cardiac-like exchanger in primary neurons in culture and assessing the effects on exchanger activity and on the ability of the neurons to handle stimuli that lead to large elevations of free $[Ca^{2+}]_i$. These studies should enable us to obtain new insights into the role of Na^+/Ca^{2+} exchangers in enabling neurons to carry out Ca^{2+} signaling operations.

CONCLUSION

The studies from several laboratories briefly discussed herein certainly suggest that the properties of some critical neuronal Ca^{2+}-regulating systems are altered in aging neurons. It is possible that such alterations underlie, at least to some degree, the enhanced vulnerability of the aging brain to various types of insults, such as excessive glutamate activation or oxidative stress as discussed in other chapters in this volume. There is little doubt that regulation of normal Ca^{2+} signaling is crucial for neuronal viability and that loss of precise regulation can lead to toxicity and cell death. However, the entire phenomenon of cell death, sometimes divided into necrotic versus apoptotic processes, is turning out to involve an enormous array of signaling events, and the role that Ca^{2+} plays in all of these has yet to be delineated. Recent developments in our understanding of the molecular nature of the Ca^{2+}-regulating proteins are certain to contribute significantly to sorting out the steps in these cascades.

It is an understatement that as molecular genetics has revealed to us the molecular nature of many heretofore poorly characterized neuronal proteins, the picture has become far more complex than had ever been envisioned. Major receptor, channel, and transport proteins, as well as peptide transmitters and hormones, have all been shown to exist in multiple forms and to be encoded by multiple genes, making the task of deciphering the dynamics of cell biology, even in postmitotic cells, incredibly challenging. On the more optimistic side, however, is the development of a remarkable convergence of information as patterns begin to appear. One case in point can be illustrated by consideration of the long-standing assertion that free radical-induced cell damage is a major mechanism in the aging process. Certainly oxidation-induced alterations in Ca^{2+}-regulating proteins could underlie the changes in Ca^{2+} transport activities just described, and new research efforts are being directed to this question. However, until recently there was little to suggest a relationship between oxidative cell damage and the signaling events that occur in the active process of cell death, that is, apoptosis. The discovery of the *bcl*-2 oncogene[50] and its role in inhibiting apoptosis have revolutionized our thinking about this sequence of events. But in terms of convergence of information, it has been very exciting to learn recently that the *bcl* protein is a "death repressor molecule functioning in an *antioxidant* pathway."[51,52] Regulation of intracellular Ca^{2+} is still likely to be a major player in any of the signaling cascades activated by excitotoxic and/or oxidative stresses, but the complexity of the cell's armamentarium for dealing with these events is only now beginning to reveal itself. It is clear we have much, much more to learn.

REFERENCES

1. FUJITA, T. 1986. Aging and calcium. Miner. Electrolyte Metab. **12:** 149-156.
2. ARMBRECHT, H. J. 1986. Age-related changes in calcium and phosphorus reuptake by rat small intestine. Biochim. Biophys. Acta **882:** 281-286.
3. GALLAGHER, J. C., B. L. RIGGS, C. M. JERBAK & C. D. ARNAUD. 1980. The effect of age on serum immunoreactive parathyroid hormone in normal and osteoporotic women. J. Lab. Clin. Med. **95:** 373-385.
4. MILLER, R. J. 1991. The control of neuronal Ca^{2+} homeostasis. Prog. Neurobiol. **37:** 255-285.
5. TSIEN, R.W. & R. Y. TSIEN. 1990. Calcium channels, stores and oscillations. Ann. Rev. Cell Biol. **6:** 715-760.
6. MILLER, R. J. & A. P. FOX. 1990. Voltage-sensitive calcium channels. *In* Intracellular Calcium Regulation. F. Bronner, Ed.: 97-138. Wiley-Liss. New York.
7. FURUICHI, T., S. YOSHIKAWA, A. MIYAWAKI, K. WADA, N. MAEDA & K. MIKOSHIBA. 1989. Primary structure and functional expression of the inositol 1,4,5-trisphosphate binding protein P400. Nature **342:** 87-89.
8. HAKAMATA, Y., J. NAKAI, H. TAKESHIMA & K. IMOTO. 1992. Primary structure and distribution of a novel ryanodine receptor/calcium release channel from rabbit brain. FEBS Lett. **312:** 229-235.
9. HEIZMANN, C. W. & W. HUNZIKER. 1991. Intracellular calcium-binding proteins: More sites than insights. Trends Biochem. Sci. **16:** 98-103.
10. SMITH, V. L., M. A. KAETZEL & J. R. DEDMAN. 1990. Stimulus response coupling: The search for extracellular mediator proteins. Cell Regul. **1:** 165-172.
11. BENNETT, M. K. & R. H. SCHELLER. 1993. The molecular machinery for secretion is conserved from yeast to neurons. Proc. Natl. Acad. Sci. USA **90:** 2559-2563.
12. BLAUSTEIN, M. P. 1988. Calcium transport and buffering in neurons. Trends Neurosci. **11:** 438-443.
13. CARAFOLI, E. 1987. Intracellular calcium homeostasis. Ann. Rev. Biochem. **56:** 395-433.
14. MCCORMACK, J. G. & R.M. DENTON. 1990. The regulation of mitochondrial function in mammalian cells by Ca^{2+} ions. Trans. Biochem. Soc. **16:** 523-527.
15. MCCORMACK, J. G., A. P. HALESTRAP & R. M. DENTON. 1990. Role of calcium in the regulation of mammalian intramitochondrial metabolism. Physiol. Rev. **70:** 391-425.
16. MICHAELIS, E. K., M. L. MICHAELIS, H. H. CHANG & T. E. KITOS. 1983. High affinity Ca^{2+}-stimulated Mg^{2+}-dependent ATPase in rat brain synaptosomes, synaptic membranes, and microsomes. J. Biol. Chem. **258:** 6101-6108.
17. LYTTON, J., M. WESTLIN & M. R. HANLEY. 1991. Thapsigargin inhibits the sarcoplasmic or endoplasmic reticulum Ca-ATPase family of calcium pumps. J. Biol. Chem. **266:** 17067-17071.
18. BERRIDGE, M. J. 1993. Inositol triphosphate and calcium signalling. Nature **361:** 315-325.
19. SORRENTINO, V. & P. VOLPE. 1993. Ryanodine receptors: How many, where, and why? Trends Pharmacol. Sci. **14:** 98-103.
20. GALIONE, A. 1992. Ca^{2+}-induced Ca^{2+} release and its modulation by cyclic ADP-ribose. Trends. Pharmacol. Sci. **13:** 304-306.

21. Burgoyne, R. D., T. R. Cheek, A. Morgan, A. J. O'Sullivan, R. B. Moreton, M. J. Berridge, A. M. Mata, J. Colyer, A. G. Lee & J. M. East. 1989. Distribution of two distinct Ca^{2+}-ATPase-like proteins and their relationships to the agonist-sensitive Ca^{2+} store in adrenal chromaffin cells. Nature **342:** 72-74.
22. Putney, J. W., Jr. 1992. Inositol phosphates and calcium entry. *In* Advances in Second Messenger and Phosphoprotein Research. J. W. Putney, Jr., Ed. Vol. **26:** 143-159. Raven Press. New York.
23. Michaelis, M. L., T. E. Kitos, E. W. Nunley, E. Lecluyse & E. K. Michaelis. 1987. Characteristics of Mg^{2+}-dependent, ATP-activated Ca^{2+} transport in synaptic and microsomal membranes and in permeabilized synaptosomes. J. Biol. Chem. **262:** 4182-4189.
24. Michaelis, M. L. & E. K. Michaelis. 1981. Ca^{++} fluxes in resealed synaptic plasma membrane vesicles. Life Sci. **28:** 37-45.
25. Hoel, G., M. L. Michaelis, W. J. Freed & J. E. Kleinman. 1990. Characterization of Na^{+}-Ca^{2+} exchange activity in plasma membrane vesicles from postmortem human brain. Neurochem. Res. **15:** 881-887.
26. Michaelis, M. L., K. Johe & T. E. Kitos. 1984. Age-dependent alterations in synaptic membrane systems for Ca^{2+} regulation. Mech. Age. Dev. **25:** 215-225.
27. Michaelis, M. L. 1989. Ca^{2+} handling systems and neuronal aging. Ann. N.Y. Acad. Sci. **568:** 89-94.
28. Carafoli, E. 1991. Calcium pump of the plasma membrane. Physiol. Rev. **71:** 129-153.
29. Shull, G. E. & J. Greeb. 1988. Molecular cloning of two isoforms of the plasma membrane Ca^{2+} transporting ATPase from rat brain. Structural and functional domains exhibit similarity to Na^{+}, K^{+} and other cation transport ATPases. J. Biol. Chem. **263:** 8646-8657.
30. Strehler, E. E., P. James, R. Fischer, R. Heim, T. Vorherr, A. Filoteo, J. T. Penniston & E. Carafoli. 1990. Peptide sequence analysis and molecular cloning reveal two calcium pump isoforms in the human erythrocyte membrane. J. Biol. Chem. **265:** 2835-2842.
31. Burk, S. E. & G. E. Shull. 1992. Structure of the rat plasma membrane Ca^{2+}-ATPase isoform 3 gene and characterization of alternative splicing and transcription products. J. Biol. Chem. **267:** 19683-19690.
32. Keeton, T. P., S. E. Burk & G. E. Shull. 1993. Alternative splicing of exons encoding the calmodulin-binding domains and C termini of plasma membrane Ca^{2+}-ATPase isoforms 1, 2, 3 and 4. J. Biol. Chem. **268:** 2740-2748.
33. Llinas, R., M. Sugimori & R. B. Silver. 1992. Microdomains of high calcium concentration in a presynaptic terminal. Science **256:** 677-679.
34. Reeves, J. P. 1990. *In* Intracellular Ca^{2+} Regulation. F. Bronner, Ed.: 305-347. Alan R. Liss, Inc., New York.
35. Blaustein, M. P. 1988. Calcium and synaptic function. *In* Calcium in Drug Action. P. F. Baker, Ed.: 275-304. Springer-Verlag. Berlin.
36. Nicoll, D. A., S. Longoni & K. D. Philipson. 1990. Molecular cloning and functional expression of the cardiac sarcolemmal Na^{+}-Ca^{2+} exchanger. Science **250:** 562-565.
37. Reilander, H., A. Achilles, U. Friedel, G. Maul, F. Lottspeich & N. J. Cook. 1992. Primary structure and functional expression of the Na/Ca,K-exchanger from bovine rod photoreceptors. EMBO J. **11:** 1689-1695.
38. Furman, I., O. Cook, J. Kasir & H. Rahamimoff. 1993. Cloning of two isoforms of the rat brain Na^{+}-Ca^{2+} exchanger gene and their functional expression in HeLa cells. FEBS Lett. **319:** 105-109.
39. Martinez-Serrano, A., P. Blanco & J. Satrústegui. 1992. Calcium binding to the cytosol and calcium extrusion mechanism in intact synaptosomes and their alterations with aging. J. Biol. Chem. **267:** 4672-4679.
40. Martinez-Serrano, A., J. Vitórica & J. Satrústegui. 1988. Cytosolic free calcium levels increase with age in rat brain synaptosomes. Neurosci. Lett. **88:** 336-342.

41. FARRAR, R. P., S. MEHDI REZAZADEH, J. L. MORRIS, J. E. DILDY, K. GRAU & S. W. LESLIE. Aging does not alter Cytosolic calcium levels of cortical synaptosomes in Fischer 344 rats. Neurosci. Lett. **100:** 319-325.
42. MARTINEZ-SERRANO, A., E. BOGONEZ, J. VITÓRICA & J. SATRÚSTEGUI. 1989. Reduction of K^+-stimulated ^{45}Ca influx in synaptosomes with age involves inactivating and non-inactivating calcium channels and is correlated with temporal modifications in protein dephosphorylation. J. Neurochem. **52:** 576-584.
43. LESLIE, S. W., L. J. CHANDLER, E. M. BARR & R. P. FARRAR. 1986. Reduced calcium uptake by rat brain mitochondria and synaptosomes in response to aging. Brain Res. **329:** 177-183.
44. VITÓRICA, J. & J. SATRÚSTOGUI. 1986. Involvement of mitochondria in the age-dependent decrease in calcium uptake in rat brain synaptosomes. Brain Res. **378:** 36-48.
45. MICHAELIS, M. L., C. T. FOSTER & C. JAYAWICKREME. 1992. Regulation of calcium levels in brain tissue from adult and aged rats. Mech. Age. Dev. **62:** 291-306.
46. SMITH, D. O. 1988. Muscle-specific decrease in presynaptic calcium dependence and clearance during neuromuscular transmission in aged rats. J. Neurophysiol. **59:** 1069-1082.
47. FIFKOVA, E. & K. CULLEN-DOCKSTADER. 1986. Calcium distribution in dendritic spines of the dentate fascia varies with age. Brain Res. **376:** 357-362.
48. MICHAELIS, M. L., E. W. NUNLEY, C. JAYAWICKREME, M. HURLBERT, S. SCHUELER & C. GUILLY. 1992. Purification of a synaptic membrane Na^+/Ca^{2+} antiporter and immunoextraction with antibodies to a 36-kDa protein. J. Neurochem. **58:** 147-157.
49. MICHAELIS, M. L., J. L. WALSH, R. PAL, M. HURLBERT, G. HOEL, K. BLAND, J. FOYE & W. H. KWONG. 1994. Immunologic localization and kinetic characterization of a Na^+/Ca^{2+} exchanger in neuronal and non-neuronal cells. Brain Res., in press.
50. HOCKENBERY, D., G. NUÑEZ, C. MILLIMAN, R. D. SCHREIBER & S. J. KORSMEYER. 1990. Bcl-2 is an inner mitochondrial membrane protein that blocks programmed cell death. Nature **348:** 334-336.
51. VEIS, D. J., C. M. SORENSON, J. R. SHUTTER & S. J. KORSMEYER. 1993. Bcl-2-deficient mice demonstrate fulminant lymphoid apoptosis, polycystic kidneys and hypopigmented hair. Cell **75:** 229-240.
52. KANE, D. J., T. A. SARAFIAN, R. ANTON, H. HAHN, E. B. GRALLA, J. S. VALENTINE, T. ÖAURD & D. E. BREDESEN. 1993. Bcl-2 inhibition of neural death: Decreased generation of reactive oxygen species. Science **262:** 1274-1277.

Calcium and Neuronal Dysfunction in Peripheral Nervous System

W. H. GISPEN AND F. P. T. HAMERS

Rudolf Magnus Institute
Department of Pharmacology
Utrecht University
Universiteitsweg 100
3594 CG Utrecht, the Netherlands

Neuronal plasticity is defined as the capacity of the neuron to adapt to a changing internal or external environment, to previous experience, or to trauma. Plasticity is an essential feature of neuronal function, and it has long been recognized that loss of neuronal plasticity may contribute to the pathogenesis of neurodegenerative and age-related brain disorders.[1] Studies in developmental neurobiology revealed the significance of specific extracellular neuronotrophic growth factors[2] and intracellular Ca^{2+} homeostasis.[3] The role of the latter is relevant to the present paper. The Ca^{2+} hypothesis suggests that neurite outgrowth during development and repair can only proceed when intracellular [Ca^{2+}] levels lie within a specific outgrowth-permissive range. Cessation of outgrowth or reduction in neuronal plasticity can be induced by signals that elevate [Ca^{2+}] above that range. Extreme excitatory imbalance in transmitter input leading to marked elevation of intracellular [Ca^{2+}] is neurotoxic and eventually may lead to cell death.[3,4] In addition, it has been proposed that age-related neuronal deficits may originate from a relatively small disturbance of intraneuronal [Ca^{2+}] homeostasis that exists a long time.[25] In view of the apparent significance of the Ca^{2+} homeostasis to neuronal plasticity during development, repair, and aging, many attempts have been made to influence neuronal repair or the provide neuroprotection by the administration of drugs that are known to affect intracellular [Ca^{2+}] levels. The role of voltage-sensitive Ca^{2+} channels is of particular interest as they also respond to prolonged toxic depolarization and contribute greatly to the rise in free $[Ca^{2+}]_i$.

There are at least three classes of voltage-sensitive Ca^{2+} channels that require a large amount of depolarization before they are activated, among which is the 1,4-dihydropyridine-sensitive L-class.[5] In many cell types, L-channel opening is long-lasting. Most of what is known about L-channel function has been derived from studies in smooth and cardiac muscle. Following the discovery of 1,4-dihydropyridine receptors in brain tissue,[6] evidence has been accumulating that blockers of the L-type channel have profound neuro- and psychopharmacological effects in animal and man.[7–10] The mechanism by which such effects are brought about is still unknown. It may be that the known vascular effects are responsible for the observed neural activity, whereas an alternative mechanism would involve binding to the neuronal L-channels. The major Ca^{2+} channel blockers used clinically are relatively devoid of CNS side effects, presumably because they do not easily pass the blood-brain barrier. The major exception is nimodipine (isopropyl(2-methoxy-ethyl) 1,4-dihydro-2,6-dimethyl-4(3-nitrophenyl)-3,5-pyridinic dicarboxylase). This L-channel blocker penetrates the blood-brain barrier relatively well,[11] and therefore this member of the 1,4-dihydropyridine family is most often used to study the effects of L-channel blockers on neuronal Ca^{2+} homeostasis and neuronal function.

As has long been recognized, neurons once formed cannot divide, and although in some areas of the nervous system neuronal precursor cells are present during adulthood, severely damaged neurons die and are usually not replaced. Hence the regenerative capacity of the nervous system is limited. Postlesion repair, if occurring at all, may occur when the damage is restricted to the neuronal branches such as dendrites and neurites. In contrast to the CNS where the "milieu intérieure" seems to hamper repair, in the peripheral nervous system successful repair and reinnervation are known to occur, albeit at a slow pace. In view of the architecture of the peripheral nervous system and its relative great exposure to noxious mechanical and metabolic influences, many peripheral nerve dysfunctions (neuropathies) are known. Commonly these neuropathies reflect neurodegenerative processes, but signs of regenerative responses are observed as well. In the present paper we review some evidence suggesting that Ca^{2+}-channel blockers such as nimodipine may counteract neurodegeneration in the peripheral nerve and thus favor postlesion repair or supply additional neuroprotection.

NEURITE OUTGROWTH AND NEUROPROTECTION

Normally, the resting free $[Ca^{2+}]$ within a neuron is maintained at 10^{-8}-10^{-7} M. Such a low free cytosolic $[Ca^{2+}]$ allows for rapid and efficient detection of transient, small increases in intracellular $[Ca^{2+}]$ as a result of transmembrane signal transduction. Apparently, many stimuli that influence neuronal functions affect intracellular $[Ca^{2+}]$ by opening voltage-sensitive or receptor-mediated plasma membrane Ca^{2+} channels or intracellular Ca^{2+} stores. The neuron is equipped with a variety of systems that buffer the rise in intracellular $[Ca^{2+}]$ and provide for Ca^{2+} homeostasis.

Kater *et al.*[12] demonstrated that action potentials and neurotransmitter action converge at the level of the membrane potential, which in large part determines growth cone intracellular $[Ca^{2+}]$. They speculate that their socalled Ca^{2+} hypothesis provides a model for a basic mechanism in the regulation of growth cone behavior. The significance of their hypothesis is that neurite outgrowth can only proceed when intracellular $[Ca^{2+}]$ levels lie within a specific outgrowth-permissive range. Cessation of outgrowth can be induced by signals that elevate Ca^{2+} levels above that range. An example is the study by Robson and Burgoyne[13] using cultured dorsal root ganglion cells from neonatal rats under highly favorable conditions: the presence of conditioned medium, nerve growth factor, and 10% serum on polylysine- and laminin-coated substrata. The addition of high $[K^+]$, veratridine, or bradykinin to the culture medium inhibited the outgrowth of the neurons by about 60%. Drugs and K^+ cause depolarization, concurrent with ion shifts across the membrane, resulting in a high intracellular Ca^{2+} level. When cells were treated with nifedipine, a dihydropyridine compound blocking L-type Ca^{2+} channels, inhibition was reversed and outgrowth returned almost to normal levels. Similarly, Bär *et al.*[14] studied whether the Ca^{2+}-antagonist nimodipine could prevent inhibition of neurite outgrowth which occurs in depolarized cultures of rat fetal spinal neurons. Spinal cord slices were depolarized in culture with 50 mM K^+. Nimodipine (0.01-10 μM) was added before depolarization. After 5 and 7 days the effect of treatment was determined by (1) blind scoring of neurite outgrowth under phase contrast, and (2) measuring neurofilament (NF) protein with an ELISA. Neurite outgrowth was markedly decreased after depolarization, but was restored to control values by nimodipine (0.1 μM). Depolarization also led to a decrease in total NF content (18%). The NF content of depolarized slices incubated with 0.1 μM nimodipine was the same as that in the controls. Thus, depolarization-induced Ca^{2+} entry into spinal neurons inhibits neurite outgrowth from spinal neurons. Low concentrations of nimodipine prevented this inhibition. As nimodipine had no effect on neurite outgrowth in control cultures, it was concluded that nimodipine does not act as a neurotrophic factor but rather as a neuroprotective agent. Indeed, a number of investigators have reported that nimodipine may protect neurons from neurotoxic damage brought about by various procedures that presumably have in common an overload of intracellular free $[Ca^{2+}]_i$. To determine if Ca^{2+} antagonists were effective in blocking neuronal death under extreme conditions, differentiated PC12 cells were preincubated with rhodamine 123, a laser dye that indirectly monitors mitochondrial activity.[15] When the cells were exposed to either mercury or low oxygen conditions, there was a dose-dependent release of dye which was subsequently quantified with a spectrofluorometer. Both nimodipine and verapamil were effective at protecting mitochondrial viability as measured by inhibition of the efflux of this dye. Nanomolar concentration of nimodipine and micromolar concentrations of verapamil elicited optimum neuroprotective effects.[15] In another study fetal serotonergic neurons were dissected from rat embryos, dissociated, and grown in tissue culture.[16] The addition of 3,4-methylenedioxymethamphetamine (MDMA, ecstasy), a drug of abuse, resulted in significant neuropathology as indicated by a reduction in ^{3}H-5HT uptake capacity, a decrease in the number of surviving 5-HT neurons, and a decrease in the length of the surviving 5-HT neurites. Glutamate at concentrations of 0.5 and 1.0 mM also produced significant inhibition of 5-HT maturation by a process reversed by MI-801, thus via NMDA receptor activation. Nimodipine protected against the effects produced by glutamate and ecstasy, suggesting that closing the L-channel allows the neuron to more effectively deal with imbalances in the free-cytoplasmic levels of Ca^{2+} regardless of the nature of the toxic increase in $[Ca^{2+}]_i$.[16] This suggestion received further support by the study of Sucher *et al.*[17] in which it was reported that L-type Ca^{2+}-channel blockers attenuate NMDA-receptor-mediated neurotoxicity of retinal ganglion cells in culture and by that of Abele *et al.*[18] showing that NMDA-receptor toxicity of cultured hippocampal pyramidal cells could be ameliorated by the L-channel blocker nitrendipine.

NIMODIPINE AND RECOVERY OF FUNCTION AFTER CRUSH LESIONS OF PERIPHERAL NERVES

The neuron is an extremely specialized and differentiated cell and is a most vulnerable cell in the mammalian central and peripheral nervous system. It is generally assumed that damage to cell bodies of neurons results in irreversible degeneration and cell death. If, however, in the peripheral nervous system the damage is restricted to dendrites or axons, then regeneration and subsequent target reinnervation are possible.

In a first series of experiments, neuronal processes (nerve fibers) in the sciatic nerve were mechanically damaged by placing a crush lesion as described in detail by De Koning *et al.*[19] The speed of recovery can be monitored by measuring primarily sensory modalities in a reflex-foot withdrawal test and the quality of recovery by measuring primarily motor function in the analysis of the free walking pattern.[20] Van der Zee *et al.*[21] reported that orally administered nimodipine (in food pellets), in a dose of either 860 or 250 ppm, significantly enhanced recovery of both sensory and motor function in the nerve crush model, reducing the period of approximately 3 weeks needed for recovery by 2-3 days and improving the functional sciatic index derived from analysis of the free walking pattern. As the animals were group-housed and food pellets were available ad libitum, no precise information on the actual amount of nimodipine that was orally ingested could be obtained. A parallel experiment using high-dose nimodipine food pellets (860 ppm) demonstrated, by the use of a noninvasive tail-cuff method to monitor tail blood pressure in a longitudinal fashion, that the neurotrophic effect of nimodipine occurred without a general vasodilatory effect (Gispen, unpublished). Whether nimodipine exerts its effect by changes in existing microvessel function or new neural microvessel growth (ref. 22 and below) or on neuronal L-channels remains to be elucidated.

Further experiments using a similar sciatic nerve crush model demonstrated that both intraperitoneally and subcutaneously administered nimodipine (in a dose-dependent manner) enhances recovery of function as measured by a free walking pattern analysis. Doses of 10 and 20 mg/kg (PEG 400) given every other day were effective, whereas 5 mg/kg was not. Recently, enhancement of peripheral nerve function following a crush lesion was studied in the caudal nerve crush model as described by Gerritsen van der Hoop *et al.*[23] In this model the major caudal nerve is crushed, and recovery of sensory function can be monitored within days following the lesion over a relatively long time as the newly formed sprouts reinnervate the long tail of the rat. Hence, in addition to revealing information on an early effect on sprouting (shortening of delay) the model will also provide information on an effect on the growth rate of neurites. Interestingly, when nimodipine was given in a dose regimen of 20 mg/kg ip per 48 hours, the delay in first measurable recovery of function was shortened (early sprouting response) and the outgrowth rate of newly formed sprouts was enhanced. In contrast nifedipine, an L-channel blocker of the 1,4-dihydropyridine family with a lesser partition coefficient in neural parenchyma,[24] when given in a same dose regimen, was without effect. Similarly, the L-channel agonist Bay K 8644 in a dose of 0.5 mg/kg ip per 48 hours had no effect on the recovery of function following a crush lesion in the caudal nerve.[25] Proper dose response curves should be made before firm conclusions. Nonetheless, these preliminary data begin to shed some light on the mechanism of action of nimodipine under these conditions. If 1,4-dihydropyridine receptors in blood vessel walls play a role, why is nifedipine not effective? If blockade of L-channels improves recovery of function, why is further opening by Bay K 8644 not detrimental to the recovery of function? The role of the L-channel in the neurotrophic effect of nimodipine during nerve repair has also been questioned by results obtained with enantiomers of nimodipine. Radioligand binding studies point to a higher potency (5-20-fold) of the (-) enantiomer of nimodipine in the blockade of the L-channel than of the (+) enantiomer.[26-28] Thus, if L-channel blockade is important in the enhancement of postlesion repair, (-) nimodipine would be expected to facilitate recovery of function more effectively than would the (+) enantiomer. However, using the foot withdrawal reflex test, it was found that at various doses tested the (+) and (-) enantiomers of nimodipine were as active as the (±) racemate with little difference between the efficacy of (+) and (-) enantiomer.[29] Therefore, it appears that a difference in L-channel antagonism efficacy does not lead to a difference in efficacy in the nerve repair process. Several explanations should be considered. First, the functional tests as used here are not sensitive enough to measure subtle differences in L-channel activity of 1,4-dihydropyridines. Second, receptors other than the membrane L-type might be at play.[27] Finally, the neurotrophic action might be brought about by a common metabolite from racemate and enantiomers not acting on L-channels.

CISPLATIN NEUROPATHY

Cisplatin is an important drug in the treatment of various malignancies, most notably those of the genitourinary tract. The drug is rated as one of the most active oncolytics available for the treatment of ovarian cancer. As with other potent cytostatics, its use is accompanied by rather unpleasant side effects. These include short-term and relatively benign phenomena as nausea and vomiting but also long-term and more threatening ones as nephrotoxicity, ototoxicity, and neurotoxicity. The latter is nowadays regarded as the major dose-limiting side effect.

Cisplatin neuropathy is characterized by numbness, tingling sensations, loss of vibratory sensation, and diminished propriocepsis. Spinal ataxia resulting in disability may ensue. The motor system does not seem to be affected at all.[30,31] The first symptoms are those of a bilateral sensory neuropathy with numbness and tingling, often in a stocking and glove distribution. As the neuropathy continues, position sense becomes progressively impaired. In severe cases this may lead to wheelchair dependency because of severe sensory ataxia. These symptoms are often accompanied by uncomfortable and painful paraesthesias. Muscle strength remains essentially the same. The neuropathy becomes more severe with increasing cumulative doses of cisplatin[32] and may worsen even after cessation of therapy.[33] After more conventional dose schedules, cisplatin neuropathy is irreversible in 30-50% of patients.

Thus far the pathogenesis of cisplatin neuropathy remains obscure. The fact that cisplatin affects the sensory peripheral nerve fibers points towards involvement of the dorsal root ganglia (DRG). The neurons herein are not protected by the blood-brain barrier, and cisplatin accumulates in the DRG as easily as in tumor tissue, whereas cisplatin levels in brain and spinal cord are at least 10-fold lower.[31] How cisplatin affects the neurons, however, is still not clear.

In two experiments we evaluated the possibility that nimodipine might prevent the occurrence of cisplatin-induced neuropathy in an animal model for cisplatin-induced neuropathy. This animal model was developed by de Koning *et al.*[34] and has extensively been used for the study of putative neuroprotective measures in this particular disease.[35–37] In the first experiment, cisplatin neuropathy was induced by intraperitoneal injection of cisplatin (1 mg/kg 2 times a week for 10 weeks); a neuropathy as evidenced by a significant decrease in the sensory nerve conduction velocity (SNCV) was present from week 7 onwards (cumulative dose of 14 mg/

kg). Oral administration (800 ppm corresponding to about 32-48 mg/kg/day) of nimodipine, however, completely prevented the cisplatin-induced decrease in SNCV. In the second experiment the animals were subjected to more intense (double doses) cisplatin treatment, and nimodipine was administered intraperitoneally (20 mg/kg every 48 hours). Again nimodipine prevented the drug-induced decrease in SNCV. We moreover outruled the possibility that nimodipine negatively influences the antitumor efficacy of cisplatin *in vivo.* From this study we concluded that nimodipine protects against the development of cisplatin-induced neuropathy, even if the neuropathy is induced by a more aggressive cisplatin treatment schedule than is normally employed.[38] Moreover, as nimodipine does not interfere with cisplatin's oncolytic efficacy, this observation opens the way for its use in a clinical setting.

DIABETIC NEUROPATHY

Diabetic neuropathy is a demonstrable disorder evident either clinically or subclinically that occurs in the setting of diabetes mellitus without there being other causes for the peripheral neuropathy. Usually it involves a combination of sensory, motor, and autonomic nerve fiber abnormalities. The prevalance of diabetic neuropathy varies between 5 and 80%[39] and increases with both age and duration of diabetes mellitus.[40] Considerable uncertainty about the pathogenesis and treatment of diabetic neuropathy exists. Currently, three working hypotheses concerning the pathogenesis can be distinguished. Firstly, the "metabolic" hypothesis suggests that chronic hyperglycemia causes activation of the polyol pathway at the cost of myoinositol availability, resulting in reduced neural Na/K-ATPase activity and consequently reduced nerve conduction velocity and function. Secondly, the high glucose levels would lead to aberrant glycation of neural proteins and hence to disturbed neuronal functions. Thirdly, chronic hyperglycemia is known to induce endoneurial pathology, resulting in decreased neural blood flow. The subsequent anoxia is thus taken as the primary course of nerve dysfunction and degeneration. Recently, a new lead in understanding the pathogenesis seems to emerge. It was reported that serum from patients with type I insulin-dependent diabetes mellitus increased L-type calcium channel activity of insulin-producing cells.[41] The subsequent increase in free $[Ca^{2+}]_i$ was associated with programmed cell death. This sequence of events could be blocked by L-type Ca^{2+} channel blockers. As peripheral neurons possess L-type calcium channels as well and diabetic nerves are known to contain increased $[Ca^{2+}]$ levels,[42] it is suggested that enhanced L-type Ca^{2+} channel activity may underlie neuronal dysfunction representing diabetic neuropathies. It may well be that a combination of the aforementioned factors underlies the development of diabetic neuropathy. In two different experimental models of diabetic neuropathy the efficacy of chronic treatment nimodipine was determined. The first model involved the single administration of streptozotocin that is known to kill the insulin-producing β-cells in the islets of Langerhans and hence to increase blood glucose levels. No insulin therapy was applied. After several weeks of chronic hyperglycemia a severe neuropathy developed as evidenced by marked slowing of motor NCV (MNCV) and H-related sensory NCV (SNCV) in peripheral nerves.[43] Recently, Kappelle *et al.*[44] reported that chronic nimodipine treatment resulted in partial protection from metabolic neurointoxication in STZ-treated rats.

Treatment consisted of the 20 mg/kg ip per 48 hours regimen and commenced on the day that the rats were made diabetic with a single injection of STZ. The values for both MNCVs and SNCVs (FIG. 2) in nimodipine-treated rats were significantly higher than those in placebo-treated diabetic rats, although at the end of the 10-week experimental period in nimodipine-treated rats the NCVs were still lower than those in age-matched control rats. A series of follow-up experiments investigated whether nimodipine treatment would also be of benefit once the neuropathy had already been established. Such beneficial effects would be of more clinical relevance than the documented protection effect of nimodipine. Interestingly there is a dose-dependent enhancement of MNCV and SNCV in diabetic rats when nimodipine treatment commenced 4 weeks following the single STZ injection and a marked neuropathy had developed. As in the crush model, 10 and 20 mg/kg ip every 48 hours are effective, whereas 5 mg/kg is not. Furthermore, 20 mg/kg of nifedipine as in the tail crush model was less effective in the diabetic neuropathy model.[45]

The second model employed the BB/Wor rat, a strain of rats that genetically develop an autoimmune disease directed against insulin-producing cells and hence resulting in chronic hyperglycemia. Insulin therapy is required to maintain the rats that develop the disease. Again longitudinal measurement of SNCV and MNCV was used to assess the development of peripheral neuropathy. Six weeks following the onset of diabetes mellitus, as monitored by blood glucose levels, a marked decrease in SNCV and MNCV had occurred. At that time point treatment with nimodipine (20 mg/kg ip every 48 hours) commenced and lasted another 6 weeks. Drug treatment resulted in gradual improvement of nerve function as indicated by significantly increased NCV values. Thus, also in this experimental model of human diabetic neuropathy chronic treatment counteracted the onset of diabetic neuropathy.

In view of these preclinical data, pilot clinical testing seems warranted and one wonders what the precise mechanism of action of nimodipine in diabetic neuropathy might be.

As just indicated, one of the presumed mechanisms in the pathogenesis of diabetic neuropathy involves the observed vascular pathology resulting in neural hypoxia and ischemia.[46,47] Indeed, a correlation was observed between the hypoxia condition and the slowing of nerve conduction velocity.[46,48,49] Hence studies were performed to determine if part of the beneficial effect of nimodipine in experimental diabetic neuropathy may be explained by a drug effect on local neural blood supply. Sciatic nerve blood flow was assessed by laser-Doppler flowmetry in both the STZ rat and the BB/Wor rat model of human diabetic neuropathy. As expected under both conditions diabetes mellitus was paralleled by a gradual decrease in nerve blood flow during the first 10 weeks following the onset of the disease. After this decline nerve blood remained relatively stable at 60-70% of the original level. In both models interventive treatment with nimodipine (20 mg/kg ip per 48 hours) beginning 4-6 weeks after the onset of the disease when a neuropathy was well established, significantly improved local nerve blood flow without affecting gross hemodynamic parameters. Drug treatment reduced the flow deficit by nearly 50%.[45,50] In experiments with BB/Wor rats, using the values of nerve blood flow and nerve condition velocity per individual rat, a linear relationship between the two parameters could be established. Data suggest that calcium antagonists such as nimodipine that exert in principle both neuronal and vascular effects may be of specific interest to the treatment of mixed pathologies such as the diabetic neuropathy. So far in none of our experiments did we observe an effect of nimodipine on metabolic parameters such as blood glucose levels. Hence, improvement in diabetic nerve function is not the result of improvement in metabolic control. The fact that nifedipine is effective at the higher dose regimens,[51] if at all, may be explained by differences in pharmacokinetic properties of the two drugs. Alternatively, L-channel blockade may not be the sole mechanism by which nimodipine affects peripheral nerve plasticity (see above).

AGING AND MOTOR COORDINATION

Age-related deficits in motor function in rats were noted by many researchers and are discussed in detail by Coper *et al.*[52] The decline in motor performance is neither abrupt nor uniform and probably not random, but rather occurs in steps and is almost systematic. These investigators suggest that this decline with age is most likely due primarily to a loss of precision and a slowing down of central mechanisms in the control of motor function.[52] In a series of experiments the age-related decline in sensorimotor function of rats was followed. As expected, old rats showed a reduction in spontaneous locomotion, a disturbed walking pattern, and an impaired performance in various motor coordination tasks.[53]

Long-term (6 months) nimodipine treatment (860 ppm in the diet), starting at the age of 24 months, counteracted the age-related impairment of sensorimotor performance in rats. The most profound effect was observed during the first 4 months of treatment. During the last 2 months of treatment differences between treated and nontreated rats became smaller.[53] In similar experiments the effect of long-term nimodipine treatment on walking patterns of aging rats was studied.[53–55] Therefore, the hindpaws were dipped into photographic developer; thereafter the rats walked on photographic paper. Analysis of footprints showed that aging rats develop abnormal walking patterns; additional footprints, rotating prints, and fuzzy prints associated with dragging legs are typical.[53] Nimodipine administered in the diet delayed the onset of abnormal footprints significantly.

At the completion of the experiment, nerve conduction velocities and the number of myelinated axons in the sciatic nerve were determined. Drug treatment had enhanced the conduction velocities and resulted in higher nerve fiber density.[56] Using a principal other experimental design De Jong *et al.*[57] found lesser effects of nimodipine on histological parameters in peripheral nerves of old rats. Nonetheless collectively the data point to improvement in sensorimotor function and motor coordination of old rats by a possible combined action of the Ca^{2+} channel blocker on central and peripheral nervous system mechanisms.

CONCLUDING REMARKS

In the present paper I review our current knowledge of the effect of nimodipine on plasticity in the rat peripheral nervous system *in vivo*. Under proper conditions nimodipine in a dose-dependent manner and following different routes of administration was shown to (1) enhance recovery of function following crush lesioning of sciatic or caudal nerve, (2) protect against cisplatin-induced peripheral neuropathy, (3) delay onset of or ameliorate an already existing experimental diabetic neuropathy, and (4) counteract the age-related decline in sensorimotor function of senescent rats. Thus, under conditions in which plasticity is necessary for the regrowth of fibers, for protection or repair during intoxication, or for compensation during old age, treatment with a member of the Ca^{2+} antagonists of the 1,4-dihydropyridine type known to readily penetrate the neural parenchyma is extremely useful. These data need further confirmation in other animal models and eventually should serve as a basis for testing in clinical studies of peripheral nerve diseases (polyneuropathies, age-related disfunction, etc.). The animal studies should bring out whether or not blockade of the L-channel (neural or vessel) is part of the neurotrophic or neuroprotective profile of nimodipine. At present, the functional and neurophysiological data obtained on the rat peripheral nerve *in vivo* are open for interpretations other than L-channel blockade.

REFERENCES

1. Gelijns, A. C., P. J. Graff, F. H. Lopes Da Silva, W. H. Gispen & F. G. I. Jennekens. 1987. Future health care applications resulting from progress in the neurosciences: The significance of neural plasticity research. Health Policy **8:** 265-276.
2. Varon, S. 1993. Factors promoting the growth cone of the nervous system. *In* Discussions in Neurosciences. FESN. Geneva, 2, no. 3.
3. Kater, S. B. & L. R. Mills. 1991. Regulation of growth cone behavior by calcium. J. Neurosci. **11:** 891-899.
4. Choi, D. W. 1988. Glutamate neurotoxicity and diseases of the nervous system. Neuron **1:** 623-634.
5. Janis, R. A. & D. J. Triggle. 1991. Drugs acting on calcium channels. *In* Calcium Channels: Their Properties, Functions, Regulation, and Clinical Relevance. L. Hurwitz, L. D. Partridge & J. K. Leach, Eds. CRC Press. Boca Raton, Ann Arbor, Boston, London.
6. Belleman, P., A. Schade & R. Towart. 1983. Dihydropyridine receptor in rat brain labeled with [^{3}H] nimodipine. Proc. Natl. Acad. Sci. USA **80:** 2356-2360.
7. Betz, E., K. Deck & F. Hoffmeister, Eds. 1985. Nimodipine; Pharmacological and Clinical Properties. Schattauer Verlag. Stuttgart.
8. Traber, J. & W. H. Gispen, Eds. 1989. Nimodipine and Central Nervous System Function: New Vistas. Schattauer Verlag. Stuttgart.
9. Scriabine, A., T. Schuurman & J. Traber. 1989. Pharmacological basis for the use of nimodipine. Ann. N.Y. Acad. Sci. **552:** 698-706.
10. Bär, P. R. D., J. Traber, T. Schuurman & W. H. Gispen. 1990. CNS and PNS effects of nimodipine. J. Neural Transm. (Suppl.) **31:** 55-71.
11. Hoffmeister, F., H. P. Bellemann, U. Benz & W. Van Den Kerckhoff. 1985. Psychotrophic actions of nimodipine. *In* Nimodipine: Pharmacological and Clinical Properties. E. Betz, K. Deck & F. Hoffmeister, Eds.: 77-99. Schattauer Verlag. Stuttgart.
12. Kater, S. B., M. P. Mattson, C. Cohan & J. Connor. 1989. Calcium regulation of the neural growth cone. Trends Neurosci. **11:** 315-321.
13. Robson, S. J. & R. D. Burgoyne. 1989. L-type calcium channels in the regulation of neurite outgrowth from rat dorsal root ganglion neurons in culture. Neurosci. Lett. **104:** 110-114.
14. Bär, P. R., G. H. Renkema, C. M. H. H. M. Veraart & W. H. Gispen. 1993. Nimodipine prevents depolarization-induced inhibition of neurite outgrowth from cultured rat spinal cord slices. Cell Calcium **14:** 293-299.

15. FAHEY, J. M., B. HEDAYAT, J. R. COOK, J. L. ISAACSON & R. B. VAN BURSKIRK. 1989. Nimodipine's ability to inhibit neuronal death due to oxygen deprivation or exposure to methylmercury and its effects on differentiation. *In* Nimodipine and Central Nervous System Function: New Vistas. J. Traber & W. H. Gispen, Eds.: 117-140. Schattauer Verlag. Stuttgart.
16. AZMITIA, E. C. 1989. Nimodipine attenuates NMDA- and MNDA-induced toxicity of cultured serotonergic neurons: Evidence for a generic model of calcium toxicity. *In* Nimodipine and Central Nervous System Function: New Vistas. J. Traber & W. H. Gispen, Eds.: 141-159. Schattauer. Stuttgart, New York.
17. SUCHER, N. J., S. Z. LEI & S. A. LIPTON. 1991. Calcium channel antagonists attenuate NMDA receptor mediated neurotoxicity of retinal ganglion cells in culture. Brain Res. **551:** 297-302.
18. ABELE, A. E., K. P. SCHOLZ, W. K. SCHOLZ & R. J. MILLER. 1990. Excitotoxicity induced by enhanced exitatory neurotransmission in cultured hippocampal pyramidal neurons. Neuron **4:** 413-419.
19. DE KONING, P., J. H. BRAKKEE & W. H. GISPEN. 1986. Methods for producing a reproducible crush in the sciatic and tibial nerve of the rat and precise testing of return of sensory function. J. Neurol. Sci. **74:** 237-246.
20. DE KONING, P. & W. H. GISPEN. 1987. ORG 2766 improves functional and electrophysiological aspects of regenerating sciatic nerve in the rat. Peptides **8:** 415-422.
21. VAN DER ZEE, C. E. E. M., T. SCHUURMAN, J. TRABER & W. H. GISPEN. 1987. Oral administration of nimodipine accelerates functional recovery following peripheral nerve damage in the rat. Neurosci. Lett. **83:** 143-148.
22. YUAN, X. Q., T. L. SMITH, D. S. PROUGH, D. S. DE WITT, *et al.* 1990. Long-term effects of nimodipine on pial microvasculature and systemic circulation in conscious rats. Am. J. Physiol. **258:** 1395-1401.
23. GERRITSEN VAN DER HOOP, R., J. H. BRAKKEE, A. C. KAPPELLE, M. SAMSON, P. DE KONING & W. H. GISPEN. 1988. A new technique to evaluate recovery after peripheral nerve damage: Neurotrophic activity of Org 2766. J. Neurosci. Meth. **26:** 111-116.
24. SCRIABINE, A. 1987. Ca^{2+} channel ligands: Comparative pharmacology. *In* Structure and Physiology of the Slow Inward Calcium Channel. J. C. Venter & D. Triggle, Eds. Alan R. Liss. New York.
25. KAPPELLE, A. C., G. BIESSELS, T. VAN BUREN, J. H. BRAKKEE & W. H. GISPEN. 1993. Influence of dihydropyridines on the recovery of sensory function after a caudal nerve crush in the rat. Rest. Neurol. Neurosci., in press.
26. TOWART, R. & S. KAZDA. 1985. Effects of the calcium antagonist nimodipine on isolated cerebral vessels. *In* Nimodipine, Pharmacological and Clinical Properties. E. Betz, K. Deck & F. Hoffmeister, Eds. Schattauer Verlag. Stuttgart.
27. GLOSSMANN, H., G. ZERNIG, I. GRAZIADEI & T. MOSHAMMER. 1989. Non L-type Ca^{2+} channel linked receptor for 1,4-dihydropyridines and phenylalkylamines. *In* Nimodipine and Central Nervous System Function. J. Traber & W. H. Gispen, Eds. Schattauer Verlag. Stuttgart.
28. GODFRAIND, T., M. WIBO, C. EGLEME & J. WAUQUAIRE. 1985. The interaction of nimodipine with calcium channels in rat isolated aorta and in human neuroblastoma cells. *In* Nimodipine, Pharmacological and Clinical Properties. E. Betz, K. Deck & F. Hoffmeister, Eds. Schattauer Verlag. Stuttgart.
29. GISPEN, W. H., R. E. SPOREL-OZAKAT, E. DUCKERS & P. M. EDWARDS. 1991. The Ca^{2+} antagonist nimodipine and recovery of function following peripheral nerve damage in the rat. Neurosci. Res. Commun. **8:** 175-184.
30. ROELOFS, R. I., W. HRUSKESKY, J. ROGIN & L. ROSENBERG. 1984. Peripheral sensory neuropathy and cisplatin chemotherapy. Neurology **34:** 934-938.
31. THOMPSON, S. W., L. E. DAVIS, M. KORNFELD, R. D. HILGERS & J. C. STANDEFER. 1984. Cisplatin neuropathy. Cancer **54:** 1269-1275.

32. Gerritsen Van Der Hoop, R., J. C. Van Houwelingen, M. E. L. Van Der Burg, W. W. Ten Bokkel Huinink & J. P. Neijt. 1990. The incidence of neuropathy in 395 ovarian cancer patients treated with or without cisplatin. Cancer **66:** 1697-1702.
33. Grundberg, S. M., S. Sonka, L. L. Stevenson & F. M. Muggia. 1989. Progressive parestesias after cessation of therapy with very high-dose cisplatin. Cancer Chemother. Pharmacol. **25:** 62-64.
34. De Koning, P., J. P. Neijt, F. G. I. Jennekens & W. H. Gispen. 1987. Evaluation of cis-diamminedichloroplatinum(II) (Cisplatin) neurotoxicity in rats. Toxicol. Appl. Pharmacol. **89:** 81-87.
35. Hamers, F. P. T., W. H. Gispen & J. P. Neijt. 1991. Neurotoxic side-effects of cisplatin. Eur. J. Cancer **27:** 372-376.
36. Hamers, F. P. T., J. H. Brakkee, E. Cavalletti, M. Tedeschi, L. Marmonti, G. Pezzoni, J. P. Neijt & W. H. Gispen. 1993. Reduced glutathione protects against cisplatin-induced neurotoxicity in rats. Cancer Res. **53:** 544-549.
37. Hamers, F. P. T., C. Pette, B. Bravenboer, C. J. Vecht, J. P. Neijt & W. H. Gispen. 1993. Cisplatin-induced neuropathy in mature rats: Effects of the melanocortin-like peptide ORG 2766. Cancer Chemother. Pharmacol. **32:** 162-166.
38. Hamers, F. P. T., R. Gerritsen Van Der Hoop, P. A. Steerenburg, J. P. Neijt & W. H. Gispen. 1991. Putative neurotrophic factors in the protection of cisplatin-induced peripheral neuropathy in rats. Toxicol. Appl. Pharmacol. **111:** 514-522.
39. Dyck, P. J., J. Janes & P. C. O'Brien. 1987. Diagnosis, staging and classification of diabetic neuropathy and association with other complications. *In* Diabetic Neuropathy. P. J. Dyck, P. K. Thomas, A. K. Asbury & A. I. Winegrad, Eds. W. B. Saunders Co. Philadelphia.
40. Young, R. J., A. J. M. Boulton, A. F. Macleod, D. R. R. Williams & P. H. Sonsken. 1993. A multicentre study of the prevalence of diabetic peripheral neuropathy in the United Kingdom hospital clinical population. Diabetologia **36:** 150-154.
41. Juntti-Berggren, L., O. Larsson, P. Rorsman, C. Ämmälä, K. Bokvist, K. Wåhlander, P. Nicotera, J. Dypbukt, S. Orrenius, A. Hallberg & P. O. Berggren. 1993. Increased activity of L-type Ca^{2+} channels exposed to serum from patients with type I diabetes. Science **261:** 86-90.
42. Lowery, J. M., J. Eichberg, A. J. Saubermann & M. Lopachin, Jr. 1990. Distribution of elements and water in peripheral nerve of streptozotocin-induced diabetic rats. Diabetes **39:** 1498-1503.
43. Van Der Zee, C. E. E. M., R. Gerritsen Van Der Hoop & W. H. Gispen. 1989. Beneficial effect of Org 277 in the treatment of peripheral neuropathy in streptozotocin-induced diabetic rats. Diabetes **38:** 225-230.
44. Kappelle, A. C., B. Bravenboer, J. Traber, D. W. Erkelens & W. H. Gispen. 1993. The Ca^{2+} antagonist nimodipine counteracts the onset of an experimental neuropathy in streptozotocin induced diabetic rats. Neurosci. Res. Commun., in press.
45. Kappelle, A. C., G. Biessels, B. Bravenboer, T. Van Buren, J. Traber, D. J. De Wildt & W. H. Gispen. 1993. Beneficial effect of the Ca^{2+} antagonist nimodipine on existing diabetic neuropathy in the BB/Wor rat. Br. J. Pharmacol., in press.
46. Low, P. A., R. R. Tuck, P. J. Dyck, J. D. Schmelzer & J. K. Yao. 1984. Prevention of some electrophysiologic and biochemical abnormalities with oxygen supplementation in experimental diabetic neuropathy. Proc. Natl. Acad. Sci. **81:** 6894-6898.
47. Dyck, P. J. 1989. Hypoxic neuropathy: Does hypoxia play a role in diabetic neuropathy? Neurology **39:** 111-118.
48. Low, P. A., R. R. Tuck & M. Takeuchi. 1987. Nerve microenvironment in diabetic neuropathy. *In* Diabetic Neuropathy. P. J. Dyck, P. K. Thomas, A. K. Asbury, A. I. Winegrad & D. Porte, Eds. Saunders. Philadelphia.
49. Hendriksen, P. J., P. L. Oey, G. H. Wieneke, A. C. Van Huffelen & W. H. Gispen. 1992. Hypoxic neuropathy versus diabetic neuropathy: An electrophysiological study in rats. J. Neurol. Sci. **110:** 99-106.

50. Kappelle, A. C., G. Biessels, T. Van Buren, D. W. Erkelens, D. J. De Wildt & W. H. Gispen. 1993. Effects of nimodipine on sciatic nerve blood flow and vasa nervorum responsiveness in the diabetic rat. Eur. J. Pharmacol., in press.
51. Cameron, N. E. & M. A. Cotter. 1993. Potential therapeutic approaches to the treatment or prevention of diabetic neuropathy: Evidence from experimental studies. Diab. Med. **10:** 593-605.
52. Coper, H., B. Janicke & G. Schulze. 1986. Biophysiological research on adaptivity across the life-span of animals. *In* Life-span Development and Behaviour. P. D. Baltes, D. L. Featherman & R. M. Lerner, Eds.: 207-232. Erlbaum. Hilsdale.
53. Schuurman, T., H. Klein, M. Beneke & J. Traber. 1987. Nimodipine and motor deficits in the aged rat. Neurosci. Res. Commun. **1:** 9-15.
54. Scriabine, A., T. Schuurman & J. Traber. 1989. Pharmacological basis for the use of nimodipine in central nervous system disorders. FASEB J. **3:** 1799-1806.
55. Schuurman, T. & J. Traber. 1989. Effects of nimodipine on behaviour of old rats. *In* Nimodipine and Central Nervous System Function: New Vistas. J. Traber & W. H. Gispen, Eds.: 195-208. Schattauer Verlag. Stuttgart.
56. Van Der Zee, C. E. E. M., T. Schuurman, R. Gerritsen Van Der Hoop, J. Traber & W. H. Gispen. 1990. Beneficial effect of nimodipine on peripheral nerve function in aged rats. Neurobiol. Aging **11:** 451-456.
57. De Jong, G. I., A. S. P. Jansen, E. Horvath, W. H. Gispen & P. G. M. Luiten. 1991. Nimodipine effect on cerebral microvessels and sciatic nerve in aging rats. Neurobiol. Aging **13:** 73-81.

Cerebrovascular, Neuronal, and Behavioral Effects of Long-Term Ca^{2+} Channel Blockade in Aging Normotensive and Hypertensive Rat Strains

P. G. M. LUITEN,[a,b] G. I. DE JONG,[a] AND T. SCHUURMAN[c]

[a]*Department of Animal Physiology*
University of Groningen
Kerklaan 30, 9751 NN Haren, the Netherlands

[c]*Institute for Neurobiology*
Troponwerke, Neurather Ring 1
5000 Cologne 80, Germany

In the search for mechanisms underlying the detrimental influence of the aging process on the functions of the nervous system, the progressive derangement of intracellular calcium homeostasis has gained increasing attention.[1-3] Changes in calcium homeostasis during aging are considered to lead to a gradual but chronic Ca^{2+} overload with serious consequences for neuronal excitation and cellular communication.[2,4]

In contrast to extensive knowledge on altered $[Ca^{2+}]$ in the cellular and extracellular components of the cardiovascular system during aging,[5] such information on cerebral vasculature is scarce. The total free $[Ca^{2+}]$ in major cardiac and body arteries show a dramatic, progressive increase in an age-dependent fashion, which is accelerated by risk factors like nicotine, diabetes, and hypertension.[6]

Like the peripheral vessels, however, the vasculature of the brain is prone to considerable pathological changes during the aging process,[7,8] which are apt to exert a profound influence on the neuronal functions of the brain.[9] Because the blood-brain barrier (BBB) limits free entry of nutrients and metabolites from the blood stream to the parenchyma, the condition and integrity of the cerebral microvascular system may have a profound impact on general neuronal activity and the nervous control of behavioral functions.[10] As such, the causal mechanism of behavioral impairment as a consequence of aging may be regarded as being one of the key questions in aging research. The importance of the BBB for brain functioning prompted us to study the consequences of the aging process with specific interest in the relation between vascular, neuronal, and behavioral parameters.

[b]Address for correspondence: Dr. P. G. M. Luiten, Department of Animal Physiology, University of Groningen, Kerklaan 30, 9751 NN Haren, the Netherlands.

In the homeostatic regulation of intracellular free calcium $[Ca^{2+}]_i$ various mechanisms are involved that interact in a complex fashion. Free intracellular Ca^{2+} is the result of a balance between receptor-, channel-, and electric potential related Ca^{2+} influxes, intracellular release mechanisms, buffering by Ca^{2+} binding proteins, intracellular uptake and storage, and efflux by ion exchange and ion pumps.[11]

In the search for tools to antagonize the chronic elevation of $[Ca^{2+}]_i$ associated with brain aging, it has become a promising strategy to limit Ca^{2+} influx by blockade of Ca^{2+} entry pathways.[12] Nimodipine, a lipophilic dihydropyridine acting as an L-type channel Ca^{2+} antagonist, easily crosses the membranes of the BBB, and this way has ready access to intra- and extracellular components of neuronal, glial, and vascular domains of the nervous system. Evidence has accumulated in recent years that blockade of Ca^{2+} influx via L-type channels with this compound exerts powerful effects on neuronal,[4] cerebrovascular,[13] and behavioral[14] alterations in aging mammals. Nimodipine was demonstrated to enhance neuronal excitability and improve behavioral and cognitive performance in senescence in several mammalian species including man.[15,16] These nimodipine findings together with our own experience with drug application on cerebrovascular condition prompted us to investigate effects of chronic application of this drug on some vascular, neuronal, and behavioral characteristics of aging normotensive rats.

The nimodipine effects in normotensive animals were extended with studying the influence of this drug in aging hypertensive stroke-prone rats. Hypertension is considered a pathological condition that during aging poses an additional risk factor for the development of cardiovascular and cerebrovascular infarctions and ischemia.[17] In that sense hypertension may be thought to accelerate the dysfunctions and pathologies attributed to the aging process. Hypertension enhances the already dramatic aging-related increase of $[Ca^{2+}]$ in the vascular tissue in several vessels of the peripheral arterial system, pointing to a common mechanism in disturbance of Ca^{2+} homeostasis. Hypertension not only affects peripheral vessels, but also threatens cerebral vasculature in senescence when stroke and edema may develop.[7,17–19] The fact that hypertension accelerates $[Ca^{2+}]$ in the vascular wall in the course of the aging process suggests that aging and hypertension share common mechanisms of altered Ca^{2+} homeostasis. For that reason it became interesting to establish the influence of the Ca^{2+} entry blocker nimodipine in aging hypertensive rats.

EFFECTS OF NIMODIPINE IN AGING NORMOTENSIVE RATS

When investigated by electronmicroscopy, a large variety of alterations can be observed in the brain of the aging mammal. In senescence changes occur in essentially all cellular and extracellular components of the brain such as nerve cells, various glial cell types, and cerebrovascular cells. As described by many investigators, these changes include the gradual accumulation of lipofuscin (FIG. 1A), excessive myelin breakdown by oligodendrocytes, neuronal degeneration (FIG. 1B), synaptic reorganization and glial hypertrophy, and degenerative changes of the vascular wall. Notably the latter category of vascular breakdown was studied in greater detail with specific interest for the progressive course of events and the influence of chronic treatment with nimodipine on this vascular degenerative process.

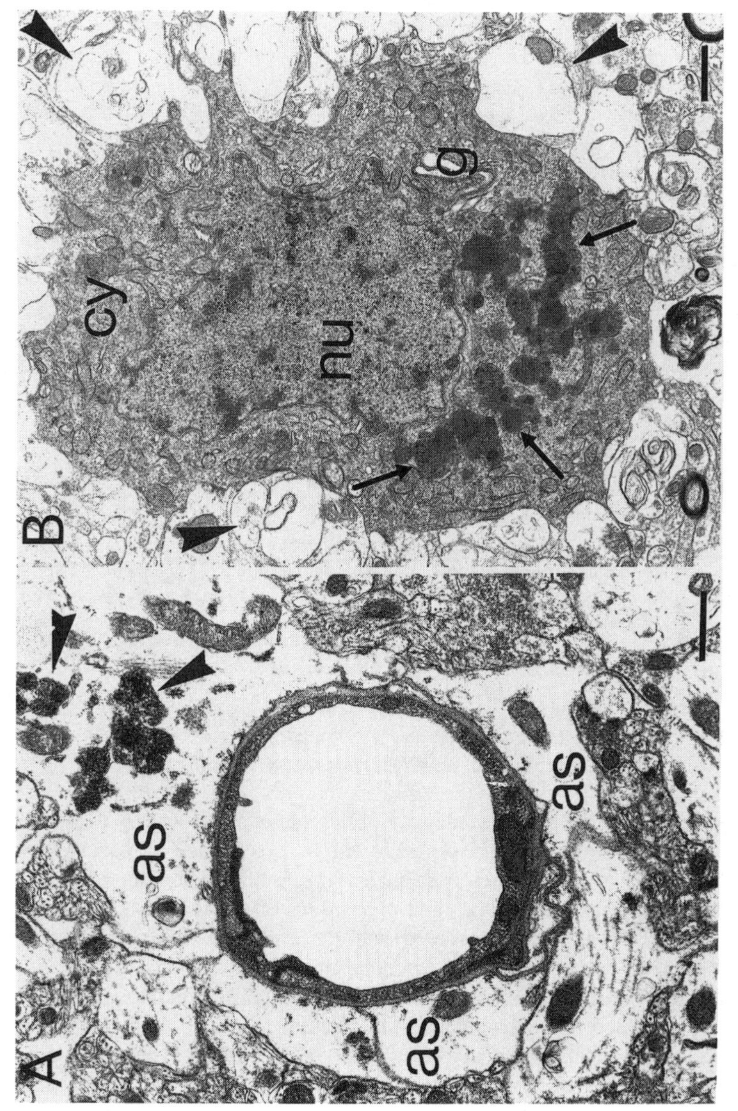

FIGURE 1. **(A)** Electron micrograph of hypertrophied astroglial endfeet (as) surrounding a microvessel in the cortex of a 30-month-old rat. *Arrowheads* point to lipofuscin granules, a common feature of aging glia and neurons. **(B)** Degenerating cell body of a pyramidal neuron in the cortex characterized by the irregular shape of cell and nucleus. The cytoplasm is electron dense and contains many lysosomes (*arrows*). Scale bar = 1 μm.

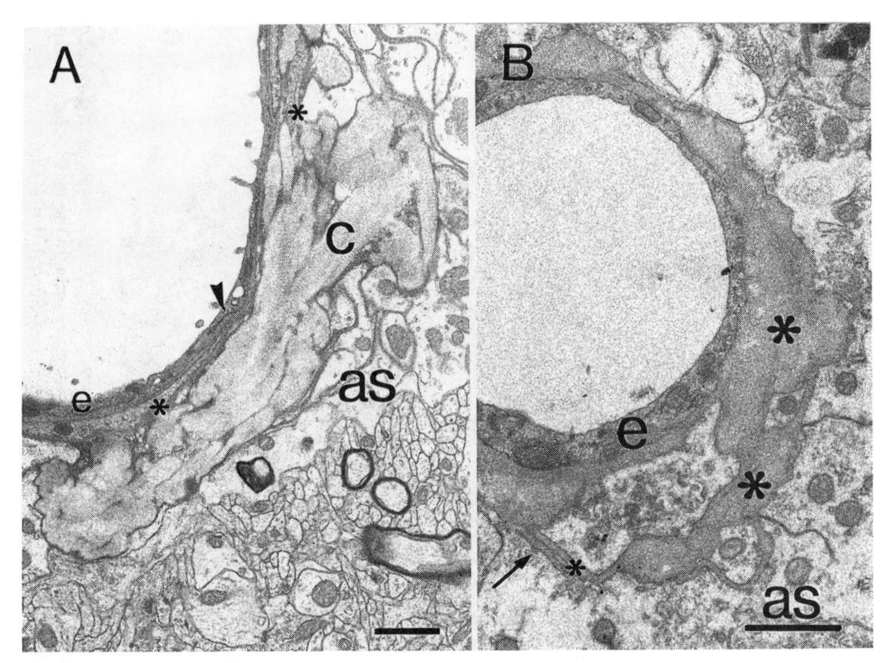

FIGURE 2. Cortex of rat aged 30 months. **(A)** Appearance of perivascular deposits consisting of a dense amount of banded collagen fibrils (c). The vascular lumen is lined by the endothelium (e) containing several large vacuoles, but with an intact tight junction. Astrocytes are hypertrophied (as). **(B)** Thickening of basement membrane (*) as a form of perivascular deposition. Scale bar = 1 μm.

Aging, Nimodipine, and Microvascular Morphology

This study was focused on the integrity of the microvessels as they comprise the anatomical substrate of the BBB. As reported in a sequence of papers, several structural alterations occur in senescence such as stages of degeneration of pericytes[20] and a series of infrequent anomalies and depositions of collagen-like and collagen-derived components in the microvascular wall designated as fibrosis and basement membrane thickening[7,8,21] (FIG. 2). It is notably the perivascular deposit, which, throughout the lifespan of the Wistar rat of approximately 32 months, is subject to a prominent influence of treatment with the calcium blocker nimodipine. In aging control animals the incidence of perivascular deposits gradually increases up to the age of 30 months and then levels off over 30 months in the parietal motor cortex.[13] During the latter final life stage the nature of the deposits changes from an obvious collagen fiber appearance to an amorphic bed of perivascular basement membrane thickening most likely a result of molecular depolymerization.[13] During the aging process we treated groups of Wistar rats chronically with an optimal dose of 1,000 ppm nimodipine administered via their daily food intake. The drug effects were studied after several treatment periods from 16-30 months, 24-30 months, and 24-

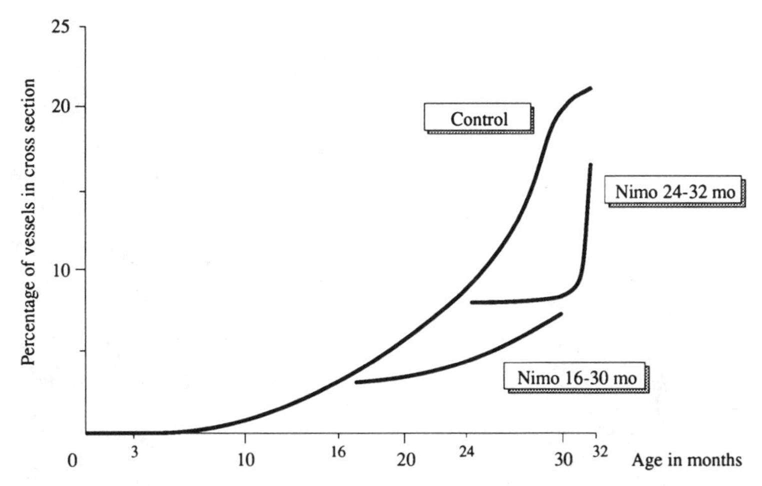

FIGURE 3. Diagram with the development of perivascular deposits during the lifespan of Wistar rats up to the age of 32 months. Note that the gradual increase during aging levels off at ages over 30 months. Treatment with the Ca^{2+} antagonist nimodipine from 16-30 months suppresses the formation of the deposits, as it does from 24-30 months. Over the age of 30 months drug treatment becomes ineffective.

32 months.[13,22] After all treatment periods the incidence of perivascular deposits was significantly suppressed except for the final life phase over 30 months. Here we observed that further treatment could no longer delay the malformations of the perivascular wall (FIG. 3). The treatment effects were proportional to the duration of treatment and the starting point of nimodipine application. More detailed analysis of the deposits in the microvascular wall revealed the collagen fibrotic nature of the perivascular malformation, which is characterized by typical banded fibrils with a periodicity of 64 nm. An important observation from a functional point of view was the apparent endothelial origin of the fibrosis[13] (FIG. 4A). In several instances collagen-like fibrils could be detected within the endothelial vacuoles or cytoplasm, whereas in the region with the thickened basement membrane often large amounts of pinocytotic vesicles occurred indicative of abnormal transport processes over the BBB (FIG. 4B). At the same time the affected microvessel was surrounded by enlarged astrocytic endfeet. The complex of aberrant microvascular structure may well be interpreted as the morphologic basis of dysfunctions of the BBB that coincide with the decreased BBB transport capacity during aging[23-25] (FIG. 5). The aforementioned observations were made on the parietal cortex as a representative of forebrain cortical structures. Sample studies on hippocampus and spinal cord indicate that the aging-related vascular decline and the nimodipine effects were basically similar in these structures of the central nervous system. The spinal effects, however, were of a lower magnitude, which may be related to the lower density of L-type Ca^{2+} channels in the brainstem and spinal cord.[26]

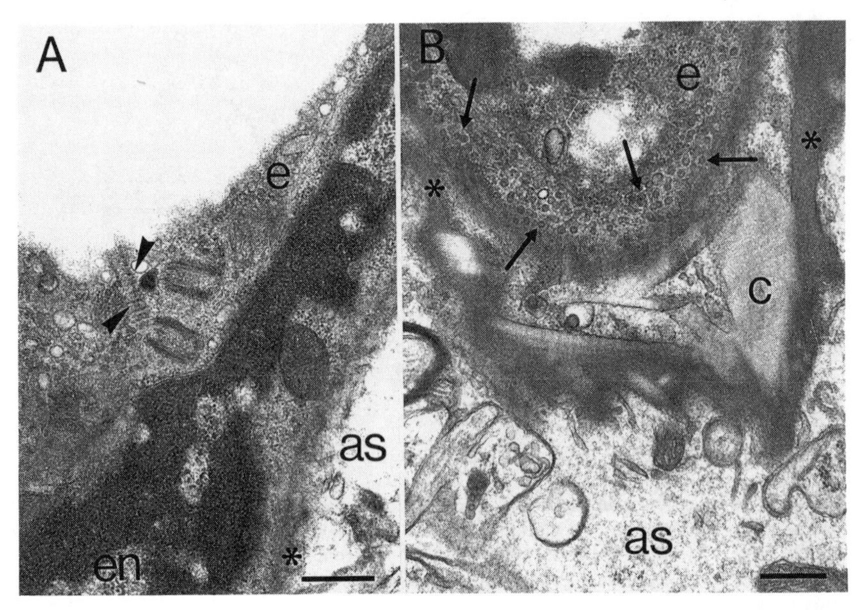

FIGURE 4. **(A)** Short collagen fibril fragment (*arrowhead*) in the cytoplasm of the endothelial cell (e). **(B)** Occurrence of large number of pinocytotic vesicles (*arrows*) in the endothelial cell of a microvessel in the cortex of an aged (30-month-old) rat. Within the basement membrane the collagen fibrils are deposited, while the astrocytic endfeet are enlarged. Scale bar = 0.5 μm.

Aging, Nimodipine, and Synaptic Ultrastructure

Apart from the vascular changes in aging animals, a variety of neuronal changes also were recorded. In more detail we established the impact of the calcium antagonist

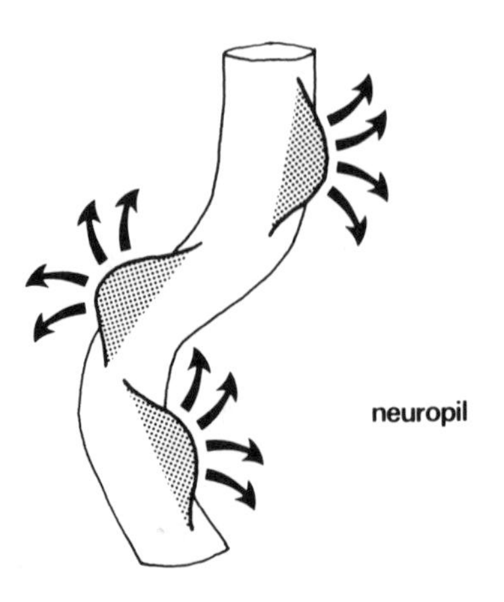

FIGURE 5. Schematic impression of local perivascular deposits in microvessel walls in the aging brain. The impact of microvascular aberrations is indicated by the *arrows* that represent dysfunction of the blood-brain barrier and its influence on the surrounding neuropil.

on the synaptic density and size exemplified on the supragranular layer of the dentate gyrus. This region was selected because of the general consensus that in this region the number of synapses is decreased in aged mammals including man in health and disease.[27–29] The synaptic changes during aging were assessed with the ethanolic-phosphotungstic acid method which selectively stains for synaptic structures for electron microscopic application. Our observations were in line with the reports of others that the synaptic density expressed as the number of synapses per cubic millimeter significantly decreased by a fourth.[30] Although the average synaptic surface did not change, the decrease in density coincided with a very significant decline in total synaptic surface area. The effect of nimodipine treatment became prominent in both density and total synaptic surface parameters, such that the decline in density was entirely prevented by the drug, while the synaptic surface was significantly higher than that of the untreated age-matched controls. In summary, these nimodipine effects may be interpreted as a long-term counteractive effect on the age-dependent decline of the synaptic structure in this part of the hippocampus.

Aging, Nimodipine, and Calcium Binding Proteins

Inasmuch as the currently used calcium antagonist exerts such a profound influence on neuronal structure and function, we included some calcium-related parameters of aging neurons. As one parameter that is of potential importance for calcium homeostatic mechanisms, we investigated the immunoreactivity (ir) of calcium-bound forms of calcium binding proteins during aging in rabbit and rat.[31] In rabbit dentate gyrus the calbindin-D28k (CaB) protein showed a striking decline during the first 12 months of the rabbit's lifespan after which the CaB-ir level remained constant up to the age of 48 months (FIG. 6). Between 48 and 60 months a second phase of CaB decline was observed when CaB-ir reached a level of about 25% of the optical density of 1-month-old animals. This age-dependent decrease of CaB-ir was essentially similar for granule cells, their dendrites in the dentate molecular layer, and the mossy fiber projection to the CA3 region. A similar trend was found for CaB-positive cells in the CA1 pyramidal cell layer, but not in the stratum oriens. In the rabbit, age had no effect on the numbers of parvalbumin-positive GABAergic neurons in the various regions of the cornu ammonis.

In the rat, CaB-ir strongly declined in senescence to a similar degree as in rabbit notably in the dentate gyrus and the neocortex (FIG. 7), but not in the CA1 of the hippocampus. As in the rabbit the number of parvalbumin-positive cells did not change in cortex and hippocampus. In rat we also investigated if the CaB and PARV pattern changes were influenced by chronic nimodipine treatment. Chronic treatment with the calcium antagonist only moderately antagonized the decline of the two calcium binding proteins investigated. Only in the cortex was the reduction of CaB attenuated, and this was significant in the parietal cortex (FIG. 7).

Aging, Nimodipine, and Open-Field Behavior

There are numerous reports on the impact of calcium channel blockers on various behavioral parameters. (See the contributions of Traber and Fannelli, this volume.)

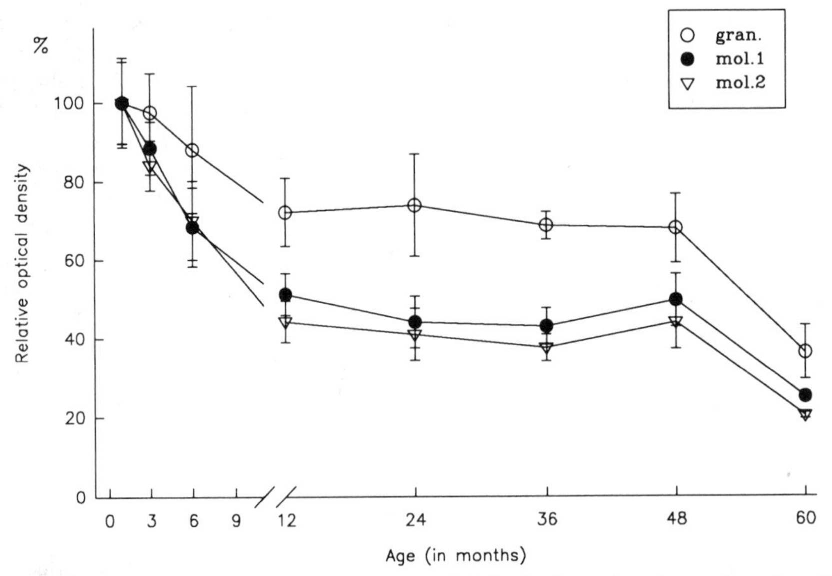

FIGURE 6. Calbindin D28k immunoreactivity (CaB-ir) in the various layers (granule cell layer, gran; inner molecular layer, mol.1; outer molecular layer, mol.2) of the dentate gyrus at eight different ages in the rabbit. Immunoreactivity was quantified with image analysis. Values were calculated as a percentage of the highest value of 1-month-old rabbit, which was set at 100%. The CaB-ir rapidly declines during the first 12 months, stabilizes up to 48 months, and is followed by a final reduction in extremely aged cases of 60 months.

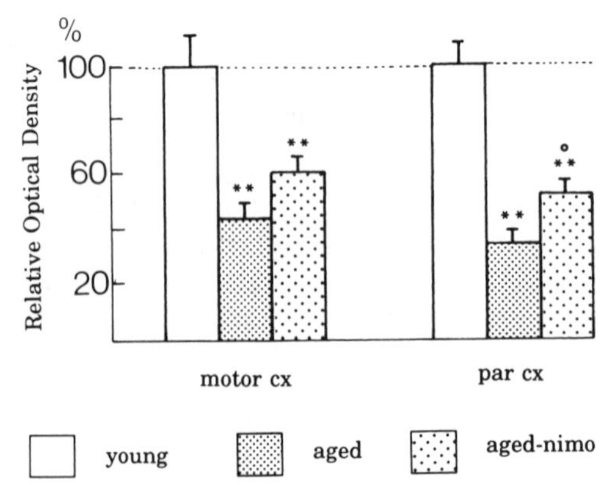

FIGURE 7. CaB immunoreactivity measured by optical density image analysis and expressed as percentages in layers II-IV of parietal and motor cortex in young (3-month-old), aged (30-month-old), and aged (24-30-month-old), nimodipine-treated Wistar rats. The sharp decrease in CaB-ir in the aged rat is slightly attenuated by nimodipine treatment, which was significant in the parietal cortex ($^{**}p < 0.01$; $^{\circ}p < 0.05$).

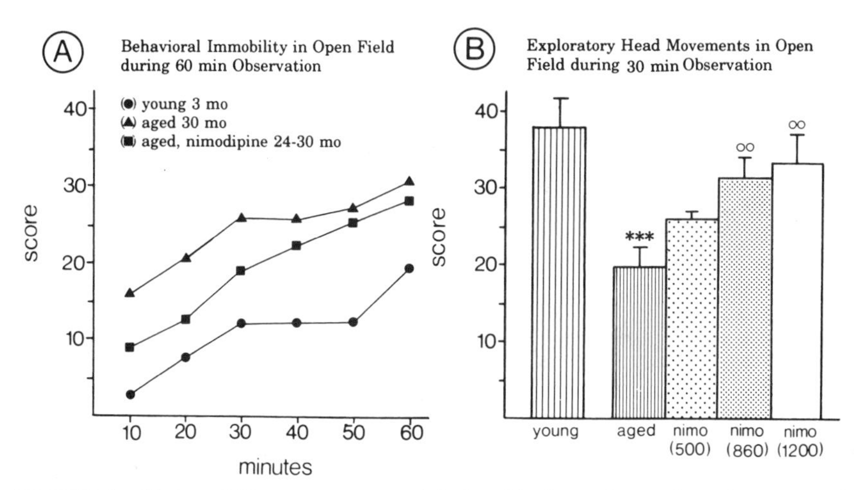

FIGURE 8. **(A)** Immobility scores measured during 60-minute observation of rats exposed to novelty-induced behavioral arousal in an open field test. Young animals show a lower level of immobility than do aged animals. The immobility of the nimodipine group (24-30 months of treatment) was significantly lower than that of the age-matched controls. **(B)** The reversed effects are seen by scores of head movements as a measure of exploratory behavior in the open field. Nimodipine effects are dose-dependent.

In short, these studies demonstrate that chronic nimodipine application improves behavioral scores in motor performance and spatial memory, facilitates and accelerates learning and conditioning, improves short-term memory in the radial maze, and attenuates the age-related decline in visual discrimination.[32,33] Also the first trials in humans with various types of dementia point to improved neuropsychological scores and a prophylactic effect on the progression of the dementing process.[15,34]

In the same animals used to investigate cerebrovascular and neuronal effects, we also assessed the behavioral scores in a novelty-induced behavioral arousal paradigm (FIG. 8). In this test the spontaneous behavior of young (3 months), aged (30 months), and aged, nimodipine-treated rats (24-30 months) was recorded during 60 minutes. Aged animals of 30 months compared to 3-month-old controls were characterized by a very significant suppression of exploration, rearing, walking, and sniffing accompanied by an equally large increase in immobility. All these age-dependent spontaneous behavioral expressions were significantly antagonized in a dose-dependent manner by chronic nimodipine application starting at the age of 24 months[16] (FIG. 8B). The observed behavioral effects of the aging process (as in the small open field test in aging) and the impact of nimodipine both corroborate the effects of this drug in motor and cognitive behaviors as reported by others in various species.[14,15,32-34]

Summary of Normotensive Brain Aging

Aging in the rodent, as in other mammalian species, reveals a wide variety of cerebrovascular, neuronal, and behavioral changes that collectively represent a slow but gradual deterioration in the cellular components of the brain, its blood supply, and BBB, leading to deficiency of basically all behavioral performance. In a series of investigations we demonstrated a significant effect of the L-type calcium channel blocker nimodipine on a number of microvascular, synaptic, and behavioral parameters. Control over calcium influx eventually will yield maintenance of microvascular function, control of the endothelium over BBB function, improvement of astrocytic and glial support, and delay of neuronal breakdown and neuronal signal transduction. The mechanism by which a compound such as nimodipine affects the aging process as yet remains a subject of study. Convincing evidence exists that aging affects several aspects of calcium-regulating homeostasis. More recent studies establish the notion that aging burdens the cell with a persistent overload of basal $[Ca]_i$ levels and delays the mechanisms that restore intracellular calcium concentrations.[35,36] Convincing electrophysiological reports[2,4] indicate that a compound such as nimodipine attenuates the prolonged afterhyperpolarization in aging hippocampal pyramidal neurons, a phenomenon that is associated with improved behavioral conditioning by nimodipine.[4]

The way in which nimodipine exerts such a powerful effect in delaying the deterioration of the microvascular wall during aging remains unclear. The aberrant microvessels in the aging rat brain are often circumvented by hypertrophied astrocytes. Because astrocytes play an important role in the maintenance of the BBB, these observations suggest that a loss of cerebrovascular integrity induces BBB alterations. More specifically, the increased number of pinocytotic vesicles in the endothelial cytoplasm in compromised microvessels points to disturbed nutrient transport over the microvascular wall. Such impaired BBB transport function has been demonstrated in the aging CNS[10,23,24,37] and clearly affects the surrounding neuropil and subsequent neuronal functioning.

On first sight, the percentage of microvessels displaying microvascular deposits may seem relatively small. However, each microvessel probably will contain several anomalies along its longitudinal surface, so that examination of microvascular cross-sections provides the relative density of these aberrations. Assuming that each microvessel in the aging brain displays loss of integrity and concomitant impaired nutrient transport along its surface, the devastating implications for neuronal functioning become clear (FIG. 5). In this light we can consider the protective effect of chronic calcium channel blockade (50% fewer microvessels with deposits in animals aged up to 30 months) as highly beneficial. Moreover, the fact that a calcium antagonist can be so effective implicates that disturbed calcium homeostasis of the microvascular wall strongly contributes to aging-related impaired neuronal functioning and consequent behavioral performance.

Most evidence points to the endothelial cell as the main origin of microvascular deterioration during aging. Not only are the number of endothelial pinocytotic vesicles increased in the aging brain, but also small collagen fibrils are present within the endothelial cytoplasm of aberrant microvessels. The cerebrovascular effects of calcium antagonists were thus far explained by vasodilatory influences on larger vessels, but never anticipated the presence of L-type calcium channels in endothelial cells. In contrast to those in the peripheral circulation, the larger arteries in the CNS are major determinants of local microvascular pressure.[38] Our studies cannot exclude that the beneficial influence of calcium channel blockade on microvascular integrity in the aging brain is secondary to alterations in the larger resistance vessels. When larger blood vessels in the brain dilate under the influence of nimodipine, the microvascular blood pressure will be reduced, possibly preventing the formation of microvascular aberrations. However, all the ultrastructural evidence for endothelial involvement in microvascular deterioration during aging and the influence of calcium channel blockade lead to an alternative hypothesis. Although most studies indicate that Ca^{2+} enters endothelial cells through nonselective cation channels,[39,40] Bossu and coworkers[41] demonstrated the presence of L-type calcium channels in isolated capillary endothelial cells. We now hypothesize that increased calcium influx through L-type calcium channels in the endothelial cell contributes to the declined microvascular condition and subsequent impaired neuronal functioning in the aging CNS (FIG. 12).

EFFECTS OF NIMODIPINE ON AGING SPONTANEOUSLY HYPERTENSIVE STROKE-PRONE RATS

General Condition and Occurrence of Stroke

High blood pressure is now commonly regarded as the most frequently occurring risk factor in vascular pathology in the peripheral circulation. Analysis of coronary and larger arterial tissue reveals high blood pressure as a major contributor of Ca^{2+} accumulation in the vascular wall notably combined with aging.[6] Little is known about the influence of elevated blood pressure and aging on Ca^{2+} contents in the cerebral vasculature. On the other hand, there is well documented evidence for the acute effects of the Ca antagonist nimodipine on the vasodilatory characteristics of the cerebral vasculature,[42] which is probably mediated via the smooth muscle cells of the larger vessels. Here we investigated the impact of the Ca^{2+} entry blocker in aging spontaneously hypertensive stroke-prone (SHR-SP) rats, in which the effects of chronic drug application were assessed on a number of vascular, physiological, neuronal, and behavioral parameters. The influence of high blood pressure on the aging process became readily obvious as the hypertensive strains showed all signs of aging-related deterioration in body condition and behavioral performance at relatively very early stages of life as compared to normotensive Wistar-Kyoto (WKY) animals. In a first set of experiments SHR-SP rats were chronically treated with nimodipine applied via daily food intake at a concentration of 1,000 ppm starting at the age of 46 weeks (immediately before the period of stroke development) up to the age of 56 weeks. During the treatment period the physical condition (fur condition and locomotor activity) gradually deteriorated. The body weight of the placebo-treated SHR-SP animals decreased from 330 to 280 g, whereas the nimodipine-treated group (SP-nimo) remained at a consistently higher level and even showed a slight increase from 323 g to an average of 354 g. No effects of treatment were found in age-matched normotensive WKY rats (TABLE 1). Furthermore, during the 10-week treatment all SP-placebo rats developed neurologic (irritability and paralysis) and histologic signs of stroke, but none of the SP-nimo cases did. In fact, because of the increasing number of lethal SP-placebo cases, the treatment period had to be terminated to maintain a sufficient number of placebo cases for statistical analysis. All of the 11 SP-nimo animals survived the experimental period, whereas 4 of 11 SP-placebo rats died with the symptoms of stroke. The incidence of stroke at the end

TABLE 1. Effect of Treatment on the Occurrence of Neurologic and Histologic Signs of Stroke, Body Weight, and Brain Weight in Spontaneously Hypertensive Stroke-Prone (SHR-RP) Animals at the Start of the Experiment (Onset) and after 10 Weeks of Treatment with Nimodipine Food (1,000 ppm) or Placebo Food

Group	Age (wk)	Rats with Stroke	Body Weight (g)	Brain Weight (g)
SHR-SP onset (5)	46	2	330 ± 15.6	2.04 ± 0.20
SHR-SP nimo (7)	56	0	354 ± 7.1**	1.81 ± 0.02
SHR-SP plac (7)	56	7	281 ± 9.3*	2.53 ± 0.11**

NOTE: Significantly different from onset group at $*p < 0.05$; $** p < 0.001$.

of the experimental period was accompanied by the presence of severe brain edema in the SP-placebo rats. The edema became apparent from the heavily swollen brain immediately after removal from the skull and was quantitatively expressed by the significant increase by 40% in brain weight in the SP-placebo versus the SP-nimo animals (TABLE 1), whereas the body weight of the SP-placebo group significantly decreased. It is important to note that the prevention of stroke by nimodipine could not be attributed to a blood pressure decrease by the drug. Nimodipine had no effect on the very high pressure of the SHR-SP animals and even appeared to stabilize the blood pressure at a value of 220 mm Hg (FIG. 9).

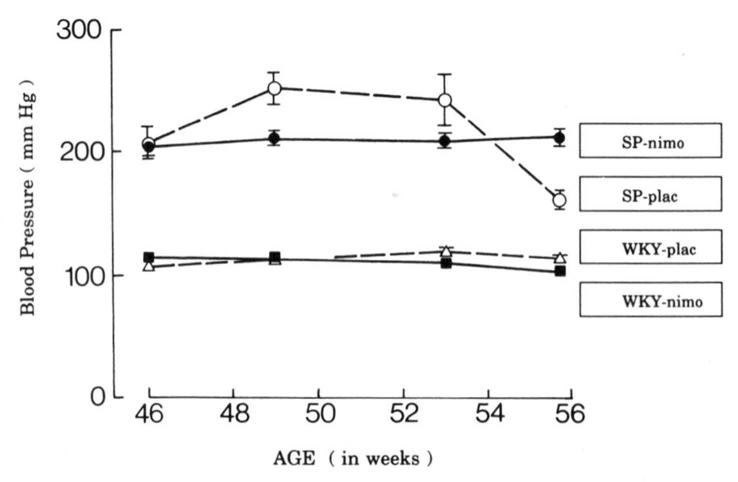

FIGURE 9. Development of blood pressure (BP) in spontaneously hypertensive stroke-prone (SHR-SP) and control Wistar Kyoto (WKY) rats treated with nimodipine or placebo food. In neither the WKY nor the SHR-SP group did nimodipine lower BP values. In the SHR-SP cases nimodipine appeared to stabilize the high BP levels.

Stroke, Hippocampal Alterations, and Behavioral Dysfunction

A major finding in the first series of experiments with SHR-SP animals was the neurochemical changes that occurred in the hippocampal cell groups in animals that had neocortical strokes. Apparently as a neuropathological consequence of cortical strokes, hippocampal pyramidal neurons revealed abnormally enhanced immunoreactivity for protein kinase Cγ, whereas the GABA synthesizing enzyme GAD and its colocalized calcium binding protein parvalbumin in the GABAergic interneurons were significantly decreased.[19] Taken together, these stroke-induced alterations were interpreted as GABAergic cell degeneration leading to abnormal disinhibition of the pyramidal cell group. The lack of inhibition or chronic overstimulation of the pyramidal cells then would yield an abnormally high level of PKCγ-ir, probably mediated by relatively high activation of G-protein-coupled receptor types. These changes in the neurochemical nature of hippocampal cell groups did not occur in the aging SHR-SP animals that were treated for 10 weeks with nimodipine and that were free of strokes. The patterns of PKCγ, parvalbumin, and GAD immunoreactivity of the SP-nimo animals were basically similar to those in the control WKY rats.

Before sacrifice the SP-nimo and SP-placebo animals were tested for their novelty-induced behavioral arousal in an open-field task. In this test condition the time the animals spent on exploration of their new environment was recorded and revealed an almost threefold higher level of activity of the SP-nimo group than of the stroke victims of the SP-placebo group.[19] It may therefore be concluded that the occurrence of stroke and the concurrent pathological condition of the hippocampal circuitry in these animals coincides with near total absence of behavioral arousal.

It remained to be determined, however, if maintenance of the normal hippocampal and behavioral features in the SP-nimo cases was the result of prevention of stroke by nimodipine or a direct neuronal effect of nimodipine in aging hypertensive conditions. In other words, what is the behavioral profile and the condition of hippocampal neuronal parameters in aging SP rats before the development of stroke. For these reasons we performed a second series of experiments in which WKY and SHR-SP rats were tested and treated with the calcium antagonist nimodipine. Animals of the second experiment revealed a somewhat lower blood pressure level of around 180 mm Hg (220 mm Hg in the first experiment) in which the symptoms of stroke started to appear at the age of 50 weeks (compared to 42 weeks in the first experimental group). Nimodipine was administered from 40-60 weeks to both WKY normotensive controls and SHR-SP rats. During this treatment period none of the WKY animals and SP-nimo rats died, whereas only 50% of the animals in the placebo SHR-SP survived (TABLE 2). None of the remaining SP-placebo rats showed neurological or histological signs of cerebrovascular strokes. This is also reflected by the similar brain weights of all SHR-SP groups and thus by the lack of brain edema formation (TABLE 2). Also the general condition and body weight of the SP-placebo rats did not decline during the observation period. As in this first experiment nimodipine did not diminish blood-pressure in normotensive WKY and hypertensive SHR-SP rats (TABLE 2).

In the experimental setup we followed the behavioral scores in the open field test over the animals' lifespan from 12-60 weeks. It was previously shown by others that SHR rats are more active in a novel environment than are their normotensive

TABLE 2. Effects of Nimodipine Treatment on the Percentage of Animals that Survive, Blood Pressure, Body Weight, and Brain Weight in Wistar Kyoto (WKY) and Spontaneously Hypertensive Stroke-Prone (SHR-SP) Animals at the Start of the Experiment (Onset) and after 20 Weeks of Treatment with Placebo Food or Nimodipine Food (1,000 ppm)

Group	Age (wk)	Survival	Rats with Stroke	Blood Pressure (mm Hg)	Body Weight (g)	Brain Weight (g)
WKY onset (6)	40	—	0	100 ± 4	428 ± 5	2.07 ± 0.03
WKY nimo (6)	60	100%	0	106 ± 3	426 ± 11	2.15 ± 0.03
WKY plac (6)	60	100%	0	111 ± 4	415 ± 6	2.19 ± 0.02
SHR-SP onset (6)	40	—	0	161 ± 7	368 ± 7	1.70 ± 0.03
SHR-SP nimo (6)	60	100%	0	184 ± 2	378 ± 4	1.96 ± 0.01
SHR-SP plac (8)	60	50%	0	179 ± 11	359 ± 3	1.90 ± 0.02

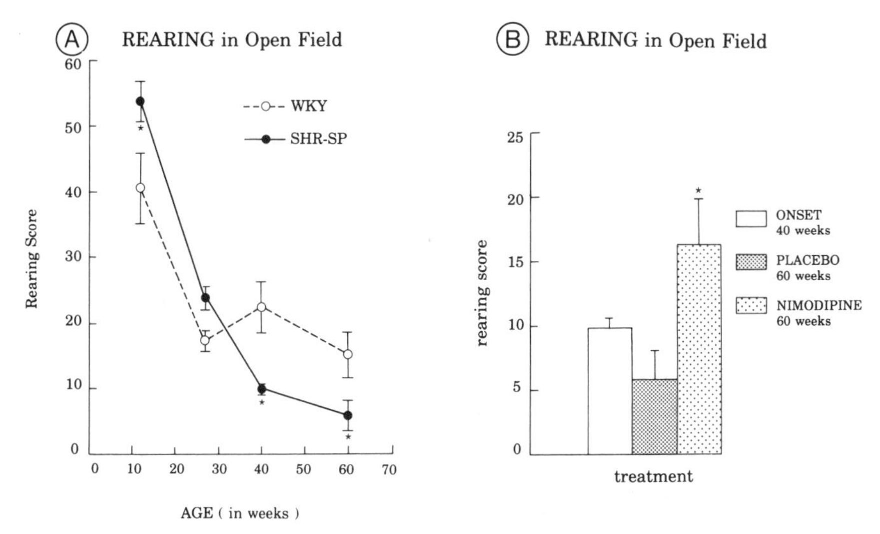

FIGURE 10. **(A)** Behavioral scores of novelty-induced exploration activity in an open field in the lifespan of SHR-SP animals compared with age-matched WKY controls. The initial hyperarousal and hyperactivity expressed by rearing of the animals are reversed to an almost lack of activity in the accelerated aging period of the SHR-SP group. **(B)** SHR-SP rats treated with nimodipine from 40-60 weeks show a significantly higher level of novelty-induced arousal than do the SHR-SP animals receiving placebo.

controls,[43,44] which corroborates our data in which young SHR-SP rats (12 weeks of age) displayed significantly more rearing behavior than did WKY controls (Fig. 10A). It is generally accepted that especially rearing as a component of exploratory behavior declines rapidly with advancing age. Despite the higher rearing scores at young age, the older SHR-SP animals (40 and 60 weeks of age) show significantly less rearing behavior in the open field than do WKY controls (Fig. 10A). These data indicate that aging-related behavioral alterations occur earlier in genetically hypertensive rats. Nimodipine treatment from 40-60 weeks of age increased the rearing scores, even when compared to that in animals aged 40 weeks (Fig. 10B). In other words, nimodipine completely counteracted the accelerated behavioral decline in aging hypertensive rats before they developed neocortical strokes.

In a recent pilot experiment we compared the performance of older nonsymptomatic SHR-SP rats (40 weeks) with age-matched WKY rats in a holeboard spatial orientation learning test that requires an intact hippocampus. The holeboard contains 16 equidistant holes in which rats were trained to learn a pattern of four baited out of 16 holes, after which the reference memory ratio (RMR) was calculated. At the age of 40 weeks the RMR of SHR-SP rats was significantly lower than that of WKY controls (Fig. 11A). The latter demonstrates that the spatial orientation capacity of aging hypertensive animals is reduced compared to that of their normotensive controls. Similar findings were described by Wyss *et al.*[45] who showed early impairment of cognitive functioning in SHR rats.

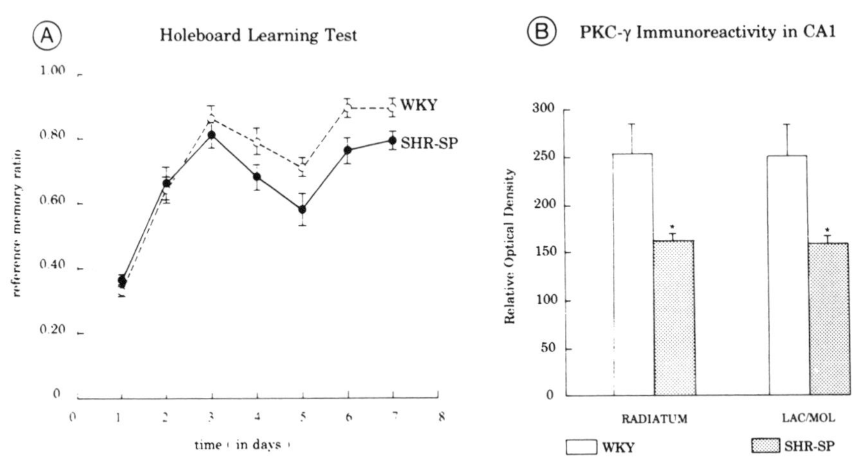

FIGURE 11. Performance of SHR-SP and age-matched WKY rats at the age of 40 weeks in the holeboard spatial learning test. In this test the animals had to learn a random spatial pattern of 4 baited holes of a total of 16 holes. Reference memory scores (visits food holes + revisits food holes/total visits food holes and non-food holes) of the WKY rats was significantly higher (MANOVA, $p < 0.05$) than those of the SHR-SP group. **(B)** Twenty-four hours after completion of the holeboard test, learning task-induced PKCγ immunoreactivity (PKC-ir) enhancement in the hippocampal CA1 area was quantified by optical density measurement. The PKC-ir response was significantly higher in the WKY than in the SHR-SP animals, consistent with their higher level of performance in the holeboard.

Previous work from our group[46] revealed that the immunoreactivity (ir) for PKCγ was enhanced in columns of the hippocampal formation of mice and rats that underwent the holeboard task. This increased PKCγ-ir was most prominent in the dendritic fields of the CA1 area and can be attributed to increased activation of CA1 pyramidal cells.[46] We examined PKCγ-ir in the hippocampus CA1 of both SHR-SP and WKY animals after holeboard training by image analysis. The relative optical density was measured in the CA1 stratum radiatum and in the lacunosum moleculare of all trained animals. In both areas the relative optical density of PKCγ-ir was significantly higher in the WKY animals (FIG. 11B). This suggests a relation between learning performance in the holeboard task and PKCγ-ir in the CA1, because the better learners (WKY animals) display a denser PKCγ-ir in the dendritic fields of the CA1. However, it remains to be established if the difference in PKCγ-ir is related to learning performance or strain differences. However, preliminary data show that PKCγ-ir does not differ between naive WKY and SHR-SP rats at 12 weeks of age (data not shown), ruling out a general strain difference. Other preliminary data show that PKCγ-ir remains stable throughout the age of 12-40 weeks in WKY rats, and declines in none-stroke SHR-SP rats. The latter suggests that decreased activation of CA1 pyramidal cells, as reflected by a reduced level of PKCγ-ir, in aging nonsymptomatic SHR-SP rats underlies the impaired learning performance of these animals.

SUMMARY AND CONCLUSIONS

The pathogenesis of essential hypertension is not fully understood, but most of the cardiovascular, metabolic, neurogenic, and humoral abnormalities are explained by dysfunctions in the control of intracellular Ca^{2+} concentrations in the cells of the vascular wall.[47,48] Most theories of disturbed calcium regulation focus on the calcium concentration within vascular smooth muscle cells.[47] The implications of hypertension for the increased calcium content of aging arteries seem to be clear, but were only studied in the peripheral circulation; hypertension prominently augments the aging-related accumulation of calcium in the vessel wall.[5]

Although the contribution of calcium overload in hypertensive cerebrovascular damage is well documented,[42] it is not clear yet if hypertension per se is the main cause of hypertension-associated calcium-dependent cerebral damage. Thus far, the hypotensive effects of most calcium antagonists were extensively described, and their efficacy in stroke prevention was proven.[19,42] Earlier studies indicated that chronic administration of nimodipine revealed a protective effect in the occurrence of strokes in SHR-SP rats, yielding a decreased mortality rate.[42] Because nimodipine did not lower the extremely high blood pressure of these animals,[19,42] the mechanisms behind such nimodipine-induced stroke prevention may be attributed to a direct cerebrovascular and/or neuronal action of nimodipine.

Hypertension is generally considered a vascular pathologic condition, and most research has been directed towards the influences of hypertension on large peripheral arteries such as the aorta and coronary artery. The influence of the CNS on the regulation of cardiovascular system and blood pressure regulation was described in detail, and the role of the CNS in hypertension also was the subject of study.[45] The increased risk of stroke in hypertensive subjects generated numerous studies on the precise nature of compromised cerebrovascular functioning under hypertensive conditions.

Few data are available on Ca^{2+} alterations in cerebral neurons during hypertension. Honda *et al.*[49] demonstrated that voltage-dependent Ca^{2+} uptake was higher in cortical synaptosomes from SHR than form normotensive animals and suggested that an important alteration in Ca^{2+} channel characteristics may occur in SHR brain synaptosomes. Although the density of L-type calcium channels was shown to be higher in the hippocampus of SHR rats,[50] others reported that the number of L-type calcium channels was significantly lower in the brain of SHR rats than WKY normotensive controls.[51] The latter data suggest that hypertension may be associated with similar alterations in neuronal calcium homeostasis as demonstrated for aging in normotensive subjects.[50,52] To date little is known about the relationship between hypertension, the cerebrovascular condition, and neuronal functioning in the CNS. A major finding in these and other studies is that the hypertensive condition strongly accelerates the neuronal and behavioral alterations commonly associated with senescence in normotensive animals. The accelerated behavioral decline of spontaneous hypertensive rats[45,53] indicates that hypertension also progressively affects neuronal functioning during aging. We anticipate that hypertension combined with aging will have a profound impact on the structure of the vascular wall of both larger and fine vessels in the brain. As such, hypertension in aging animals may considerably contribute to impaired neuronal functioning in such animals. We hypothesize that endothelial cell

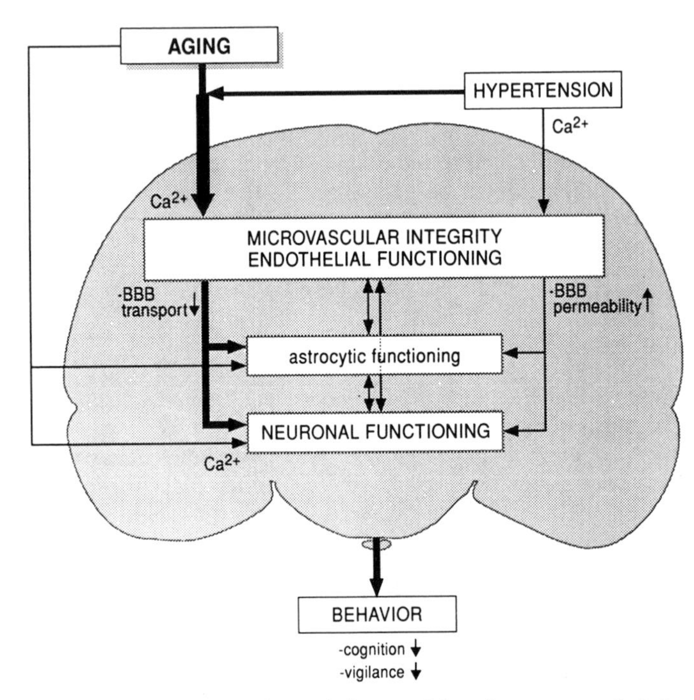

FIGURE 12. Survey diagram depicting the influence of the aging process and of a hypertensive condition on microvascular integrity and function of the endothelial lining of the microvascular wall. Aging and hypertension apparently mediated by Ca^{2+}-dependent mechanisms directly affect the vascular condition and functioning of the blood-brain barrier. This way aging may indirectly influence glial and neuronal functioning. The aging process naturally exerts direct Ca^{2+}-mediated effects on the glial and neuronal condition as well. Together aging and hypertension have a profound impact on all behavioral functions regulated by the nervous system.

dysfunction (due to disturbed calcium homeostasis) is a major causal factor for the accelerated decline in CNS functioning in aging hypertensive subjects (Fig. 12). This way the efficacy of the calcium channel blocker nimodipine in stroke prevention can possibly be explained by maintenance of endothelial integrity and condition in aging hypertensive rats. The current results also allow the conclusion that maintenance of calcium balance, without reduction of blood pressure, is a fruitful strategy in therapeutic treatment of essential hypertension as a risk factor for brain dysfunction and stroke.

REFERENCES

1. Khachaturian, Z. S. 1984. Towards theories of brain aging. *In* Handbook of Studies on Psychiatry and Old Age. Key and Burrows, Eds.: 7-31. Elsevier. Amsterdam.
2. Landfield, P. W. 1989. Calcium homeostasis in brain aging and Alzheimer's disease. *In* Diagnosis and Treatment of Senile Dementia. Bergener and Reisberg, Eds.: 276-287. Springer. Berlin.

3. LANDFIELD, P. W., O. THIBAULT, M. L. MAZZANTI, N. M. PORTER & D. S. KERR. 1992. Mechanisms of neuronal death in brain aging and Alzheimer's disease: Role of endocrine-mediated calcium dyshomeostasis, J. Neurobiol. **23:** 1247-1260.
4. MOYER, J. R., L. T. THOMPSON, J. P. BLACK & J. F. DISTERHOFT. 1992. Nimodipine increases excitability of rabbit CA1 pyramidal neurons in an age- and concentration-dependent manner. J. Neurophysiol. **68:** 2100-2109.
5. FLECKENSTEIN, A., M. FREY, J. ZORN & G. FLECKENSTEIN-GRÜN. 1990. Calcium, a neglected key factor in hypertension and arteriosclerosis. Experimental vasoprotection with calcium antagonists or ACE inhibitors. *In* Hypertension: Pathophysiology, diagnosis and management. J. H. Laragh & B. M. Brenner, Eds.: 471-509. Raven Press. New York.
6. FLECKENSTEIN-GRUN, G. & A. FLECKENSTEIN. 1991. Calcium, a neglected key factor in arteriosclerosis. The pathogenetic role of arterial calcium overload and its prevention by calcium antagonists. Ann. Med. **23:** 589-599.
7. KNOX, C. A., R. D. YATES, I. CHEN & P. M. KLARA. 1980. Effects of aging on the structure and permeability characteristics on cerebrovasculature in normotensive and hypertensive strains of rats. Acta Neuropathol. **51:** 1-13.
8. TOPPLE, A., E. FIFKOVA & K. CULLEN-DOCHSTADER. 1990. Effect of age on blood vessels and neurovascular appositions in the rat dentate fascia. Neurobiol Aging **11:** 371-380.
9. RAVENS, J. R. 1970. Vascular changes in the human senile brain. Adv. Neurol. **20:** 487-501.
10. MOORADIAN, A. D. 1988. Effect of aging on the blood-brain barrier. Neurobiol. Aging **9:** 31-39.
11. RASMUSSEN, H., P. BARRETT, J. SMALLWOOD, W. BOLLAG & C. ISALES. 1990. Calcium ion as intracellular messenger and toxin. Environm. Health Persp. **84:** 17-25.
12. SCRIABINE, A. 1990. Pharmacology overview: Nimodipine in CNS indications. *In* Nimodipine. Pharmacological and Clinical Results in Cerebral Ischemia. A. Scriabine, G. M. Teasdale, D. Tettenborn & W. Young, Eds.: 1-7. Springer. Berlin.
13. DE JONG, G. I., J. TRABER & P. G. M. LUITEN. 1992. Formation of cerebrovascular anomalies in the aging rat is delayed by chronic nimodipine application. Mech. Ageing Dev. **64:** 255-272.
14. SCHUURMAN, T. & J. TRABER. 1989. Effects of nimodipine on behavior of old rats. *In* Nimodipine and Central Nervous System Functioning. J. Traber and W. H. Gispen, Eds.: 195-208. Schattauer, Stuttgart.
15. TOLLEFSON, G. D. 1990. Short-term effects of the calcium channel blocker nimodipine (Bay-e-9736) in the management of primary degenerative dementia. Biol. Psychiatry **27:** 1133-1142.
16. DE JONG, G. I., C. NYAKAS, T. SCHUURMAN & P. G. M. LUITEN. 1993. Aging-related alterations in behavioral activation and cerebrovascular integrity in rats are dose-dependently influenced by nimodipine. Neurosci. Res. Comm. **12:** 1-8.
17. PHILIPS, S. J. & J. P. WHISNANT. 1990. Hypertension and stroke. *In* Hypertension: Pathophysiology, Diagnosis and Management. J. H. Laragh & B. M. Brenner, Eds.: 417-431. Raven Press. New York.
18. TAGAMI, M., Y. NARA, A. KUBOTA, H. FUJINO & Y. YAMORI. 1990. Ultrastructural changes in cerebral pericytes and astrocytes of stroke-prone spontaneously hypertensive rats. Stroke **21:** 1064-1071.
19. LUITEN, P. G. M., G. I DE JONG, E. A. VAN DER ZEE, M. BRAAKSMA, F. W. MAES, T. SCHUURMAN & C. NYAKAS. Neuroprotection by chronic nimodipine treatment in aging hypertensive stroke prone rats. Drugs Dev. **2:** 183-191.
20. CASEY, M. A. & M. L. FELDMAN. 1985. Aging in the rat medial nucleus of the trapezoid body. III. Alterations in capillaries. Neurobiol. Aging **6:** 39-46.
21. DE JONG, G. I., H. DE WEERD, T. SCHUURMAN, J. TRABER & P. G. M. LUITEN. 1990. Microvascular changes in aged rat forebrain. Effects of chronic nimodipine treatment. Neurobiol. Aging **11:** 381-389.

22. De Jong, G. I., A. S. P. Jansen, E. Horváth, W. H. Gispen & P. G. M. Luiten. 1991. Nimodipine effects on cerebral microvessels and sciatic nerve in aging rats. Neurobiol. Aging **13:** 73-81.
23. Mooradian, A. D., A. M. Morin, L. J. Cipp & H. C. Haspel. 1991. Glucose transport is reduced in the blood-brain barrier of aged rats. Brain Res. **551:** 145-149.
24. Samuels, S., I. Fish, S. A. Schwartz & U. Hochgeschwender. 1983. Age related changes in blood-to-brain amino acid transport and incorporation into brain protein. Neurochem. Res. **8:** 167-177.
25. Mooradian, A. D. 1988. Blood-brain barrier transport of choline is reduced in the aged rat. Brain Res. **440:** 328-332.
26. Cortés, R., P. Supavilai, M. Karobath & J. M. Palacios. 1984. Calcium antagonist binding sites in the rat brain: Quantitative autoradiographic using the 1,4-dihydropyridines [^{3}H]PN200-110 and [^{3}H]PY108-068. J. Neural Transm. **60:** 169-197.
27. DeToledo-Morrell, L., Y. Geinisman & F. Morrell. 1988. Age-dependent alterations in hippocampal synaptic plasticity: Relation to memory disorders. Neurobiol. Aging **8:** 409-416.
28. Bertoni-Freddari, C., C. Giuli, C. Pieri & D. Paci. 1986. Quantitative investigation of the morphological plasticity of synaptic junctions in rat dentate gyrus during aging. Brain Res. **366:** 187-192.
29. Bertoni-Freddari, C., P. Fattoretti, T. Casoli, W. Meier-Ruge & J. Ulrich. 1990. Morphological adaptive response of the synaptic junctional zones in the human dentate gyrus during aging and Alzheimer's disease. Brain Res. **517:** 69-75.
30. De Jong, G. I., B. Buwalda, T. Schuurman & P. G. M. Luiten. 1992. Synaptic plasticity in the dentate gyrus of aged rats is altered after chronic nimodipine application. Brain Res. **596:** 345-348.
31. Heizman, C. W. & K. Braun. 1992. Changes in Ca^{2+}-binding proteins in human neurodegenerative disorders. TINS **15:** 259-264.
32. Deyo, R. A., K. T. Straube & J. F. Disterhoft. 1989. Nimodipine facilitates associative learning in aging rabbits. Science **243:** 809-811.
33. McMonagle-Strucko, K. & R. J. Fanelli. 1993. Enhanced acquisition of reversal training in a spatial learning task in rats treated with chronic nimoidipine. Pharmacol. Biochem. Behav. **44:** 827-835.
34. Tedeshi, D. 1991. Calcium regulation in brain aging by nimodipine: A multicenter trial in Italy. Current Ther. Res. **50:** 553-563.
35. Kirischuk, S., N. Pronchuk & A. Verkhratsky. 1992. Measurements of intracellular calcium in sensory neurons of adult and old rats. Neuroscience **50:** 947-951.
36. Michaelis, M. L., C. T. Foster & C. Jayawickreme. 1992. Regulation of calcium levels in brain tissue from adult and aged rats. Mech. Aging Dev. **62:** 291-306.
37. Banks, W. A. & A. J. Kastin. 1985. Aging and the blood-brain barrier: Changes in the carrier-mediated transport of peptides in rats. Neurosci. Lett. **61:** 171-175.
38. Faraci, F. M. & D. D. Heistad. 1990. Regulation of large cerebral arteries and cerebral microvascular pressure. Circ. Res. **66:** 8-17.
39. Popp, R., J. Hoyer, J. Meyer, H.-J. Galla & H. Gögelein. 1992. Stretch-activated nonselective cation channels in the antiluminal membrane of porcine cerebral capillaries. J. Physiol. **454:** 435-449.
40. Himmel, H. M., A. R. Whorton & H. C. Strauss. 1993. Intracellular calcium, currents, and stimulus coupling in endothelial cells. Hypertension **21:** 112-127.
41. Bossu, J. L., A. Elhamdani & A. Feltz. 1992. Voltage-dependent calcium entry in confluent bovine capillary endothelial cells. FEBS Lett. **299:** 239-242.
42. Kazda, S., B. Garthoff, H. P. Krause & K. Schlossmann. 1982. Cerebrovascular effects of the calcium antagonistic dihydropyridine derivative nimodipine in animal experiments. Arzneim. Forsch. **32:** 331-338.
43. Knardahl, S. & T. Sagvolden. 1979. Open-field behavior of spontaneously hypertensive rats. Behav. Neural Biol. **27:** 187-200.

44. DANYSZ, W., A. PLAZNIK, O. PUCILOWSKI, M. PLEWAKO, M. OBERSZTYN & W. KOSTOWSKI. 1983. Behavioral studies in spontaneously hypertensive rats. Behav. Neural Biol. **39:** 22-29.
45. WYSS, J. M., G. FISK & T. VAN GROEN. 1992. Impaired learning and memory in mature spontaneously hypertensive rats. Brain Res. **592:** 135-140.
46. VAN DER ZEE, E. A., J. C. COMPAAN, M. DE BOER & P. G. M. LUITEN. 1992. Changes in PKCγ immunoreactivity in mouse hippocampus induced by spatial discrimination learning. J. Neurosci. **12:** 4808-4815.
47. WADSWORTH, R. M. 1990. Calcium and vascular reactivity in aging and hypertension. J. Hypertens. **8:** 975-983.
48. NOJIMA, H. 1990. Role of molecular genetics in the understanding of the pathogenesis of hypertension. Gerontology **36:** 31-41.
49. HONDA, H., T. SHIBUYA & B. SALAFSKY. 1990. Brain synaptosomal Ca^{2+} uptake: Comparison of Sprague-Dawley, Wistar Kyoto and spontaneously hypertensive rats. Comp. Biochem. Physiol. **95:** 555-558.
50. HUGUET, F., A. HUCHET, P. GERARD & G. NARCISSE. 1987. Characterization of dihydropyridine binding sites in the rat brain: Hypertension and age-dependent modulation of [^{3}H](+)-PN200-100 binding. Brain Res. **412:** 125-130.
51. GALETTI, F., A. RITLEDGE, V. KROGH & D. J. TRIGGLE. 1991. Age related changes in Ca^{2+} channels in spontaneously hypertensive rats. Gen. Pharmacol. **22:** 173-176.
52. HARA, H., H. ONODERA, H. KATO & K. KOGURE. 1992. Effects of aging on signal transduction and transduction systems in the gerbil brain: Morphological and autoradiographic studies. Neuroscience **46:** 475-488.
53. MORI, S., S. IBAYASHI, M. KATO & M. FUJISHIMA. 1993. Aging of brain function in spontaneously hypertensive rats: Radial maze learning and local cerebral glucose utilization. J. Cereb. Blood Flow Metab. **13:** S397.

Comparison of Structural Synaptic Modifications Induced by Long-Term Potentiation in the Hippocampal Dentate Gyrus of Young Adult and Aged Rats

YURI GEINISMAN,[a,b] LEYLA DETOLEDO-MORRELL,[c,d] AND FRANK MORRELL[c]

[a]Department of Cell and Molecular Biology
Northwestern University Medical School
and
Departments of [c]Neurological Sciences
and [d]Psychology
Rush Medical College
Chicago, Illinois 60611

The calcium hypothesis of brain aging[1–3] suggests that an increase in the concentration of free Ca^{2+} occurring in the brain with advancing chronological age may represent a common determinant of age-related changes in neuronal structure and function, leading eventually to neuronal death. This hypothesis implies that an age-related memory decline, which is a cardinal sign of normal aging, may result from alterations in calcium-dependent processes underlying learning and memory. One calcium-dependent phenomenon that relates at least to some forms of memory[4,5] is hippocampal long-term potentiation (LTP). The essence of LTP is a persistent enhancement of synaptic responses which can be elicited by brief, repetitive stimulation of presynaptic fibers with high-frequency trains of electrical pulses.[6,7] Because the basis for information storage in the brain is believed to be an augmented efficacy of synaptic transmission resulting from repetitive activation, LTP is widely regarded as a synaptic model of memory.[8]

A growing body of evidence indicates that calcium plays an important role in the induction, expression, and maintenance of LTP. Being a prerequisite for LTP induction, a transient rise in postsynaptic calcium levels also appears to be required both for triggering biochemical events that subserve the expression of LTP as well as for controlling LTP maintenance.[9–13] As predicted by the calcium hypothesis of brain aging, the process of aging has been demonstrated to affect some properties of hippocampal LTP, most notably its duration. In young animals, enhancement of synaptic efficacy, which defines LTP, is extremely durable and may last

[b] Address for correspondence: Dr. Yuri Geinisman, Department of Cell and Molecular Biology, Northwestern University Medical School, 303 East Chicago Avenue, Chicago, Illinois 60611.

weeks.[7,14–16] Although aged animals can be potentiated to the same degree as young ones by perforant path stimulation, they cannot retain the potentiated synaptic response in the dentate gyrus for a long time and lose it much more rapidly than do young adults.[14,17–19]

The reason for this age-related deficiency in LTP retention is unknown. The long-lasting nature of the LTP phenomenon suggests that the maintenance of LTP may depend on structural modifications of the activated synaptic population. In fact, previous electron microscopic studies showed that the induction of LTP is followed by structural synaptic alterations.[20–24] Especially interesting are the observations regarding an LTP-induced increase in the numerical density of certain morphological types of synapses per unit tissue area or volume of a potentiated synaptic field. Such a change has been shown to involve synapses on dendritic shafts, concave and double-headed spines in the hippocampal dentate gyrus,[25,26] as well as synaptic contacts on dendritic branches and stubby (neckless) spines in hippocampal field CA1.[27–30]

These results, however, are difficult to interpret because they were obtained with the aid of conventional methods for synapse quantitation that provide estimates of synaptic numerical density biased by uncontrollable factors. In our earlier studies, unbiased stereological techniques were used to ascertain whether LTP is accompanied by changes in synaptic numbers.[31,32] The induction of LTP by high-frequency stimulation of the medial perforant path was indeed followed by an increase in the number of synapses in the hippocampal dentate gyrus of either young[31] or aged[32] rats. This modification was found to be characteristic of only one subtype of axospinous synapse, the postsynaptic density (PSD) of which consists of discrete segments.

Recently, three-dimensional reconstructions of synapses revealed that most synaptic contacts with segmented PSDs exhibit multiple, completely partitioned transmission zones.[33] It was postulated that these axospinous junctions represent synaptic contacts of an unusually high efficacy and thus may play a special role in various forms of synaptic plasticity including LTP.[33] Our previous LTP study of *young* animals demonstrated that only those synapses that have multiple, completely partitioned transmission zones are significantly increased in numbers as a consequence of potentiating stimulation.[34] The present study was designed to determine if *aged* animals exhibit a similar kind of structural synaptic plasticity following the induction of LTP.

The results to be described were obtained by reexamining the material of our previous LTP experiment on aged animals.[32] Data on young potentiated animals were published elsewhere[34] and are presented here for comparison purposes.

MORPHOLOGICAL CLASSIFICATION OF SYNAPTIC CONTACTS

Synapses were examined in electron micrographs of serial sections through the molecular layer of the hippocampal dentate gyrus. In aged (28-month-old) rats of the Fischer-344 strain, the synaptic population of this area was composed of the same morphological varieties of synapses which were observed earlier in young adult (5-month-old) animals.[31,33] Two major categories of synaptic junction are represented by axodendritic synapses involving dendritic shafts and axospinous ones involving dendritic spines. The latter category may be further divided into *perforated* synaptic contacts exhibiting a discontinuous PSD profile in at least one serial section and *nonperforated* ones showing continuous PSD profiles in all consecutive sections.

Perforated synapses are morphologically heterogeneous, and they were previously categorized on the basis of the configuration of their PSDs.[31,32,35–40] Analysis of these synaptic contacts with the aid of three-dimensional reconstructions suggested another classification of perforated axospinous synapses which takes into account not only the PSD shape, but also the presence or absence of spine partitions.[33] According to this classification, the most numerous subtype of perforated axospinous junctions is represented by *synapses with multiple (2-4), completely partitioned transmission zones.* Electron micrographs of consecutive serial sections through such a synaptic contact are shown in FIGURE 1. In some sections, profiles of the presynaptic axon terminal are notched by a finger-like spine extension (FIG. 1g and h) known as a spinule.[41,42] The spinule base is interposed between two separate PSD profiles (FIG. 1h). More proximally in the postsynaptic element, a typical spinule profile is replaced by a band of the spine cytoplasm that interconnects the opposite walls of the spine head cavity and borders two distinct PSD profiles (FIG. 1i).

A three-dimensional reconstruction of this synapse demonstrates that the presynaptic axon terminal has two protrusions at its distal end (FIG. 2a, arrows). The surface of the postsynaptic spine head contacting the axon terminal is concave, and a cavity of the postsynaptic element is subdivided into two compartments (FIG. 2b) filled by corresponding protrusions of the presynaptic bouton. Separating the spine head compartments and axon terminal protrusions is a complete spine partition (FIG. 2b, arrow). It provides a barrier between two discrete *transmission zones,* each one being formed presynaptically by a separate axon terminal protrusion and delineated postsynaptically by a separate PSD segment.

Spine partitions are also found in two other subtypes of perforated synapses. One of these is characterized by a sectional partition emanating from the sector of the postsynaptic surface limited by two arms of the PSD horseshoe (FIG. 3c). Synaptic junctions of another subtype exhibit a focal spine partition restricted to a perforation in the fenestrated PSD (FIG. 3b). The partitioned subtypes of perforated synapses are complemented by their nonpartitioned counterparts. Although the latter also exhibit a segmented (FIG. 3e), horseshoe-shaped (FIG. 3f), or fenestrated (FIG. 3g) PSD, they lack spine partitions.

MODEL OF STRUCTURAL REARRANGEMENTS UNDERLYING SYNAPTIC PLASTICITY ASSOCIATED WITH LONG-TERM POTENTIATION

Perforated axospinous synapses have been implicated in synaptic plasticity.[21,31,32,36,43–45] The existing models of structural alterations that underlie synaptic plasticity postulate that perforated axospinous synapses with a segmented PSD are formed from nonperforated ones through the stages of synaptic contacts with a fenestrated and horseshoe-shaped PSD and then split into nonperforated axospinous junctions.[36,43,45] Although the central concept of these models is that of synapse division, no experimental evidence supports it.

The existence of the various synaptic subtypes just described has suggested a novel model of structural synaptic plasticity.[33] Structural intermediates in synaptic plasticity are hypothesized to include all subtypes of perforated axospinous synapses as well as atypical nonperforated ones (FIG. 3). Although the latter are characterized by a nonperforated PSD, they are indistinguishable from perforated synaptic contacts with regard to such basic morphological characteristics as large dimensions, a complex configuration, and the presence of a spine apparatus.

A cascade of structural modifications leading to long-lasting enhancement of synaptic efficacy is postulated to culminate in the formation of additional synapses with multiple, completely partitioned transmission zones. These synapses may evolve from atypical nonperforated ones through the stages of various perforated axospinous junctions as shown in FIGURE 3 (from "a" through "b" and "c" or through "g", "f" and "e" to "d"). Synaptic contacts that have multiple transmission zones separated by complete partitions appear to be designed as specialized elements of an exceptionally high efficacy. Their distinctive feature is that they exhibit two to four transmission zones instead of only one, as is usual. If discrete axon terminal protrusions in a single synaptic contact of this subtype represent separate sites of transmitter release, a larger total volume of a neurotransmitter may be released during transmission of an impulse. Although a single synaptic vesicle releases enough transmitter molecules to saturate postsynaptic receptors at a central synapse,[24] the compartmentalization of multiple transmission zones by complete spine partitions may prevent the saturation and facilitate the action of multiple transmitter quanta. This would require each release site to be associated with a separate postsynaptic receptor cluster localized within a distinct PSD segment. Coincidentally, the PSD is the zone of highest concentration of receptors for neurotransmitters,[46,47] and the total PSD area along the postsynaptic membrane is largest in those synaptic contacts that exhibit a PSD consisting of separate segments.[32] These peculiarities of synapses with multiple, completely partitioned transmission zones may be indicative of their unusually high strength.

CHANGES IN SYNAPTIC NUMBERS AFTER LTP INDUCTION IN AGED RATS

The electrophysiological experiment with LTP induction was described in detail earlier.[32] Briefly, aged rats were chronically implanted with bipolar stimulating electrodes into the right medial perforant path and recording electrodes into the hilus of the ipsilateral dentate gyrus. The animals were assigned to the following three groups: (1) potentiated rats were stimulated (with fifteen 20-ms bursts of 400 Hz delivered at 0.2 Hz) on each of 4 consecutive days and sacrificed 1 hour after the fourth stimulation; (2) coulombic controls were stimulated at a low frequency (0.2 Hz) that does not elicit LTP and were matched with respective potentiated animals according to the total amount of current; and (3) unstimulated controls received the same handling as did rats from the other two groups.

Synaptic contacts were examined in the middle molecular layer (MML) of the dentate gyrus where virtually all perforated axospinous synapses are formed by perforant path axons[36] which were stimulated during LTP induction. As a control, synaptic contacts were additionally analyzed in the dentate inner molecular layer (IML) which was not directly stimulated in this experiment. Because neither axoden-

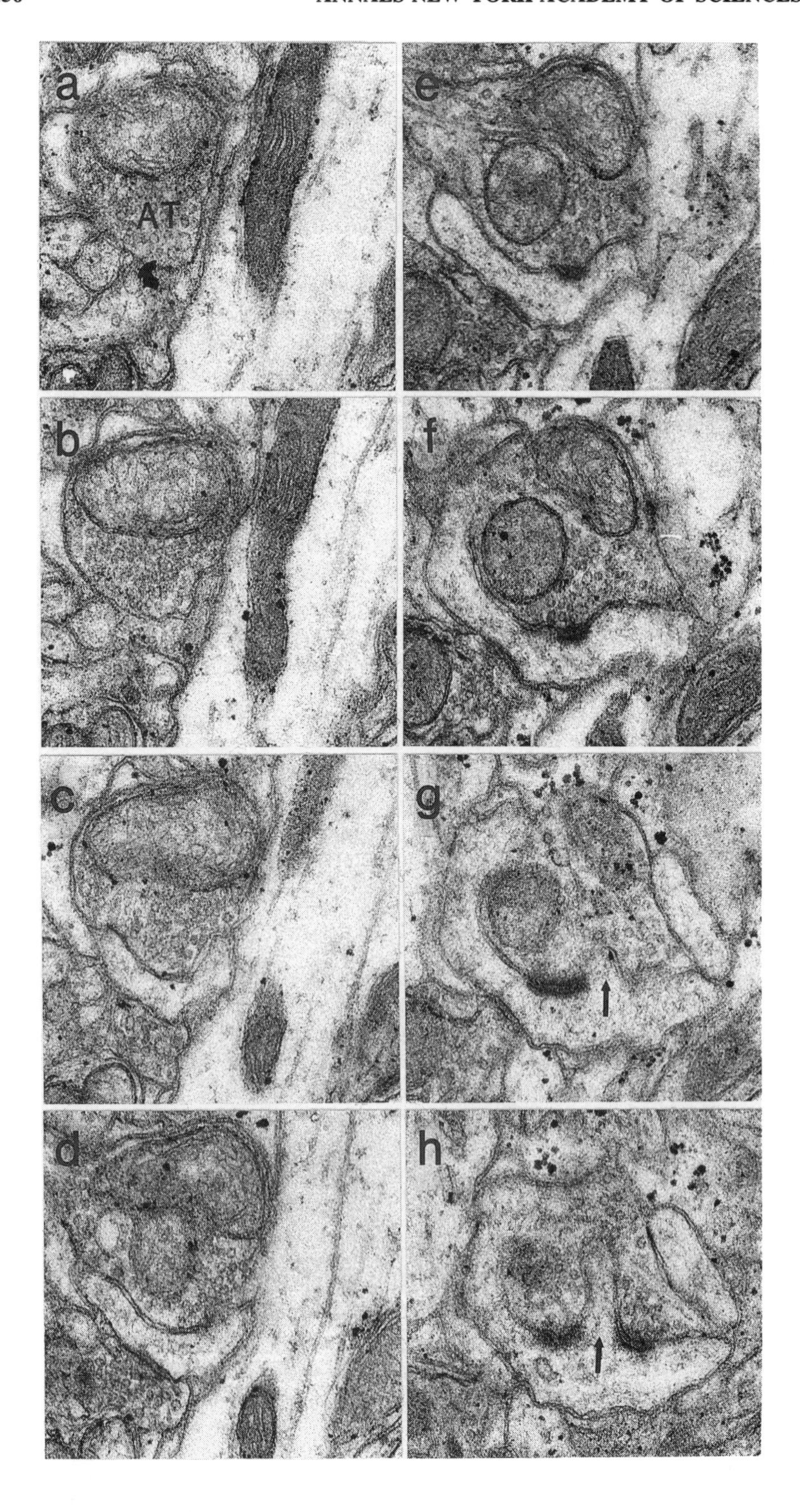
a
AT
b
c
d
e
f
g
h

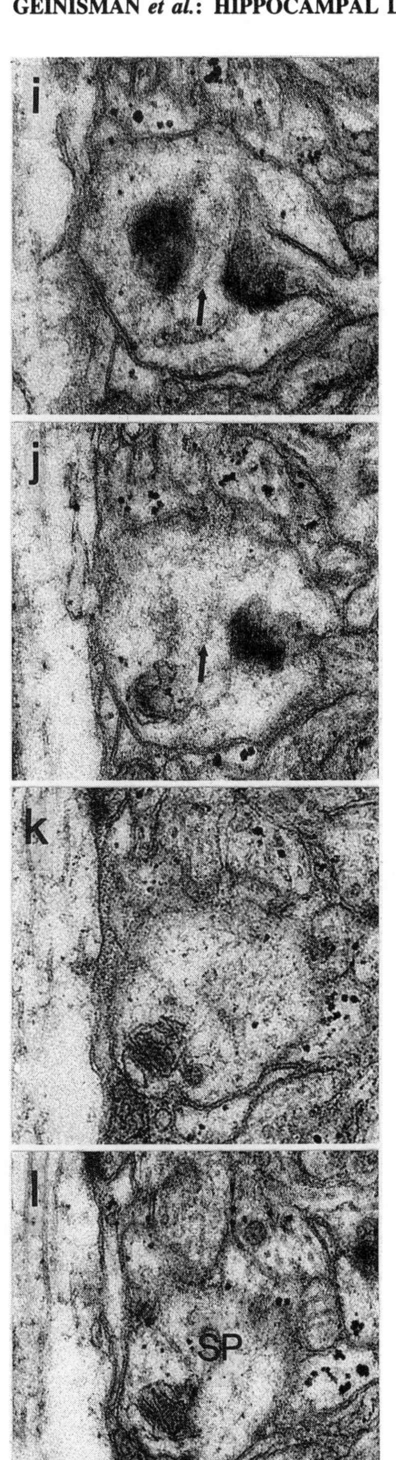

FIGURE 1. Electron micrographs of serial ultrathin sections (**a-l**) demonstrating an axospinous synapse with two completely partitioned transmission zones. All profiles of the presynaptic axon terminal in contact with the postsynaptic dendritic spine are shown. (The remaining axon terminal profiles are illustrated in FIG. 2a.) The axon terminal (labeled AT in **a**) contains synaptic vesicles and mitochondria. The spine head (labeled SP in **l**), all profiles of which are presented, is consecutively sectioned from its distal end (**c**) to the proximal one (**l**). The proximal end of the spine head is continuous with a short neck emanating from a parent dendrite (not shown). In one section containing electron dense PSD profiles (**i**), the spine head exhibits a centrally placed band of cytoplasm (*arrow*) interconnecting its opposite walls. In sections through more distal portions of the postsynaptic element (**g, h**), this band of spine cytoplasm is seen in the form of a fingerlike extension or spinule (*arrow*) that arises from a spine head wall and invaginates an axon terminal profile. Calibration bar = 0.25 μm.

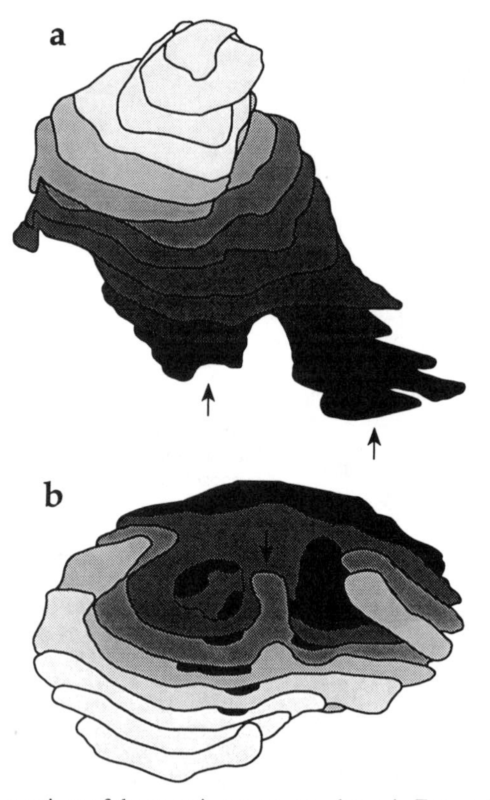

FIGURE 2. Reconstructions of the axospinous synapse shown in FIGURE 1. (**a**) Three-dimensional reconstruction of the presynaptic axon terminal demonstrating that two separate protrusions (*arrows*) emanate from its distal end. Image has been rotated by 310 degrees around the *x* axis. (**b**) Three-dimensional reconstruction of the postsynaptic spine head exhibiting a partition (*arrow*). This partition divides a spine head cavity into two compartments or pits, each one being fitted by a corresponding protrusion of the presynaptic axon terminal. PSD segments delineating two separate transmission zones are placed on each side of the partition. The image has been rotated by 50 degrees around the *x* axis.

dritic nor typical nonperforated axospinous synapses were increased in numbers after the induction of LTP in young adult or aged rats,[31,32] the present analysis was limited only to perforated axospinous junctions and atypical nonperforated ones. The existence of spine partitions was established by inspecting perforated synapses in electron micrographs of serial sections. In some cases, it was necessary to perform computer-aided three-dimensional reconstructions of synapses for establishing if a spine partition completely separated transmission zones or was incomplete. The PSD configuration was assessed by two-dimensional reconstructions of PSD plates along the postsynaptic membrane.

The number of synapses per neuron was differentially estimated for each synaptic subtype with the aid of the unbiased stereological disector technique,[48,49] according

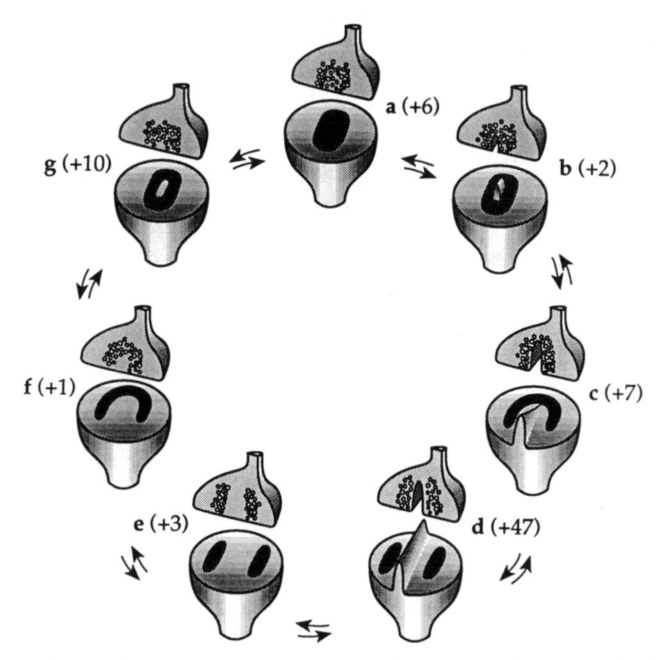

FIGURE 3. Diagram illustrating the proposed model of structural synaptic plasticity as described in the text. Shown schematically are the following structural intermediates in synaptic plasticity: (**a**) atypical nonperforated axospinous synapses, the subtypes of perforated axospinous junctions that have (**b**) a focal spine partition and fenestrated PSD, (**c**) a sectional partition and horseshoe-shaped PSD, or (**d**) a complete partition(s) and segmented PSD as well as nonpartitioned counterparts of these synaptic subtypes characterized by (**e**) segmented, (**f**) horseshoe-shaped, or (**g**) fenestrated PSDs. Values (in parentheses) represent differences in the number of MML synapses per neuron between control (data obtained from unstimulated and coulombic controls were combined) and potentiated animals. The only statistically significant difference detected was the increase (+47) in the number of synapses with complete spine partitions and segmented PSDs.

to our established protocol.[31,32,34] The results showed that only synapses with multiple, completely partitioned transmission zones were significantly increased in numbers in the MML of aged potentiated rats as compared with their unstimulated or coulombic controls (FIG. 4). This change did not involve any other synaptic subtype (FIG. 4), and it was observed only in the MML, but not in the IML (data are not shown for the latter).

It is necessary to note that the number of spine partitions per synapse was not altered as a result of LTP induction. In the MML of both control and potentiated animals, partitioned synaptic contacts with a fenestrated or horseshoe-shaped PSD exhibited only a single spine partition. The mean number (±SEM) of complete partitions per MML synapse with multiple transmission zones was estimated to be 1.17 ± 0.04 for unstimulated controls, 1.20 ± 0.04 for coulombic controls, and 1.17 ± 0.03 for potentiated animals.

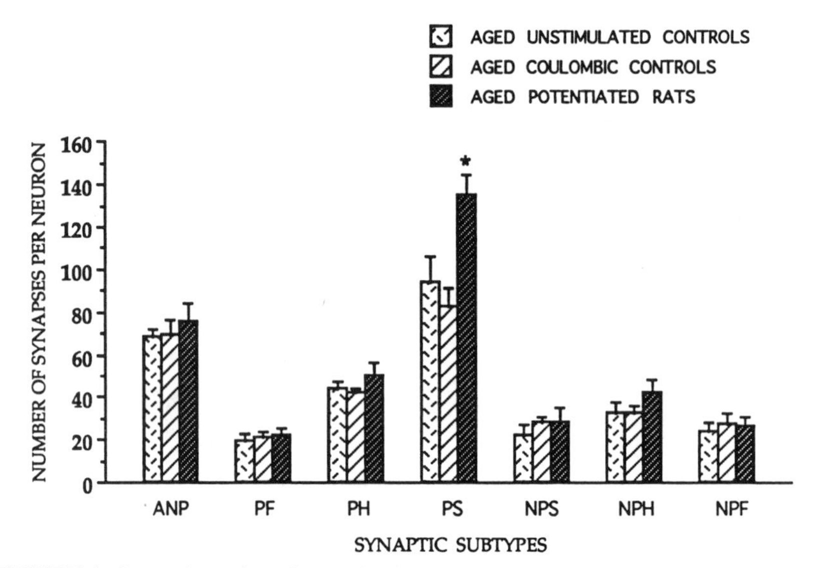

FIGURE 4. Comparison of aged potentiated and control rats with respect to the number of synapses per neuron in the MML. Data are means ± SEM. Estimates of the mean number per neuron were differentially obtained for atypical nonperforated (ANP) synaptic junctions, partitioned fenestrated (PF), horseshoe-shaped (PH), and segmented (PS) synapses as well as for nonpartitioned segmented (NPS), horseshoe-shaped (NPH), and fenestrated (NPF) synaptic contacts. A statistically significant change in potentiated animals as compared with controls is indicated by an *asterisk.*

COMPARISON OF AGED AND YOUNG RATS WITH RESPECT TO CHANGES IN SYNAPTIC NUMBERS AFTER THE INDUCTION OF LTP

It seemed interesting to compare the results of the present work with those of our previous LTP study[34] of young adult animals. Analysis of electrophysiological data indicated that young and old rats under comparison did not differ significantly in terms of the extent of potentiation observed. In aged animals, at a 1-hour time interval after the fourth high-frequency stimulation, the slope of the extracellularly recorded excitatory postsynaptic potential (EPSP) was potentiated by 69 ± 22%, the population spike amplitude by 704 ± 172%, and the input-output function (population spike/EPSP slope) by 365 ± 57%. Equivalent values for the young were 74 ± 23, 873 ± 261, and 441 ± 103%, respectively.

In accordance with the electrophysiological data, the induction of LTP in animals of both ages was followed by the same structural synaptic modification. The number of synapses with multiple, completely partitioned transmission zones was significantly increased in the MML of both young and old potentiated rats relative to their respective controls (TABLE 1). The extent of change was comparable in animals of the two different chronological ages (TABLE 1).

Senescent potentiated rats, however, had significantly fewer synapses characterized by multiple, completely partitioned transmission zones in the MML than did

TABLE 1. Increase in the Number of Synapses with Multiple, Completely Partitioned Transmission Zones in Young and Aged Potentiated Rats[a]

Triplet of Rats	A. Unstimulated Controls	B. Coulombic Controls		C. Potentiated Animals		
	n/N	n/N	ΔB-A	n/N	ΔC-A	ΔC-B
		Young Adult Rats				
1	114	111		218		
2	98	82		160		
3	207	66		227		
4	160	179		173		
5	93	161		248		
6	147	104		136		
7	134	183		194		
Group mean	136	127	−6.6%	194	+42.6%*	+52.8%**
± SEM	± 15	± 18		± 15		
		Aged Rats				
1	107	58		116		
2	148	115		154		
3	81	112		103		
4	66	78		138		
5	103	79		107		
6	105	79		176		
7	46	53		148		
Group mean	94	82	−12.8%	135	+43.6%*	+64.6%***
± SEM	± 12	± 9		± 10		

[a] Designations: n/N = number of synapses per neuron in the middle molecular layer of the dentate gyrus; Δ = difference between group means; *p <0.05,**p<0.02, and ***p<0.005, two-tailed randomization test for two independent samples.

young ones (TABLE 1). Examination of other subtypes of axospinous junctions in the MML demonstrated that this was the only significant difference between potentiated rats of the two ages (FIG. 5).

Additionally, it was possible to combine the data obtained from unstimulated and coulombic controls, because there was no significant difference in synaptic numbers between the two control groups at each chronological age studied. Comparison of such combined values showed that the mean number of MML synapses having multiple, completely partitioned transmission zones was 88 per neuron for aged controls versus 132 for young ones, the age-dependent synaptic loss being equal to −33.3%. Although these synapses were increased in numbers to 135 per neuron in aged rats and to 194 in young adults after LTP induction (TABLE 1), the difference in synaptic numbers between potentiated animals of the two ages (−30.4%) was of the same magnitude as that between old and young controls.

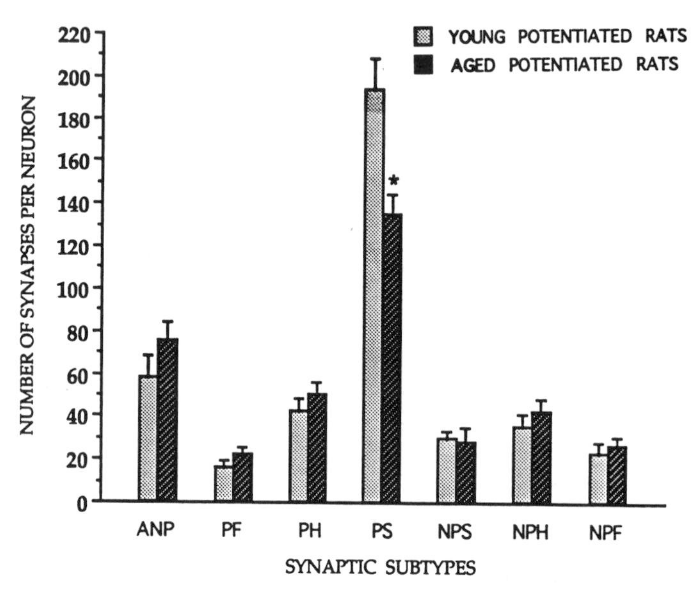

FIGURE 5. Comparison of young and aged potentiated rats with respect to the number of synapses per neuron in the MML. Designations are the same as in FIGURE 4.

FUNCTIONAL SIGNIFICANCE OF OBTAINED MORPHOLOGICAL DATA

A major finding of the present study is that the induction of LTP in the hippocampal dentate gyrus of aged rats is followed by an increase in the number of synapses with multiple, completely partitioned transmission zones (FIG. 4, TABLE 1). This structural alteration is highly selective because it involves only one particular synaptic subtype. Moreover, it occurs only in the terminal synaptic field of stimulated axons (MML), but not in an immediately adjacent one (IML) which was not directly activated by potentiating stimulation. Such topographical specificity strongly suggests that the increase in the number of synapses with multiple, completely partitioned transmission zones is indeed associated with LTP inasmuch as it is restricted, as is the LTP phenomenon itself,[50–53] to the site of termination of tetanized axons.

The structural synaptic modification that occurs in aged rats as a consequence of LTP is predicted by the proposed model of structural synaptic plasticity.[33] This model also implies, however, that the observed morphological change may be due to a remodeling of preexisting axospinous junctions involving the assembly of complete spine partitions which split a single transmission zone into multiple ones. Therefore, the LTP-induced increase in the number of synapses with multiple, completely partitioned transmission zones should have been accompanied by a corresponding loss of other synaptic subtypes which are also postulated to be structural intermediates in synaptic plasticity. Although such a correspondence was not observed (FIG. 3), it is difficult to disregard the possibility that the proposed synaptic remodeling did take place, but was undetected in this study examining just one, relatively late time point following the establishment of LTP. Further studies of earlier time points are necessary to clarify the issue of whether the described morphological change results from a restructuring of pre-existing synapses or from synaptogenesis. If this change is detected at very short poststimulation periods (e.g., 2-10 min), which are insufficient for synaptogenesis to be completed, then it can only be due to synaptic remodeling.

Comparison of relative changes in synaptic numbers in young and aged potentiated rats as opposed to their respective controls shows a similar pattern of LTP-induced synaptic restructuring. In either young or old animals, only those synapses that are distinguished by multiple, completely partitioned transmission zones were markedly and significantly increased in numbers in the potentiated synaptic field. Moreover, the magnitude of the relative increases in synaptic numbers was similar in young and old rats (TABLE 1). The equal extent of the structural synaptic change in animals of the two different chronological ages may explain why senescent rats can be potentiated to the same degree as young ones. Activation of about 400 axospinous synapses formed by perforant path fibers is sufficient to evoke a response of a dentate granule cell.[54] Following LTP induction via perforant path stimulation, each granule cell acquires some 60 or 50 additional axospinous synapses with multiple, completely partitioned transmission zones in young or aged rats, respectively. Such a substantial and selective increase in the number of presumably most efficacious synaptic contacts may represent a structural modification which is required for a long-lasting enhancement of synaptic responses regardless of age.

Comparison of animals of two different chronological ages with respect to the absolute number of synaptic contacts per neuron, however, reveals striking age-related differences. The complement of MML synapses with multiple, completely partitioned transmission zones is significantly lower in aged potentiated rats than in their young adult counterparts (TABLE 1). The same difference in this particular synaptic subtype is also found between aged and young controls. These data suggest that the maintenance of synapses having multiple, completely partitioned transmission zones may be impaired during aging. Such an age-related deficit in the maintenance of a particular synaptic subtype, which appears to be essential for synaptic plasticity associated with LTP, may account for the impaired LTP retention characteristic of senescent rats.

REFERENCES

1. KHACHATURIAN, Z. S. 1987. Hypothesis on the regulation of cytosol calcium concentration and the aging brain. Neurobiol. Aging **8:** 345-346.
2. KHACHATURIAN, Z. S. 1989. The role of calcium regulation in brain aging: Reexamination of a hypothesis. Aging **1:** 17-34.
3. LANDFIELD, P. W. 1987. "Increased calcium current" hypothesis of brain aging. Neurobiol. Aging **8:** 346-347.
4. MORRIS, R. G. M., S. DAVIS & S. P. BUTCHER. 1990. Hippocampal synaptic plasticity and NMDA receptors: A role in information storage? Phi. Trans. R. Soc. Lond. (Biol.) **329:** 187-204.
5. DOYÈRE, V. & S. LAROCHE. 1992. Linear relationship between the maintenance of hippocampal long-term potentiation and retention of an associative memory. Hippocampus **2:** 29-38.

6. BLISS, T. V. P. & T. LØMO. 1973. Long-lasting potentiation of synaptic transmission in the dentate area of the anaesthetized rabbit following stimulation of the perforant path. J. Physiosl. (Lond.) **232:** 331-356.
7. BLISS, T. V. P. & A. GARDNER-MEDWIN. 1973. Long-lasting potentiation of synaptic transmission in the dentate area of the unanaesthetized rabbit following stimulation of the perforant path. J. Physiol. (Lond.) **232:** 357-374.
8. BLISS, T. V. P. & G. L. COLLINGRIDGE. 1993. A synaptic model of memory: Long-term potentiation in the hippocampus. Nature **361:** 31-39.
9. LYNCH, G., M. KESSLER, A. ARAI & J. LARSON. 1990. The nature and causes of hippocampal long-term potentiation. Prog. Brain Res. **83:** 233-250.
10. MALENKA, R. C. & R. A. NICOLL. 1990. Intracellular signals and LTP. Sem. Neurosci. **2:** 335-344.
11. MADISON, D. V., R. C. MALENKA & R. A. NICOLL. 1991. Mechanisms underlying long-term potentiation of synaptic transmission. Ann. Rev. Neurosci. **14:** 379-397.
12. MALENKA, R. C. 1991. The role of postsynaptic calcium in the induction of long-term potentiation. Molec. Neurobiol. **5:** 289-295.
13. COLLEY, P. A. & A. ROUTTENBERG. 1993. Long-term potentiation as synaptic dialogue. Brain Res. Rev. **18:** 115-122.
14. BARNES, C. A. 1979. Memory deficits associated with senescence: A neurophysiological and behavioral study in the rat. J. Comp. Physiol. Psychol. **93:** 74-104.
15. RACINE, R. G., N. W. MILGRAM & S. HAFNER. 1983. Long-term potentiation phenomena in the rat limbic forebrain. Brain Res. **260:** 217-231.
16. STAUBLI, U. & G. LYNCH. 1987. Stable hippocampal long-term potentiation elicited by "theta" pattern stimulation. Brain Res. **435:** 227-234.
17. BARNES, C. A. & B. L. MCNAUGHTON. 1980. Spatial memory and hippocampal synaptic plasticity in senescent and middle-aged rats. *In* Psychobiology of Aging: Problems and Perspectives. D. Stein, Ed.: 253-272. Elsevier. Amsterdam.
18. BARNES, C. A. & B. L. MCNAUGHTON. 1985. An age comparison of the rates of acquisition and forgetting of spatial information in relation to long-term enhancement of hippocampal synapses. Behav. Neurosci. **99:** 1040-1048.
19. DETOLEDO-MORRELL, L., Y. GEINISMAN & F. MORRELL. 1988. Age-dependent alterations in hippocampal synaptic plasticity: Relation to memory disorders. Neurobiol. Aging **9:** 581-590.
20. DESMOND, N. L. & W. B. LEVY. 1988. Anatomy of associative long-term synaptic modification. *In* Long-Term Potentiation: From Biophysics to Behavior. P. W. Landfield & S. A. Deadwyler, Eds.: 265-305. Liss. New York.
21. GEINISMAN, Y., L. DETOLEDO-MORRELL & F. MORRELL. 1991. Structural synaptic substrates of kindling and long-term potentiation. *In* Kindling and Synaptic Plasticity: The Legacy of Graham Goddard. F. Morell, Ed.: 124-159. Birkhauser. Boston, MA.
22. WALLACE, C., N. HAWRYLAK & W. T. GREENOUGH. 1991. Studies of synaptic structural modifications after long-term potentiation and kindling: Context for a molecular morphology. *In* Long-Term Potentiation: A Debate of Current Issues. M. Baudry & J. L. Davis, Eds.: 189-232. MIT Press. Cambridge, MA.
23. BAILEY, C. G. & E. R. KANDEL. 1993. Structural changes accompanying memory storage. Annu. Rev. Physiol. **55:** 397-426.
24. LISMAN, J. E. & K. M. HARRIS. 1993. Quantal analysis and synaptic anatomy-integrating two views of hippocampal plasticity. Trends Neurosci. **16:** 141-147.
25. DESMOND, N. L. & W. B. LEVY. 1986. Changes in the numerical density of synaptic contacts with long-term potentiation in the hippocampal dentate gyrus. J. Comp. Neurol. **253:** 466-475.
26. WENZEL, J. & H. MATTHIES. 1985. Morphological changes in the hippocampal formation accompanying memory formation and long-term potentiation. *In* Memory Systems of the Brain. N. M. Weinberger, J. L. McGaugh & G. Lynch, Eds.: 151-170. Guilford Press, New York.

27. LEE, K., F. SCHOTTLER, M. OLIVER & G. LYNCH. 1980. Brief bursts of high-frequency stimulation produce two types of structural change in rat hippocampus. J. Neurophysiol. **44:** 247-258.
28. LEE, K., M. OLIVER, F. SCHOTTLER & G. LYNCH. 1981. Electron microscopic studies of brain slices: The effects of high-frequency stimulation on dendritic ultrastructure. *In* Electrophysiology of Isolated Mammalian CNS Preparations. G. A. Kerkut & H. V. Wheal, Eds.: 189-211. Academic Press. New York.
29. CHANG, F.-L. & W. T. GREENOUGH. 1984. Transient and enduring morphological correlates of synaptic activity and efficacy change in the rat hippocampal slice. Brain Res. **309:** 35-46.
30. CHANG, F.-L. F., K. R. ISSAKS & W. T. GREENOUGH. 1991. Synapse formation occurs in association with the induction of long-term potentiation in two-year-old rat hippocampus *in vitro.* Neurobiol. Aging **12:** 517-522.
31. GEINISMAN, Y., L. DETOLEDO-MORRELL & F. MORRELL. 1991. Induction of long-term potentiation is associated with an increase in the number of axospinous synapses with segmented postsynaptic densities. Brain Res. **566:** 77-88.
32. GEINISMAN, Y., L. DETOLEDO-MORRELL, F. MORRELL, I. S. PERSINA & M. ROSSI. 1992. Structural synaptic plasticity associated with the induction of long-term potentiation is preserved in the dentate gyrus of aged rats. Hippocampus **2:** 445-456.
33. GEINISMAN, Y. 1993. Perforated axospinous synapses with multiple, completely partitioned transmission zones: Probable structural intermediates in synaptic plasticity. Hippocampus **3:** 417-434.
34. GEINISMAN, Y., L. DETOLEDO-MORRELL, F. MORRELL, R. E. HELLER, M. ROSSI & R. F. PARSHALL. 1993. Structural synaptic correlate of long-term potentiation: Formation of axospinous synapses with multiple, completely partitioned transmission zones. Hippocampus **3:** 435-446.
35. PETERS, A. & I. R. KAISERMAN-ABRAMOF. 1969. The small pyramidal neuron of the rat cerebral cortex. The synapses upon dendritic spines. Z. Zellforsch. **100:** 487-506.
36. NEITO-SAMPEDRO, M., S. W. HOFF & C. W. COTMAN. 1982. Perforated postsynaptic densities: Probable intermediates in synapse turnover. Proc. Natl. Acad. Sci. USA **79:** 5718-5722.
37. COHEN, R. S. & P. SIEKEVITZ. 1983. Form of the postsynaptic density. A serial section study. J. Cell Biol. **78:** 36-46.
38. SPACEK, J. & M. HARTMANN. 1983. Three-dimensional analysis of dendritic spines. I. Quantitative observations related to dendritic spine and synaptic morphology in cerebral and cerebellar cortices. Anat. Embryol. **167:** 289-310.
39. JONES, D. G. & P. K. S. CALVERLEY. 1991. Perforated and non-perforated synapses in rat neocortex: Three-dimensional reconstructions. Brain Res. **556:** 247-258.
40. GEINISMAN, Y., F. MORRELL & L. DETOLEDO-MORRELL. 1987. Axospinous synapses with segmented postsynaptic densities: A morphologically distinct synaptic subtype contributing to the number of profiles of "perforated" synapses visualized in random sections. Brain Res. **423:** 179-188.
41. WESTRUM, L. E. & T. W. BLACKSTAD. 1962. An electron microscopic study of the stratum radiatum of the rat hippocampus (regio inferior, CA 1) with particular emphasis on synaptology. J. Comp. Neurol. **119:** 281-309.
42. TARRANT, S. B. & A. ROUTTENBERG. 1977. The synaptic spinule in the dendritic spine: Electron microscopic study of the hippocampal dentate gyrus. Tissue & Cell **9:** 461-473.
43. CARLIN, P. K. & P. SIEKEVITZ. 1983. Plasticity in the central nervous system: Do synapses divide? Proc. Natl. Acad. Sci. USA **80:** 3517-3521.
44. CALVERLEY, P. K. S. & D. G. JONES. 1990. Contribution of dendritic spines and perforated synapses to synaptic plasticity. Brain Res. Rev. **15:** 215-249.

45. DYSON, S. E. & D. G. JONES. 1984. Synaptic remodelling during development and maturation: Junction differentiation and slitting as a mechanisms of modifying connectivity. Dev. Brain Res. **13:** 125-137.
46. COTMAN, C. W. & P. T. KELLY. 1980. Macromolecular architecture of CNS synapses. *In* The Cell Surface and Neuronal Function. C. W. Cotman, G. Poste & G. L. Nicolson, Eds.: 506-533. Elsevier. Amsterdam.
47. SIEKEVITZ, P. 1985. The postsynaptic density: A possible role in long-lasting effects in the central nervous system. Proc. Natl. Acad. Sci. USA **82:** 3494-3498.
48. STERIO, D. C. 1984. The unbiased estimation of number and sizes of arbitrary particles using the disector. J. Microsc. (Lond.) **134:** 127-136.
49. BRÆNDGAARD, H. & H. J. G. GUNDERSEN. 1986. The impact of recent stereological advances on quantitative studies of the nervous system. J. Neurosci. Meth. **18:** 39-78.
50. BLISS, T. V. P., A. R. GARDNER-MEDWIN & T. LØMO. 1973. Synaptic plasticity in the hippocampal formation. *In* Macromolecules and Behaviour. G. B. Ansell & P. B. Bradley, Eds.: 193-203. MacMillan. London.
51. ANDERSEN, P., S. H. SUNDBERG, O. SVEEN & H. WINGSTROM. 1977. Specific long-lasting potentiation of synaptic transmission in hippocampal slices. Nature **266:** 736-737.
52. LYNCH, G., T. DUNWIDDIE & V. GRIBKOFF. 1977. Heterosynaptic depression: A postsynaptic correlate of long-term potentiation. Nature **266:** 737-739.
53. MCNAUGHTON, B. L. & C. A. BARNES. 1977. Physiological identification and analysis of dentate granule cell responses to stimulation of the medial and lateral perforant pathways in the rat. J. Comp. Neurol. **175:** 439-454.
54. MCNAUGHTON, B. L., C. A. BARNES & P. ANDERSEN. 1981. Synaptic efficacy and EPSP summation in granule cells of rat fascia dentata studied *in vitro*. J. Neurophysiol. **46:** 952-966.

Calcium Antagonists in Aging Brain

T. SCHUURMAN[a] AND J. TRABER

Institute for Neurobiology
Troponwerke
Cologne, Germany

Calcium antagonists have a 20-year history in the therapy of hypertension. However, their potential indications are not limited to cardiovascular diseases. Preclinical and clinical evidence is accumulating that calcium antagonists may be useful in the treatment of various disorders of the central nervous system (CNS). Efficacy of calcium antagonists have been shown in animal models of epilepsy, cerebral ischemia, depression, and dementia.[1-3] Results from initial clinical studies up to now are in agreement with the outcome of animal experiments. However, more clinical data are needed to definitely assess the therapeutic value of selected calcium antagonists in the treatment of brain diseases.

The effects of calcium antagonists in the brain are determined by several factors. First, the availability of the drug to the brain is a prerequisite for direct effects of that drug on nervous tissue. Most known calcium antagonists do not cross the blood brain barrier, whereas others do so. Second, the molecular target (the subtype of calcium channel) and the intrinsic activity of the substance at its target(s) are important for the drug-induced CNS effects. The biological functions of the various subtypes of calcium channels (voltage-dependent L-, N-, T-, and P-channels and different receptor-operated calcium channels) differ. Consequently, pharmacological manipulations of different calcium channel subtypes may lead to different CNS effects. Third, the state of the system may determine the effect of the drug. For example, the ability of dihydropyridine calcium antagonists such as nifedipine and nimodipine to block calcium currents is stronger if the cell membrane is depolarized.[4] Finally, the tissue distribution of calcium antagonists, their interaction with systems other than calcium channels, and other factors contribute to the pharmacological profile of a calcium antagonist.

In the following, some crucial neuropharmacological studies with the dihydropyridine calcium antagonist nimodipine are reviewed. Nimodipine was selected, because no other calcium antagonist has been studied so extensively with respect to its effects on the brain. Only the experiments that provide a rationale for the use of this tool substance in disorders related to brain aging and dementia will be discussed. For other possible CNS indications the reader is referred to other reviews.[1,5]

CALCIUM ANTAGONISTIC AND NEUROPROTECTIVE ACTIVITY OF NIMODIPINE IN NEURONS

Nimodipine is different from many other calcium antagonists because it passes the blood brain barrier rapidly, as was shown for tritiated nimodipine administered

[a] Present address: Wageningen Agricultural University, Department of Animal Husbandry, Section Ethology, P.O. Box 338, 6700 A H Wageningen, the Netherlands.

to animals.[6] In the brain, nimodipine binds with high affinity (K_d = 1.1 nMolar in displacement experiments) and high specificity to the dihydropyridine receptors of the L-type calcium channel.[7–9] Functional, electrophysiological studies showed that nimodipine at nanomolar concentrations blocks calcium currents through L-type channels in neurons.[10–12] Similarly, nimodipine at low nanomolar concentrations blocks K^+-induced elevations of intracellular calcium in hippocampal neurons from newborn rats, as shown with calcium imaging techniques.[3,13] By its calcium antagonistic action nimodipine interferes with many cellular processes such as neuronal excitability, release of neurotransmitters, axonal transport, and activity of calcium-dependent enzymes.[1] Under pathological conditions associated with high intracellular calcium concentrations, nimodipine may protect neurons against cell damage or cell death. Indeed, nimodipine antagonizes the toxic effects of mercury or low oxygen concentrations in PC12 cells.[14] Glutamate toxicity in rat retinal ganglion cells can be reduced by micromolar concentrations of this calcium antagonist.[15] Low nanomolar concentrations of nimodipine prevent the toxic effects of the combined administration of AMPA and Zn^{2+} in cortical neurons from fetal rat brain.[16] The increase in intracellular calcium concentration and neuronal death induced by the human immunodeficiency virus (HIV) coat protein gp 120 can be successfully prevented by 100 nM nimodipine (ref. 17 and this volume). Recently, it was demonstrated that nimodipine at low micromolar concentrations reduces the beta-amyloid peptide-induced toxicity to mouse cortical neurons (ref. 18 and this volume). Acute neuroprotective effects of nimodipine have also been shown in various animal models of brain ischemia.[19–21]

FUNCTIONAL EFFECTS OF NIMODIPINE IN THE BRAINS OF AGING ANIMALS

Brain aging is associated with a reduction or even the loss of various functions such as learning and memory capability, diurnal rhythms, social behavior, and sensorimotor functions. This is observed not only in humans but also in many animal species including rodents.[22] Attempts to improve one or more of these old age-related deficiencies by pharmacological treatments have not been successful until now. However, a number of animal studies suggest that the calcium antagonist nimodipine may be a drug candidate in the treatment of some behavioral symptoms that occur during aging.

We trained 16-month-old, male Wistar rats to escape from a water-filled tank. The animals learned to swim around vertical barriers in the maze to reach the exit staircase. The old rats were subjected one time a day to this navigational task. The number of errors (swimming in the wrong direction away from the exit) and the time needed to reach the exit were recorded for each animal. Old Wistar rats perform less well than do young conspecifics in this maze.[22] The daily oral administration of nimodipine 10 mg/kg of body weight 45 minutes before each swim trial significantly reduced the number of errors and the escape latency times of 16-month-old rats compared to vehicle-treated rats of the same age (FIG. 1). In other maze experiments using 26-month-old Wistar rats similar beneficial effects of nimodipine administration on maze learning were demonstrated.[23,24]

The promnestic effects of nimodipine are not restricted to rats. Deyo *et al.*[25] and Disterhoft *et al.* (ref. 26 and this volume) showed that nimodipine improves the

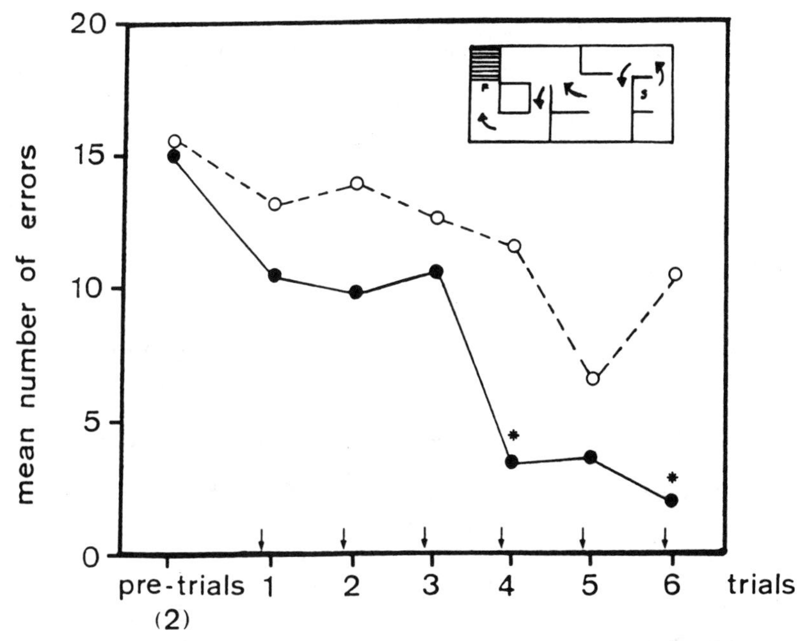

FIGURE 1. Acquisition curves of vehicle-treated (○----○; n = 11) and nimodipine-treated (●——●; n = 12) rats tested in a water labyrinth.

acquisition of an eyeblink response in aging rabbits. Sandin *et al.*[27] investigated the effect of nimodipine on cognitive performance in 28-year-old female rhesus monkeys subjected to a delayed response task. Doses of nimodipine 3 and 9 mg/kg of body weight given orally improved the working memory of these aged nonhuman primates.

Besides learning and memory deficits disturbances of sensorimotor function are common in aged subjects. Pathologic conditions in the CNS as well as the peripheral nervous system (PNS) contribute to age-related deficits of sensorimotor behavior.[22–24,30–33]

A series of experiments investigated if the long-term administration of the calcium antagonist nimodipine in aging male Wistar rats could slow down the deterioration of sensorimotor behavior. Sensorimotor performance was measured by applying a battery of simple tests described by Gage *et al.*[28] Among these tests were balance rod tests, a suspended wire, and a pole climbing test. Rats subjected to a balance rod test were placed with all 4 feet in the middle of the elevated horizontal rod. The time they could maintain balance before falling off was measured. A comparison of the balancing times of rats belonging to different age categories (3-24 months) proved that this bridge test is very useful in the assessment of age-related disturbances of sensorimotor function.[22–24] Balance rod and other sensorimotor tests were applied in pharmacological experiments in which 24-month-old male Wistar rats were fed nimodipine-containing food pellets (concentrations of 300, 860, and 1,200 ppm) until they reached the age of 28 months. During the 4-month treatment period the aging

TABLE 1. Balancing Times of Aging Nimodipine-Treated Male Rats and Nontreated Control Rats on an Elevated Horizontal Rod (width 5 cm)

Age (mo.)	Duration of Treatment (wk.)	Median Latency(s) before Falling-off		Significance
		Controls (n = 26)	Nimodipine (n = 26)	
24	0	84	84	...
25	4	64	92	0.01
26.5	10	48	93	0.001
28	16	52	80	0.01

animals were repeatedly subjected to balance rod and other sensorimotor tests. The performance of the drug-treated animals was compared to that of control animals of the same age fed with drug-free rat chow.

TABLE 1 shows that the performance of control animals subjected repeatedly to the bridge test gradually became worse in the course of the experiment. The balancing times of these rats were reduced by about 40% between 24 and 28 months of age. In contrast, the balancing times of aging rats treated with nimodipine (860 ppm in the food) did not decrease during 4 months of observation. Apparently, drug treatment prevented the deterioration of this type of sensorimotor behavior for at least 4 months. However, long-term treatment with nimodipine cannot slow down the age-related deterioration of sensorimotor behavior until the end of life. In other longitudinal experiments the treatment period was extended to 6 or 9 months. In these experiments the surviving animals were 30 and 33 months old at the end of the experiment. Sensorimotor scores of 30-33-month-old animals treated chronically with nimodipine did not differ from those of rats that served as drug-free controls of the same age.[23,24,29] Apparently, long-term treatment with the calcium antagonist can delay or slow down the worsening of sensorimotor function in male rats up to the age of about 30 months. Beyond that age the beneficial effects of drug treatment are probably passed or overruled by the increasing pathological processes of (brain) aging.

To further study sensorimotor deficits in the aging rat and the effects of long-term nimodipine treatment thereupon, we applied a less known procedure that analyzes walking patterns. In short, the hindpaws of a rat were dipped into photographic developer. Thereafter, the rat was placed in a narrow corridor (length 75 cm) with access to a darkened goal box. The bottom of the corridor was covered with a strip of photographic paper. After the rat walked through the corridor into the dark compartment, prints of the hindpaws appeared on the paper. A comparison of the footprints of young 3-month-old animals with those of 24-30-month-old rats led to the definition of the following "pathologies": (a) fuzzy prints as a result of lateral rotation of one or both hindfeet after placement on the floor (exorotation); (b) fuzzy prints resulting from the lack of lifting the foot between steps (dragging foot) on one or both sides; and (c) small additional prints between the regular steps.

In longitudinal studies walking patterns of aging rats that received nimodipine in rat food between 24 and 30 months of age were compared to those of nontreated

TABLE 2. Occurrence of Abnormal Walking Patterns in Aging Male Rats: Effects of Nimodipine Application

Age (mo.)	Duration of Treatment (wk.)	Percentage of Rats with Pathological Prints	
		Nondrugged Controls ($n = 26$)	Nimodipine-Treated ($n = 26$)
24	0	24	19
25	4	52	13
26	8	64	13
27	12	92	32
28	16	100	46

rats of the same age. Long-term treatment with nimodipine (860 and 1,200 ppm in the chow) delayed the onset of abnormal walking patterns as assessed in the footprint test (TABLE 2 and FIG. 2). In accordance with the previous notion that chronic nimodipine treatment cannot improve balancing times of very old animals, the walking patterns of drug-treated and control animals aged more than 30 months also did not differ.[23,24,29]

Detailed histological examinations of various brain regions of 30-33-month-old rats treated for 6-9 months with nimodipine proved that long-term treatment with this calcium antagonist can prevent or at least delay the loss of synapses in some brain regions.[30] Furthermore, the integrity of the cerebral microvasculature was maintained under conditions of chronic drug treatment (refs. 31 and 32 and this volume). Finally, histological examinations of sciatic nerves of 28-month-old rats treated for 4 months with nimodipine suggest that this drug can prevent the loss of small sensory and motor fibers in the aging PNS.[33] In accordance with this observation, nerve conduction velocities were significantly higher in the sciatic nerves of drug-treated rats than in those of nontreated samples. These histological and electrophysiological findings are part of the morphological and physiological base for the behavioral effects of long-term nimodipine administration. The results of the experiments reviewed here strongly suggest that long-term treatment with nimodipine has antidegenerative effects in the PNS and CNS.

CONCLUSION

The dihydropyridine calcium antagonist nimodipine passes the blood brain barrier and binds with high affinity to the dihydropyridine binding sites of L-type calcium channels. Both electrophysiological and calcium imaging experiments showed that nimodipine at nanomolar concentrations blocks calcium currents in neurons. This calcium antagonist effect is most probably the mechanism by which nimodipine exerts neuroprotective effects in cell culture and animal experiments.

Nimodipine has promnestic effects in various animal models of cognitive dysfunction such as aging rodents. Long-term nimodipine treatment has beneficial effects

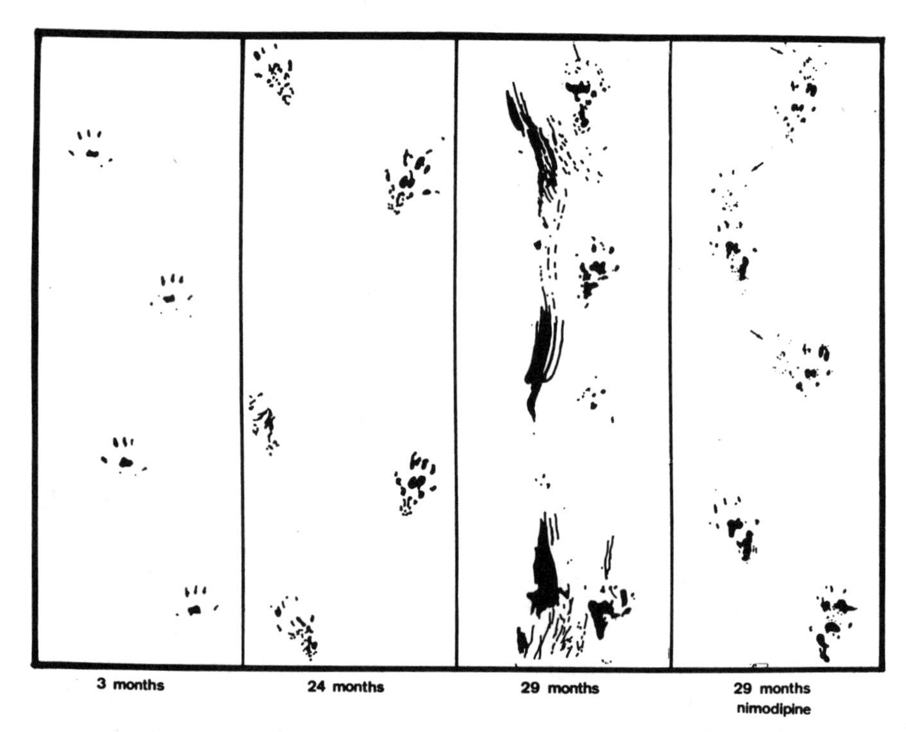

FIGURE 2. Prints of the hind paws of rats. In the right panel, prints of a 29-month-old rat treated for 5 months with nimodipine (860 ppm in the food) are represented.

on the sensorimotor behavior of aging rats. This can best be explained by neuroprotective or antidegenerative actions of the drug. These and other data (see also this volume) make nimodipine an interesting drug candidate for the treatment of behavioral disturbances occurring in conditions of old age and dementia.

REFERENCES

1. Scriabine, A., T. Schuurman & J. Traber. 1989. Pharmacological basis for the use of nimodipine in central nervous system disorders. FASEB J. **3:** 1799-1806.
2. Czyrak, A., E. Mogilnicka & J. Maj. 1989. Dihydropyridine channel antagonists as antidepressant in mice and rats. Neuropharmacology **28:** 229-233.
3. De Jonge, M., A. Friedel & J. De Vrij. 1993. CNS pharmacology of nimodipine: Antidepressant effects, drug discrimination, and Ca^{2+} imaging. *In* Ca^{2+} Antagonists in the CNS. Drugs in Development. Volume 2. A. Scriabine, R. A. Janis & D. J. Triggle, Eds.: 165-174. Neva Press. Connecticut.
4. Cohen, C. J. & R. T. McCarthy. 1987. Nimodipine block of calcium channels in rat anterior pituitary cells. J. Physiol. **387:** 195-225.
5. Betz, E., K. Deck & F. Hoffmeister (Eds). 1987. Nimodipine: Pharmacological and clinical properties. Schattauer. Stuttgart.

6. VAN DEN KERCKHOFF, W. & L. R. DREWES. 1985. Transfer of the Ca-antagonists nifedipine and nimodipine across the blood-brain barrier and their regional distribution *in vivo*. J. Cereb. Blood Flow Metab. **5:** 459-460.
7. BELLEMAN, P., D. FERRY, F. LÜBBECKE & H. GLOSSMAN. 1982. ^{3}H-nimodipine and ^{3}H-nitrendipine as tools to directly identify the sites of action of 1,4-dihydropyridine calcium-antagonists in guinea pig tissues. Arzneim. Forsch/Drug Res **32:** 361-363.
8. PEROUTKA, S. J. & G. S. ALLEN. 1983. Calcium channel antagonists binding sites labeled by ^{3}H-nimodipine in human brain. J. Neurosurg. **59:** 933-937.
9. DOMPERT, W. U. & J. TRABER. 1984. Binding sites for dihydropyridine calcium-antagonists. *In* Calcium Antagonists and Cardiovascular Disease. L. H. Opie, Ed.: 175-179. Raven Press. New York.
10. FRY, H. K. & R. T. MCCARTHY. 1989. Nimodipine block of L-type calcium channels in freshly dispersed rabbit DRG neurons. Soc. Neurosci. Abstr. **14:** 134.
11. MCCARTHY, R. T. 1989. Nimodipine block of L-type calcium channels in dorsal root ganglion cells. *In* Nimodipine and Central Nervous System Function: New Vistas. J. Traber & W. H. Gispen, Eds.: 35-49. Schattauer. Stuttgart, New York.
12. MCCARTHY, R. T. & P. E. TAN PIENGO. 1993. Nimodipine block of L-type calcium channels in rat hippocampal neurons. *In* Ca^{2+} Antagonists in the CNS. Drugs in Development. Volume 2. A. Scriabine, R. A. Janis & D. J. Triggle, Eds.: 71-77. Neva Press. Connecticut.
13. CHISHOLM, J. C., E. J. HUNNICUTT, JR. & J. N. DAVIS. 1993. Neuronal $[Ca^{2+}]_i$ responses to different classes of calcium-lowering agents. *In* Ca^{2+} Antagonists in the CNS. Drugs in Development. Volume 2. A. Scriabine, R. A. Janis & D. J. Triggle, Eds.: 117-125. Neva Press, Connecticut.
14. FAHEY, J. M., B. HEDAYAT, J. R. COOK, R. L. ISAACSON & R. G. VAN BUSKIRK. 1989. Nimodipine's ability to inhibit neuronal death due to oxygen deprivation or exposure to methylmercury and its effects on differentiation. *In* Nimodipine and Central Nervous System Function: New Vistas. J. Traber & W. H. Gispen, Eds.: 177-140. Schattauer. Stuttgart.
15. SUCHER, N. J., S. Z. LEI & S. A. LIPTON. 1991. Calcium channel antagonists attenuate NMDA receptor-mediated neurotoxicity in retinal ganglion cells in culture. Brain Res. **551:** 297-302.
16. FREUND, W. D. & S. REDDIG. 1994. AMPA/Zn^+-induced neurotoxicity in rat primary cortical cultures: Involvement of L-type calcium channels. Brain Res. **654:** 257-264.
17. DREYER, E. B., P. K. KAISER, J. T. OFFERMAN & S. A. LIPTON. 1990. HIV-1 protein neurotoxicity prevented by calcium antagonists. Science **248:** 364-367.
18. WEISS, J. H., C. J. PIKE & C. W. COTMAN. 1994. Ca^{2+} channel blockers attenuate β-amyloid peptide toxicity to cortical neurons in culture. J. Neurochem. **62:** 372-375.
19. UEMATSU, D., J. H. GREENBERG, W. F. HICKEY & M. REIVICH. 1989. Nimodipine attenuates both increase in cytosolic free calcium and histologic damage following focal cerebral ischemia and reperfusion in cats. Stroke **20:** 1531-1537.
20. MEYER, F. B., R. E. ANDERSON, T. L. YAKS & T. M. SUNDT, JR. 1986. Effect of nimodipine on intracellular brain pH, cortical blood flow and EEG in experimental focal cerebral ischemia. J. Neurosurg. **64:** 617-626.
21. HADLEY, M. M., J. M. ZABRAMSKI, R. F. SPETZLER, D. RIGAMONTI, M. S. FIFIELD & P. C. JOHNSON. 1989. The efficacy of intravenous nimodipine in the treatment of focal cerebral ischemia in a primate model. Neurosurgery **25:** 63-70.
22. SCHUURMAN, T., E. HORVÀTH, D. G. SPENCER, JR. & J. TRABER. 1986. Old rats: An animal model for senile dementia. *In* Senile Dementias: Early Detection. A. Bès, J. Cahn, R. Cahn, S. Hoyer, J. P. Marc-Vergnes & H. M. Wisniewski, Eds.: 629-630. John Libbey Eurotext. London-Paris.
23. SCHUURMAN, T. & J. TRABER. 1989. Effects of nimodipine on behavior of old rats. *In* Nimodipine and Central Nervous System Function: New Vistas. J. Traber & W. H. Gispen, Eds.: 195-208. Schattauer. Stuttgart.

24. SCHUURMAN, T. & J. TRABER. 1989. Old rats as an animal model for senile dementia: behavioral effects of nimodipine. *In* Diagnosis and Treatment of Senile Dementia. M. Bergener & B. Reisberg, Eds.: 295-307. Springer. Berlin-Heidelberg-New York.
25. DEYO, R., K. T. STRAUBE & J. F. DISTERHOFT. 1989. Nimodipine facilitates conditioning of the eye-blink response in aging rabbits. Science **243:** 809-811.
26. DISTERHOFT, J. F., R. A. DEYO, J. BLACK, M. DE JONGE, K. T. STRAUBE & L. T. THOMPSON. 1989. Associative learning in aging rabbits is facilitated by nimodipine. *In* Nimodipine and Central Nervous System Function. J. Traber & W. H. Gispen, Eds.: 209-235. Schattauer. Stuttgart, New York.
27. SANDIN, M., S. JASMIN & T. E. LEVERE. 1990. Aging and cognition: Facilitation of recent memory in aged nonhuman primates by nimodipine. Neurobiol. Aging **11:** 573-575.
28. GAGE, F. H., S. B. DUNNETT & A. BJÖRKLUND. 1984. Spatial learning and motor deficits in aged rats. Neurobiol. Aging **5:** 43-48.
29. SCHUURMAN, T., H. KLEIN, M. BENEKE & J. TRABER. 1987. Nimodipine and motor deficits in the aged rat. Neurosci. Res. Commun. **1:** 9-16.
30. DE JONG, G. I., B. BUWALDA, T. SCHUURMAN & P. G. M. LUITEN. 1992. Synaptic plasticity in the dentate gyrus of aged rats is altered after chronic nimodipine application. Brain Res. **596:** 345-348.
31. DE JONG, G. I., H. DE WEERD, T. SCHUURMAN, J. TRABER & P. G. M. LUITEN. 1990. Microvascular changes in aged rat forebrain. Effects of chronic nimodipine treatment. Neurobiol. Aging **11:** 381-389.
32. DE JONG, G. I., A. S. P. JANSEN, E. HORVÁTH, W. H. GISPEN & P. G. M. LUITEN. 1991. Nimodipine effects on cerebral microvessels and sciatic nerve in aging rats. Neurobiol. Aging **13:** 73-81.
33. VAN DER ZEE, C. E. E. M., T. SCHUURMAN, R. GERRITSEN VAN DER HOOP, J. TRABER & W. H. GISPEN. 1990. Beneficial effect of nimodipine on peripheral nerve function in aged rats. Neurobiol. Aging **11:** 451-456.

Subject Index

Index of Contributors